$$\left[-\frac{\hbar^2\nabla^2}{2m_e} + U(\mathbf{r})\right]\Phi = i\hbar\frac{\partial\Phi}{\partial t} \qquad a_B = \frac{4\pi\varepsilon_o\hbar^2}{m_e q^2}$$

$$\frac{\quad}{8\pi\varepsilon_o a_B}$$

$$j = \mu_n n \frac{dE_{Fn}}{dx} + \mu_p p \frac{dE_{Fp}}{dx}$$

$$L_{Dn} = \left(\frac{\varepsilon V_{th}}{qN_D}\right)^{1/2} \qquad x_{dn} = \left[\frac{2\varepsilon(V_{Bi} - V)}{qN_D}\right]^{1/2}$$

$$D_a = \frac{\mu_p p_n D_n + \mu_n n_n D_p}{\mu_n n_n + \mu_p p_n} \qquad \mu_a = \frac{\mu_n \mu_p (n_n - p_n)}{\mu_n n_n + \mu_p p_n}$$

$$D_d \frac{\partial^2 p_n}{\partial x^2} - \mu_a F \frac{\partial p_n}{\partial x} + G - R = 0$$

$$\beta^{-1} = \frac{D_p W^2 I_c}{qD_n^2 N_{de} x_e S} + \frac{D_p N_{ab} W}{D_n N_{de} x_e} + \frac{W\, n_i}{\mu_n F_{np} n_{bo}\, \tau_{rec}\, [WI_c/(qD_n n_{bo} S)]^{(1 - 1/m_{re})}} + \frac{W^2}{2L_{nb}}$$

$$I_{dsat} = \beta V_{sl}^2 \frac{(1 + 2\beta R_s V_{gst} + V_{gst}^2/V_{sl}^2)^{1/2} - 1 - \beta R_s V_{gst}}{1 - \beta^2 R_s^2 V_{sl}^2}$$

$$I = I_L + I_s\left[1 - \exp\left(\frac{V + R_s I}{m_{id} V_{th}}\right)\right] - V/R_{sh}$$

$$\omega = -kv(F_o) - i\left[\frac{1}{\tau_{md}(F_o)} + D_n k^2\right]$$

$$D = m/(\pi\hbar^2)$$

$$k_o d_o = n\pi - \arcsin\{\gamma E/[V_o + (\gamma - 1)E]\}^{1/2} - \arcsin\{\gamma E/[V_o + V_a + (\gamma - 1)E]\}^{1/2}$$

Physics of Semiconductor Devices

PRENTICE HALL SERIES IN SOLID STATE PHYSICAL ELECTRONICS

Nick Holonyak, Jr., Editor

Cheo, *Fiber Optics: Devices and Systems 2E*
Haus, *Waves and Fields in Optoelectronics*
Hess, *Advanced Theory of Semiconductor Devices*
Pulfrey and Tarr, *Introduction to Microelectronic Devices*
Shur, *Physics of Semiconductor Devices*
Soclof, *Analog Integrated Circuits*
Soclof, *Applications of Analog Integrated Circuits*
Streetman, *Solid State Electronic Devices 3E*
Verdeyen, *Laser Electronics 2E*
Wolfe/Holonyak/Stillman, *Physical Properties of Semiconductors*

Physics of Semiconductor Devices

MICHAEL SHUR

PRENTICE HALL, Englewood Cliffs, New Jersey 07632

Editorial/production supervision and
interior design: ***bookworks***
Cover design: George Cornell
Manufacturing buyer: Robert Anderson

Prentice Hall Series in Solid State Physical Electronics
Nick Holonyak, Jr., Editor

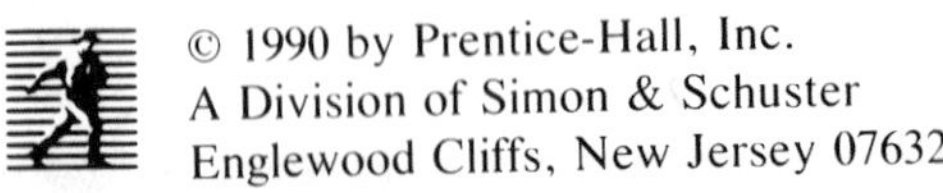

Printed in the United States of America

10 9 8 7 6 5 4 3 2 1

ISBN 0-13-666496-2

Prentice-Hall International (UK) Limited, *London*
Prentice-Hall of Australia Pty. Limited, *Sydney*
Prentice-Hall Canada Inc., *Toronto*
Prentice-Hall Hispanoamericana, S.A., *Mexico*
Prentice-Hall of India Private Limited, *New Delhi*
Prentice-Hall of Japan, Inc., *Tokyo*
Simon & Schuster Asia Pte. Ltd., *Singapore*
Editora Prentice-Hall do Brasil, Ltda., *Rio de Janeiro*

To students of electrical engineering who enter this field
with dedication and vigor, who challenge their professors
and advisors, and who are our hope for a better future.

Contents

List of Symbols

a_1, a_2, a_3	primitive base vectors
a_B	Bohr radius
A	active layer thickness
A*	Richardson constant
A_o	width of space charge region
A_d	depletion layer width of Schottky contact
$A_{dS,D}$	depletion layer width of Schottky contact at source/drain end of the gate
B	magnetic field
BV_{cb}	collector-base breakdown voltage
BV_{ce}	collector-emitter breakdown voltage
c	normalizing constant
C	energy gap produced by the ionic potential
C_d	capacitance of the depletion layer
C_{dom}	domain capacitance
C_{geom}	geometric capacitance
C_{gd}	internal gate-to-drain capacitance

C_{gs}	internal gate-to-source capacitance
D_a	ambipolar diffusion coefficient
D_n	electron diffusion coefficient
D_p	hole diffusion coefficient
D_{xy}	diffusion coefficient tensor
E	energy
E_1	acoustic deformation potential
E_a	acceptor level energy
E_B	Bohr energy
E_{ba}	energy gap between the bonding and antibonding state
E_c	energy at bottom of conduction band
E_d	donor level energy
E_{ij}	energy of intervalley scattering phonon
E_F	Fermi energy level
E_g	energy gap
E_h	energy gap produced by the covalent part of the potential
E_i	carrier critical ionization energy
E_n	electron energy
E_o	energy of optical phonon
E_{on}	energy of non-polar optical phonon
E_v	energy at ceiling of valence band
f	Fermi-Dirac distribution (occupation) function
f_n	Fermi-Dirac electron occupation function
f_p	Fermi-Dirac hole occupation function
F	electric field
F_b	built in electric field in base
F_{br}	breakdown field strength
F_m	maximum electric field
F_p	peak electric field
F_s	velocity saturation field
F_s	force constant in lattice vibrations
g_a	acceptor degeneracy factor
g_d	donor degeneracy factor
g_m	transconductance
G	generation rate of electron-hole pairs
h	Planck constant

$\hbar$	reduced Planck constant
I	total current
I_b	base current
I_{br}	base recombination current
I_c	collector current
I_{CBO}	common base collector saturation current
I_{CEO}	common emitter collector saturation current
I_e	emitter current
I_{gc}	generation current in depletion region of collector-base junction
I_i	intercept current
I_{nb}	base electron current
I_{nc}	collector electron current
I_{ne}	emitter electron current
I_{pe}	emitter hole current
I_{pc}	collector hole current
I_{re}	recombination current in depletion region of emitter-base junction
I_t	tunneling current
j	current density
j_1	small signal diode current density
j_F	total forward current density
j_{gen}	generation current density
j_n	electron current density
j_p	hole current density
j_R	reverse current density
j_s	diode saturation current
$\mathbf{k}$	wave vector
$\mathbf{K}_1, \mathbf{K}_2, \mathbf{K}_3$	primitive vectors of the reciprocal lattice
k_B	Boltzmann constant
k_F	Fermi wave vector
K_i	heat conductivity of electron gas in i-th valley
κ_s	low frequency dielectric constant
κ_0	high frequency dielectric constant
l	orbital quantum number
L	Lorentz number
L	device length

L	gate length of a field effect transistor
L_{bn}	diffusion length of electrons in the base region
L_D	Debye radius
L_{cp}	diffusion length of holes in the collector region
L_{ep}	diffusion length of holes in the emitter region
L_n	electron diffusion length
L_p	hole diffusion length
L_s	length of carrier velocity saturation region under the gate of a field effect transistor
L_T	transfer length
m	magnetic quantum number
m_{dn}	effective mass of the density of states' electrons
m_{dp}	effective mass of the density of states' holes
m_e	free electron mass
m_i	ideality factor
m_l	longitudinal effective electron mass
m_{lp}	effective mass for light holes
m_n	conductivity electron effective mass
m_{pn}	reduced effective mass
m_r	factor in recombination current exponent
m_t	transverse effective electron mass
M	collector multiplication factor
M	atomic mass
M_n	multiplication factor for electrons
M_p	multiplication factor for holes
n	principal quantum number
n	electron concentration
n_b	electron concentration in the base
n_{bo}	equilibrium electron concentration in the base
n_{bc}	electron concentration in the base at the collector-base junction
n_{be}	electron concentration in the base at the emitter-base junction
n_i	intrinsic carrier concentration
n_{ib}	intrinsic carrier concentration in the base region
n_{ic}	intrinsic carrier concentration in the collector region
n_{ie}	intrinsic carrier concentration in the emitter region
n_{po}	equilibrium concentration of electrons in p-region

N_a	concentration of acceptors
N_{ab}	acceptor concentration in the base
N_c	effective density of states in the conduction band
N_d	donor concentration
N_{dc}	donor concentration in the collector
N_{de}	donor concentration in the emitter
N_I	impurity concentration
N_s	density of surface states
N_{sub}	substrate doping density
N_v	effective density of states in the valence band
$\mathbf{p}$	momentum
p_b	hole concentration in the base
p_c	hole concentration in the collector
p_{co}	equilibrium hole concentration in the collector
p_e	hole concentration in the emitter
p_{eo}	equilibrium hole concentration in the emitter
$\mathbf{p}_i$	electron momentum in the i-th valley
p_{no}	equilibrium concentration of holes in the n region
q	elementary charge
$\mathbf{q}_i$	heat flow vector
Q_b	Gummel number
Q_d	charge in depletion layer
r	distance
R	recombination rate
R_b	bulk resistance
R_c	contact resistance
R_{chc}	channel sheet resistance
R_{end}	end resistance
R_p	resistance of interconnect wires
$\mathbf{R}$	space vector
R_{ch}	channel resistance
R_d	series drain resistance
R_g	series gate resistance
R_{sh}	drain-to-source shunt resistance
$\mathbf{R}_{k,l,m}$	coordinates of points belonging to the crystal lattice
R_s	series source resistance
s	spin

S	device cross-section
t	time
t_s	electron free flight time
T	temperature
T_e	effective electron temperature
T_{po}	Einstein temperature
u	sound velocity
u_k	Bloch amplitude
U	potential energy
U_m	amplitude of atomic displacement
U_M	atom velocity
U_R	generation-recombination rate
v_d	average drift velocity
V	crystal potential
V_A	Early voltage
V_{bi}	built-in voltage
V_{bisb}	built-in potential of channel-substrate junction
V_{eb}	emitter-base voltage
V_{cb}	collector-base voltage
V_{ch}	channel potential
V_{ds}	intrinsic drain-to-source voltage
V_g	gate potential
$V_{gs} = V_g - V_s$	intrinsic gate-to-source voltage
$V_{gd} = V_g - V_d$	intrinsic gate-to-drain voltage
$V_{po} = V_{bi} - V_T$	ideal pinch-off voltage
$V_{s,d}$	channel potential at source/drain ends
V_T	threshold voltage
V_{th}	thermal voltage
W	base width of bipolar junction transistor
W	gate width of field effect transistor
$W(\mathbf{k},\mathbf{k}')$	transition probability
W_i	kinetic energy density
x,y,z	space coordinates
x_c	width of the collector region
x_e	width of the emitter region
x_d	width of depletion layer
X_m	work function of metal

X_s	work function of semiconductor
Y	small-signal impedance
Z	atomic number
α	common-base current gain
α_T	base transport factor
β	common-emitter short-circuit current gain
γ	emitter injection efficiency of a bipolar junction transistor
δ_i	average length of electron travel in electric field
δ_{ox}	thickness of interfacial layer
ε_o	permittivity of vacuum
ε_s	static dielectric permittivity
ε_∞	high frequency dielectric constant
ϕ_b	barrier height
ϕ_o	neutral level
λ	scattering rate
λ_b	characteristic length of the exponential doping profile in the base
λ_e	intervalley scattering rate between equivalent valleys
μ	low field mobility
μ_a	ambipolar mobility
μ_n	electron mobility
μ_p	hole mobility
ρ	space charge density
σ_n	electron capture cross-section
σ_p	hole capture cross-section
τ	relaxation time
τ_{ac}	acoustic scattering relaxation time
τ_A	Auger recombination lifetime
τ_F	effective minority carrier lifetime for forward current
τ_{gen}	effective generation time
τ_{ii}	ionized impurity scattering relaxation time
τ_{ni}	neutral impurity scattering relaxation time
τ_{npo}	non-polar optical scattering relaxation time
τ_{pe}	piezoelectric scattering relaxation time
τ_{po}	polar optical scattering relaxation time

τ_{rec}	effective recombination time in emitter-base depletion region
τ_R	effective minority carrier lifetime for reverse current
ψ	wave function
ψ_a	antibonding orbital
ψ_b	bonding orbital
ψ_k	Bloch wave function
ω	frequency
ω_a	frequency of acoustic lattice vibrations
ω_o	frequency of optical lattice vibrations

Preface

The rapid development of semiconductor devices and integrated circuits has been accompanied by an enormous increase of information in the field of semiconductor physics and electronics. New ideas, new theories, new models, new devices, and new circuits have not only led to numerous practical applications but have also created opportunities for further and, perhaps, even more exciting developments. To work in this rapidly growing field is a challenge that attracts and inspires many researchers and students.

This book is intended to serve as a text for a three-quarter or two-semester sequence of courses on semiconductor devices for first year graduate students and qualified seniors. Some background in solid state physics and quantum mechanics may be helpful (but not required) for the students using this book. In addition to material typically found in textbooks on semiconductor devices, this book describes new important developments, such as amorphous silicon, compound semiconductor technologies, and novel heterostructure transistors. Theories and models presented in the book are implemented in microcomputer programs that make a "toolbox" for modeling and simulation of semiconductor devices. Appendices include information on semiconductor parameters. These device models and material parameters allow students to solve practical problems related to analysis, design, and characterization of different semiconductor devices. This book includes nearly 150 of such problems—from simple to advanced—with a detailed solution manual available for instructors. The book also gives many references that can serve as a material for further reading. These features should make this book useful for engineers and researchers working on semiconductor devices and also for students encountering this exciting field for the first time.

Chapter 1 starts from a brief discussion of semiconductor physics that introduces Schrödinger's equation, atomic states, chemical bonds, crystal structure, energy bands, semiconductor statistics, transport properties, and basic semiconductor equations. In addition, Chapter 1 includes more advanced topics, such as the Boltzmann transport equation, Monte Carlo simulation, and high electric field transport. These topics may be omitted from a typical course on semiconductor materials and devices, and the corresponding sections, 1-13, 1-14, and 1-15 (marked by asterisks in the Table of Contents) can be used as material for further reading.

Chapter 2 deals with the semiconductor junctions and contacts which are present in every semiconductor device. I have also included a section describing heterojunctions formed at the interfaces of dissimilar semiconductor materials. Because heterojunction devices have become extremely important for a variety of different applications—from light sources to ultrafast switching and microwave devices—the reader will find this section especially beneficial in understanding new devices emerging from research laboratories. Section 2-8-2 (marked by an asterisk in the Table of Contents) includes a more detailed analysis of avalanche breakdown than may be required for a typical course on physics of semiconductor devices. This Section can be used as a material for further study.

Chapter 3 describes bipolar junction transistors. In addition to the conventional material, I have included a description of the Gummel-Poon model and a section on heterojunction bipolar transistors. The Gummel-Poon model is required for realistic modeling of a bipolar junction transistor and is widely used in popular circuit simulators, such as SPICE developed at Berkeley. Heterojunction bipolar transistors have the potential to become one of the fastest solid-state technologies, both in the analog and digital worlds.

Chapter 4 treats field-effect transistors. The silicon field-effect transistor, considered in this chapter, is a work horse of modern electronics. Chapter 4 also describes compound semiconductor devices, such as gallium arsenide field-effect transistors, as well as amorphous silicon Thin Film Transistors (TFTs). Amorphous silicon TFTs have emerged as a very promising technology for driving flat screen displays and for applications in electronic copiers and printers.

Chapter 5 deals with photonic devices—solar cells, light-emitting diodes, semiconductor lasers, and integrated optoelectronic circuits. In particular, I discuss amorphous silicon solar-cell technology which has become the most practical solar cell technology because of its combination of high solar energy conversion efficiency with relatively low fabrication and material costs.

Chapter 6 covers microwave diodes—the most powerful solid-state sources of microwave energy. Some of these diodes utilize the negative differential resistance found in gallium arsenide in high electric fields. The physical mechanisms that lead to negative differential resistance are also discussed in Chapter 6.

Finally, Chapter 7 deals with new device concepts, such as ballistic (collisionless) transport in short semiconductor structures, so-called energy band engineering, Hot Electron Transistors, superlattice devices, resonant tunneling devices, etc.

As George Bernard Shaw once remarked, all professions are conspiracies against laity. With this quote in mind, I include a detailed list of symbols.

Appendices A1 through A27 contain material parameters of a dozen important semiconductor materials, important formulas, typical device parameters, the description of popular device and circuit simulation programs, and the Periodic Table.

If this book is used as a text for a three-quarter sequence of courses, the material may be divided as follows: The first course covers semiconductor physics, material properties, and p-n junctions. The second course deals with Schottky barriers, ohmic contacts, heterojunctions, and bipolar junction transistors. The third course is devoted to field effects transistors, photonic devices, and novel device concepts. For a two-semester sequence the first course may cover semiconductor physics, material properties, p-n junctions, Schottky barriers, ohmic contacts, heterojunctions, and principle of operation of bipolar junction transistors, with the remaining material taught in the second course.

SOFTWARE MANUAL TO ACCOMPANY THIS TEXT

Many excellent programs have been developed for simulation and modeling of semiconductor devices. Some are distributed to the public by Universities (most notably by Berkeley and Stanford) for a charge ranging from several hundred to several thousand dollars, some are proprietary and developed by big companies, such as IBM or AT&T, for their internal use, some are available from commercial software houses. Most of these programs run on mainframes or minicomputers and take time to learn. By their nature they are geared towards departmental or company use.

By contrast, numerous program modules accompanying this book are written in BASIC for an IBM PC or compatible microcomputer. They range from the programs describing the crystal vibrations to the programs simulating current-voltage characteristics of a Si MOSFET (Metal-Oxide-Semiconductor Field Effect Transistor) and are based on the concepts and models discussed in this book. BASIC was chosen because it is bundled with most IBM PC's or compatibles and is widely available to students. Hence, they have the option of modifying the source code or writing their own programs. Many of the figures that appear in the text have been generated using this software. The reader will find a caption, (see Program Description Section) beneath each of these figures.

All programs available as a supplement to this book are written as subroutines for a general purpose graphics program—PLOTF—(meaning "Plot a Function"). The manual that accompanies the $5\frac{1}{4}''$ diskette contains a **PLOTF User's**

Guide, as well as a **Program Description Section.** PLOTF allows a user to plot one or several functions for different values of parameters along with data points entered from a keyboard or stored on a diskette. Simulation programs used as subroutines of PLOTF may be easily changed by the user. All variables and parameters as well as the limitations and applicability of the models used in the programs are described in the Program Description Section.

The software and manual are available in a shrinkwrapped package at a nominal cost through Prentice Hall.

Professors may obtain a free examination copy of the **Software and Manual to accompany PHYSICS OF SEMICONDUCTOR DEVICES ISBN 0-13-666587-X** by contacting their local Prentice Hall representative. For those professors who would like to require purchase of this supplement for use in their courses, this package may be ordered in quantity simultaneously with the text, through the bookstore. Alternatively, should professors want to make the purchase of this software package optional to the student, there is a tear-out order card located in the back of the text that may be completed and returned to Prentice Hall.

My experience with this software in a classroom environment showed that a small number of students preferred to write their own programs using spreadsheets such as Lotus 1-2-3 or Microsoft Excell, etc. On the other hand many students have chosen to use this software in many different courses and even in their research.

The development of the PLOTF subroutines used in this book was partially supported by the WOKSAPE project at the University of Minnesota. This project (sponsored by the IBM corporation) supported the development of computer-aided instruction software.

In writing this book, I was inspired by a genuine interest in the subject and by enthusiasm from students who have taken my courses on semiconductor devices at the University of Minnesota. My graduate and undergraduate students, Young Byun, Jun-ho Baek, Byoung Moon, Michael Norman, Tim Bianchi, Bill Broker, John Leighton, Trung Dung, and others helped me with proofreading. I am especially grateful to Professor Michael Melloch of Purdue University and Professor Christine Maziar of the University of Texas at Austin, who read the manuscript and made many thoughtful comments and suggestions, and to Professor Yannis Tsividis of Columbia University who made many useful suggestions regarding MOSFET models. I am grateful to my editor, Ms. Elizabeth Kaster, for her patience and help. Finally, I would like to thank my wife, Dr. Paulina Shur, for her indispensable support and for her help in developing the software.

Michael Shur
Charlottesville
Virginia

1

Basic Semiconductor Physics

1-1. INTRODUCTION

Modern semiconductor electronics requires faster and faster devices operating at smaller and smaller amounts of power. This requires the scaling down of typical device sizes. Advances in fabrication technology have led to a shrinking of the minimum device feature size from about 20 microns in the early sixties to submicron dimensions in the late eighties. In shorter devices, electrons take less time to travel across the device, leading to higher speeds and operating frequencies. In addition, a smaller active device volume translates into lower operating power.

Today, most semiconductor devices are made of silicon. However, submicron devices made of compound semiconductors such as gallium arsenide successfully compete for applications in microwave and ultrafast digital circuits. As the dimensions of semiconductor devices shrink and more complicated and exotic compound semiconductor materials are used in electronic circuits, the physics involved in understanding device behavior becomes more complicated and more exciting. In novel device structures, the dimensions are so small that quantum effects become important or even dominant. At the same time, the number of charge carriers in a semiconductor device is very large and even a basic device represents a very complicated system. Modeling, or even qualitative understanding of, such a system presents a formidable challenge. In this sense, physics of semiconductor devices differs from more established "classical" engineering courses, such as electromagnetic fields or circuit theory. The material presented here is not as firmly established and is somewhat in a state of flux. Kirchhoff's current law will never change. However, even basic semiconductor equations that have been used for decades to analyze semiconductor devices have to be questioned and revised when applied to very small devices.

This chapter starts with a brief discussion of quantum mechanics and Schrödinger's equation, followed by the discussion of atomic states, chemical bonds, crystal structure, energy bands, semiconductor statistics, transport properties, and basic semiconductor equations. In addition, it includes more advanced topics, such as the Boltzmann transport equation, Monte Carlo simulation, and equations describing short semiconductor devices in high electric fields. These topics are usually omitted from a typical course on semiconductor materials and devices, and the corresponding sections, 1-13, 1-14, and 1-15 (marked by asterisks in the Table of Contents) can be used as material for further reading. However, this material is useful for a deeper understanding of advanced devices and device concepts (topics considered in Chapter 7).

The material in this chapter will prepare students for the analysis of semiconductor devices given in subsequent chapters of this book. At the same time, material constants and other information that should make this chapter useful as a reference for a semiconductor researcher or engineer is included. The chapter is written in such a way that no special background in quantum mechanics or solid-state physics is required, although a student lacking such a background may have to accept some of the material in this chapter on trust. Solving the numerous problems accompanying this chapter is a must for such a student. Many references are provided for further study for readers who may be interested in delving deeper into the subject.

1-2. QUANTUM MECHANICAL CONCEPTS AND ATOMIC STATES

Our understanding of semiconductor materials and devices is based on quantum mechanical concepts. The quantum theory started from the pioneering work of Max Planck, who explained the energy distribution of blackbody radiation. A blackbody is defined as an object that absorbs all incoming radiation at all frequencies. As was first established by Kirchhoff (1824–1887), the spectrum of the blackbody is independent of the material it is made of. A fairly good realization of a blackbody is a small hole into a cavity trapping and absorbing all incoming radiation (see Fig. 1-2-1). In 1900 Planck showed that the energy distribution of blackbody radiation can only be explained if one assumes that this radiation is emitted and absorbed as discrete energy quanta-photons. The energy of a photon, E, is proportional to the frequency of light, ω,

$$E = \hbar\omega \tag{1-2-1}$$

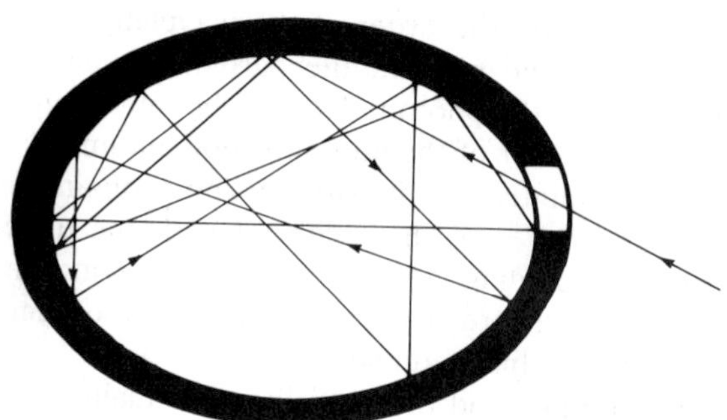

Fig. 1-2-1. Possible realization of a black body using a cavity with a small opening, trapping, and absorbing all incoming radiation. This figure schematically shows reflections of an incoming ray of light.

where $\hbar = h/2\pi$ is the reduced Planck constant and $h = 6.62 \times 10^{-34}$ J is the Planck constant.

Albert Einstein was the first to show that each photon has a momentum

$$\mathrm{p} = E/\mathrm{c} \tag{1-2-2}$$

where **c** is the velocity of light. The direction of this momentum coincides with the direction of the propagation of the light wave.

Hence, light has not only wavelike but also particlelike properties. The most important quantum mechanical concept states that all matter, including electrons, behaves like both particles and waves. Particles can be characterized by a *wave function*, $\psi(x,y,z)$, such that the probability, P, of finding a particle in an incremental volume $dx\ dy\ dz$ is equal to $|\psi(x,y,z)|^2\ dx\ dy\ dz$. Hence, $\psi(x,y,z)$ is an amplitude of the probability density of finding the particle in a given point of space. For a particle in free space, the wave function is a plane wave; i.e., $\psi(x,y,z)$ is proportional to exp $[i(k_x x + k_y y + k_z z)]$, where k_x, k_y, and k_z are components of the wave vector, **k**, related to the wavelength $\lambda = 2\pi/|k|$. In 1924 de Broglie predicted that the wavelength associated with a free particle (called the *de Broglie wavelength*) is given by

$$\lambda = h/p \tag{1-2-3}$$

Here p is momentum ($p = mv$, where m is the mass of a particle and v is the particle velocity). Wavelike but also particlelike duality is reflected in the *uncertainty principle*, discovered by Werner Heisenberg in 1927. Heisenberg, who in 1932 received the Nobel Prize for his work, stated that the product of the uncertainties of a particle's momentum and of its coordinates, Δp and Δx, respectively, is always larger than $\hbar/2$:

$$\Delta p \Delta x > \hbar/2$$

This principle can be understood if we consider attempts to determine the position of a small particle, such as an electron, by shining light at it. For us to "see" this particle, it must reflect at least one photon. Even then we will know the position of the particle within an uncertainty on the order of the wavelength of such a photon, λ. During the reflection of the photon, the particle momentum may change because of its interaction with the photon. Hence, the uncertainty of the particle's momentum will be on the order of the photon's momentum, $\hbar k = 2\pi\hbar/\lambda$. Hence, the product of the uncertainties of the particle's position and momentum will be on the order of $\lambda(2\pi\hbar/\lambda) \sim \hbar$, in agreement with the uncertainty principle. Many great scientists, including Albert Einstein, have attempted to devise some other measurement procedure that can circumvent the uncertainty principle. However, none have been able to do so.

It is very difficult to fully comprehend the particle-wave duality of nature: particles having wavelike properties or waves having particlelike properties. This duality defies conventional images. But although we cannot quite visualize this particle-wave duality, numerous experimental data have proven to us that it exists. Here Emily Dickinson's poem "Chartless" comes to mind:

> I never saw a moor,
> I never saw the sea,
> Yet know I how a heather looks
> And what a wave must be.
> I never spoke with God,
> Nor visited in Heaven,
> Yet certain am I of the spot,
> As if the chart were given.

In the same way, we know that quantum mechanical duality exists and can be used in the interpretation of physical phenomena.

In a semiconductor at room temperature a typical velocity of random electronic motion, v_T, is on the order of 10^5 m/s. The mass of a free electron is $m_e = 9.11 \times 10^{-31}$ kg, so that $p = m_e v_T \approx 9 \times 10^{-26}$ kgm/s, $k = p/\hbar \approx 8.6 \times 10^8$ m^{-1}, and $\lambda = 2\pi/k \approx 7.3 \times 10^{-9}$ m $= 73$ Å. Hence, λ may become comparable to the dimensions of small semiconductor devices (as small nowadays as 200 Å in laboratories and 800 Å in advanced practical devices). In principle, even macroscopic objects have an associated de Broglie wavelength. However, it is so small (for example, $\lambda \approx 10^{-38}$ m for a 2000 kg car moving with the velocity $v = 55$ miles per hour) that it does not have any practical importance.

An electron wave function satisfies the Schrödinger wave equation,

$$\left[-\frac{\hbar^2 \nabla^2}{2m_e} + U(\mathbf{r})\right]\Phi = i\hbar\frac{\partial \Phi}{\partial t} \tag{1-2-4}$$

where $\hbar = h/2\pi$ is the reduced Planck constant, m_e is the electron mass, $\mathbf{r}$ is the space vector, and U is the potential energy. We can search for the solution of this equation in the following form:

$$\Phi(\mathbf{r}, t) = \psi(\mathbf{r})f(t) \tag{1-2-5}$$

Substituting eq. (1-2-5) into eq. (1-2-4) and dividing by $\psi(\mathbf{r})\, f(t)$ we obtain

$$-\frac{\hbar^2\, \nabla^2\, \psi}{2m_e\, \psi} + U(\mathbf{r}) = i\hbar\, \frac{\partial f}{f\, \partial t} \tag{1-2-6}$$

The left-hand side of this equation depends only on $\mathbf{r}$, while the right-hand side of this equation depends only on t. Hence, they both must be equal to a constant independent of either $\mathbf{r}$ or t. Denoting this constant as E, we find

$$-\frac{\hbar^2\, \nabla^2\, \psi}{2m_e} + U(\mathbf{r})\, \psi = E\, \psi \tag{1-2-7}$$

and

$$i\hbar\, \frac{\partial f}{\partial t} = Ef \tag{1-2-8}$$

The solution of eq. (1-2-8) is given by

$$f = \exp(-iEt/\hbar) \tag{1-2-9}$$

Equation (1-2-7) is called the *time-independent Schrödinger equation*. We consider the solution of this equation for three important cases: free space, an infinitely deep one-dimensional potential well, and for the simplest atomic system—a hydrogen atom. The hydrogen atom is especially important for us, as many properties of atomic systems may be illustrated using this simplest atom as an example.

In a free space the potential $U = 0$ and the time-independent Schrödinger equation becomes

$$-\frac{\hbar^2\, \nabla^2\, \psi}{2m_e} = E\psi \tag{1-2-10}$$

Let us consider the electron motion only in one direction, x, so that $\nabla^2 = d^2/dx^2$. The solution of this second-order linear differential equation is given by

$$\psi = A\, \exp(ikx) + B\, \exp(-ikx) \tag{1-2-11}$$

where the wave vector, k, is expressed as

$$\mathrm{k} = (2m_e E/\hbar^2)^{1/2} \tag{1-2-12}$$

The wavelength, $\lambda = 2\pi/k$, and the electron momentum, $p = m_e v$, are linked by the de Broglie relation (eq. (1-2-3)). Using eq. (1-2-3) we obtain

$$p = \hbar k \tag{1-2-13}$$

From this equation and eq.(1-2-12) we find

$$E = p^2/2m_e = m_e v^2/2 \tag{1-2-14}$$

Hence, the separation constant E turns out to be the electron energy.

Substituting eq. (1-2-9) and eq. (1-2-11) into eq. (1-2-5) we obtain

$$\Phi = A \exp[i(kx - \omega t)] + B \exp[i(-kx - \omega t)] \tag{1-2-15}$$

where we used eq. (1-2-1) to express the electron energy in terms of the corresponding electron frequency

$$\omega = E/\hbar \tag{1-2-16}$$

Hence, as was previously mentioned, the free electron motion is described by a plane wave. The frequency of this wave is proportional to the electron energy.

Let us now consider an infinitely deep one-dimensional potential well. In this case

$$\begin{aligned} U &= \infty \qquad \text{for } x < 0 \\ U &= 0 \qquad \text{for } 0 < x < a \\ U &= \infty \qquad \text{for } x > a \end{aligned} \tag{1-2-17}$$

where a is the width of the potential well.

Within the potential well the solution for ψ is still given by eq. (1-2-11). However, the solution in the regions, where $U = \infty$, is $\psi = 0$. Hence, the boundary conditions for the potential well are

$$\begin{aligned} \psi &= 0 \qquad \text{for } x = 0 \\ \psi &= 0 \qquad \text{for } x = a \end{aligned} \tag{1-2-18}$$

We search for the solution in the form that coincides with the real part of the wave function given by eq.(1-12-11), which can be rewritten as

$$\psi = A_1 \sin(kx + \delta) \tag{1-2-19}$$

where $k = (2m_e E)^{1/2}/\hbar$. The condition $\psi = 0$ for $x = 0$ requires $\delta = 0$, whereas the condition $\psi = 0$ for $x = a$ leads to the requirement $ka = \pi n$, i.e.

$$E = E_n = \frac{\pi^2 \hbar^2}{2m_e a^2} n^2 \tag{1-2-20}$$

where $n = 1, 2, 3, 4, 5, \ldots$ (any positive integer number). As ψ is the probability amplitude—i.e., $|\psi|^2$ is the probability density—the normalization constant, A_1, can be found from

$$\int_0^a |A_1 \sin[(\pi n/a)x]|^2 \, dx = 1 \tag{1-2-21}$$

This condition means that the particle is localized within the potential well, so that the probability of finding the particle in the potential well is equal to unity. Using eq. (1-2-21), we obtain $A_1 = (2/a)^{1/2}$.

The most interesting result is that the energy of a particle in the potential well can have only discrete (quantized) values. These values for a 100 Å potential

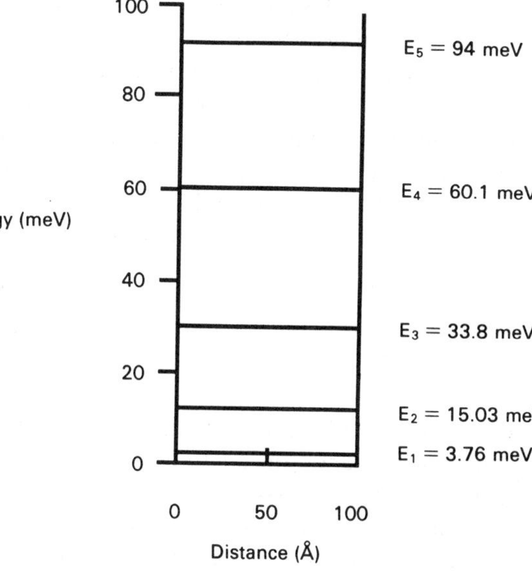

Fig. 1-2-2. Lowest energy levels for an infinitely deep potential well. The energy levels are calculated using eq. (1-2-20) for $m_e = 9.11 \times 10^{-31}$ kg and $a = 100$ Å.

well are shown in Fig. 1-2-2. A similar analysis can be performed for two-dimensional and three-dimensional potential wells (see Problem 1-2-1).

Let us now consider a hydrogen atom that has only one electron and a positive nucleus. The energy levels and wave functions for an electron in a hydrogen atom are found from the solution of the time-independent Schrödinger equation (see eq. (1-2-7)) with the Coulombic potential

$$U(\mathbf{r}) = -\frac{q^2}{4\pi\varepsilon_o|\mathbf{r}|} \tag{1-2-22}$$

Here q is the electronic charge and ε_o is the permittivity of vacuum.

For this potential, the solutions of eq. (1-2-7) that approach zero at large r are found to exist when and only when the electron energy E is given by

$$E_n = -\frac{E_\mathrm{B}}{n^2} \tag{1-2-23}$$

where

$$n = 1, 2, 3, 4, \ldots \tag{1-2-24}$$

is the principal quantum number,

$$E_B = \frac{q^2}{8\pi\varepsilon_o a_\mathrm{B}} \tag{1-2-25}$$

is called the Bohr energy ($E_\mathrm{B} = 13.6$ eV $= 2.18 \times 10^{-18}$ J), and

$$a_B = \frac{4\pi\varepsilon_o\hbar^2}{m_e q^2} \tag{1-2-26}$$

is called the Bohr radius (a_B = 0.52917 Å). The coulomb potential and the first several energy levels for the hydrogen atom are shown in Fig. 1-2-3.

Besides the principal quantum number, n, there are three more quantum numbers that characterize the electronic states in a hydrogen atom: the orbital quantum number, l, the magnetic quantum number, m, and the spin, S. The angular dependence of ψ is determined by the orbital quantum number, l, and the magnetic quantum number, m. The analysis of the Schrödinger equation for the hydrogen atom shows that these numbers can accept the following values:

$$l = 0, 1, 2, \ldots, n - 1 \tag{1-2-27}$$

$$m = -l, -l + 1, \ldots, l - 1, l \tag{1-2-28}$$

$$S = \pm\tfrac{1}{2} \tag{1-2-29}$$

Each electronic state in a hydrogen atom can be characterized by this set of four quantum numbers. The wave function depends on all four quantum numbers. However, the electron energy depends only on n (see eq. (1-2-23)).

The simplest possible approximation in the treatment of many-electron atoms is to assume that the wave functions are the same as those for the hydrogen atom, with the modification that the nuclear charge is not q but Zq, where Z is the atomic number equal to the number of protons in the nucleus. According to the Pauli exclusion principle, no two electrons can occupy the same state, that is, have the same four quantum numbers (including the spin). As two values of spin are possible ($S = \pm\frac{1}{2}$), two electrons may have the same n, l, and m numbers. The electronic structure of complex atoms may be understood in terms of filling up

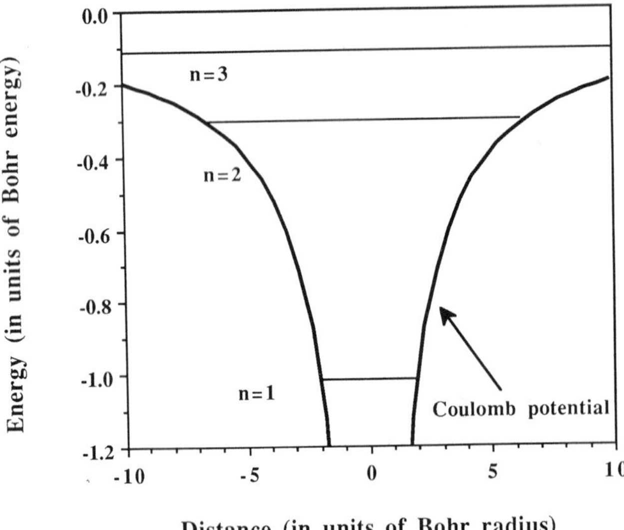

Fig. 1-2-3. Coulomb potential and the first several energy levels of the hydrogen atom (in units of Bohr energy) vs. distance (in units of Bohr radius).

higher and higher energy levels, taking into account that the chemical properties are determined mainly by the outermost (valence) electrons.

The average distance of an electron from the nucleus depends on n. The innermost "shell" of electrons consists of those in the $n=1$ states, the next shell corresponds to the $n=2$ states, etc. These shells are sometimes identified by capital letters starting with K:

$$n = 1, 2, 3, 4, \ldots$$

$$\text{shell} = \text{K, L, M, N}, \ldots$$

The energy levels in many-electron atoms depend not only on n (as is the case for a hydrogen atom) but also on l (due to the screening of the nuclear potential by the electrons). Therefore, electrons are subdivided into "subshells" according to the value of l. These subshells are often labeled by lowercase letters:

$$l = 0, 1, 2, 3, 4, 5, \ldots$$

$$\text{subshell} = \text{s, p, d, f, g, h}, \ldots$$

For a given l there are $2l+1$ values of m, each corresponding to a distinct state, and for each of these there are two possible values of the spin quantum number. The levels in different subshells are given in Table 1-2-1.

For silicon (Si), germanium (Ge), gallium (Ga), and arsenic (As)—elements that are particularly important for semiconductor electronics—the valence configurations are given by

Si (atomic number 14):	core + $3s^2 3p^2$
Ge (atomic number 32):	core + $4s^2 4p^2$
Ga (atomic number 31):	core + $4s^2 4p^1$
As (atomic number 33):	core + $4s^2 4p^3$

TABLE 1-2-1. SHELLS AND SUBSHELLS

Shell	n l	m	Spectroscopic notation	Number of levels
K	1 0	0	1s	2
L	2 0	0	2s	2
	2 1	$0, \pm 1$	2p	6
M	3 0	0	3s	2
	3 1	$0, \pm 1$	3p	6
	3 2	$0, \pm 1, \pm 2$	3d	10
N	4 0	0	4s	2
	4 1	$0, \pm 1$	4p	6
	4 2	$0, \pm 1, \pm 2$	4d	10
	4 3	$0, \pm 1, \pm 2, \pm 3$	4f	14

Here "core" means all inner shells; the first digit in the second term is the principal quantum number of the outer (valence) shell—3 for Si, 4 for Ge, Ga, and As. Superscripts denote the number of electrons in the subshells. As the valence configurations of Si and Ge are very similar (in both cases we have two s electrons and two p electrons), we can expect that their physical and chemical properties should be similar too. Indeed, both elements are semiconductors. If we combine Ga and As into a GaAs compound, each atom, on the average, will also have two s electrons and two p electrons. GaAs is another important semiconductor material. In a similar way, we can form many semiconductor compounds by combining other elements of column III of the periodic table (having three valence electrons, two s electrons, and one p electron) with elements from column V (having 5 valence electrons, as, for example, As—two s electrons and three p electrons). Some examples of these III-V compounds are GaAs, InAs, InP, AlAs, GaP, AlP, InSb, BN, AlN, GaSb, and GaN. Elements from the second and sixth columns of the periodic table (having two and six valence electrons, respectively) can also be combined to form II-VI compound semiconductors. CdS, ZnS, CdTe, and CdSe are examples of II-VI compound semiconductors. Moreover, many III-V and II-VI compound semiconductors may form solid-state solutions, such as $Al_xGa_{1-x}As$, where x is a molar fraction of Al. (Such materials are called *ternary compounds*.) By varying x from 0 to 1 one can change properties of $Al_xGa_{1-x}As$ from those of GaAs to those of AlAs. This particular compound is especially important because the lattice constants of GaAs and AlAs are very close (5.6533 Å and 5.6605 Å at 300 K for GaAs and AlAs, respectively). That is why AlGaAs can be easily grown on GaAs, forming a heterostructure, i.e., a structure including two different semiconductor materials in intimate contact. Heterostructures have found numerous and increasingly important applications in novel semiconductor devices and circuits, which are considered in Chapters 3, 4, 5, and 7. Other important examples of ternary compounds include $In_xGa_{1-x}As$, $GaIn_xP_{1-x}$, and $Al_xIn_{1-x}As$. Moreover, *quarternary compounds*, such as $In_xGa_{1-x}As_yP_{1-y}$ can also be formed. This "material engineering" approach allows us to design semiconductor materials with desired properties. Many of these new materials, however, demand further study and refinement before they find applications in the new and exciting high-performance devices of the future.

1-3. CHEMICAL BONDS

The most stable electron configurations correspond to complete s and p subshells of an outer (valence) shell. Indeed, in the periodic system, elements having completely filled s and p valence subshells correspond to inert gases. For example,

$$\text{Ne} \quad 1s^2 2s^2 2p^6$$

$$\text{Ar} \quad 1s^2 2s^2 2p^6 3s^2 3p^6$$

$$\text{Kr} \quad \text{core} + 4s^2 4p^6$$

$$\text{Xe} \quad \text{core} + 5s^2 5p^6$$

$$\text{Rn} \quad \text{core} + 6s^2 6p^6$$

Bonds in a crystal are formed in such a way that neighboring atoms share their valence electrons, satisfying (on the average) the valence requirements, i.e., having completed s and p subshells of the valence shell. For example, in Si or Ge there are four valence electrons per atom:

$$\text{Si} \quad \text{core} + 3s^2 3p^2$$

$$\text{Ge} \quad \text{core} + 4s^2 4p^2$$

In a silicon crystal, a silicon atom forms four bonds with four other silicon atoms (four nearest neighbors) and shares two valence electrons with each of them. Hence, it shares eight valence electrons with all four nearest neighbors. In other words, in a silicon or germanium crystal, each atom is tetrahedrally coordinated (see Fig. 1-3-1) in order to share eight electrons (two electrons per bond), corresponding to complete p and s valence subshells. These bonds can be presented as a linear combination of atomic wave functions (atomic orbitals)—s- and p-type atomic orbitals. One s orbital and three p orbitals, having two electrons with opposite spins each, form the outer subshell of a silicon atom. The s-type atomic orbital has a spherical symmetry. The angular dependence of the three p-type atomic orbitals may be visualized as linear combinations of three perpendicular lobes, as shown in Fig. 1-3-2. Four bonds can be formed using linear combinations of these atomic orbitals for each atom:

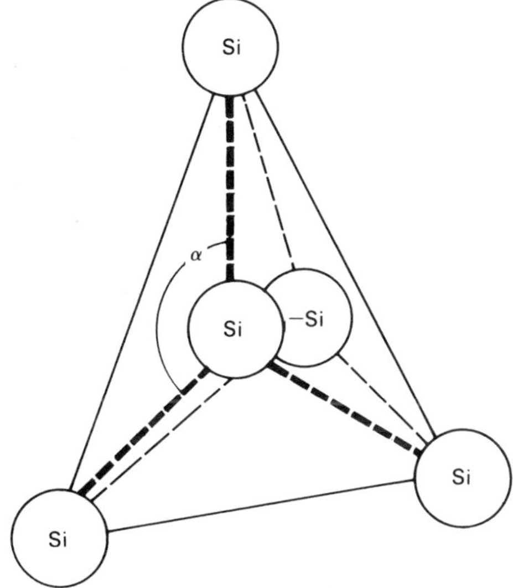

Fig. 1-3-1. Tetrahedral atomic configuration of silicon atoms. α is the angle between the bonds.

$$\psi_1 = \tfrac{1}{2}(|s\rangle + |p_x\rangle + |p_y\rangle + |p_z\rangle) \tag{1-3-1}$$

$$\psi_2 = \tfrac{1}{2}(|s\rangle + |p_x\rangle - |p_y\rangle - |p_z\rangle) \tag{1-3-2}$$

$$\psi_3 = \tfrac{1}{2}(|s\rangle - |p_x\rangle + |p_y\rangle - |p_z\rangle) \tag{1-3-3}$$

$$\psi_4 = \tfrac{1}{2}(|s\rangle - |p_x\rangle - |p_y\rangle + |p_z\rangle) \tag{1-3-4}$$

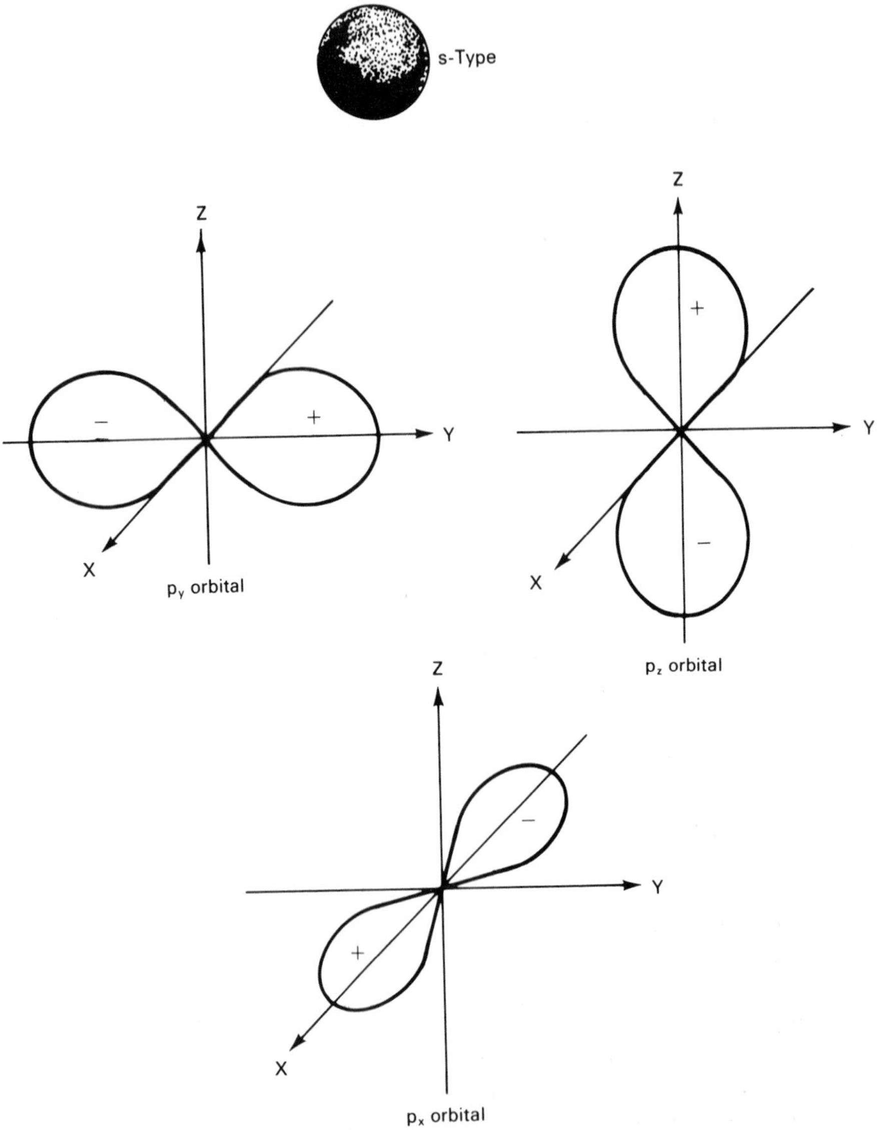

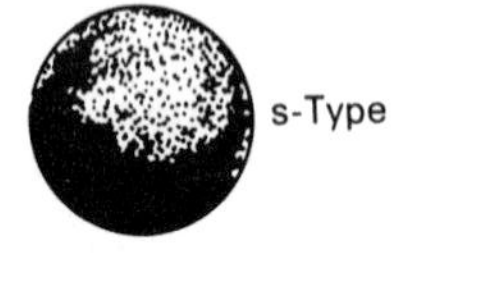

p_y orbital

p_z orbital

p_x orbital

Fig. 1-3-2. Schematic representation of the angular dependence of *s*-type and *p*-type atomic orbitals.

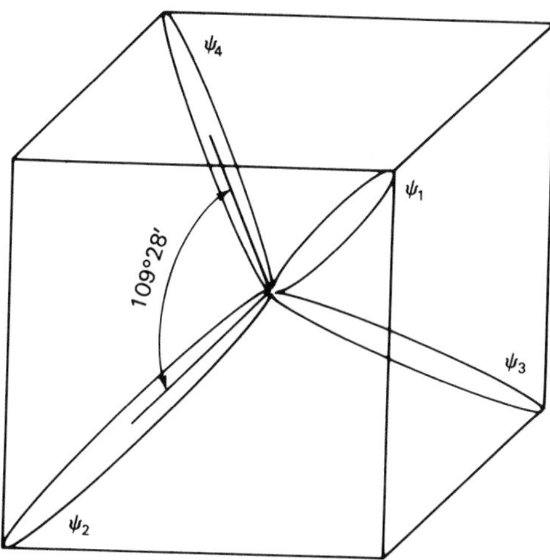

Fig. 1-3-3. Hybridized orbitals ψ_1, ψ_2, ψ_3, and ψ_4.

where $|s>$, $|p_x>$, $|p_y>$, and $|p_z>$ denote the atomic orbitals. These functions are called *hybridized orbitals*. They have maximum values in the directions shown in Fig. 1-3-3. Comparison of Fig. 1-3-2 with Fig. 1-3-3 clearly illustrates the significance of the hybridized orbitals in the bond formation.

Hybridized orbitals do not correspond to the ground state of an atom; atomic orbitals do. Their energy is of the order of 5 to 10 eV higher. This energy is recovered in a crystal, as a result of the interaction between the atoms. The resulting bonding energy level is lower than the ground atomic state by an amount of the order of 1 eV/valence electron. This gain in energy, caused by the formation of a crystal, is called cohesion energy.

In a solid, hybridized (or directed) orbitals are combined into bonding, antibonding, and nonbonding orbitals forming covalent bonds. A bonding orbital, ψ_b, consists of two directed orbitals associated with the nearest atoms and combined in phase. Hence, ψ_b is large in the bonding region between the atoms, as shown in Fig. 1-3-4a. An antibonding orbital, ψ_a, is similar to the bonding orbital, but the phase between the directed orbitals is reversed. Hence, ψ_a has a node in the bonding region (see Fig. 1-3-4b). A nonbonding orbital is centered on one atom and does not have a directional character.

The bonding state has the lowest energy because the absolute value of the wave function is larger (compared with the antibonding orbital) in the center of the bond, where the potential is lower owing to the overlap of the coulomb potential between the nearest ion cores (see Fig. 1-3-4c). For the antibonding state, the wave function is zero at the center of the bond. Hence, there is no energy gain owing to the potential lowering caused by the overlap of the coulomb potential, and the antibonding state has the largest energy. Usually the bonding and nonbonding states are occupied and the antibonding states are empty.

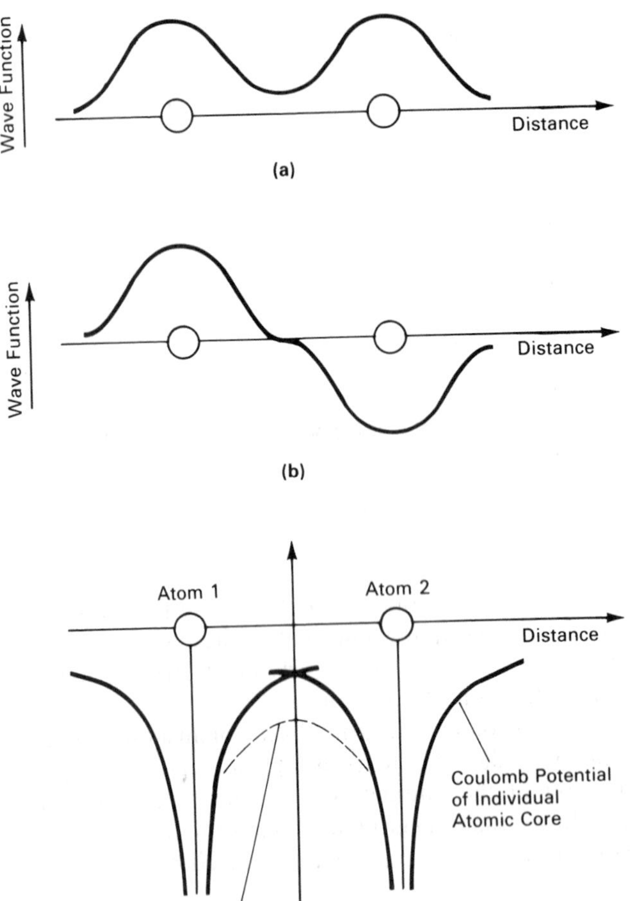

Fig. 1-3-4. (a) Bonding, (b) antibonding orbitals, and (c) decrease of potential in the middle of the bond due to overlap of the ion core coulombic potentials of the nearest atoms.

If all atoms in a crystal are identical (as in Si or Ge, for example), the bonding orbital between the nearest neighbors 1 and 2 has the following form

$$\psi_b \sim \psi_1 + \psi_2 \tag{1-3-5}$$

where ψ_1 and ψ_2 are hybridized orbitals. Because of symmetry, electrons shared in a bond must spend, on the average, the same time on each atom. This corresponds to a symmetrical distribution of the electron density and to a purely homopolar (covalent) bond. In a compound semiconductor, such as GaAs, the bonding orbital includes the weighting factor λ,

$$\psi_b \sim \psi_1 + \lambda\psi_2 \tag{1-3-6}$$

TABLE 1-3-1. ENERGIES E_h, C, AND E_{ba} AND IONICITY f_i IN BINARY TETRAHEDRALLY COORDINATED CRYSTALS (from Phillips 1973)

Crystal	E_h (eV)	C (eV)	E_{ba} (eV)	f_i
C	13.5	0	13.5	0
Si	4.77	0	4.77	0
Ge	4.31	0	4.31	0
Sn	3.06	0	3.06	0
BAs	6.55	0.38	6.56	0.002
BP	7.44	0.68	7.47	0.006
BeTe	4.54	2.05	4.98	0.169
SiC	8.27	3.85	9.12	0.177
AlSb	3.53	2.07	4.14	0.250
BN	13.1	7.71	15.2	0.256
GaSb	3.55	2.10	4.12	0.261
BeSe	5.65	3.36	6.57	0.261
AlAs	4.38	2.67	5.14	0.274
BeS	6.31	3.99	7.47	0.286
AlP	4.72	3.14	5.67	0.307
GaAs	4.32	2.90	5.20	0.310
InSb	3.08	2.10	3.73	0.321
GaP	4.73	3.30	5.75	0.327
InAs	3.67	2.74	4.58	0.357
InP	3.93	3.34	5.16	0.421
AlN	8.17	7.30	11.0	0.449
GaN	7.64	7.64	10.8	0.500
MgTe	3.20	3.58	4.80	0.554
InN	5.93	6.78	8.99	0.578
BeO	11.5	13.9	18.0	0.602
ZnTe	3.59	4.48	5.74	0.609
ZnO	7.33	9.30	11.8	0.616
ZnS	4.82	6.20	7.85	0.623
ZnSe	4.29	5.60	7.05	0.630
HgTe	2.92	4.0	5.0	0.65
HgSe	3.43	5.0	6.1	0.68
CdS	3.97	5.90	7.11	0.685
CuI	3.66	5.50	6.61	0.692
CdSe	3.61	5.50	6.58	0.699
CdTe	3.08	4.90	5.79	0.717
CuBr	4.14	6.90	8.05	0.735
CuCl	4.83	8.30	9.60	0.746
CuF	8.73	15.8	18.1	0.766
AgI	3.09	5.70	6.48	0.770
MgS	3.71	7.10	8.01	0.786
MgSe	3.31	6.41	7.22	0.790
HgS	3.76	7.3	8.3	0.79

which may be interpreted in terms of fractions of time f_1 and f_2 the bonding electron spends on the first and the second atom, respectively:

$$f_1 = \frac{1}{1 + \lambda^2} \quad \text{and} \quad f_2 = \frac{\lambda^2}{1 + \lambda^2} \tag{1-3-7}$$

Electrons in the bonding states spend a greater fraction of time on the anions (negatively charged atoms). This situation corresponds to a partially heteropolar (partially ionic) bond.

In covalent crystals the difference in the energies of bonding and antibonding states depends on the bond length. In ionic compounds there is also a contribution that depends on the difference of atomic potentials. According to Phillips (1973), this can be taken into account by presenting the crystal potential as the sum of the symmetric potential (as in a covalent crystal) and antisymmetric potential (as in a purely ionic crystal) and using the following empirical equation:

$$E_{ba}^2 = E_h^2 + C^2 \tag{1-3-8}$$

Here E_{ba} is the energy gap between the bonding and the antibonding state, E_h is the energy gap produced by the covalent (symmetrical) part of the crystal potential, and C is the magnitude of the energy gap produced by the antisymmetric (ionic) potential. The fractional ionicity and covalency of the bond are then defined as

$$f_i = C^2/E_{ba}^2 \tag{1-3-9}$$

$$f_c = E_h^2/E_{ba}^2 \tag{1-3-10}$$

Values of E_h, C, E_{ba}, and f_i for binary tetrahedrally coordinated crystals are given in Table 1-3-1 (from Phillips 1973) as shown on page 15.

Many properties of semiconductors, such as their crystalline structure, cohesive energy, elastic constants, and vibrational spectra, are strongly dependent on the degree of the ionicity of the bonds (see Problem 1-3-1).

1-4. SOLID–STATE STRUCTURE

There are three different types of solids: crystalline, polycrystalline, and amorphous (see Fig. 1-4-1). In a crystal, atoms are arranged into a regular, periodically repeated structure that extends throughout the whole sample. The atoms are said to have *long-range order*. Slices of a silicon crystal—round crystalline silicon wafers, as large as 6 inches in diameter or more—are used for integrated circuit fabrication. Polycrystalline material is composed of many small crystals, or grains, of somewhat irregular size. These small crystals are separated by grain boundaries. Doped polycrystalline silicon (polysilicon) is widely used as a gate material in silicon transistors. In an amorphous solid, the interatomic distance is more or less the same as in the crystalline material of the same substance. The bond angles are also similar, but a long-range order is absent. Amorphous mate-

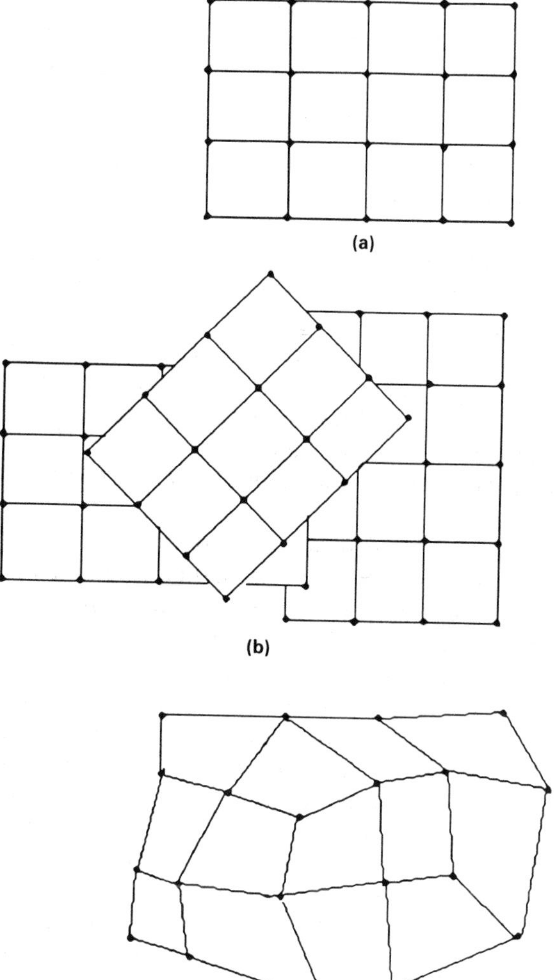

Fig. 1-4-1. Schematic representation of atomic structure in single crystal (a), polycrystalline (b), and amorphous material (c).

rial is said to have *short-range order.* Amorphous silicon (more precisely, an alloy of amorphous silicon with hydrogen and other similar amorphous alloys) may be inexpensively deposited as a thin film over very large areas (as large as 2 feet by 4 feet or even larger). Therefore, this material has found important applications in photovoltaic technology and in large-area integrated circuits used in flat displays, printers, copiers, scanners, and imagers.

A crystal structure can be visualized using the concept of the *crystal lattice,* which is a three-dimensional array of periodically located ideal points in space.

This array can be defined by the primitive basis vectors **a, b,** and **c,** which are three independent shortest vectors connecting lattice sites. In other words, the coordinates of all points belonging to the crystal lattice are given by vectors

$$\mathbf{R}_{k,l,m} = k\mathbf{a} + l\mathbf{b} + m\mathbf{c} \tag{1-4-1}$$

where k, l, and m are integers. A crystal structure is formed by placing an identical group of atoms (called a basis) into the same position with respect to each point in a crystal lattice.

Primitive vectors form a parallelepiped, which is called a *primitive cell.* In Fig. 1-4-2 three cubic lattices—simple cubic, body-centered cubic, and face-centered cubic—are shown with their corresponding primitive vectors. As can be seen from this figure, it is more convenient to represent these cubic lattices by larger cubic cells (unit cells) than by their primitive cells. Gallium (Ga) crystallizes in a simple cubic structure that can be visualized by putting a gallium atom at each point of a simple cubic lattice. Sodium (Na) and tungsten (W) are examples of crystals with body-centered cubic lattices. Aluminum (Al) and gold (Au) have a face-centered cubic structure.

Another way to characterize a crystal structure is to consider the *Wigner-Seitz primitive cell.* This cell is formed by bisecting all lattice vectors emanating from a lattice site by planes and choosing the smallest volume formed by these intersecting planes. Wigner-Seitz primitive cells for simple cubic, face-centered cubic, and body-centered cubic lattices are shown in Fig. 1-4-3.

The most important semiconductor material is silicon (Si). As was discussed in Section 1-3, each silicon atom forms four bonds with its nearest neighbors. This corresponds to the *tetrahedral configuration*, in which each silicon atom is located at the center of the tetrahedron formed by four silicon atoms (see Fig. 1-3-1). In a silicon crystal this tetrahedral bond configuration is repeated, forming the same crystal structure as in a diamond crystal (shown in Fig. 1-4-4). This structure (called the diamond structure) is formed by two interpenetrating face-centered cubic sublattices of atoms, shifted with respect to each other by one fourth of the body diagonal. Perhaps we should call this structure the silicon structure instead, because we can argue that silicon is much more important for our civilization than diamond. (Another important semiconductor—germanium [Ge]—has exactly the same crystal configuration as well.)

Si and Ge are *elemental semiconductors. Compound semiconductors* containing more than one element (such as GaAs, InP, and ZnSe) have also found numerous applications. As we discussed in Section 1-2, most of the compound semiconductors "simulate" silicon by having the same average number of valence electrons per atom. Therefore, they may have a chemical structure $A_{III}B_{V}$, where A_{III} represents an element from the third column of the periodic table (three valence electrons) and B_{V} represents an element from the fifth column of the periodic table (five valence electrons). Examples are GaAs, InP, InSb, GaSb, GaP, and AlAs. Another chemical formula for compound semiconductors is $A_{II}B_{VI}$. Again, the total number of valance electrons is eight. Examples are ZnSe, ZnTe, and ZnS.

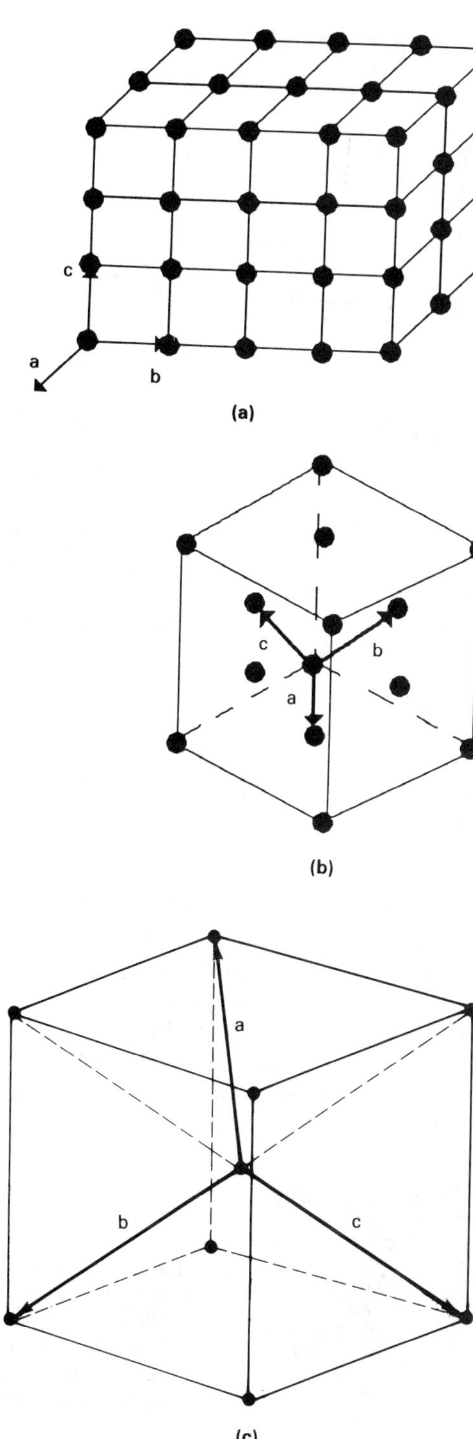

Fig. 1-4-2. Crystal cubic lattices and primitive vectors: (a) simple cubic lattice, the unit cell is a simple cube with atoms located at the corners; (b) face-centered cubic lattice, the unit cell is a face-centered cube with atoms located at the corners of the cube and in the centers of the cube faces; and (c) body-centered cubic lattice, the unit cell is a body-centered cube with atoms located at the corners of the cube and in the center of the cube.

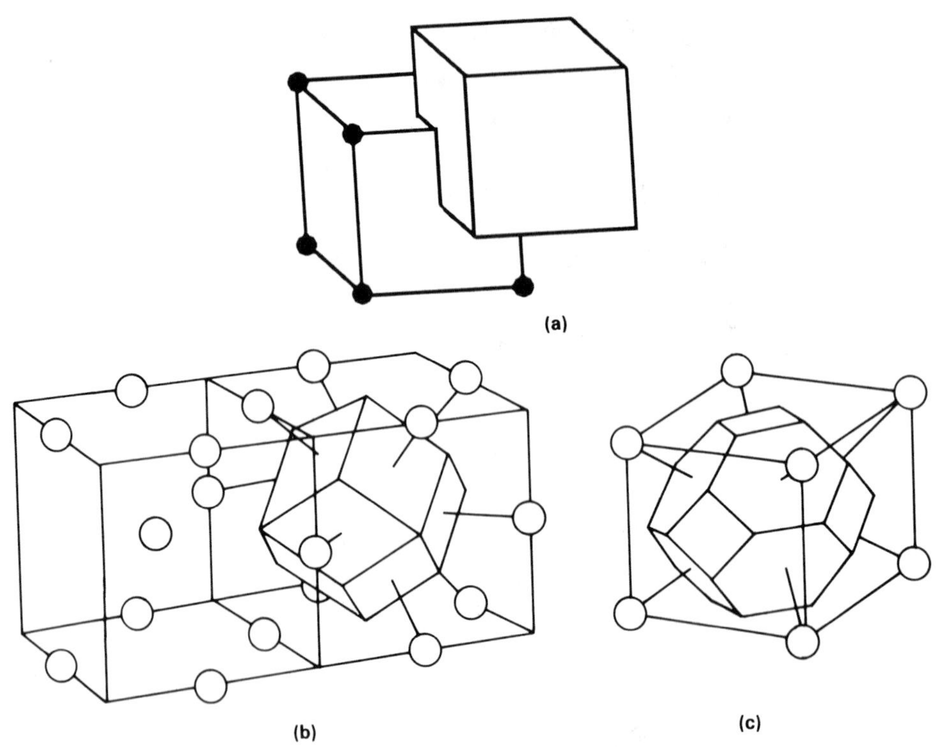

Fig. 1-4-3. Wigner-Seitz primitive cells for (a) simple, (b) face-centered, and (c) body-centered cubic lattices.

In compound semiconductors the chemical bond between the nearest neighbors is partially heteropolar. Nevertheless, in many such compounds this bond is still more or less covalent, leading to the tetrahedral bond configuration similar to that in a silicon crystal (see Fig. 1-4-5). As a consequence, most III-V compounds crystallize in the *zinc blende crystal structure* (see Fig. 1-4-6), which is very similar to the diamond structure. The primitive cell of the zinc blende structure contains two atoms A and B that are repeated in space, with each species forming

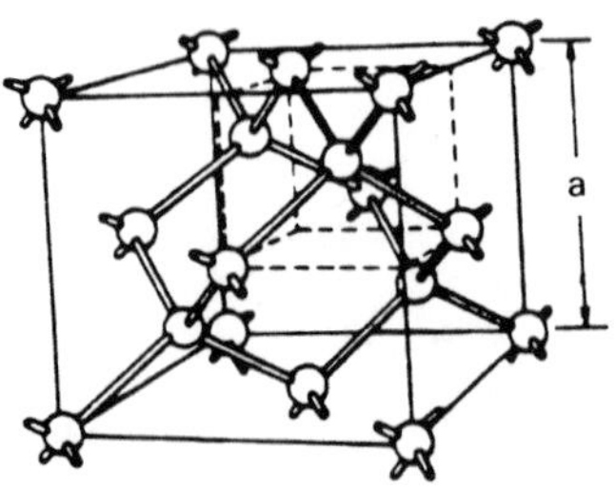

Fig. 1-4-4. Diamond crystal structure (from S. M. Sze, *Physics of Semiconductor Devices*, John Wiley & Sons, New York (1981)). Lattice constant is 3.56 Å for carbon, 5.43 Å for silicon, and 5.66 Å for germanium.

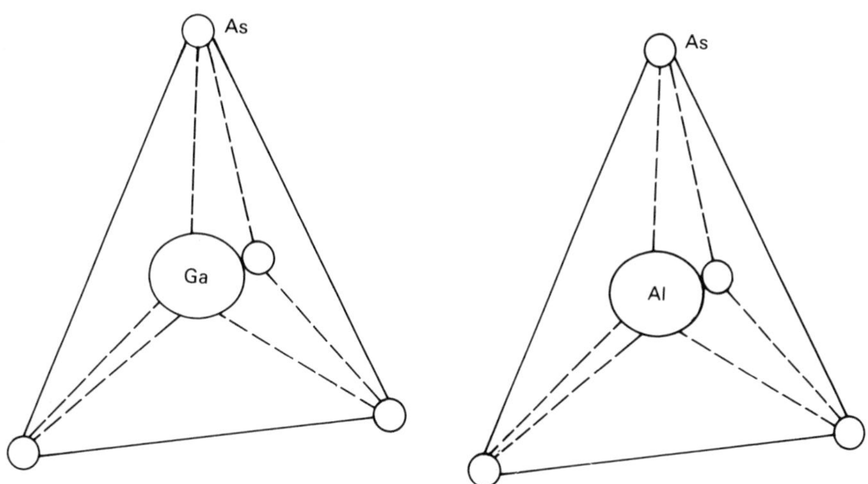

Fig. 1-4-5. Tetrahedral atomic configuration in GaAs and AlAs. (Compare with Fig. 1-3-1.)

a face-centered cubic lattice. It can be described as mutually penetrating face-centered cubic (fcc) lattices of element A and element B (gallium [Ga] and arsenic [As] in the case of GaAs) shifted relative to each other by a quarter of the body diagonal of the unit cell cube.

If atoms A and B are identical, this crystal structure reduces to a diamond structure. The diamond structure has an inversion symmetry. This means that if all the atoms of the crystal are moved in such a way that their space coordinates, x, y, and z (counted with respect to the point called the center of the inversion) are changed to $-x$, $-y$, and $-z$, the crystal will remain unchanged. The center of inversion for the diamond structure is at the midpoint between two nearest atoms. The zinc blende structure lacks such inversion symmetry.

Crystallographic directions and crystal planes are specified using a set of three integer numbers called *Miller indices*. A direction in a crystal is specified by a set of three integers u, v, and w defining a vector $u\mathbf{a}_1 + v\mathbf{a}_2 + w\mathbf{a}_3$, which points along the given direction. Here $\mathbf{a}_1$, $\mathbf{a}_2$, and $\mathbf{a}_3$ are the unit vectors (forming a unit

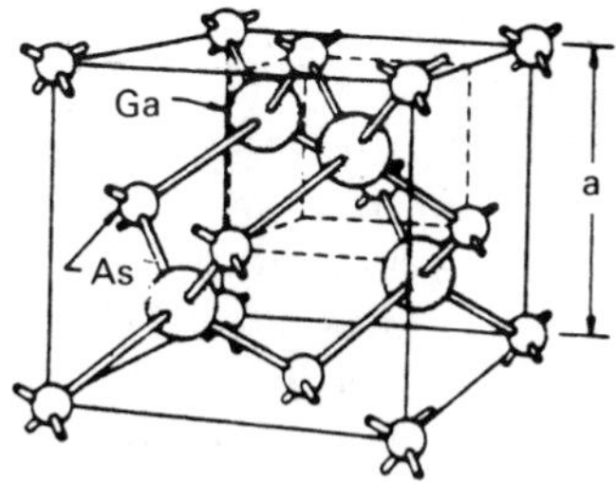

Fig. 1-4-6. Zinc blende crystal structure (from S. M. Sze, *Physics of Semiconductor Devices*, John Wiley & Sons, New York (1981)).

cell). It is clear that vectors $u\mathbf{a}_1 + v\mathbf{a}_2 + w\mathbf{a}_3$ and $k(u\mathbf{a}_1 + v\mathbf{a}_2 + w\mathbf{a}_3)$, where k is an arbitrary integer, point in the same direction. Hence, the choice of u, v, and w specifying the given direction is not unique. It is convenient to choose integers u, v, and w that have no common integral divisor. The common way to denote directions in a crystal is to enclose these integers, u, v, and w, in square brackets. Examples are [100], [111], and [110]. A negative integer is represented by placing a bar above the integer (for example, $[\bar{1}\,\bar{1}\,\bar{1}]$; see Fig. 1-4-7a). Many directions in a crystal are equivalent because of crystal symmetry. Crystal properties along these equivalent directions are exactly the same. Examples are directions [100], [001], and [010] in a cubic crystal, such as silicon. A set of all equivalent directions is denoted as ⟨uvw⟩. For instance, symbol ⟨110⟩ for a cubic lattice stands for all 12 directions along the face diagonals.

A crystal plane is specified using intercepts with the crystallographic axes pointing in directions $\mathbf{a}_1$, $\mathbf{a}_2$, and $\mathbf{a}_3$. These intercepts are expressed as vectors $n_1\mathbf{a}_1$, $n_2\mathbf{a}_2$, and $n_3\mathbf{a}_3$. Then integer numbers u, v, and w that have no common integral divisor and are proportional to $1/n_1$, $1/n_2$, and $1/n_3$ are found. These numbers are used to denote the crystal plane, using a symbol (uvw). The reason for using such a notation is that in the very important case of a cubic crystal (such as silicon or gallium arsenide) plane (uvw) is perpendicular to the direction $[uvw]$. A set of equivalent planes is denoted as $\{uvw\}$. Indices $\{uwv\}$ and $[uwv]$ are called Miller indices. Some important planes and directions for a cubic crystal are shown in Fig.1-4-7b.

As can be seen from Fig. 1-4-4 and Fig. 1-4-6, the bonds between the nearest Si atoms in silicon and between the nearest Ga and As atoms in gallium arsenide are in the ⟨111⟩ directions. The closest equivalent atoms in gallium arsenide (Ga and Ga or As and As) are in the ⟨110⟩ directions. Usually GaAs films are grown in direction [100] and GaAs wafers are cleaved along (011) and $(01\bar{1})$ planes, as shown in Fig. 1-4-8. Silicon and germanium crystals break (cleave) easily along {111} planes. This difference is related to the difference between purely covalent bonds in elemental semiconductors and partially heteropolar bonds in gallium arsenide.

Because of the lack of inversion symmetry in GaAs, directions [111] and $[\bar{1}\,\bar{1}\,\bar{1}]$ are not equivalent. Usually symbol [111] denotes the direction from Ga to the nearest As whereas $[\bar{1}\,\bar{1}\,\bar{1}]$ corresponds to the opposite direction. The surface of the gallium arsenide crystal cleaved along the (111) plane consists either of Ga atoms, which have three bonds with the crystal, or of As atoms, which have just one bond with the crystal. The opposite is true for the $(\bar{1}\,\bar{1}\,\bar{1})$ plane. The (111) plane is called the Ga plane and the $(\bar{1}\,\bar{1}\,\bar{1})$ plane is called the As plane. Fig. 1-4-9 shows the three basic crystal planes for GaAs. Each As atom on the (100) surface has two bonds with Ga atoms from the layer below. Two other bonds are free. The (110) plane contains the same number of Ga and As atoms. Each atom has one bond with the layer below. Atoms on the (111) surface have three bonds with the As atoms from the layer below. The fourth bond is free. These free bonds may be responsible for the surface electronic states that strongly affect the behavior of semiconductor devices fabricated from GaAs.

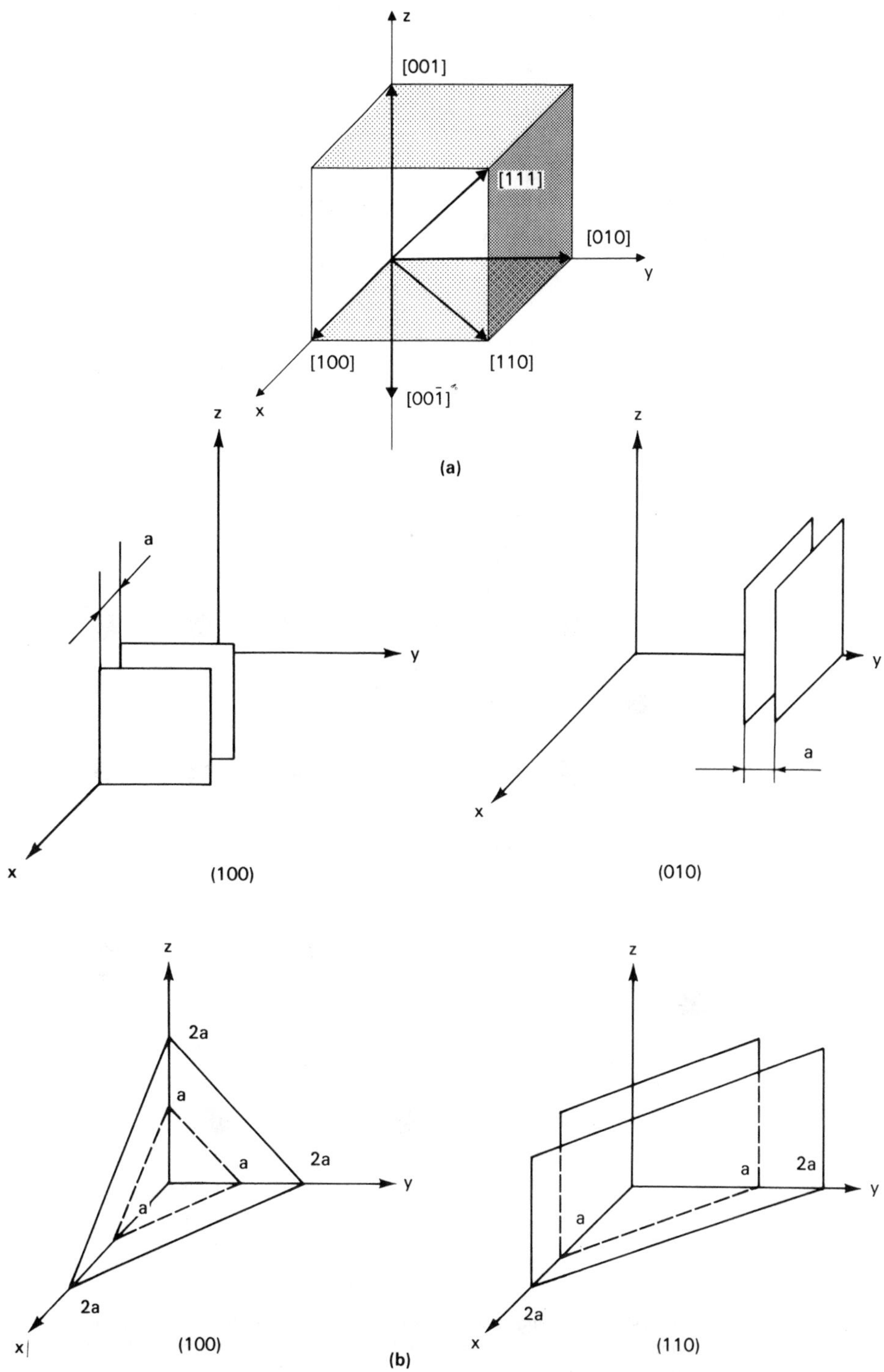

Fig. 1-4-7. (a) Important directions in cubic crystal, (b) Important cubic planes.

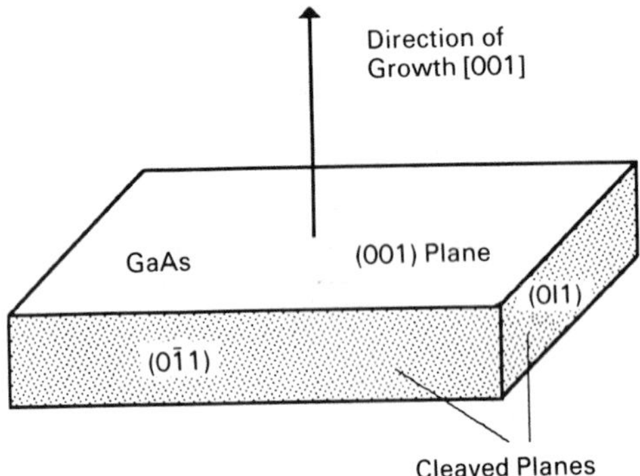

Fig. 1-4-8. Cleaved planes and direction of growth for GaAs sample.

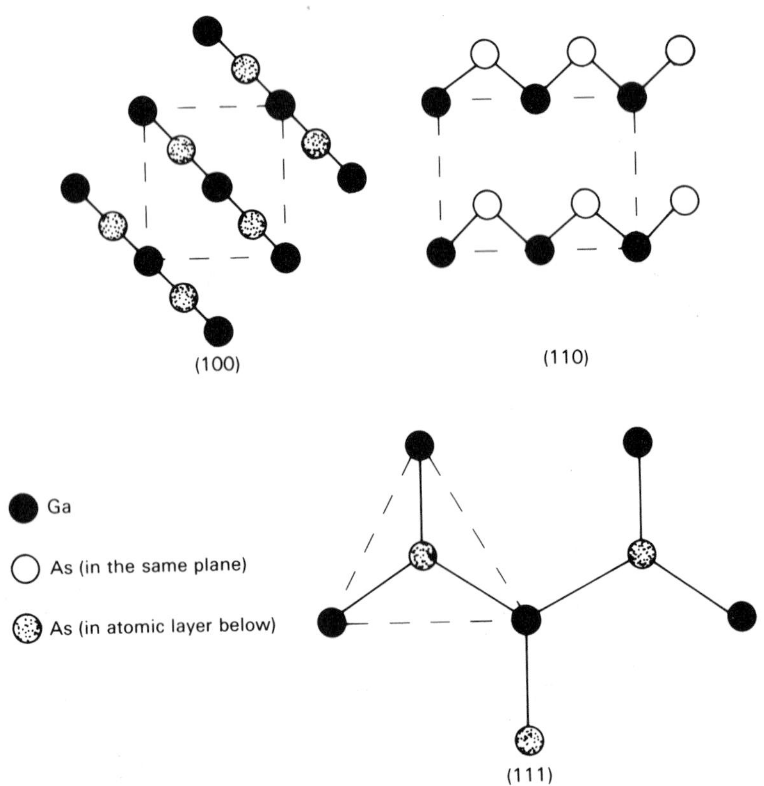

Fig. 1-4-9. Gallium arsenide crystal planes. Solid lines correspond to crystal bonds; dashed lines correspond to the intercepts of atomic planes with a unit cell. (a) (100), (b) (110), (c) (111).

The distance between the nearest neighbors in Si, Ge, and III-V compounds may be found as the sum of the atomic radii. These radii, the distances between the nearest neighbors, and lattice constants of several important semiconductors with diamond or zinc blende structure are given in Tables 1-4-1 and 1-4-2. (For a cubic lattice the lattice constant is equal to the side of the unit cell cube.) Once again we notice that the lattice constants of GaAs and AlAs are very close indeed. Another important example of compound semiconductors with "matched" lattice constants is the CdTe-HgTe pair. Hence, this combination can also be used to produce a ternary compound, $Hg_xCd_{1-x}Te$ (where the mole fraction of mercury, x, may vary from zero to 1), or to grow a CdTe-HgTe heterostructure. Another way to match lattice constants is to use a heterostructure consisting of a ternary compound and a binary compound. The lattice constant, a_{ter}, of a ternary compound $A_xC_{1-x}B$ varies roughly linearly with composition,

$$a_{ter} \approx a_{bin1}\, x + a_{bin2}\,(1 - x) \tag{1-4-2}$$

where a_{bin1} is the lattice constant of the binary compound AB, and a_{bin2} is the lattice constant of the binary compound CB. Hence, for $In_{0.47}Ga_{0.53}As$, for example, we find from eq. (1-4-2) and Table 1-4-2 that $a_{ter} \approx 5.84$ Å, which matches quite well the lattice constant of InP ($a = 5.86$ Å). As a consequence of such a good match, $In_{0.47}Ga_{0.53}As$ can be grown on InP substrates.

TABLE 1-4-1. COVALENT (TETRAHEDRAL) RADII

Element	Tetrahedral radius (Å)
Al	1.26
As	1.18
B	0.88
Bi	1.46
C	0.77
Ga	1.26
Cd	1.48
Ge	1.22
Hg	1.48
In	1.44
Mn	1.27
N	0.70
P	1.10
Sb	1.36
Si	1.17
Sn	1.40
Te	1.47
Zn	1.31

TABLE 1-4-2. LATTICE CONSTANTS

Material	Lattice constant, a (Å) at 25 °C	Distance between nearest neighbors, $a\sqrt{3}/4$ (Å)	Sum of covalent radii (Å)
Si	5.434	2.353	2.34
Ge	5.657	2.450	2.44
A_3B_5			
AlAs	5.612	2.430	2.440
AlP	5.451	2.360	2.360
AlSb	5.136	2.224	2.620
BAs	4.776	2.068	2.060
BN	3.615	1.565	1.580
BP	4.538	1.965	1.980
BSb	5.170	2.239	2.240
GaAs	5.653	2.448	2.440
GaP	5.451	2.360	2.360
GaSb	6.095	2.639	2.620
InAs	6.058	2.623	2.620
InP	5.867	2.540	2.540
InSb	6.479	2.805	2.800
A_2B_6			
CdTe	6.482	2.807	2.950
HgS	5.841		
HgSe	6.084		
HgTe	6.462	2.798	2.950
ZnS	5.415		
ZnSe	5.653		
ZnTe	6.101	2.642	2.780

1-5. BAND STRUCTURE

As was shown in Section 1-1 (see eq. (1-2-11) and Problem 1-2-1), the electron wave function $\psi_{\mathbf{k}}$ of a free electron is given by a plane wave

$$\psi_{\mathbf{k}} = C\, e^{i\mathbf{k}\cdot\mathbf{r}} \tag{1-5-1}$$

where $\mathbf{k} = \mathbf{P}/\hbar$ is the wave vector, $\mathbf{P}$ is the momentum, $\mathbf{r}$ is the space vector, and C is a normalizing constant. The energy of a free electron is given by

$$E = \frac{p^2}{2m_e} = \frac{\hbar^2 k^2}{2m_e} \tag{1-5-2}$$

In a crystal, the electronic wave function may be presented as a *Bloch wave*, i.e., as the plane wave modulated by a periodic function,

$$\psi_{\mathbf{k}}(\mathbf{r}) = e^{i\mathbf{k}\cdot\mathbf{r}}\, u_{\mathbf{k}}(\mathbf{r}) \tag{1-5-3}$$

where the function $u_{\mathbf{k}}(\mathbf{r})$ (called the Bloch amplitude) has the same spatial periodicity as the crystal lattice. Equation (1-5-3) is a direct consequence of crystal symmetry. Indeed, crystal symmetry requires that the crystal potentials $U(\mathbf{r})$ and $U(\mathbf{r} + \mathbf{a})$ and the probability densities $|\psi(\mathbf{r})|^2$ and $|\psi(\mathbf{r} + \mathbf{a})|^2$ must be equal for any lattice vector $\mathbf{a}$ of the crystal. This means that $\psi(\mathbf{r})$ and $\psi(\mathbf{r} + \mathbf{a})$ may only differ by a constant A, such that $|A|^2 = 1$:

$$A\psi(\mathbf{r}) = \psi(\mathbf{r} + \mathbf{a}) \tag{1-5-4}$$

The constant A can be expressed as

$$A = \exp(i\mathbf{k} \cdot \mathbf{a}) \tag{1-5-5}$$

where $\mathbf{k}$ is the wave vector. Hence,

$$\psi(\mathbf{r}) = \exp(-i\mathbf{k} \cdot \mathbf{a})\, \psi(\mathbf{r} + \mathbf{a}) \tag{1-5-6}$$

or

$$\psi(\mathbf{r}) = \exp(i\mathbf{k} \cdot \mathbf{r})\exp[-i\mathbf{k} \cdot (\mathbf{r}+\mathbf{a})]\, \psi(\mathbf{r} + \mathbf{a}) \tag{1-5-7}$$

Comparing this result with eq. (1-5-3), we identify the Bloch function as

$$u_{\mathbf{k}}(\mathbf{r}) = \exp[-i\mathbf{k} \cdot (\mathbf{r}+\mathbf{a})]\, \psi(\mathbf{r} + \mathbf{a}) \tag{1-5-8}$$

which has the periodicity of the crystal lattice (as can be easily proved using eq. (1-5-6)). Indeed,

$$u_{\mathbf{k}}(\mathbf{r} + \mathbf{b}) = \exp[-i\mathbf{k} \cdot (\mathbf{r} + \mathbf{b} + \mathbf{a})]\, \psi(\mathbf{r} + \mathbf{b} + \mathbf{a}) =$$

$$= \exp[-i\mathbf{k} \cdot (\mathbf{r} + \mathbf{b} + \mathbf{a})]\exp(i\mathbf{k} \cdot \mathbf{b})\, \psi(\mathbf{r} + \mathbf{a}) = u_{\mathbf{k}}(\mathbf{r})$$

where $\mathbf{b}$ is an arbitrary lattice vector of the crystal.
Hence, eq. (1-5-7) can be rewritten as

$$\psi_{\mathbf{k}}(\mathbf{r}) = e^{i\mathbf{k}\cdot\mathbf{r}}\, u_{\mathbf{k}}(\mathbf{r}) \tag{1-5-9}$$

where we added subscript $\mathbf{k}$ to the wave function ψ to emphasize that different electronic states in a crystal are characterized by different values of the wave vector $\mathbf{k}$.

The wave functions $\psi_{\mathbf{k}}$ are found from the solution of the Schrödinger equation:

$$-\frac{\hbar^2}{2m_e} \nabla^2 \psi_{\mathbf{k}} + U(\mathbf{r})\psi_{\mathbf{k}} = E\psi_{\mathbf{k}} \tag{1-5-10}$$

where E is the electron energy.

To determine possible values of the wave vector $\mathbf{k}$ we should introduce the concept of a *reciprocal lattice*. The reciprocal lattice is defined as a set of points in $\mathbf{k}$-space corresponding to the tips of vectors $\mathbf{K}$ given by

$$\exp(i\mathbf{K} \cdot \mathbf{a}) = 1 \tag{1-5-11}$$

where $\mathbf{a} = n_1\mathbf{a}_1 + n_2\mathbf{a}_2 + n_3\mathbf{a}_3$ are lattice vectors of the crystal, $\mathbf{a}_1$, $\mathbf{a}_2$, and $\mathbf{a}_3$ being the primitive vectors of the crystal lattice, and n_1, n_2, and n_3 being integers. In the simple case of a one-dimensional lattice with the lattice constant a_1, eq. (1-5-11) becomes

$$\exp(iKn_1a_1) = 1 \tag{1-5-12}$$

or

$$a_1/\lambda = m \tag{1-5-13}$$

where $\lambda = 2\pi/K$ is the wavelength and m is an integer. Hence, **K** is a wave vector such that the distance between two crystal planes (separated by distance a) is equal to an integer number of corresponding wavelengths, i.e., to $2\pi m/K$. As we will discuss presently, this condition and eq. (1-5-11) are important for determining the values of **k** that correspond to a constructive interference of electron wave functions reflected from the crystal planes.

To satisfy eq. (1-5-11), the primitive vectors of the reciprocal and real lattices $\mathbf{K}_i$ and $\mathbf{a}_j$ must satisfy the following conditions:

$$\mathbf{K}_i \cdot \mathbf{a}_j = 2\pi\delta_{ij} \tag{1-5-14}$$

Based on eq. (1-5-14), one can show that

$$\mathbf{K}_1 = 2\pi\frac{\mathbf{a}_2 \times \mathbf{a}_3}{\mathbf{a}_1 \cdot (\mathbf{a}_2 \times \mathbf{a}_3)} \tag{1-5-15}$$

$$\mathbf{K}_2 = 2\pi\frac{\mathbf{a}_3 \times \mathbf{a}_1}{\mathbf{a}_1 \cdot (\mathbf{a}_2 \times \mathbf{a}_3)} \tag{1-5-16}$$

$$\mathbf{K}_3 = 2\pi\frac{\mathbf{a}_1 \times \mathbf{a}_2}{\mathbf{a}_1 \cdot (\mathbf{a}_2 \times \mathbf{a}_3)} \tag{1-5-17}$$

Eqs. (1-5-15) to (1-5-17) allow us to find the primitive vectors of the reciprocal lattice.

Just as the direct crystal lattice can be reproduced by repeating one primitive cell in space (with an atomic basis), a primitive cell in the reciprocal space contains complete information about the reciprocal lattice. All points in the reciprocal lattice may be obtained from just one primitive cell by adding different vectors of the reciprocal lattice. Usually such a primitive cell is chosen as the Wigner-Seitz cell of a reciprocal lattice and is called the first *Brillouin zone*. The Wigner-Seitz cell for the direct lattice was defined in Section 1-4. This cell is formed by bisecting all lattice vectors emanating from a lattice site by planes and choosing the smallest volume formed by these intersecting planes. Hence, we can obtain the first Brillouin zone by constructing planes that normally bisect the reciprocal lattice vectors emanating from the origin in the **k**-space. All values of **k** in the first Brillouin zone are such that subtracting any vector of the reciprocal lattice from a vector **k** inside the zone does not give another vector also inside this zone. As shown below, physically significant and unique solutions of the Schrödinger equation for an electron in a crystal may be assigned to the wave vectors **k** in the first Brillouin zone.

Indeed, let us consider two wave functions $\psi_{\mathbf{k}}(\mathbf{r})$ and $\psi_{\mathbf{k}+\mathbf{K}}(\mathbf{r})$, where $\mathbf{K}$ is a vector of a reciprocal lattice:

$$\psi_{\mathbf{k}}(\mathbf{r}) = \exp(i\mathbf{k} \cdot \mathbf{r})u_{\mathbf{k}}(\mathbf{r})$$

$$\psi_{\mathbf{k}+\mathbf{K}}(\mathbf{r}) = \exp[i(\mathbf{k}+\mathbf{K}) \cdot \mathbf{r}]u_{\mathbf{k}+\mathbf{K}}(\mathbf{r}) = \exp(i\mathbf{k} \cdot \mathbf{r})u_{\mathbf{k}+\mathbf{K}}(\mathbf{r})\exp(i\mathbf{K} \cdot \mathbf{r})$$

Functions $u_{\mathbf{k}}(\mathbf{r})$, $u_{\mathbf{k}+\mathbf{K}}(\mathbf{r})$ and $\exp(i\mathbf{K} \cdot \mathbf{r})$ are periodic functions of $\mathbf{r}$ with the period of a lattice potential. Hence,

$$\psi_{\mathbf{k}+\mathbf{K}}(\mathbf{r}) = \exp(i\mathbf{k} \cdot \mathbf{r})u'_{\mathbf{k}}(\mathbf{r}) \tag{1-5-18}$$

where $u'_{\mathbf{k}}(\mathbf{r}) = u_{\mathbf{k}+\mathbf{K}}(\mathbf{r})\exp(i\mathbf{K} \cdot \mathbf{r})$ is also a periodic function of $\mathbf{r}$ with the period of a lattice potential. To find this function, we must substitute (1-5-18) into the Schrödinger equation (1-5-10). For the same energy, E, we will then obtain exactly the same solution for $u'_{\mathbf{k}}(\mathbf{r})$ as for $u_{\mathbf{k}}(\mathbf{r})$. Hence,

$$u'_{\mathbf{k}}(\mathbf{r}) = u_{\mathbf{k}}(\mathbf{r})$$

$$\psi_{\mathbf{k}+\mathbf{K}}(\mathbf{r}) = \psi_{\mathbf{k}}(\mathbf{r})$$

This proves that all physically significant solutions can be labeled using the wave vectors within the first Brillouin zone because it contains all values of **k,** such that no two vectors can be obtained from each other by subtracting a vector of the reciprocal lattice.

The boundaries of the first, second, third, and higher-order Brillouin zones may also be defined as regions bounded by the *Bragg planes* in the reciprocal space. The Bragg planes determine the conditions for reflections of electron wave functions from the crystal planes. To find the equation describing the Bragg planes, let us consider the reflection of the Bloch wave with wave vector $\mathbf{k}_b$ from crystal planes separated by distance a (see Fig. 1-5-1a). Such a reflection occurs when the waves reflected from each crystal plane add up with the same phase (this is called constructive interference). For this to happen, the phase difference between the waves reflected from adjacent planes, $2a \cos \phi$, must be equal to an integer number of wavelengths, $n\lambda$, where $\lambda = 2\pi/|\mathbf{k}_b|$. Thus we obtain the condition

$$2a \cos \phi = n\lambda \tag{1-5-19}$$

Taking into account that $\cos \phi = \mathbf{k}_b \cdot \mathbf{K}/(k_b K)$, we can rewrite eq. (1-5-19) as

$$2\mathbf{k}_b \cdot \mathbf{K} = K^2 \tag{1-5-20}$$

where $\mathbf{K}$ is a vector of the reciprocal lattice ($K=2\pi n/a$). For a given vector of the reciprocal lattice, $\mathbf{K}$, all waves with wave vectors satisfying eq. (1-5-20) will be reflected from the given set of crystal planes separated by vector $\mathbf{a}$, as shown in Fig. 1-5-1a. Such a reflection is called Bragg reflection, and the plane in **k**-space formed by tips of vectors $\mathbf{k}_b$ satisfying eq. (1-5-20) is called the Bragg plane (see Fig. 1-5-1b). When the wave vector, $\mathbf{k}_b$, of the incident electron wave satisfies eq. (1-5-20), the component of $\mathbf{k}$ in the direction of $\mathbf{K}$ is equal to $(1/2)K$. The compo-

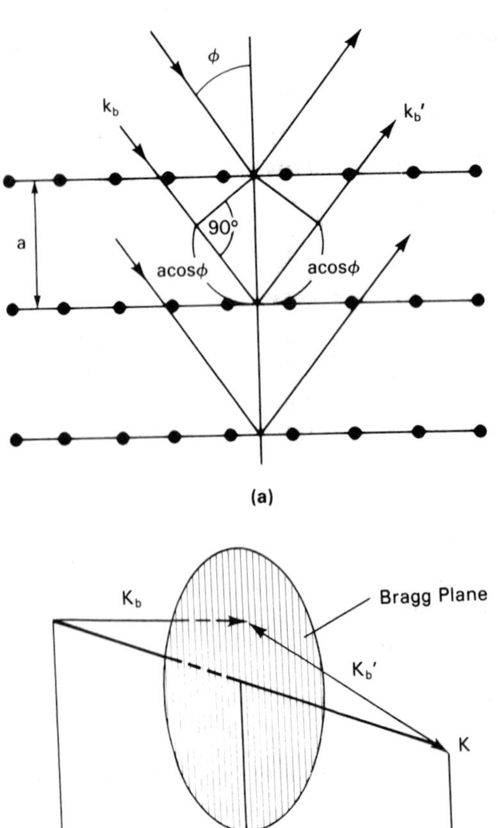

Fig. 1-5-1. (a) Electron wave reflection from crystal planes; (b) Bragg plane in reciprocal lattice. This plane is perpendicular to and bisects the vector of the reciprocal lattice **K**. Component of $\mathbf{k}_b$ along **K** is equal to $K/2$. $\mathbf{k}_b'$ is the wave vector of the electronic wave function reflected from the crystal planes separated by vectors of the crystal lattice **a**, such that $\exp(i\mathbf{K} \cdot \mathbf{a}) = 1$ (see text).

nent of the wave vector of the reflected wave in the direction of **K** is equal to $-1/2K$. The component of the wave vector parallel to the Bragg plane does not change. Hence, the total change in the wave vector is equal to **K**. As was shown previously, this leads to the constructive interference of the incident electron wave function with the electron wave functions reflected from the crystal planes separated by vectors **a,** satisfying eq. (1-5-11):

$$\exp(i\mathbf{K} \cdot \mathbf{a}) = 1$$

The Bragg planes determine which electron wave functions, i.e., which electron states, are strongly affected by reflections from crystal planes. These wave functions with wave vectors $\mathbf{k}_b$ determined by eq. (1-5-20) are more strongly affected by the crystal potential than wave functions with other values of $\mathbf{k}$. The Bragg planes form the boundaries of the Brillouin zones (see Problem 1-5-10).

We should point out that the shapes of the second-, third-, and higher-order Brillouin are rather complicated. For this reason the energy states of a crystal are usually labeled using the wave vectors in the first Brillouin zone.

The first Brillouin zone for the face-centered cubic, diamond, and zinc blende structures is the same (see Fig. 1-5-2). This is not surprising because all three structures can be formed from the same face-centered cubic lattice placing a different atomic basis into the lattice nodes. The electron energy is a function of k_x, k_y, and k_z that varies within the first Brillouin zone. The four-dimensional plot, required to reproduce such a dependence, defies our imagination. The closest we can come to it is to reproduce three dimensional surfaces of equal energy in the $\mathbf{k}$-space. The alternative is to study different cross sections of this plot, varying only a certain component of k (for example, k_x) and keeping other components (k_y and k_z) constant. Fortunately, most important semiconductor properties are determined by regions in the first Brillouin zone corresponding to lines and points of symmetry. These symmetry points and lines of the Brillouin zone are marked on Fig. 1-5-2 (see also Table 1-5-1). A detailed study of the symmetry of electronic states in the zinc blende structure was done by Dresselhaus (1955) and Parmenter (1955). In silicon, germanium, diamond, and compound III-V and II-IV semiconductors, points Γ (in the center of the first Brillouin zone), X, and L, and direction Δ (see Table 1-5-1 and Fig. 1-5-2) are the most important regions. The dependencies of the electronic energy on the wave vector (energy bands) are usually calculated for the lines of symmetry in the first Brillouin zone.

Calculations of energy bands are based on the solution of the Schrödinger equation (1-5-10). Such calculations are difficult and computer-intensive. They

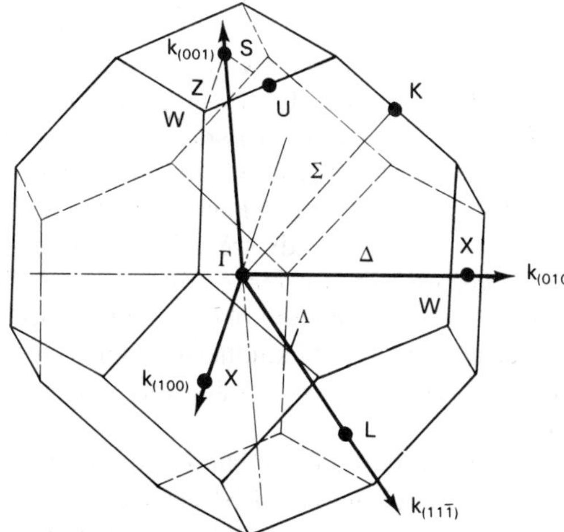

Fig. 1-5-2. The first Brillouin zone for the face-centered cubic, diamond, and zinc blende structures. Also shown are symmetry points and directions of the first Brillouin zone.

TABLE 1-5-1. SYMMETRY POINTS OF THE BRILLOUIN ZONE OF THE ZINC BLENDE STRUCTURE (see Fig. 1-5-2)

Number of equiv. K vectors	Point	Symmetry operations	Point groups
1	$\Gamma(000)$	E, $8C_3$, $3C_2$, $6S_4$, $6C_s$	T_d
3	$X\frac{2\pi}{a}(100)$	E, $3C_2$, $2S_4$, 2σ	D_{2d}
4	$L\frac{2\pi}{a}\left(\frac{1}{2}\frac{1}{2}\frac{1}{2}\right)$	E, $2C_3$, 3σ	C_{3v}
6	$W\frac{2\pi}{a}\left(1\,0\,\frac{1}{2}\right)$	E, $2S_4$, C_2	S_4
6	$\Delta\,(K00)$	E, 2σ, C_2	C_{2v}
4	$\Lambda\,(KKK)$	E, $2C_3$, 3σ	C_{3v}
12	$\Sigma(K0K)$	E, σ	C_s
12	$K\frac{2\pi}{a}\left(\frac{3}{4}\,0\,\frac{3}{4}\right)$	E, C_2	C_s

are frequently done by starting from a known solution for a simpler system and using perturbation theory to introduce corrections for a given crystal. In particular, we can point out two complementary approaches. In the first approach, an electron wave function of the crystal is represented as a linear combination of the atomic orbitals, and the crystal potential is considered as a perturbation. The band structure is then determined using the perturbation theory. This method is called the Linear Combination of Atomic Orbitals (LCAO) method (see, for example, Harrison 1980). This approach is illustrated by Fig. 1-5-3, which shows how energy levels of isolated atoms split into energy bands when atoms are combined into a crystal. According to the Pauli exclusion principle, only two electrons (with different spins) may occupy an atomic energy level with a given set of quantum numbers n, l, and m (see Section 1-1). Accordingly, $2N$ electrons may occupy an energy band. Just as in an atom, where the lowest energy levels are filled and higher energy levels are empty, the lowest energy bands in a crystal are filled and higher energy bands are empty. The bands of allowed energy states are separated by forbidden ranges of energy states. Empty bands are called conduction bands, and bands filled by valence electrons are called valence bands. The highest filled valence band and the lowest empty conduction band play the most important role in determining electronic properties of a semiconductor. Generally speaking (with a notable exception discussed below), these conduction and valence bands are separated by the energy gap, E_g.

A second approach to the calculation of energy-band structure is used in the *nearly free electron model*. In this model unperturbed electron wave functions are assumed to be given by plane waves (just as in free space; see eq. (1-5-1)). The

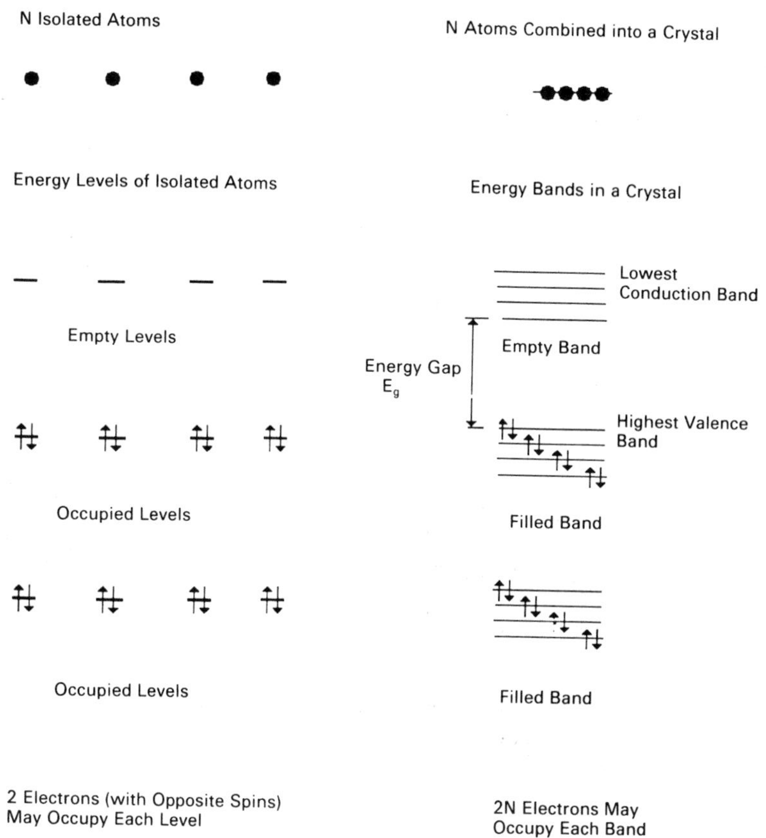

Fig. 1-5-3. Splitting of energy levels of isolated atoms into energy bands in a crystal. Arrows pointing up and down represent electrons with different values of spin ($S = \pm\frac{1}{2}$).

periodic crystal potential is regarded as a small perturbation. In this model, a nearly parabolic E vs. $\mathbf{k}$ curve is obtained with energy gaps at the boundaries of the Brillouin zones where Bragg reflections of the electron wave functions from the crystal planes strongly affect the electronic states (see Fig. 1-5-4). The resulting E vs. $\mathbf{k}$ curve can be "folded" back into the first Brillouin zone. This simple picture provides a good qualitative description of the band structure. In quantitative analysis, however, there are problems related to very large values of the crystal Coulomb potential near the centers of the ion cores and its rapid variation in this region. This requires the representation of wave functions as sums of plane waves. A very large number of plane waves must be used in order to approach the correct solution. To some extent, the method can be improved if care is taken in the selection of plane waves of proper symmetry properties.

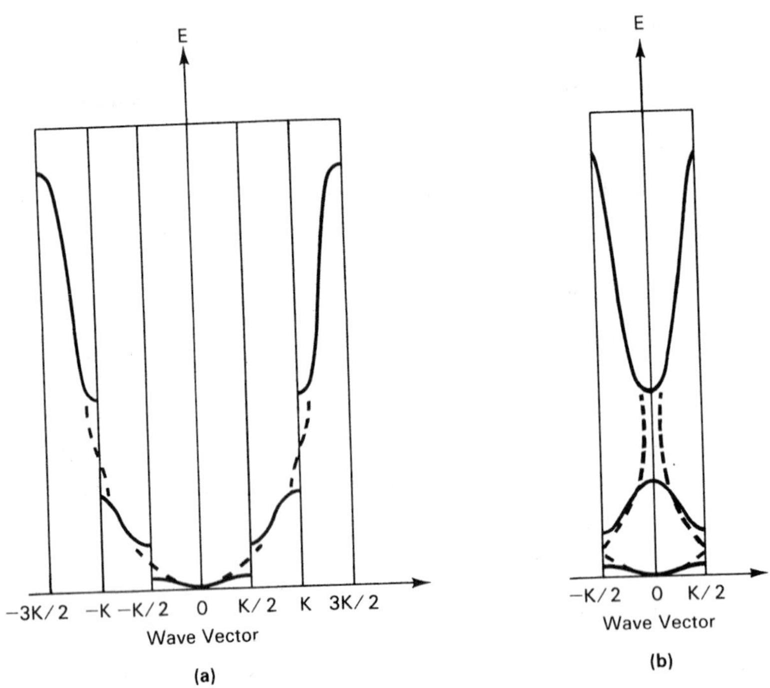

Fig. 1-5-4. Energy band diagram for one-dimensional lattice in a nearly free electron model. (a) Extended bands; (b) Bands "folded" into the first Brillouin zone K = $2\pi/a$ where a is the lattice constant.

More accurate calculations of the band structure are based on the *pseudopotential method* (see, for example, Heine 1970 and Chelikowsky and Cohen 1976). The idea of this approach is to use a model potential that leads to the same energy levels as the real potential but not to the same wave functions. The pseudopotential has a form that allows improved convergence of the series of plane waves that represent the electron pseudowave functions. In many cases it is convenient to choose the pseudopotential to be a constant within the ion core (see Fig. 1-5-5). The parameters of the pseudopotential can be determined from the spectroscopic data for the individual atoms. Hence, this method allows us to predict the energy bands of a crystal based on the experimental data obtained for individual atoms.

The calculated energy bands for Ge, Si, and GaAs are shown in Fig. 1-5-6 (from Chelikowsky and Cohen 1976). The most important features of the band structures shown in Fig. 1-5-6 are represented by the minima of the lowest conduction band and by the maxima of the highest valence band because the states with energies much higher than the minima of the conduction band are empty and the states with energies well below the maxima of the valence band are completely filled. The model that summarizes the most important features of the band struc-

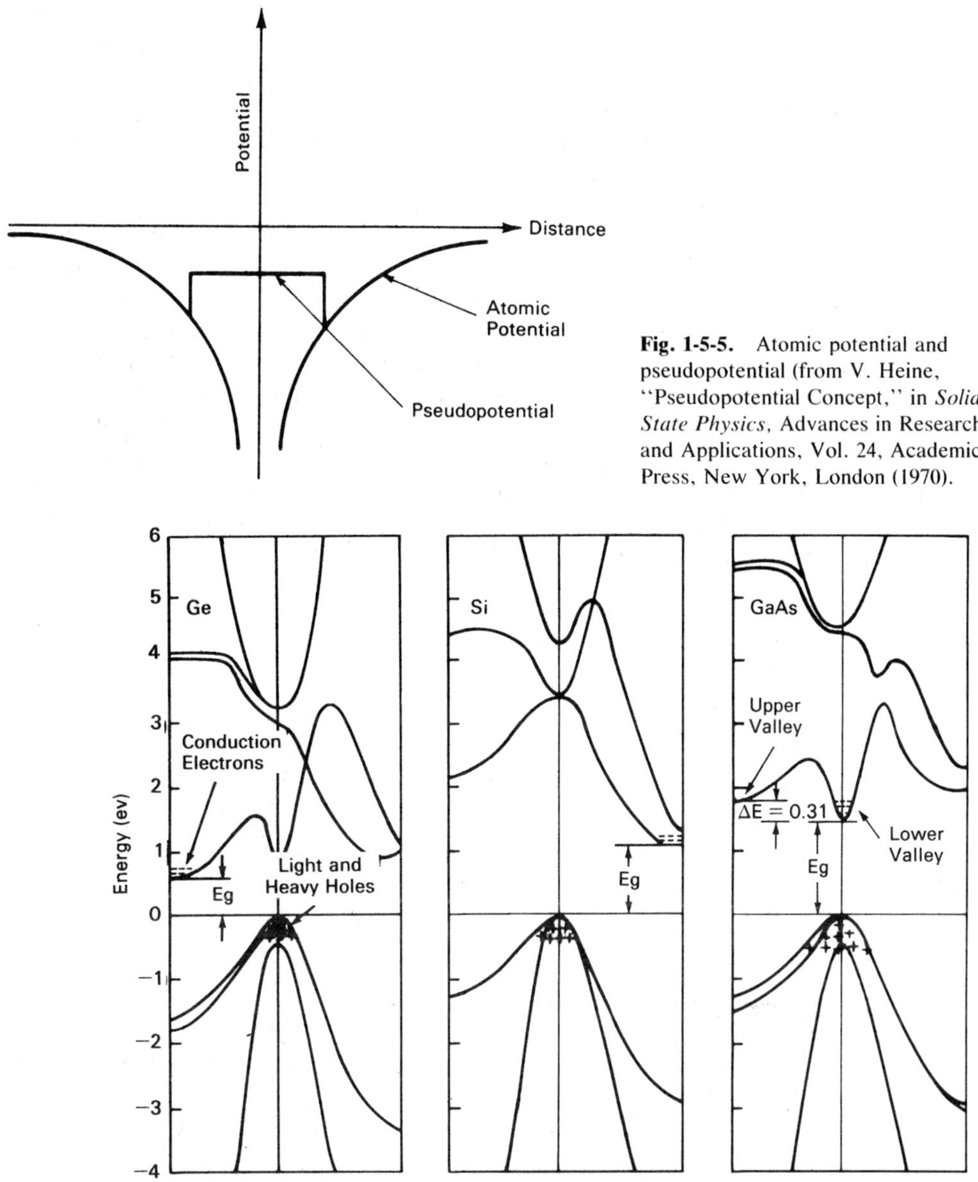

Fig. 1-5-5. Atomic potential and pseudopotential (from V. Heine, "Pseudopotential Concept," in *Solid State Physics,* Advances in Research and Applications, Vol. 24, Academic Press, New York, London (1970).

Fig. 1-5-6. Computed energy bands for silicon, germanium, and gallium arsenide (from J. R. Chelikowsky and M. L. Cohen, "Nonlocal Pseudopotential Calculations for the Electronic Structure of Eleven Diamond and Zinc-Blend Semiconductors," *Phys. Rev.,* B14, p. 556 (1976), and S. M. Sze, *Physics of Semiconductor Devices,* John Wiley & Sons, New York (1981).

ture of the cubic semiconductors is shown in Fig. 1-5-7 (from Jacoboni and Reggiani 1979). In all important cubic semiconductors, the top of the highest filled (valence) band corresponds to the Γ point of the first Brillouin zone (see Fig. 1-5-2), that is, to point $k = (0,0,0)$. The bands shown in Fig. 1-5-7a are typical for Si, Ge, III-V, and most II-VI semiconductors where the top valence band and the lowest conduction band are separated by an energy gap. The tops of the valence bands are located at point Γ; two of these bands (light and heavy holes) have the same energy at this point (i.e., they are degenerate), and the third one is split (it is called a split-off band). The lowest minimum of the conduction band in different semiconductors corresponds to different points of the first Brillouin zone. For example, in germanium, the L minimum (corresponding to the point (1,1,1) of the first Brillouin zone) is the lowest. The lowest minimum of the conduction band in silicon is located along the Δ axis (corresponding to the direction $(k,0,0)$ of the first Brillouin zone), close to the X-point (corresponding to the point (1,0,0) of the first Brillouin zone). In GaAs, the lowest minimum of the conduction band is at point Γ(0,0,0), i.e., at the same value of the wave vector **k** as the top of the valence band. Such a semiconductor is called a direct gap semiconductor. Silicon and germanium are indirect gap semiconductors.

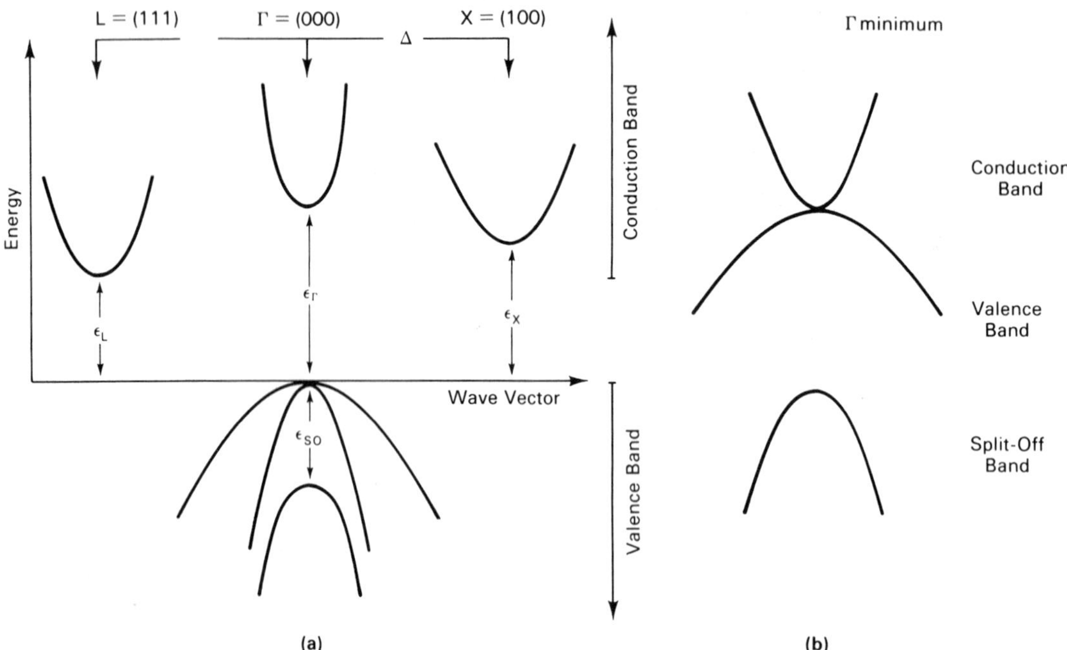

Fig. 1-5-7. Important minima of the conduction band and maxima of the valence band in cubic semiconductors (from C. Jacoboni and L. Reggiani, Bulk Hot-Electron Properties of Cubic Semiconductors, *Advances in Physics,* Vol. 28, No. 4, pp 493–553, (1979). The values of energies E_Γ, E_X and E_L for different cubic semiconductors are given in Table 1-5-2.

The band structure of Fig. 1-5-7b occurs in gapless semiconductors such as gray tin and HgTe. There is no energy gap! This leads to very interesting and unusual properties of these materials. Using, for example, CdHgTe alloys, one may create materials with narrow energy gaps, to register interband transitions caused by infrared photons and thus make sensitive infrared detectors.

Near the minima of the conduction band the energy-versus-wave vector dependencies, such as shown in Fig. 1-5-6, can be approximated by one of the following functions:

$$E(k) = \hbar^2 k^2/(2m^*) \text{ (spherical)} \tag{1-5-21}$$

$$E(k) = \hbar^2/2\ (k_x^2/m_x + k_y^2/m_y + k_z^2/m_z) \text{ (ellipsoidal)} \tag{1-5-22}$$

Wave vector **k** and energy E in eqs. (1-5-21) to (1-5-22) are measured from the corresponding minima of the energy band. Equation (1-5-21) represents a parabolic dependence of energy on the wave vector, similar to that of an electron in free space, $E(k) = \hbar^2 k^2/(2m_e)$. However, the value of m^* (which is called the effective mass) can be very different from a free electron mass, m_e. For example, in gallium arsenide $m^* \approx 0.067\ m_e$. This difference is caused by the periodic crystal potential. Equation (1-5-21) corresponds to a band with a single scalar effective mass and spherical surfaces of equal energy. It is an appropriate model for the Γ minimum of the conduction band. A similar equation,

$$E(k) = E_1 - \hbar^2 k^2/(2m_p^*) \text{ (spherical)} \tag{1-5-23}$$

can be used for the split-off valence band. Equation (1-5-21) is also the simplest possible model of the band structure and is frequently used for crude estimates of transport properties. The similarity with the free electron motion is very useful for providing an insight into the physics of electronic motion. When eq. (1-5-21) is applicable, one can write the Schrödinger equation for a semiconductor crystal in the same form as eq. (1-2-7), simply substituting m_e by m^*, provided that potential U varies slowly over distances on the order of the lattice constant. For example, we can find energy levels related to a positive charge introduced into the semiconductor solving the Schrödinger equation with the Coulombic potential given by eq. (1-2-22) (we just have to substitute ε_o by $\kappa\varepsilon_o$, where κ is the dielectric constant of the semiconductor, and m_e by m^*). Another example is the problem of electron energy levels in a thin layer of GaAs imbedded into an AlGaAs crystal. Using the effective mass concept, one can reduce this problem to the solution of the Schrödinger equation for a one-dimensional potential well. The importance of the effective-mass concept is related to the fact that in many cases electrons and holes occupy energy states in the vicinity of minima and maxima, shown in Fig. 1-5-7. On the other hand, even a casual look at Fig. 1-5-6 is enough to convince us that this concept should be used with caution, especially in high electric fields when electron and hole energies can be very high.

Equation (1-5-22) represents a band with ellipsoidal surfaces of equal energy. It can be used to describe the L and X minima of the conduction band. In this case the approximation using a single effective mass, m^*, is not adequate because the effective mass depends on the direction of the wave vector, **k**. As a consequence, the effective mass tensor, $\mathbf{m}_{ij}$, has to be defined. The components of the effective mass tensor are related to the curvature of E vs. **k** dependence:

$$\mathbf{m}_{ij} = \hbar^2/(\partial^2 E/\partial k_i \partial k_j) \tag{1-5-24}$$

Here $i, j = x, y, z$. This definition can be understood by using a Taylor series expansion of $E(k)$ in the vicinity of a conduction band minimum, $E = E_o$, $k = k_o$, where $\partial E/\partial k = 0$:

$$E = E_o + (1/2) \sum_{i,j} \partial^2 E/\partial k_i \partial k_j \,(k_i - k_{oi})\,(k_j - k_{oj}) \tag{1-5-25}$$

This equation can be rewritten as

$$E = E_o + (1/2)\hbar^2 \sum_{i,j} (k_i - k_{oi})\,(k_j - k_{oj})/m_{ij} \tag{1-5-26}$$

where components of the effective mass tensor, $\mathbf{m}_{ij}$, are defined by eq. (1-5-24).

As a result of the rotational symmetry of the ellipsoids of equal energy for the L and X minima of a conduction band (see Figs. 1-5-8 and 1-5-9), eq. (1-5-22) can be rewritten in a simpler form:

$$E(k) = \frac{\hbar^2}{2}\left(\frac{k_l^2}{m_l} + \frac{k_t^2}{m_t}\right) \tag{1-5-27}$$

where $1/m_l$ and $1/m_t$ are the longitudinal and transverse components of the inverse effective mass tensor, and k_l and k_t are the longitudinal and transverse components of the wave vector.

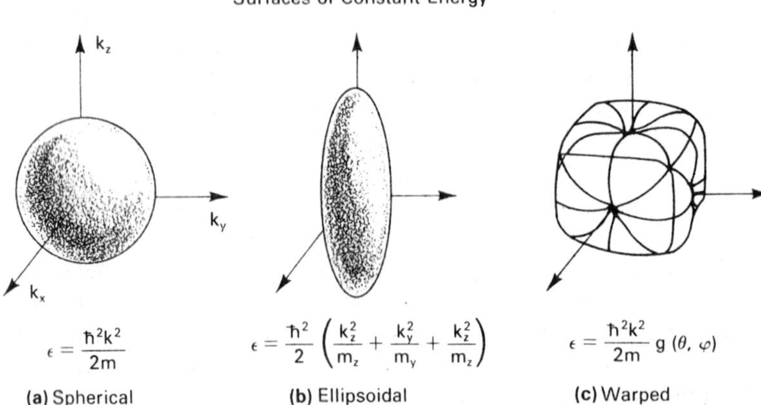

Fig. 1-5-8. Surfaces of equal energy in **k**-space (from C. Jacoboni and L. Reggiani, Bulk Hot-Electron Properties of Cubic Semiconductors, *Advances in Physics*, Vol. 28, No. 4, pp 493–553, (1979). (a) Spherical equal energy surface, (b) ellipsoidal equal energy surface, (c) "warped" equal energy surfaces.

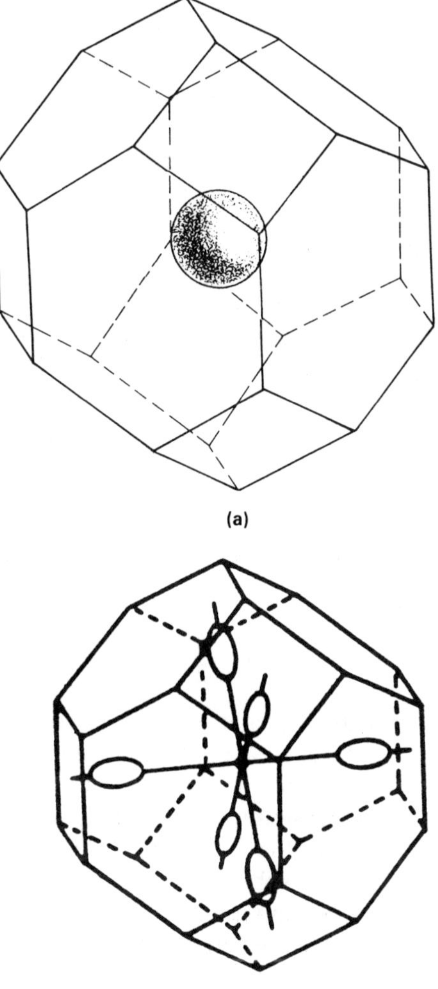

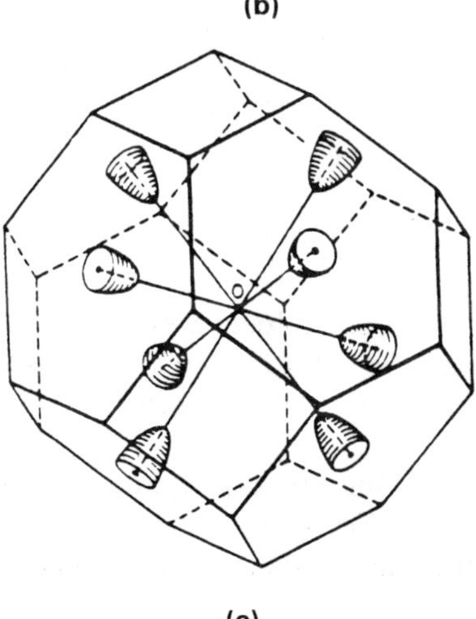

Fig. 1-5-9. Constant energy surfaces in the first Brillouin zone: (a) spherical constant energy surface for the lowest (Γ) minimum of the conduction band in GaAs, (b) constant energy surfaces for the lowest minima of the conduction band in Si. There are six ellipsoids along ⟨100⟩ axes with the centers of the ellipsoids located at about three quarters of the distance from the Brillouin zone center to the Brillouin zone boundary, (c) constant energy for the lowest minima of the conduction band in Ge (from Allen Nussbaum, *Semiconductor Device Physics*, © 1962 p. 207. Reprinted by permission of Prentice-Hall, Inc., Englewood Cliffs, NJ.) There are eight ellipsoids of revolution along ⟨111⟩ axes, and the Brillouin zone boundaries are in the middle of the ellipsoids.

The dependence of energy on the wave vector near the top of the valence band was calculated by Dresselhaus et al. (1955):

$$E(k) = E_V - \hbar^2 k^2 g(k)/2m^* \tag{1-5-28}$$

where

$$g = A_g \pm [B_g^2 + C_g^2\,(k_x^2k_y^2/k^4 + k_x^2k_z^2/k^4 + k_y^2k_z^2/k^4)]^{1/2} \tag{1-5-29}$$

Here A_g, B_g, and C_g are constants. Plus and minus signs in eq. (1-5-26) correspond to the heavy and light holes, respectively. These bands are called warped bands because of the "warped" surface of equal energy (see Fig. 1-5-8c).

More realistic models should take into account the deviation of the E vs. $\mathbf{k}$ relationship from parabolicity. In the simplest way it can be done by substituting for E in eq. (1-5-21) with

$$\gamma(E) = E(1 + \alpha E) \tag{1-5-30}$$

For example, for the minimum of the conduction band in the direct gap semiconductors such as GaAs we have

$$E(1 + \alpha E) = \frac{\hbar^2 k^2}{2m_n} \tag{1-5-31}$$

Eq. (1-5-31) takes into account the increase of the electron effective mass with energy, which can be noticed as the decrease in the curvature of E vs. $\mathbf{k}$ dependencies with an increase in E in Fig. 1-5-6.

Equations describing the dispersion relationships in the vicinity of the symmetry points may be derived using the *k-p method*, first developed by Kane (1957). In this method, eq. (1-5-3) is substituted into the Schrödinger equation. The resulting equation for the Bloch amplitude $u_{\mathbf{k}}(r)$ near the symmetry points of the Brillouin zone contains a term $\mathbf{k} \cdot \mathbf{p}$, where $\mathbf{p}$ is the momentum operator. This term is considered as a perturbation. (For this reason the approach is called the k-p method.) The solution of the equation for the Bloch amplitude $u_{\mathbf{k}}(r)$ is then found based on the assumption that $u_{\mathbf{k}}(r)$ can be presented as a linear combination of the wave functions corresponding to the top valence bands and the bottom conduction band. In particular, this method yields the following approximate expression for the nonparabolicity constant:

$$\alpha = \frac{1}{E_g}\left(1 - \frac{m_n}{m_e}\right)^2 \tag{1-5-32}$$

Here E_g is the energy gap, m_n is the effective electron mass, and m_e is the free electron mass (see, for example, Blakemore 1982). Band gaps and effective masses of some cubic semiconductors are given in Table 1-5-2 (after Jacoboni and Reggiani 1979).

All parameters characterizing the band structure are temperature dependent. According to Blakemore (1982), the temperature dependencies of energy gaps can be described by the following phenomenological equation:

$$E_g(T) = E_{go} - \alpha_{temp}T^2/(T + \beta_{temp}) \quad (1\text{-}5\text{-}33)$$

where the values of the energy gap E_{go} at $T = 0$ and coefficients α_{temp} and β_{temp} are given for Si, Ge, and GaAs in Table 1-5-3 (see also Fig. 1-5-10). For GaAs it is also useful to know that the energy gaps between the top of the valence band and the bottom of L and X minima, E_L and E_X, vary with temperature. The phenomenological equations describing these dependencies for GaAs are

$$E_L(T) = 1.815 - 6.05 \times 10^{-4}\, T^2/(T + 204)\ (\text{eV}) \quad (1\text{-}5\text{-}34)$$

and

$$E_X(T) = 1.981 - 4.60 \times 10^{-4}\, T^2/(T + 204)\ (\text{eV}) \quad (1\text{-}5\text{-}35)$$

where temperature T is in degrees Kelvin (from Blakemore 1982).

TABLE 1-5-2. BAND GAPS AND EFFECTIVE MASSES OF SOME CUBIC SEMICONDUCTORS (after Jacoboni and Reggiani 1979)

	E_Γ (eV)	E_L (eV)	E_Δ (eV)	E_{so} (eV)	Electron effective masses			Hole effective masses	
					m_l	m	m_t	m_{hh}	m_{hl}
Si	4.08	1.87	1.13	0.04	0.98	—	0.19	.53	.16
Ge	0.89	0.76	0.96	0.29	1.64	—	0.082	.35	.043
AlP	3.3	3.0	2.1	0.05	—	—	—	.63	.2
AlAs	2.95	2.67	2.16	0.28	2.0	—	—	.76	.15
AlSb	2.5	2.39	1.6	0.75	1.64	—	0.23	.94	.14
GaP	2.24	2.75	2.38	0.08	1.12	—	0.22	.79	.14
GaAs	1.42	1.71	1.90	0.34	—	0.067	—	.62	.074
GaSb	0.67	1.07	1.30	0.77	—	0.045	—	.49	.046
InP	1.26	2.0	2.3	0.13	—	0.080	—	.85	.089
InAs	0.35	1.45	2.14	0.38	—	0.023	—	.6	.027
InSb	0.17	1.5	2.0	0.81	—	0.014	—	.47	.015
ZnS	3.8	5.3	5.2	0.07	—	0.28	—	—	—
ZnSe	2.9	4.5	4.5	0.43	—	0.14	—	—	—
ZnTe	2.56	3.64	4.26	0.92	—	0.18	—	—	—
CdTe	1.80	3.40	4.32	0.91	—	0.096	—	—	—

TABLE 1-5-3. TEMPERATURE DEPENDENCE OF THE ENERGY GAPS (from Blakemore 1982)

Semiconductor	E_{go} (eV)	α_{temp} (eV/K^2)	β_{temp} (K)
Si	1.170	4.73×10^{-4}	636
Ge	0.7437	4.77×10^{-4}	235
GaAs	1.519	5.405×10^{-4}	204

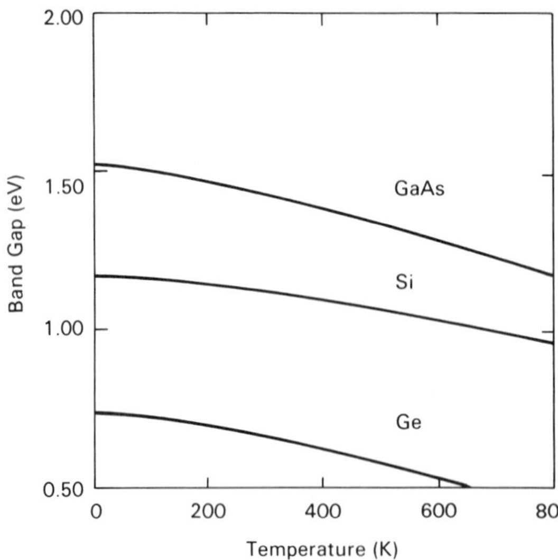

Fig. 1-5-10. Temperature dependence of energy gaps for silicon, germanium, and gallium arsenide. This plot is generated by the program PLOTF with subroutine PEGNI (see the subroutine listing in the Program Description Section).

1-6. ELECTRONS AND HOLES: SEMICONDUCTOR STATISTICS

Let us consider a crystal cube with dimensions L along the x-, y-, and z-axes and demand that electron wave functions

$$\psi_{\mathbf{k}}(\mathbf{r}) = e^{i\mathbf{k}\cdot\mathbf{r}}\, u_{\mathbf{k}}(\mathbf{r}) \tag{1-6-1}$$

(see eq. (1-5-3)) vanish at the crystal boundaries. This condition requires that

$$k_x L = 2\pi n_1 \tag{1-6-2}$$

$$k_y L = 2\pi n_2 \tag{1-6-3}$$

$$k_z L = 2\pi n_3 \tag{1-6-4}$$

where n_1, n_2, and n_3 are integers. Hence, the difference between the two closest allowed k_x values (or the difference between the two closest allowed k_y or k_z values) is $2\pi/L$ and each allowed value of k with coordinates k_x, k_y, and k_z occupies volume $(2\pi/L)^3$ in the reciprocal space. According to the Pauli exclusion principle applied to a crystal, only two electrons (with opposite spins) may have the same wave vector, **k.** If we use an extended band scheme (shown in Fig. 1-5-4a) and start filling energy states with valence electrons (two valence electrons per each state with a given value of **k**) from the bottom up, we will fill a certain volume (a sphere with radius $\mathbf{k}_{Fv}$, the Fermi wave vector of valence electrons) in the reciprocal space. The volume of the sphere, V_{Fv}, is equal to $4\pi\mathbf{k}_{Fv}{}^3/3$. The total number of states, N_{states}, inside the sphere is given by

$$N_{\text{states}} = V_F/(2\pi/L)^3 = \mathbf{k}_{Fv}{}^3 L^3/(6\pi^2) \tag{1-6-5}$$

The total number of valence electrons in the crystal, N_{val}, is equal to $2N_{states}$ (because two electrons with different spins share each occupied state). Hence,

$$2k_{Fv}^3 L^3/(6\pi^2) = N_{val} \tag{1-6-6}$$

and

$$\mathbf{k}_{Fv} = (3\pi^2 n_{val})^{1/3} \tag{1-6-7}$$

where $n_{val} = N_{val}/L^3$ is the volume concentration of the valence electrons (four per atom for silicon, germanium, GaAs, and related compounds).

In this calculation we assumed that all states with wave vectors smaller than k_F are filled and all states with wave vectors higher than $\mathbf{k}_F$ are empty. This is only true when the temperature tends to zero. For a more realistic case of a finite temperature, T, the probability of having an energy state occupied is given by the Fermi-Dirac occupation function, f_n:

$$f_n = \frac{1}{1 + \exp\left(\frac{E_n - E_F}{k_B T}\right)} \tag{1-6-8}$$

Here $E_n(\mathbf{k})$ is the energy of electrons and E_F is the Fermi level. [A detailed discussion of the Fermi-Dirac occupation function is given, for example, by Blokhintsev (1964)]. Later we will discuss the factors determining the position of the Fermi level with respect to the energy bands in a crystal. For now, let us simply state that in semiconductors, the Fermi level is located either in the energy gap between the conduction band and valence band, or in the conduction band, relatively close to the bottom (typically within 100 to 200 meV or so), or in the valence band, relatively close to the top. Function f_n is plotted in Fig. 1-6-1 for different values of temperature. As can be seen from this figure, all states below approximately $E_F - 3k_BT$ are essentially filled ($f \simeq 1$ for such energies), all states above approximately $E_F + 3k_BT$ are practically empty ($f << 1$ for such energies), and $f_n(E_F) = 1/2$. When $E_n - E_F >> k_BT$, the Fermi function may be approximated as

$$f_n \approx \exp\left(\frac{E_F - E_n}{k_B T}\right) \tag{1-6-8a}$$

If we take into account the finite probability of occupation of energy states, the expression for the concentration of electrons, under the conditions of thermal equilibrium, is given by

$$n = \frac{2}{(2\pi)^3} \int f_n[E_n(\mathbf{k})]\, d\mathbf{k} \tag{1-6-9}$$

If we use the energy bands folded into the first Brillouin zone (see Fig. 1-5-4b), the integration in eq. (1-6-9) is over the first Brillouin zone. If we want to find the concentration of all valence electrons, $E_n(\mathbf{k})$ should be taken as multivalued function of k that includes all branches (see Fig. 1-5-4b). If we would like to find the

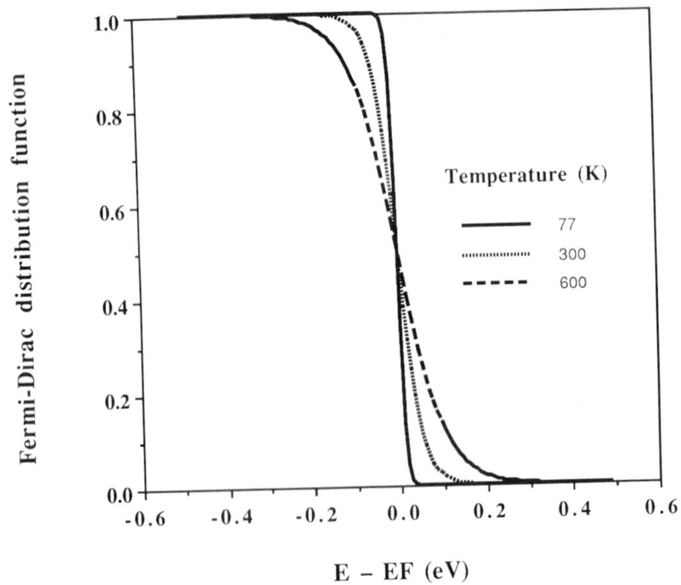

Fig. 1-6-1. Fermi-Dirac occupation function for different values of temperature. This plot is generated by program PLOTF with subroutine PFD (see the subroutine listing in the Program Description Section).

concentration of electrons in the conduction band alone, we should only consider one branch of $E_n(\mathbf{k})$ corresponding to the conduction band.

Let us consider the concentration of electrons in the conduction band. If the Fermi level, E_F, is in the conduction band and temperature is low, $f_n \approx 1$ for filled states with wave vectors smaller than

$$\mathbf{k}_F = [2m_n(E_F - E_c)]^{1/2}/\hbar$$

and $f_n \approx 0$ for empty states with wave vectors larger than k_F. The wave vector k_F is called the Fermi vector of the conduction band electrons. A semiconductor with the Fermi level, E_F, such that $E_F - E_c > 3k_BT$ is called a degenerate semiconductor. (In a similar way, a p-type semiconductor is called a degenerate semiconductor if $E_v - E_F > 3k_BT$.) At low temperatures, in an n-type degenerate semiconductor with an isotropic (independent of direction) effective mass, electrons occupy the sphere of the allowed states in the first Brillouin zone, called the Fermi sphere (see Fig. 1-6-2).

For a degenerate semiconductor with the Fermi level in the conduction band, the integral in eq. (1-6-9) is simply the volume of the Fermi sphere in **k**-space with radius k_F, i.e., $4\pi k_F^3/3$, and we obtain the following expression for k_F:

$$k_F = (3\pi^2 n)^{1/3} \tag{1-6-10}$$

where n is the concentration of electrons in the conduction band (compare with eq. (1-6-7)).

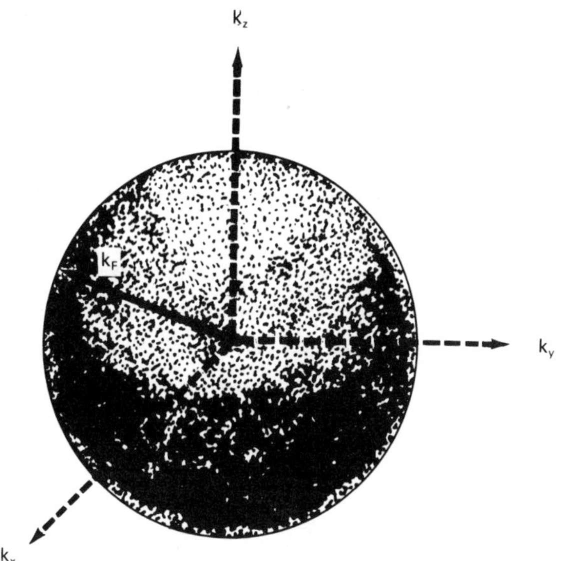

Fig. 1-6-2. Fermi sphere.

The electron concentration in the conduction band is a more important number than the total concentration of valence electrons. Indeed, there are many unoccupied levels in the conduction band, so that electrons in the conduction band can change energy (and, hence, acquire velocity) in an electric field, leading to an electric current. Most of the electronic states (those not in the conduction band and not near the top of the highest filled valence band) remain occupied when the electric field is applied to the sample. Hence, they do not contribute to the electric current.

To describe the contribution of mostly filled states in the valence band to the electronic conduction, we introduce the important concept of a *hole*. This concept is illustrated by Fig. 1-6-3, which shows a line of classical dancers. In the beginning there is a space at the left side of the stage, and ballerinas jump one by one into the empty space to the left. What we see can be interpreted as a motion of this empty space on stage from the left to the right, rather than the jumps of many ballerinas. In a similar way, we can follow the motion of an unoccupied state in a valence band. It can be visualized in real space as a motion of an empty space on a bond in a crystal lattice that propagates when it is filled by electrons from neighboring bonds (see Fig. 1-6-4).

A hole can be represented as a positive "particle" (corresponding to the absence of a negative electron), and the occupation Fermi-Dirac function for a hole can be calculated as

$$f_p(E) = 1 - f_n(E), \qquad (1\text{-}6\text{-}11)$$

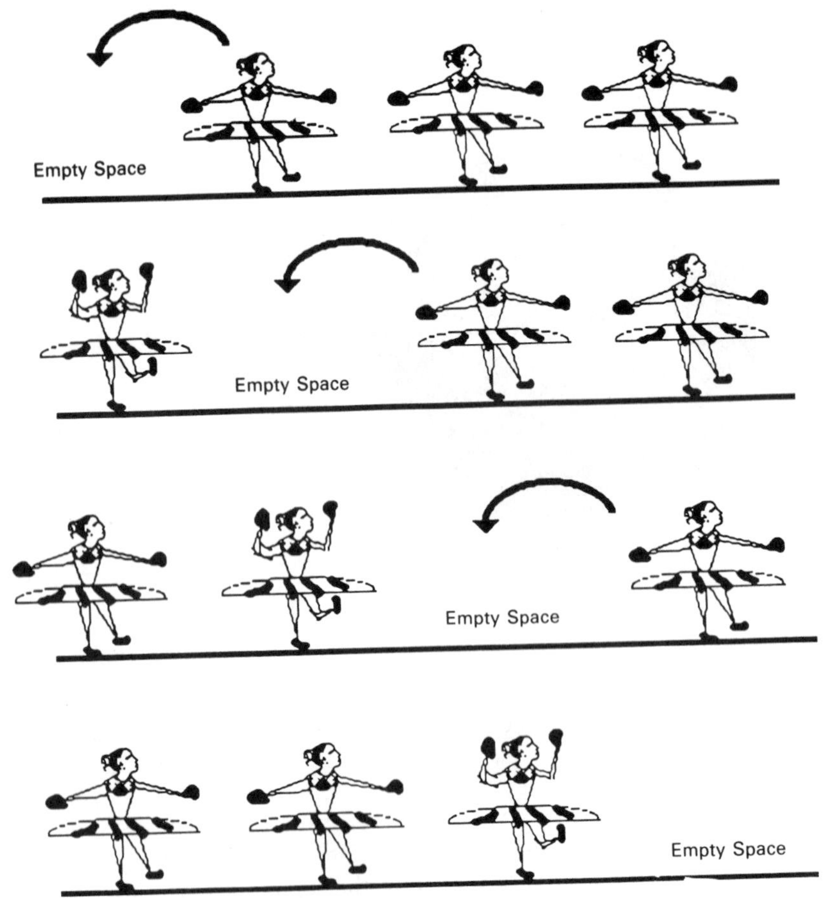

Fig. 1-6-3. Concept of a *hole*.

Hence

$$f_p = 1 - \frac{1}{1 + \exp\left(\dfrac{E - E_F}{k_B T}\right)} = \frac{1}{1 + \exp\left(\dfrac{E_F - E}{k_B T}\right)} \tag{1-6-12}$$

The hole concentration in the valence band can be found similarly to the electron concentration in the conduction band:

$$p = \frac{2}{(2\pi)^3} \int f_p \, [E_p(\mathbf{k})] \, d\mathbf{k} \tag{1-6-13}$$

The integration in eqs. (1-6-9) and (1-6-13) is over the first Brillouin zone. In the simplest case of the parabolic bands,

$$E_n(\mathbf{k}) = E_c + \frac{\hbar^2 k^2}{2m_n} \tag{1-6-14}$$

and

$$E_p(\mathbf{k}) = E_v - \frac{\hbar^2 k^2}{2m_p} \tag{1-6-15}$$

where E_c and E_v are the bottom of the conduction band and the ceiling of the valence band, respectively. The substitution of eq. (1-6-14) into eq. (1-6-9) yields

$$n = N_c\, F_{1/2}\,(\eta_n) \tag{1-6-16}$$

where

$$N_c = 2\left(\frac{m_n k_B T}{2\pi\hbar^2}\right)^{3/2} \tag{1-6-17}$$

is called the effective density of states for the conduction band,

$$\eta_n = (E_F - E_c)/k_B T \tag{1-6-18}$$

and

$$F_{1/2}(\eta) = (2/\sqrt{\pi})\int_0^\infty \frac{x^{1/2}\,dx}{[1 + \exp(x - \eta)]} \tag{1-6-19}$$

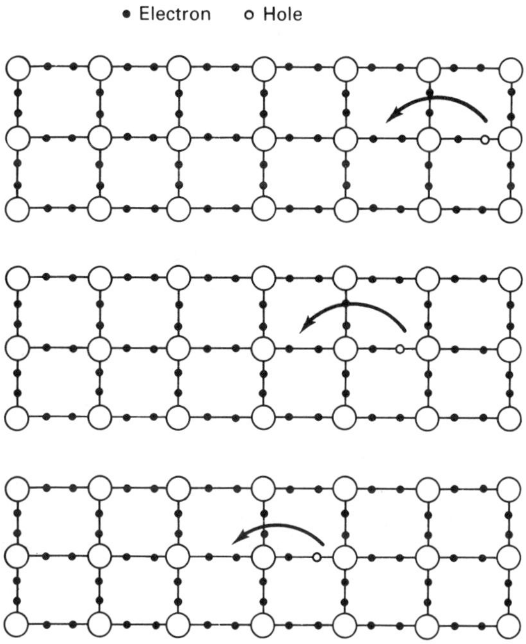

• Electron ○ Hole

Fig. 1-6-4. Propagation of a hole in a crystal lattice.

is the Fermi integral. As shown in Appendix 21, when $\eta << -1$

$$F_{1/2}(\eta) \approx \exp(\eta) \tag{A21-3}$$

When $\eta >> 1$

$$F_{1/2}(\eta) \approx \frac{4\eta^{3/2}}{3\sqrt{\pi}} \tag{A21-4}$$

For $-10 < \eta < 10$ the Fermi integral $F_{1/2}(\eta)$ can be interpolated by the following expression (see Fig. A21-1):

$$F_{1/2}(\eta) = \exp(-0.32881 + 0.74041\eta - 0.045417\eta^2 - 8.797 \times 10^{-4}\,\eta^3 + 1.5117 \times 10^{-4}\eta^4)$$

In a similar fashion, we find that for holes,

$$p = N_v\, F_{1/2}\,(\eta_p) \tag{1-6-20}$$

where

$$N_v = 2\left(\frac{m_p k_B T}{2\pi\hbar^2}\right)^{3/2} \tag{1-6-21}$$

is the effective density of states for the valence band,

$$\eta_p = \frac{E_v - E_F}{k_B T} \tag{1-6-22}$$

In semiconductors, such as Si, Ge, and GaAs, dependencies of $E_n(\mathbf{k})$ and $E_p(\mathbf{k})$ are more complicated. In particular, in Si and Ge the electronic effective mass is anisotropic, so that an effective mass tensor should be introduced. Also, there are several equivalent minima in the conduction band (see Section 1-5). In this case, the effective mass, m_n, in eq. (1-6-17) should be replaced by the density-of-states effective mass,

$$m_{dn} = Z^{2/3}\,(m_x m_y m_z)^{1/3} \tag{1-6-23}$$

where Z is the number of the equivalent minima (see Table 1-5-1).

In nearly all cubic semiconductors there are "light" and "heavy" holes (see Section 1-5) and, as a consequence, the effective mass, m_p, in eq. (1-6-12) should be replaced by the density of states effective mass, m_{dp}, that is shown to be given by

$$m_{dp} = (m_{ph}{}^{3/2} + m_{pl}{}^{3/2})^{2/3} \tag{1-6-24}$$

When the Fermi level is in the energy gap and is separated by more than several thermal energies from the edges of the conduction and valence bands ($\eta_n < -3$, $\eta_p < -3$), a semiconductor is called nondegenerate. In this important case

$$F_{1/2}(\eta) \approx \exp(\eta) \tag{1-6-25}$$

so that

$$n = N_c \exp\left(\frac{E_F - E_c}{k_B T}\right) \quad (1\text{-}6\text{-}26)$$

$$p = N_v \exp\left(\frac{E_v - E_F}{k_B T}\right) \quad (1\text{-}6\text{-}27)$$

In the opposite limiting case when the Fermi level enters the conduction or the valence band, a semiconductor is called a *degenerate semiconductor.* In this case

$$F_{1/2}(\eta) = \frac{4}{3\sqrt{\pi}}\,\eta^{3/2} \quad (1\text{-}6\text{-}28)$$

and

$$E_F - E_c \approx \frac{\hbar^2 k_F^2}{2m_{dn}} \quad (1\text{-}6\text{-}29)$$

and

$$n \approx \frac{1}{3\pi^2}\left[\frac{2m_{dn}\,(E_F - E_c)}{\hbar^2}\right]^{3/2} \quad (1\text{-}6\text{-}30)$$

Equation (1-6-30) follows from eq. (1-6-10).

Similarly, when the Fermi level is in the valence band ($\eta_p > 3$), we find that

$$p \approx \frac{1}{3\pi^2}\left[\frac{2m_{dp}\,(E_v - E_F)}{\hbar^2}\right]^{3/2} \quad (1\text{-}6\text{-}31)$$

For the intermediate values of η_n and η_p we have to use eqs. (1-6-16) and (1-6-20). The dependence of $F_{1/2}$ on η is shown in Fig. A21-1.

For a nondegenerate semiconductor, we find from eqs.(1-6-26) and (1-6-27) that

$$np = N_v N_c \exp(-E_g/k_B T) \quad (1\text{-}6\text{-}32)$$

where $E_g = E_c - E_v$ is the energy gap. This equation can be rewritten as

$$np = n_i^2 \quad (1\text{-}6\text{-}33)$$

where

$$n_i = (N_v N_c)^{1/2} \exp[-E_g/(2k_B T)] \quad (1\text{-}6\text{-}34)$$

is called the intrinsic carrier concentration.

As can be seen from eqs. (1-6-32) and (1-6-33), for a nondegenerate semiconductor the *np* product is independent of the position of the Fermi level and is determined by densities of states in the valence and conduction bands, energy gap, and temperature.

Equations (1-6-9) and (1-6-13) can be rewritten as

$$n = \int_0^{E_{tn}} (dn/dE)dE = \int_0^{E_{tn}} f_n(E)\, g_n(E)\, dE \tag{1-6-35}$$

$$p = \int_{E_{bp}}^{0} (dp/dE)dE = \int_{E_{bp}}^{0} f_p(E)\, g_p(E)\, dE \tag{1-6-36}$$

where E_{tn} and E_{bp} are energies corresponding to the highest energy in the conduction band and to the lowest energy in the valence band, respectively, and $g_n(E)$ and $g_p(E)$ are densities of states in the conduction and valence band per unit energy, respectively (these densities of states differ from effective densities of states, N_c and N_v). Comparing eq. (1-6-35) with eqs. (1-6-16) and (1-6-19), we find that

$$g_n(E) = (2/\pi^{1/2})N_c\,(E - E_c)^{1/2}/(k_BT)^{3/2} \tag{1-6-37}$$

or

$$g_n(E) = (2^{1/2}/\pi^2)(m_n/\hbar^2)^{3/2}\,(E - E_c)^{1/2} \tag{1-6-38}$$

Similarly, for holes

$$g_p(E) = (2^{1/2}/\pi^2)(m_p/\hbar^2)^{3/2}\,(E_v - E)^{1/2} \tag{1-6-39}$$

Equation (1-6-38) can be also derived by counting the number of states with wave vectors **k** between **k** and **k** + d**k**. The corresponding volume in **k**-space is equal to $4\pi k^2 dk$. The density of allowed states is equal to the number of allowed points in this volume in **k** space times two (the factor of 2 takes into account two possible values of spin). The density of allowed points in **k**-space is $V/(2\pi)^3$. Here V is the crystal volume (see eqs. (1-6-2) to (1-6-4) and related discussion). Hence, the total number of states with values of k between k and $k + dk$ is

$$dN = 8\pi V k^2\, dk/(2\pi)^3 \tag{1-6-40}$$

Taking into account that

$$k^2 = 2m_n(E - E_c)/\hbar^2 \tag{1-6-41}$$

and

$$2k\, dk = 2m_n dE/\hbar^2 \tag{1-6-42}$$

we obtain

$$\begin{aligned} dN &= 8\pi V[2m_n(E - E_c)/\hbar^2]\,(2m_n dE/\hbar^2)\;/\;\{2\,[2m_n(E - E_c)/\hbar^2]^{1/2}(2\pi)^3\} \\ &= g_n(E)\, dE \end{aligned} \tag{1-6-43}$$

For the unit volume crystal this equation leads to eq. (1-6-38).

Qualitative dependencies of the density of states, $g_n(E)$, the distribution function, $f_n(E)$, and the electron density, dn/dE, on energy for nondegenerate and degenerate semiconductors are shown in Fig. 1-6-5.

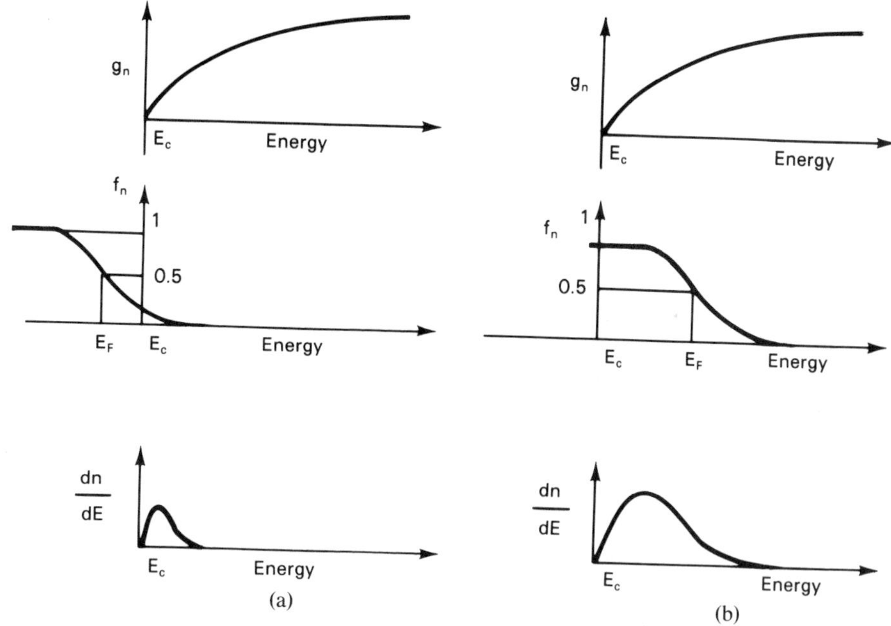

Fig. 1-6-5. The density of states $g_n(E)$, distribution function $f_n(E)$, and the electron density per unit energy dn/dE vs. energy for nondegenerate (a) and degenerate (b) semiconductors.

As will be discussed in Chapter 7, *quantum well structures* have found important applications in novel semiconductor devices. In such structures a thin region of a narrow-gap semiconductor is sandwiched between layers of a wide-band-gap semiconductor, as shown in Fig. 1-6-6. If the narrow-gap semiconductor layer is thin enough, the motion of carriers in the direction perpendicular to the heterointerfaces is quantized. The lowest energy levels can be estimated using eq. (1-2-20):

$$E_n - E_c = \frac{\pi^2 \hbar^2}{2 m_n a^2} n^2 \tag{1-6-44}$$

(as long as E_n is well below the bottom of the conduction band in the wide-band material). Here n is the quantum number labeling the levels, and a is the thickness of the quantum well. For the quantization, the difference between the levels should be much larger than the thermal energy, $k_B T$, i.e.,

$$\frac{\pi^2 \hbar^2}{2 m_n a^2} >> k_B T \tag{1-6-45}$$

Using this condition, we find that, for example, for GaAs, where $m_n/m_e \approx 0.067$, the levels are quantized at room temperature ($T = 300$ K) when $a << 147$ Å.

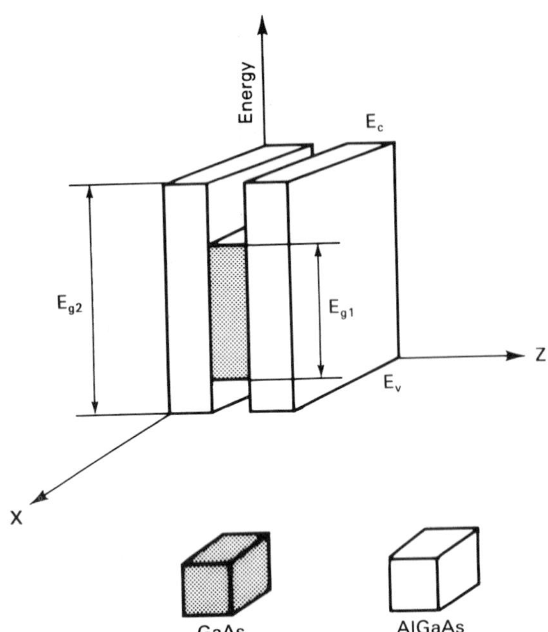

Fig. 1-6-6. Quantum well structure, where a thin region of narrow gap semiconductor (such as GaAs) is sandwiched between layers of wide band semiconductor (such as AlGaAs).

In the direction parallel to the heterointerfaces, the wave function can be described as a two-dimensional Bloch function with the dispersion relation for the conduction band

$$E - E_n = [\hbar^2/(2m_n)](k_x^2 + k_y^2) \tag{1-6-46}$$

(notice the absence of a k_z component; the motion in the z direction is quantized, so that the wave function in the z direction is completely different from the plane wave proportional to exp $(ik_z z)$). In other words, each quantum level given by eq. (1-6-44) corresponds to an energy subband as shown in Fig. 1-6-7.

The density of states for each subband can be found using an approach similar to that used previously for a three-dimensional density of states, i.e., by counting the number of states with wave vectors **k** between **k** and **k** + d**k**. The corresponding area in **k**-space is equal to $2\pi k\, dk$. The density of allowed states is equal to the number of allowed values of k in this area in **k** space times two (the factor of 2 takes into account two possible values of spin). The density of allowed points in **k**-space for the unit-size sample is $1/(2\pi)^2$. Hence, the total number of states with values of **k** between **k** and **k** + d**k** is

$$dN = 4\pi k\, dk/(2\pi)^2 \tag{1-6-47}$$

Taking into account that

$$k^2 = 2m_n(E - E_n)/\hbar^2 \tag{1-6-48}$$

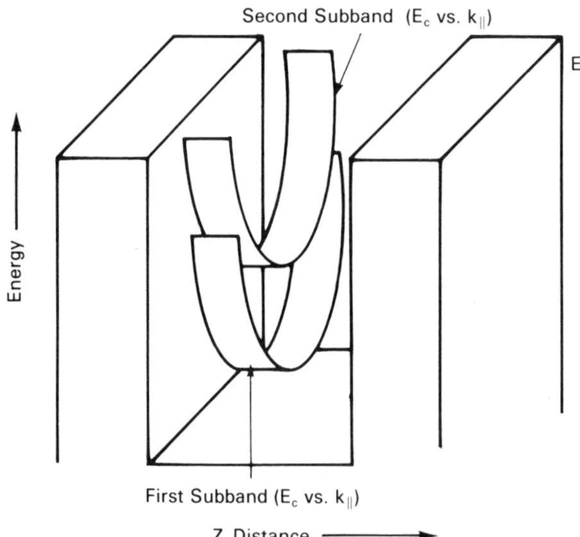

Fig. 1-6-7. Energy subbands in quantum well structure.

and

$$2k\,dk = 2m_n dE/\hbar^2 \tag{1-6-49}$$

we obtain

$$\begin{aligned} dN &= 4\pi[2m_n(E - E_n)/\hbar^2]^{1/2}(2m_n dE/\hbar^2) \,/\, \{2[2m_n(E - E_n)/\hbar^2]^{1/2}(2\pi)^2\} \\ &= D\,dE \end{aligned} \tag{1-6-50}$$

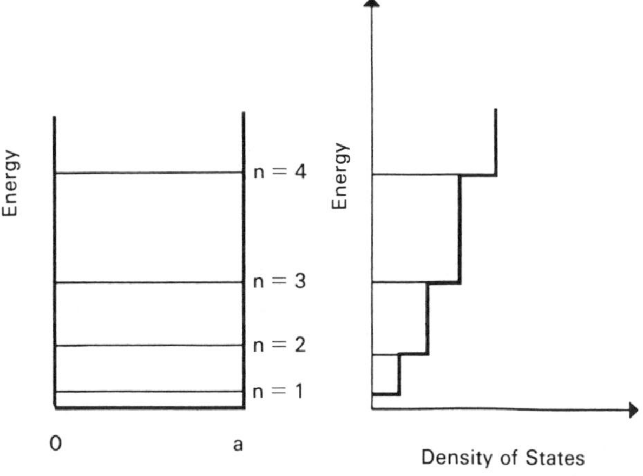

Fig. 1-6-8. Density of states for quantum well structure.

where the density of states, D, for one subband is given by

$$D = m_n/(\pi\hbar^2) \tag{1-6-51}$$

As can be seen from Fig. 1-6-7, the states of the first (bottom) subband overlap with the states of the second subband for energies larger than the second (from the bottom) energy level, and so on. As a consequence, the overall density of states has a staircase shape, as shown in Fig. 1-6-8.

1-7. INTRINSIC, EXTRINSIC, AND COMPENSATED SEMICONDUCTORS

To determine n and p from eqs. (1-6-16) and (1-6-20) one needs to know the position of the Fermi level, E_F. It is found from the requirement of the electric neutrality for a semiconductor. This condition follows from a very general principle of the conservation of the electric charge. The electric charges cannot be created or destroyed, only separated. Under equilibrium conditions, a semiconductor sample is neutral. For a pure (intrinsic) semiconductor, this means that the concentration of negatively charged electrons must be equal to the concentration of positively charged holes:

$$p = n = n_i \tag{1-7-1}$$

Using eq. (1-6-34) for the intrinsic concentration of carriers, n_i, for a nondegenerate semiconductor

$$n = p = n_i = (N_c N_v)^{1/2} \exp\left(-\frac{E_g}{2k_BT}\right) \tag{1-7-2}$$

and substituting eqs. (1-6-17), (1-6-21), (1-6-26), and (1-6-27) into the equation $n = p$, we find the position of the Fermi level in pure (undoped) semiconductor. Such a semiconductor is called an intrinsic semiconductor, and the corresponding Fermi level is called the intrinsic Fermi level, $E_F = E_i$,

$$E_i = \frac{E_c + E_v}{2} + \frac{3}{4} k_BT \ln \frac{m_{dp}}{m_{dn}} \tag{1-7-3}$$

The intrinsic concentration of carriers for Si, Ge, and GaAs is shown as a function of temperature in Fig. 1-7-1.

The concentrations of electrons and holes, the condition of neutrality (see eq. (1-7-1)), and the position of the Fermi level may all be changed by doping, i.e., by introducing impurities into a semiconductor. Doped semiconductors are called *extrinsic semiconductors*. To achieve n-type doping (which supplies electrons to the conduction band), one introduces impurity atoms with a larger number of valence electrons than the host atoms that they substitute. For example, impurity atoms belonging to the fifth column of the periodic table (and, hence, having five valence electrons) will act as n-type dopants in silicon, which belongs to the fourth

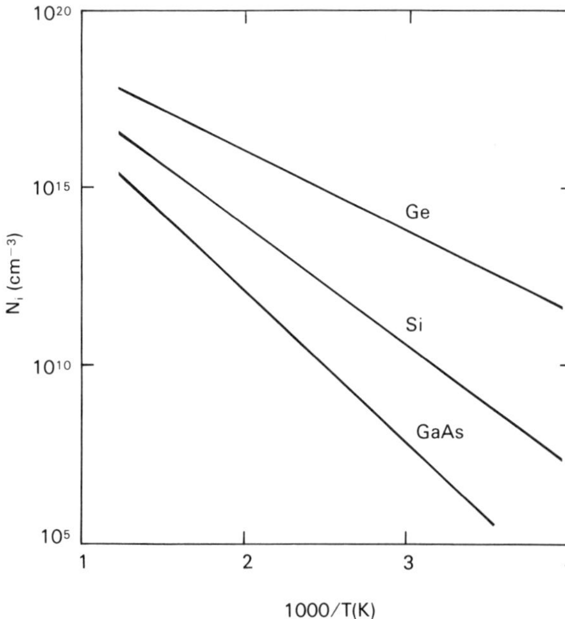

Fig. 1-7-1. Intrinsic carrier concentrations for silicon, germanium, and gallium arsenide. The temperature dependence of energy gaps used in the calculation is given by eq. (1-6-15). This plot is generated by program PLOTF with subroutines PEGNI (see the subroutine listing in the Program Description Section).

column of the periodic table and has four valence electrons. Antimonide (Sb), phosphorus (P), and arsenic (As) act as n-type dopants in silicon. The n-type dopants are called donors. They are neutral when occupied by electrons and become positively charged when they "donate" their excess valence electrons to the conduction band.

Atoms belonging to the third column of the periodic table act as p-type dopants (acceptors) in silicon. Examples are boron (B), aluminum (Al), gallium (Ga), and indium (In). Acceptor impurities are negatively charged when they "accept" extra electrons from the valence band. This means that they are negatively charged when filled and neutral when empty. If both donors and acceptors are present in a semiconductor, the semiconductor is called compensated. In other words, the donors and acceptors donate and accept electrons from each other rather than from the valence and conduction bands. They compensate each other in their impact on changing the free carrier concentration.

In an energy band scheme, dopants are characterized by the discrete energy levels which they introduce. The energy levels of *shallow donors* are located below and close to the bottom of a conduction band. The energy levels of shallow acceptors are above and close to the top of a valence band. This means that a very small energy (in many cases less than a thermal energy k_BT) is needed to ionize a shallow impurity and supply an electron from a shallow donor into a conduction band, or a hole from a shallow acceptor into a valence band.

Fig. 1-7-2 illustrates the concept of doping by showing a phosphorus atom (a donor) and a boron atom (an acceptor) incorporated into silicon, as well as corre-

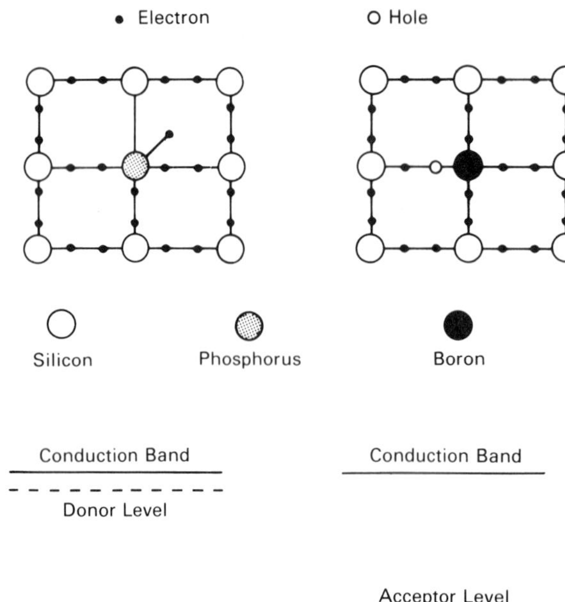

Fig. 1-7-2. Schematic diagram of a silicon crystal structure with a silicon atom replaced by a phosphorus atom and a boron atom.

sponding energy band diagrams. Ionization energies for different impurities in Ge, Si, and GaAs are shown in Fig. 1-7-3 (from Sze 1981). As can be seen from the figure, only a few impurities behave as shallow donors or acceptors even at room temperature ($k_BT/q \approx 0.02584$ eV). Some impurity atoms, *deep donors* and *deep acceptors*, create "deep" levels, which are separated from the conduction or valence band edges by more than several k_BT/q even at room temperature. Also, certain impurity atoms (such as Si in GaAs) may behave either as donors or acceptors, depending on which site (Ga or As) they occupy. Some impurities, such as copper or gold, give rise to multiple levels (see Fig. 1-7-3). Such impurities are said to be amphoteric, and their behavior is quite complicated.

The ratio of the concentrations of the filled (N_{d^0}) and empty (N_{d^+}) donors is given by

$$\frac{N_{d^0}}{N_{d^+}} = g_d \exp\left(\frac{E_F - E_d}{k_BT}\right) \tag{1-7-4}$$

where E_d is the donor level energy, E_F is the Fermi level, and g_d is the degeneracy factor. In the simplest case $g_d = 2$ (because of two possible values for the electron spin), and E_d can be calculated using the equation for the ground state of a hydrogen-atom-like impurity. This means that the difference between the bottom of the conduction band and the donor energy level is estimated as the Bohr energy (see eq. (1-2-25)), where the free electron mass m is substituted by the electron

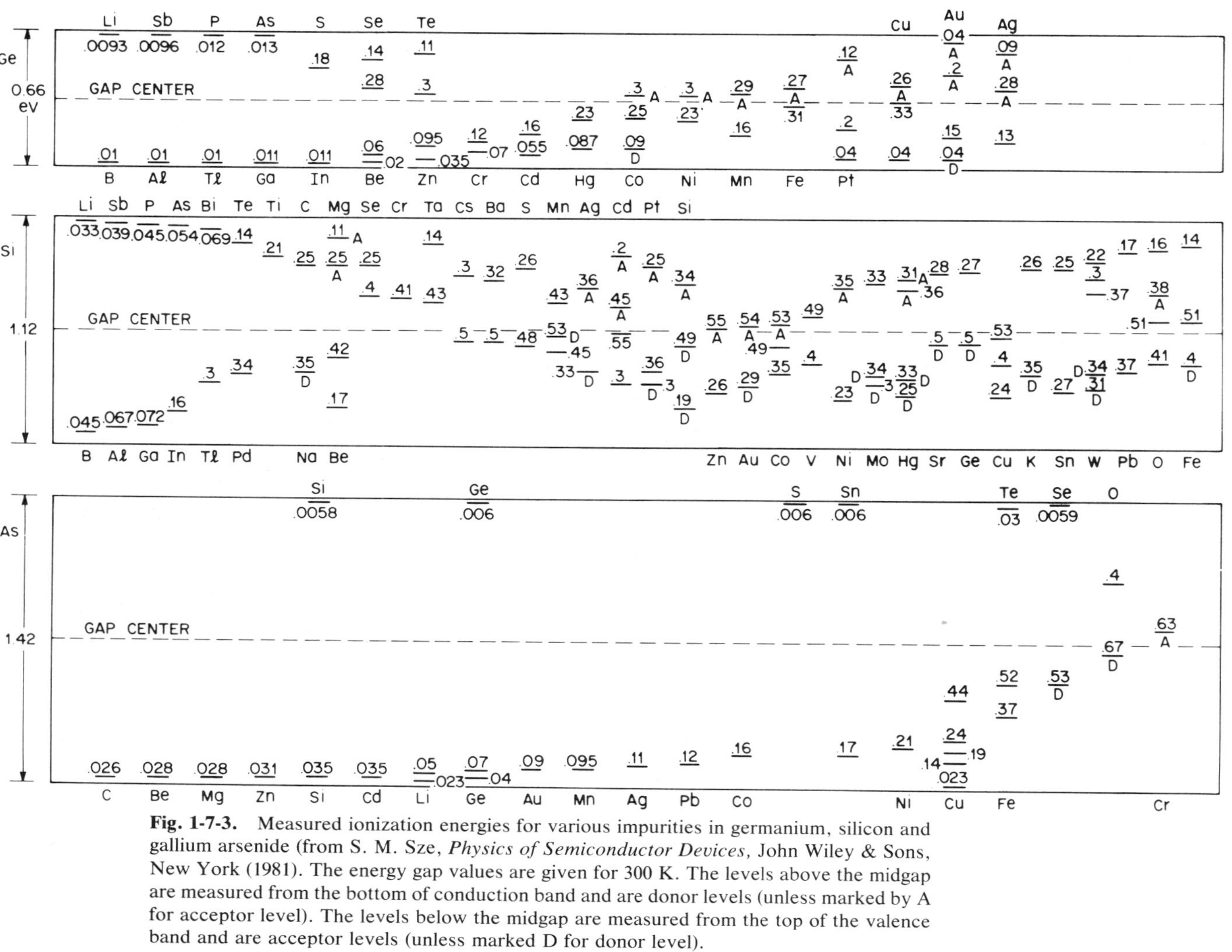

Fig. 1-7-3. Measured ionization energies for various impurities in germanium, silicon and gallium arsenide (from S. M. Sze, *Physics of Semiconductor Devices*, John Wiley & Sons, New York (1981). The energy gap values are given for 300 K. The levels above the midgap are measured from the bottom of conduction band and are donor levels (unless marked by A for acceptor level). The levels below the midgap are measured from the top of the valence band and are acceptor levels (unless marked D for donor level).

effective mass, m_n, and the free space permittivity, ε_o, is substituted by the static dielectric permittivity, ε_s, of the semiconductor crystal:

$$E_c - E_d \approx \left(\frac{\varepsilon_o}{\varepsilon_s}\right)^2 \left(\frac{m_n}{m_e}\right) \frac{q^4 m_e}{32\pi^2 \varepsilon_o^2 \hbar^2} \tag{1-7-5}$$

or

$$E_c - E_d \text{ (eV)} \approx 13.6 \left(\frac{\varepsilon_o}{\varepsilon_s}\right)^2 \left(\frac{m_n}{m_e}\right) \tag{1-7-5a}$$

In fact, the experimental values of E_d and g_d may be quite different from these simple estimates.

The Bohr radius of an electron on a shallow donor is given by

$$a_{Bd} = \frac{4\pi\varepsilon_o \hbar^2}{m_e q^2} (\varepsilon_s/\varepsilon_o)(m_e/m_n) \tag{1-7-6}$$

(compare with eq. (1-2-26) for the Bohr radius $a_B = 4\pi\varepsilon_o\hbar^2/(m_e q^2) = 0.52917$ Å). Coulomb forces in semiconductor are $(\varepsilon_s/\varepsilon_o)$ times smaller than in vacuum, and in many semiconductors the electron effective mass is smaller than the electron free mass. Hence, $E_c - E_d$ is typically much smaller than E_B, and a_{Bd} is much greater than a_B. For example, in GaAs $\varepsilon_s/\varepsilon_o = 13.1$ and $m_e/m_n = 14.9$. Hence, $a_{Bd} \approx 103$ Å, and the electronic impurity state extends over many lattice constants (the lattice constant of GaAs is 5.6533 Å at room temperature). This is illustrated by Fig. 1-7-4, which compares a circle with Bohr radius for a shallow donor in GaAs with a lattice constant of GaAs.

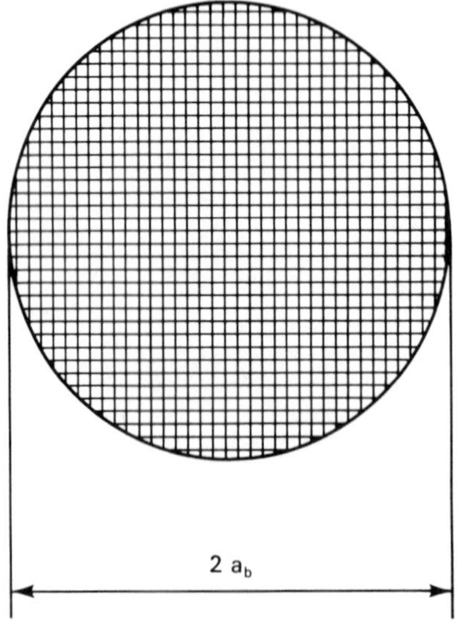

Fig. 1-7-4. The Bohr radius for a shallow donor in GaAs. Each square represents a unit cell of GaAs.

Similarly, for acceptor levels we have

$$\frac{N_{a^-}}{N_{a^o}} = \frac{1}{g_a} \exp\left(\frac{E_F - E_a}{k_B T}\right) \tag{1-7-7}$$

where E_a is the acceptor level energy. In the simplest case of a shallow hydrogen-like acceptor level,

$$E_a - E_v \approx \left(\frac{\varepsilon_o}{\varepsilon_s}\right)^2 \left(\frac{m_p}{m_e}\right) \frac{q^4 m_e}{32\pi^2 \varepsilon_o^2 \hbar^2} \tag{1-7-8}$$

and $g_a = 4$ (a factor of 2 in g_a is due to the electron spin and another factor of 2 is due to the double degeneracy of the valence band in the cubic semiconductors). Again, more reliable values of E_a and g_a should be found from the experimental data (see Fig. 1-7-3).

The position of the Fermi level, E_F, can be found from the condition of neutrality

$$p + \sum_j Z_j N_j - n = 0 \tag{1-7-9}$$

where Z_j is the charge (in units of the electronic charge) of an impurity of type j and N_j is the concentration of such impurities. When a semiconductor is doped by shallow donors with concentration N_d, we find from eq. (1-7-9) that

$$n = p + N_{d^+} \tag{1-7-10}$$

This equation may be solved together with eq. (1-6-32), which may be rewritten as

$$np = n_i^2 \tag{1-7-11}$$

yielding the concentration of electrons, n_n, in the n-type semiconductor

$$n_n = \frac{\sqrt{(N_{d^+})^2 + 4n_i^2} + N_{d^+}}{2}. \tag{1-7-12}$$

For shallow donors $N_{d^+} \approx N_d$, and eq. (1-7-12) may be simplified:

$$n_n \approx \frac{\sqrt{(N_d)^2 + 4n_i^2} + N_d}{2}. \tag{1-7-12a}$$

In a general case

$$N_{d^+} = \frac{N_d}{1 + g_d \exp\left(\frac{E_F - E_d}{k_B T}\right)}. \tag{1-7-13}$$

(This equation is obtained from eq. 1-7-4, taking into account that $N_d = N_{d^+} + N_{d^o}$.) Substituting eq. (1-6-26) and eq. (1-7-13) into eq. (1-7-9) and assuming $n >> p$ we obtain

$$N_c \exp(\eta) = \frac{N_d}{1 + g_d \exp(\eta) \exp\left(\frac{E_c - E_d}{k_B T}\right)} \tag{1-7-14}$$

where $\eta = \dfrac{E_F - E_c}{k_B T}$. Equation (1-7-14) is a quadratic equation with respect to $X = \exp(\eta)$. Solving eq. (1-7-14) we find

$$E_F = E_D + k_B T \ln \left\{ \frac{1}{2g_d} \sqrt{1 + 4g_d \frac{N_d}{N_c} \left[\exp\left(\frac{E_c - E_d}{k_B T} \right) - 1 \right]} \right\} \tag{1-7-15}$$

For $T \to 0$,

$$N_c << N_d \exp\left(\frac{E_c - E_d}{k_B T} \right)$$

and eq. (1-7-15) reduces to

$$E_F = \frac{E_c + E_d}{2} + \frac{k_B T}{2} \ln \frac{N_d}{g_d N_c(T)} \tag{1-7-16}$$

The electron concentration may now be found from eq. (1-6-26):

$$n = N_c \exp\left(\frac{E_F - E_c}{k_B T} \right)$$

The concentration of holes in the n-type material is found from eq. (1-6-32):

$$p_n = \frac{n_i^2}{n_n}$$

In many practical cases the temperature is high enough so that $N_d \approx N_{d^+} >> n_i$ in an n-type semiconductor, and the equations for carrier concentrations and for the Fermi level are reduced to

$$n_n \approx N_d \tag{1-7-17}$$

$$p_n \approx \frac{n_i^2}{N_d} \tag{1-7-18}$$

$$E_F \approx E_c - k_B T \ln \frac{N_c}{N_d} \tag{1-7-19}$$

When $T \to 0$,

$$N_c << N_d \exp\left(\frac{E_c - E_d}{k_B T} \right) \tag{1-7-20}$$

and we find from eq. (1-7-15) that

$$E_F = \frac{E_c + E_d}{2} + \frac{k_B T}{2} \ln \frac{N_d}{g_d N_c} \tag{1-7-21}$$

and the electron concentration is given by

$$n \approx \left(\frac{N_d N_c}{g_d} \right)^{1/2} \exp\left(\frac{E_d - E_c}{2k_B T} \right) \tag{1-7-22}$$

At very high temperatures $n \approx p \approx n_i >> N_d$. Hence, as temperature increases, the electron concentration increases exponentially at very low temperatures with the activation energy $(E_c - E_d)/2$, then becomes nearly independent of temperature in the extrinsic temperature range ($n \approx N_{d^+} \approx N_d$), and then rises again at very high temperatures with the activation energy $E_g/2$ ($n \approx p \approx n_i$). We should mention, however, that this analysis is valid for the nondegenerate case only.

In fact, at very low temperatures, the experimentally observed activation energy is equal to $E_c - E_d$ (and not to $(E_c - E_d)/2$ as may be expected from the foregoing analysis), becoming equal to $(E_c - E_d)/2$ at somewhat higher temperatures. The explanation of this fact is related to residual acceptor (compensating) impurities always present in any n-type material. Even though $N_d > N_a$, at very low temperatures $N_a >> n >> p$ and the neutrality condition eq. (1-7-9) becomes

$$N_{d^+} = N_{a^-} \tag{1-7-23}$$

which describes the situation when all acceptors are filled by electrons supplied by donors. Using eq. (1-7-13) we can rewrite eq. (1-7-23) as

$$N_a \approx N_{a^-} = \frac{N_d}{1 + g_d \exp\left(\dfrac{E_F - E_d}{k_B T}\right)} \tag{1-7-24}$$

yielding

$$n = N_c \exp\left(\frac{E_F - E_c}{k_B T}\right) = \frac{N_c}{g_d}\frac{N_d - N_a}{N_a} \exp\left(-\frac{E_c - E_d}{k_B T}\right) \tag{1-7-25}$$

Hence, the activation energy in this case is equal to $E_c - E_d$, in agreement with the experimental data.

This result may also be explained in the following way. If a semiconductor is partially compensated by acceptors with concentration N_a, at very low temperatures the concentration of electrons on donors is $N_d - N_a$. At very low temperatures, the Fermi-Dirac distribution function

$$f = 1/(1 + \exp[(E - E_F)/k_B T])$$

is very close to a step function of energy, and $f(E_F) = 1/2$ (see Fig. 1-6-1). Thus, the Fermi level must coincide with the donor level. Hence, at very low temperatures the activation energy is equal to $E_c - E_d$. However, eq. (1-7-24) is valid only when $n << N_a$. As soon as the temperature increases slightly, this condition is violated, and the activation energy becomes equal to $(E_c - E_d)/2$.

Similar arguments are valid for a p-type semiconductor. In Fig. 1-7-5 the electric conductivity of compensated p-Si is shown as a function of temperature (Seeger 1985). As can be seen from the figure at $T \rightarrow 0$ the activation energy is 46 meV. At slightly higher temperatures the activation energy becomes equal to 23 meV. This is consistent with the value of $E_a - E_v = 46$ meV.

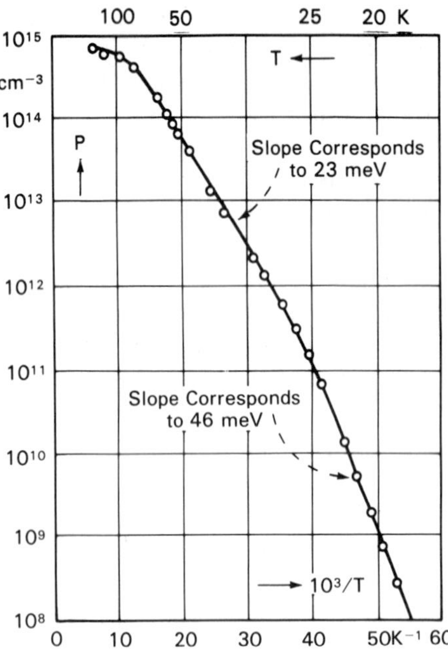

Fig. 1-7-5. Hole concentration in *p*-Si vs. $1/T$ (from K. Seeger, *Semiconductor Physics, An Introduction,* Series in Solid-State Sciences, vol. 40, 3d ed. Springer-Verlag, Berlin, Heidelberg, New York (1985). Acceptor density is 7.4×10^{14} cm^{-3}, donor density, 10^{11} cm^{-3}, and acceptor activation, energy 46 meV.

In general, for a p-type semiconductor, the charge neutrality condition is given by

$$n + N_{a^-} = p \tag{1-7-26}$$

Performing calculations similar to those for the n-type material, we find that

$$p = \frac{\sqrt{N_{a^-}^2 + 4\,n_i^2} + N_{a^-}}{2} \tag{1-7-27}$$

For shallow acceptors $N_{a^-} \approx N_a$, and in general,

$$N_{a^-} = \frac{N_a}{1 + g_a \exp\left(\dfrac{E_a - E_F}{k_B T}\right)} \tag{1-7-28}$$

At relatively high temperatures $N_{a^-} \approx N_a >> n_i$, and

$$p_p \approx N_a \tag{1-7-29}$$

$$n_p \approx \frac{n_i^2}{N_a} \tag{1-7-30}$$

$$E_F \approx E_v + k_B T \ln \frac{N_v}{N_a} \tag{1-7-31}$$

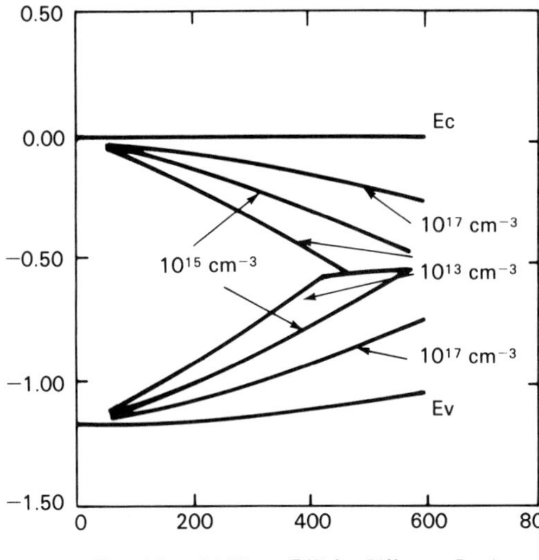

Fig. 1-7-6. Fermi level vs. temperature for silicon doped with different concentrations of shallow donors (phosphorus) and shallow acceptors (boron). Donor ionization energy is 4.5 meV, *g*-factor 2 and acceptor ionization energy is 4.5 meV, *g*-factor 4. This plot is generated by program PLOTF with subroutines PEFNP (see the subroutine listing in the Program Description Section).

The position of the Fermi level for different concentrations of shallow impurities in silicon is shown in Fig. 1-7-6. The temperature dependence of the carrier concentration is shown in Fig. 1-7-7.

The foregoing analysis is valid only for nondegenerate semiconductors where

$$F_{1/2}(\eta) \approx e^{\eta} \tag{1-7-32}$$

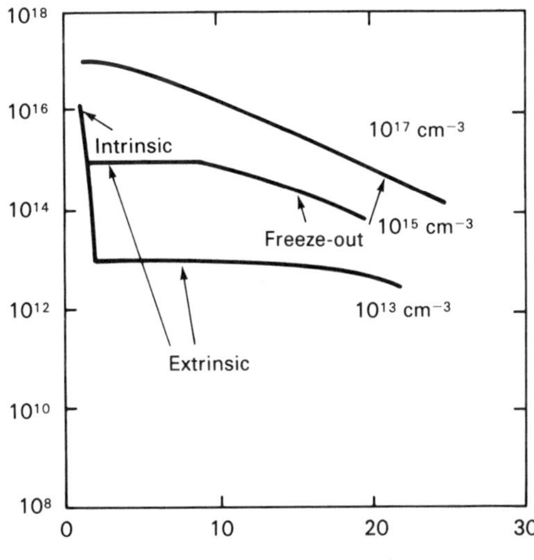

Fig. 1-7-7. Approximate calculation of electron concentration in *n*-type silicon for different doping levels. Donor activation energy is 4.5 meV, *g*-factor 2. This plot is generated by program PLOTF with subroutines PEFNP (see the subroutine listing in the Program Description Section).

(see eq. (1-6-25) and Appendix 21). In general, eq. (1-7-14), for example, should be replaced by

$$N_c F_{1/2}(\eta) = \frac{N_d}{1 + g_d \exp(\eta)\exp\left(\dfrac{E_c - E_d}{k_B T}\right)} \tag{1-7-33}$$

In the case of shallow ionized donors,

$$n = N_d \tag{1-7-34}$$

and the position of the Fermi level is found from

$$F_{1/2}(\eta) = \frac{N_d}{N_c} \tag{1-7-35}$$

For $\eta >> 1$

$$F_{1/2}(\eta) \approx \frac{4\eta^{3/2}}{3\sqrt{\pi}} \tag{1-7-36}$$

In this case, from eqs. (1-7-36) and (1-6-16) we find that

$$E_F - E_c \approx \frac{\hbar^2}{2m_n}(3\pi^2 N_d)^{2/3} \tag{1-7-37}$$

This result may also be obtained in a simple way by counting the number of allowed states in the Fermi sphere, as was discussed in Section 1-5.

Practically all semiconductors, even those of extreme purity, contain many different impurities. Both donors and acceptors are present, and the conductivity type (p or n) is determined by the larger concentration of ionized impurities. This situation is illustrated by Fig. 1-7-8a, which represents the situation when $N_a > N_d$. In this case

$$N_{a\text{eff}} \approx N_a - N_d \tag{1-7-38}$$

In the opposite case when $N_d > N_a$,

$$N_{d\text{eff}} \approx N_d - N_a \tag{1-7-39}$$

(see Fig. 1-7-8b). We can now use the equations describing the extrinsic carrier concentration and derived in this section if we substitute N_d by $N_{d\text{eff}}$ or N_a by $N_{a\text{eff}}$.

As was mentioned previously, a semiconductor doped with donors and acceptors is called a *compensated semiconductor* because the effect of one dopant species is "compensated" by the other dopant.

One consequence of compensation—the change in the activation energy of the sample conductivity from half of the dopant level energy to the full dopant level energy at very low temperatures—has already been mentioned. It can be clearly seen from Fig. 1-7-5.

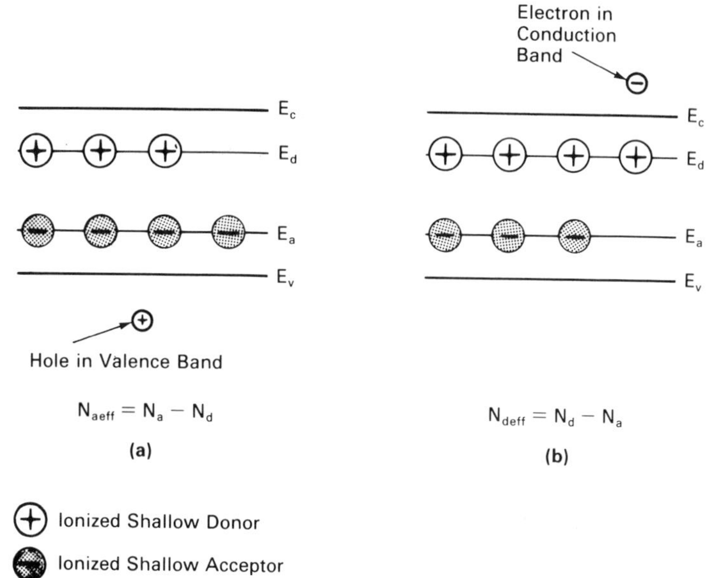

Fig. 1-7-8. Schematic band diagram for compensated semiconductor: $N_a > N_d$, (b) $N_a < N_d$. E_d is the donor level and E_a the acceptor level, E_c is the bottom of the conduction band and E_v is the top of the valence band.

The value of

$$K = \begin{cases} N_a/N_d & \text{for} \quad N_a < N_d \\ N_d/N_a & \text{for} \quad N_a > N_d \end{cases} \tag{1-7-40}$$

is called the compensation ratio. As was mentioned previously, practically all semiconductor samples are compensated to some extent, but in most cases the term "compensated semiconductor" is used when the value of K is not negligibly small (that is, $K \approx 0.05$ or larger).

As can be seen from eqs. (1-7-38) and (1-7-39) compensation may be used to decrease the effective doping density and, hence, the sample conductivity. Let us consider the situation when we have a concentration of donor impurities that we cannot reduce for technological reasons. We may be able to reduce the sample conductivity by adding a controlled amount of deep acceptors. Another, more practical way is to add more deep acceptors than uncontrollable shallow donors. The deep acceptors should not be ionized in the temperature range of interest. In this case all donors will give up electrons to the deep acceptor impurities and the overall sample conductivity may become quite low. This approach is used to produce high resistivity chromium-doped (compensated) gallium arsenide substrates. The GaAs semi-insulating substrate resistivity may be as high as 10^8 ohm-cm.

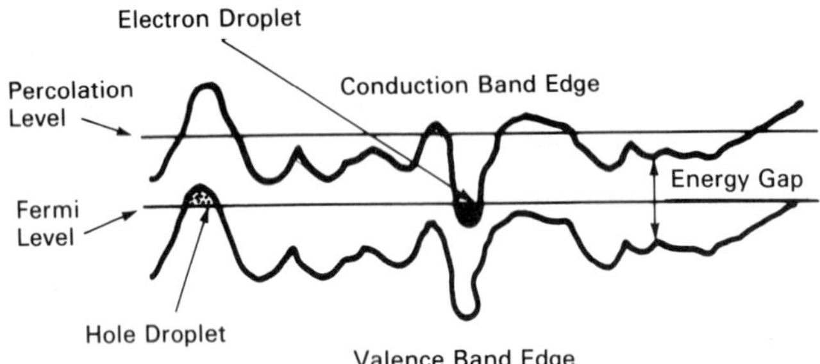

Fig. 1-7-9. Variation of energy bands with position in strongly compensated semiconductor. Fluctuations of the electric potential are created by fluctuations in the acceptor and donor densities.

A very interesting situation may occur in a semiconductor sample when the acceptor and donor concentrations, N_a and N_d, are both large but nearly equal (compensation ratio, K, is close to 1). At first glance such a semiconductor will behave almost like an intrinsic semiconductor. In fact, the incorporation of dopants into a semiconductor material is a statistical process, and the local densities of acceptors and donors fluctuate with position. When N_a and N_d are large, such fluctuations become important, leading to large variations of the electric potential from point to point. The band diagram may look like a set of peaks and valleys randomly distributed in space (see Fig. 1-7-9). Strongly compensated semiconductors have many interesting and unusual properties (see Shklovkii and Efros 1984). For example, as can be seen from Fig. 1-7-9, in parts of a highly compensated semiconductor sample the Fermi level may be in the conduction band, and in other places the Fermi level may be in the valence band. This means that electron and hole gas "droplets" will form, and in certain ways, this material may behave (at low temperatures) as an insulator with small, highly conductive metal balls. Also, in such heavily compensated material, conduction-band electrons with low energies are prevented from traveling across the sample by the peaks of the conduction band. Conduction-band electrons with high energies can travel across, over, or around the peaks of the conduction band. The critical energy separating these low and high energy electrons is called a percolation level. The position of the percolation level (with respect to the Fermi level) determines the conductivity of a strongly compensated sample.

1-8. LATTICE VIBRATIONS

So far we have considered electrons and holes in an ideal crystal with perfect symmetry and with atoms located in strictly fixed positions. In fact, defects, impurities, dislocations, and crystal boundaries may strongly affect the electronic motion. And, what is especially important, atoms in a crystal lattice are vibrating in perpetual thermal motion.

This random atomic motion (lattice vibrations) may be represented as a superposition of very many plane waves of atomic displacements:

$$U_M = \Sigma\, U_n \tag{1-8-1}$$

where U_M is the displacement of an atom, and terms

$$U_n = U_{no} \cdot \exp\{i[\mathbf{q} \cdot \mathbf{r} - \omega(\mathbf{q})t]\} \tag{1-8-2}$$

represents plane waves (called normal modes) of atomic vibrations. The frequencies of these waves are dependent on the relative displacement of atoms (i.e., on the wave length, λ) and on the direction of the wave propagation (i.e., on the direction of the wave vector, $\mathbf{q}$). Because of the crystal symmetry, we have to consider, once again , only the values of $\mathbf{q}$ belonging to the first Brillouin zone (see Fig. (1-6-2)). This is illustrated by Fig. 1-8-1 where we show an atomic chain and two sinusoidal waves of atomic displacements with wave vectors $q_1 = 3\pi/(2a)$ and $q_2 = \pi/(2a)$, respectively, where a is the lattice constant. As can be seen from the figure the values of the displacements are the same at the atomic sites (the only points where it matters). The difference $q_1 - q_2 = \pi/a$ is equal to a vector of the reciprocal lattice for this atomic chain.

The dependencies of ω on $\mathbf{q}$ may be understood by analyzing lattice vibrations of a simple atomic chain (see Fig. 1-8-2). We consider such a chain with two kinds of atoms with atomic masses, M_A and M_B, at each lattice site, n, connected by springs with the force constant F_s. We further assume that atoms may only be displaced in the direction along the chain. If the displacement of atom A located

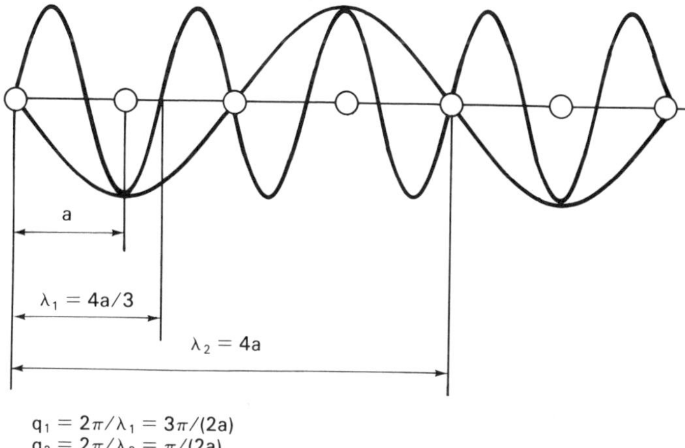

Fig. 1-8-1. Atomic chain and two sinusoidal waves of atomic displacements with wave vectors $q_1 = 3\pi/(2a)$ and $q_2 = \pi/(2a)$.

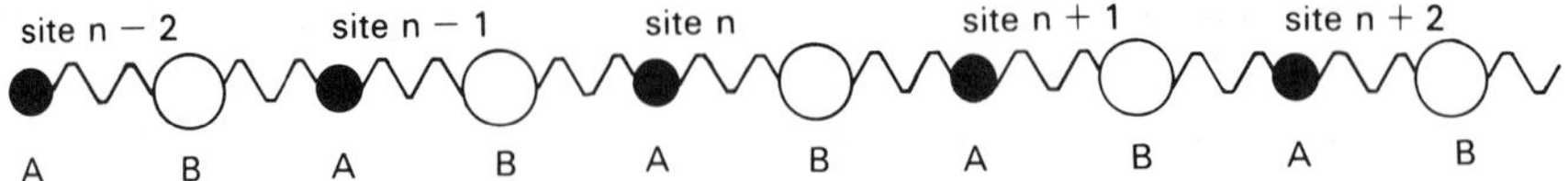

Fig. 1-8-2. Linear atomic chain connected with springs.

at the lattice site n is equal to $U_{A,n}$ and the displacement of the adjacent atom B at the lattice site n is equal to $U_{B,n}$, the force caused by the deformation of the spring, connecting these two atoms and acting on the atom A, is given by $-F_s(U_{A,n} - U_{B,n})$. The negative sign means that the direction of this force is opposite to the displacement, $(U_{A,n} - U_{B,n})$. The equations of motion for atoms of each kind are given by

$$M_A \ddot{U}_{A,n} = -F_s(2U_{A,n} - U_{B,n} - U_{B,n+1}) \tag{1-8-3}$$

$$M_B \ddot{U}_{B,n} = -F_s(2U_{B,n} - U_{A,n} - U_{A,n-1}) \tag{1-8-4}$$

Substituting the trial solutions

$$U_{A,n} = U_A \exp(\text{iqna}) \tag{1-8-5}$$

and

$$U_{B,n} = U_B \exp(\text{iqna}) \tag{1-8-6}$$

into eqs. (1-8-3) and (1-8-4), we obtain

$$M_A \ddot{U}_A = -F_s[\,2U_A - U_B\,(1 + e^{iqa})] \tag{1-8-7}$$

$$M_B \ddot{U}_B = -F_s[\,2U_B - U_A\,(1 + e^{-iqa})] \tag{1-8-8}$$

Substituting

$$U_A = U_{Ao} \cdot \exp(-i\omega t) \tag{1-8-9}$$

$$U_B = U_{Bo} \cdot \exp(-i\omega t) \tag{1-8-10}$$

into eqs. (1-8-7) and (1-8-8), we obtain the following equation determining frequency ω:

$$\begin{vmatrix} 2F_s - M_1\omega^2 & -F_s\,(1 + e^{iqa}) \\ -F_s\,(1 + e^{-iqa}) & 2F_s - M_2\omega^2 \end{vmatrix} = 0 \tag{1-8-11}$$

This equation has the following two solutions:

$$\omega_a^2 = \frac{F_s}{M}\left[1 - \left(1 - \frac{4M\,\sin^2(qa/2)}{M_A + M_B}\right)^{1/2}\right] \tag{1-8-12}$$

$$\omega_o^2 = \frac{F_s}{M}\left[1 + \left(1 - \frac{4M\,\sin^2(qa/2)}{M_A + M_B}\right)^{1/2}\right] \tag{1-8-13}$$

Here $M = (1/M_A + 1/M_B)^{-1}$ is called the reduced atomic mass.

The first branch (ω_a) corresponds to *acoustic vibrations*. For this acoustic branch,

$$\omega_a \approx qa \sqrt{F_s/2(M_A + M_B)} \tag{1-8-14}$$

for $qa << 1$. At small $qa << 1$, this branch is identical to sound waves. The long-wave (small q) acoustic lattice vibrations are such that the adjacent atoms move nearly in phase (see Fig. 1-8-3a).

The second branch (ω_o) corresponds to *optical vibrations*. For this optical branch,

$$\omega_o \approx \sqrt{2F_s/M} \tag{1-8-15}$$

for $qa << 1$, i.e., ω_o becomes a constant when $qa << 1$. In optical vibrations with $qa << 1$, the adjacent atoms move in the opposite directions (see Fig. 1-8-3b).

The existence of optical branches is related to the difference in the atoms comprising a crystal. For example, when M_A becomes equal to M_B for the linear atomic chain, the optical branch becomes equivalent to the acoustic branch folded into a smaller first Brillouin zone, which is one half of the first Brillouin zone for the atomic chain with two identical atoms. Indeed, when $M_A = M_B$, the lattice constant in our model becomes equal to $a/2$ instead of a. This increases the dimensions of the first Brillouin zone by a factor of 2. (See Fig. 1-8-4, where vibration spectra of a linear atomic chain are shown for different ratios of M_1/M_2.)

The vibrational spectrum of a three-dimensional crystal is more complicated. In the three-dimensional case we have not two but three S branches, where S is the number of atoms in the primitive cell. Three branches among them are acoustic branches. For some crystal directions, along the lines of symmetry, two of the acoustic branches in a cubic crystal have the same dependence of ω vs. q

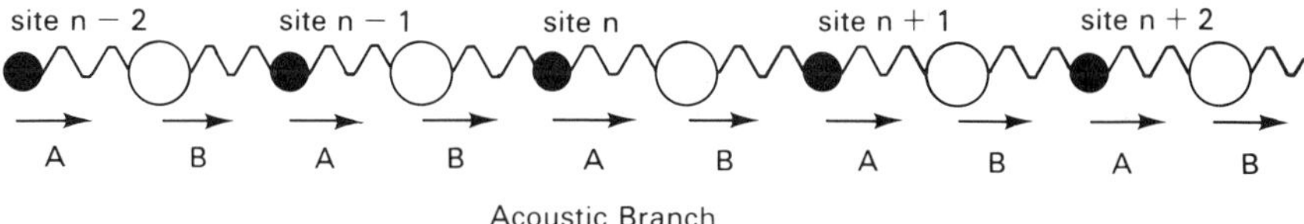

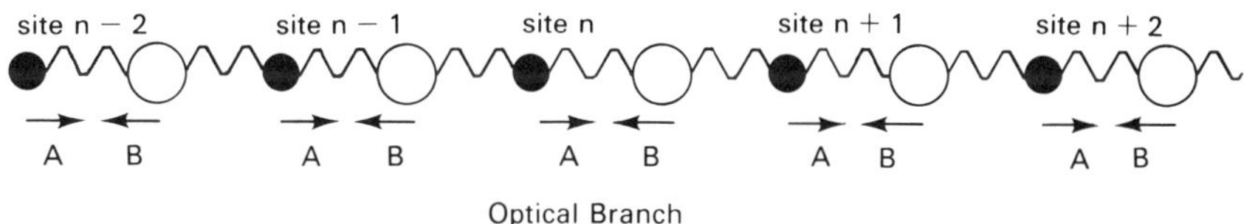

Fig. 1-8-3. Relative atomic displacements in linear atomic chain (shown by arrows): (a) acoustic mode, (b) optical mode.

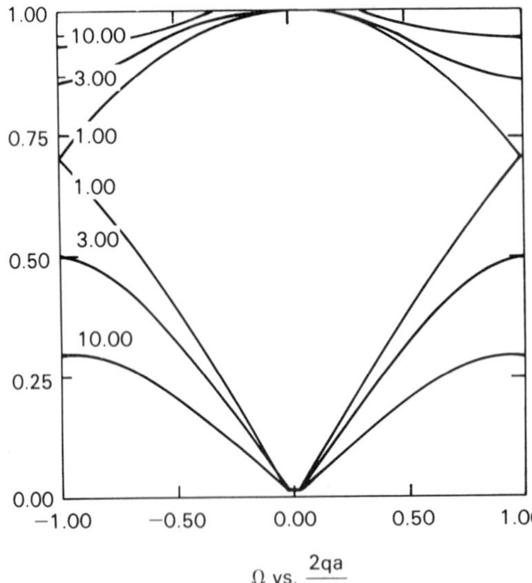

Fig. 1-8-4. Frequencies of lattice vibrations of one dimensional atomic chain vs. wave vector for different M_2/M_1 ratios (shown next to the curves). The curves were generated using program PLOTF with subroutines PCHAIN (see Program Description Section).

(i.e., they are degenerate). At small values of q these two acoustic branches correspond to the transverse sound waves:

$$\omega_{ta} \approx c_{st} q \tag{1-8-16}$$

where c_{st} is the transverse sound velocity. The third acoustic branch corresponds to the longitudinal sound waves for $qa << 1$:

$$\omega_{la} \approx c_{sl} q \tag{1-8-17}$$

Here c_{sl} is the longitudinal sound velocity. Each type of lattice vibration is associated with different phonons—the energy quanta of lattice vibrations.

In many cubic crystals there are only one longitudinal and two degenerate transverse optical branches of lattice vibrations (see Fig. 1-8-5 from Johnson and Cochran 1962 showing these dependencies, called dispersion relations, for a GaAs crystal).

For covalent crystals, such as silicon or germanium, the frequency ω_{to} of transverse optical phonons at $q \to 0$ is equal to the frequency of the longitudinal optical phonons, ω_{lo}. In partially heteropolar crystals, such as gallium arsenide, longitudinal vibrations are accompanied by a periodic variation of the dipole moment caused by a relative displacement of two atomic sublattices of opposite charge. This increases the effective force constant, and the frequency of the longitudinal mode is higher than the frequency of the transverse mode that does not involve such a variation in the dipole moment. The change in the dipole moment of charged atomic sublattices contributes to the semiconductor dielectric permittivity, ε_o, at frequencies smaller than the frequency of the optical mode

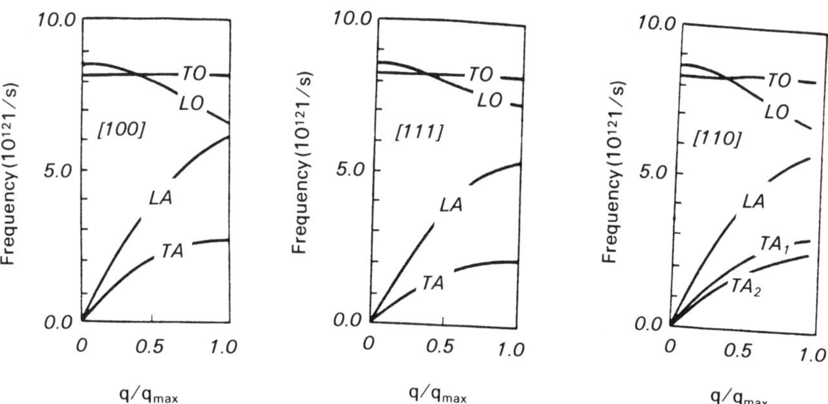

Fig. 1-8-5. Calculated ω vs. q curves for GaAs (from Johnson and Cochran [1962]); q_{max} corresponds to the boundary of the first Brillouin zone.

because at such frequencies the atomic sublattices respond to the variation of an external electric field. Typically, the frequencies of optical vibrations correspond to the infrared range (i.e., light with such a frequency is an infrared light). At much higher frequencies (such as frequencies of visible light), the atomic sublattices do not respond to the variation of an external electric field. The dielectric constant, ε_∞, at such frequencies is determined by electronic motion alone and is smaller than ε_o. As was first shown by Liddane, Sachs, and Teller, static and high frequency dielectric constants, ε_o and ε_∞ and frequencies of optical vibrations, ω_{lo} and ω_{to}, are related as follows:

$$\frac{\omega_{lo}}{\omega_{to}} = \left(\frac{\varepsilon_o}{\varepsilon_\infty}\right)^{1/2} \tag{1-8-18}$$

This equation is known as the *Liddard-Saks-Teller relationship*.

The sound velocities and the frequencies of the optical vibrations for different semiconductors are given in Table 1-8-1 (M. Haas et al. 1959).

This highly simplified description of lattice vibrations may be obtained based on the assumption that all atoms vibrate with the same frequency, ω, so that the atomic displacements are given by

$$U_M = U_{Mo} \sin \omega t \tag{1-8-19}$$

This simplified model is called the *Einstein model*. As can be seen from Fig. 1-8-4, this model may be adequate for optical branches of lattice vibrations in crystals with very different atomic masses.

The average energy of these vibrations may be estimated as

$$\frac{MC_M^2}{2} \approx \frac{3k_BT}{2} \tag{1-8-20}$$

TABLE 1-8-1. FREQUENCIES OF THE OPTICAL VIBRATIONS FOR DIFFERENT SEMICONDUCTORS (after M. Haas et al. 1962)

Semiconductor	Transverse optical frequency ω_t (cm^{-1})	Longitudinal optical frequency ω_l (cm^{-1})
AlSb	318	345
GaAs	373.9	297.3
GaSb	230.5	240.3
InP	307	351
InAs	218.9	243.3
InSb	184.7	197.0

where M is the reduced atomic mass, C_M is the effective atomic velocity, and T is the lattice temperature. The atomic velocity for a vibration with frequency ω is given by

$$C_M = \omega U_M \tag{1-8-21}$$

where

$$\omega = \sqrt{2F_s/M} \tag{1-8-22}$$

is the characteristic frequency of the lattice vibrations, F_s is the force constant, and U_M is the amplitude of the relative atomic displacement. Hence,

$$U_M \approx \left(\frac{3k_BT}{M\omega^2}\right)^{1/2} = \left(\frac{3k_BT}{2F_s}\right)^{1/2} \tag{1-8-23}$$

Of course, in a crystal U_M should be much smaller than the lattice constant, a. When T increases so much that U_M becomes equal to a considerable fraction of the lattice constant (say 15% to 20%), the crystal melts down.

Another simplified model of lattice vibrations is the *Debye model*, which assumes that the frequency of the lattice vibrations is proportional to the wave vector, q:

$$\omega \approx c_s q \tag{1-8-24}$$

where c_s is the velocity of sound. This model may be adequate for an approximate description of acoustic branches, and it is often used for evaluating scattering of electrons by acoustic vibrations. Wave vector q in eq. (1-8-24) varies from zero to the maximum value, q_D, where q_D is called the *Debye wave vector*. The Debye wave vector is chosen in such a way that the total number of the modes of the lattice vibrations for a crystal of a given size is equal to the number of allowed values of q in the sphere of radius q_D. One can show (see Problem 1-8-2) that

$$q_D = (6\pi^2/V_c)^{1/3} \tag{1-8-25}$$

where V_c is the volume of the primitive cell. The frequency

$$\omega_D = c_s q_D \quad (1\text{-}8\text{-}26)$$

is called the *Debye frequency*, and temperature

$$\Theta_D = \hbar\omega_D / k_B \quad (1\text{-}8\text{-}27)$$

is called the *Debye temperature*. The Debye temperature is an important material parameter used in the calculation of transport properties of a semiconductor.

1-9. ELECTRON AND HOLE MOBILITIES AND DRIFT VELOCITIES

In low electric fields, the carrier drift velocity, $\mathbf{v}$, is proportional to the electric field, $\mathbf{F}$,

$$\mathbf{v} = \mu\mathbf{F} \quad (1\text{-}9\text{-}1)$$

The coefficient of proportionality, μ, is called the *low field mobility*.

This equation can be obtained from the second law of motion for an electron moving in an electric field:

$$m_n \, d\mathbf{v}/dt = q\mathbf{F} - m_n\mathbf{v}/\tau \quad (1\text{-}9\text{-}2)$$

Here m_n is the electron effective mass. The first term in the right-hand side of eq. (1-9-2) represents the electron acceleration by the electric field; the second term describes collisions caused by lattice vibrations, impurities, and crystal imperfections. This term limits the electron drift velocity and electron drift momentum. Therefore, the time constant, τ, is called the *momentum relaxation time*. Usually, τ is of the order of 10^{-12} to 10^{-14} s. At low frequencies, the left-hand side of eq. (1-9-2) is small compared with either term in the right-hand side of this equation, so that

$$q\mathbf{F} \approx m_n\mathbf{v}/\tau \quad (1\text{-}9\text{-}3)$$

and, hence,

$$\mathbf{v} = q\tau\mathbf{F}/m_n \quad (1\text{-}9\text{-}4)$$

In low electric fields, τ and m_n are independent of the electric field, so that $\mathbf{v} = \mu\mathbf{F}$, where

$$\mu = q\tau/m_n \quad (1\text{-}9\text{-}5)$$

is the low field mobility. We should emphasize that the linear dependence of the electron drift velocity does not hold in high electric fields when electrons may gain a considerable energy from the electric field. This process can be described by the following equation:

$$dE/dt = q\mathbf{F} \cdot \mathbf{v} - (E - E_o)/\tau_E \quad (1\text{-}9\text{-}6)$$

Here E is the electron energy, $E_o = 3k_BT/(2q)$ is the electron energy under the conditions of the thermal equilibrium, and τ_E is the effective energy relaxation time. Usually the values of τ_E are of the order of 10^{-11} to 10^{-13} s. In a steady state, when $dE/dt = 0$,

$$E = E_o + q\mathbf{F} \cdot \mathbf{v}\, \tau_E \tag{1-9-6a}$$

When the electric field, F, is small, $E \approx E_o$. However, in high electric fields the electron energy can greatly exceed E_o. In this case, electrons are called *hot electrons*. For hot electrons, τ, τ_E, and m_n depend on the electric field, and the electron drift velocity is no longer proportional to the electric field. As a matter of fact, in many cases the drift velocity in high electric fields becomes nearly independent of the electric field. This trend is illustrated by Fig. 1-9-1, in which are shown the dependencies of the electron drift velocities for several important semiconductor compounds. As can be seen from Fig. 1-9-1, in many semiconductors, such as GaAs, InP, and InGaAs, the electron velocity in a certain range of electric fields may actually decrease with the increase of the electric field. (This important effect is considered in Chapter 6.) As can be also seen from this figure, compound semiconductors have a potential for higher speed operation than silicon because electrons in these materials may move faster. This will be discussed in more detail in Chapters 4 and 7.

A more rigorous derivation of eq. (1-9-5) is based on the Boltzmann equation (see Section 1-13). This equation and the Monte-Carlo method (considered in Section 1-14) allow us to calculate the dependence of the drift velocity on the electric field for different semiconductor materials.

The foregoing discussion assumes that the electron drift velocity depends on the electric field alone, so that this velocity should be the same in a very long and

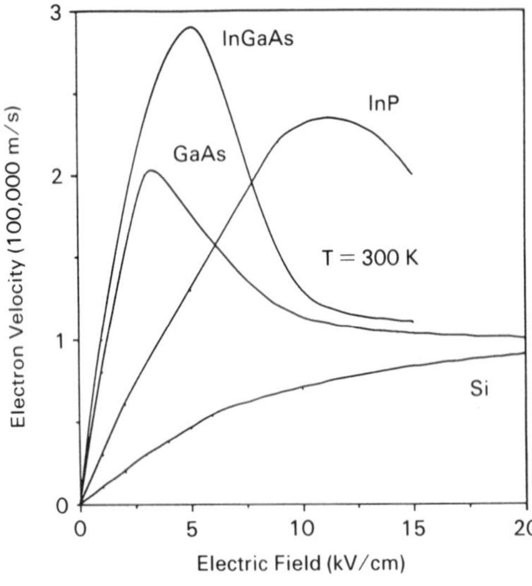

Fig. 1-9-1. Electron drift velocities in silicon and in several compound semiconductors at room temperature. The curve for $In_xGa_{1-x}As$ is for $x = 0.53$.

in a very short sample, provided that the electric field is the same. However, this is only true for relatively long samples when transit times $T = L/v$ and $T_T = L/v_T$ are much greater than the momentum and energy relaxation times. Here L is the sample length and

$$v_T = (3k_BT/m_n)^{1/2} \tag{1-9-7}$$

is the average electron thermal velocity. These conditions are frequently violated in modern short semiconductor devices. In the limited case of a very short device, the electron transit time may become so small that electrons will not have time to experience any collisions during the transit. Such a mode of electron transport is called *ballistic transport* (see Shur and Eastman 1979 and Section 7-1). Basic semiconductor equations describing electron transport in short devices are considered in Section 1-15.

The low field mobility, μ, is determined by electron collisions with phonons and impurities. These collision processes are called *scattering mechanisms*. The momentum relaxation time, τ, can be approximately expressed as

$$\frac{1}{\tau} = \frac{1}{\tau_{\text{ii}}} + \frac{1}{\tau_{\text{ni}}} + \frac{1}{\tau_{\text{ac}}} + \frac{1}{\tau_{\text{npo}}} + \frac{1}{\tau_{\text{po}}} + \frac{1}{\tau_{\text{pe}}} + \cdots \tag{1-9-8}$$

where terms in the right-hand side represent relaxation times due to different scattering processes such as ionized impurity scattering (τ_{ii}), neutral impurity scattering (τ_{ni}), acoustic (or deformation potential) scattering (τ_{ac}), nonpolar optical scattering (τ_{npo}), polar optical scattering (τ_{po}), and piezoelectric scattering (τ_{pe}). The last two scattering mechanisms are present only in partially heteropolar crystals, such as GaAs. Typically, only two or so scattering mechanisms are dominant for given values of temperature and impurity concentration. In nonpolar semiconductors, such as silicon and germanium, scattering by acoustic phonons and ionized impurities determine the electron and hole mobilities in low electric fields. Scattering by nonpolar optical phonons becomes dominant in these materials in high electric fields, leading to the saturation of the electron and hole velocities. Scattering by polar optical phonons and ionized impurity scattering dominate for electrons and holes in GaAs in low electric fields. Scattering by acoustic phonons (also called deformation potential scattering) and scattering caused by piezoelectric properties (piezoelectric scattering) become important in pure GaAs samples at low temperatures. The relative contributions of the three major scattering mechanisms in GaAs are illustrated in Fig. 1-9-2 (from Stillman et al. 1970 and Blakemore 1982), where the electron mobility is shown as a function of temperature for three samples with different doping densities. Polar-optical scattering and intervalley scattering dominate in large electric fields in GaAs, leading to the negative differential mobility (see Fig. 1-9-1, Section 1-14, and Chapter 6).

Relaxation times determined by different scattering mechanisms are considered, for example, in Seeger 1979. Formulas that allow one to calculate momentum relaxation times for different scattering mechanisms and, hence, low-field

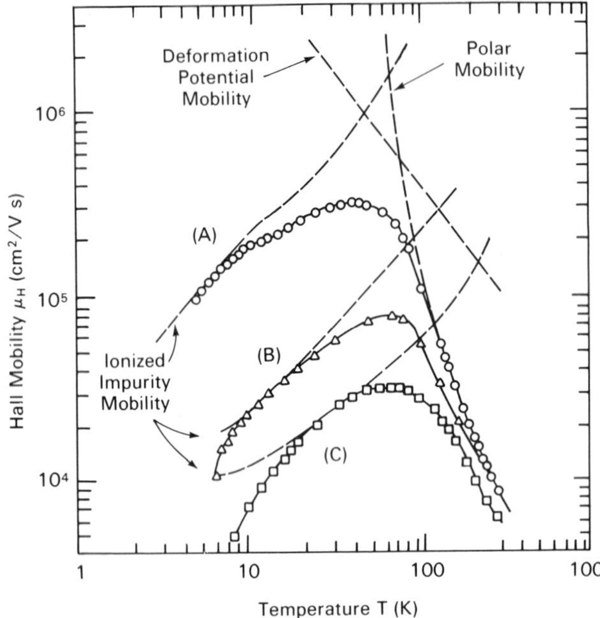

Fig. 1-9-2. Temperature dependence of the electron Hall mobility μ_H (for $B = 5kG$) for three N-type GaAs samples (after Stillman et al. [1970]. The estimated donor densities of (a) 5×10^{13} cm^{-3}, (b) 10^{15} cm^{-3}, and (c) 5×10^{15} cm^{-3} for the three samples so identified, with $(N_a/N_d) \approx 0.3$ to 0.4 in each case. Expected contributions of three major processes towards the scattering are shown. (Reproduced from the review paper by J. S. Blakemore, "Semiconductor and Other Major Properties of GaAs," *J. Appl. Phys.*, 53, no. 10, pp. R123–R181, Oct. (1982).

mobilities for electrons and holes are summarized in Appendix 23. In Figs. 1-9-3 to 1-9-5 we show dependencies of electron and hole mobilities in silicon on temperature and doping density.

The electron and hole drift velocities in high electric fields depend on temperature and on the concentration of charged impurities, $N_T = N_A + N_D$. According to Caugley and Thomas (1967), Arora et al. (1982), Sze (1981), and Yu and Dutton (1985), the electron velocity in bulk silicon, $v_n(F)$, can be approximated by

$$v_n = \frac{\mu_n F}{[1 + (\mu_n F/v_s)^2]^{1/2}} \tag{1-9-9}$$

where

$$\mu_{\mathrm{n}}(N_T,T) = \mu_{\mathrm{mn}} + \frac{\mu_{\mathrm{on}}}{1 + (N_T/N_{cn})\nu} \tag{1-9-10}$$

$$\mu_{\mathrm{mn}} = 88(T/300)^{-0.57}\ (\mathrm{cm^2/Vs}) \tag{1-9-11}$$

$$\mu_{\mathrm{on}} = 7.4 \times 10^8\ T^{-2.33}\ (\mathrm{cm^2/Vs}) \tag{1-9-12}$$

$$\nu = 0.88\ (T/300)^{-0.146} \tag{1-9-13}$$

$$N_{cn} = 1.26 \times 10^{17}\ (T/300)^{2.4}\ (\mathrm{cm^{-3}}) \tag{1-9-14}$$

The saturation velocity, v_s, for electrons and holes is given by (see Sze 1981 and Yu and Dutton 1985)

$$v_s = 2.4 \times 10^5/[1 + 0.8 \exp (T/600)]\ (\mathrm{m/s}) \tag{1-9-15}$$

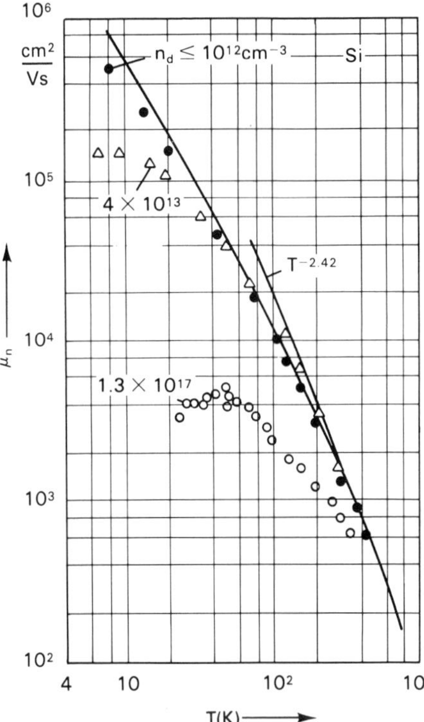

Fig. 1-9-3. Measured and calculated electron mobility in silicon versus temperature for different doping levels (Reprinted with permission from C. Jacoboni and L. Reggiani, "Bulk Hot-Electron Properties of Cubic Semiconductors," *Advances in Physics*, 28, no. 4, pp. 493–553 (1979), Pergamon Press p/c). The solid line shows calculations and the symbols represent measured data. Also shown is a straight line corresponding to $T^{-2.42}$ temperature dependence.

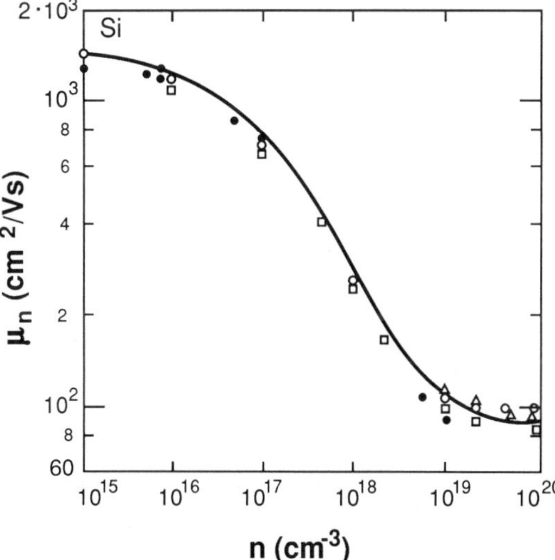

Fig. 1-9-4. Measured and calculated electron mobility in silicon versus doping at room temperature (Reprinted with permission from T. I. Tosic, D. A. Tjapkin, and M. M. Jevtic, *Solid State Electron.*, 24, p. 577 (1981) Pergamon Press p/c). The solid line shows calculations and the symbols represent measured data.

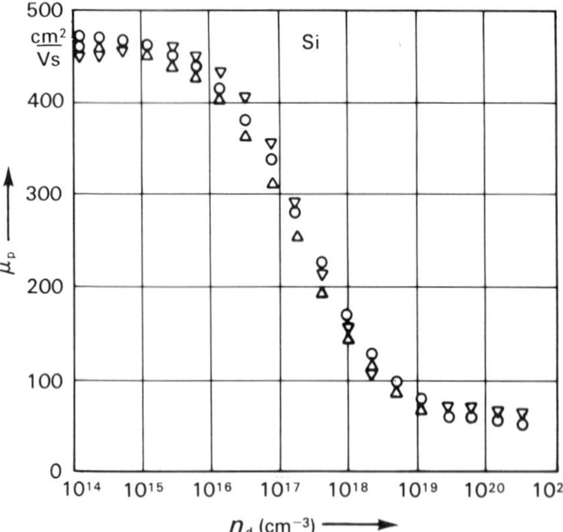

Fig. 1-9-5. Measured hole mobility in silicon versus doping at room temperature (Reprinted with permission from H. S. Bennett, *Solid State Electron.*, 26, p. 1157 (1983) Pergamon Press p/c).

According to Caugley and Thomas (1967), Arora et al. (1982), Sze (1981), and Yu and Dutton (1985), the hole velocity in silicon, $v_p(F)$, is given by

$$v_p = \frac{\mu_p F}{1 + (\mu_p F / v_s)} \tag{1-9-16}$$

where

$$\mu_p(N_T, T) = \mu_{mp} + \frac{\mu_{op}}{1 + (N_T/N_{cn})\nu} \tag{1-9-17}$$

$$\mu_{mp} = 54\ (T/300)^{-0.57}\ (\text{cm}^2/\text{Vs}) \tag{1-9-18}$$

$$\mu_{op} = 1.36 \times 10^8\ T^{-2.33}\ (\text{cm}^2/\text{Vs}) \tag{1-9-19}$$

$$\text{N}_{cp} = 2.35 \times 10^{17}\ (T/300)^{2.4}\ (\text{cm}^{-3}) \tag{1-9-20}$$

$$\nu = 0.88\ (T/300)^{-0.146} \tag{1-9-21}$$

As can be seen, the exponent ν is the same for holes as for electrons. The saturation velocity for holes is also approximately the same as for electrons.

The foregoing expressions for the hole and the electron velocity are obtained for majority carriers in bulk silicon. The mobility and velocity of minority carriers may be quite different because of the electron-hole scattering. Nevertheless, the same equations are frequently used for minority carriers as well because of the lack of sufficient experimental data (see Yu and Dutton 1985).

1-10. HALL EFFECT AND MAGNETORESISTANCE

Standard experimental techniques of measuring mobilities include the studies of transport properties in a magnetic field. Let us consider a p-type semiconductor sample placed into a magnetic field as shown in Fig. 1-10-1. When voltage, V, is applied to the sample, holes move from contact 2 to contact 1 with velocity

$$v_p = \mu_p F = \mu_p V/L \tag{1-10-1}$$

where F is the electric field, L is the sample length, and μ_p is the hole mobility. The magnetic field, B, exerts a force $\mathbf{f}_L = q\mathbf{v}_p\mathbf{x}\mathbf{B}$ (called the Lorentz force), acting on the holes and deflecting them toward side 3 as shown in Fig. 1-10-1. As a consequence, holes accumulate on this side and create a net positive charge there. Side 4 gets depleted by holes and is negatively charged. The magnitude of these charges is such that the electric field F_H, created by the charges, exactly counterbalances the Lorentz force:

$$F_H = v_p B = |f_L|/q \tag{1-10-2}$$

(Here we assume that the magnetic field is small so that $\mu_p B << 1$.) Hence, the voltage difference

$$V_H = v_p B d \tag{1-10-3}$$

develops between contacts 3 and 4.

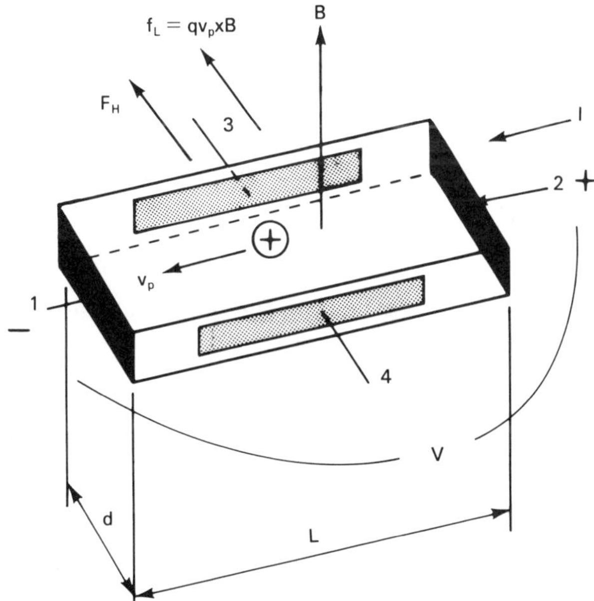

Fig. 1-10-1. Hall effect in *p*-type semiconductor sample placed into magnetic field.

This effect is called the *Hall effect*, contacts 3 and 4 are called the *Hall contacts*, and electric field F_H and voltage V_H are called the *Hall electric field* and the *Hall voltage*, respectively.

The electric current density

$$j = qp\,\mu_p F \tag{1-10-4}$$

and, hence

$$F_H = Bj/(qp) \tag{1-10-5}$$

The Hall voltage, $V_H = F_H d$, is given by

$$V_H = RBI/t \tag{1-10-6}$$

where $I = jtd$ is the electric current, and t is the sample thickness.

$$R = 1/(qp) \tag{1-10-7}$$

is called the *Hall constant*. Hence, the mobility of holes can be found as

$$\mu_p = \sigma_p\,R \tag{1-10-8}$$

where

$$\sigma_p = j/F = qp\mu_p \tag{1-10-9}$$

is the conductivity of the sample.

This approach allows us to understand the physics of the Hall effect. However, it is not totally accurate. Indeed, we have assumed that all holes in the sample move with the same velocity, v_p. In fact, this velocity is the average drift velocity superimposed on a random thermal motion. This means that some holes move slower and some move faster. The Hall electric field is the same for all the holes, so that it counterbalances the Lorentz forces acting on individual holes with different velocities only on the average. A more accurate analysis of this problem shows that eqs. (1-10-7) and (1-10-8) have to be modified as follows:

$$R = r_H/(qp) \tag{1-10-7a}$$

$$\mu_{Hp} = \sigma_p R \tag{1-10-8a}$$

where r_H is called the *Hall factor* and μ_{Hp} is called the *Hall mobility of holes*

$$\mu_{Hp} = r_H\mu_p \tag{1-10-9a}$$

The value of r_H depends on temperature, doping, magnetic field, and other factors. When the relaxation time approximation (used in Section 1-10 for the mobility calculation) is valid,

$$r_H = \langle\tau^2\rangle/\langle\tau\rangle^2 \tag{1-10-10}$$

(see, for example, Seeger 1982), where for nondegenerate carrier distribution,

$$\langle\tau\rangle = 4/(3\sqrt{\pi}) \int_0^\infty \tau(E)E^{3/2} \exp\left(-\frac{E}{k_BT}\right) dE/(k_BT)^{5/2} \tag{1-10-11}$$

$\tau(E)$ is the momentum relaxation time, which is a function of the electron energy, E,

$$\langle \tau^2 \rangle = 4/(3\sqrt{\pi}) \int_0^\infty \tau^2 E^{3/2} \exp\left(-\frac{E}{k_B T}\right) dE/(k_B T)^{5/2} \qquad (1\text{-}10\text{-}12)$$

In a particular case when the momentum relaxation time τ may be presented as

$$\tau = \tau_o \left(\frac{E}{k_B T}\right)^{-S} \qquad (1\text{-}10\text{-}13)$$

where τ_o and S are constants, the integrals in eqs. (1-10-11) and (1-10-12) can be expressed through a gamma function, Γ:

$$\langle \tau \rangle = 4/(3\sqrt{\pi})\tau_o\, \Gamma(5/2 - S) \qquad (1\text{-}10\text{-}14)$$

$$\langle \tau^2 \rangle = 4/(3\sqrt{\pi})\tau_o^2\, \Gamma(5/2 - 2S) \qquad (1\text{-}10\text{-}15)$$

(The values of the gamma function are given in Table 1-10-1.) The Hall factor for holes in silicon as a function of temperature is shown in Fig. 1-10-2 (from Szmilovicz and Madarasz 1983).

Similar expressions can be derived for an n-type sample:

$$R = -r_H/(qn) \qquad (1\text{-}10\text{-}16)$$

$$\mu_{Hn} = \sigma_n R \qquad (1\text{-}10\text{-}17)$$

TABLE 1-10-1. GAMMA FUNCTION (from Shur 1987)

Definition of Γ function:

$$\Gamma(x) = \int_0^\infty t^{x-1} \, exp(-t) \, dt$$

x	$\Gamma(x)$
1	1.00
1.25	0.906
1.5	0.886
1.75	0.919
2	1.00

For positive integer values of x

$$\Gamma(x) = (x - 1)!$$

The values of $\Gamma(x)$ for $x < 1$ and for $x > 2$ may be evaluated using the following formulas:

$$\Gamma(x) = \Gamma(x + 1)/x$$

$$\Gamma(x) = (x - 1)\Gamma(x - 1)$$

Example: $\Gamma(2.5) = 1.5\Gamma(1.5) = 1.33$

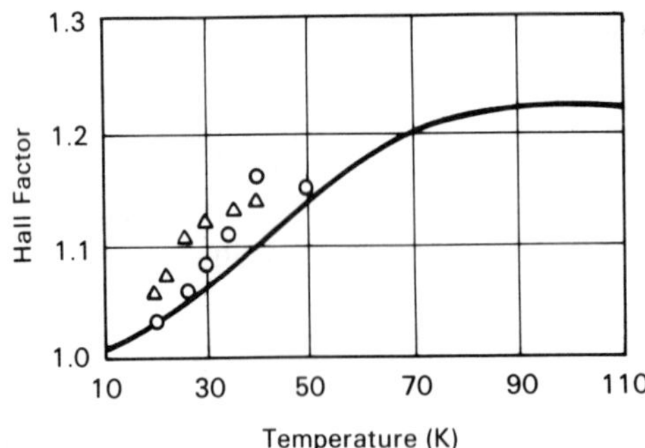

Fig. 1-10-2. Hall factor for holes in silicon and temperature (from F. Szmilovicz and F. L. Madarasz, *Phys. Rev*, B27, p. 2605 (1983). © 1983 IEEE.

where μ_{Hn} is the Hall mobility of electrons,

$$\mu_{Hn} = r_H \mu_n \tag{1-10-18}$$

The Hall factor for electrons in GaAs as a function of temperature is shown in Fig. 1-10-3 (from Rode 1975).

An interesting problem is to consider the Hall effect in the intrinsic semiconductor. It may be shown (see Problem P1-10-1) that the Hall constant, R, in this case is given by

$$R = -\frac{1}{n_i}\frac{\mu_n/\mu_p - 1}{\mu_n/\mu_p + 1} \tag{1-10-19}$$

where n_i is the intrinsic carrier concentration.

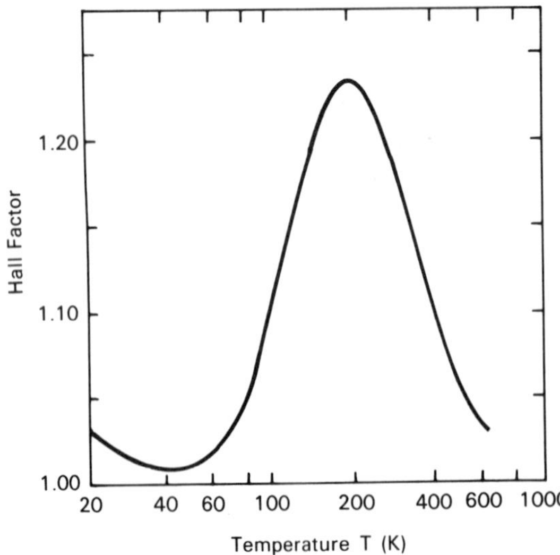

Fig. 1-10-3. Hall factor for electrons in GaAs in weak magnetic field as function of temperature (from D.L. Rode, *Physica Status Solidi*, 55, p. 687 (1975).

The Hall effect is used in *Hall-effect sensors* that generate an electrical signal (the Hall voltage) in response to a changing magnetic field. Hall sensors are used in computer keyboards, where pushing a key moves a permanent magnet close to a Hall-effect sensor. Such a sensor is simply a piece of an n-type semiconductor with four contacts, as shown in Fig. 1-10-1. (In practical devices, however, all contacts are made on top of a silicon wafer to make the fabrication process compatible with integrated-circuit technology.) The movement of the magnet creates a magnetic field of the order of several hundred gauss. A typical sensitivity of a Hall sensor (defined as a ratio of the Hall voltage over the magnetic field) is of the order of tens of microvolts per gauss. Hence, pushing a computer keyboard key produces a Hall voltage of the order of several tens of millivolts which is then amplified to a desired level.

The Hall electric field develops as a result of the carrier flux in the direction perpendicular to the current flow. This perpendicular motion is caused by the magnetic field when it is turned on. At first, the carriers move under the angle

$$\theta = \tan^{-1}\left(\frac{v_y}{v_x}\right) = \tan^{-1}\frac{\mu v_x B}{v_x} = \tan^{-1}(\mu B) \qquad (1\text{-}10\text{-}20)$$

called the *Hall angle*. Here v_y and v_x are components of the carrier velocity in the x and y directions (see Fig. 1-10-4). Then carriers accumulate at one of the sample side surfaces, depleting the other side and building up the Hall electric field. The Hall electric field compensates the Lorentz force and stops the further motion of carriers in the direction perpendicular to the current flow.

This implies, however, that the sample dimensions should satisfy the requirement $L >> d$ for the Hall electric field to appear. In the opposite limiting case, $d << L$, most of the carriers moving under the Hall angle arrive at the contact, not at the sample side, as illustrated by Fig. 1-10-4b, and hence, the Hall electric field is almost totally shorted by the contact. As a result, carriers travel a longer distance,

$$L_H = L(1 + \tan^2\theta)^{1/2} \qquad (1\text{-}10\text{-}21)$$

and the component of their velocity in the direction of the applied electric field decreases owing to the component of the Lorentz force related to the component of the electron velocity parallel to the contacts. Assuming a zero Hall field we find

$$\mathbf{v} = \mu(\mathbf{F} + \mathbf{vxB}) \qquad (1\text{-}10\text{-}22)$$

where μ is a low field mobility. Hence,

$$v_x = \mu(F_x + v_y B) \qquad (1\text{-}10\text{-}23)$$

$$v_y = -\mu v_x B \qquad (1\text{-}10\text{-}24)$$

(because $F_y = 0$). Substituting eq. (1-10-24) into eq. (1-10-23) we find

$$v_x = \frac{\mu F_x}{1 + \mu^2 B^2} \qquad (1\text{-}10\text{-}25)$$

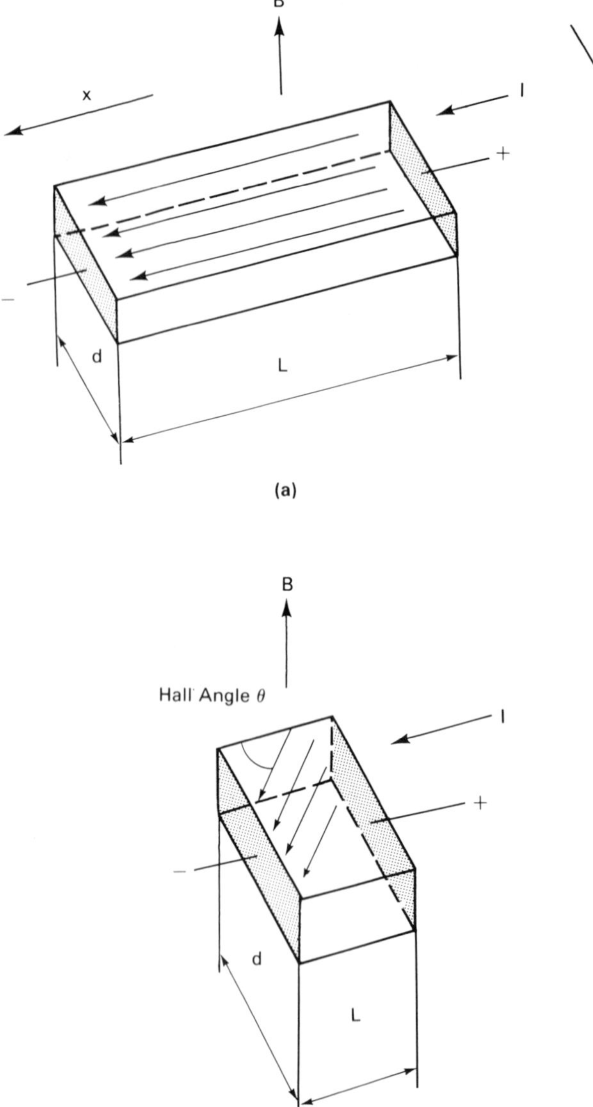

Fig. 1-10-4. Streamlines of electric current in semiconductor samples with $d << L$ (a) and $d >> L$ (b).

Eq. (1-10-25) means that the effective low-field mobility, μ_B, for a sample with $L/d << 1$ in the magnetic field B is given by

$$\mu_B = \frac{\mu}{1 + \mu^2 B^2} \tag{1-10-26}$$

The related increase in the low-field resistivity, $\Delta\rho$, in the magnetic field (geometric magnetoresistance) is given by

$$\frac{\Delta\rho}{\rho_o} = (\mu B)^2 \tag{1-10-27}$$

Here

$$\rho_o = \frac{1}{qn\mu} \tag{1-10-28}$$

is the low-field resistivity, and n is the carrier concentration.

Measuring the geometric magnetoresistance allows one to measure the low-field mobility, μ, (see, for example, Jervis and Johnson 1970). Equation (1-10-27) for the geometric magnetoresistance is valid when the Hall field is completely shorted by the contacts. This may only be achieved when the ratio L/d tends to infinity. Experimentally, such conditions can be realized using a *Corbino disk configuration* (see Fig. 1-10-5). Figure 1-10-6 shows how the geometric magnetoresistance varies depending on the shape of the sample. The expressions that allow one to calculate the geometric magnetoresistance for arbitrary values of L/d are given, for example, by Madelung (1964).

In addition to the geometric magnetoresistance a *physical magnetoresistance* is also observed in semiconductors, even when $L/d \to \infty$. One of the reasons for this magnetoresistance is the difference in carrier velocities related to random thermal motion. The Hall electric field compensates the Lorentz force acting on an average carrier. Faster carriers are slowed down as a result of the magnetic field; slower carriers are accelerated. However, the faster carriers are slowed down relatively more than the slower carriers are accelerated. As a conse-

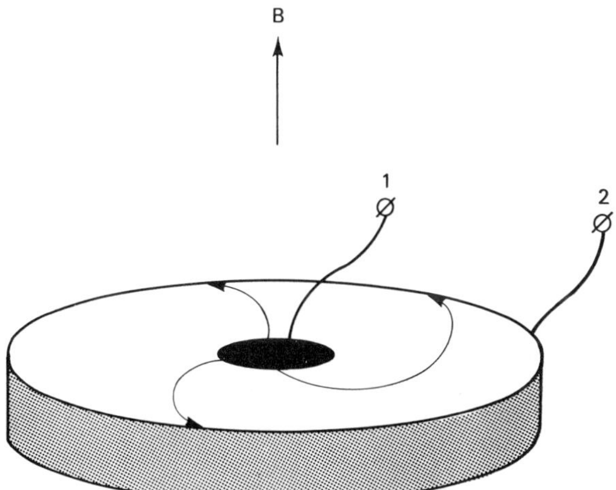

Fig. 1-10-5. Streamlines of electric current in Corbino disk.

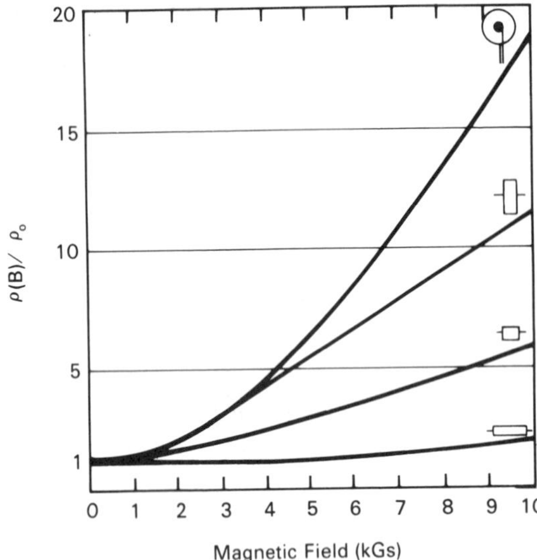

Fig. 1-10-6. Measured geometric magnetoresistance in four InSb samples of different shapes (from Weiss and Welker [1963] and O. Madelung, *Physics of III-V Compounds*, John Wiley & Sons, New York (1964)). Sample shapes are shown near the curves.

quence, there is some increase in the sample resistance. The magnitude of this resistance increase depends on dominant scattering mechanisms. In the particular case when the dependence of the momentum relaxation time on energy may be approximated as

$$\tau = \tau_o \left(\frac{E}{k_B T}\right)^{-S}$$

the physical magnetoresistance is given by

$$\frac{\Delta\rho}{\rho_o} = \left\{ \left[\frac{\Gamma^2(5/2)\Gamma(5/2 - 3S)}{\Gamma^3(5/2 - S)}\right] \left(\frac{\mu_n^3 n + \mu_p^3 p}{\mu_n n + \mu_p p}\right) \right. \tag{1-10-29}$$

$$\left. - \left[\frac{\Gamma^2(5/2)\Gamma(5/2 - 2S)}{\Gamma^2(5/2 - S)}\right]^2 \left(\frac{\mu_n^2 n - \mu_p^2 p}{\mu_n n + \mu_p p}\right)^2 \right\} B_z^2$$

(see Smith 1979). For $n >> p$, $\Delta\rho/\rho \sim (\mu_n B)^2$, and for $p >> n$, $\Delta\rho/\rho \sim (\mu_p B)^2$. However, when $L/d << 1$ the physical magnetoresistance is much smaller than the geometric magnetoresistance.

1-11. SEMICONDUCTOR EQUATIONS BASED ON THE FIELD DEPENDENT VELOCITY AND DIFFUSION

In a low electric field the current densities for electrons and holes are given by

$$\mathbf{j}_n = q(n\mu_n \mathbf{F} + D_n \nabla n) \tag{1-11-1}$$

$$\mathbf{j}_p = q(p\mu_p \mathbf{F} - D_p \nabla p) \tag{1-11-2}$$

The first and second terms in parenthesis in the right-hand-sides of eqs. (1-11-1) and (1-11-2) represent drift and diffusion current densities, respectively. As was discussed in Section 1-9, the low-field mobilities, μ_n and μ_p, are related to the momentum relaxation times and effective masses as follows:

$$\mu_n = \frac{q\tau_n}{m_n} \tag{1-11-3}$$

$$\mu_p = \frac{q\tau_p}{m_p} \tag{1-11-4}$$

Here τ_n, τ_p, m_n, and m_p are momentum relaxation times and effective masses of electrons and holes, respectively. In low electric fields, the diffusion coefficients are related to the mobilities via the Einstein relationship,

$$D_n = \frac{\mu_n k_B T}{q} \tag{1-11-5}$$

$$D_p = \frac{\mu_p k_B T}{q} \tag{1-11-6}$$

Equations (1-11-5) and (1-11-6) may be derived as follows. Let us consider, for simplicity, a one-dimensional problem. Furthermore, let us assume that we have an n-type sample with nonuniform carrier concentration (that may be related to a nonuniform doping) and no electric current ($j = 0$). Then from eq. (1-11-1) we find that

$$0 = n\mu_n F + D_n \partial n/\partial x \tag{1-11-7}$$

The electric field $F = -\partial V/\partial x$ where V is the electric potential, and hence,

$$qF = \partial E_c/\partial x \tag{1-11-8}$$

where $E_c = -qV$ is the bottom of the conduction band. The carrier concentration, n, is given by

$$n = N_c \exp[(E_F - E_c)/k_B T] \tag{1-11-9}$$

where N_c is the effective density of states in the conduction band and E_F is the Fermi level. Hence,

$$\partial n/\partial x = -N_c \exp[(E_F - E_c)/(k_B T)](k_B T)^{-1} \partial E_c/\partial x \tag{1-11-10}$$

or

$$\partial n/\partial x = -n(k_B T)^{-1}\, \partial E_c/\partial x \tag{1-11-11}$$

Substituting eqs. (1-11-8) and (1-11-11) into eq. (1-11-7), we obtain $D_n = \mu_n k_B T/q$.

As can be seen from this derivation, eqs. (1-11-5) and (1-11-6) are valid for nondegenerate samples (because eq. (1-11-9) is only valid for nondegenerate samples). At high carrier concentrations this relationship has to be modified. For the electron diffusion coefficient we have (see Smith 1978):

$$D_n = \frac{2\left(\frac{\mu_n k_B T}{q}\right) F_{1/2}\left(\frac{E_F - E_c}{k_B T}\right)}{F_{-1/2}\left(\frac{E_F - E_c}{k_B T}\right)} \tag{1-11-5a}$$

where $F_{1/2}$ and $F_{-1/2}$ are Fermi integrals (see Appendix 21). For $n < N_c$, where N_c is the effective density of states in the conduction band, this equation may be approximated as (Kroemer 1978)

$$D_n = \frac{\mu_n k_B T}{q}\left[1 + 0.35355\left(\frac{n}{N_c}\right) - 9.9 \times 10^{-3}\left(\frac{n}{N_c}\right)^2 + 4.45 \times 10^{-4}\left(\frac{n}{N_c}\right)^3 + \cdots\right] \tag{1-11-5b}$$

In a high electric field, carriers are heated by the electric field, so that the carrier energy becomes larger than the average thermal energy, $3k_BT/2$, where T is the sample temperature. This changes the conditions for scattering. For example, scattering by ionized impurities becomes less important as the carriers traveling at higher speed spend less time in the vicinity of a scattering center, where their motion is changed by the scattering potential. On the other hand, the scattering involving emission of phonons becomes more important, as the probability of a carrier to have enough energy to emit a phonon increases with energy. As a consequence, the electron and hole velocities are no longer proportional to the electric field where the electric field is high (see Fig. 1-9-1). The diffusion coefficients also become dependent on the electric field (see Section 1.14).

The following phenomenological equations are frequently used in device modeling to describe the electron and hole transport in both low and high electric fields:

$$\mathbf{j}_n = q[-n\mathbf{v}_n(\mathbf{F}) + D_n(\mathbf{F})\nabla n] \tag{1-11-12}$$

$$\mathbf{j}_p = q[p\mathbf{v}_p(\mathbf{F}) - D_p(\mathbf{F})\nabla p] \tag{1-11-13}$$

Here $\mathbf{v}_n(\mathbf{F})$,$\mathbf{v}_p(\mathbf{F})$, $D_n(\mathbf{F})$, and $D_p(\mathbf{F})$ are assumed to be the same functions of electric field as computed or measured for the uniform sample under the steady-state conditions. In a low electric field

$$\mathbf{v}_n = -\mu_n \mathbf{F}$$

$$\mathbf{v}_p = \mu_p \mathbf{F}$$

and eqs. (1-11-12) and (1-11-13) reduce to eqs. (1-11-1) and (1-11-2). The critical analysis of the applicability of equations (1-11-12) and (1-11-13) in a high electric field was given by Blotekjaer (1970). These equations are not derived from the first principles. They may lead to considerable errors in describing the hot electron behavior.

At high frequencies comparable to the inverse energy relaxation time (which is of the order of 2 ps for electrons in the central minimum of the conduction band in GaAs) the velocity and diffusion do not instantaneously follow the variations of the electric field. Therefore, the effective differential mobility, for example, becomes frequency dependent (see Fig. 1-11-1), and eqs. (1-11-12) and (1-11-13) do not apply (Rees 1969).

The advantage of using eqs. (1-11-12) and (1-11-13) or even similar equations with field-independent diffusion coefficients is the relative simplicity of the analysis, which allows one to achieve some insight into the device physics. Several different models that have been introduced as a substitution for eqs. (1-11-12) and (1-11-13) for an approximate analysis of electron and hole transport in short semiconductor devices at high frequencies are considered in Section 1-15.

Equations (1-11-12) and (1-11-13) (or eqs. (1-11-1) and (1-11-2) in a low electric field) should be solved together with the Poisson equation,

$$\nabla \cdot \mathbf{F} = \frac{\rho}{\varepsilon} \tag{1-11-14}$$

and continuity equations,

$$\frac{\partial n}{\partial t} = \frac{1}{q} \nabla \cdot \mathbf{j}_n + G - R \tag{1-11-15}$$

$$\frac{\partial p}{\partial t} = -\frac{1}{q} \nabla \cdot \mathbf{j}_p + G - R \tag{1-11-16}$$

Here ε is the dielectric permittivity,

$$\rho = q(N_D - N_A - n + p) + \rho_T \tag{1-11-17}$$

is the space charge density, N_D is the concentration of the ionized donors, N_A is the concentration of the ionized acceptors, ρ_T is the charge density of the recombination centers and traps, G is the generation rate of electron-hole pairs (due to the light excitation or impact ionization), and R is the recombination rate. Different

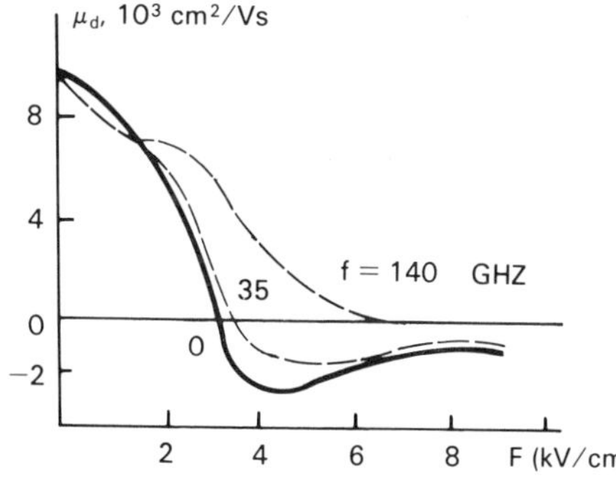

Fig. 1-11-1. Electron differential mobility in GaAs at different frequencies. μ_d is a real part of the differential mobility. (Reprinted with permission from H. D. Rees, "Hot Electron Effects at Microwave Frequencies in GaAs," *Solid State Comm.*, 7, no. 2, pp. 267–269 (1969), Pergamon Press.

recombination mechanisms and the detailed theory that allows us to calculate ρ_T and R are considered in Section 1-12.

Let us now consider the application of these basic semiconductor equations to a nearly neutral device region that may contain both electrons and holes with concentrations n_n and p_n that are larger than equilibrium concentrations n_{no} and p_{no}. An example would be a piece of an n-type semiconductor in which extra carriers are generated by light. For simplicity, we consider a one-dimensional steady-state situation in which $\partial n/\partial t = 0$ and $\partial p/\partial t = 0$ and the continuity equations (1-11-15) and (1-11-16) can be rewritten as

$$D_n \frac{\partial^2 n_n}{\partial x^2} + \mu_n F \frac{\partial n_n}{\partial x} + \mu_n n_n \frac{\partial F}{\partial x} + G - R = 0 \qquad (1\text{-}11\text{-}18)$$

$$D_p \frac{\partial^2 p_n}{\partial x^2} - \mu_p F \frac{\partial p_n}{\partial x} - \mu_p p_n \frac{\partial F}{\partial x} + G - R = 0 \qquad (1\text{-}11\text{-}19)$$

We assume that the semiconductor is almost neutral (i.e., that the space charge density is nearly zero), so that

$$n_n - n_{no} \cong p_n - p_{no} \qquad (1\text{-}11\text{-}20)$$

Let us now exclude the $\partial F/\partial x$ term from the continuity equations (1-11-18) and (1-11-19). Multiplying eq. (1-11-18) by $\mu_p p_n$ and eq. (1-11-19) by $\mu_n n_n$ and adding the resulting equations we obtain

$$\mu_p p_n D_n \frac{\partial^2 n_n}{\partial x^2} + \mu_n n_n D_p \frac{\partial^2 p_n}{\partial x^2} + \mu_n \mu_p p_n F \frac{\partial n_n}{\partial x}$$
$$- \mu_n \mu_p n_n F \frac{\partial p_n}{\partial x} + (G - R)\mu_n n_n + (G - R)\mu_p p_n = 0 \qquad (1\text{-}11\text{-}21)$$

The quasineutrality condition (eq. (1-11-20)) yields

$$\frac{\partial n_n}{\partial x} \cong \frac{\partial p_n}{\partial x} \qquad (1\text{-}11\text{-}22)$$

and

$$\frac{\partial^2 n_n}{\partial x^2} \cong \frac{\partial^2 p_n}{\partial x^2} \qquad (1\text{-}11\text{-}23)$$

Using eqs. (1-11-22) and (1-11-23) we find from eq. (1-11-21) that

$$D_a \frac{\partial^2 p_n}{\partial x^2} - \mu_a F \frac{\partial p_n}{\partial x} + G - R = 0 \qquad (1\text{-}11\text{-}24)$$

where

$$D_a = \frac{\mu_p p_n D_n + \mu_n n_n D_p}{\mu_n n_n + \mu_p p_n} \qquad (1\text{-}11\text{-}25)$$

is called the *ambipolar diffusion coefficient* and

$$\mu_a = \frac{\mu_n \mu_p (n_n - p_n)}{\mu_n n_n + \mu_p p_n} \tag{1-11-26}$$

is called the *ambipolar mobility*. When $n_n >> p_n$ we obtain

$$D_a \cong D_p$$

$$\mu_n \cong \mu_p$$

so that eq. (1-11-24) reduces to

$$D_p \frac{\partial^2 p_n}{\partial x^2} - \mu_p F \frac{\partial p_n}{\partial x} + G - R = 0 \tag{1-11-27}$$

Equation (1-11-27) is the continuity equation for minority carriers (holes) in an n-type sample. This equation will be extremely useful for the analysis of different semiconductor devices, such as p-n junctions (see Chapter 2), bipolar junction transistors (see Chapter 3), and solar cells (see Chapter 5).

Another useful semiconductor equation may be derived from equations given previously when the trapped charge ρ_T in eq. (1-11-17) may be neglected. Indeed, subtracting eq. (1-11-15) from eq. (1-11-16) we obtain

$$q \frac{\partial}{\partial t}(p - n) + \nabla \cdot (\mathbf{j}_n + \mathbf{j}_p) = 0 \tag{1-11-28}$$

From eq. (1-11-17), neglecting ρ_T, we find that

$$q \frac{\partial}{\partial t}(p - n) = \frac{\partial \rho}{\partial t} \tag{1-11-29}$$

Substituting this expression for $\frac{\partial}{\partial t}(p - n)$ into eq. (1-11-29) and using Poisson's equation we obtain

$$\varepsilon \frac{\partial}{\partial t} \nabla \cdot \mathbf{F} + \nabla \cdot (\mathbf{j}_n + \mathbf{j}_p) = 0 \tag{1-11-30}$$

Integration of eq. (1-11-30) over the space coordinates leads to

$$\mathbf{j}(t) = \mathbf{j}_n + \mathbf{j}_p + \varepsilon \frac{\partial \mathbf{F}}{\partial t} \tag{1-11-31}$$

where $\mathbf{j}(t)$ is the total current density $\left(\text{including the displacement current } \varepsilon \frac{\partial \mathbf{F}}{\partial t}\right)$; and

$$I = \int \mathbf{j}\, ds \tag{1-11-32}$$

where I is the current in the circuit and s is the sample cross section. In a one-dimensional case

$$j = \frac{I}{S} \tag{1-11-33}$$

for a sample with a constant cross section.

In low electric fields when the Einstein relationship is valid, the electron current density may be expressed through the electron quasi-Fermi level. The concept of the electron and hole quasi-Fermi levels, E_{Fn} and E_{Fp}, is considered in Section 1-12. In low electric fields for nondegenerate semiconductors, the electron and hole quasi-Fermi levels are defined by the following equations:

$$n = N_c \exp\left(\frac{E_{Fn} - E_c}{k_B T}\right)$$

$$p = N \exp\left(\frac{E_v - E_{Fp}}{k_B T}\right)$$

(In a zero electric field E_{Fn} and E_{Fp} coincide with the Fermi level, E_F.) Using these definitions, we find that

$$\begin{aligned} j_n &= q\mu_n nF + qD_n \frac{\partial n}{\partial x} \\ &= q\mu_n n \frac{1}{q}\frac{\partial E_c}{\partial x} + qD_n \frac{\partial}{\partial x}\left[N_c \cdot \exp\left(\frac{E_{Fn} - E_c}{k_B T}\right)\right] \\ &= \mu_n n \frac{\partial E_{Fn}}{\partial x} \end{aligned} \tag{1-11-34}$$

In a similar fashion we find that

$$j_p = \mu_p p \frac{\partial E_{Fp}}{\partial x} \tag{1-11-35}$$

Modern semiconductor devices may have very short dimensions (as small as 0.1 μm or less). Typical operating voltages are usually of the order of several volts, so that internal electric fields are extremely large (10^4 to 10^5 V/cm and higher). These devices may operate at very high frequencies (up to 100 GHz and higher). As was mentioned previously, eqs. (1-11-12) and (1-11-13) are not valid under such conditions. Within certain limitations they still may provide some useful insight into the device operation; however, more accurate analysis requires either a numerical solution of the Boltzmann transport equation (see Section 1-13) together with Poisson's equation or, for very small devices, even a direct solution of the Schrödinger equation. The difficulties in such a "brute force" approach are related not only to the large computational resources required for the calculations but also to the lack of detailed information about numerous material parameters needed for a realistic description of the high-field transport. Therefore, several simple (but less accurate) approaches for dealing with high-field transport have been proposed. Some of these approaches are discussed in Section 1-15.

1-12. QUASI-FERMI LEVELS: GENERATION AND RECOMBINATION OF CARRIERS

In Section 1-6 we considered concentrations of electrons and holes under the conditions of thermal equilibrium when the probability of occupancy of an electronic state is given by the Fermi-Dirac distribution function. However, the distribution function changes quite dramatically when a high electric field is applied to a semiconductor sample (see Section 1-14, where we describe distribution functions in high electric fields computed using the Monte-Carlo method). Under such nonequilibrium conditions, the electron and hole concentrations are no longer related by

$$pn = n_i^2 \tag{1-12-1}$$

(as for the equilibrium nondegenerate case, see eq. (1-6-33)) and the very concept of a Fermi level is no longer applicable. Nonequilibrium conditions are also created by generating extra electron-hole pairs in a semiconductor by absorbed light. Photons with energies greater than the energy gap may promote electrons from the valence into the conduction band, generating electron-hole pairs. This process is illustrated by Fig. 1-12-1.

Under such nonequilibrium conditions, it may still be useful to represent the distribution functions for electrons and holes, f_n and f_p, as

$$f_n = \frac{1}{1 + \exp\left(\dfrac{E - E_{Fn}}{k_B T}\right)} \tag{1-12-2}$$

$$f_p = \frac{1}{1 + \exp\left(\dfrac{E_{Fp} - E}{k_B T}\right)} \tag{1-12-3}$$

(compare with eqs. (1-6-8) and (1-6-12)). Equations (1-12-2) and (1-12-3) may be considered definitions of E_{Fp} and E_{Fn} that are called *electron* and *hole quasi-Fermi levels* (sometimes they are also called Imrefs—Fermi spelled backward). Under equilibrium conditions, $E_{Fp} = E_{Fn} = E_F$. However, under nonequilibrium condi-

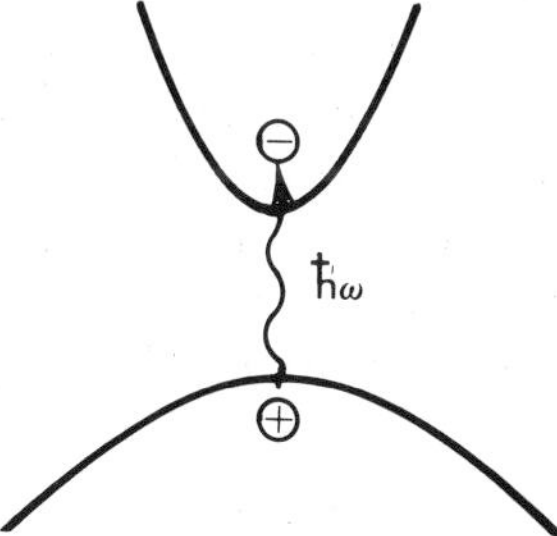

Fig. 1-12-1. Generation of electron-hole pair by photon.

tions, E_{Fp} is not equal to E_{Fn} and they both may be functions of coordinates. Actually, the difference $E_{Fn} - E_{Fp}$ can be used as the measure of a deviation from equilibrium. In the nondegenerate case, we obtain from eqs. (1-12-2) and (1-12-3)

$$f_n \approx \exp\left(\frac{E_{Fn} - E}{k_B T}\right) \tag{1-12-4}$$

$$f_p \approx \exp\left(\frac{E - E_{Fp}}{k_B T}\right) \tag{1-12-5}$$

Substituting these expressions into eqs. (1-6-9) and (1-6-13) and performing the integration, we obtain

$$n = N_c \exp\left(\frac{E_{Fn} - E_c}{k_B T}\right) \tag{1-12-6}$$

$$p = N_v \exp\left(\frac{E_v - E_{Fp}}{k_B T}\right) \tag{1-12-7}$$

(see Section 1-11). In other words, in this case the quasi-Fermi levels are proportional to the logs of electron and hole concentrations.

We can try to extend this concept for the case in which an applied electric field causes a substantial increase in average energies of random motion of electrons or holes by introducing effective electron and hole temperatures, T_e and T_p. The effective electron temperature, T_e, equals $2E/3k_B$, where E is the electron energy (see Section 1-9). In this case eqs. (1-12-2) and (1-12-3) become

$$f_n = \frac{1}{1 + \exp\left(\dfrac{E - E_{Fn}}{k_B T_e}\right)} \tag{1-12-8}$$

$$f_p = \frac{1}{1 + \exp\left(\dfrac{E_{Fp} - E}{k_B T_p}\right)} \tag{1-12-9}$$

However, the concept of effective electron (or hole) temperature may not be very accurate (see Section 1-14 for the related discussion).

The concept of quasi-Fermi levels is very useful because carrier concentrations in a practical semiconductor device may vary as functions of position or bias by many orders of magnitude, whereas the quasi-Fermi levels change within the energy gap or close to the bottom of the conduction band or the top of the valence band. This variation is much easier to visualize.

Let us now consider an example. Let us say that we shine light on the n-type GaAs sample with doping density N_d. The light is uniformly absorbed and produces electron-hole pairs with density P. Hence, the electron concentration in the sample is equal to

$$n \approx P + N_d \tag{1-12-10}$$

and the hole concentration is equal to

$$p \approx P + n_i^2/N_d \tag{1-12-11}$$

The electron and hole quasi-Fermi levels, calculated using eqs. (1-12-6), (1-12-7), (1-12-10), and (1-12-11), are shown in Fig. 1-12-2. As can be seen from Fig. 1-12-2, when P varies from 10^8 to 10^{17} cm^{-3}, E_{Fn} varies from 1.25 to 1.35 eV and E_{Fp} varies from 0.7 eV to approximately 0.15 eV.

Electron-hole pairs generated in a semiconductor recombine. Recombination processes become more intensive as the electron-hole concentration increases. Steady-state values of carrier concentration are reached when the generation rate, G, is balanced by the recombination rate, R:

$$G = R \tag{1-12-12}$$

Recombination processes include direct (band-to-band) radiative recombination, radiative band-to-impurity recombination, nonradiative recombination via impurity (trap) levels, and surface recombination. The radiative band-to-band recom-

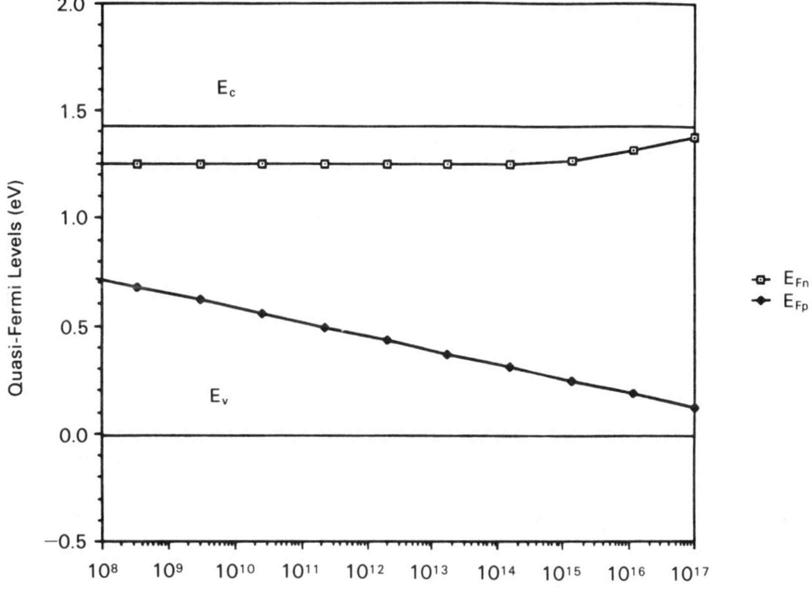

Fig. 1-12-2. Electron and hole quasi-Fermi levels vs. concentration of light-generated electron-hole pairs for n-type GaAs a sample. Parameters used in the calculation are $T = 300$ K, $Nd = 10^{15}$ cm^{-3}, $N_c = 4.7 \times 10^{17}$ cm^{-3}, $N_v = 7 \times 10^{18}$ cm^{-3}, and $n_i = 1.79 \times 10^6$ cm^{-3}.

bination rate is proportional to an np product. For a nondegenerate semiconductor:

$$R = G_{th}\, np/n_i^2 \tag{1-12-13}$$

where G_{th} is the thermal generation rate (so that at equilibrium $np = n_i^2$ as expected). The expression for G_{th} was first derived by van Roosbroeck and Shockley (1954), who showed that

$$G_{th} = 32\pi^2(k_BT/h)^4 \int \xi(\nu) n_r^3 x^3\, dx/[\exp(x) - 1] \tag{1-12-14}$$

where ν is frequency, n_r is the refraction index, $x = h\nu/(k_BT)$, $\xi(x) = hc\alpha(x)/(4\pi k_BTn_rx)$, c is the velocity of light in a vacuum, and α is the absorption coefficient. Absorption coefficients for different semiconductors will be considered in Section 5-3. The constant in front of the integral in eq. (1-12-14) is equal to $1.785 \times 10^{22}\,[T(\mathrm{K})/300]^4$ (cm^{-3}/s). Equation (1-12-13) may be rewritten as

$$R = C_r\, np \tag{1-12-15}$$

In the steady state, we find that

$$G = C_rnp = C_r(n_o + \Delta n)(p_o + \Delta p) \tag{1-12-16}$$

where $\Delta n = \Delta p$ are concentrations of extra electrons and holes, and n_o and p_o are equilibrium concentrations of electrons and holes ($n_o\, p_o = n_i^2$). When the generation of the electron-hole pairs is caused by light, the generation rate, G, is proportional to the light intensity, I. For example, consider an n-type semiconductor where under equilibrium conditions $n_o = N_d$, where N_d is the concentration of shallow donors and $p_o = n_i^2/N_d$. At low light intensities when $\Delta n << N_d$ but $\Delta p \approx \Delta n >> n_i^2/N_d$, we find from eq. (1-12-16) that

$$\Delta n = G\tau_r \tag{1-12-17}$$

where $\tau_r = 1/(C_rN_d)$ is called the *radiative band-to-band recombination lifetime*. When the light intensity is small, Δn is proportional to G, and hence, to I. At high intensities, when $\Delta n >> n_o, p_o$, $\Delta n \approx \Delta p$, $G \approx C_r\Delta n\Delta p = C_r\,\Delta n^2$, and Δn is proportional to $\sqrt{G}$, and hence, to $\sqrt{I}$.

In the practical light-emitting semiconductor devices described, for example, in Section 5-3, radiative band-to-impurity recombination (illustrated by Fig. 1-12-3b) is often more important than radiative band-to-band recombination (see Fig. 1-12-3a). The radiative band-to-impurity recombination lifetime is given by

$$\tau_r = 1/(B_rN_A) \tag{1-12-18}$$

where B_r is the radiative recombination coefficient and N_A is the concentration of impurities involved in this recombination process (see Goodfellow et al. 1985).

In many cases the dominant recombination mechanism is recombination via traps. The theory of this recombination process was developed by Shockley and Read (1952) and later by Sah et al. (1957). Four electron and hole transitions involved in this recombination mechanism are shown in Fig. 1-12-4. When an

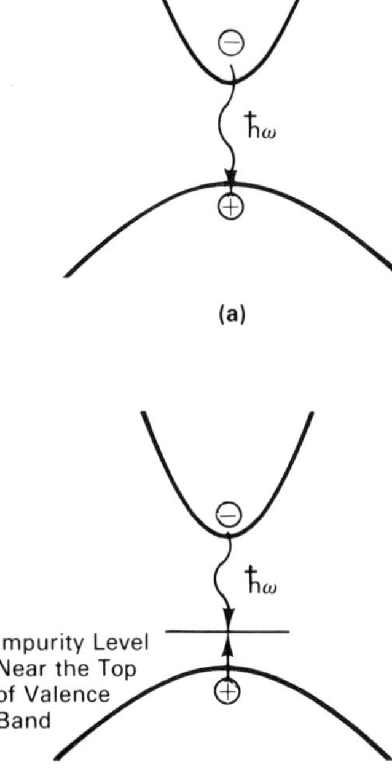

Fig. 1-12-3. Mechanisms of radiative recombination: (a) band-to-band recombination and (b) band-to-impurity recombination.

electron is captured by an empty trap and then a hole is captured by the trap filled by the electron, the electron-hole pair recombines. Inverse processes are the emission of an electron from a filled trap into the conduction band and the emission of a hole from an empty trap into the valence band. The rate of the electron capture by the traps, R_{nc}, is proportional to the number of electrons and to the number of empty traps, $(1 - f_t)N_t$:

$$R_{nc} = C_n n(1 - f_t)N_t \qquad (1\text{-}12\text{-}19)$$

Here f_t is the occupancy function of the trap level. The coefficient C_n may be presented as

$$C_n = \sigma_n v_{thn} \qquad (1\text{-}12\text{-}20)$$

where σ_n is called the *capture cross section* for electrons,

$$v_{thn} = (3k_B T/m_n)^{1/2} \qquad (1\text{-}12\text{-}21)$$

is the electron thermal velocity, and m_n is the electron effective mass.

The rate of the electron emission from the traps, R_{ne}, is proportional to the number of filled traps, $f_t N_t$:

$$R_{ne} = e_n f_t N_t \qquad (1\text{-}12\text{-}22)$$

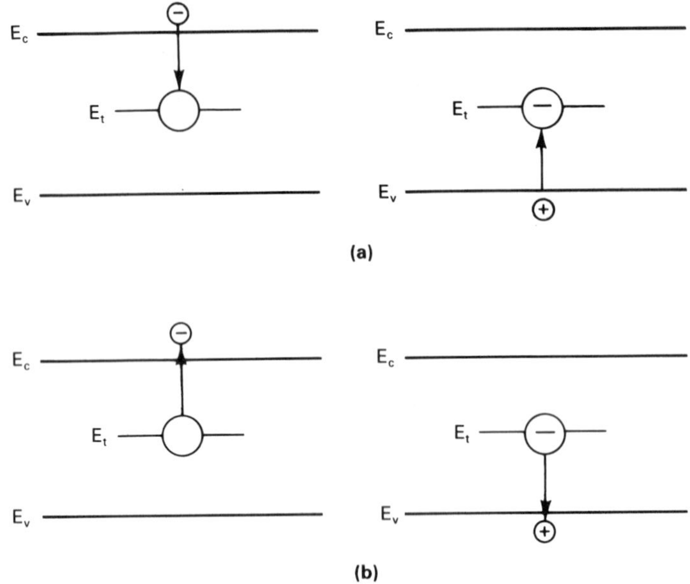

Fig. 1-12-4. Electron and hole transitions involved in recombination via traps. (a) The electron captured by empty trap and hole captured by filled trap result in recombination of electron-hole pair. (b) Inverse processes: The electron is generated from filled trap into conduction band and the hole is generated from empty trap into valence band.

Under the equilibrium conditions,

$$R_{nc} = R_{ne} \tag{1-12-23}$$

and hence,

$$C_n n_o = e_n f_{to}/(1 - f_{to}) \tag{1-12-24}$$

where

$$n_o = N_c \exp[(E_F - E_c)/(k_B T)] \tag{1-12-25}$$

is the equilibrium electron concentration, and f_{to} is the equilibrium occupancy of the trap level. The ratio $f_{to}/(1 - f_{to})$ can be found using the Fermi-Dirac occupation function:

$$f_{to}/(1 - f_{to}) = \exp[-(E_t - E_F)/(k_B T)] \tag{1-12-26}$$

Here E_t is the energy of the trap level. Hence, from eq. (1-12-24) we find that

$$e_n = n_t C_n \tag{1-12-27}$$

where

$$n_t = N_c \exp[(E_t - E_c)/(k_B T)] \tag{1-12-28}$$

The difference between the electron capture and electron emission rates is given by

$$R_n = R_{nc} - R_{ne} = C_n N_t[(1 - f_t)n - f_t n_t] \tag{1-12-29}$$

A similar derivation yields the following expression for the difference between the hole capture and the hole emission rates:

$$R_p = R_{pc} - R_{pe} = C_p N_t[f_t p - (1 - f_t)p_t] \tag{1-12-30}$$

where

$$C_p = \sigma_p \mathrm{v}_{\mathrm{thp}} \tag{1-12-31}$$

σ_p is the capture cross section for holes,

$$v_{thp} = (3k_B T/m_p)^{1/2} \tag{1-12-32}$$

is the hole thermal velocity, and m_p is the hole effective mass. Under steady-state conditions there is no net accumulation of charge and, hence, electrons and holes must recombine in pairs. Thus,

$$R_p = R_n = R \tag{1-12-33}$$

where R is the recombination rate. The occupation function, f_t, may be found from the condition $R_p = R_n$:

$$f_t = \frac{nC_n + p_t C_p}{C_n(n + n_t) + C_p(p + p_t)} \tag{1-12-34}$$

Substituting this expression into eq. (1-12-29) we find that

$$R = \frac{pn - n_i^2}{\tau_{pl}\,(n + n_t) + \tau_{nl}\,(p + p_t)} \tag{1-12-35}$$

Here n_i is the intrinsic concentration, τ_{nl} and τ_{pl} are electron and hole lifetimes:

$$\tau_{nl} = 1/(v_{thn}\sigma_n N_t) \tag{1-12-36}$$

$$\tau_{pl} = 1/(v_{thp}\sigma_p N_t) \tag{1-12-37}$$

In particular, when electrons are minority carriers ($n << p \approx N_A$, where N_A is the concentration of shallow ionized acceptors, $p >> p_t$, $p >> n_t$), eq. (1-12-35) reduces to

$$R = \frac{n - n_o}{\tau_{nl}} \tag{1-12-38}$$

where $n_o = n_i^2/N_D$. When holes are minority carriers ($p << n \approx N_D$, where N_D is the concentration of shallow ionized donors, $n >> n_t$, $n >> p_t$), eq. (1-12-35) reduces to

$$R = \frac{p - p_o}{\tau_{pl}} \tag{1-12-39}$$

where $p_o = n_i^2/N_A$. Electron lifetime, τ_{nl}, in p-type silicon samples, and hole lifetime, τ_{pl}, in n-type silicon samples, are shown as functions of carrier concentration in Figures 1-12-5a and 1-12-5b. The carrier concentration is determined by doping. The decrease of lifetime with doping at low doping levels may be explained by higher trap concentrations in doped samples. If the trap concentration is proportional to the concentration of dopants, we should expect $\tau \sim 1/N_D$. However, as can be seen from Fig. 1-12-5b, at relatively high doping levels τ_{pl} decreases with doping concentration faster than $1/N_D$. The reason is that a different recombination mechanism, called *Auger recombination*, becomes important at very high doping levels. In this recombination mechanism electron and hole recombine without involving trap levels, and the released energy (of the order of the energy gap) is transferred to another carrier (a hole in p-type material and an electron in n-type material). This recombination mechanism is illustrated by Fig. 1-12-6. Such a process is an inverse process to an impact ionization mechanism via which an energetic carrier causes the generation of an electron-hole pair. Because two electrons (in n-type material) or two holes (in p-type material) are involved in the Auger recombination process, the recombination lifetime associated with the Auger recombination is inversely proportional to the square of the majority carrier concentration. For p-type material,

$$\tau_{nl} = 1/(G_p N_A^2) \tag{1-12-40}$$

For n-type material,

$$\tau_{pl} = 1/(G_n N_D^2) \tag{1-12-41}$$

For silicon, $G_p = 9.9 \times 10^{-32}$ cm^6/s, and $G_n = 2.28 \times 10^{-31}$ cm^6/s (see Dziewior and Schmid 1982).

In many semiconductor devices the recombination rate is very high near the surface, where extra defects and traps increase the recombination rate. As a consequence, the diffusion flux of minority carriers at the surface is determined by the surface recombination processes. For example, when minority carriers are holes this surface recombination can be described by the following equation:

$$D_p \partial p_n/\partial x|_{x=0} = -S_p[p_n(x = 0) - p_{no}] \tag{1-12-42}$$

where D_p is the hole diffusion coefficient, p_n is the hole concentration, $p_{no} = n_i^2/N_D$ is the equilibrium hole concentration, $x = 0$ corresponds to the surface of the sample,

$$S_p = \sigma_p v_{thp} N_{st} \tag{1-12-43}$$

is the surface recombination rate, and N_{st} is the surface density of the surface traps. Equation (1-12-42) should be used as one of the two boundary conditions required for solving continuity equation (1-11-27) (see Problem 1-12-2).

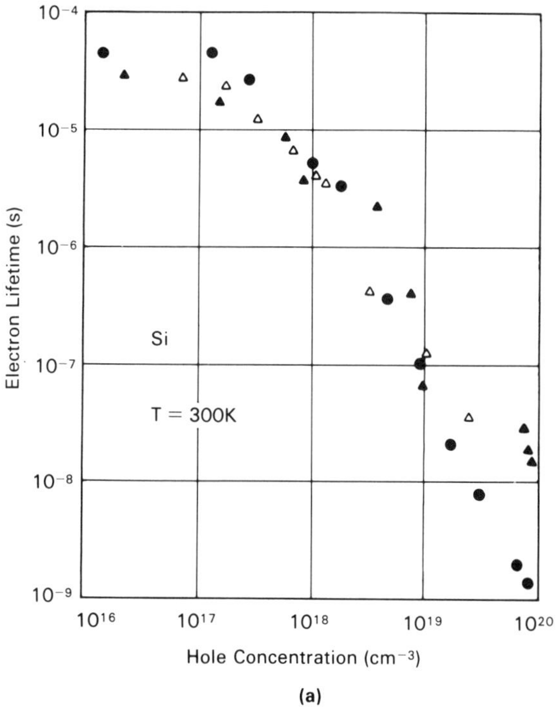

(a)

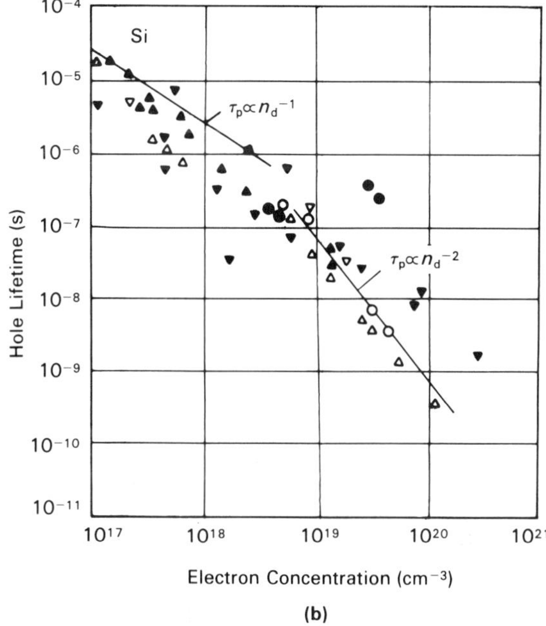

(b)

Fig. 1-12-5. (a) Electron lifetime in *p*-type silicon samples and (b) hole lifetime in *n*-type silicon samples vs. doping concentration (Reprinted with permission from M. S. Tyagi and R. van Overstraaten, *Solid State Elecron.*, 26, p. 577 (1983) and J. G. Fossum, R. P. Mertens, D. S. Lee, and J. F. Nijs, *Solid State Electron.*, 26, p. 569 (1983) Pergamon Press p/c, respectively).

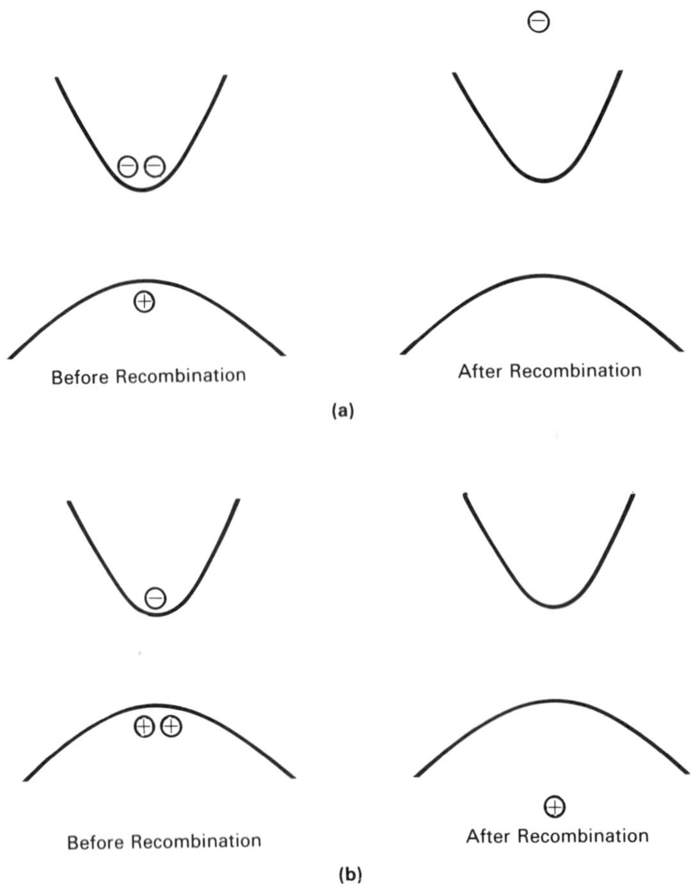

Fig. 1-12-6. Auger recombination: (a) n-type sample and (b) p-type sample.

*1-13. BOLTZMANN TRANSPORT EQUATION AND SCATTERING RATES

Electrons in a crystal with a given band structure may be thought of as a collection of particles with a given relationship between the energy E and the wave vector $\mathbf{k}$ (see Section 1-5).

In a six-dimensional phase space $(\mathbf{r},\mathbf{k})$, the motion of each particle can be represented by a moving point. The equation of motion for the particle is given by

$$\hbar\dot{\mathbf{k}} = q(\mathbf{F} + \mathbf{v} \times \mathbf{B}) \tag{1-13-1}$$

where $\mathbf{v}$ is the velocity, $\mathbf{F}$ is the electric field, q is the electronic charge, and $\mathbf{B}$ is the magnetic field. The dot (.) corresponds to the differentiation with respect to time. This equation can be understood by referring to an electron in free space, where the electron momentum is

$$\mathbf{p} = m_n\mathbf{v} = \hbar\mathbf{k} \tag{1-13-2}$$

Here velocity $\dot{\mathbf{v}} = \mathbf{r}$ (see eq. (1-1-13)). (In a crystal $\hbar\mathbf{k}$ is called quasi momentum because there are many equivalent values of the wave vector that differ by a vector of a reciprocal lattice.) The right-hand side of eq. (1-13-1) represents the force acting on an electron from external electric and magnetic fields. Hence, eq. (1-13-1) represents the second law of motion. In a free space the kinetic energy is given by $E = m_n v^2/2$ (or $E = \hbar k^2/(2m_n)$). In a crystal the $E(\mathbf{k})$ relation may be far more complex (see Section 1-5) and

$$\mathbf{v} = \frac{1}{\hbar}\nabla_{\mathbf{k}} E(\mathbf{k}) \tag{1-13-3}$$

Equation (1-13-1) describes free (ballistic) motion of electrons in a crystal. However, this ballistic motion is interrupted by collisions with impurities, quanta of lattice vibrations (phonons), different defects, etc. In most cases, the collisions are considered to be instantaneous events changing the electron wave vector. If a collision is instantaneous, it may be visualized as the particle disappearing and instantaneously reappearing in a different point of the phase space.

The average occupancy $f(\mathbf{k},\mathbf{r},t)$ of a point in the phase space is called the *distribution function*. The distribution function may change as a result of the scattering, the flow of electrons in real space determined by their velocity, $\mathbf{v}$, and the flow of electrons in $\mathbf{k}$-space, which is determined by the time derivative, $\dot{\mathbf{k}}$. The equation for the distribution function, f, is called the *Boltzmann transport equation:*

$$\frac{\partial f}{\partial t} + \mathbf{v}\cdot\nabla_{\mathbf{r}}f + \dot{\mathbf{k}}\cdot\nabla_{\mathbf{k}}f = \left(\frac{\partial f}{\partial t}\right)_{coll} \tag{1-13-4}$$

The Boltzmann transport equation may be interpreted as the continuity equation for the distribution function. It simply states that the change of the distribution function with time (represented by the first term in the left-hand side of eq. (1-13-4)) is determined by the flow of electrons in real space (the second term in the left-hand side of eq. (1-13-4)), by the flow of electrons in the $\mathbf{k}$-space (the third term in the left-hand side of eq. (1-13-4)), and by collisions represented by the right-hand side of the Boltzmann transport equation. As was noted previously, this equation is based on the assumption that the collisions are instantaneous. Also, for the Boltzmann equation to be valid the external fields should be sufficiently small and band-to-band transitions should be absent.

The term in the right-hand part of eq. (1-13-4) is called the *collision integral* because it can be calculated integrating probabilities of transitions:

$$\left(\frac{\partial f}{\partial t}\right)_{\mathrm{coll}} = \int [W(\mathbf{k}',\mathbf{k})f_{\mathbf{k}'}(1 - f_{\mathbf{k}}) - W(\mathbf{k},\mathbf{k}')f_{\mathbf{k}}(1 - f_{\mathbf{k}'})]\, dV_{\mathbf{k}'} \tag{1-13-5}$$

Here the integration is over the first Brillouin zone, and $W(\mathbf{k},\mathbf{k}')\, dV_{\mathbf{k}'}dt$ is the conditional probability of the transition from the state $\mathbf{k}$ to a state $\mathbf{k}'$ in $dV_{\mathbf{k}'}$ in time dt given that an electron is initially in state $\mathbf{k}$ and the state $\mathbf{k}'$ is empty. Hence, the first term in the integral in the left-hand side of eq. (1-13-5) gives the rate of increase in $f_{\mathbf{k}}(r,t)$ due to transitions from all other states. The second term in the integral in the left-hand side of eq. (1-13-5) gives the rate of decrease in $f_{\mathbf{k}}(r,t)$ due to transitions into all other states.

The principle of detailed balance requires that the transition probability be symmetric for scattering processes that conserve electron energy, i.e.,

$$W(\mathbf{k},\mathbf{k}') = W(\mathbf{k}',\mathbf{k}) \tag{1-13-6}$$

For such processes the collision integral can be simplified:

$$\left(\frac{\partial f}{\partial t}\right)_{\text{coll}} = \int W(\mathbf{k},\mathbf{k}')(f'_{\mathbf{k}'} - f_{\mathbf{k}})\, dV_{\mathbf{k}} \tag{1-13-7}$$

In the uniform equilibrium state the Boltzmann equation reduces to

$$\left(\frac{\partial f}{\partial t}\right)_{\text{coll}} = 0 \tag{1-13-8}$$

It may be shown that the solution of this equation is

$$f_o = \frac{1}{\exp\left[(E_{\mathbf{k}} - E_F)/k_BT\right] + 1} \tag{1-13-9}$$

i.e., the Fermi-Dirac occupation function. Here E_F is the Fermi level and T is temperature.

If the transition probabilities are independent of $\mathbf{r}$ and t, eq. (1-13-8) is also valid for a nonuniform equilibrium state. An example of such a system would be a p-n junction or nonuniformly doped semiconductor with zero electric current (no external voltage applied).

The general nonuniform-equilibrium solution of the Boltzmann equation is given by

$$f = \frac{1}{\exp\left[(E_{\mathbf{k}} - E_F + qV)/k_BT\right] + 1} \tag{1-13-10}$$

Here V is the electric potential.

In an external electric field, the Boltzmann equation has to be solved in order to determine the distribution function. Probabilities $W(\mathbf{k},\mathbf{k}')$ are calculated using the so-called golden rule of quantum mechanics,

$$W(\mathbf{k},\mathbf{k}') = \frac{2\pi}{\hbar} |H|^2\, \delta(E' - E) \tag{1-13-11}$$

where E' and E are the initial and final energies of the crystal (i.e., energies before and after the transition), and $|H|^2$ is the squared matrix element that depends on a particular scattering mechanism. For scattering processes that do not involve a change in **k** by a vector of reciprocal lattice called (non-umklapp processes),

$$|H|^2 = AG(\mathbf{k},\mathbf{k}') \tag{1-13-12}$$

where A depends on a particular scattering mechanism and $G(\mathbf{k},\mathbf{k}')$ is the overlap factor,

$$G(\mathbf{k},\mathbf{k}') = \left| \int_{\text{cell}} u_{\mathbf{k}}(\mathbf{r}) u_{\mathbf{k}'}(\mathbf{r}) d\mathbf{r} \right|^2 \tag{1-13-13}$$

between the periodic parts $u_{\mathbf{k}}(\mathbf{r})$ and $u_{\mathbf{k}'}(\mathbf{r})$ of the Bloch wave functions of the initial and final states.

The expressions for constant A in the squared matrix element (see eq. (1-13-12) for different scattering mechanisms are given, for example, by Jacoboni and Reggiani (1979). The most important scattering mechanisms in semiconductors are phonon scattering (i.e., scattering by lattice vibrations) and ionized impurity scattering. Long-wave phonons (acoustic and optical) are involved in the intravalley scattering (such as electron scattering in one valley of a conduction band), and phonons with wave vectors close to the boundary of the first Brillouin zone are involved in the intervalley transitions (such as transitions between different valleys in silicon; see Fig. 1-5-7).

Scattering by polar optical phonons in partially heteropolar crystals (like gallium arsenide) is caused by the electric field associated with relative periodic displacement of oppositely charged atomic sublattices (such as gallium and arsenic sublattices in Gatts). Scattering by acoustic and nonpolar phonons (in homopolar crystals like silicon) is caused by the modulation of energy bands by the lattice vibrations due to the crystal deformation (these scattering mechanisms are sometimes referred to as deformation potential scattering).

A convenient way to characterize the relative strength of the different scattering mechanisms is to compute the total scattering rate,

$$\lambda(\mathbf{k}) = \int W(\mathbf{k},\mathbf{k}')\, dV_{\mathbf{k}'} \tag{1-13-14}$$

which is the probability per unit time that an electron with the wave vector **k** will be scattered. Here the integration is over the Brillouin zone.

The expressions for the electron scattering rates obtained using eqs. (1-13-11) to (1-13-14) and eq. (1-13-1) are given in Appendix 22 (from Ruch and Fawcett 1970).

Polar optical scattering (as well as all other scattering mechanisms involving phonons) is actually a superposition of two scattering processes involving emission and absorption of a phonon, respectively. The scattering rates for processes with absorption and emission of a polar optical phonon for 300 K and 77 K are shown in Figs. 1-13-1 and 1-13-2. Polar optical scattering with emission of a polar optical phonon becomes effective when the electron energy exceeds the energy of a polar optical phonon. We also notice that scattering rates for the process with

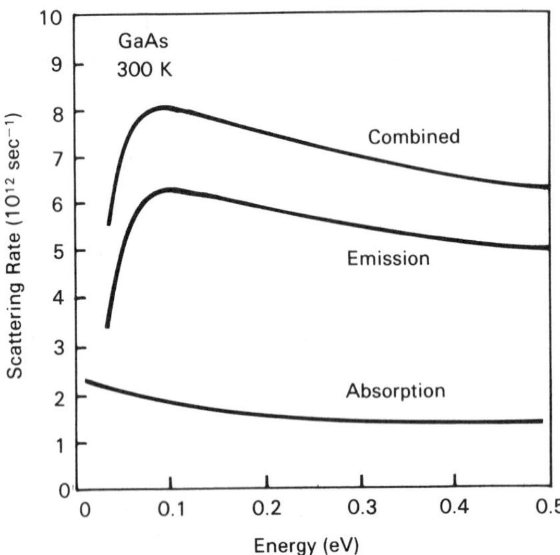

Fig. 1-13-1. Energy dependence of the polar-optical scattering rate for electrons (emission, absorption, and combined) in the (000) minimum of GaAs at 300 K for a parabolic band (from M. Shur, *GaAs Devices and Circuits*, Plenum Publishing, New York (1987). Parameters used in the calculation are the same as used by Fawcett, Boardman, and Swain [1970].

absorption of an optical phonon are very small at cryogenic temperatures, when there are very few phonons available. At the same time, scattering rates for the process with emission of an optical phonon are not very dependent on temperature. A similar qualitative trend also applies to scattering by nonpolar optical phonons. This comment will later provide a clue for an explanation of temperature dependence of electron velocity in small and large electric fields. In particular, we can expect that the electron velocity in high electric fields is much less temperature dependent than a low-field mobility.

In addition to the magnitude of scattering rates, it is important to know the probability distribution of the wave vectors of the final state (i.e., the states after scattering) and the dependence of this probability distribution on the wave vector of the initial state (the state before scattering). For polar optical scattering, the probability of scattering under a small angle, Θ, between the wave vectors of initial and final states is much higher. This is illustrated by Fig. 1-13-3, which shows the probability of scattering under angle Θ as a function of $\cos\Theta$ computed using eqs. given in Appendix 22. In other words, polar optical scattering is a "forward" scattering—scattered electrons tend to move approximately in the same direction.

Once the scattering mechanisms are established and the corresponding scattering rates are calculated, the Boltzmann transport equation has to be solved to yield the distribution function. The average drift velocity, $\mathbf{v}_d$, the average electron energy, E, etc., can then be found using the integration over the first Brillouin zone. For example,

$$v = \int v_{\mathbf{k}} f_{\mathbf{k}}\, d\mathbf{k} \tag{1-13-15}$$

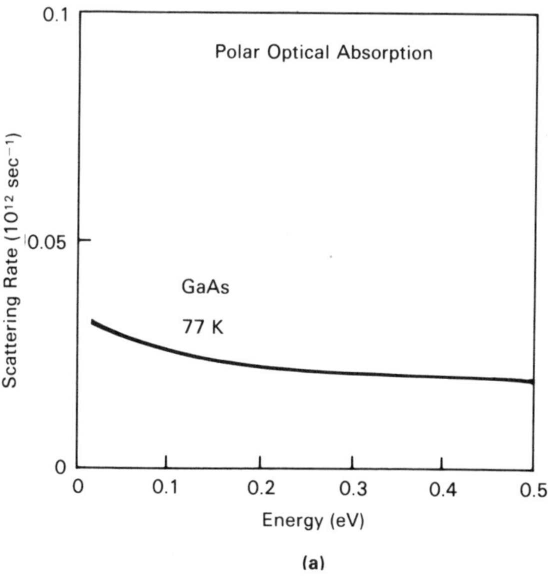

(a)

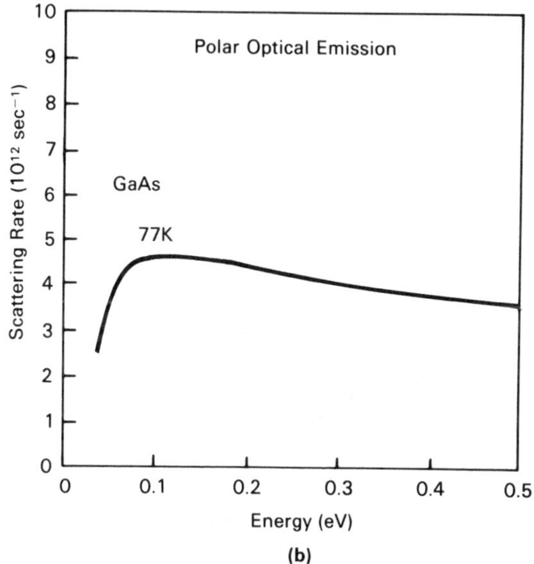

(b)

Fig. 1-13-2. Energy dependence for electrons of the polar optical scattering rate in the (000) minimum of GaAs at 77 K for a parabolic band (from M. Shur, *GaAs Devices and Circuits*, Plenum Publishing, New York (1987). (a) absorption and (b) emission. Parameters used in the calculation are the same as used by Fawcett, Boardman, and Swain [1970].

In the next Section we will consider a very popular numerical technique—the Monte Carlo method—for simulating electron transport in semiconductors in high electric fields without directly solving the Boltzmann transport equation.

In low electric fields, the Boltzmann equation may be solved analytically because the distribution function is not very different from the equilibrium Fermi-

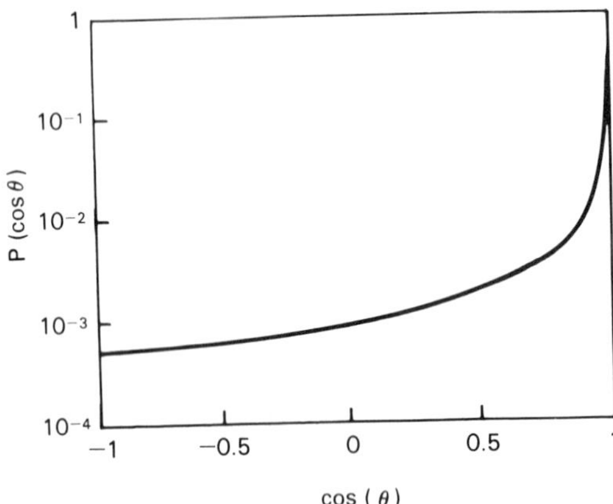

Fig. 1-13-3. Angular probability P(Θ) for electrons for polar optical scattering vs. cos Θ in the (000) valley of GaAs at 300 K for the electron energy 0.4 eV (for a parabolic band), from W. Fawcett, D. A. Boardman, and S. Swain, "Monte Carlo Determination of Electron Transport Properties in Gallium Arsenide," *J. Phys. Chem. Solids*, 31, pp 1963–1970 (1970).

Dirac distribution function. For elastic scattering (such as impurity, acoustic, or piezoelectric scattering) one may use a relaxation time approximation to solve the Boltzmann equation in low electric fields. According to this approximation the distribution function is presented as

$$f = f_o + f_1 \tag{1-13-16}$$

where f_o is the equilibrium Fermi-Dirac distribution function and f_1 is a small correction caused by the applied electric field. The collision integral in the Boltzmann equation is presented as $-f_1/\tau$, where τ is the momentum relaxation time. This relaxation time is assumed to be a function of the wave vector **k**, or in isotropic media, a function of the electron energy, E.

For a long uniform sample under steady-state conditions, the first two terms in the left-hand side of the Boltzmann equation (1-13-4) are equal to zero. If a magnetic field is zero, the time derivative of **k** is equal to $q\mathbf{F}/\hbar$, where **F** is the constant electric field applied to the sample. The gradient of the distribution function in the **k**-space in the third term of the left-hand side of the Boltzmann equation (1-13-4) can be presented as

$$\nabla_{\mathbf{k}} f = \partial f/\partial E \ \nabla_{\mathbf{k}} E = \partial f/\partial E (1/\hbar) \mathbf{v}_{\mathbf{k}} \tag{1-13-17}$$

Here $\mathbf{v}_{\mathbf{k}}$ is the electron velocity corresponding to the wave vector **k**:

$$\mathbf{v}_{\mathbf{k}} = \hbar \mathbf{k}/m_n$$

where m_n is the effective mass (see eq. (1-13-2)). Hence, the Boltzmann equation (1-13-4) can be rewritten as

$$q \frac{df_o}{dE} \mathbf{v}_{\mathbf{k}} \cdot \mathbf{F} = -\frac{f_1}{\tau} \tag{1-13-18}$$

and hence,

$$f_1 = -q\frac{df_o}{dE}(\mathbf{v_k}\cdot\mathbf{F})\tau \tag{1-13-19}$$

Once f_1 is known, the electron drift velocity, $\mathbf{v}$, can be calculated as

$$\mathbf{v} = \frac{\int \mathbf{v_k}\,[f_o + f_1(\mathbf{k})]\,d\mathbf{k}}{\int f(\mathbf{k})\,d\mathbf{k}} \tag{1-13-20}$$

Here $d\mathbf{k}$ is the infinitesimal volume in the $\mathbf{k}$-space and the integration is over the first Brillouin zone. From eq. (1-6-9) we find that

$$\int f(\mathbf{k})\,d\mathbf{k} = 4\pi^3 n \tag{1-13-21}$$

where n is the electron concentration in the conduction band. Furthermore, the integral $\int \mathbf{v_k} f_o\,d\mathbf{k}$ is proportional to an average electron velocity in zero electric field, which is zero. Hence,

$$\mathbf{v} = \frac{\int \mathbf{v_k} f_1(\mathbf{k})\,d\mathbf{k}}{4\pi^3 n} \tag{1-13-22}$$

For the simple case of nondegenerate statistics and parabolic spherical bands,

$$f_o = \exp\left(-\frac{E - E_F}{k_B T}\right) \tag{1-13-23}$$

(see eq. (1-6-8a)). The Fermi level can be expressed in terms of the electron concentration, n. For a nondegenerate case we have

$$n = N_c \exp\left(-\frac{E_c - E_F}{k_B T}\right) \tag{1-13-24}$$

where N_c is the density of states in the conduction band and E_c is the energy corresponding to the bottom of the conduction band (see eq. (1-6-26)). Hence,

$$f_o = \frac{n}{N_c}\exp\left(-\frac{E - E_c}{k_B T}\right) \tag{1-13-25}$$

and

$$\partial f_o/\partial E = -\frac{n}{N_c k_B T}\exp\left(-\frac{E - E_c}{k_B T}\right) \tag{1-13-26}$$

For simple parabolic bands, $E = \hbar^2 k^2/(2m_n)$, and

$$v_{\mathbf{k}} = (2E/m_n)^{1/2} \tag{1-13-27}$$

The dot product $\mathbf{v_k}\cdot\mathbf{F}$ is given by

$$\mathbf{v_k}\cdot\mathbf{F} = (2E/m_n)^{1/2}\,F\cos\theta \tag{1-13-28}$$

where θ is the angle between $\mathbf{k}$ and $\mathbf{F}$. Substituting eqs. (1-13-26) to (1-13-28) into eq. (1-13-19) we find that

$$f_1 = \frac{qn}{N_c k_B T} \exp\left(-\frac{E - E_c}{k_B T}\right)(2E/m_n)^{1/2} F\tau \cos\theta \tag{1-13-29}$$

Hence, the integral in the numerator of the right-hand side of eq. (1-13-22) and **v** are proportional to the applied electric field, **F**, when **F** is small, as expected (see eq. (1-9-1)).

Let us now evaluate the integral in eq. 1-13-22 using a spherical coordinate system with axis z in the direction of the electric field (see Fig. 1-13-4). The incremental volume in the **k**-space, $d\mathbf{k}$, may be presented as

$$d\mathbf{k} = k^2\, dk\, d\phi \sin\theta\, d\theta \tag{1-13-30}$$

Angle ϕ varies from $-\pi$ to π ($-\pi < \phi \leq \pi$), and angle θ varies from 0 to π ($0 \leq \theta \leq \pi$).

Substituting $k^2 = 2m_n E/\hbar^2$ and $dk = [m_n/(2\hbar^2 E)]^{1/2}\, dE$ into eq. (1-13-30) we find that

$$d\mathbf{k} = m_n^{3/2} E^{1/2}\, dE\, d\phi \sin\theta\, d\theta/(2^{1/2}\hbar^3) \tag{1-13-31}$$

We are now ready to evaluate the integral in eq. (1-13-22). First, we notice that **v** must have the same direction as **F** because of the symmetry of the problem. Hence, the component of **v** in the direction of **F** can be found as

$$v = \frac{\int v_{\mathbf{k}} \cos\theta\, f_1(\mathbf{k})\, d\mathbf{k}}{4\pi^3 n} \tag{1-13-32}$$

The integrand in eq. (1-13-32) is independent of ϕ, and hence, the integration with respect to ϕ can be carried out as follows:

$$\int_{-\pi}^{\pi} d\phi = 2\pi \tag{1-13-33}$$

Next we substitute eqs. (1-13-27), (1-13-29), and (1-13-31) into eq. (1-13-32) and carry out the integration. Collecting terms containing θ ($\cos\theta$ from the integrand in eq. (1-13-32), $\cos\theta$ from f_1, and $\sin\theta\, d\theta$ from dk) we find that

$$\int_0^{\pi} \cos^2\theta \sin\theta\, d\theta = \int_{-1}^{1} \cos^2\theta\, d\cos\theta = 2/3 \tag{1-13-34}$$

Using eqs. (1-13-33) and (1-13-34) and eq. (1-6-17) for the effective density of states, N_c, we obtain the expression for v that coincides with eq. (1-9-1):

$$v = \mu F \tag{1-13-35}$$

$$\mu = \frac{q\langle\tau\rangle}{m_n} \tag{1-13-36}$$

and

$$\langle\tau\rangle = 4/(3\sqrt{\pi}) \int_0^{\infty} \tau(E) E^{3/2} \exp\left(-\frac{E}{k_B T}\right) dE/(k_B T)^{5/2} \tag{1-13-37}$$

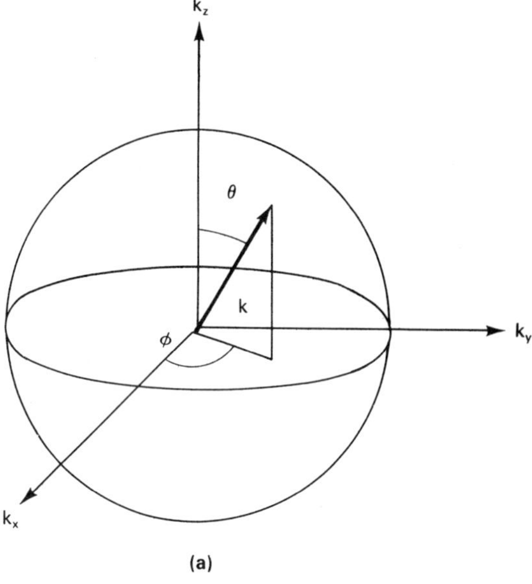

(a)

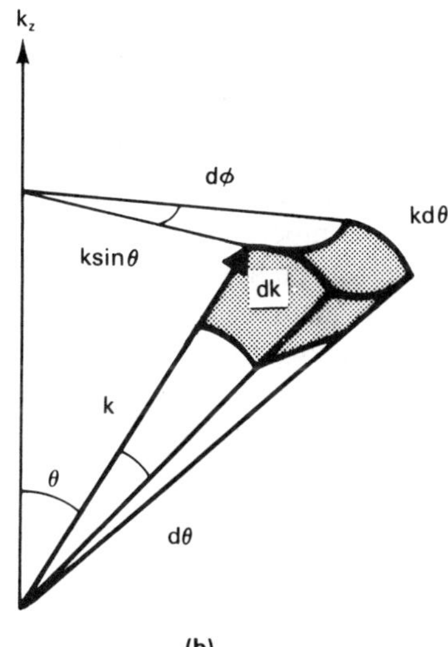

(b)

Fig. 1-13-4. (a) Spherical coordinate system and (b) incremental volume in spherical coordinate system.

This equation is valid for a nondegenerate electron gas. However, the foregoing derivation may be repeated for an arbitrary statistics to yield (see Seeger 1982)

$$\langle \tau \rangle = -(2/3)\,\frac{k_B T \int_0^\infty \tau(E) E^{1/2} \partial f_o / \partial E \, dE}{\int_0^\infty E^{1/2} f_o \, dE} \tag{1-13-38}$$

In many cases the momentum relaxation time, τ, may be presented as

$$\tau = \tau_o E^{-S} \tag{1-13-39}$$

where τ_o and S are constants. In this case the integrals in eq. (1-13-37) and (1-13-38) can be expressed through a gamma function, Γ:

$$\langle \tau \rangle = 4/(3\sqrt{\pi})\tau_o \, \Gamma(5/2 - S) \tag{1-13-37a}$$

$$\langle \tau \rangle = 4/(3\sqrt{\pi})\tau_o \, \Gamma(5/2 - S) F_{1/2-S}(\eta_n)/F_{1/2}(\eta_n) \tag{1-13-38a}$$

where the Γ function is defined as

$$\Gamma(x) = \int_0^\infty t^{x-1} \exp(-t)\, dt \tag{1-13-40}$$

The values of the Γ function are given in Table 1-10-1. The expressions for momentum relation times τ and for low-field mobilities calculated for different scattering mechanisms using eqs. (1-13-36) and (1-13-37) are given in Appendix 23.

*1-14. MONTE CARLO SIMULATION

The Monte Carlo method has become a standard numerical method of simulating electron and hole transport in semiconductors. It is based on the approach developed by Kurosawa (1966). The idea of this technique is to simulate the electron motion in **k**-space using random numbers to account for the random nature of this motion. We consider free electron flights that are interrupted by the scattering processes and resumed again and again, each time from a new starting point in the **k**-space. If we observe a single electron for a sufficiently long time, the distribution of times the electron spends in the vicinity of different points in **k**-space will reproduce the shape of the distribution function, $f(\mathbf{k})$. This process has been shown to be equivalent to the solution of the Boltzmann equation. Hence, the Monte Carlo method is a way to solve the Boltzmann transport equation without even writing this equation on a piece of paper.

During free flights (between the scattering events) the electron wave vector changes at the rate determined by the electric field, $\mathbf{F}$:

$$\mathbf{k}(t) = \mathbf{k}_o + \frac{q\mathbf{F}}{\hbar} t \tag{1-14-1}$$

where $\mathbf{k}_o$ is the initial value of the wave vector. This equation is obtained by choosing $\mathbf{B} = 0$ in eq. (1-13-1) and integrating the resulting equation with respect to time. In the Monte Carlo method, the moments of time, when the scattering events occur, can be determined using computer-generated random numbers.

The frequency of scattering events depends on the total scattering rate,

$$\lambda(\mathbf{k}) = \sum_{i=1}^{n} \lambda_i(\mathbf{k}) \tag{1-14-2}$$

Here $\lambda_i(\mathbf{k})$ are scattering rates for different scattering mechanisms (see Appendix 22).

The total scattering rate $\lambda(\mathbf{k})$ is a complicated function of $\mathbf{k}(t)$. This makes it difficult to simulate the probability distribution of scattering events. This difficulty can be avoided by introducing an additional fictitious scattering process, which does not change the wave vector. The scattering probability of this process (called a self-scattering process) is defined as

$$W_o(\mathbf{k},\mathbf{k}') = \lambda_o(\mathbf{k})\delta(\mathbf{k} - \mathbf{k}') \tag{1-14-3}$$

The function $\lambda_o(\mathbf{k})$ is chosen in such a way that

$$\lambda_o(\mathbf{k}) + \lambda(\mathbf{k}) = \Gamma \tag{1-14-4}$$

where the total scattering rate, Γ, is constant and $\lambda_o(\mathbf{k})$ is positive. Now we have to find the probability of electron scattering events for the constant scattering rate, Γ.

To find the probability for an electron to scatter during the time interval between t and $t + dt$, we divide t into very small equal time intervals, Δt. The probability for an electron to experience scattering during an interval of time Δt is given by

$$\Delta P = \Gamma \Delta t \tag{1-14-5}$$

Hence, the probability not to be scattered during this period of time is equal to

$$1 - \Delta P = 1 - \Gamma \Delta t \tag{1-14-6}$$

The probability, P_{not}, for an electron not to experience scattering during time t is given by

$$P_{\text{not}} = (1 - \Gamma\Delta t)(1 - \Gamma\Delta t)(1 - \Gamma\Delta t) \ldots (1 - \Gamma\Delta t) \tag{1-14-7}$$

with N terms in the right-hand side of eq. (1-14-7), where $N = t/\Delta t$. Taking a log of both sides of eq. (1-14-7) we obtain

$$\ln (P_{\text{not}}) = \ln (1 - \Gamma\Delta t) + \ln (1 - \Gamma\Delta t) + \ldots + \ln (1 - \Gamma\Delta t) \tag{1-14-8}$$

If the interval Δt is chosen small enough, log functions in the right-hand side of eq. (1-14-8) can be expanded into the Taylor series, so that

$$\ln (P_{\text{not}}) = -\Gamma\Delta t - \Gamma\Delta t - \ldots - \Gamma\Delta t \tag{1-14-9}$$

or

$$\ln (P_{\text{not}}) = -\Gamma t \tag{1-14-10}$$

The total probability for an electron not to scatter during the time interval between 0 and t is then given by

$$P_{\text{not}} = e^{-\Gamma t} \tag{1-14-11}$$

The probability for an electron to scatter during the time interval between 0 and t, P_{yes}, is equal to

$$P_{\text{yes}} = 1 - P_{\text{not}} = 1 - e^{-\Gamma t} \tag{1-14-12}$$

and the probability for an electron to scatter during the time interval between t and $t + dt$, $P(t)\,dt = (dP_{\text{yes}}/dt)\,dt$, can be found by differentiating eq. (1-14-12):

$$P(t)\,dt = \Gamma e^{-\Gamma t}\,dt \tag{1-14-13}$$

Here t is the time passed from the previous scattering event. When t tends to infinity, the probability that the scattering even will occur tends to 1,

$$\int_0^\infty P(t)\,dt = 1 \tag{1-14-14}$$

When $t << \Gamma$,

$$P(t)\,dt \simeq \Gamma\,dt \tag{1-14-15}$$

The time of the free flight, t_s, can now be related to the random number r distributed with equal probability between 0 and 1:

$$r = \int_o^{t_s} P(t)\,dt \tag{1-14-16}$$

or

$$r = 1 - \exp(-\Gamma t_s), \tag{1-14-17}$$

$$t_s = -\frac{1}{\Gamma}\ln(1 - r) \tag{1-14-18}$$

Indeed, such a choice of t_s yields the probability distribution P that satisfies eq. (1-14-13).

After the time t_s of the free flight has been determined, it is necessary to establish which scattering process was responsible for terminating the flight. Since the probability of the free flight termination by the given scattering process is proportional to its scattering rate, $\lambda_i(\mathbf{k})$, and since $\Sigma\,\lambda_i = \Gamma$, this can be done by generating a random number s, distributed with equal probability between 0 and Γ, and testing inequalities

$$S < \sum_{i=o}^{m} \lambda_i(\mathbf{k}) \tag{1-14-19}$$

for $m = 0, 1, 2, \ldots, n$. Here m is the total number of the real scattering mechanisms. If the inequality is satisfied for $m = 0$, the self-scattering process is selected. If it is satisfied for $m = 1$, the first scattering mechanism is selected, and so on. This process is illustrated by Fig. 1-14-1, which shows a casino dealer throw-

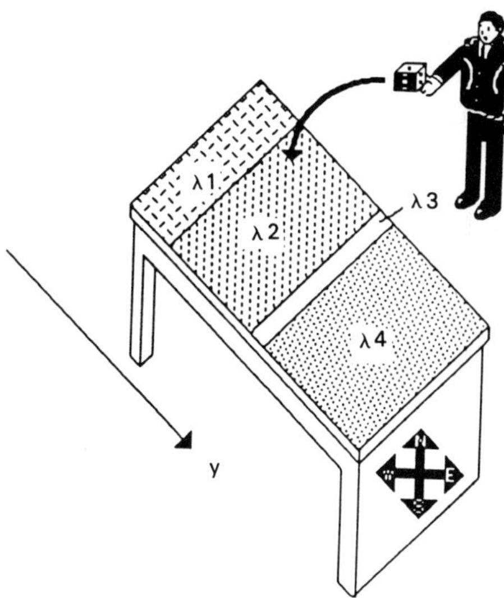

Fig. 1-14-1. Monte Carlo process. The table width is Γ. The dealer throws dice randomly at the table. Hence, the coordinate of the dice, s, is a random number uniformly distributed between 0 and Γ. Events 1, 2, 3, and 4 are selected when the dice lands on stripes 1, 2, 3, and 4, respectively. The width of the i-th stripe is proportional to the scattering rate, λ_i, of the i-th process.

ing a dice at a casino table. The table width is Γ. The dealer throws the dice randomly. The y-coordinate of the dice, s, counted from the edge of the table is a random number uniformly distributed between 0 and Γ. Events 1, 2, 3, and 4 are selected depending on which stripe the dice lands on the table. The width of the ith stripe is proportional to the scattering rate, λ_i, of the ith process. Hence, for many scattering events the different scattering mechanisms are selected proportionally to their scattering rates.

The next step is to determine the final state after the scattering. As mentioned in Section 1-13, we assume that collisions are instantaneous. As a result of scattering, the electron disappears from point $\mathbf{k}$ of the phase space (the initial state) and instantaneously reappears in point $\mathbf{k}'$ (the final state). This new position $\mathbf{k}'$ is a starting point for the next free flight, i.e., it replaces $\mathbf{k}_o$ in eq. (1-14-1).

For the self-scattering process, the final state is known because this process does not change wave vector $\mathbf{k}$ (see eq. (1-14-3)). (It is clearly advantageous to choose $\lambda_o(\mathbf{k})$ and, hence, Γ as small as possible—keeping, however, $\lambda_o(\mathbf{k}) > 0$—to minimize the frequency of the selection of the self-scattering process.)

For all real scattering processes, additional random numbers have to be generated to simulate the probability distribution of the final states. For randomizing scattering events, such as acoustic scattering in a parabolic valley and intervalley scattering, all states on the energy surface corresponding to the final state are equally probable (see Appendix 22). Hence, the probability that the angle between $\mathbf{k}'$ and the direction of the electric field $\mathbf{F}$ is θ and is proportional to the number of states in a circle with radius $|\mathbf{k}'| \sin\theta$ (see Fig. 1-14-2).

$$P(\theta)\, d\theta = \frac{1}{2} \sin\theta \, d\theta \qquad (1\text{-}14\text{-}20)$$

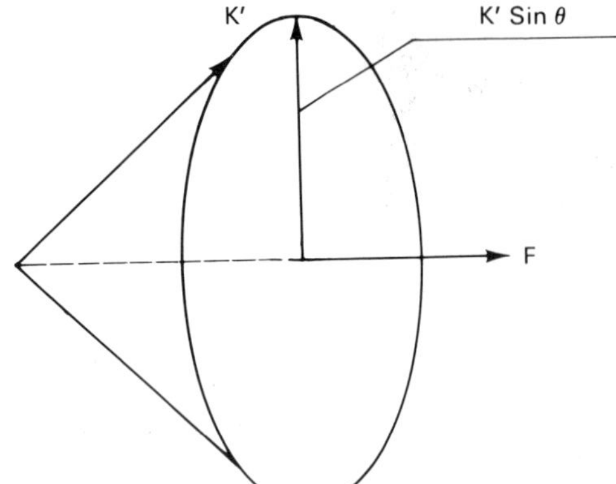

Fig. 1-14-2. Selection of a final state. The number of final states with the same energy ($\mathbf{k}'$ = constant) and with the same angle θ between $\mathbf{k}'$ and the electric field $\mathbf{F}$ is proportional to the length of the circle ($2\pi\mathbf{k}' \sin \theta$).

Here θ varies between 0 and π and $\frac{1}{2}$ is a normalizing constant. A random number, S_1, uniformly distributed between 0 and 1,

$$S_1 = \int_0^\theta P(\theta)\, d\theta \tag{1-14-21}$$

is selected to determine θ. It reproduces the probability distribution of the final states given by eq. (1-14-20). Substituting eq. (1-14-20) into eq. (1-14-21), we find that

$$S_1 = \frac{1}{2}(1 - \cos \theta) \tag{1-14-22}$$

For the acoustic scattering in a nonparabolic valley (such as Γ minimum in GaAs) and for the polar optical scattering a more complicated procedure has to be adopted (see Fawcett et al. 1970).

In a particular case of gallium arsenide, for example, three sets of valleys have to be taken into account (Γ, L, and X minima see Fig. 1-5-7). To determine the distribution functions in the Γ, L, and X valleys, histograms are set up in (k_z, k_ρ) space where k_z is parallel and k_ρ is perpendicular to the electric field. Counts proportional to the time the electron spends in each cell of the histograms are recorded. The distribution function is proportional to these histograms. The distribution function, $f_\mathbf{k}$, may be presented as

$$f_\mathbf{k} = f_{\mathbf{k}s}(E) + f_{\mathbf{kF}}(\theta)$$

where $f_{\mathbf{k}s}(E)$ is independent of the angle, θ, between the electric field and the wave vector, $\mathbf{k}$ ($f_{\mathbf{k}s}(E)$ is called the symmetrical part of the distribution function) and $f_{\mathbf{kF}}(\theta)$ is dependent on θ.

The symmetrical part of the distribution function in Γ and L valleys of GaAs for a different value of the electric field is shown in Fig. 1-14-3 (see Pozhela and

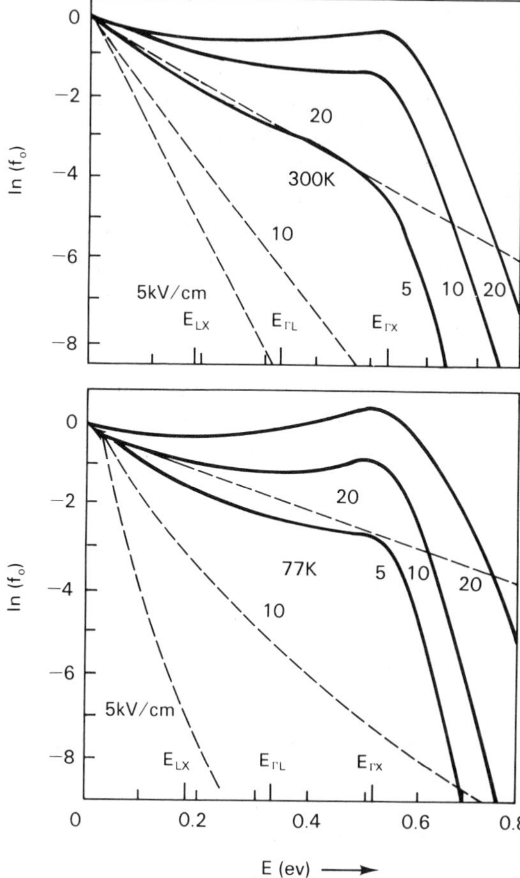

Fig. 1-14-3. Energy dependence of spherically symmetric part of the electron distribution function in Γ (solid curves) and L (dashed curves) valleys at 300 and 77 K (from Pozhela and Reklaitis [1980]). Electric field strengths are indicated in the figure. Impurity scattering is not included. Also shown are energies $E_{\Gamma X} = E_\Gamma - E_X$, $E_{GL} = E_\Gamma - E_L$, and $E_{LX} = E_X - E_L$.

Reklaitis 1980). The situation depicted in this figure corresponds to a nondegenerate sample, so that at zero electric field the distribution function is given by

$$f_{\mathbf{k}} = \exp\,[(E_F - E)/(k_B T)] = (n/N_c)\exp\,([-\,E)/(k_B T)]$$

(see eqs. (1-6-8a) and (1-6-26)). This means that the dependence of ln ($f_{\mathbf{k}}$) on E should be a straight line with the slope of $1/k_BT$ (q/k_BT if E is in eV). It is interesting to note that the dependence of the log of the symmetrical part of the distribution function in L valleys, $f_{\mathbf{k}sL}$, at 300 K in applied electric fields of 5 kV/cm, 10 kV/cm, and 20 kV/cm may be still fairly well approximated by straight lines with smaller slopes (see Fig. 1-14-3a). This means that this function may be presented as

$$f_{\mathbf{k}sL} \sim \exp([-\,E/(k_B T_e)] \qquad (1\text{-}14\text{-}23)$$

where T_e is called the electron temperature. From the slopes of the dashed lines shown in Fig. 1-14-3a we find that $T_e \approx 420$ K at 5 kV/cm, $T_e \approx 657$ K at 5 kV/cm, and $T_e \approx 1550$ K at 20 kV/cm. However, as can be seen from Fig. 1-14-3b, such an approximation is considerably less accurate for the symmetrical part of the distribution function in L valleys, $f_{\mathbf{k}sL}$, at 77 K. The distribution function in the Γ valley is drastically different from the Maxwell-Boltzmann distribution with the effective electron temperature T_e. Moreover, the distribution function actually becomes inverted in electric fields higher than 10 kV/cm, so that in a certain range the distribution function actually increases with energy, in sharp disagreement with what may be expected from eq. (1-14-23). The peak in the distribution function corresponds to the energy close to the difference of the bottoms of X and Γ valleys.

Some quantities, such as the drift velocity, for example, are calculated directly from the initial and final values of **k** for each free flight. The average drift velocity, v_d, in valley j, for example, is given by (Fawcett et al. 1970)

$$v_j = \frac{1}{\hbar k_j} \sum (E_{fn} - E_i) \tag{1-14-24}$$

where E_{fn} and E_i are the energies of the final and initial state for each free flight, and k_j is the total length of the electron trajectory in **k**-space in the valley:

$$k_j = \frac{qFT_j}{\hbar} \tag{1-14-25}$$

where T_j is the total time the electron spends in the valley. The summation in eq. (1-14-24) is over all free flights in valley j.

The Monte Carlo algorithm is summarized in Table 1-14-1.

To determine when the number of simulated scattering events is sufficiently large for an accurate simulation, the simulation time is split into a number of successive time intervals of equal duration. The electron drift velocity (or other quantity of interest) is computed for each of them. The mean value and standard deviation of the electron are then found from these partial "measurements."

TABLE 1-14-1. MONTE CARLO ALGORITHM (from M. S. Shur, *GaAs Devices and Circuits*, Plenum, New York (1987)).

Generate random number r and determine the time of free flight.
Record time the electron spends in each cell of the **k**-space during the free flight, drift velocity, mean energy, etc.
Generate random numbers and the final state and determine which scattering process has occurred.
Repeat until the desired number of scattering events is reached.
Calculate the distribution function, drift velocity, mean energy, etc.

The Monte Carlo technique is also used to determine the diffusion coefficient tensor, D_{ik}, that determines diffusion current (proportional to space derivatives of carrier concentration):

$$j_{\text{diff}} = q \sum D_{\text{ik}} \, \partial n/\partial x_i \tag{1-14-26}$$

Here j_{diff} is the diffusion current density, n is the electron concentration, and i, k = 1,2,3 ($x_1 = x$, $x_2 = y$, and $x_3 = z$). In the one-dimensional case, the diffusion tensor becomes a scalar quantity (diffusion coefficient $D = \mu k_B T/q$; see Section 1-11).

The diffusion process is related to a random thermal motion. It may be illustrated using a simple one-dimensional model that assumes that carriers may move in random to the left or right, with equal probability. Let us consider two adjacent incremental regions with different numbers of electrons: more electrons in the left region and fewer in the right region (see Fig. 1-14-4). On the average, half of the electrons in each region will move to the left and half will move to the right. This means that more electrons move from the left region to the right region than vice versa, thus creating a diffusion flux that tends to equalize the electron concentration. This flux is proportional to the gradient of the electron concentration, n.

The components of the diffusion tensor can be related to random electron trajectories simulated by the Monte Carlo method. If x and y are the distances traveled during interval t along the x and y directions, respectively, the diffusion tensor is given by

$$\begin{aligned} D_{xy} &= \frac{1}{2}\frac{d}{dt}\langle(x - \langle x\rangle)(y - \langle y\rangle)\rangle \\ &= \frac{1}{2}\frac{d}{dt}(\langle xy\rangle - \langle x\rangle\langle y\rangle) \end{aligned} \tag{1-14-27}$$

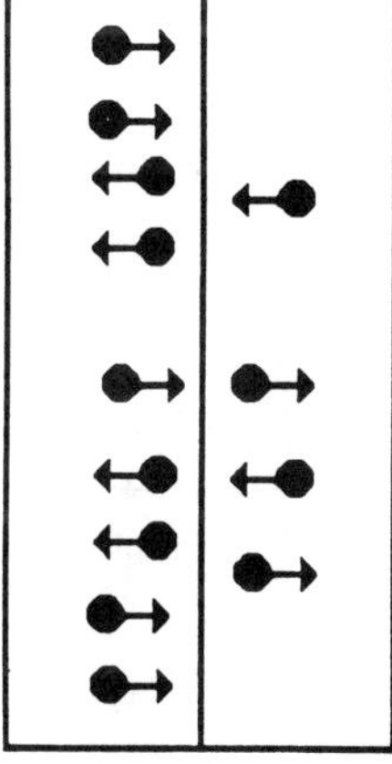

Fig. 1-14-4. Illustration of diffusion process.

where ⟨⟩ indicates the average over all time intervals (see, for example, Pozhela and Reklaitis 1980).

A typical electron trajectory in **k**-space and real space simulated by the Monte Carlo method for a two-dimensional case is shown in Fig. 1-14-5 (from Reggiani 1985). The electric field is applied in the x direction and heavy lines parallel to the k_x-axis in Fig. 1-14-5a correspond to free flights. Thin lines represent instantaneous changes of the wave vector resulting from scattering. Eight free flights are represented in Fig. 1-14-5a and b. In real space each trajectory is a parabolic segment with breaks in the trajectory slopes appearing when scattering events occur.

It takes a fairly large number of simulated scattering events to reproduce electron transport in a semiconductor. This is illustrated by Fig. 1-14-6, which shows the calculated electron drift velocity in an electric field for a GaAs sample as a function of simulated electron time of flight. As we see from the figure, several nanoseconds of electron flight time are required to reproduce the velocity with a reasonable accuracy. An average time between collisions is only of the

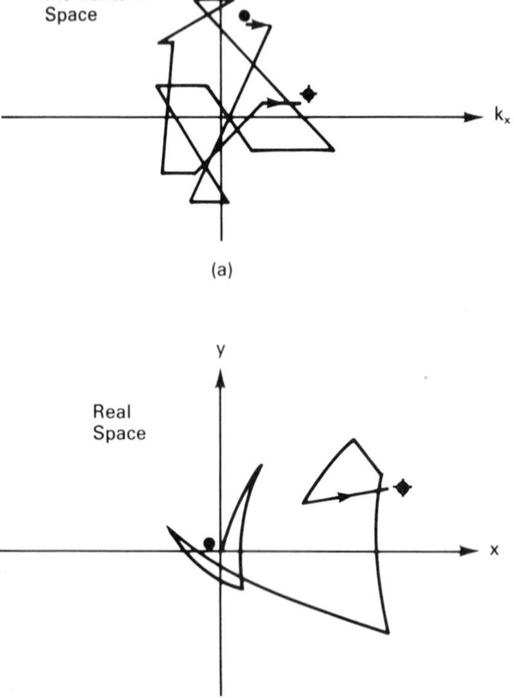

Fig. 1-14-5. Electron trajectory in two dimensional **k**-space (a) and two-dimensional real space (b) simulated by Monte-Carlo technique for eight scattering events (From L. Reggiani, "Hot Electron Transport in Semiconductors," *Topics in Physics*, vol. 58, ed. L. Reggiani, Springer-Verlag, Berlin, p. 7 (1985). * is the starting point and ϕ is the final point. Circles and stars denote intervalley transitions, points denote the self-scattering events.

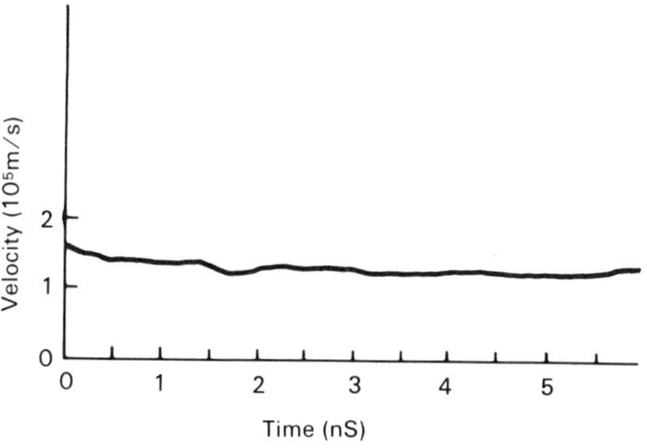

Fig. 1-14-6. The average drift velocity vs. simulation time for GaAs at 300 K (from Yu. K. Pozhela, *Plasma and Current Unstabilities in Semiconductors*, Nauka (1977). Electric field 1.5 kV/cm was turned on at $t = 0$.

order of 10^{-13} s. Hence, this simulation time corresponds to many tens of thousands of simulated scattering events.

Fig. 1-14-7 shows measured and calculated drift velocity and diffusion coefficient in silicon at 77 K as a function of electric field (from Reggiani 1985). Measured and calculated dependencies of electron drift velocity in silicon at 300 K are shown in Fig. 1-14-8 (from Canali et al. 1985). Similar dependencies for GaAs are shown in Fig. 1-14-9 (from Pozhela and Reklaitis 1980). As can be seen from

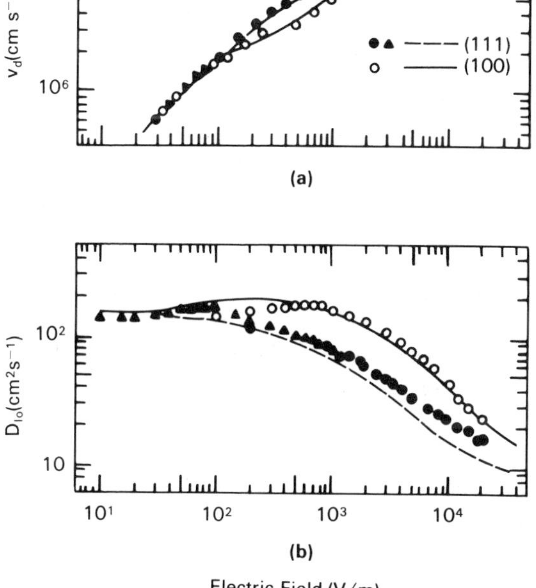

Fig. 1-14-7. (a) The measured and calculated drift velocity and (b) longitudinal diffusion coefficient in silicon at 77 K vs. electric field (from L. Reggiani, "Hot Electron Transport in Semiconductors," *Topics in Physics*, vol. 58, ed. L. Reggiani, Springer-Verlag, Berlin, p. 7 (1985). Solid and dotted lines show the results of Monte-Carlo simulation for different orientations of electric field. Symbols represent measured data.

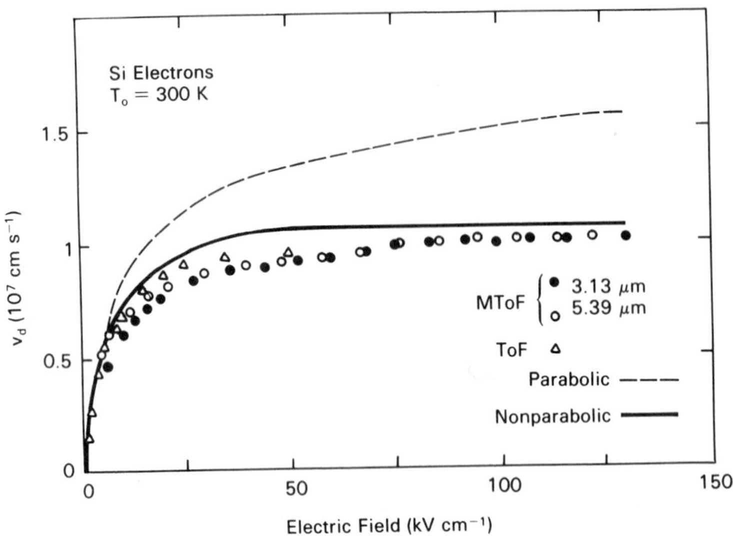

Fig. 1-14-8. Measured and calculated drift velocity in silicon at 300 K vs. electric field (from C. Canali, F. Nava & L. Reggiani, "Hot Electron Transport in Semiconductors," *Topics in Physics*, vol. 58, ed. L. Reggiani, Springer-Verlag, Berlin, p. 87 (1985). Measured data is from Smith et al [1980]). The electric field is in ⟨111⟩ direction.

all these figures the agreement between the Monte Carlo results and measured data is quite good. We have to keep in mind, however, that parameters characterizing scattering mechanisms are not well known and have to be adjusted to provide a good fit to experimental data.

The Monte Carlo technique can also be used to describe the transient effects and investigate the electron transport in small semiconductor devices. For such applications the Monte Carlo simulation is performed not just for one but for thousands of electrons with the initial wave vectors $\mathbf{k}_o$ distributed according to the initial equilibrium distribution. First, an approximate distribution of an electric field is chosen for a given device geometry. Then electron trajectories in this electric field are calculated, the electron charge distribution is determined, and Poisson's equation is solved in order to find an updated electric field distribution. The process is repeated again and again. The number of simulated collisions for each electron can be relatively small, but the number of particles involved in the simulation can be quite large.

For example, Hesto et al. (1984) simulated the distribution of 5000 electrons in a gallium arsenide metal-semiconductor field effect transistor. Five thousand electrons is a large number, but it is still small compared with an actual number of electrons in a semiconductor device. To some extent, the computational require-

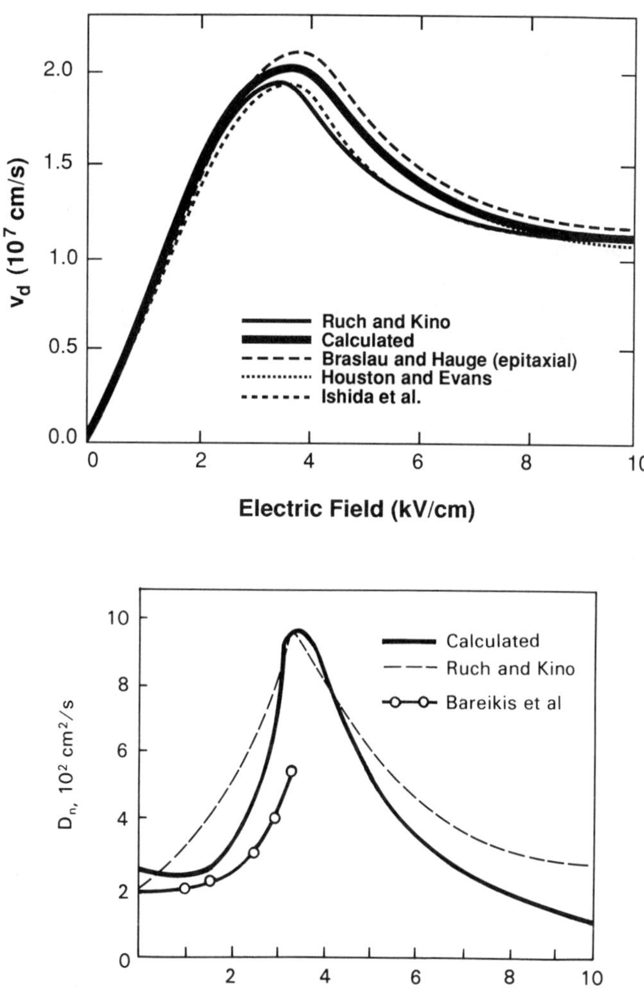

Fig. 1-14-9. (a) Measured and calculated drift velocity and (b) longitudinal diffusion coefficient in GaAs at 300 K vs. electric field (Reprinted with permission from Yu. K. Pozhela and A. Reklaitis, "Electron Transport Properties in GaAs at High Electric Fields," *Solid-State Electronics*, 23, pp. 927-933 (1980) Pergamon Press p/c).

ments may be made less stringent by assigning to each simulated electron a multiple charge so that each simulated "super" electron represents several electrons in the actual device. Such calculations are still very involved and consume enormous amounts of computer time. However, they give a valuable insight into complicated device physics that cannot be obtained any other way.

*1-15. PHENOMENOLOGICAL TRANSPORT EQUATIONS FOR HIGH ELECTRIC FIELDS

As was pointed out in Section 1-11, conventional semiconductor equations based on field dependent velocity and diffusion are inadequate for modeling small semiconductor devices. Some insight into the physics of such devices is provided by the Monte Carlo method described in Section 1-14. However, approximate analytic equations can be extremely useful for the analysis of the device behavior. This is related not just to the large amounts of computer time required for Monte Carlo calculations but, more important, to the lack of detailed and accurate information about numerous scattering parameters and details of the band structure required for accurate Monte Carlo simulations. Traditionally, the electron temperature approximation has been used for deriving approximate equations describing hot electrons. Indeed, as was discussed in Section 1-14, the symmetrical part of the distribution function for electrons in L valleys of the conduction band of GaAs at 300 K may be approximated as

$$f = C\exp(-E/k_B T_e) \tag{1-15-1}$$

We also noticed that this was a very poor approximation, for example, for electrons in the Γ valley of the conduction band of GaAs (see Fig. 1-14-3).

Equation (1-15-1) is valid when the electrons exchange energy through the electron-electron collisions at a faster rate than they lose it through electron-phonon scattering. At even higher electron concentrations the electron-electron collision redistributes both electron energy and momentum, leading to the *displaced Maxwellian distribution function,*

$$f(\mathbf{p}) = c\exp\left(-\left[\frac{(\mathbf{p} - \mathbf{p}_o)^2}{2m_n k_B T_e}\right]\right) \tag{1-15-2}$$

where $\mathbf{p}$ is momentum, $\mathbf{p}_o$ is the drift momentum, and T_e is the effective electron temperature.

For this approximation to be valid the average time between the electron-electron collisions should be much smaller than the momentum relaxation time. If the number of electrons in the conduction band is determined by the concentration of the ionized donors, this is not true and, therefore, the displaced Maxwell distribution function can be considered only as a crude approximation.

If, however, this distribution function is assumed, phenomenological transport equations can be derived from the Boltzmann equation by substituting eq. (1-15-2) into the Boltzmann equation (see Blotekjaer 1970). The resulting equations are fairly complicated and may be inaccurate if the electron distribution function is very non-Maxwellian (as is the distribution function for electrons in the Γ valley of the conduction band in GaAs in high electric fields). In this section we describe two simple models based on the results of steady-state Monte Carlo calculations and the application of these results to short semiconductor devices under non-equilibrium conditions. These models are empirical and can only be justified by comparing the results of the calculations based on these models with more rigorous calculations and experimental data. This is the price to pay for their relative simplicity and ease of use.

The first model (proposed by Shur [1976] and further developed by Carnez et al. [1980] and Cappy et al. [1980]) can be called the *energy-dependent relaxation time model.* This model describes the electron transport in short samples or in rapidly varying electric fields based on the results of the Monte Carlo simulations for long samples under steady-state conditions. When the diffusion is neglected, the equations of this model are given by

$$\frac{dm(E)\mathbf{v}}{dt} = -q\mathbf{F} - \frac{m\mathbf{v}}{\tau_p(E)} \tag{1-15-3}$$

$$\frac{dE}{dt} = \frac{\mathbf{j}\cdot\mathbf{F}}{n} - \frac{E - E_o}{\tau_n(E)} \tag{1-15-4}$$

$$\mathbf{j} = -qn\mathbf{v} \tag{1-15-5}$$

Here E is the average electronic energy, τ_p is the effective momentum relaxation time, τ_E is the effective energy relaxation time, $E_o = 3/2\, k_B T_o$ where T_o is the lattice temperature, and $m(E)$ is the effective mass of electrons. $m(E)$, $\tau_p(E)$, and $\tau_E(E)$ are determined from the steady-state Monte Carlo calculations from the requirement that in the steady state eqs. (1-15-16) to (1-15-18) give the same results as the Monte Carlo calculations. This requirement leads to the following expressions:

$$\tau_p(E) = \left\{\frac{m[F(E)]\, v[F(E)]}{qF(E)}\right\}_{\text{steady state}} \tag{1-15-6}$$

$$\tau_E(E) = \frac{E - E_o}{q\,\{F(E)\, v[F(E)]\}_{\text{steady state}}} \tag{1-15-7}$$

where the steady-state curves are taken from the Monte Carlo equations. This empirical approach has been shown to be in very good agreement with the direct Monte Carlo simulation for nonequilibrium conditions (see Shur 1976, Carnez et al. 1980, Cappy et al. 1980, and Fig. 1-15-1). The agreement is very good not only for GaAs but also for InP and $Ga_{.47}In_{.53}As$ samples (see Cappy et al. 1980).

The dependencies of the steady-state velocity, energy, and effective mass on the electric field deduced from the Monte Carlo simulation and used in this model are shown in Figs. 1-15-2, 1-15-3, and 1-15-4 for samples with doping densities of 10^{17} cm^{-3} and 3×10^{17} cm^{-3} (see Carnez et al. 1980). These curves have been used for the simulation of GaAs field effect transistors (see Carnez et al. 1980).

In short semiconductor devices the diffusion effects may be very important. These effects have been incorporated into this model by using the following equation for the electric current density:

$$\mathbf{j} = -qn\mathbf{v} + qD(E)\nabla n \tag{1-15-8}$$

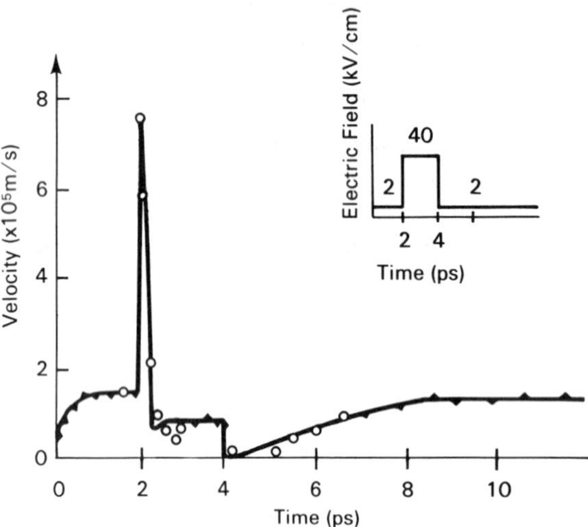

Fig. 1-15-1. Electron velocity vs. time for applied electric field step (A. Cappy, B. Carnez, R. Fauquembergues, G. Salmer, and E. Constant, *Comparative Potential Performance of Si, GaAs, GaInAs, InAs Submicrometer-Gate FET's*, IEEE Trans. on Electron Devices, ED-27, No. 11, p. 2158, Nov. (1980)). Lines are the solution to phenomenological equations based on the relaxation time approximation points (see Eqs. (1-15-1) - (1-15-3)) in the Monte-Carlo simulation.

where the diffusion coefficient, $D(E)$, is determined from steady-state Monte Carlo simulations.

A different model was proposed by Thornber (1982), who suggested a generalized current equation for short semiconductor devices. For a one-dimensional case the proposed equation is

$$j = qn(x,t)[\mu(F)F + W(F)\partial F/\partial x + B(F)\partial F/\partial t] + qD(F)\partial n/\partial x + qA(F)\partial n/\partial t \tag{1-15-9}$$

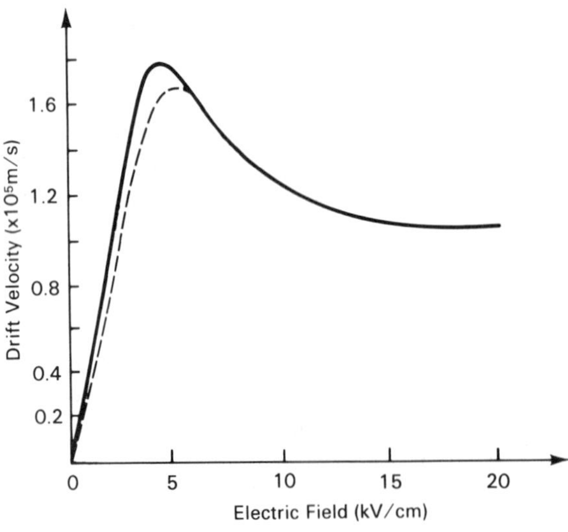

Fig. 1-15-2. Steady state drift velocity vs. electric field (A. Cappy, B. Carnez, R. Fauquembergues, G. Salmer, and E. Constant, *Comparative Potential Performance of Si, GaAs, GaInAs, InAs Submicrometer-Gate FET's*, IEEE Trans. on Electron Devices, ED-27, No. 11, p. 2158, Nov. (1980)). The solid line is $N_d = 10^{17}$ cm^{-3}, and the dashed line is $N_d = 3 \times 10^{17}$ cm^{-3}.

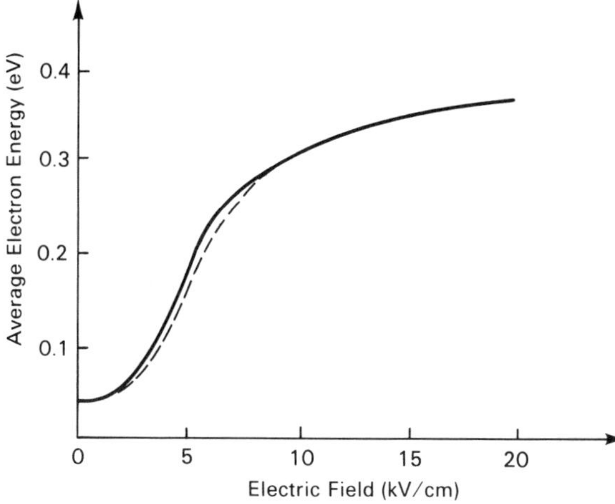

Fig. 1-15-3. Steady state average electron energy vs. electric field (A. Cappy, B. Carnez, R. Fauquembergues, G. Salmer, and E. Constant, *Comparative Potential Performance of Si, GaAs, GaInAs, InAs Submicrometer-Gate FET's,* IEEE Trans. on Electron Devices, ED-27, No. 11, p. 2158, Nov. (1980)). The solid line is $N_d = 10^{17}$ cm^{-3} and the dashed line is $N_d = 3 \times 10^{17}$ cm^{-3}.

where $\mu(F)$, $W(F)$, $B(F)$, $D(F)$, and $A(F)$ are mobility, gradient coefficient, rate coefficient, diffusion coefficient, and relaxation coefficient, respectively. Two new coefficients—rate and relaxation coefficients—were calculated by Kizialyalli and Hess (1987) using the Monte Carlo technique.

More theoretical and experimental work is required to evaluate these models and, perhaps, develop better analytic approaches to complicated physical processes occurring in short semiconductor devices. In this area of physics, experiments and technology are well ahead of theory.

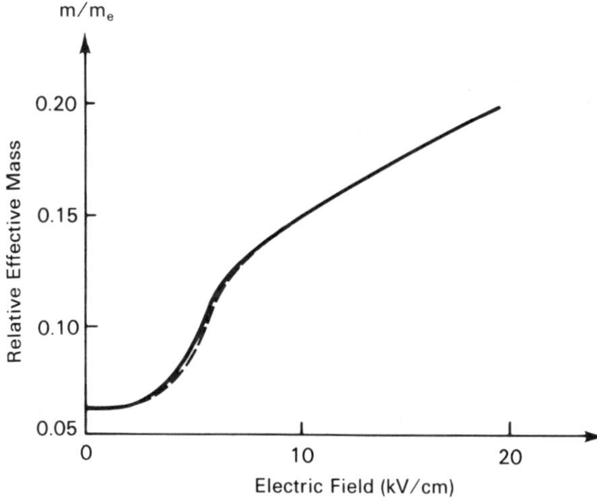

Fig. 1-15-4. Steady state effective mass ratio vs. electric field (A. Cappy, B. Carnez, R. Fauquembergues, G. Salmer, and E. Constant, *Comparative Potential Performance of Si, GaAs, GaInAs, InAs Submicrometer-Gate FET's,* IEEE Trans. on Electron Devices, ED-27, No. 11, p. 2158, Nov. (1980)). The solid line is $N_d = 10^{17}$ cm^{-3} and the dashed line is $N_d = 3 \times 10^{17}$ cm^{-3}.

REFERENCES

N. D. Arora, J. R. Hauser, and D. J. Rulson, *IEEE Trans. Electron Devices,* ED-29, p. 292 (1982).

N. W. Ashcroft and N. D. Mermin, *Solid State Physics,* Holt, Rinehart, and Winston, Philadelphia (1976).

K. Ashida, M. Inoue, J. Shirafuji, and Y. Inuisi, *J. Phys. Soc. Japan,* 37, p. 408 (1974).

Y. Awano, K. Tomizawa, N. Hashizume, and M. Kawashima, "Monte Carlo Simulation of a Submicron Sized GaAs n^+-i(n)-n^+ Diode," *Electronics Letters,* 18, no. 3, pp. 133–134, Feb. (1982).

G. Baccarani, C. Jacoboni, and A. Mazzone, *A. M. Solid State Electron.,* 20, p. 5 (1977).

H. S. Bennett, *Solid State Electron.,* 26, p. 1157 (1983).

S. Blakemore, "Semiconductor and Other Major Properties of GaAs," *J. Appl. Phys.,* 53, no. 10, pp. R123–R181, Oct. (1982).

D. I. Blokhintsev, *Principles of Quantum Mechanics,* Allyn and Bacon, Boston (1964).

K. Blotekjaer, "Transport Equations for Two-Valley Semiconductors," *IEEE Trans. Electron. Devices,* ED-17, no. 1, pp. 38–47, Jan. (1970).

K. Blotekjaer and E. B. Lunde, "Collision Integrals for Displaced Maxwellian Distributions," *Physica Status Solidi,* 35, p. 581 (1969).

R. Bosch and H. W. Thim, *IEEE Transactions on Electron. Devices,* ED-21, no. 1, pp. 16–25, Jan. (1974).

N. Braslau and P. S. Hauge, *IEEE Trans. Electron. Dev.,* ED-17, p. 616 (1970).

H. Brooks, *Advances in Electronics and Electron Physics,* vol. 7, ed. L. Marton, Academic Press, New York, p. 85 (1955).

P. N. Butcher, W. Fawcett, and C. Hilsum, "A Simple Analysis of Stable Domain Propagation in the Gunn Effect," *Brit. J. Appl. Phys.,* 17, no. 7, pp. 841–850 (1966).

C. Canali, F. Nava, and L. Reggiani, "Hot Electron Transport in Semiconductors," in *Topics in Physics,* vol. 58, ed. L. Reggiani, Springer-Verlag, Berlin, p. 87 (1985).

A. Cappy, B. Carnez, R. Fauquembergues, G. Salmer, and E. Constant, "Comparative Potential Performance of Si, GaAs, GaInAs, InAs Submicrometer-Gate FET's," *IEEE Trans. on Electron. Devices,* ED-27, no. 11, p. 2158, Nov. (1980).

B. Carnez, A. Cappy, A. Kaszinski, E. Constant, and G. Salmer, "Modeling of Submicron Gate Field-Effect Transistor Including Effects of Non-Stationary Electron Dynamics," *J. Appl. Phys.,* 51, no. 1 (1980).

D. M. Caugley and R. E. Thomas, *Proc. IEEE,* 55, pp. 2192–2193 (1967).

J. R. Chelikowsky and M. L. Cohen, "Nonlocal Pseudopotential Calculations for the Electronic Structure of Eleven Diamond and Zinc-Blende Semiconductors," *Phys. Rev.,* B14, p. 556 (1976).

E. Conwell, V. F. Weisskopf, *Phys. Rev.,* vol. 77, p. 388 (1950).

W. R. Curtice and Y.-H. Yun, "A Temperature Model for the GaAs MESFET," *IEEE Trans. Electron. Devices,* ED-28, no. 8, pp. 954–962, Aug. (1981).

V. L. Dalal, *Appl. Phys. Lett.,* 16, no. 12, pp. 489–491 (1970).

G. DRESSELHAUS, *Phys. Rev.,* 100, p. 580 (1955).

G. DRESSELHAUS, A. F. KIP, and C. KITTEL, *Phys. Rev.,* 98, p. 368 (1955).

J. DZIEWIOR and W. SCHMID, *Appl. Phys. Lett.,* 31, p. 346 (1982).

W. FAWCETT, D. A. BOARDMAN, and S. SWAIN, "Monte Carlo Determination of Electron Transport Properties in Gallium Arsenide," *J. Phys. Chem. Solids,* 31, pp. 1963–1970 (1970).

J. G. FOSSUM, R. P. MERTENS, D. S. LEE, and J. F. NIJS, *Solid State Electron.,* 26, p. 569 (1983).

R. C. GOODFELLOW, B. T. DEBNEY, G. J. REES, and J. BUUS, *IEEE Trans. Electron. Devices,* ED-32, no. 12, p. 2562, Dec. (1985).

M. HAAS et al., *J. Phys. Chem. Solids,* 8, p. 282 (1959).

W. A. HARRISON, *Electronic Structure and Properties of Solids,* W. H. Freeman, San Francisco (1980).

J. W. HARRISON and J. R. KAUSER, *J. Appl. Phys.,* 47, p. 292 (1976).

V. HEINE, "Pseudopotential Concept," *Solid State Physics,* Advances in Research and Applications, vol. 24, Academic Press, New York, London (1970).

P. HESTO, J. F. PONE, R. CASTAGNÉ, and J. L. PELOUARD, *The Physics of Submicron Structures,* ed. H. L. Grubin, K. Hess, G. J. Iafrate, and D. K. Ferry, Plenum Press, New York, p. 101 (1984).

A. HOUSTON and A. G. R. EVANS, *Solid State Electron.,* 20, p. 197 (1977).

C. JACOBONI, C. CANALI, G. OTTAVIANI, and A. QUARANTA, *Solid State Electron.,* 20, p. 77 (1977).

C. JACOBONI and L. REGGIANI, "Bulk Hot-Electron Properties of Cubic Semiconductors," *Advances in Physics,* 28, no. 4, pp. 493–553 (1979).

T. R. JERVIS and E. F. JOHNSON, "Geometric Magnetoresistance and Hall Mobility in Gunn Effect Devices," *Solid State Electron.,* 13, pp. 181–189 (1970).

F. A. JOHNSON and W. COCHRAN, *Proc. Int. Conf. Phys. Semicond.* (Exeter 1962), London, p. 498 (1962).

E. O. KANE, *J. Phys. Chem. Solids,* 1, p. 249 (1957).

C. KITTEL, *Introduction to Solid State Physics,* John Wiley & Sons (1976).

I. C. KIZIALYALLI and K. HESS, *IEEE Trans. Electron. Devices,* ED-34, no. 11, p. 2353–2354, Nov. (1987).

H. KROEMER, "The Einstein Relation for Degenerate Carrier Concentration," *IEEE Trans. Electron. Dev.,* ED-25, p. 850 (1978).

T. KUROSAWA, *J. Phys. Soc. Jap. Suppl.,* 21, p. 424 (1966).

C. P. LEE, R. ZUCCA, and B. M. WELCH, "Orientation Effect on Planar GaAs Schottky Barrier Field Effect Transistors," *Appl. Phys. Lett.,* 37, no. 3, pp. 311–314, Aug. (1980).

K. LEE, M. S. SHUR, T. J. DRUMMOND, and H. MORKOÇ, *J. Appl. Phys.,* 54, p. 2093 (1983).

O. MADELUNG, *Physics of III-V Compounds,* John Wiley & Sons, New York (1964).

A. NUSSBAUM, *Semiconductor Device Physics,* Prentice-Hall, Englewood Cliffs, N.J. (1962).

R. H. PARMENTER, *Phys. Rev.,* 100, p. 573 (1955).

J. C. PHILLIPS, *Bonds and Bands in Semiconductors,* Academic Press, New York (1973).

YU. K. POZHELA, *Plasma and Current Unstabilities in Semiconductors,* in Russian, Nauka, Moscow (1977).

YU K. POZHELA and A. REKLAITIS, "Electron Transport Properties in GaAs at High Electric Fields," *Solid-State Electronics,* 23, pp. 927–933 (1980).

H. D. REES, "Hot Electron Effects at Microwave Frequencies in GaAs," *Solid State Comm.,* 7, no. 2, pp. 267–269 (1969).

L. REGGIANI, "Hot Electron Transport in Semiconductors," in *Topics in Physics,* vol. 58, ed. L. Reggiani, Springer-Verlag, Berlin, p. 7 (1985).

D. L. RODE, *Phys. Rev.,* B2, no. 4, pp. 1012–1024, Aug. (1970).

D. L. RODE, *Semiconductors and Semimetals,* ed. by R. K. Willardson and A. C. Beer, Academic Press, New York, vol. 10, p. 1 (1975).

W. VAN ROOSBROECK and W. SHOCKLEY, *Phys. Rev.,* 94, p. 1558 (1954).

J. G. RUCH and W. FAWCETT, "Temperature Dependence of the Transport Properties of Gallium Arsenide Determined by a Monte Carlo Method," *J. Appl. Phys.,* 41, no. 9, pp. 3843–3849, Aug. (1970).

G. RUCH and G. S. KINO, *Phys. Rev.,* 174, p. 921 (1968).

C. T. SAH, R. N. NOYCE, and W. SHOCKLEY, "Carrier Generation and Recombination in p-n Junction and p-n Junction Characteristics," *Proc. IRE,* 45, p. 1228 (1957).

K. SEEGER, *Semiconductor Physics, An Introduction,* Series in Solid-State Sciences, vol. 40, 3d ed. Springer-Verlag, Berlin, Heidelberg, New York (1985).

B. I. SHKLOVKIĬ and A. L. EFROS, *Electronic Properties of Doped Semiconductors,* Springer Series in Solid State Sciences, vol. 45, Springer-Verlag (1984).

M. S. SHUR, "Influence of Non-Uniform Field Distribution on Frequency Limits of GaAs Field-Effect Transistors," *Electron Letters,* 12, no. 23, pp. 615–616, Nov. (1976).

M. S. SHUR and L. F. EASTMAN, "Ballistic Transport in Semiconductors at Low-Temperatures for Low Power High Speed Logic," *IEEE Trans. Electron. Devices,* ED-26, no. 11, pp. 1677–1683, Nov. (1979).

M. S. SHUR, C. HYUN, and M. HACK, "New High Field-Effect Mobility Regimes of Amorphous Silicon Alloy Thin-Film Transistor Operation," *J. of Applied Phys,* 59, no. 7, p. 2488 (1986).

P. M. SMITH, M. INOUE, and J. FREY, *Appl. Phys. Lett.,* 39, p. 797 (1980).

R. A. SMITH, *Semiconductors,* 2d ed., Cambridge University Press, Cambridge, London, New York, Melbourne, Sydney (1978).

G. E. STILLMAN, C. M. WOLFE, and J. O. DIMMOCK, *J. Phys. Chem. Solids,* 31, p. 1199 (1970).

S. M. SZE, *Physics of Semiconductor Devices,* John Wiley & Sons, New York (1981).

F. SZMILOVICZ and F. L. MADARASZ, *Phys. Rev,* B27, p. 2605 (1983).

K. K. THORNBER, *IEEE Electron. Device Letters,* EDL-3, no. 3, pp. 69–71 (1982).

T. I. TOSIC, D. A. TJAPKIN, and M. M. JEVTIC, *Solid State Electron.,* 24, p. 577 (1981).

M. S. TYAGI and R. VAN OVERSTRAATEN, *Solid State Electron.,* 26, p. 577 (1983).

V. VAREIKIS, A. GALDIKAS, R. MILISYTE, and V. VIKTORAVICIUS, *Program and Papers of 5th Int. Conf. on Noise in Physical Systems,* p. 212, 13–16 March, Bad Nauheim, W. Germany (1978).

H. WEISS and H. WELKER, *Zs. Phys.*, 138, p. 322 (1963).

Z. YU and R. W. DUTTON, *Sedan III: A General Electronic Material Device Analysis Program,* program manual, Stanford University, July (1985).

J. M. ZIMAN, *Electrons and Phonons,* Oxford University Press, London (1960).

PROBLEMS

1-2-1. **(a)** Find energy levels and a wave function for an infinitely deep two-dimensional potential well such that

$$U = 0 \qquad \text{for } 0 < x < a \text{ and } 0 < y < b$$

$$U = \infty \qquad \text{outside of this region}$$

(b) Find energy levels and a wave function for an infinitely deep three-dimensional potential well such that

$$U = 0 \qquad \text{for } 0 < x < a,\ 0 < y < b, \text{ and } 0 < z < c$$

$$U = \infty \qquad \text{outside of this region}$$

1-2-2. Write down and compare electronic configurations for C, Si, Ge, Ga, Al, P, As, N, and Ne. Example: C ($1s^2 2s^2 2p^2$). $1s^2$ means that for $n=1$ there are two s electrons ($l=1$), $2s^2$ means that for $n=2$ there are two s electrons ($l=1$), $2p^2$ means that for $n=2$ there are two p electrons ($l=2$).
Hint: Use the periodic system.

1-2-3. A particle on a string is described by the following equation of motion

$$m\, dx^2/dt^2 + kx = 0$$

where m is the mass of the particle and k is the force constant of the string. This equation can be rewritten as

$$dx^2/dt^2 + \omega^2 x = 0$$

where $\omega = (k/m)^{1/2}$ is the frequency of oscillations (such a system is an example of a linear, or harmonic, oscillator).
The average value of the energy, $\langle E \rangle$, of a harmonic oscillator is given by $\langle E \rangle = \langle p^2 \rangle/(2m) + m\,\omega^2 \langle x^2 \rangle/2$, where $\langle x^2 \rangle$ and $\langle p^2 \rangle$ are the average values of x^2 and p^2. Use the uncertainty principle to determine the low bound of $\langle E \rangle$.
Hint: $\langle p^2 \rangle \geq (\Delta p^2)$ and $\langle x^2 \rangle \geq (\Delta x^2)$.

1-2-4. Find energy levels for a particle with mass m in the one-dimensional potential well shown in Fig. P1-2-4 with potential $U(x)$ given by

$$\mathrm{U} = \mathrm{U}_1 \qquad \text{for x} \leq 0$$

$$\mathrm{U} = 0 \qquad \text{for } 0 < \text{x} < \text{d}$$

$$\mathrm{U} = \mathrm{U}_2 \qquad \text{for x} \geq \text{d}$$

Also, consider the case of the symmetrical well ($U_2 = U_1$) and prove that such a well has at least one discrete energy level.

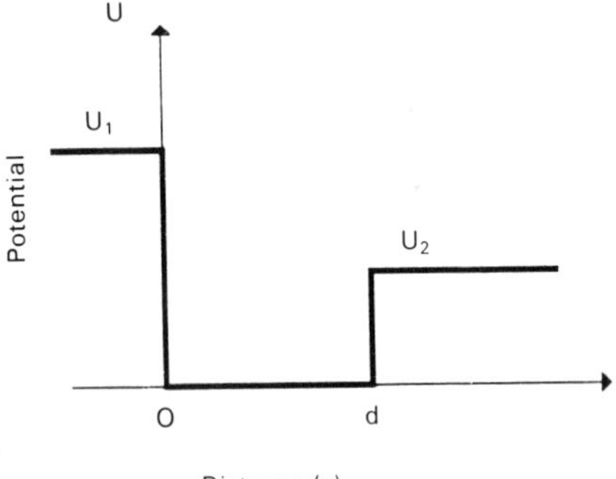

Fig. P1-2-4

1-3-1. Piezoelectricity is the effect of creating an electric field in a crystal by applying stress. Fig. P1-3-1 shows the dependence of the piezoelectric constant (which determines the strength of a piezoelectric effect) in different semiconductor compounds as a function of the bond ionicity. Why is there a strong correlation between the bond ionicity and piezoelectricity?

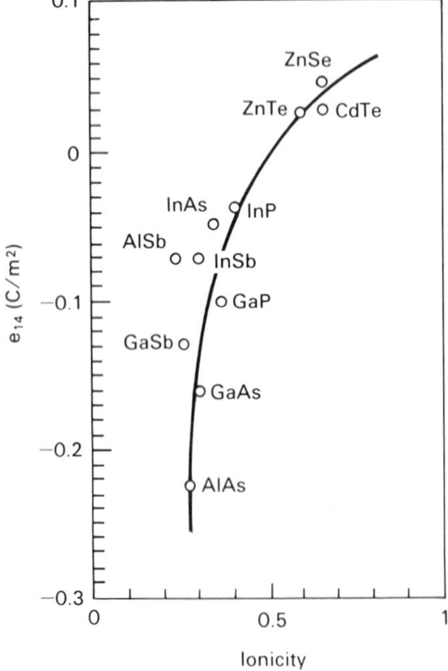

Fig. P1-3-1. Plots of e_{14} versus Phillips' ionicity for several III-V and II-VI compounds. (After Adachi. *J. Appl. Phys.*, 58, no. 3, August 1985.)

1-4-1. Assume that a lattice constant $a = 5$ Å. Find the volume of a primitive cell for simple cubic, fcc, and bcc crystal lattices.

1-4-2. **(a)** What is the number of nearest neighbors for simple cubic, fcc, bcc, and diamond crystal lattices?

(b) What is the number of second nearest neighbors for simple cubic, fcc, bcc, and diamond crystal lattices?

1-4-3. Sketch primitive lattice vectors and identify the basis atoms for the diamond crystal structure.

1-4-4. Show the (211) plane and the [211] direction in a cubic crystal lattice.

1-4-5. **(a)** What are the Miller indices for the plane shown in Fig. P1-4-5a and for the plane shown in Fig. P1-4-5b?

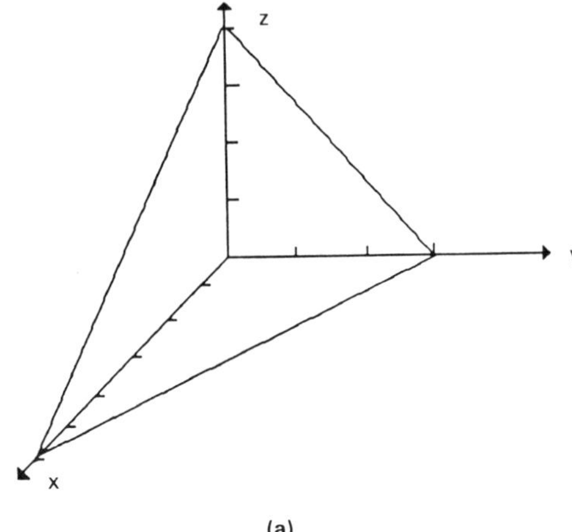

(a) **Fig. P1-4-5a**

(b) What are the Miller indices for the lines formed by the intersection of the crystal plane shown in Fig. P1-4-5b with planes [001], [010], and [100]?

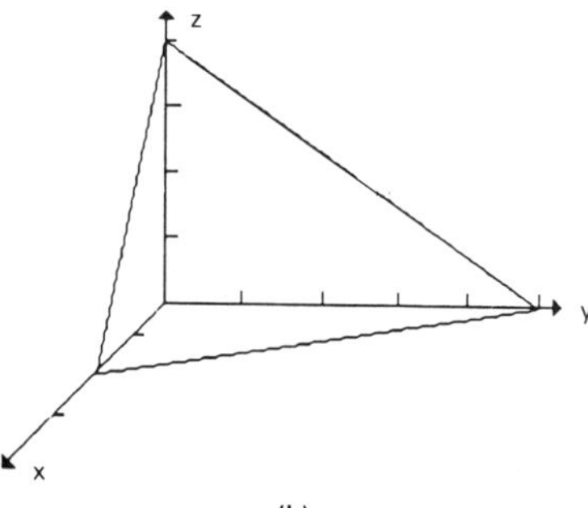

(b) **Fig. P1-4-5b**

1-4-6. Find angle α between two nearest bonds in the diamond lattice (see diagram).

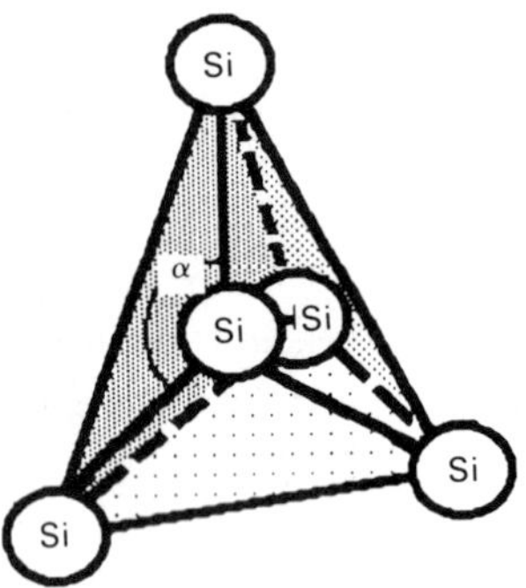

Fig. P1-4-6

1-4-7. Covalent radii of In and Sb are approximately 1.44 Å and 1.36 Å, respectively. Find the lattice constant of InSb and the volume of the primitive cell.
Hint: The lattice constant is the length of the side of the cube of the unit cell.

1-5-1. Find primitive vectors of the reciprocal lattice for simple cubic, bcc, and fcc lattices.

1-5-2. Calculate the Fermi vector of valence electrons for Si, Ge, and GaAs. Densities of Si, Ge, and GaAs are 2.33, 5.33, and 5.32 g/cm^3, respectively. Lattice constants of Si, Ge, and GaAs are 5.43, 5.64, and 5.65 Å, respectively.

1-5-3. Using eqs.(1-5-27) and (1-5-28) plot energy E (in eV) versus wave vector $\mathbf{k}$ for direction [001] of the reciprocal lattice between $k=0$ and the boundary of the Brillouin zone for GaAs. Compare the resulting plot with the parabolic dependence for the same effective mass. The lattice constant of GaAs is 5.65 Å. The electron effective mass in the central valley of the conduction band $m = 0.067m_e$, where $m_e = 9.11 \times 10^{-31}$ kg is a free electron mass, and the energy gap E_g = 1.42 eV.

1-5-4. The difference in the energies of the low valley (Γ valley) in GaAs and upper valleys is approximately 0.3 eV. The effective electron mass is 0.067 m_e. Assuming that electrons reaching this energy in the Γ valley transfer into the upper valleys (where the density of states is much larger) calculate the maximum velocity, $v_{\Gamma\max}$, the maximum momentum, $P_{\Gamma\max}$, and the maximum wave vector, $\mathbf{k}_{\Gamma\max}$ of electrons in the Γ valley. Find $v_{\Gamma\max}/v_{th}$, $\mathbf{k}_{\Gamma\max}a/2\pi$, and $\mathbf{k}_{\Gamma\max}\lambda$ where a is the lattice constant, v_{th} is the thermal velocity at 300 K, and λ is the de Broglie wavelength.

1-5-5. Consider an atomic ring of six atoms. The Schrödinger equation for this chain may be written in the matrix form as

$$(E_i - E)\psi_i - \sum_{k \neq i} J_{ik}\,\psi_k = 0 \qquad \text{(P1-5-5-1)}$$

where E_i are diagonal matrix elements, J_{ik} are exchange integrals, and ψ_i is the wave function corresponding to the electron residing on the atom i. Assuming that only the exchange integrals between the nearest neighbors are equal to J and the rest of the exchange integrals are equal to zero, and that the diagonal elements E_i are the same as E_o, calculate the energy spectrum of the chain.

Hint: Seek the solution of the Schrödinger equation, the wave junction ψ, in the following form

$$\psi_j = ae^{i\theta j} \tag{P1-5-5-2}$$

and use the periodic boundary condition, $\psi_{j+6} = \psi_j$.

1-5-6. Determine and sketch first Brillouin zones for simple cubic, fcc, and bcc lattices. **Hint:** Define the lattice vectors , compute $\mathbf{K}_1$, $\mathbf{K}_2$, and $\mathbf{K}_3$, then sketch in the $\mathbf{k}_x$, $\mathbf{k}_y$, $\mathbf{k}_z$ system.

1-5-7. Consider a one-dimensional crystal lattice with a separation between the atoms of 2 Å. Assume that the electron effective mass is equal to 10^{-30} kg. At what smallest kinetic energy will electrons experience Bragg reflection?
Hint: For Bragg reflection the wave vector, $\mathbf{k_b}$, should satisfy the following condition: $2\mathbf{k_b} \cdot \mathbf{K} = K^2$, where $\mathbf{K}$ is the reciprocal lattice vector and K is it's magnitude.

1-5-8. The dependence of energy on the wave vector for the Γ minimum of the conduction band in GaAs may be approximated by

$$E(1 + \alpha E) = \hbar^2 k^2/(2m_n)$$

where m_n is the effective mass for $E = 0$, $\mathbf{k}$ is the wave vector, and α is a constant. Calculate the dependence of the effective mass, m_n, on energy.

1-5-9. A Wigner-Seitz primitive cell is formed by bisecting all lattice vectors emanating from a lattice site by planes and choosing the smallest volume formed by these intersecting planes. For a two-dimensional lattice a Wigner-Seitz primitive cell is formed by bisecting all lattice vectors emanating from a lattice site by perpendicular lines and choosing the smallest area formed by these intersecting lines.

(a) Find the Wigner-Seitz primitive unit cell for the direct two-dimensional square centered lattice shown in Fig. P1-5-9a.

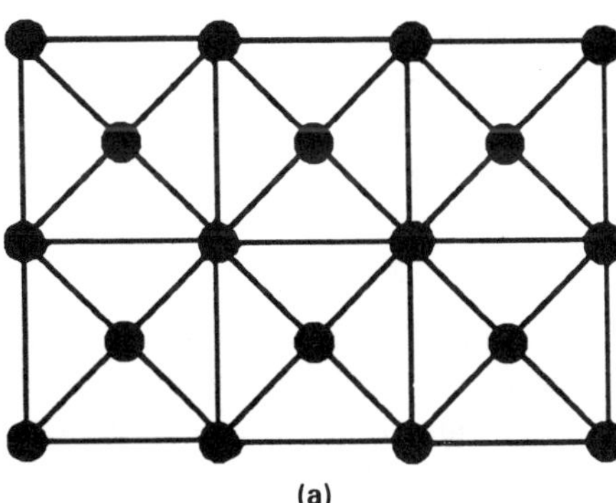

Fig. P1-5-9a. Two-dimensional square centered lattice.

(b) Find the first Brillouin zone for the direct two-dimensional square centered lattice shown in Fig. P1-5-9a. Show dimensions. Assume that the distance between the nearest neighbors in the direct lattice is c Å.

1-5-10. Prove (for a two-dimensional square lattice) that the Bragg planes form the boundaries of the Brillouin zones.

1-5-11. Find the area of the first Brillouin zone for the direct two-dimensional square centered lattice shown in Fig. P1-5-11. Assume that the distance between the nearest neighbors in the direct lattice is 3 Å.

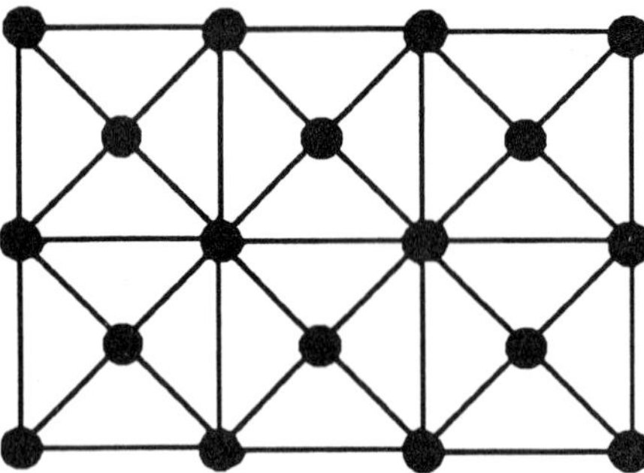

Fig. P1-5-11. Two-dimensional square centered lattice.

1-6-1. The degeneracy of the electron gas in a semiconductor becomes noticeable when the electron concentration becomes larger than the effective density of states, N_c. Using eqs.(1-6-16) and (1-6-28) find at what ratio of n/N_c the condition of degeneracy (which is $E_F - E_c > 3k_BT$) is satisfied. Here $E_F - E_c$ is the difference between the Fermi level and the bottom of the conduction band, n is the electron concentration, and N_c is the effective density of states in the conduction band.

1-6-2. A *pn* product for a nondegenerate semiconductor in equilibrium is given by

$$np = N_c N_v \exp(-E_g/k_BT)$$

Here N_c and N_v are the effective densities of states in the conduction and valence band, respectively, E_g is the energy gap, k_B is the Boltzmann constant, and T is temperature. Ionized donor density is N_D. Derive an expression for the *pn* product for the degenerate n-type semiconductor in terms of N_D, N_c, N_v, k_BT, and E_g.

1-6-3. Assume that the maximum donor concentration can be achieved when donor atoms effectively "touch" each other (assuming that they can be regarded as spheres with radii

$$a_{Bd} = a_B(\varepsilon_s/\varepsilon_o)(m_e/m_n)$$

where $a_B \approx 0.53$ Å is the Bohr radius.) Assuming that for GaAs $(\varepsilon_s/\varepsilon_o) = 12.9$ and $(m_n/m_e) = 0.067$, estimate the maximum doping concentration.
Hint: Assume that at the maximum doping level donors are as closely packed as possible, even though this is not necessarily a very realistic assumption.

1-6-4. Consider two-dimensional band structure with the dispersion relation for the conduction band

$$E = [\hbar^2/(2m_e)](k_x^2 + k_y^2)$$

(no k_z-component). (This is a fairly realistic model for electron levels in heterostructure quantum wells.)
Derive the expression linking the density of electrons in the conduction band to the position of the Fermi level, E_F, and to temperature T.

(a) Assume nondegenerate distribution of electrons in the conduction band.

(b) Assume degenerate distribution of electrons in the conduction band, i.e., assume that $E_F - E_c >> k_BT$, where E_c is the bottom of the conduction band.

(c) Derive this expression for an arbitrary position of the Fermi level with respect to the bottom of the conduction band, E_c.

1-6-5. An effective density of states in the conduction band of silicon is equal to 2.8×10^{19} cm^{-3} (at room temperature). The density of states effective mass for silicon is 1.182 m_e at 300 K and 1.077 m_o at 77 K, where m_e is the free electron rest mass. Consider a sample doped at 10^{16} cm^{-3} by shallow ionized donors. Find an expression for the number of electrons per unit energy, dn/dE, in the conduction band as a function of energy in this sample and plot it at T = 77 K and T =300K.

1-6-6. Consider a one-dimensional quantum wire, where electron motion in two directions is quantized and in one direction electrons are free to move. Assuming that the effective electron mass is equal to m_n and that the dispersion relation for the electron energy in a subband is parabolic ($E = \hbar^2k^2/(2m_n)$), find the density of states, $\Omega(E)$, in the subband as a function of energy E.
Hint: $dN = \Omega(E)\, dE$, where dN is the number of states in the energy interval between E and $E + dE$.

1-7-1. Conductivity of a silicon sample in the thermal equilibrium is given by

$$\sigma = q(n\mu_n + p\mu_p)$$

Assume that the sample temperature is 300 K. The intrinsic carrier concentration of Si at this temperature is about 1.5×10^{10} cm^{-3}. $\mu_n = 0.1$ m^2/Vs, $\mu_p = 0.03$ m^2/Vs. The sample is doped by donors with the donor concentration 10^{16} cm^{-3} and then compensated by acceptors. What values of the acceptor concentration will yield the resistivity of 1 Ωcm?

1-7-2. Consider a silicon sample in the thermal equilibrium at room temperature (300K). The intrinsic carrier concentration of Si at this temperature is about 1.5×10^{10} cm^{-3}. Plot the concentration of electrons and holes versus acceptor doping in the range from 10^{13} cm^{-3} to 10^{17} cm^{-3} for the following shallow-donor concentrations:

(a) 10^{15} cm^{-3}

(b) 10^{16} cm^{-3}

1-7-3. Find the relationship between the Fermi level, E_F, and the electron concentration, n, for a two-dimensional degenerate electron gas. The electron effective mass is equal to m_n.

1-7-4. Find the relationship between the Fermi level, E_F, and the electron concentration, n, for a one-dimensional degenerate electron gas. The electron effective mass is equal to m_n.

1-8-1. Consider a linear chain of 11 atoms. The atomic mass is equal to M. Assume that the ends of the chain are fixed and that the force F_{ik} acting on an atom i displaced from an equilibrium position from atom k is given by

$$F_{ik} = -\kappa(x_i - x_k)\,\delta_{|i-k|,1}$$

Calculate the frequencies of the atomic vibrations.

1-8-2. A simplified model of lattice vibrations called the Debye model that assumes that the frequency of the lattice vibrations is proportional to the wave vector, $\mathbf{q}$:

$$\omega \approx c_s \mathbf{q}$$

where c_s is the velocity of sound. Wave vector $\mathbf{q}$ in this equation varies from zero to the maximum value, q_D, where q_D is called the Debye wave vector. The Debye wave vector is chosen in such a way that the total number of the modes of the lattice vibrations for a crystal of a given size is equal to the number of allowed values of q in the sphere of radius q_D. Show (using a simple cubic lattice as an example) that

$$q_D = (6\pi^2/V_c)^{1/3}$$

where V_c is the unit cell volume.

1-9-1. An electron mobility in a semiconductor with charged impurities may be increased if additional electrons are induced into the semiconductor. These electrons "screen" the scattering potential of a charged impurity.

(a) Why is mobility not decreased because of additional electron-electron scatterings as more electrons are induced into the semiconductor?

(b) Calculate the electron mobility, limited by impurity scattering, for silicon as a function of the electron concentration in silicon in the range from 10^{16} to 10^{18} cm^{-3} for fixed densities of the charged impurities $N_{dA} = 10^{16}\ \text{cm}^{-3}$ and $N_{dA} = 10^{18}\ \text{cm}^{-3}$ and $T = 300\text{K}$.

Hint: Use equations given in Appendix 23.

1-10-1. The Hall effect can be used to determine the type of carriers present in a material. Show that the Hall constant, R, for an intrinsic semiconductor is given by

$$R = -\frac{1}{qn_i}\frac{\mu_n/\mu_p - 1}{\mu_n/\mu_p + 1} \tag{1-11-19}$$

where n_i is the intrinsic carrier concentration, and μ_n and μ_p are electron and hole mobility, respectively.

Hints: Average velocities for holes and electrons in the y-direction, v_{yp} and v_{yn}, perpendicular to both magnetic field, B_z, (in the z direction) and bias electric field, F_x, in the x direction are approximately as follows:

$$v_{yn} = -\mu_n F_y - \mu_n^2 B_z F_x$$

$$v_{yp} = \mu_p F_y - \mu_p^2 B_z F_x$$

Derive an expression for the y component of the current density, j_y in terms of n, p, and these velocities, then use the condition that $j_y = 0$.

1-12-1. Calculate the dependence of electron and hole quasi-Fermi levels in silicon doped at $10^{16}\ \text{cm}^{-3}$ at 300 K versus the density of electron-hole pairs, P, generated by light for $10^9\ \text{cm}^{-3} < P < 10^{17}\ \text{cm}^{-3}$.

Assume the effective densities of states in the conduction and valence bands, $N_c = 2.8 \times 10^{19}\ \text{cm}^{-3}$ and $N_v = 1.04 \times 10^{19}\ \text{cm}^{-3}$, and intrinsic carrier concentration, $n_i = 1.45 \times 10^{10}\ \text{cm}^{-3}$.

1-12-2. Consider a n-type semiconductor sample, shown in Fig. P1-12-2. The ultraviolet light shining at surface is absorbed in a very thin layer near the surface, creating the hole concentration at the surface $p(0) = G_L t_{\text{eff}}$. The backside contact has a surface recombination rate S_p. Calculate and plot the steady-state hole distribution in the sample for the values of S_p = 1000 cm/s, 100,000 cm/s, 10^7 cm/s. Neglect the intrinsic concentration of holes in the sample. Use the following values of parameters:

Hole diffusion coefficient $D_p = 30$ cm²/Vs

Hole lifetime $\tau_p = 10^{-7}$ s

Device length $L = 10\ \mu$m

$t_{\text{eff}} = 10^{-8}$ s

$G_L = 10^{24}$ 1/cm^{-3} s

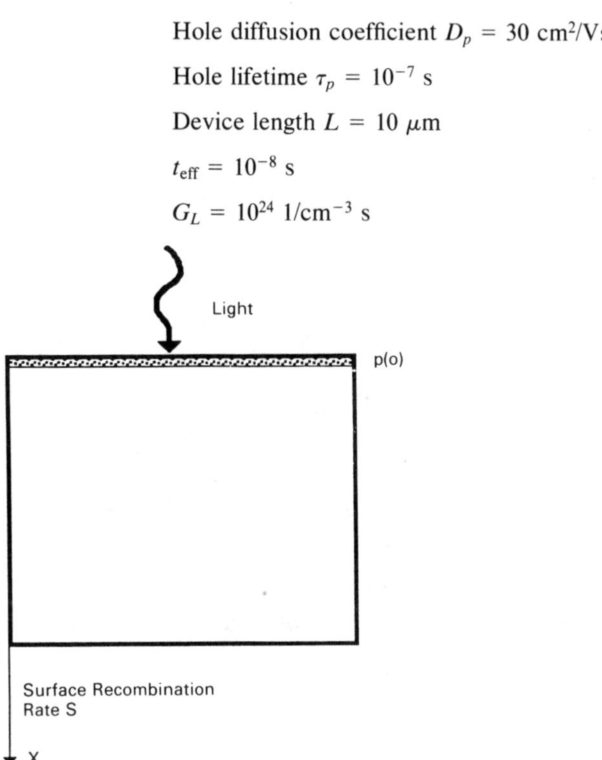

Fig. P1-12-2

1-14-1. Simulate an electron trajectory in **k**-space for a square lattice. Assume that the initial state is $k_x = 0$, $k_y = 0$. The electric field $F_x = F_y = 10^7$ V/m. The average time $1/\Gamma$ between collisions is 10^{-13} s. Assume that collisions are random, so that the times of free fights can be related to random numbers by eq. (1-14-8). The direction of **k** after collision is randomized. Simulate the trajectory for 5, 10, and 100 free flights.

2

p-n Junctions, Schottky Barrier Junctions, Heterojunctions, and Ohmic Contacts

2-1. INTRODUCTION

Contacts—gradual or abrupt—between two different materials, such as semiconductors with different doping, semiconductor and metal, semiconductor and insulator, semiconductor regions with different composition (heterojunctions), are primary building blocks for all semiconductor devices. In this chapter we consider the basic concepts describing such junctions. For p-n junctions or abrupt metal-semiconductor contacts these concepts are well established and understood. Our understanding of heterojunctions and of *ohmic contacts* is more qualitative. Ohmic contacts are used to attach external voltage or current sources and other circuit elements and devices to a semiconductor device. Ideally, they should have as little effect as possible on the device performance. In practice, the effect of ohmic contacts is very often important. At the same time, the fabrication of ohmic contacts is still more of an art than a science. This chapter starts from the discussion of p-n junctions and metal-semiconductor junctions. Then we try to apply the same basic concepts to ohmic contacts and heterojunctions.

2-2. p-n JUNCTION UNDER ZERO BIAS CONDITIONS

A standard approach to understanding the behavior of a semiconductor device is to plot and analyze its energy band diagram that shows the bottom of the conduction band, the top of the valence band, and the Fermi level as functions of distance. The energy band diagrams of n- and p-type semiconductors are shown in Figs. 2-2-1a and 2-2-1b. In the n-type material the Fermi level is closer to the bottom of the conduction band, whereas in the p-type material it is closer to the top of the valence band. Let us now consider what happens if these n- and p-type semiconductors are brought in contact. According to eqs. (1-11-34) and (1-11-35), electric current will flow if the Fermi level is not constant throughout the whole system:

$$j = \mu_n n \frac{dE_{Fn}}{dx} + \mu_p p \frac{dE_{Fp}}{dx} \tag{2-2-1}$$

Hence, under equilibrium conditions (when no external bias is applied), the Fermi level must be constant throughout the whole sample. The requirement of the constancy of the Fermi level leads to the energy band diagram shown in Fig. 2-

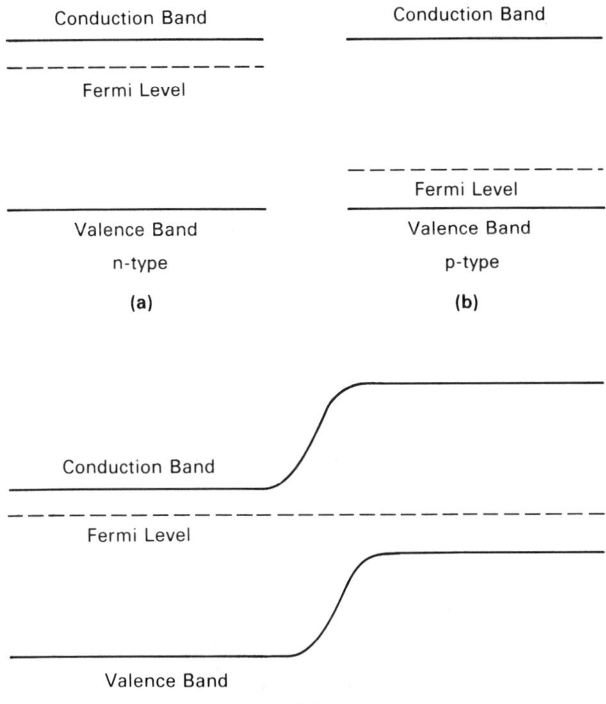

Fig. 2-2-1. (a) Band diagrams of *n*-type semiconductor (b), *p*-type semiconductor, and (c) *p-n* junction.

2-1c. The concentration of carriers can be related to the position of the Fermi level. For a nondegenerate semiconductor we have

$$n = N_c \exp \frac{E_F - E_c}{k_B T} \tag{2-2-2}$$

$$p = N_v \exp \frac{E_v - E_F}{k_B T} \tag{2-2-3}$$

(see eqs. (1-6-26) and (1-6-27)). Hence, in the transition region, where the energy bands are bent, the concentration of electrons is much smaller than the concentration of electrons in the n region, where $n \cong N_D$ and the energy bands are flat. In fact, as may be expected, the concentration of electrons throughout the transition region varies from $n = n_{no} = N_D$ in the n-type region to $n = n_{po} = n_i^2/N_A$ in the p-type region, where n_{no} and n_{po} are equilibrium electron concentrations in the n and p regions respectively (see Fig. 2-2-2). As can be seen from eqs. (2-2-2) and (2-2-3), the electron and hole concentrations change by a factor of exp(1) when the band bending changes by $V_{th} = k_B T/q$ (or 25.8 mV at room temperature). The total band bending is typically several tenths of a volt or more (depending on doping

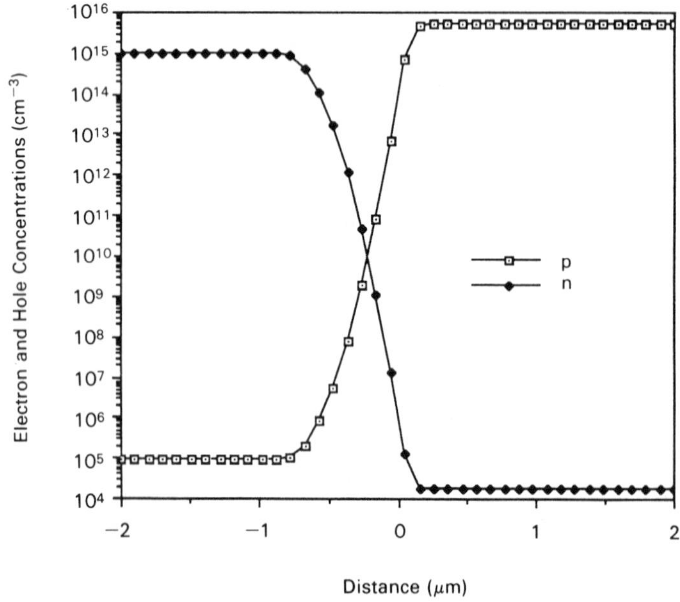

Fig. 2-2-2. Electron and hole concentration profiles for a silicon *p-n* junction. The curves were generated using program PLOTF.BAS with subroutines PPNJN-.BAS (see the listing in the Program Description Section). Parameters used in the calculation: energy gap $E_g = 1.12$ eV, $N_c = 3.22 \times 10^{19}\text{cm}^{-3}$, $N_v = 1.83 \times 10^{19}\text{cm}^{-3}$, $\varepsilon = 1.054 \times 10^{-10}$F/m.

levels, the energy gap, and effective densities of states in the conduction and valence bands), i.e., much greater than $k_B T/q$. As a consequence, almost in the entire transition region

$$n << N_D$$
$$p << N_A \tag{2-2-4}$$

and, hence, the charge densities are

$$\rho_n \cong qN_D \tag{2-2-5}$$

in the n-type section of the transition region and

$$\rho_p \cong -qN_A \tag{2-2-6}$$

in the p-type section of the transition region. Here N_D and N_A are the concentrations of shallow (completely ionized) donors and acceptors, and we assume that $n = N_D$ and $p = N_A$ in the n and p sections, respectively. Equations (2-2-5) and (2-2-6) correspond to the *depletion approximation*. The transition region is frequently called the depletion region or space charge region.

Please note that the depletion approximation does not mean that the depletion region is devoid of carriers. As can be seen from Fig. 2-2-2, not only are there electrons and holes in the depletion region but their concentrations are such that

$$n >> n_{po} \tag{2-2-7a}$$

and

$$p >> p_{no} \tag{2-2-7b}$$

The depletion approximation merely states that the concentrations of electrons and holes in the depletion region, n and p, are much smaller than N_D and N_A, respectively, and may be neglected when the charge densities are evaluated (see eqs. (2-2-5) and (2-2-6)). They cannot be neglected, however, in the evaluation of the current flow through a p-n junction.

The depletion approximation is not valid in boundary layers between the neutral sections and the depletion regions. However, these layers are relatively thin as long as the total band bending—i.e., the difference between the bottoms of the conduction band in the p and n regions is much greater than the thermal voltage $V_{th} = k_B T/q$.

The bottom of the conduction band and the top of the valence band correspond to the potential energies of electrons and holes, respectively. Consequently,

$$E_c = -q\phi + \text{const} \tag{2-2-8}$$

$$E_v = -q\phi - E_g + \text{const} \tag{2-2-9}$$

where q is the electronic charge, $E_g = E_c - E_v$ is the energy gap, and ϕ is the electric potential. Choosing $E_c = 0$, $\phi = 0$ in the n-type region far from the junction we find that

$$n = N_D \exp[q\phi(x)/k_BT] = N_D \exp[-E_c(x)/k_BT] \tag{2-2-10}$$

$$p = N_A \exp\{-q[V_{bi} + \phi(x)]/k_BT\} = N_A \exp\{-q[V_{bi} - E_c(x)/q]/k_BT\} \tag{2-2-11}$$

Here x is coordinate, and

$$V_{bi} = V_{th} \ln \frac{N_D N_A}{n_i^2} \tag{2-2-12}$$

is called the *built-in voltage*. (V_{bi} is equal to $-\phi(\infty)$, where $\phi(\infty)$ is the potential of the p region far from the junction.) Equations (2-2-10) to (2-2-12) are derived from the condition that far from the junction the carrier concentrations have their equilibrium values:

$$n_{no} \cong N_D \tag{2-2-13}$$

$$p_{no} \cong \frac{n_i^2}{N_D} \tag{2-2-14}$$

$$p_{po} \cong N_A \tag{2-2-15}$$

$$n_{po} \cong \frac{n_i^2}{N_A} \tag{2-2-16}$$

(see eqs. (1-7-17), (1-7-18), (1-7-29), and (1-7-30)). Equation (2-2-12) may also be derived from the requirement of the constancy of the Fermi level across the p-n junction. The Fermi level in the n and p sections of the device is given by

$$E_F - E_{cn} = k_BT \ln \frac{N_D}{N_c} \tag{2-2-17}$$

$$E_F - E_{cp} = -E_g - k_BT \ln \frac{N_A}{N_v} \tag{2-2-18}$$

where E_{cn} and E_{cp} are the bottom of the conduction band in the n and p sections of the device far from the junction, respectively. (For our choice of reference $E_{cn} = 0$, and $E_{cp} = V_{bi}/q$). Hence,

$$qV_{bi} = E_{cp} - E_{cn} = E_g + k_BT \ln \frac{N_A N_D}{N_c N_v} \tag{2-2-19}$$

which coincides with eq. (2-2-12) because

$$n_i^2 = N_c N_v \exp[-E_g/(k_BT)] \tag{2-2-20}$$

(see eq. (1-6-34)). Equation (2-2-19) shows that the built-in voltage is primarily determined by the energy gap of a semiconductor (see also Fig. 2-2-3).

We conclude that there is a potential variation across the p-n junction even when the external bias is zero. This voltage drop is caused by the difference in doping. However, the p-n junction cannot be used as an electric power supply if

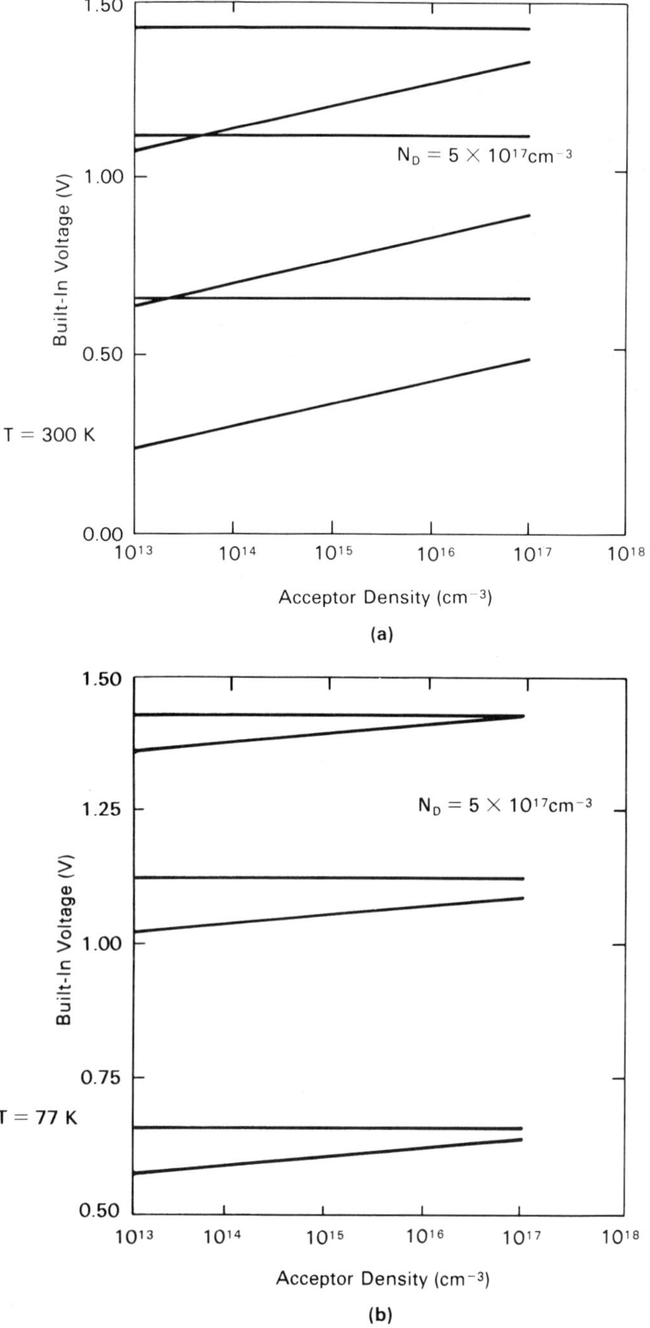

Fig. 2-2-3. Built-in voltage vs. acceptor density for n^+-p junctions. Energy gaps (horizontal lines) are shown for comparison. $N_D = 5 \times 10^{17}$ cm^{-3}. (a) T = 300K; (b) T = 77K. The curves were generated using program PLOTF.BAS with subroutines PVBI.BAS (see Program Description Section.)

there is no continuous separation of the electric charges generated by an external source, such as illumination by light. Indeed, if we try to make a continuous circuit by making, for example, the n-type region long and bending it into a ring to contact the p section on the other side (see Fig. 2-2-4), the voltage drops at the two junctions will cancel each other, so that the total voltage drop along the loop will be zero. One can establish that the same conclusion remains true even if one inserts into this loop other semiconductors or conductors. However, if we shine light on a p-n junction creating electron-hole pairs, they may be separated by the built-in electric field in the junction, producing an electric current. This effect is utilized in different photonic devices, such as solar cells and photodetectors (see Chapter 5).

The distribution of the electric potential in the junction is determined from Poisson's equation:

$$\frac{d^2\phi}{dx^2} = -\frac{\rho}{\varepsilon} \tag{2-2-21}$$

where ε is the dielectric permittivity, x is the space coordinate, and

$$\rho = q(p - n + N_D - N_A) \tag{2-2-22}$$

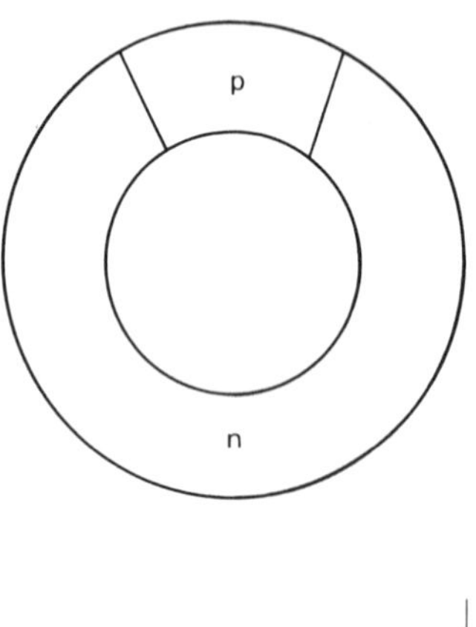

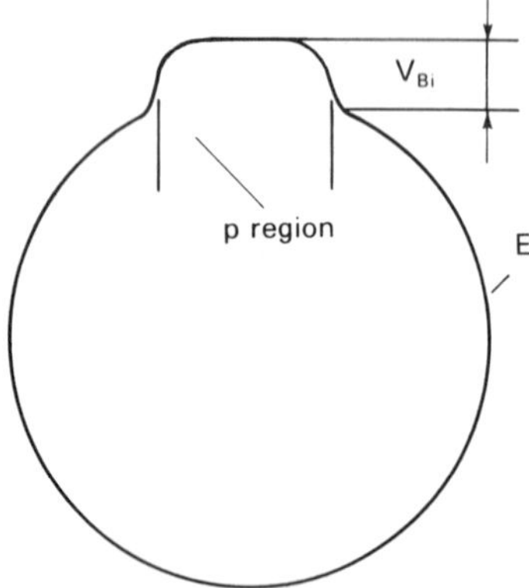

Fig. 2-2-4. Ring shaped *p-n* junction and bottom of conduction band in the junction as a function of the position along the ring. The overall change in E_c (and in the electric potential) around the loop is zero.

is the space charge density. According to the depletion approximation, the space charge density in the space charge region is given by eqs. (2-2-5) and (2-2-6).

Using the depletion approximation, we can rewrite Poisson's equation as follows:

$$\frac{d^2\phi}{dx^2} = \begin{cases} -\dfrac{qN_D}{\varepsilon} & -x_n < x < 0 \\[2ex] \dfrac{qN_A}{\varepsilon} & 0 < x < x_p \end{cases} \tag{2-2-23}$$

Here x = 0 corresponds to the boundary between n- and p-type regions. The charge distribution determining the right-hand side of eq. (2-2-23) is shown in Fig. 2-2-5.

The electric field,

$$F = -d\phi/dx = (dE_c/dx)/q \tag{2-2-24}$$

is given by the first integral of eq. (2-2-23),

$$F = \begin{cases} F_m\left(1 + \dfrac{x}{x_n}\right) & -x_n \leqslant x \leqslant 0 \\[2ex] F_m\left(1 - \dfrac{x}{x_p}\right) & 0 \leqslant x \leqslant x_p \end{cases} \tag{2-2-25}$$

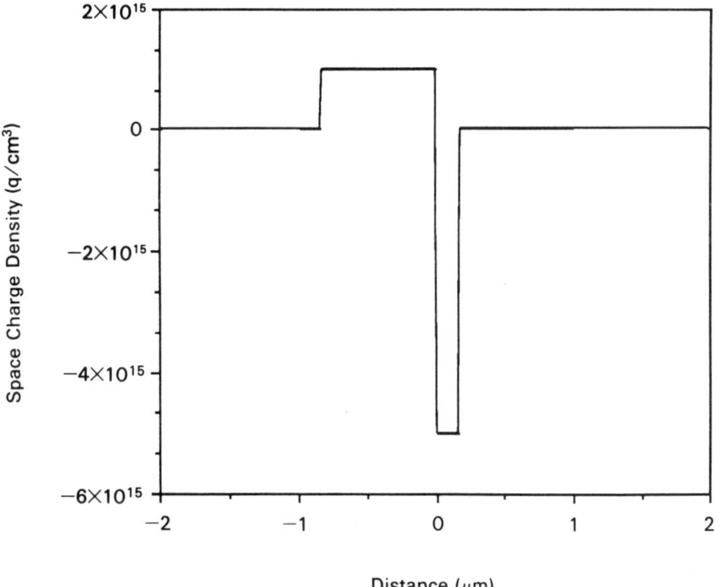

Fig. 2-2-5. Charge density profile (in units of electronic charge per cm³) for a silicon *p-n* junction. Rectangular profile corresponds to the depletion approximation (see Eqs. 2-2-5 and 2-2-6).

(see Fig. 2-2-6). The maximum electric field, F_m, in the junction (reached at the p-n interface, i.e., at $x = 0$) is given by

$$F_m = \frac{qN_D x_n}{\varepsilon} = \frac{qN_A x_p}{\varepsilon} \tag{2-2-26}$$

The potential distribution is found by integrating eq. (2-2-25),

$$\phi = \begin{cases} -\dfrac{qN_D(x + x_n)^2}{2\varepsilon} & -x_n < x < 0 \\ -V_{bi} + \dfrac{qN_A(x_p - x)^2}{2\varepsilon} & 0 < x < x_p \end{cases} \tag{2-2-27}$$

where we used the boundary condition

$$\phi\,(-x_n) = 0 \tag{2-2-28}$$

A potential distribution for a silicon p-n junction at zero external bias is shown in Fig. 2-2-7. Equation (2-2-27) and eqs. (2-2-10) and (2-2-11) can be used to calculate approximate profiles of electron and hole concentrations in the space charge region. Then a more accurate potential profile can be computed by integrating Poisson's equation, where the charges of mobile electrons and holes are included into the space charge density:

$$\frac{d^2\phi}{dx^2} = \begin{cases} -\dfrac{q(N_D + p - n)}{\varepsilon} & -x_n < x < 0 \\ -\dfrac{q(-N_A + p - n)}{\varepsilon} & 0 < x < x_p \end{cases} \tag{2-2-29}$$

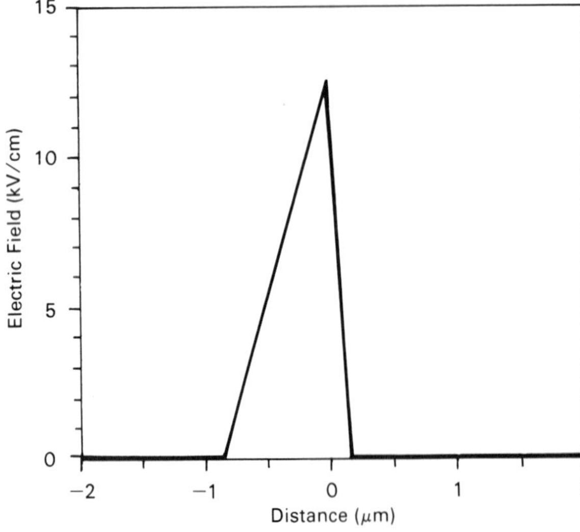

Fig. 2-2-6. Field distribution in a silicon *p-n* junction at zero bias. The field profile was calculated using program PLOTF.BAS with subroutines PPNJN.BAS (see Program Description Section). The calculation is based on the depletion approximation. $x = 0$ corresponds to the boundary between *p* and *n* regions.

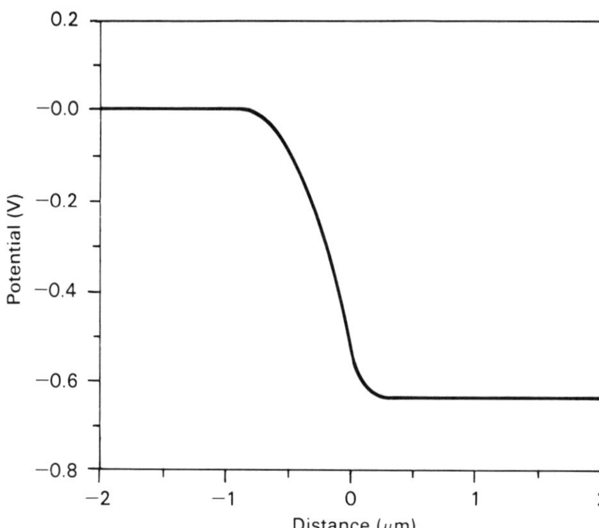

Fig. 2-2-7. Potential distribution in a silicon *p-n* junction at zero bias. The potential reference ($\phi = 0$) was chosen in the *n*-type region far from the junction. The potential profile was calculated using program PLOTF.BAS with subroutines PPNJN.BAS (see Program Description Section). The calculation is based on the depletion approximation. $x = 0$ corresponds to the boundary between *p* and *n* regions.

Then more accurate dependencies of n and p on distance, x, can be calculated by substituting the new values of ϕ into eqs. (2-2-10) and (2-2-11). This iterative process may be continued until we achieve a desired accuracy.

From eq. (2-2-26) we find that

$$\frac{x_n}{x_p} = \frac{N_A}{N_D} \tag{2-2-30}$$

and from eq. (2-2-27) we obtain

$$\frac{qN_D x_n^2}{2\varepsilon} + \frac{qN_A x_p^2}{2\varepsilon} = V_{bi} \tag{2-2-31}$$

Hence,

$$x_n = \left[\frac{2\varepsilon V_{bi}}{qN_D\left(1 + \dfrac{N_D}{N_A}\right)}\right]^{1/2} \tag{2-2-32}$$

and

$$x_p = \left[\frac{2\varepsilon V_{bi}}{qN_A\left(1 + \dfrac{N_A}{N_D}\right)}\right]^{1/2} \tag{2-2-33}$$

(see Fig. 2-2-8). These equations for x_n and x_p show that for a one-sided n^+-p junction, where $N_D >> N_A$, $x_p >> x_n$, so that the electric field distribution looks like a right-angle triangle.

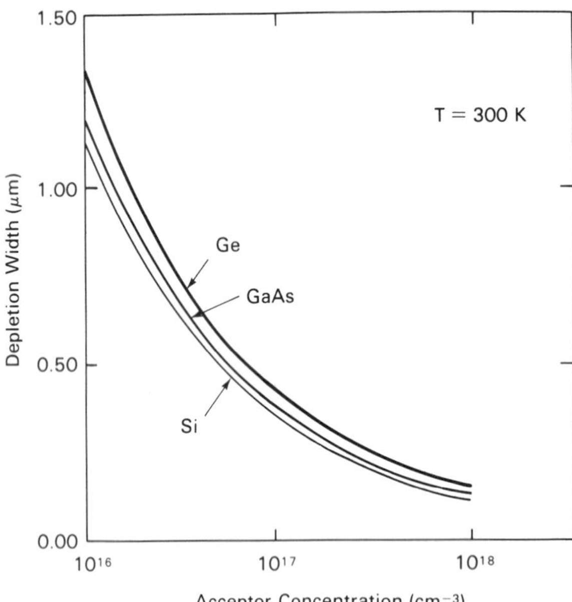

Fig. 2-2-8. Depletion width of n^+-p junction for GaAs, Si, and Ge vs. the acceptor density at zero external bias. The donor density in the n^+ region is 5×10^{17} cm^{-3}. T = 300K. The curves are calculated using program PLOTF-.BAS with subroutines PVBI.BAS (see Program Description Section). The bottom curve is for Si, the middle curve is for GaAs, and the top curve is for Ge.

As was mentioned previously, the depletion approximation breaks down in thin boundary layers between the depletion layer and neutral layers, where the carrier concentrations are comparable with the dopant concentrations. In the boundary layer in the n-type region we can rewrite Poisson's equation as follows:

$$\frac{d^2\phi}{dx^2} = -\frac{qN_D}{\varepsilon} + \frac{qN_D}{\varepsilon}\exp\left(\frac{\phi}{V_{th}}\right) \tag{2-2-34}$$

where the thermal voltage $V_{th} = k_B T/q$. (Equation (2-2-34) is derived using eqs. (2-2-10), (2-2-21), and (2-2-22).) When $n \cong N_D$ ($\phi << V_{th}$) the second term in the right-hand side of eq. (2-2-34) can be expanded in the Taylor series ($\exp(\phi/V_{th}) \approx 1 + \phi/V_{th}$) to yield

$$\frac{d^2\phi}{dx^2} \cong \frac{qN_D\phi}{\varepsilon V_{th}} = \frac{\phi}{L_{Dn}^2} \tag{2-2-35}$$

where the characteristic scale of the potential variation,

$$L_{Dn} = \left(\frac{\varepsilon V_{th}}{qN_D}\right)^{1/2} \tag{2-2-36}$$

is called the *Debye radius*. The width of the boundary layer is of the order of a few Debye radii. This is confirmed by numerical calculations of the potential distribution. Similarly, for the p-type region the width of the boundary region is determined by the Debye radius

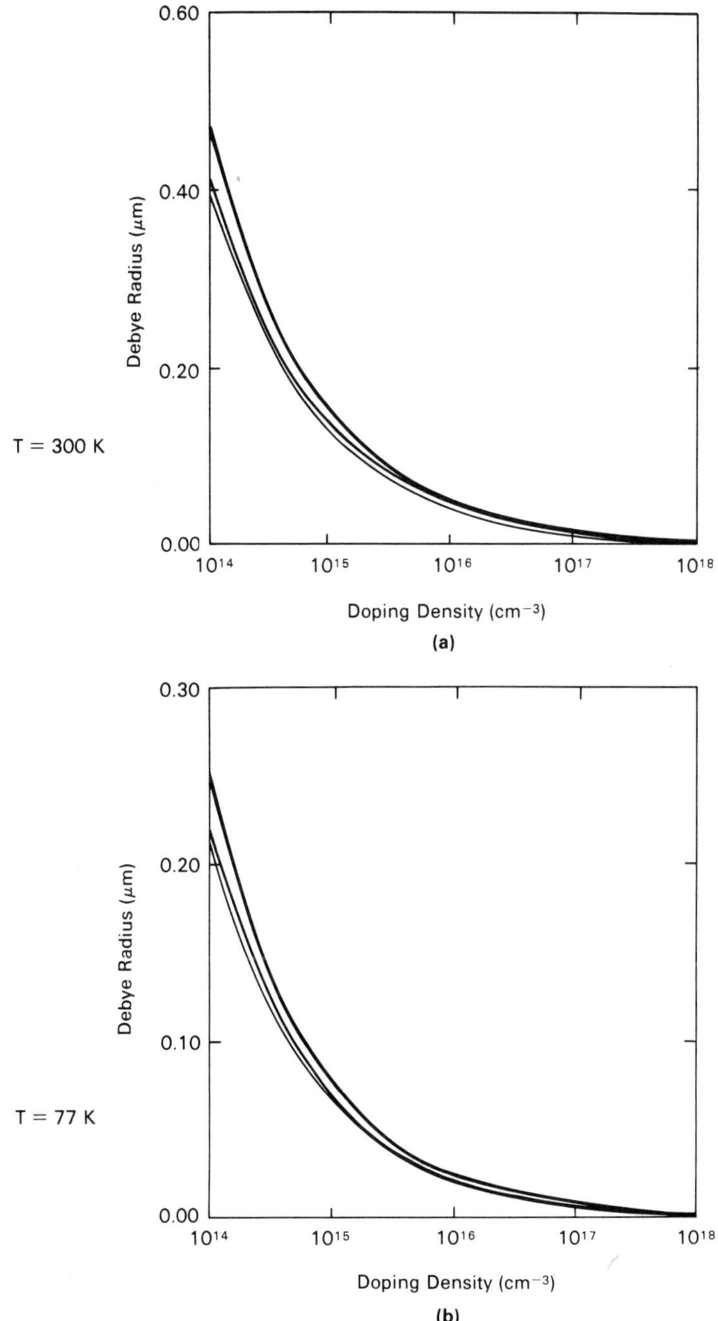

Fig. 2-2-9. Debye radii in GaAs, Si, and Ge and doping density. (a) T = 300K; (b) T = 77K. The curves are calculated using program PLOTF.BAS with subroutines PVBI.BAS (see Program Description Section). The bottom curve is for Si, the middle curve is for GaAs, and the top curve is for Ge.

$$L_{dp} = \left(\frac{\varepsilon V_{th}}{qN_A}\right)^{1/2} \tag{2-2-37}$$

Debye radii for Si, GaAs, and Ge at 77 and 300 K are shown in Fig. 2-2-9 as functions of doping densities. For a one-sided n$^+$-p junction, $x_p >> x_n$ and

$$x_p \cong \left(\frac{2\varepsilon V_{bi}}{qN_A}\right)^{1/2} \tag{2-2-38}$$

(see eq. (2-2-31)). Equation (2-2-38) may also be rewritten as

$$x_p = L_{Dp}\left(\frac{2V_{bi}}{V_{th}}\right)^{1/2} \tag{2-2-39}$$

As $V_{bi} \approx 0.5$ V for Ge, 0.9V for Si, and 1.3 V for GaAs (see Fig. 2-2-3a), $x_p/L_{Dp} \cong 6$ for Ge, 8 for Si, and 10 for GaAs at 300 K. At 77 K these ratios are approximately two times higher. The same ratios are obtained for x_n/L_{Dn} for a p$^+$-n junction.

2-3. CURRENT-VOLTAGE CHARACTERISTICS OF AN IDEAL p-n JUNCTION (THE DIODE EQUATION)

The most resistive portion of a p-n junction is a space charge (depletion) region. When a negative voltage is applied to the p region of a p-n junction (with respect to the n region), the holes and electrons are drawn away from the junction, increasing the width of the depletion region and, hence, the resistance of the device. This situation corresponds to the reverse bias. The voltage of the opposite polarity (forward bias-positive potential applied to the p region) pushes the electrons and holes toward the junction (see Fig. 2-3-1). This decreases the width of the depletion region and the device resistance.

Under the reverse bias and small forward voltages the resistance of the space-charge region is much greater than the resistance of the neutral sections of the device and, as a consequence, almost all applied voltage is dropped across the depletion region where the electric field is very high. At zero bias there is an exact balance between the drift component of the electric current proportional to the electric field and the compensating diffusion current. The total current is zero, and the Fermi level is constant. When external voltage is applied to a p-n junction this balance between the drift and diffusion components of the electric current is violated. However, for small currents, the total current is still much smaller than the drift and diffusion components of the current in the depletion region evaluated separately. This means that the carrier distribution is still close to the equilibrium state, and as a consequence, the electron and hole quasi-Fermi levels remain nearly constant throughout the depletion region (see Fig. 2-3-2). Indeed, numerical simulations of p-n junctions at reverse bias and low forward biases show that the concentrations of electrons and holes in the depletion region are approxi-

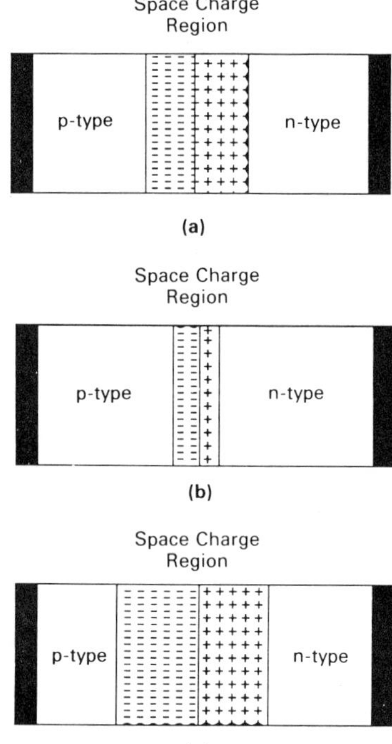

Fig. 2-3-1. Depletion region in *p-n* junction under zero bias (a), forward bias (b), and reverse bias (c).

mately given by

$$n = N_c \cdot \exp\left(\frac{E_{Fn} - E_c}{k_B T}\right) \tag{2-3-1}$$

$$p = N_v \cdot \exp\left(\frac{E_v - E_{Fp}}{k_B T}\right) \tag{2-3-2}$$

where electron and hole quasi-Fermi levels E_{Fn} and E_{Fp} remain practically at the same level as the equilibrium quasi-Fermi levels in the n-type and p-type regions,

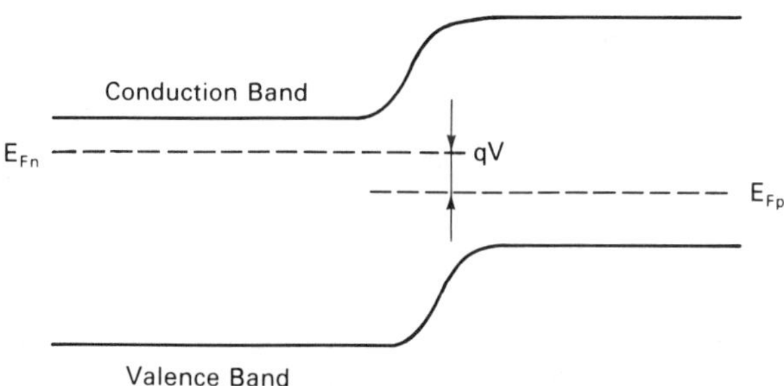

Fig. 2-3-2. Quasi-Fermi levels in *p-n* junction under forward bias.

respectively. The applied forward voltage reduces the potential barrier between the p-type and n-type regions from the built-in voltage, V_{bi}, to $V_{bi} - V$ (assuming that all applied voltage drops across the space charge region with a negligible voltage drops across neutral regions). Hence, the difference between E_{Fn} and E_{Fn} in the depletion region is equal to qV (see Fig. 2-3-2).

$$qV = E_{Fn} - E_{Fp} \tag{2-3-3}$$

From eqs. (2-3-1) to (2-3-3), we obtain that in the depletion region

$$pn = n_i^2 \exp\left(\frac{E_{Fn} - E_{Fp}}{k_B T}\right) \tag{2-3-4}$$

or

$$pn = n_i^2 \exp\left(\frac{qV}{k_B T}\right) \tag{2-3-4a}$$

Let us now consider the boundary between the depletion region and the neutral region at the n-side of the junction. At this boundary $n \cong N_D$. Hence, from eq. (2-3-4a) we obtain

$$p \cong \frac{n_i^2}{N_D} \exp\left(\frac{qV}{k_B T}\right) = p_{no} \exp\left(\frac{qV}{k_B T}\right) \tag{2-3-5}$$

Similarly, at the boundary between the depletion and neutral regions at the p-side of the junction, we find that

$$n \cong \frac{n_i^2}{N_A} \exp\left(\frac{qV}{k_B T}\right) = n_{po} \exp\left(\frac{qV}{k_B T}\right) \tag{2-3-6}$$

Here n_{po} and p_{no} are the concentrations of the minority carriers in the neutral regions at zero applied bias. Equations (2-3-5) and (2-3-6) mean that the concentrations of the minority carriers at the boundaries of the depletion region increase exponentially with the increase in the forward bias (and decrease exponentially with the increase in the reverse bias).

Let us now consider the neutral regions. As was shown in Section 1-11, the concentration of holes in the n-type neutral region can be found by solving the continuity equation

$$D_p \frac{\partial^2 p_n}{\partial x^2} - \mu_p F \frac{\partial p_n}{\partial x} + G - R = 0 \tag{2-3-7}$$

(see eq. (1-11-27)). In the case of a p-n junction the generation rate, G, is zero and the net recombination rate may be expressed as

$$U_R = G - R = -\frac{p_n - p_{no}}{\tau_p} \tag{2-3-8}$$

where τ_p is the ambipolar lifetime (coinciding with the hole lifetime for the low injection level when $p_n << n_n$). The second term in eq. (2-3-7) is small, as the electric field is very small in the neutral region. Hence, we obtain

$$D_p \frac{\partial^2 p_n}{\partial x^2} - \frac{p_n - p_{no}}{\tau_p} = 0 \tag{2-3-9}$$

For a long neutral region ($x_n < x < \infty$), eq. (2-3-9) has to be solved with two boundary conditions:

$$p_n(x_n) = p_{no} \exp\left(\frac{qV}{k_B T}\right) \tag{2-3-10}$$

(see eq. (2-3-5)) and

$$p_n(x \to \infty) = p_{no} \tag{2-3-11}$$

(Here x_n is the boundary between the depletion region and the neutral n-type region.) A general solution of eq. (2-3-9) is given by

$$p_n - p_{no} = A \exp[(x - x_n)/L_p] + B \exp[-(x - x_n)/L_p] \tag{2-3-12}$$

where constants A and B are to be determined from the boundary conditions. Using eqs. (2-3-10) and (2-3-11), we find from eq. (2-3-12) that

$$p_n - p_{no} = p_{no}\left[\left(\exp \frac{qV}{k_B T}\right) - 1\right]\exp[-(x - x_n)/L_p] \tag{2-3-12a}$$

where

$$L_p = (D_p \tau_p)^{1/2} \tag{2-3-13}$$

is the hole diffusion length. The hole current density in the neutral n-type region is primarily the diffusion current density:

$$j_p \cong j_{pD} = -qD_p \frac{\partial p_n}{\partial x} = \frac{qD_p p_{no}}{L_p}\left[\exp\left(\frac{qV}{k_B T}\right) - 1\right] \exp[-(x - x_n)/L_p] \tag{2-3-14}$$

It is a function of distance. The electron diffusion current is given by

$$j_{nD} = qD_n \frac{\partial n_n}{\partial x} \cong qD_n \frac{\partial p_n}{\partial x} = -\frac{qD_n p_{no}}{L_p}\left[\exp\left(\frac{qV}{k_B T}\right) - 1\right] \exp[-(x - x_n)/L_p] \tag{2-3-15}$$

Equation (2-3-15) follows from the quasi-neutrality condition, $n_n - n_{no} \approx p_n - p_{no}$. From eqs. (2-3-14) and (2-3-15) we find that

$$j_{nD} = -\frac{D_n}{D_p} j_{pD} \tag{2-3-16}$$

so that the total diffusion current

$$j_D = j_{pD} + j_{nD} = \frac{qp_{no}}{L_p}(D_p - D_n)\left[\exp\left(\frac{qV}{k_B T}\right) - 1)\right] \exp\left(-\frac{x - x_n}{L_p}\right) \tag{2-3-17}$$

Let us now consider the boundary between the neutral and depletion regions under forward bias. As we approach this boundary, the concentration of the majority carriers increases in order to compensate for the charge of the minority carriers. In the space charge region, however, the concentration of the majority carriers drops, whereas the concentration of the minority carriers continues to increase (compare with Fig. 2-2-2). Clearly, in the depletion region the quasi neutrality is violated. The points where

$$\frac{\partial n_n}{\partial x} = 0 \tag{2-3-18}$$

and

$$\frac{\partial p_p}{\partial x} = 0 \tag{2-3-19}$$

are fairly close to the the boundaries of the depletion region. Equation (2-3-17) is only valid in the neutral n-type region up to these boundaries. Beyond the points determined by eqs. (2-3-18) and (2-3-19) the minority and majority carrier concentrations vary very rapidly with distance.

The hole and electron diffusion current densities (and hence total electron and hole current densities) vary with distance because of the recombination. The characteristic scales of this variation are the hole diffusion length ($L_p = (D_p\tau_p)^{1/2}$) for the n-type region and the electron diffusion length ($L_n = (D_n\tau_n)^{1/2}$) for the p-type region. Typically,

$$L_n >> x_n + x_p \tag{2-3-20}$$

and

$$L_p >> x_n + x_p \tag{2-3-21}$$

In this case relatively little recombination occurs in the depletion region. That is why the ratio of the hole and electron current densities throughout the depletion region remains nearly constant. Therefore, the current density through the p-n junction is given by

$$j = j_p|_{x=x_n} + j_n|_{x=x_p} \tag{2-3-22}$$

Hence, from eq. (2-3-14) and from a similar equation for the electrons in the p-type neutral region we find

$$j = j_s\left[\exp\left(\frac{qV}{k_BT}\right) - 1\right] \tag{2-3-23}$$

where the *diode saturation current density* is

$$j_s = \frac{qD_pp_{no}}{L_p} + \frac{qD_nn_{po}}{L_n} \tag{2-3-24}$$

Equation (2-3-23) is called the *Shockley equation* or the *diode equation*. This equation is in good agreement with the experimental data for germanium diodes at low current densities but does not describe very well the current-voltage characteristics of silicon or gallium arsenide p-n junctions at low forward biases (see Figs. 2-3-3c and 2-3-3d).

In most practical semiconductor diodes the lengths, X_n and X_p, of the n-type and p-type regions are comparable or smaller than the diffusion length of minority carriers. In this case the continuity equation (2-3-9) has to be solved with the

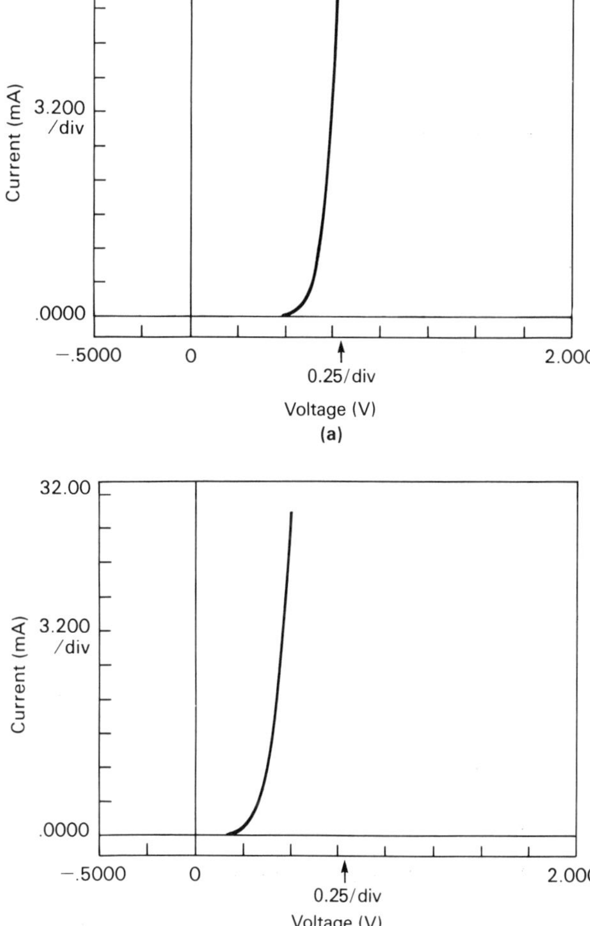

Fig. 2-3-3. Current-voltage characteristics of silicon and germanium *p-n* diodes: (a) I-Vs for silicon *p-n* diode (linear scale), (b) I-Vs for germanium *p-n* diode (linear scale), (c) I-Vs for germanium *p-n* diode (semi-log scale), and (d) I-Vs for silicon *p-n* diode (semi-log scale). The curves were measured using an HP4541 semiconductor parameter analyzer.

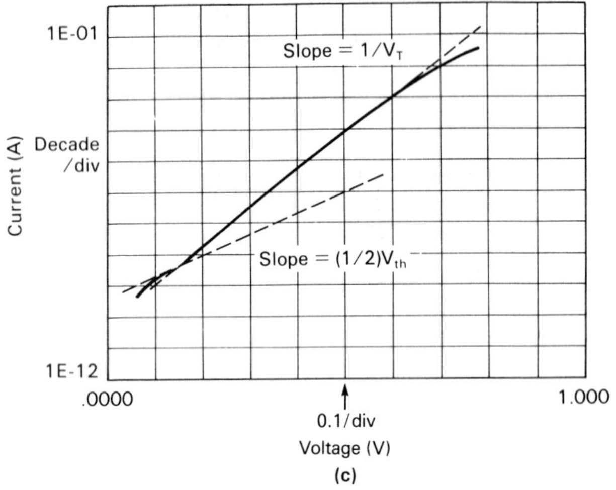

(c)

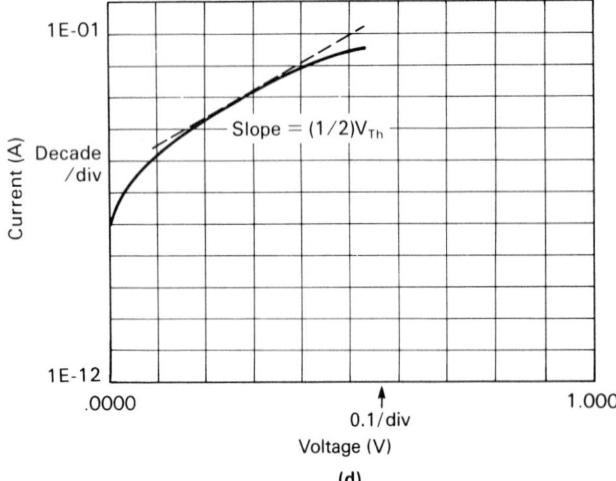

(d)

Fig. 2-3-3. (Cont.)

boundary conditions

$$p_n(x_n) = p_{no} \exp\left(\frac{qV}{k_B T}\right) \tag{2-3-25}$$

$$p_n(x = X_n) = p_{no} \tag{2-3-26}$$

Using eqs. (2-3-25) and (2-3-26), we find from eq. (2-3-12) that

$$p_n - p_{no} = p_{no}\left[\exp\left(\frac{qV}{k_B T}\right) - 1\right](a_1 \exp[(x - x_n)/L_p] + b_1 \exp[-(x - x_n)/L_p]) \tag{2-3-27}$$

where

$$a_1 = -\exp[-(X_n - x_n)/L_p]/\{2\sinh[(X_n - x_n)/L_p]\}$$

$$b_1 = \exp[(X_n - x_n)/L_p]/\{2\sinh[(X_n - x_n)/L_p]\}$$

The current density, j_p, is then given by

$$j_p = -qD_p\partial p/\partial x|_{x=x_n} = (qD_p p_{no}/L_p)[\exp(qV/k_BT) - 1]\coth[(X_n - x_n)/L_p] \quad (2\text{-}3\text{-}28)$$

A similar expression can be derived for the electron current density in the p region at the boundary between the neutral and depletion regions.

It is also interesting to consider the case of a short p^+-n diode such that the width of the n section, X_n, is much smaller than the hole diffusion length, L_p. Such a solution can be found from eq. (2-3-27) using a Taylor series expansion:

$$p_n(x) = p_{no}[\exp(qV/k_BT) - 1]\frac{X_n - x}{X_n - x_n} + p_{no} \quad (2\text{-}3\text{-}27a)$$

It corresponds to the linear distribution of the minority carriers in the n region (see Fig. 2-3-4). The current density, j, is then given by

$$j = -qD_p\partial p/\partial x|_{x=x_n} = qD_p p_{no}[\exp(qV/k_BT) - 1]/(X_n - x_n) \quad (2\text{-}3\text{-}29)$$

This expression is similar to eq. (2-3-23) for a long p^+-n junction, where L_p is substituted by $X_n - x_n$. The problem of a short p^+-n diode is very useful for the consideration of the carrier distributions in a bipolar junction transistor (see Chapter 3).

In order to provide a good agreement with experimental data the theory of a p-n junction should take into account several important nonideal effects, such as

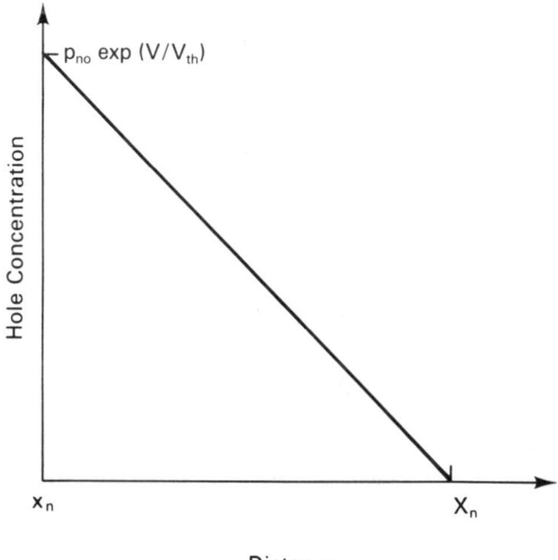

Fig. 2-3-4. Hole concentration in the n-type region of a short p^+-n diode vs. distance.

the recombination and generation of carriers within the depletion region, the voltage drops across the neutral regions (which may become important at relatively large current densities), and the effect of the parasitic series resistances. In addition, the contribution of the surface conduction may be important at low current densities, the effects of the junction breakdown occur at large reverse voltages, and the carrier tunneling drastically changes the diode current-voltage characteristics in devices with high doping densities. Nevertheless, this simple model (called a low-level injection model) gives some insight into the physics of a p-n junction. The important results are illustrated by Figs. 2-3-5 to 2-3-8, where we show carrier concentration, electric field, electric potential, and current profiles for a silicon p-n junction under different bias conditions. For this particular diode, the diffusion lengths of minority carriers, $L_p = 20\ \mu m$ and $L_n = 40\ \mu m$, are much greater than the length of n-type and p-type regions, $X_n = 2\ \mu m$ and $X_p = 2 \mu m$. As a consequence, the distributions of minority carriers at forward bias (V = 0.57 V) are practically linear with distance, as can be seen in Fig. 2-3-5. To the first order, the increase in the majority carrier concentrations under forward

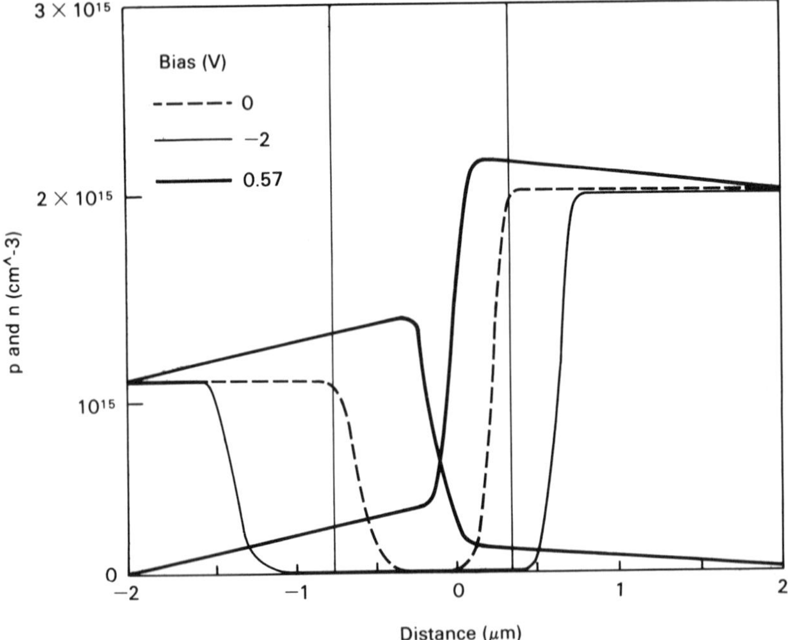

Fig. 2-3-5. Electron and hole concentrations vs. distance for silicon *p-n* junction. Numbers near the curves represent the bias (in volts). The negative sign means reverse bias. The doping level in the *p* section is $2 \times 10^{15}\text{cm}^{-3}$; the doping level in the *n* section is 10^{15}cm^{-3}. The curves are generated using program PLOTF with subroutines PPNJN.BAS (see Program Description Section).

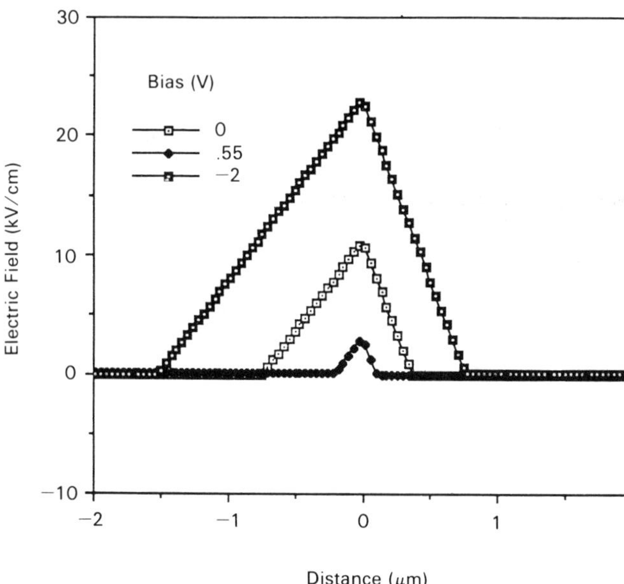

Fig. 2-3-6. Electric field vs. distance for silicon *p-n* junction. Numbers near the curves represent the bias (in Volts). The negative sign means reverse bias. The doping level in the *p* section is $2 \times 10^{15}\text{cm}^{-3}$; the doping level in the *n* section is 10^{15}cm^{-3}. The curves are generated using program PLOTF with subroutines PPNJN.BAS (see Program Description Section).

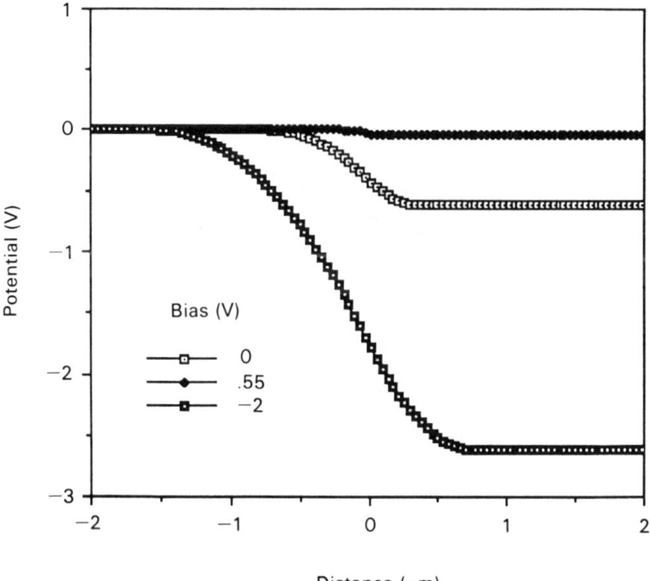

Fig. 2-3-7. Electric potential vs. distance for silicon *p-n* junction. Numbers near the curves represent the bias (in volts). The negative sign means reverse bias. The doping level in the *p* section is $2 \times 10^{15}\text{cm}^{-3}$; the doping level in the *n* section is 10^{15}cm^{-3}. The curves are generated using program PLOTF with subroutines PPNJN.BAS (see Program Description Section).

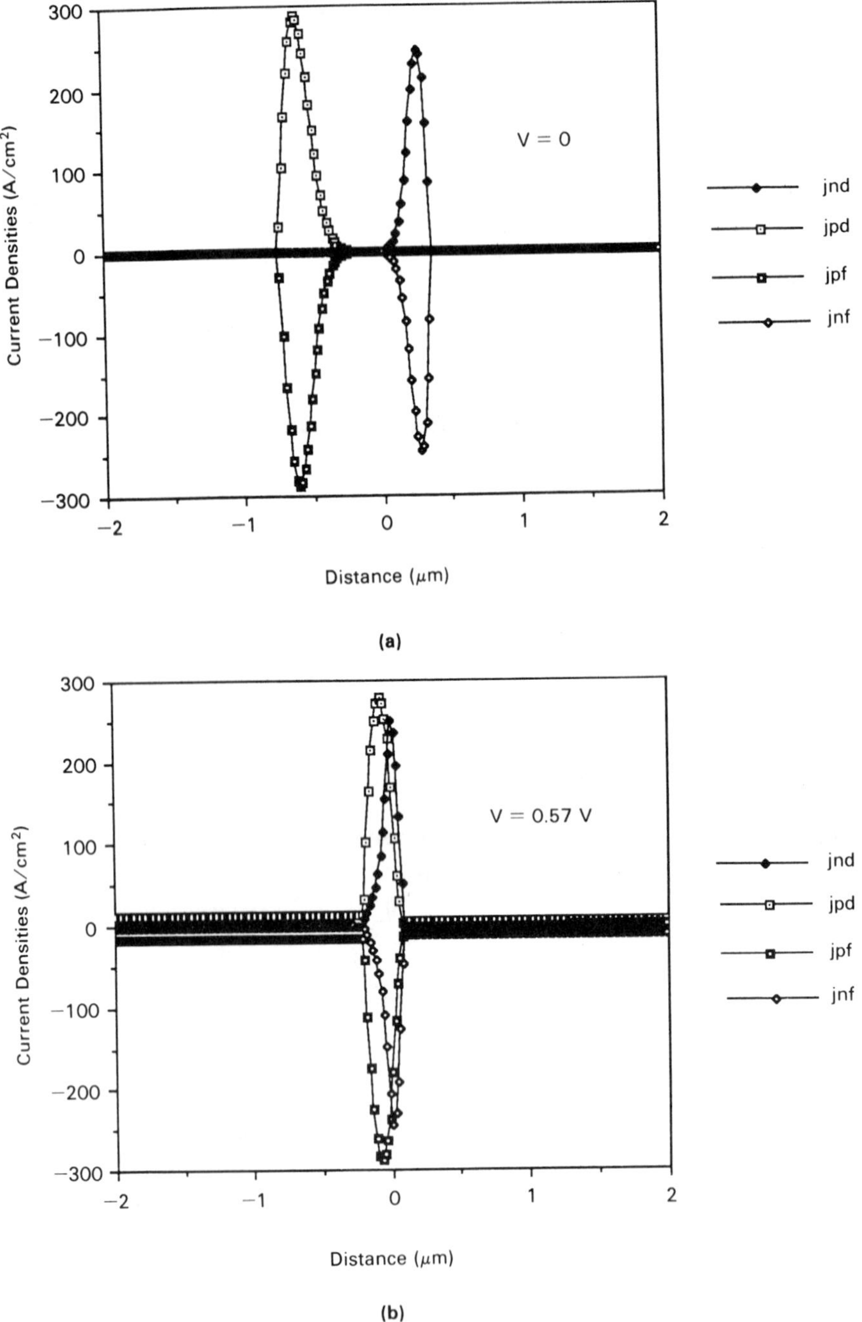

Fig. 2-3-8. Electron and hole drift and diffusion current components vs. distance for silicon *p-n* junction. The doping level in the *p* section is $2 \times 10^{15}\text{cm}^{-3}$; the doping level in the *n* section is 10^{15}cm^{-3}. The curves are generated using program PLOTF with subroutines PPNJN.BAS (see Program Description Section). (a) Bias voltage is 0 V. (b) Bias voltage is 0.57 V.

bias reproduce the profiles of minority carriers, so that quasi-neutrality is approximately preserved outside the depletion region. Also marked in Fig. 2-3-5 is the depletion width under zero bias. As can be clearly seen from Fig. 2-3-5, the depletion width decreases with forward bias and increases with the reverse bias. Figure 2-3-6 shows how the electric field in the space charge region decreases with the forward bias and decreases with the reverse bias as well. As illustrated by Fig. 2-3-7, this is accompanied by a commensurate change in the band bending.

The distributions of diffusion and drift electron and hole current densities are shown in Fig. 2-3-8 for zero bias and forward bias (V= 0.57 V). At zero bias diffusion and drift components of the current in the depletion region exactly cancel each other. We notice that the drift current densities have the largest absolute values at values of the electric field smaller than the maximum field. The reason for that is the rapid decrease in values of the carrier concentrations close to the boundary between p and n regions. Under the forward bias, the depletion region shrinks, and the drift components of the current densities within the depletion region become smaller than the diffusion components. However, as was mentioned in the beginning of this section, the total current is still much smaller than the drift and diffusion components of the current taken separately. The current profiles in the neutral regions are not clearly reproduced in Fig. 2-3-8b because of these differences in scale. However, the sum of all four current components (drift and diffusion components for electrons and holes, respectively) is constant throughout the device and is equal to the total current density.

2-4. GENERATION AND RECOMBINATION CURRENTS

The generation and recombination currents depend on the concentrations, distribution, and energy levels of traps in the depletion region. Traps associated with different impurities are always present in any semiconductor material. The simplest model describing the generation and recombination currents is based on the assumption that there is just one type of traps that dominate recombination and generation processes and are uniformly distributed across the device. In fact, traps may be nonuniformly distributed in a semiconductor device, and more than one type of trap may be involved in recombination and generation processes.

For a single type of uniformly distributed traps the net recombination rate, $R - G_{th}$ (where G_{th} is the thermal generation rate), is given by

$$R - G_{th} = \frac{pn - n_i^2}{\tau_{pl}(n + n_t) + \tau_{nl}(p + p_t)} \tag{2-4-1}$$

(see Section 1-12). Here n_i is the intrinsic concentration and τ_{nl} and τ_{pl} are electron and hole lifetimes:

$$\tau_{nl} = 1/(v_{thn}\sigma_n N_t) \tag{2-4-2}$$

$$\tau_{pl} = 1/(v_{thp}\sigma_p N_t) \tag{2-4-3}$$

where

$$v_{thn} = (3k_BT/m_n)^{1/2} \quad (2\text{-}4\text{-}4)$$

$$v_{thp} = (3k_BT/m_p)^{1/2} \quad (2\text{-}4\text{-}5)$$

are electron and hole thermal velocities, σ_n and σ_p are capture cross sections, N_t is the concentration of the recombination centers,

$$n_t = n_i \cdot \exp\left(\frac{E_t - E_i}{k_BT}\right) \quad (2\text{-}4\text{-}6)$$

$$p_t = n_i \cdot \exp\left(\frac{E_i - E_t}{k_BT}\right) \quad (2\text{-}4\text{-}7)$$

E_t is the recombination level, and E_i is the intrinsic Fermi level. The analysis of eq. (2-4-1) shows that the most effective recombination centers are located close to the middle of the energy gap and, hence, close to E_i. The recombination rate, R, is represented by the term in the right-hand side of eq. (2-4-1), which is proportional to the pn product. The second term in the right-hand side of eq. (2-4-1), which is proportional to n_i^2, represents the thermal generation of electron-hole pairs, G_{th}.

In the neutral n region of a p-n junction,

$$\tau_{pl}n \approx \tau_{pl}N_D \quad (2\text{-}4\text{-}8)$$

is much larger than all other terms in the denominator of eq. (2-4-1). Hence, the net recombination rate in the neutral n region is given by

$$R - G_{th} = (p - n_i^2/N_D)/\tau_{pl} = (p_n - p_{no})/\tau_{pl} \quad (2\text{-}4\text{-}9)$$

Equation (2-4-9) coincides with the expression for the net recombination rate $(R - G)$ used in the continuity equation for holes in the neutral n region (see eq. (2-3-8)).

According to this equation

$$p_n = p_{no} \quad (2\text{-}4\text{-}10)$$

under the equilibrium conditions, as expected for the neutral n region.

In a similar way one can show that in the neutral p region

$$R - G_{th} = (n - n_i^2/N_A)/t_{nl} = (n_p - n_{po})/\tau_{nl} \quad (2\text{-}4\text{-}11)$$

However, within the depletion region the relationships between different terms in the numerator and the denominator of eq. (2-4-1) are completely different and bias dependent, as both p and n are exponentially dependent on the bias voltage. Under the reverse bias in the depletion region p, $n << n_i$, so that

$$R - G_{th} = \frac{-n_i^2}{\tau_{pl}(n + n_t) + \tau_{nl}(p + p_t)} \quad (2\text{-}4\text{-}12)$$

Equation (2-4-12) may be rewritten as

$$R - G_{th} = -n_i/\tau_{gen} \quad (2\text{-}4\text{-}13)$$

where

$$\tau_{gen} = \tau_{pl} \exp[(E_t - E_i)/k_B T] + \tau_{nl} \exp[(E_i - E_t)/k_B T] \qquad (2\text{-}4\text{-}14)$$

(To obtain eq. (2-4-14) we neglected n and p in the denominator of eq. (2-4-12) compared with n_t and p_t.) Equation (2-4-13) describes the net thermal generation of carriers from the recombination centers. There is practically no recombination in the depletion region under reverse bias because this region has very few carriers. This generation process described by eq. (2-4-13) tries to restore the equilibrium value of the pn product (which is n_i^2), but the generated carriers are swept away, leading to the generation current density j_{gen} given by

$$j_{gen} = q \int_0^{x_d} |R - G_{th}| \, dx = q n_i x_d / \tau_{gen} \qquad (2\text{-}4\text{-}15)$$

For an abrupt junction the depletion width, x_d and, hence, j_{gen} are proportional to $(V_{bi} - V)^{1/2}$, where V_{bi} is the built-in voltage and V is the bias voltage (counting forward bias as positive).

The total reverse current density, j_R, is given by

$$j_R = j_s + j_{gen} \qquad (2\text{-}4\text{-}16)$$

where the diffusion component, j_s, is given by eq. (2-3-24), which may be rewritten as

$$j_s = [qD_p/(N_D L_p) + qD_n/(N_A L_n)] n_i^2 \qquad (2\text{-}4\text{-}17)$$

As j_s is proportional to n_i^2 and j_{gen} to n_i, the generation current is dominant when n_i is small enough. Practically, this is always the case for GaAs and for Si at room temperature and below.

Under the forward bias conditions,

$$pn = n_i^2 \exp(qV/k_B T) \qquad (2\text{-}4\text{-}18)$$

and

$$R - G_{th} = \frac{n_i^2[\exp(qV/k_B T) - 1]}{\tau_{pl}(n + n_t) + \tau_{nl}(p + p_t)} \qquad (2\text{-}4\text{-}19)$$

where V is the forward bias voltage. If the lifetimes τ_{pl} and τ_{nl} are assumed to be equal,

$$\tau_{pl} = \tau_{nl} = \tau_{rec} = 1/(\sigma v_{th} N_t)$$

(see eqs. (2-4-2) and (2-4-3)) and if we consider the trap level, E_t, to be coinciding with the intrinsic Fermi level, E_i, eq. (2-4-19) may be simplified:

$$R - G_{th} = \frac{n_i^2 \exp(qV/k_B T)}{\tau_{rec}\,(n + p + 2n_i)} \qquad (2\text{-}4\text{-}20)$$

Here we also assumed that the forward bias is sufficiently large, so that $\exp(qV/k_B T) >> 1$. Figure 2-4-1 shows how electric field, potential, and electron and hole concentrations vary with distance (for a symmetrical junction where

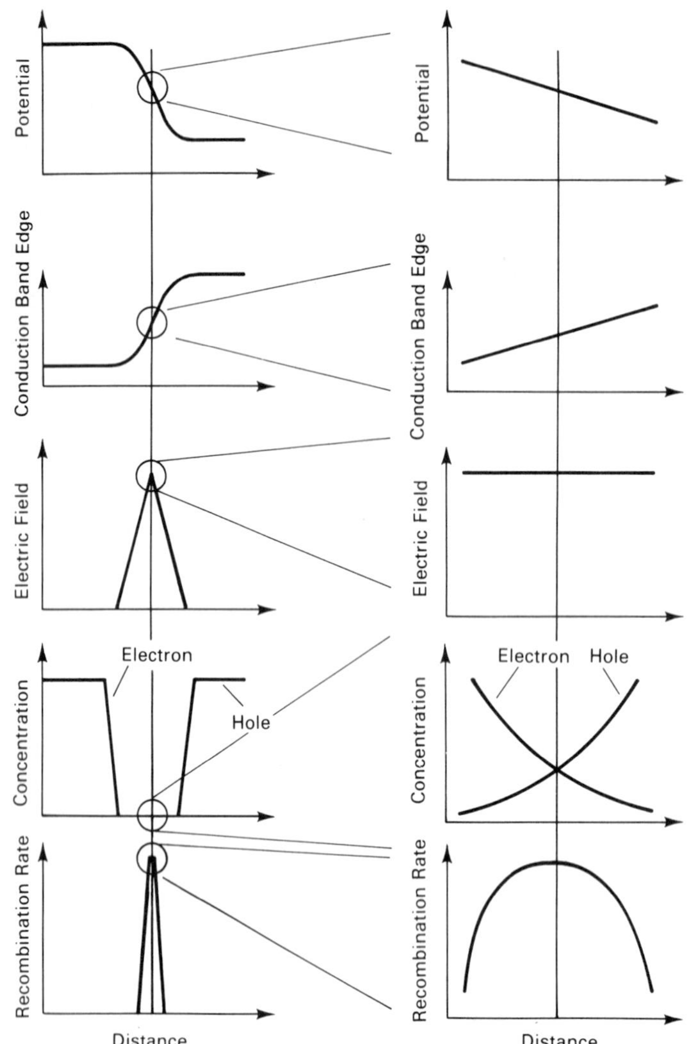

Fig. 2-4-1. Electric field, potential, and electron and hole concentrations in the depletion region of a symmetrical junction ($N_D = N_A$) vs. distance. The right half of the figure shows blown-up profiles in the vicinity of the point where $p = n$.

$N_{\mathrm{D}} = N_{\mathrm{A}}$). The right half of this figure shows these profiles expanded in the vicinity of the point where $p = n$. As can be seen from Fig. 2-4-1, n is a rapidly decreasing function of distance and p is a rapidly increasing function of distance. Hence, function $U_{\mathrm{R}} = R - G_{\mathrm{th}}$ is small close to the boundary between the depletion region and the neutral section of the n-type region, where n is large. Function U_{R} is also small close to the boundary between the depletion region and the neutral section of the p-type region, where p is large. This function reaches the maximum value at the point in the space charge region where $n = p$. Indeed, eq. (2-4-20) may be re-written as

$$U_R = \frac{n_i^2 \exp(qV/k_BT)}{\tau_{rec}\{N_D \exp(-E_c/k_BT) + (n_i^2/N_D)\exp[(E_c + qV)/k_BT] + 2n_i\}} \tag{2-4-21}$$

where $E_c(x)$ is the conduction band edge in the depletion region and we choose $E_c = 0$ in the n-type region far from the junction. The maximum value of U_R is reached at the point where $dU_R/dE_c = 0$. From this condition we find that the maximum value of U_R is reached at the point $x = x_o$, where $n = p$, as expected. The pn product in the depletion region is given by

$$np = n_i^2 \exp[qV/(k_BT)] \tag{2-4-22}$$

(see eq. (2-3-4a)). Hence at $x = x_o$

$$n = p = n_i \exp[qV/(2k_BT)] \tag{2-4-23}$$

Substituting n and p from eq. (2-4-23) into eq. (2-4-20) and neglecting the term $2n_i$ in the denominator of the right-hand-side of this equation, we obtain

$$U_{Rmax} \approx n_i \exp[qV/(2k_BT)]/(2\tau_{rec}) \tag{2-4-24}$$

The recombination current density is given by

$$j_{rec} = q\int_0^{x_d} U_R\, dx \tag{2-4-25}$$

The largest contribution to the integral in eq. (2-4-25) comes from the region near the point x_o, where $n = p$. Hence, this integral can be estimated by expanding U_R into the Taylor series in the vicinity of the point x_o and assuming that

$$\begin{aligned} U_R &\approx U_{Rmax} + 0.5\, U''_{Ro}(x - x_o)^2 \quad \text{for} \quad x_o - \Delta x \le x \le x_o + \Delta x \\ U_R &\approx 0 \qquad\qquad\qquad\qquad\qquad\; \text{for} \quad x_o - \Delta x < x \text{ and } x > x_o + \Delta x \end{aligned} \tag{2-4-26}$$

where U''_{Ro} is the second derivative of the recombination rate, U_R, with respect to x for $x = x_o$ ($U''_{Ro} < 0$), $\Delta x = (2U_{Rmax}/|U''_{Ro}|)^{1/2}$. Substituting eq. (2-4-26) into eq. (2-4-25) and evaluating the integral, we obtain

$$j_{rec} \approx q[2\pi(U_{Rmax})^3/|U''_{Ro}|]^{1/2} \tag{2-4-27}$$

(see Zeldovich and Myshkis 1967, p. 84). Here we consider the case when $N_D < N_A$. If the $N_D < N_A$, point x_o, where $n = p$, lies in the n-type region of the diode.† Using eq. (2-4-22) and the equation relating n to E_c (see eq. (2-2-10)), we find that the value of $E_{co} = E_c(x_o)$ is given by

$$E_{co} = -qV/2 + (k_BT/q)\ln(N_D/n_i) \tag{2-4-28}$$

† If $N_D > N_A$, point x_o, where $n = p$, lies in the p-type region of the diode. However, the results of our derivation will remain quite similar (see Problem 2-4-4).

(see Shur 1988). This equation can be rewritten as

$$E_{co} = q(V_{bin} - V/2) \tag{2-4-29}$$

where

$$V_{bin} = (k_B T/q)\ln(N_D/n_i) \tag{2-4-30}$$

is the built-in voltage drop in the depletion region of the n-type section of the diode. The dependence of E_c on x is given by

$$E_c = q^2 N_D x^2/(2\varepsilon) \tag{2-4-31}$$

where x is counted from the boundary between the neutral and depletion regions in the n-type section of the diode (see eqs. (2-2-27) and (2-2-8)). Hence,

$$x_o = [2\varepsilon E_{co}/(q^2 N_D)]^{1/2} \tag{2-4-32}$$

or

$$x_o = [2\varepsilon(V_{bin} - V/2)/(qN_D)]^{1/2} \tag{2-4-33}$$

The expression for U_R in the vicinity of point $x = x_o$ can be written as

$$U_R = \frac{2U_{Rmax}}{\exp(-u) + \exp(u)} \tag{2-4-34}$$

where $u = (E_c - E_{co})/k_B T$. The second derivative, U''_{Ro}, is given by

$$U''_{Ro} = -\frac{U_{Rmax}\,(dE_c/dx)^2|_{x=x_o}}{(k_B T)^2} \tag{2-4-35}$$

where

$$(dE_c/dx)|_{x=x_o} = q^2 N_D x_o/\varepsilon \tag{2-4-36}$$

(see eq. (2-2-8) and eq.(2-2-27)). Substituting eq. (2-4-33) into eq. (2-4-36) we find that

$$(dE_c/dx)|_{x=x_o} = [2q^3 N_D(V_{bin} - V/2)/\varepsilon]^{1/2} \tag{2-4-37}$$

and from eqs. (2-4-27), (2-4-24), (2-4-35), and (2-4-37) we obtain

$$j_{rec} = j_{recs}\exp[qV/(2k_B T)] \tag{2-4-38}$$

where

$$j_{recs} = q(\pi/2)^{1/2}(k_B T/q)n_i/(\tau_{rec}F_{np}) \tag{2-4-39}$$

and F_{np} is the electric field at $x = x_o$:

$$F_{np} = [qN_D(2V_{bin} - V)/\varepsilon]^{1/2} \tag{2-4-40}$$

For p-n diodes with $N_D > N_A$ eq. (2-4-40) should be replaced by

$$F_{np} = [qN_A(2V_{bip} - V)/\varepsilon]^{1/2} \tag{2-4-41}$$

where

$$V_{\text{bip}} = (k_B T/q)\ln(N_A/n_i) \tag{2-4-42}$$

For a symmetrical junction (where $N_D = N_A$) $2V_{\text{bin}} = V_{\text{bi}}$ and eq. (2-4-39) reduces to

$$j_{\text{recs}} = q(\pi/2)^{1/2}\,(k_B T/q)n_i/(\tau_{\text{rec}}F_{\text{max}}) \tag{2-4-39a}$$

This equation coincides within a factor of $(\pi/2)^{1/2} \approx 1.25$ with the approximate equation derived for the recombination current in a symmetrical junction by van der Ziel (1976). However, eq. (2-4-39) is different from equations used in many textbooks (see, for example, Grove 1967), where it is assumed that the recombination current may be estimated by multiplying U_{Rmax} by the total width of the depletion region, x_d. Such an approach overestimates j_{recs} by a large factor that is equal to $2^{-1/2}(V_{\text{bi}} - V)/V_{\text{th}}$ for a symmetrical junction (where $N_D = N_A$).

In the foregoing derivation we made an assumption that $E_t = E_i$ and that $\tau_{nl} = \tau_{pl} = \tau_{\text{rec}}$. These assumptions may not be valid in practical devices, and as a consequence, exponential dependence of the recombination current on the bias voltage may be different from the dependence predicted by eq. (2-4-38). The following empirical expression may be used instead:

$$j_{\text{rec}} = j_{\text{recs}} \exp[qV/(m_r k_B T)] \tag{2-4-38a}$$

where factor M_r may differ from 2. The total forward current density may then be found as

$$j_F = j_s \exp(qV/k_B T) + j_{\text{recs}} \exp(qV/m_r k_B T) \tag{2-4-42}$$

However, it is frequently more convenient to use an empirical formula,

$$j_F = j_{\text{seff}} \exp(qV/n_{\text{id}} k_B T) \tag{2-4-43}$$

where n_{id} is called an *ideality factor*. The deviation of n_{id} from 1 may be considered an important measure of the recombination current.

As j_s is proportional to n_i^2 and j_{recs} is proportional to n_i, the recombination current is dominant when n_i is small enough. Practically, this is always the case for relatively small applied voltages for GaAs and for Si at room temperature or at lower temperatures. At large voltages the diffusion component is dominant because it is proportional to $\exp(qV/k_B T)$, whereas the recombination current density increases with the forward bias only as $\exp[qV/(m_r k_B T)]$, where $m_r \approx 2$ (see Fig. 2-3-3).

2-5. DEPLETION CAPACITANCE

At reverse bias, zero bias, or small forward bias the capacitance of a p-n diode is primarily determined by the capacitance of the depletion layer:

$$C_d = \varepsilon S/x_d \tag{2-5-1}$$

where ε is the dielectric permittivity, S is the device cross section, and

$$x_d = x_n + x_p \tag{2-5-2}$$

is the width of the depletion layer. At zero bias x_n and x_p are given by eqs. (2-2-32) and (2-2-33). When an external bias V is applied these equations become

$$x_n = \left[\frac{2\varepsilon(V_{bi} - V)}{qN_D(1 + N_D/N_A)}\right]^{1/2} \tag{2-5-3}$$

$$x_p = \left[\frac{2\varepsilon(V_{bi} - V)}{qN_A(1 + N_A/N_D)}\right]^{1/2} \tag{2-5-4}$$

Here N_A and N_D are donor and acceptor densities, and V_{bi} is the built-in voltage (see eq. (2-2-12)). Substituting eqs. (2-5-3) and (2-5-4) into eq. (2-5-2) and the result into eq. (2-5-1), we obtain

$$C_d = \varepsilon S/x_d = S\left[\frac{q\varepsilon N_{eff}}{2(V_{bi} - V)}\right]^{1/2} \tag{2-5-5}$$

where

$$N_{eff} = \{[N_D(1 + N_D/N_A)]^{1/2} + [N_A(1 + N_A/N_D)]^{1/2}\}^2 \tag{2-5-6}$$

As can be seen from eqs. (2-5-1) to (2-5-5), the depletion capacitance C_d is inversely proportional to $(V_{bi} - V)^{1/2}$ in the case of the abrupt junction that was considered previously. For an arbitrary doping profile the depletion capacitance may be found as

$$C_d = dQ_d/dV \tag{2-5-7}$$

where Q_d is the charge in the depletion layer in the n-type section of the device. In particular, for the linearly graded profile shown in Fig. 2-5-1 we find that

$$C_d = \varepsilon S/x_d = S\left[\frac{qa\varepsilon^2}{12(V_{bi} - V)}\right]^{1/3} \tag{2-5-8}$$

(see Problem 2-5-2). The built-in voltage for the linearly graded junction is found by substituting the values of the doping concentration at $x = \pm x_d/2$ into eq. (2-2-12):

$$V_{bi} = 2k_BT/q \ln(ax_d/2n_i) \tag{2-5-9}$$

The values of the depletion capacitance per unit area for an abrupt p^+-n silicon junction are shown in Fig. 2-5-2.

Capacitance-voltage characteristics are used to determine a doping profile. Let us consider a p^+-n diode with an arbitrary doping of the n-type section as an example. The electric field in the n-type section is described by Poisson's equation:

$$dF/dx = qN_D(x)/\varepsilon \tag{2-5-10}$$

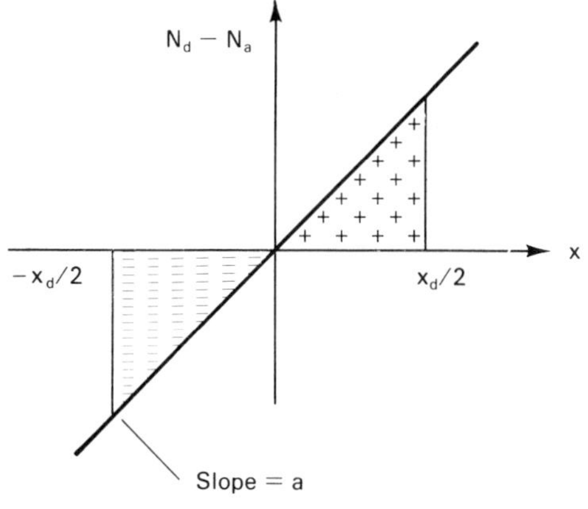

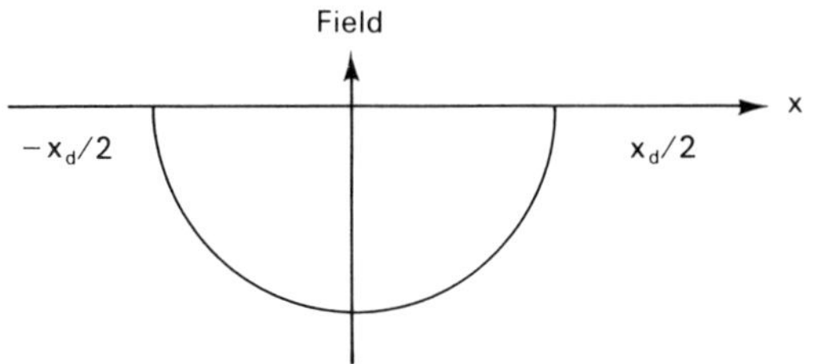

Fig. 2-5-1. Doping and electric field distributions for a linearly graded junction.

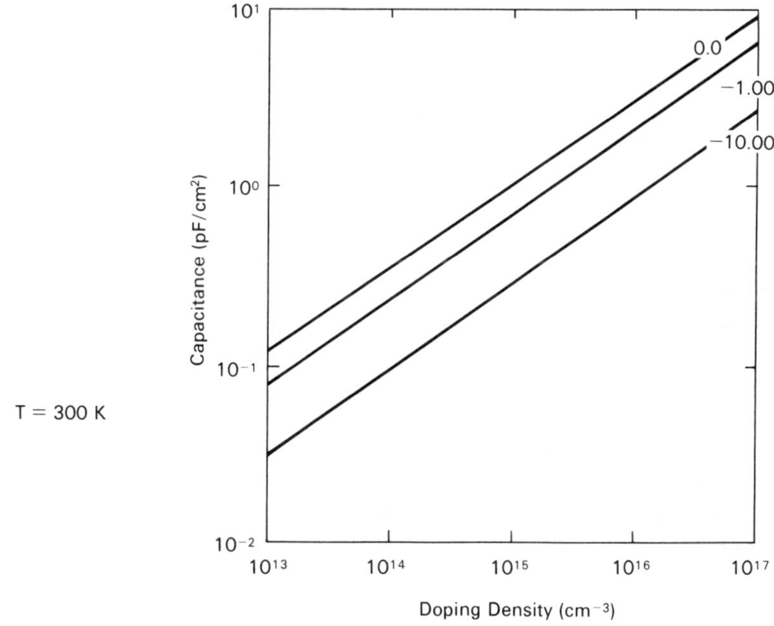

Fig. 2-5-2. Depletion capacitance vs. doping for a p^+-n Si diode. Numbers near the curves represent the bias (in volts). The negative sign means reverse bias. The doping level in the p^+ section is $2 \times 10^{18}\text{cm}^{-3}$. The curves are generated using program PLOTF with subroutines PPNCD.BAS (see Program Description Section).

Multiplying eq. (2-5-10) by x and integrating by parts yields

$$V_{bi} - V = (q/\varepsilon)\int_0^{x_d} N_D(x)x\,dx \tag{2-5-11}$$

On the other hand, the charge in the depletion region of the n-type section is given by

$$Q_d = q\int_0^{x_d} N_D(x)\,dx \tag{2-5-12}$$

Differentiating eqs. (2-5-11) and (2-5-12) with respect to x_d we find that

$$dV/dx_d = -(q/\varepsilon)N_D(x_d)x_d \tag{2-5-13}$$

$$dQ_d/dx_d = qN_D(x_d) \tag{2-5-14}$$

Hence,

$$C_d = S(dQ_d/dx_d)/(d|V|/dx_d) = \varepsilon S/x_d \tag{2-5-15}$$

This is the same expression that we used for the abrupt junction. As the foregoing derivation shows, it is applicable for an arbitrary doping profile. Differentiating eqs. (2-5-15) with respect to V and using eq. (2-5-13), we obtain

$$N_D(x_d) = C_d^3/[q\varepsilon S^2(dC_d/dV)] \tag{2-5-16}$$

or

$$N_D(x_d) = 2/[q\varepsilon S^2 d(1/C_d^2)/dV] \tag{2-5-17}$$

Equation (2-5-17) may be used to deduce the doping profile from the measured capacitance-voltage characteristics at reverse and small forward biases.

2-6. DIFFUSION CAPACITANCE AND EQUIVALENT CIRCUIT OF A p–n DIODE

Besides the depletion capacitance considered in Section 2-5, a capacitance related to the charge of the minority carriers—*diffusion capacitance*—becomes important or even dominant under forward bias conditions. In order to find this capacitance we will calculate the device impedance using the continuity equations for the minority carriers derived in Section 2-3. These equations have to be modified to include $\partial p/\partial t$ and $\partial n/\partial t$ terms so that eq. (2-3-9) becomes

$$D_p\frac{\partial^2 p_n}{\partial x^2} - \frac{p_n - p_{no}}{\tau_p} = \frac{\partial p_n}{\partial t} \tag{2-6-1}$$

We assume that the applied voltage and current density are given by

$$V(t) = V_o + V_1\exp(i\omega t) \tag{2-6-2}$$

$$j(t) = j_o + j_1\exp(i\omega t) \tag{2-6-3}$$

where $V_1 << V_o$ and $j_1 << j_o$, and seek the solution of eq. (2-6-1) as

$$p_n(x, t) = p_{ns}(x) + p_{n1}(x)\exp(i\omega t) \tag{2-6-4}$$

where $p_{ns}(x)$ is the solution for the steady state continuity equation (2-3-17) and the second term in the right-hand side represents the time dependent component. Substituting eq. (2-6-4) into eq. (2-6-1) and taking into account eq. (2-3-9), we obtain

$$D_p \frac{\partial^2 p_{n1}}{\partial x^2} - \frac{p_{n1}}{\tau_p} = i\omega p_{n1} \tag{2-6-5}$$

The boundary conditions for eq. (2-6-5) are

$$p_{n1}\ (x \to \infty) = 0 \tag{2-6-6}$$

$$p_{n1}\ (x = 0) = (p_{no} q V_1 / k_B T)\exp(q V_o / k_B T) \tag{2-6-7}$$

where $x = 0$ corresponds to the boundary of the depletion region. (Equation (2-6-7) is obtained by substituting eq. (2-6-2) into eq. (2-3-10) to yield

$$p_n\ (x = 0) = p_{no}\exp[q(V_o + V_1)/k_B T] \tag{2-6-8}$$

and expanding the exponent in the right-hand side of eq. (2-6-8) into the Taylor series.)

Equation (2-6-5) coincides with continuity equation (2-3-9) for $p_n - p_{no}$ if the lifetime τ_p is replaced by $\tau_p/(1 + i\omega\, \tau_p)$. Hence, using the relationship between p_n and the hole component of the diode saturation current, we find the component of the small-signal hole current density j_{1p}:

$$j_{1p} = [qD_p p_{no}(1 + i\omega\tau_p)^{1/2}/L_p]\exp[(V_o + V_1)/V_{th}] \tag{2-6-9}$$
$$\sim (V_1/V_{th})[qD_p p_{no}(1 + i\omega\tau_p)^{1/2}/L_p]\exp(V_o/V_{th})$$

A similar expression may be obtained for the electron component of j_{1n} so that

$$j_1 = j_{1p} + j_{1n} = \tag{2-6-10}$$
$$(V_1/V_{th})q[D_p p_{no}(1 + i\omega\tau_p)^{1/2}/L_p + D_n n_{po}(1 + i\omega\tau_n)^{1/2}/L_n]\exp(V_o/V_{th})$$

The small-signal admittance, Y, defined as

$$Y = j_1 S/V_1 \tag{2-6-11}$$

may be found from eq. (2-6-10). At relatively low frequencies when $\omega\tau_p << 1$ and $\omega\tau_n << 1$, we obtain from eqs. (2-6-10) and (2-6-11)

$$Y = G_d + i\omega C_{dif} \tag{2-6-12}$$

where

$$G_d = 1/R_d = j_o S/V_{th} \tag{2-6-13}$$

and

$$C_{dif} = \left(\frac{qL_p p_{no}}{2} + \frac{qL_n n_{po}}{2}\right) S\exp(V_o/V_{th})/V_{th} \tag{2-6-14}$$

For a p^+-n diode $p_{no} >> n_{po}$ and eq. (2-6-14) may be simplified to

$$C_{dif} = j_o\tau_p S/(2V_{th}) = G_d\tau_p/2 \tag{2-6-15}$$

This equation shows that a characteristic time constant, $C_{dif}R_d$ of a forward biased p-n junction is of the order of the recombination time, as may be expected.

The equivalent circuit of a p-n junction is shown in Fig. 2-6-1. In addition to the depletion and diffusion capacitances, C_d and C_{dif}, and the differential resistance of the p-n junction, R_d, it includes a series resistance, R_s (consisting of the resistance of the neutral semiconductor regions between the depletion region and the contacts and contact resistances), as well as a parasitic inductance, L_s, and the geometric capacitance of the sample

$$C_{geom} = \varepsilon S/L \tag{2-6-16}$$

where L is the sample length. Calculated voltage dependencies of the differential resistance and capacitance for a Si p-n junction are shown in Fig. 2-6-2.

The dependencies shown in Fig. 2-6-2 are calculated for the limiting case $\omega \to 0$. Practical measurements of a small-signal junction capacitance are usually done at a finite frequency (typically 2 MHz or so). The results of such a measurement can be dramatically different from the curves shown in Fig. 2-6-2. The reason for that is the effect of the parasitic inductance, L_s (see Fig. 2-6-1). At large forward bias, the diffusion capacitance, C_{dif}, is much greater than the depletion capacitance, C_d, and geometric capacitance, C_{geom}. Hence, to first order, the measured impedance is given by

$$Z_{meas} \approx R_s + R_d/(1 + i\omega C_{dif}R_d) + i\omega L_s \tag{2-6-17}$$

When differential resistance, R_d, becomes small enough at large forward bias, the second term in the right-hand side of eq. (2-1-17) becomes small, and reactance actually becomes inductive. A typical shape of measured small-signal capacitance curves is shown in Fig. 2-6-3. (Negative values of the capacitance in the figure correspond to the inductive response.) This, of course, introduces an additional complication, especially when one needs to estimate an equivalent capacitance, C_{eq}, of a p-n junction under forward bias for large signal applications.

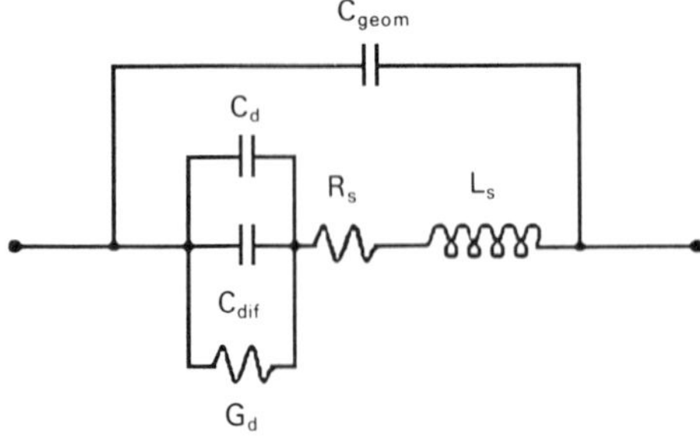

Fig. 2-6-1. Equivalent circuit of a *p-n* junction.

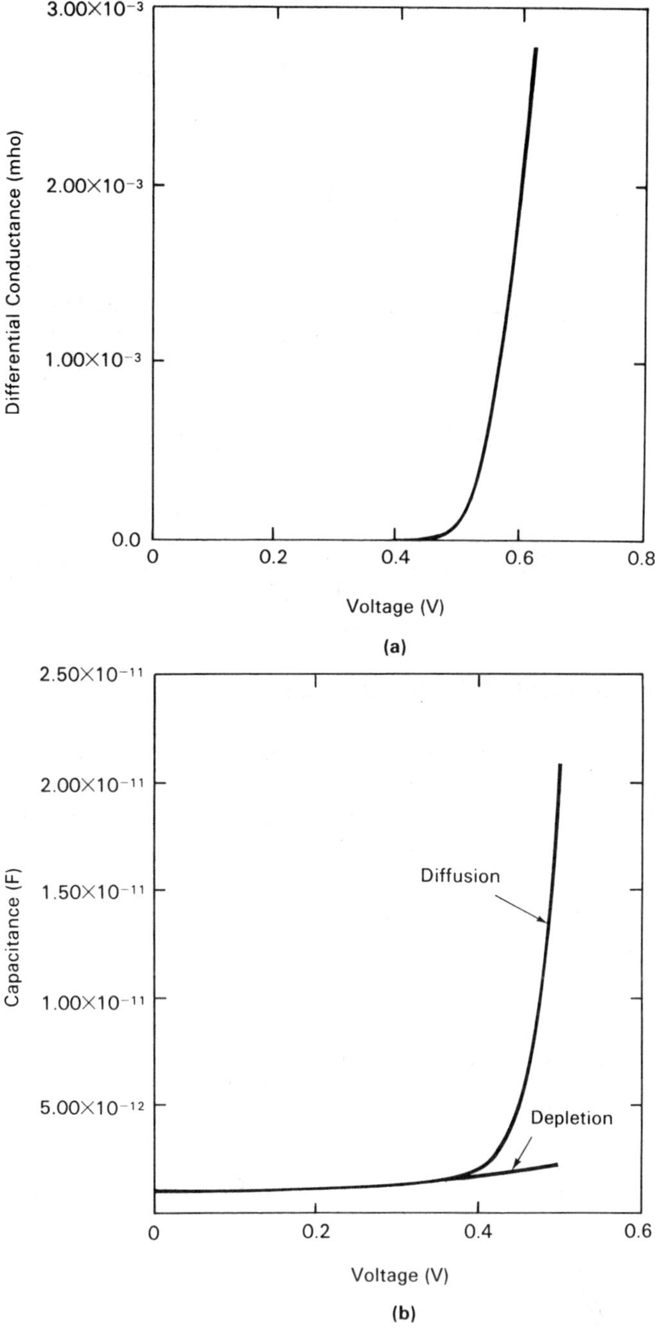

Fig. 2-6-2. (a) Voltage dependencies of the differential conductance and (b) capacitance of a Si *p-n* junction at zero frequency. The curves were generated using program PLOTF.BAS with subroutines PPEQJN.BAS (see the listing in the Program Description Section).

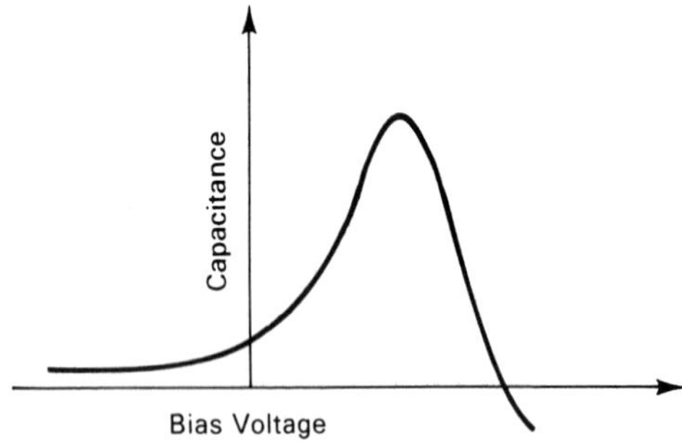

Fig. 2-6-3. Typical shape of small signal capacitance curves measured at finite frequency (usually 2 MHz).

Frequently, the following (purely empirical) estimate is used under such conditions:

$$C_{\text{eq}} \approx 4C_{\text{d}}(0) \tag{2-6-18}$$

You can probably use eq. (2-6-18) as a rule of thumb, keeping in mind that this is just a crude order-of-magnitude estimate.

2-7. TUNNELING AND TUNNEL DIODES

As we established in Section 2-3, the current through a p-n diode is expected to rise exponentially with the forward bias. However, if both the p and n regions are so heavily doped that they become degenerate, the current-voltage characteristic changes dramatically and actually has a region of a negative differential resistance (see Fig. 2-7-1). Such a p-n diode, with degenerate p and n regions is called a *tunnel diode*, and this behavior is caused by the quantum mechanical phenomenon of *tunneling*.

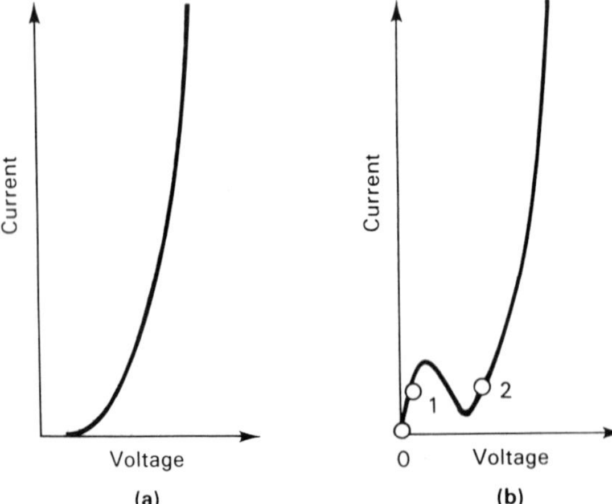

Fig. 2-7-1. (a) Current-voltage characteristics of conventional *p-n* diode and (b) tunnel diode. Points 0, 1, and 2 shown in Fig. 2-7-1b correspond to different conditions for tunneling discussed below.

In order to understand this phenomenon, let us consider a thought experiment of putting an electron with given kinetic energy and momentum into a potential box, as shown in Fig. 2-7-2a. In other words, we assume that an uncertainty in the momentum value, Δp, is zero. Let us assume that the electron kinetic energy is smaller than the barrier height. Hence, according to classical physics and, perhaps, common sense, the electron must remain in the box. However, this would mean that the uncertainty in the electron position, Δx, is equal to the size of the box, d, and, hence, that $\Delta p \Delta x = 0$, in violation of the Heizenberg uncertainty principle,

$$\Delta p \Delta x > \hbar/2 \tag{2-7-1}$$

(see Section 1-2). Therefore, there must be a finite probability of finding the electron outside the potential box. In other words, the electron may tunnel through the potential barriers bounding the box. The insert in Fig. 2-7-2b shows a similar achievement for a fish in a fishbowl. It is very unlikely to happen, however, because tunneling is a quantum phenomena. For tunneling to occur, the de

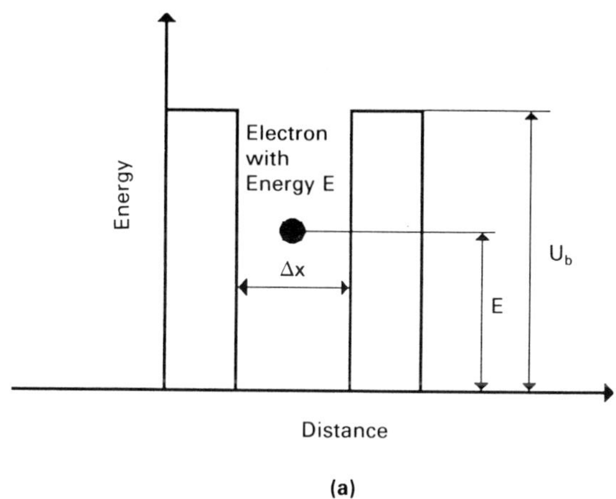

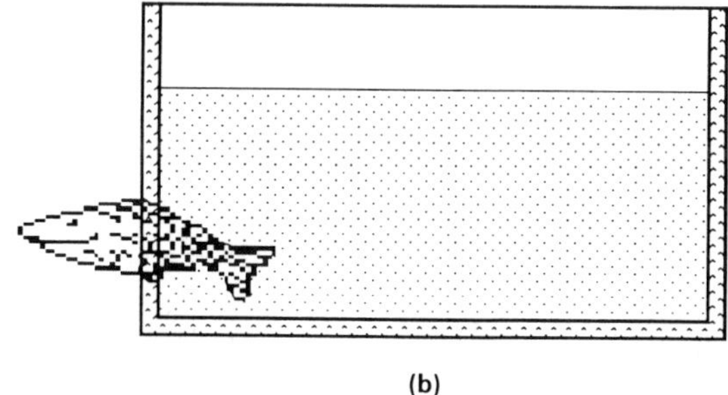

Fig. 2-7-2. Concept of tunneling: (a) Electron in potential box formed by two potential barriers and (b) Fish trying to tunnel out of a fishbowl.

Broglie wavelength of the object should be comparable with the width of the potential barrier (see Section 1-2 for a brief discussion of the de Broglie wavelength).

The quantitative analysis of tunneling should be based on the solution of the Schrödinger equation,

$$-\frac{\hbar^2\nabla^2 \psi}{2m_e} + U(\mathbf{r})\psi = E\psi \tag{2-7-2}$$

(see eq. (1-2-7)). We will first solve the Schrödinger equation for the simplest potential barrier shown in Fig. 2-7-2 and then apply the results to describe tunneling through potential barriers of arbitrary shapes. For the simplest potential barrier shown in Fig. 2-7-3a we can rewrite eq. (2-7-2) as follows:

$$\begin{aligned} d^2\psi/dx^2 + k^2\psi &= 0 \quad \text{for } x < 0 \\ d^2\psi/dx^2 + k_b^2\psi &= 0 \quad \text{for } 0 \le x \le d_b \\ d^2\psi/dx^2 + k^2\psi &= 0 \quad \text{for x} > d_b \end{aligned} \tag{2-7-3}$$

Here

$$k^2 = 2m_e E/\hbar^2 \tag{2-7-4}$$

$$k_b^2 = n_b^2 k^2 \tag{2-7-5}$$

$$n_b^2 = (E - U_b)/E \tag{2-7-6}$$

The solution of the Schrödinger equation (eq. (2-7-3)) in each region is given by

$$\psi = A_l \exp(ikx) + B_l \exp(-ikx) \quad \text{for } x < 0 \tag{2-7-7}$$

$$\psi = A_b \exp(ikx) + B_b \exp(-ikx) \quad \text{for } 0 \le x \le d_b \tag{2-7-8}$$

$$\psi = A_r \exp(ikx) + B_r \exp(-ikx) \quad \text{for } x > d_b \tag{2-7-9}$$

as was discussed in Section 1-2 (see eq. (1-2-11)). Six constants, A_l, B_l, A_d, B_d, A_r, and B_r have to be determined from the four requirements of the continuity of the wave function and its first derivative at the region boundaries,

$$\begin{aligned} \psi(-0) &= \psi(+0) \\ d\psi/dx(-0) &= d\psi/dx(+0) \\ \psi(d_d - 0) &= \psi(d_d + 0) \\ d\psi/dx(d_d - 0) &= d\psi/dx(d_d + 0) \end{aligned} \tag{2-7-10}$$

and two other conditions discussed shortly.

Let us consider an electron propagation from the left to the right. Then the term $A_l \exp(ikx)$ in eq. (2-7-9) describes the incoming wave function, and the term $B_l \exp(ikx)$ describes the reflected wave function. The term $A_r \exp(ikx)$ describes the transmitted wave function. Because we have limited ourselves to the case

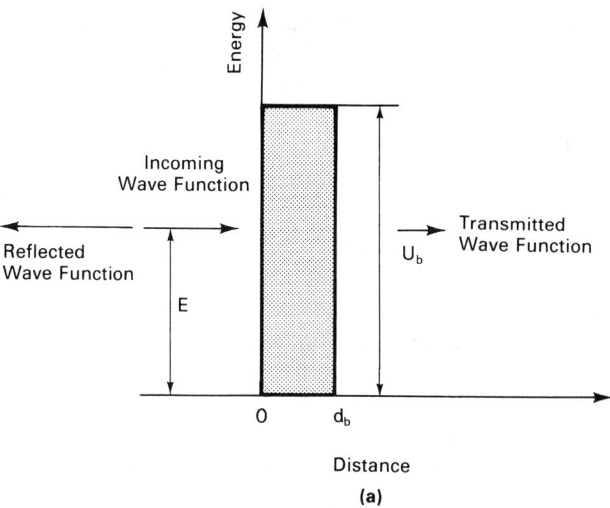

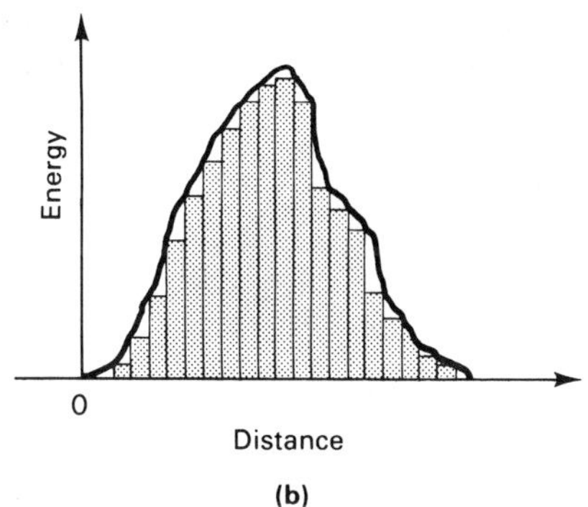

Fig. 2-7-3. Electron tunneling across potential barriers: (a) Simple rectangular barrier. Arrows represents incoming, reflected, and transmitted wave functions. Incoming electrons are coming from the left to the right. (b) Barrier of complex shape presented as a superposition of thin rectangular barriers.

when electrons propagate from the left to the right we must choose

$$B_r = 0 \tag{2-7-11}$$

The sixth condition needed in addition to eqs. (2-7-10) and (2-7-11) to determine all six constants may be obtained from the requirement that

$$\int_0^\infty |\psi^2|\, dx = 1 \tag{2-7-12}$$

which means that the probability of finding the electron somewhere is equal to unity (the normalization condition). It is more useful, however, to find the transmission coefficient for the barrier, which is defined as

$$D = |A_r|^2/|A_l|^2 \tag{2-7-13}$$

If we have a flux of incoming electrons, D gives the ratio of the flux of the electrons transmitted across the barrier to the flux of the incoming electrons. To determine D, five conditions given by eqs. (2-7-10) and (2-7-11) are enough, and we do not have to use eq. (2-7-12).

We can also define the reflection coefficient

$$R = |B_l|^2/|A_l|^2 \tag{2-7-14}$$

As the number of the electrons must be conserved,

$$D + R = 1 \tag{2-7-15}$$

It is convenient to choose $A_l = 1$. Then after somewhat tedious but fairly straightforward derivation we obtain

$$A_r = \frac{4n_b \exp(-ikd_b)}{(1 + n_b)^2 \exp(-ikn_b d_b) - (1 - n_b^2)\exp(ikn_b d_b)} \tag{2-7-16}$$

$$B_l = \frac{[\exp(-ikn_b d_b) - \exp(ikn_b d_b)](1 - n_b^2)}{(1 + n_b)^2 \exp(-ikn_b d_b) - (1 - n_b^2)\exp(ikn_b d_b)} \tag{2-7-17}$$

(see Blokhintsev 1964 and Problem 2-7-1). When $E < U_b$, n_b^2 is negative and n_b is imaginary (see eq. (2-7-6)). In the case when $k|n_b|d_b << 1$, we find from eq. (2-7-16) that

$$D \approx D_b \exp\{-2[2m_e(U_b - E)]^{1/2} d_b/\hbar\} \tag{2-7-18}$$

where

$$D_b = 16|n_b|/(1 + |n_b|^2)^2 \tag{2-7-19}$$

Following Blokhintsev (1964), we can now obtain an approximate expression for D for a barrier of an arbitrary shape, considering such a barrier as a superposition of very thin rectangular barriers (see Fig. 2-7-3b). Then for each thin barrier we can use eq. (2-7-18), and the total transmission coefficient, D, will be given by the product of transmission coefficients for all the rectangular barriers:

$$D = D_{b1} D_{b2} \ldots D_{bn} \tag{2-7-20}$$

Hence, we obtain

$$D \approx D_o \exp \int_0^d \{-2[2m_e(U_b - E)]^{1/2}/\hbar\}\, dx \tag{2-7-21}$$

Using eq. (2-7-18) and taking a typical value of D_o to be of the order of unity, we can clearly show that the fish in Fig. 2-7-2 does not stand a chance to tunnel through the fishbowl (see Problem 2-7-2). However, if we consider an electron in gallium arsenide ($m_n \approx 0.067 m_e = 6.1 \times 10^{-32}$ kg), and the barrier with the barrier height of $U_b - E = 1$ eV and the barrier width of 20 Å, we find that the tunneling probability is quite high (see Problem 2-7-3).

It is also interesting to consider the case when $E > U_b$. Under such conditions, all classical particles will go over the barrier and none will be reflected. However, based on eq. (2-7-17), we see that quantum mechanics predicts that even in this case $D < 1$ and $R > 0$.

We are now ready to consider the mechanism of tunneling in a tunnel diode (see Fig. 2-7-4). Let us consider the energy band diagram of a diode with degenerate p and n regions. As we can see from the figure, empty states are available for the electrons in the valence band. For an electron to tunnel from the conduction band in the n region into the valence band in the p region, there must be an occupied state in the conduction band and the empty state in the valence band with the same energy. To get to these empty states, electrons at the Fermi level have to tunnel across the potential barrier shown in the figure. (Holes may tunnel across the barrier as well. However, in most cases holes are heavier and, hence, tunneling probability for holes is much smaller.) In the limit $T \to 0$ there are very few electrons in the conduction band available for tunneling into the valence band because all filled states in the conduction band are below the Fermi level, and all empty states in the valence band are above the Fermi level (see Fig. 2-7-4a). Therefore, the tunneling current is zero. This situation corresponds to point 0 in Fig. 2-7-1b. However, when a relatively small bias V is applied, the quasi-Fermi levels are split by the amount equal to qV. Now there is a band of empty states

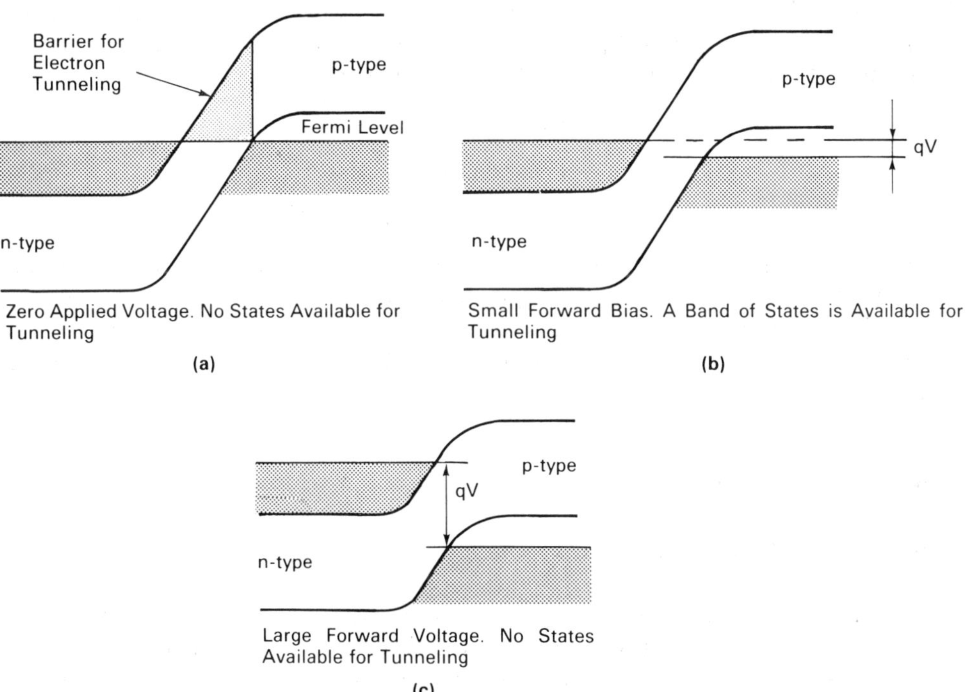

Fig. 2-7-4. Mechanism of tunneling in *p-n* junction.

available for electrons to tunnel from the conduction band in the n region (see Fig. 2-7-4b). The tunneling current can now flow. This current may be much larger than the normal diode forward current. This situation corresponds to point 1 in Fig. 2-7-1b. When the forward bias is increased even more, the band diagram becomes as shown in Fig. 2-7-4c. Under such conditions, no states are available for direct tunneling. The tunneling current drops, and the subsequent rise in the diode current is related to a normal diode mechanism. This situation corresponds to point 2 in Fig. 2-7-1b. Based on this discussion we expect that the current, I, in a tunnel diode is given by

$$I = I_{tun} + I_{diode} \tag{2-7-22}$$

where I_{diode} is the normal diode current component and I_{tun} is the tunneling current. The tunneling current can be found by calculating the transmission coefficients, taking into account electron occupation functions and integrating over all electronic states, using eq. (2-7-21). In fact, however, there is an additional tunneling current that becomes important at large forward voltages (called the excess current), so that

$$I = I_{tun} + I_{diode} + I_{excess} \tag{2-7-23}$$

This excess current component is related to the indirect tunneling via states present in the energy gap. These processes typically involve the emission of phonons. A calculation of I_{tun} and I_{excess} leads to the following expression for the current-voltage characteristics of a tunnel diode (see Sze 1981, p. 529):

$$I_{tun} \approx I_p(V/V_p)\exp(1 - V/V_p) \tag{2-7-24}$$

$$I_{excess} \approx I_v \exp[A_e(V - V_v)] \tag{2-7-25}$$

$$I_{diode} \approx I_o\{\exp[qV/(n_{id}k_BT)] - 1] \tag{2-7-26}$$

where I_p, V_p, I_v, and V_v are the peak and valley currents and voltages, I_o is the diode saturation current, n_{id} is the diode ideality factor, and A_e is a constant. At $T = 300$ K, typical values of the peak and valley voltages, V_p and V_v, are about 200 mV and 600 mV for GaAs tunnel diodes and about 100 mV and 300 mV for Ge tunnel diodes, respectively.

2-8. JUNCTION BREAKDOWN

2-8-1. Breakdown Mechanisms

When a reverse bias applied to a p-n junction exceeds some critical value, the reverse current rises rapidly with the further increase in the applied reverse bias voltage (see Fig. 2-8-1). Two basic mechanisms—avalanche or tunneling breakdown—may be responsible for this rapid rise in current. Avalanche breakdown is caused by an impact ionization process. During this process an electron (or a

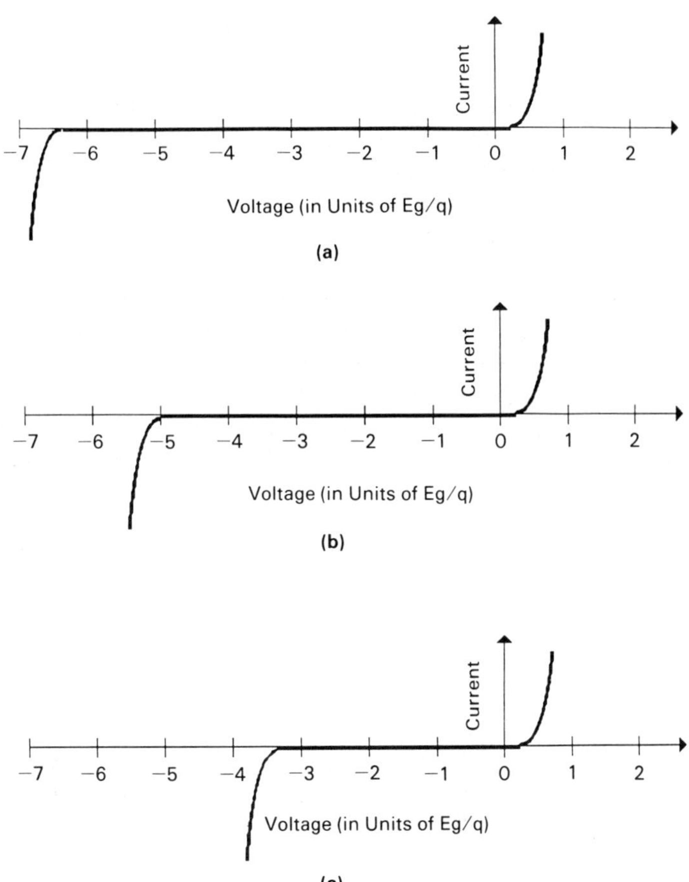

Fig. 2-8-1. Current-voltage characteristics of p^+-n diodes: (a) Relatively low doped n region. Breakdown is caused by avalanche breakdown. (b) Intermediate level of doping in the n region. Both tunneling and avalanche mechanisms may play a role. (c) Relatively high doped, degenerate n region. Breakdown is caused by tunneling.

hole) gains so much energy from the electric field that it initiates a transition of another electron from a valence band into a conduction band. Newly created carriers are, in turn, accelerated by the electric field and create new electron-hole pairs via the impact ionization process. If the applied voltage is high enough, this will lead to an uncontrolled rise of current (until the current is either limited by an external load or the sample is destroyed). A crude estimate of the critical voltage, V_{abr}, causing the avalanche breakdown may be obtained by assuming that the avalanche breakdown occurs when the electric field in the reverse biased p-n junction exceeds a certain critical field for an impact ionization process. For a p^+-n junction, this model leads to the following estimate:

$$V_{abr} = \frac{\varepsilon F_{br}^{\ 2}}{2qN_D} \tag{2-8-1}$$

The breakdown field, F_{br}, at 300 K is of the order of 100 kV/cm for germanium, 300 kV/cm for silicon, 400 kV/cm for GaAs, and 2300 kV/cm for silicon carbide. Hence, V_{abr} can be quite large for diodes with low-doped n regions. In some devices V_{abr} can exceed two thousand volts.

Tunneling breakdown usually occurs in fairly highly doped semiconductors when the maximum electric field in the depletion layer approaches values of the order of 10^6 V/cm. Under such conditions the width of the depletion layer is so narrow that electrons may tunnel from occupied states in the valence band of the p-type region into empty states of the conduction band in the n-type region. This process is illustrated by Fig. 2-8-2. The expression for the tunneling current can be derived using the equation for a barrier transmission coefficient derived in Section 2-7 (see eq. (2-7-21)). This leads to the following expression for the tunneling current, I_t, (see Moll 1964):

$$I_t = \frac{(2m_{pn})^{1/2} q^{5/2} F_{av} V S}{h^2 E_g^{\ 1/2}]} \exp\left[-\frac{8\pi(2m_{pn})^{1/2} q^{1/2} E_g^{\ 3/2}}{3hF_{av}}\right] \tag{2-8-2}$$

where

$$F_{av} = F_m/2 = \left[\frac{q(V_{bi} - V)N_A N_D}{2\varepsilon(N_A + N_D)}\right]^{1/2}$$

is the average electric field in the junction, F_m is the maximum electric field in the junction, E_g is the energy gap (in eV), V_{bi} is the built-in voltage (which may be assumed approximately equal to E_g for highly doped p-n junctions), V is the

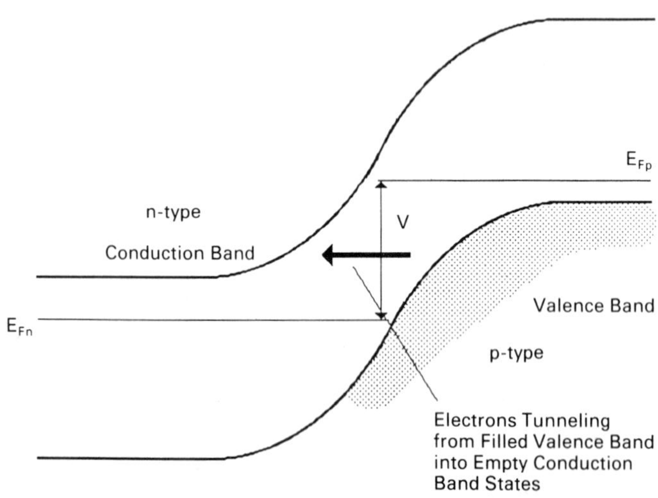

Fig. 2-8-2. Band diagram of a tunneling diode.

applied voltage, N_A and N_D are acceptor and donor concentrations, m_{pn} is the reduced effective mass

$$m_{pn} = 2(1/m_n + 1/m_{lp})^{-1} \tag{2-8-3}$$

m_n is the conduction band effective mass, and m_{lp} is the effective mass for "light" holes. The effective masses are $m_{lh} = 0.044\, m_e$, $m_{lp} = 0.16\, m_e$, and $m_{lp} = 0.082\, m_e$ for Ge, Si, and GaAs, respectively. The effective mass, m_n, is $0.067\, m_e$ for GaAs. For Si and Ge, m_n may be estimated as

$$m_n = (m_{nl}/3 + 2m_{nt}/3) \tag{2-8-4}$$

where $m_{nl} = 0.98\, m_e$, $m_{nt} = 0.19\, m_e$ for Si, and $m_{nl} = 1.64\, m_e$, $m_{nt} = 0.082\, m_e$ for Ge. (Strictly speaking, an appropriate effective mass may depend on the direction of tunneling).

The exponential rise of the tunneling current in highly doped p-n junctions with the reverse voltage leads to the tunneling breakdown. The critical voltage of the tunneling breakdown, V_{brt}, may be estimated by assuming that

$$I_t(V_{brt}) \sim 10\, j_s S \tag{2-8-5}$$

Here j_s is the diffusion saturation current density. According to Sze (1981), the critical voltage of the tunneling breakdown is usually less than approximately $4E_g/q$. When the breakdown voltage is higher than $6E_g/q$ the breakdown is typically caused by the avalanche. Both mechanisms may play a role for breakdown voltages between $4E_g/q$ and $6E_g/q$ (see Fig. 2-8-1).

In p-n diodes with a large power dissipation a *thermal breakdown* may lead to the runaway increase in the reverse current and to an S-type negative differential resistance. This mechanism is especially important in devices made from relatively narrow gap materials (such as Ge), where the device temperature may increase appreciably even at relatively low current densities.

*2-8-2. Impact Ionization and Avalanche Breakdown

Impact ionization may occur when a carrier has enough energy to initiate a transition of an electron from a valence band into a conduction band. The smallest energy required for such a process, consistent with energy and momentum conservation, is called the *threshold energy*. The threshold condition occurs when

$$\nabla_{\mathbf{k}} E_1(\mathbf{k}_1) = \nabla_{\mathbf{k}} E_2(\mathbf{k}_2) = \nabla_{\mathbf{k}} E_3(\mathbf{k}_3) \tag{2-8-6}$$

$$E_i(\mathbf{k}_i) = E_1(\mathbf{k}_1) + E_2(\mathbf{k}_2) - E_3(\mathbf{k}_3) \tag{2-8-7}$$

$$\mathbf{k}_i = \mathbf{k}_1 + \mathbf{k}_2 - \mathbf{k}_3 \tag{2-8-8}$$

where E_i and $\mathbf{k}_i$ are the energy and the wave vector of the hot electron causing the impact ionization in the initial state, E_1 and $\mathbf{k}_1$ are the energy and the wave vector of the hot carrier in the final state, E_3 and $\mathbf{k}_3$ are the energy and the wave vector of the electron in the valence band, and E_2 and $\mathbf{k}_2$ are the energy and the wave vector

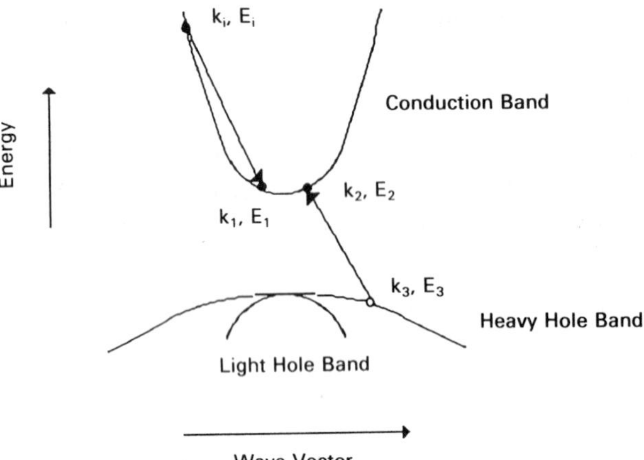

Fig. 2-8-3. Impact ionization electron and hole transitions.

of the state it is promoted to as a result of the impact ionization (see Pearsal et al. 1978; see Fig. 2-8-3).

Equations (2-8-7) and (2-8-8) describe the conservation of the energy and momentum. Equation (2-8-6) states that the group velocities of the three particles should be equal. This condition corresponds to a sharp peak in the probability of a transition corresponding to the impact ionization. If it is not fulfilled, the impact ionization event may be still possible, but it has a much smaller probability. These three equations correspond to quite severe restraints on values of **k** that may be involved in impact ionization transitions. As a consequence, the threshold energy for the impact ionization, E_{ti}, for some materials may be substantially larger than the energy gap, E_g, and may depend on crystal orientation. For example, in GaAs $E_{ti} = 2.05$ eV for the impact ionization by electrons propagating in the $\langle 100 \rangle$ direction, $E_{ti} = 2.01$ eV for the impact ionization by electrons propagating in the $\langle 110 \rangle$ direction, $E_{ti} = 1.81$ eV for the impact ionization by holes propagating in the $\langle 100 \rangle$ direction, and $E_{ti} = 1.58$ eV for the impact ionization by holes propagating in the $\langle 110 \rangle$ and $\langle 111 \rangle$ directions. Electrons moving in the $\langle 111 \rangle$ direction cannot cause the impact ionization in GaAs. (This data is taken from Pearsal et al. 1978). In silicon, the critical energy of impact ionization is closer to the energy gap.

The impact ionization is characterized by the ionization rates α_i and β_i for electrons and holes, which are defined as probabilities of impact ionization per unit length.

The ionization rate is roughly proportional to the number of electrons with energies higher than the critical energy E_i. Only a small fraction of electrons have a sufficient energy because the average electron energy is typically much smaller and is limited by optical phonon scattering. To reach energy E_i an electron (or a hole) has to travel distance

$$d_F = \frac{E_i}{qF} \tag{2-8-9}$$

without collisions. The probability of such an event is

$$P_d = \exp\left[-\frac{d_F}{\lambda}\right] \tag{2-8-10}$$

where λ is the mean free path.

If the carrier reaches the critical energy E_i it may cause the impact ionization or may scatter without causing the ionization. Let us denote as r^{-1} the probability that the carrier with $E > E_i$ causes the impact ionization. The probability that an electron creates an electron hole pair is given by

$$W_i = \frac{1}{r}\exp\left[-\frac{E_i}{qF\lambda}\right] \tag{2-8-11}$$

A phonon generation decreases the electron energy and impedes the impact ionization. The probability that the electron generates a phonon rather than creates an electron hole pair is given by

$$W_{ph} = 1 - W_i \tag{2-8-12}$$

Hence, the number of phonons, $N = W_{ph}/W_i$, generated per each electron-hole pair is given by

$$N = \frac{1}{W_i} - 1 \cong \frac{1}{W_i} \tag{2-8-13}$$

and the total energy that the carrier obtains from the electric field prior to the impact ionization event is

$$E_i^* = E_i + \frac{\hbar\omega_o}{W_i} \tag{2-8-14}$$

where $\hbar\omega_o$ is the optical phonon energy. As a result, the average length an electron travels in the electric field before causing the ionization is

$$\delta_i = \frac{E_i^*}{qF} \tag{2-8-15}$$

and, hence, the ionization rate by electrons is

$$\alpha_i = \frac{1}{\delta_i} = \frac{qF}{r\hbar\omega_o}\exp\left[-\frac{E_i}{qF\lambda}\right] \tag{2-8-16}$$

Here we have taken into account that typically $E_i << E_i^* \approx \hbar\omega_o/W_i$.

Somewhat more realistic numerical calculation of the ionization rate was done by Baraff (1964). His qualitative conclusions are in agreement with eq. (2-8-16). However, most experimental results can be described more accurately using the following empirical equation for electrons,

$$\alpha_i = \alpha_o \exp[-(F_{on}/F)^{m_{bn}}] \tag{2-8-17}$$

and a similar equation for holes:

$$\beta_i = \beta_o \exp[-(F_{op}/F)^{m_{bp}}] \tag{2-8-18}$$

Experimental curves α_i and β_i vs. $1/F$ for GaAs are shown in Fig. 2-8-4 for three orientations of the electric field. The values of α_o, β_o, and m obtained by curve fitting the experimental results using eqs. (2-8-17) and (2-8-18) are given in Table 2-8-1 (after Lee and Sze 1980). An average error of such a fit is less than 0.5%.

Electrons and holes created as a result of the impact ionization move in opposite directions, changing the electron and hole current distributions. Let us consider the fluxes of electrons and holes passing through an infinitesimal region of a uniform semiconductor sample (see Fig. 2-8-5). The electron flux,

$$\phi_n = -\frac{j_n}{q} \tag{2-8-19}$$

is equal to the number of electrons passing through the unit cross section in one second. Here j_n is the electron current density. The hole flux, ϕ_p, is given by

$$\phi_p = \frac{j_p}{q} \tag{2-8-20}$$

where j_p is the hole current density. Passing an incremental length dx, each electron creates on the average $\alpha_i\, dx$ electron-hole pairs and each hole creates $\beta_i\, dx$ electron-hole pairs. As a result, the electron and hole current densities, j_n and j_p, should satisfy the following differential equations:

$$\frac{dj_n}{dx} = \alpha_i j_n + \beta_i j_p \tag{2-8-21}$$

TABLE 2-8-1. IONIZATION RATES FOR GaAs AT T = 300 K

$\alpha_i \beta_i$	Orientation	⟨100⟩	⟨100⟩	⟨111⟩
	α_o (cm^{-1})	9.12×10^4	2.19×10^6	7.76×10^4
α_i	F_{on}(V/cm)	4.77×10^3	2.95×10^6	4.45×10^5
	m_{bn}	3.48	1	6.91
	β_o(cm^{-1})	3.47×10^6	2.47×10^6	6.31×10^6
β_i	F_{op} (V/cm)	2.18×10^6	2.27×10^6	2.31×10^6
	m_{bp}	1	1	1
Field range	(V/cm)	3.13×10^5 to 4.76×10^5	3.13×10^5 to 4.76×10^5	3.33×10^5 to 5.56×10^5

(from M. H. Lee and S. M. Sze, "Orientation Dependence of Breakdown and Voltage in GaAs," *Solid State Electronics,* 23, p. 1007 (1980).)

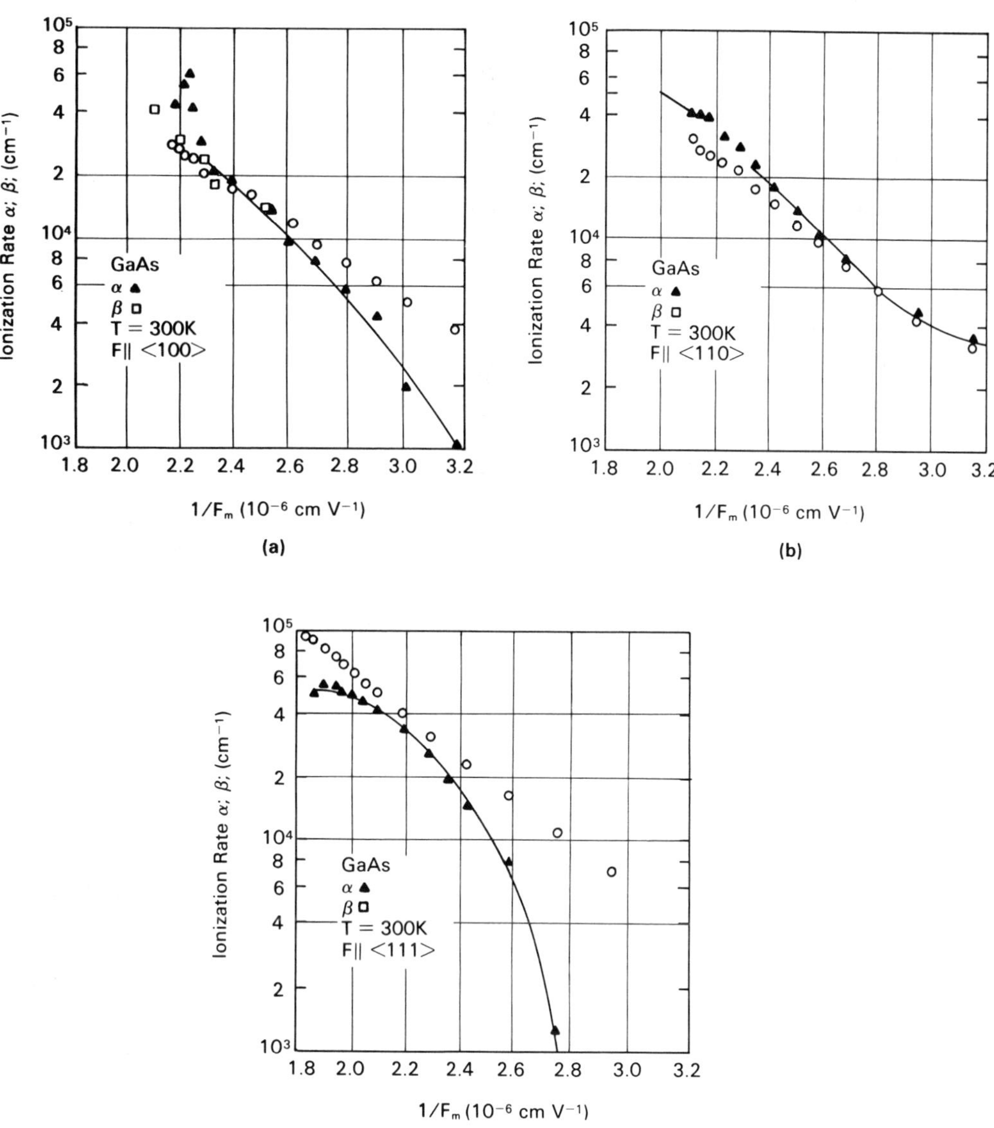

Fig. 2-8-4. Experimental curves α_i and β_i vs. $1/F$ for GaAs for three orientations of the electric field (after M. H. Lee and S. M. Sze, "Orientation Dependence of Breakdown and Voltage in GaAs" (1980)).

and

$$\frac{dj_p}{dx} = -\beta_i j_p - \alpha_i j_n \tag{2-8-22}$$

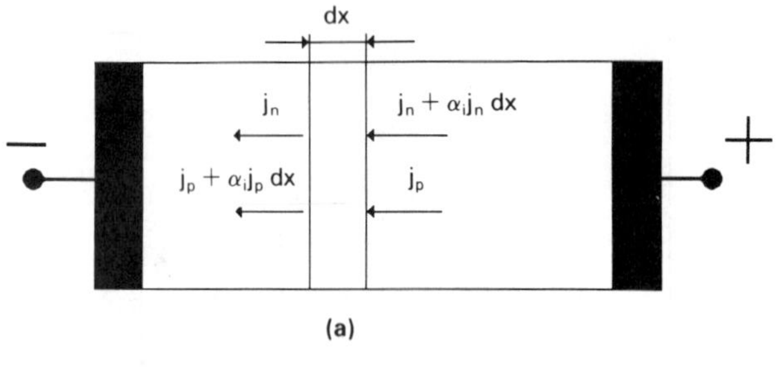

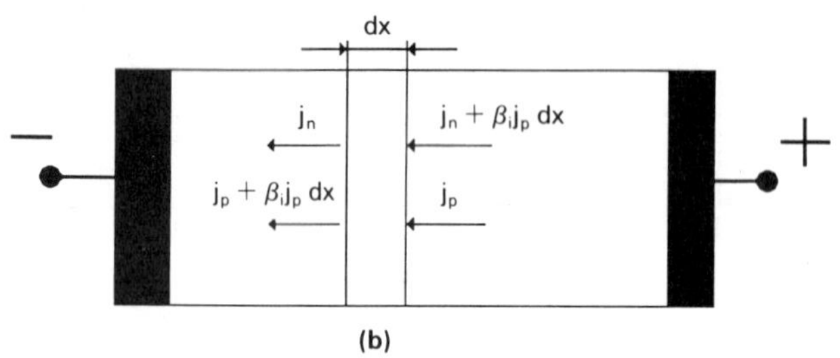

Fig. 2-8-5. Incremental increases in electron and hole current densities, j_n and j_p, caused by the impact ionization by (a) electrons and (b) holes.

(see Fig. 2-8-5). From eqs. (2-8-21) and (2-8-22) we find that

$$\frac{dj_p}{dx} = -\frac{dj_n}{dx}$$

and the total current density

$$j = j_n + j_p \tag{2-8-23}$$

is constant. The qualitative current distributions under impact ionization conditions are shown in Fig. 2-8-6. If the impact ionization is caused by electrons, we can define the multiplication factor for electrons injected at the cathode:

$$M_n = \frac{j_n(L)}{j_n(0)} \tag{2-8-24}$$

where L is the sample length. If the impact ionization is caused by holes injected at the anode, we can define a multiplication factor for holes:

$$M_p = \frac{j_p(0)}{j_p(L)} \tag{2-8-25}$$

The breakdown voltage is defined as the voltage at which the multiplication rate (M_n or M_p) becomes infinite. Hence, to calculate the breakdown voltage, we have to express M_n and M_p in terms of α_i and β_i using eqs. (2-8-21) and (2-8-22).

Using eq. (2-8-23) we can rewrite eq. (2-8-21) as

$$\frac{dj_n}{dx} - (\alpha_i - \beta_i)j_n = -(\alpha_i - \beta_i)j + \alpha_i j \tag{2-8-26}$$

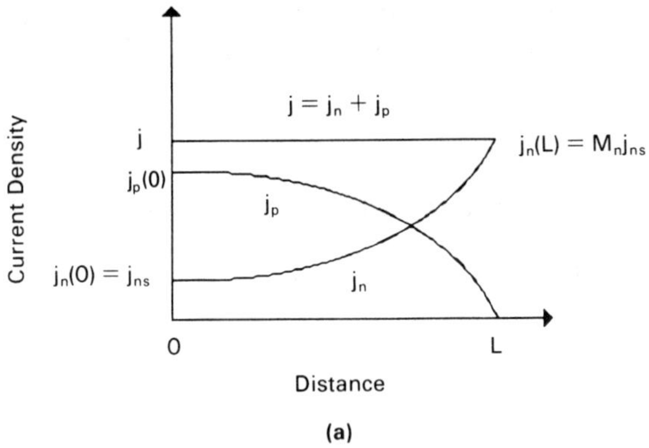

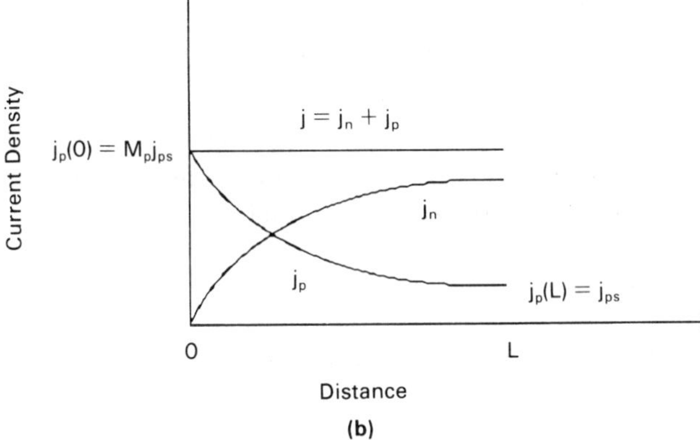

Fig. 2-8-6. Current densities as functions of distance for the impact ionization caused by (a) the electron injection and (b) the hole injection (b).

Multiplying eq. (2-8-26) by the integrating factor $\exp[-\int_0^x (\alpha_i - \beta_i)\,dx']$ and integrating the resulting equation from 0 to L, we find that

$$1 - \frac{1}{M_n} = \int_0^L \alpha_i \exp\left[-\int_0^x (\alpha_i - \beta_i)\,dx'\right] dx \tag{2-8-27}$$

In a similar way, we obtain from eqs. (2-8-22) and (2-8-23)

$$1 - \frac{1}{M_p} = \int_0^L \beta_i \exp\left[-\int_0^x (\alpha_i - \beta_i)\,dx'\right] dx \tag{2-8-28}$$

The condition of the avalanche breakdown follows from eq. (2-8-27):

$$\int_0^L \alpha_i \exp\left[-\int_0^x (\alpha_i - \beta_i)\,dx'\right] dx = 1 \tag{2-8-29}$$

It can also be shown that the same eq. (2-8-29) can be derived from eq. (2-8-28) assuming that $M_p \to \infty$. Hence, the breakdown condition does not depend on which type of carrier initiated the breakdown.

Let us now consider several simple cases. If the impact ionization is dominated by one carrier ($\beta << \alpha$, for example), we find from eq. (2-8-27) that

$$M_n = \exp\left(\int_0^L \alpha_i \, dx\right) \tag{2-8-30}$$

Equation (2-8-30) is derived as follows. We can rewrite eq. (2-8-29) for $\alpha_i >> \beta_i$ as

$$1 - \frac{1}{M_n} = \int_0^L (df/dx)\exp(-f) \, dx' \tag{2-8-31}$$

where

$$f = \int_0^x \alpha_i \, dx'$$

The integral in the right-hand side of eq. (2-8-31) can be evaluated as

$$\int (df/dx)\exp(-f) \, dx' = \int \exp(-f) \, df = -\exp(-f) + \text{const}$$

leading to eq. (2-8-30).

Equation (2-8-30) means that, in the limit $\beta \to 0$, M_n is always finite and, strictly speaking, there is no avalanche breakdown. If we further assume that M_n is small (low applied voltages), we find from eq. (2-8-30) that

$$M_n \approx 1 + \int_0^L \alpha_i \, dx \tag{2-8-32}$$

Let us now consider a different limiting case, in which $\alpha_i = \beta_i$. In this case we obtain from eq. (2-8-27)

$$M_n = \frac{1}{1 - \int_0^L \alpha_i \, dx} \tag{2-8-33}$$

Hence, the breakdown voltage for $\alpha_i = \beta_i$ can be found from the condition

$$\int_0^L \alpha_i \, dx = 1 \tag{2-8-34}$$

that corresponds to $M_n \to \infty$.

In the foregoing we considered the situation in which the impact ionization is either caused by the electron current injected at the negative terminal (cathode) or by the hole current injected at the positive terminal (anode). When the injected current is composed of both electrons and holes,

$$I_s = k_i I_{ns} + (1 - k_i) I_{ps} \tag{2-8-35}$$

the multiplication rate is given by

$$M = k_i M_p + (1 - k_i) M_n \tag{2-8-36}$$

(see Lee et al. 1964).

If the ionization rates are known, the breakdown voltage can be calculated from eq. (2-8-27) assuming $M_n \to \infty$. The results of such a calculation for GaAs p^+-n junctions (Lee and Sze 1980) are compared with the experimental results of Sze and Gibbons (1966) in Fig. 2-8-7. For comparison we also show the breakdown voltage V_{br} calculated using an elementary model that assumes that the breakdown occurs when the maximum field in the junction exceeds some critical value, F_{br}. In the frame of this model, the breakdown voltage is given by eq. (2-8-1)

$$V_{br} = \frac{\varepsilon F_{br}^2}{2qN_D} \tag{2-8-37}$$

In this calculation we assumed that F_{br} = 393 kV/cm. As can be seen from the figure, the simple model may be quite adequate for relatively low doping levels ($N_D < 10^{16}$ cm^{-3}) but becomes increasingly inaccurate at higher doping. For a linear graded junction the breakdown voltage calculated in the frame of this elementary model is given by

$$V_{br} = \frac{4F_{br}^{3/2}}{3}\left(\frac{2\varepsilon}{a_d q}\right)^{1/2} \tag{2-8-38}$$

where a_d is the doping gradient.

In Fig. 2-8-8 we show the breakdown voltage as a function of doping for one-sided abrupt GaP, GaAs, Si, and Ge p-n junctions (after Sze and Gibbons 1966).

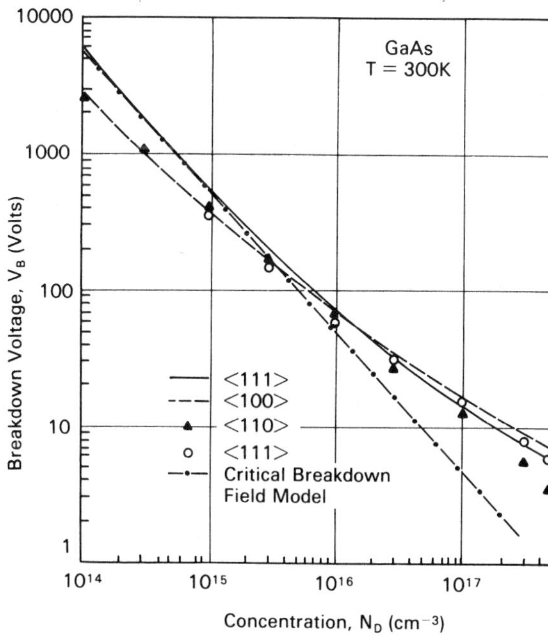

Fig. 2-8-7. Breakdown voltage vs. doping for GaAs p^+-n junctions (after M. H. Lee and S. M. Sze, "Orientation Dependence of Breakdown and Voltage in GaAs" (1980)).

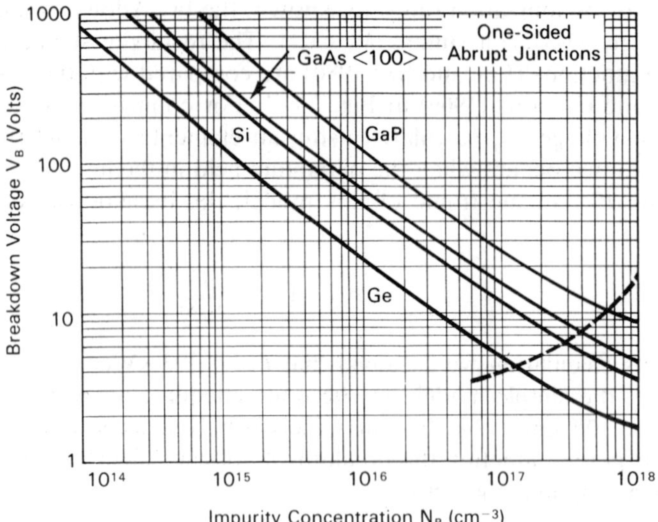

Fig. 2-8-8. Breakdown voltage as a function of doping for one-sided abrupt GaP, GaAs, Si, and Ge *p-n* junctions (after S. M. Sze and G. Gibbons, *Appl. Phys. Lect.*, 8, p. 111 (1966). The tunneling mechanism is dominant for the doping levels higher than corresponding to the dashed line in the low right corner of the figure.

2-9. SCHOTTKY BARRIERS

The first semiconductor device was a rectifier made by using a metal whisker contacting a piece of semiconductor material. Rectifying properties of such a contact are caused by an electrostatic potential barrier (Schottky barrier) that exists at the boundary between a semiconductor and a metal. To understand the reason for the existence of such a barrier let us consider boundaries of an n-type semiconductor with vacuum (see Fig. 2-9-1). The potential energy of electrons inside the crystal is smaller because the electrons are attracted by the positive ions of the crystal lattice. The energy difference, X_{so}, between the bottom of the conduction band, E_c, and the vacuum energy level is called the *electron affinity*. Owing to the thermal motion, some electrons have energy higher than $E_c + X_{so}$ and may leave the crystal. Let us calculate this flux assuming that electrons inside the crystal have a Maxwell-Boltzmann distribution function

$$f(E_n) = \exp[(E_F - E_n)/(k_B T)] \tag{2-9-1}$$

where E_F is the Fermi level, and E_n is the electron energy (see eq. (1-6-8a)). Using eq. (2-9-1) we can find the velocity distribution function f_v such that $f_v\, dv_x\, dv_y\, dv_z$ is the probability of an electron having velocity v with components between v_x, v_y, v_z and $v_x + dv_x$, $v_y + dv_y$, $v_z + dv_z$. Indeed, choosing the bottom of the conduction

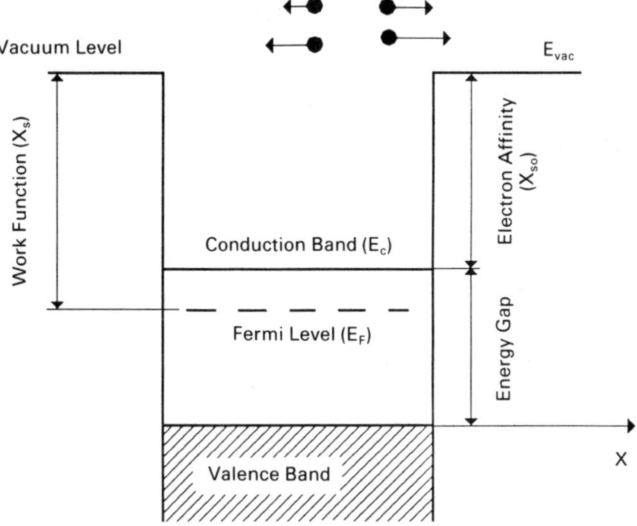

Fig. 2-9-1. Potential distribution at the semiconductor surface. E_c is the bottom of the conduction band; E_v, the top of the valence band; E_F, the Fermi level; X_s, the work function; and X_{so}, the electron affinity.

band as a reference point ($E_c = 0$ so that $E = m_n(v_x^2 + v_y^2 + v_z^2)/2$) we obtain from eq. (2-9-1)

$$f_v = A_v \exp[-m_n(v_x^2 + v_y^2 + v_z^2)/(2k_B T)] \tag{2-9-2}$$

The normalization constant, A_v, is found from the condition

$$\int_{-\infty}^{\infty} \int_{-\infty}^{\infty} \int_{-\infty}^{\infty} f_v \, dv_x \, dv_y \, dv_z = 1 \tag{2-9-3}$$

Using the following relation (that can be found in any table of integrals),

$$\int_{-\infty}^{\infty} \exp(-ax^2) \, dx = (\pi/a)^{1/2} \tag{2-9-4}$$

we obtain from eq. (2-9-3)

$$f_v = [m_n/(2\pi k_B T)]^{3/2} \exp[-m(v_x^2 + v_y^2 + v_z^2)/(2k_B T)] \tag{2-9-5}$$

We can also rewrite this expression as

$$f_v = f_{vx} f_{vy} f_{vz} \tag{2-9-6}$$

where f_{vx}, f_{vy}, and f_{vz} are the distribution functions for velocity components v_x, v_y, and v_z, respectively:

$$f_{vx} = [m_n/(2\pi k_B T)]^{1/2} \exp[-mv_x^2/(2k_B T)] \tag{2-9-7}$$

$$f_{vy} = [m_n/(2\pi k_B T)]^{1/2} \exp[-mv_y^2/(2k_B T)] \tag{2-9-8}$$

$$f_{vz} = [m_n/(2\pi k_B T)]^{1/2} \exp[-mv_z^2/(2k_B T)] \tag{2-9-9}$$

We can now find an average velocity $\langle v_x \rangle$ of electrons moving in the direction x perpendicular to the interface (see Fig. 2-9-1):

$$\langle v_x \rangle = \int_0^\infty v_x f_{vx}\, dv_x =$$

$$= [m_n/(2\pi k_B T)]^{1/2} \int_0^\infty v_x \exp[-mv_x^2/(2k_B T)]\, dv_x \tag{2-9-10}$$

Equation (2-9-10) can be rewritten as

$$\langle v_x \rangle = [2k_B T/(\pi m_n)]^{1/2} \int_0^\infty u \exp(-u^2)\, du \tag{2-9-11}$$

The integral in the right-hand side of eq. (2-9-11) is equal to $\frac{1}{2}$. Hence,

$$\langle v_x \rangle = [k_B T/(2\pi m_n)]^{1/2} \tag{2-9-12}$$

Let us now calculate the average electron velocity,

$$\langle v \rangle = \int_0^\infty v f_v\, d^3v = [m_n/(2\pi k_B T)]^{3/2} \int_0^\infty v \exp[-mv^2/(2k_B T)] 4\pi v^2\, dv$$

$$= 2[8k_B T/(\pi m_n)]^{1/2} \int_0^\infty u \exp(-u^2) u^2\, du$$

Hence,

$$\langle v \rangle = [8k_B T/(\pi m_n)]^{1/2} \int_0^\infty t \exp(-t)\, dt \tag{2-9-13}$$

In this derivation we take into account that $d^3v = 4\pi v^2\, dv$. The integral in the right-hand side of eq. (2-9-13) can be found in any table of integrals or can be easily evaluated using integration by parts. It is equal to unity. Hence, the average electron thermal velocity is given by

$$\langle v \rangle = [8k_B T/(\pi m_n)]^{1/2} \tag{2-9-14}$$

The electron current density, j_s, corresponding to the flux of electrons out of the crystal in the positive direction x can be calculated as follows:

$$j_s = q \int_{E_{vac}}^\infty v_x\, (dn/dE)\, dE \tag{2-9-15}$$

where

$$dn/dE = [4\pi(2m_n)^{3/2}/h^3]\, (E - E_c)^{1/2} \exp[(E_F - E)/k_B T] \tag{2-9-16}$$

(see eqs. (1-6-35) and (1-6-37)),

$$E - E_c = m_n(v_x^2 + v_y^2 + v_z^2)/2 = m_n v^2/2$$

$$dE = m_n v \, dv$$

$$4\pi v^2 \, dv = dv_x \, dv_y \, dv_z$$

and E_{vac} is the vacuum level. Now we should substitute these equations into eq. (2-9-15) and perform the integration in eq. (2-9-15) with respect to v_y from minus infinity to infinity, with respect to v_z from minus infinity to infinity, and with respect to v_x from v_{min} to infinity, where $v_{min} = (2X_{so}/m_n)^{1/2}$ is the minimum value of v_x required for an electron escaping the crystal. Using the definition of the work function

$$X_s = E_{vac} - E_F \tag{2-9-17}$$

we finally obtain the expression for the current density, j_s, of electrons leaving the sample (see Problem 2-9-2):

$$j_s = A^* T^2 \exp\left(-\frac{X_s}{k_B T}\right) \tag{2-9-18}$$

where

$$A^* = \frac{q m_n \, k_B{}^2}{2\pi^2 \hbar^3} \approx 120(m_n/m_e) \; (A/cm^2/K^2) \tag{2-9-19}$$

is called the *Richardson constant*. Taking into account eqs. (2-9-12) and (2-9-14), we can rewrite eq. (2-9-18) as

$$j_s = q n_{xs} \langle v_x \rangle = q n_{xs} \langle v \rangle / 4 \tag{2-9-18a}$$

where n_{xs} is the concentration of electrons with energies higher than X_s.

This effect of electrons escaping a crystal is called *thermionic emission*. The escaping electrons leave the unbalanced positive charge inside the crystal. The resulting electric field leads to band bending near the surface. The positive charge attracts electrons back, thus establishing the thermodynamic equilibrium. The simple potential distribution shown in Fig. 2-9-1 does not take the band bending into account. Therefore, it is inaccurate.

When a metal and a semiconductor are placed close to each other (forming what is called a *Schottky diode*) there are initially two competing electron currents: from the semiconductor to the metal (see eq. (2-9-18)) and from the metal to the semiconductor:

$$j_m = \frac{q m_n (k_B T)^2}{2\pi^2 \hbar^3} \exp\left(-\frac{X_m}{k_B T}\right) \tag{2-9-20}$$

where X_m is the work function of the metal. If $X_m > X_s$, $j_m < j_s$, then the metal will be charged negatively and the semiconductor will be charged positively with the resulting potential difference

$$V_{bi} = X_m - X_s \tag{2-9-21}$$

This potential difference, V_{bi}, is called the *built-in voltage*. It corresponds to the barrier height,

$$\phi_b = X_m - X_{so} \tag{2-9-22}$$

The negative charge in the metal is practically localized at the surface atomic layer (due to a very large free electron density). The positive charge density in the semiconductor is limited by the concentration of ionized donors, and the space charge region extends into the semiconductor. A simplified energy diagram of a metal-semiconductor barrier corresponding to this model is shown in Fig. 2-9-2.

In practice, Schottky barrier heights are quite different from those predicted by eq. (2-9-22) and shown in Fig. 2-9-2. The Schottky barrier height is only weakly dependent on X_m (it increases by 0.1 to 0.3 eV when X_m increases by 1 to 2 eV). Bardeen (1947) developed the model explaining this difference by the effects of the surface states at the boundary between the semiconductor and a thin oxide layer that is almost always present at the surface. The oxide is so thin that electrons can easily tunnel through. The surface states that change the barrier height are continuously distributed in energy within the energy gap. They are characterized by a "neutral" level ϕ_o such that the states below ϕ_o are neutral when filled by electrons and the states above ϕ_o are neutral when empty (see Fig. 2-9-3). It can be shown that in this case the barrier height is given by

$$\phi_b = \gamma_s(X_m - X_{so}) + (1 - \gamma_s)(E_g - \phi_o) - \gamma_s(\varepsilon_s/\varepsilon_i)F_m\delta_{ox} \tag{2-9-23}$$

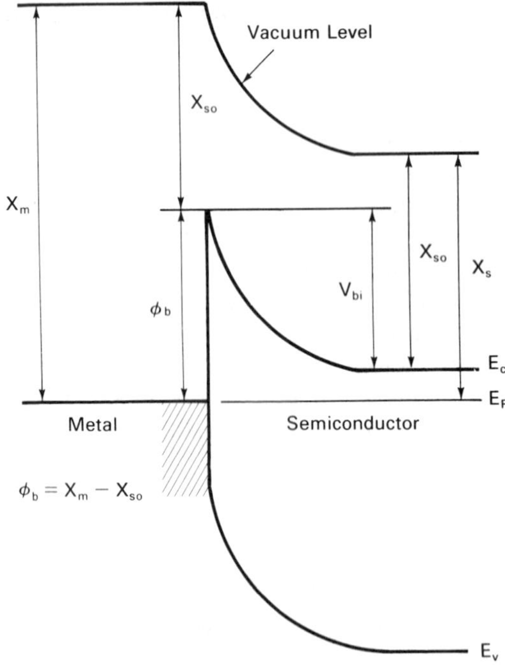

Fig. 2-9-2. Simplified energy band diagram of metal-semiconductor barrier.

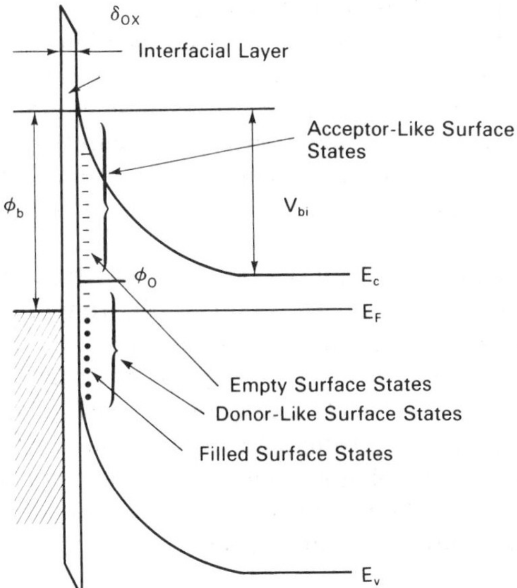

Fig. 2-9-3. Surface states at metal-semiconductor boundary.

where

$$\gamma_s = \frac{\varepsilon_i}{\varepsilon_i + qN_s\delta_{ox}} \tag{2-9-24}$$

E_g is the energy gap, ε_i is the permittivity of the interfacial layer, δ_{ox} is the thickness of the interfacial layer, N_s is the density of the surface states, and

$$F_m = \left(\frac{2qN_D V_{bi}}{\varepsilon_s}\right)^{1/2} \tag{2-9-25}$$

$$qV_{bi} = \phi_b - E_c + E_F \tag{2-9-26}$$

(see Rhoderick 1977). The last term in the right-hand part of eq. (2-9-23) is proportional to the voltage drop across the interfacial layer, which is small in most cases. Hence, eq. (2-9-23) reduces to eq. (2-9-22) when the density of the surface states is zero. In the opposite limiting case when $N_s \to \infty$, $\gamma_s \to 0$ and

$$\phi_b = E_g - \phi_o \tag{2-9-27}$$

Indeed, in this case the Fermi level in the semiconductor must coincide with the neutral level because any deviation of the Fermi level from this position will result in an infinitely large charge at the interface.

Bardeen's model is in better agreement with experimental data than the simplistic model that assumes that the barrier height is given by eq. (2-9-22). Still, it cannot explain many properties of the Schottky barrier diodes, nor can it explain the mechanism of the surface states formation. Spicer et al. (1979) related the formation of the surface states to defects formed during metal deposition (the unified defect model). Tersoff (1984) proposed that the Schottky barrier heights (as well as band discontinuities at heterointerfaces; see Section 2-12) are controlled by electrons tunneling from one material into another, forming an interfacial dipole. Even though a detailed and accurate understanding of Schottky barrier formation still remains a challenge, many properties of Schottky barriers may be understood independently of the exact mechanism determining the barrier height. In other words, we simply determine the effective barrier height from experimental data.

The barrier height, ϕ_b, can be determined experimentally from the current-voltage characteristics of Schottky diodes (see Section 2-10), from internal photoemission measurements, and from capacitance measurements. The internal photoemission method is, perhaps, the most direct way to measure the Schottky barrier height. In this technique, the light shining on a semiconductor with a Schottky barrier contact generates electrons and leads to an electric current, called *photocurrent*. The ratio of the photocurrent over the number of absorbed photons, called *yield,* is measured as a function of the light frequency. The yield, Y, is related to the Schottky barrier height, ϕ_b, and to the frequency of the absorbed radiation, ν, as

$$Y = \text{const}\,(h\nu - \phi_b)^2 \tag{2-9-28}$$

if $h\nu > \phi_b + 3k_BT$ and $\phi_b + E_F >> h\nu$ where E_F is the Fermi level of the metal counted from the bottom of the metal conduction band (see, for example, Eizenberg et al. 1986 for details). In Fig. 2-9-4 we show the results of the internal photoemission experiment for determining the Schottky barrier heights on $Al_xGa_{1-x}As$ with different values of x (from Eizenberg et al. 1986). Generally there is a correlation between the energy gap and Schottky barrier height (the Schottky barrier height is close to approximately $\frac{2}{3}$ of the energy gap for Schottky barriers on n-type semiconductors). This correlation is illustrated by Fig. 2-9-5, in

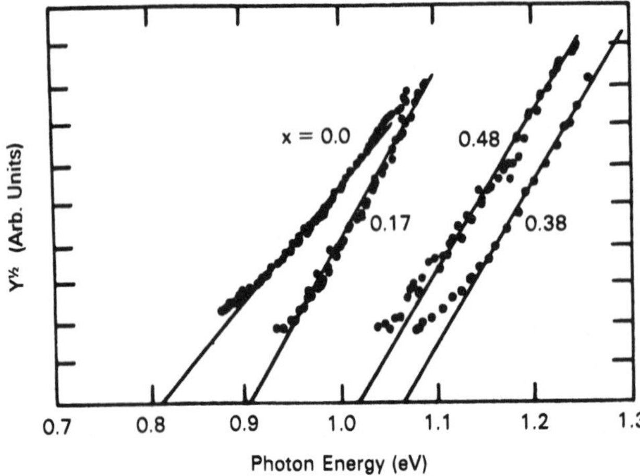

Fig. 2-9-4. Internal photoemission experiment for determining the Schottky barrier heights on $Al_xGa_{1-x}As$ with different values of x (from M. Eizenberg, M. Heiblum, M. I. Nathan, N. Braslau, and P. M. Mooney, "Barrier-Heights and Electrical Properties of Intimate Metal-AlGaAs Junctions," *J. Appl. Physics.*, pp. 1516–1522 (1987).

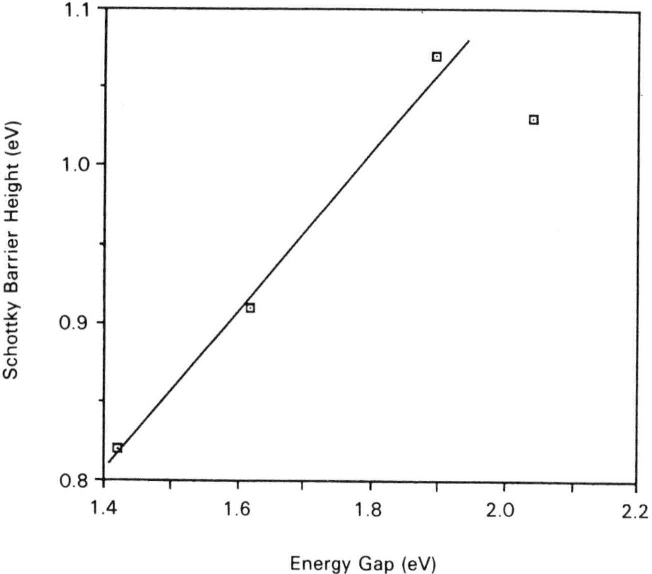

Fig. 2-9-5. Schottky barrier heights determined from the data in Fig. 2-9-6 vs. an energy gap of $Al_xGa_{1-x}As$.

which we plot the Schottky barrier heights determined from the data in Fig. 2-9-4 vs. the energy gap of $Al_xGa_{1-x}As$.

Once the Schottky barrier height is known, the variation of the space charge, electric field, and potential in the semiconductor space charge region can be found using the depletion approximation:

$$\rho = qN_D \tag{2-9-29}$$

$$F = -\frac{qN_D(A_o - x)}{\varepsilon_s} \tag{2-9-30}$$

$$\phi = -\frac{qN_D}{2\varepsilon_s}(A_o - x)^2 = -V_{bi}\left(1 - \frac{x}{A_o}\right)^2 \tag{2-9-31}$$

(see Fig. 2-9-6). Here

$$A_o = \left[\frac{2\varepsilon_s(V_{bi} - V)}{qN_D}\right]^{1/2} \tag{2-9-32}$$

is the width of the space charge region, ε_s is the dielectric permittivity of the semiconductor, and the built-in voltage, V_{bi}, is given by eq. (2-9-26). The shape of the depletion region under reverse bias and small forward bias may be obtained by substituting V_{bi} with $V_{bi} - V$, where V is the applied voltage (see Fig. 2-9-7). Here a positive V corresponds to the forward bias.

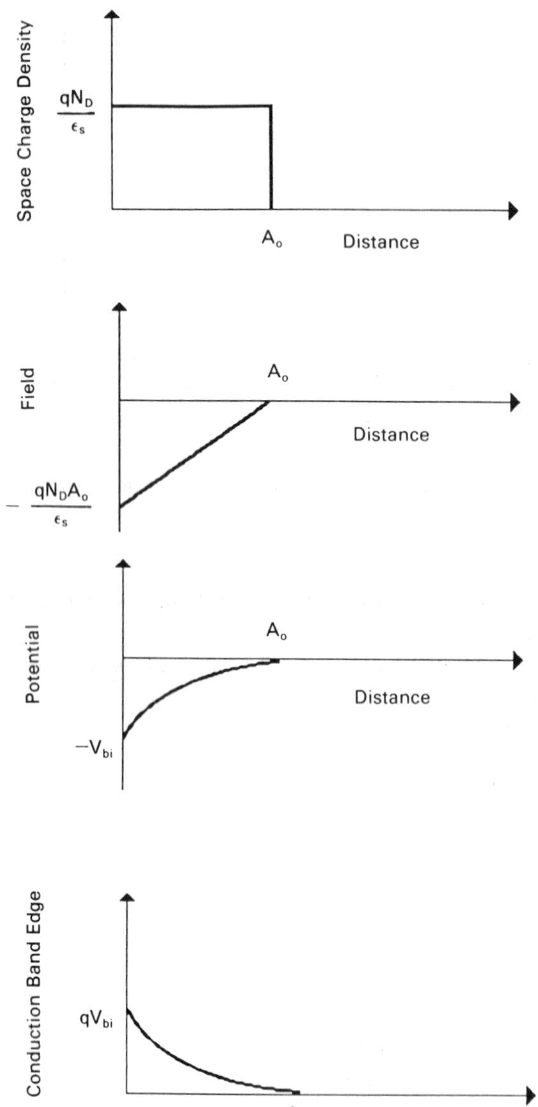

Fig. 2-9-6. Space charge, electric field, potential, and conduction band edge in depletion region of Schottky diode.

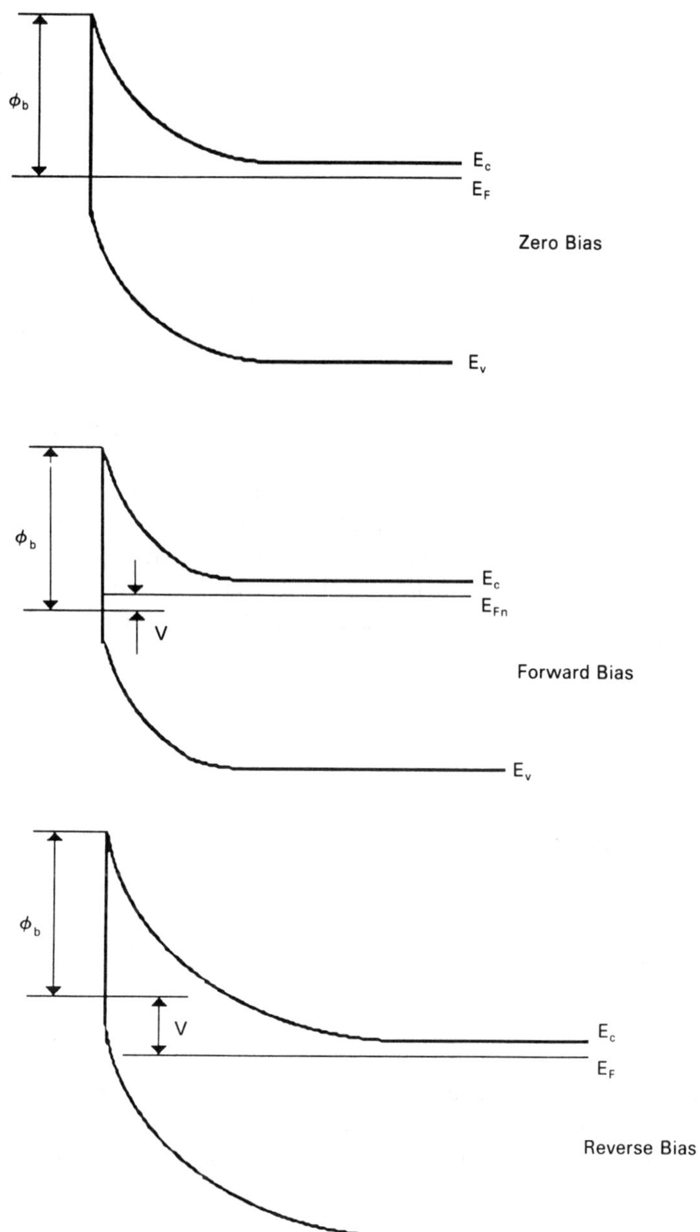

Fig. 2-9-7. Energy band diagram of Schottky barrier under zero, forward, and reverse bias.

2-10. CURRENT-VOLTAGE CHARACTERISTICS OF SCHOTTKY DIODES

2-10-1. Thermionic Emission Model

The thermionic model of the electron transport in Schottky barriers, considered in Section 2-9, is valid when the interface barrier presents an important impediment to the current flow. The situation here is very similar to that in a p-n junction (see Section 2-3). Under reverse bias or relatively small forward bias, the electron quasi-Fermi level in the depletion region near the interface remains practically constant with distance. Just as in a p-n junction under similar conditions, drift and diffusion current components in the depletion region are both much greater than their sum, which is the total current. This behavior of the electron quasi-Fermi level illustrated by Fig. 2-9-7 is confirmed by numerical simulations.

The band diagrams shown in Fig. 2-9-7 imply that an external voltage drop is primarily across the metal-semiconductor interface. In low mobility and relatively low doped semiconductors the current through the Schottky barrier may be limited more by processes of diffusion and drift in the space charge region rather than by the barrier at the metal-semiconductor interface. Under such conditions, the electron quasi-Fermi level will vary in the depletion region, and the thermionic model will not apply. The quantitative criterion of the validity of the thermionic model is obtained by comparing the mean free path of electrons, λ, with the distance

$$d_T = k_B T/(qF_{max}) \tag{2-10-1}$$

over which the potential in the depletion region near the interface decreases by $k_B T/q$ (see Rhoderick 1977). Here F_{max} is the maximum electric field at the metal semiconductor interface,

$$F_{max} = \left(\frac{2qN_D V_{bi}}{\varepsilon_s}\right)^{1/2} \tag{2-10-2}$$

and the mean free path is given by

$$\lambda = \frac{\mu_n}{q}\sqrt{3k_B T m_n} \tag{2-10-3}$$

where μ_n is a low-field electron mobility and m_n is the effective mass. (Equation (2-10-3) is obtained by substituting the thermal velocity $v_T = (3k_B T/m_n)^{1/2}$ and the momentum relaxation time $\tau = \mu_n m_n/q$ into the equation for the mean free path, $\lambda = v_T\tau$.) The thermionic model is valid if

$$\lambda > d_T \tag{2-10-4}$$

In this case electrons near the metal-semiconductor interface that have sufficient energy to go over the barrier have a fair chance to do so before experiencing scattering. Of course, the momentum relaxation time of these high-energy electrons may be quite different from the momentum relaxation time in low electric field, τ, that we related to the low-field mobility. Therefore, eq. (2-10-4) is a fairly crude criterion. Nevertheless, it gives us some idea about conditions of the validity of the thermionic model.

Substituting eqs. (2-10-1) and (2-10-3) into eq. (2-10-4) we find that

$$N_D \mu_n^2 > (k_B T/q)\varepsilon_s/(6V_{bi} m_n) \qquad (2\text{-}10\text{-}5)$$

or

$$N_D(\text{cm}^{-3})\mu_n^2(\text{cm}^2/\text{Vs}) > 4.19 \times 10^{18}(T/300)(\varepsilon_s/\varepsilon_o)(m_e/m_n)/V_{bi} \qquad (2\text{-}10\text{-}6)$$

For Si and GaAs at room temperature inequality (2-10-6) is fulfilled for N_D greater than approximately 10^{12} cm^{-3}.

When $\lambda << d_T$, the position of the quasi-Fermi level in the depletion region becomes dependent on distance, and the thermionic model is no longer applicable. This situation corresponds to the *diffusion model* (see Rhoderick 1977). It occurs, for example, in Schottky barriers on amorphous silicon where the electron mobility is very small ($\mu < 10$ cm^2/Vs).

When the thermionic model is valid, the current-voltage characteristics of the Schottky diode may be determined by calculating the electronic fluxes in and out of the semiconductor at the metal-semiconductor interface. The electronic flux, $J_{sm} = j_s/q$, out of the semiconductor may be derived using eq. (2-9-18) and substituting the work function X_s by the effective barrier height, $\phi_b - V$ (see Fig. 2-9-7). The flux in the opposite direction, J_{ms} (from the metal into the semiconductor) is independent of the applied voltage (assuming that the barrier height is independent of the applied voltage; see Fig. 2-9-7 which shows that the barrier for the electrons in the metal is equal to ϕ_b for zero, forward, and reverse biases). The total flux, J, must be equal to zero at $V = 0$. Hence,

$$J_{ms} = J_{sm} \qquad (V = 0) \qquad (2\text{-}10\text{-}7)$$

and we obtain the following expression for the electric current density:

$$j = j_{ss}\left[\exp\left(\frac{qV}{k_B T}\right) - 1\right] \qquad (2\text{-}10\text{-}8)$$

where

$$j_{ss} = A^* \, T^2 \exp\left[-\frac{q\phi_b}{k_B T}\right] \qquad (2\text{-}10\text{-}9)$$

The Richardson constant,

$$A^* = \alpha \frac{m_n q k_B^2}{2\pi\hbar^3} \approx 120\, \alpha(m_n/m_e)\ (\text{A/cm}^2\ \text{K}^2) \qquad (2\text{-}10\text{-}10)$$

now includes an empirical factor, α, that accounts for deviations from the simple theory considered in Section 2-9. According to Crowell and Sze (1966), $\alpha \approx 0.5$. This equation is valid for spherical surfaces of equal energy and parabolic bands (which is approximately true for the conduction band in GaAs where $m_n \approx 0.067\ m_e$). For ellipsoidal surfaces of equal energy (as in the conduction band of silicon), A^* depends on the direction of the the current flow. According to Crowell (1965), m_n in eq. (2-10-10) should be replaced by

$$m^* = (l_\theta^2 m_y m_z + m_\theta^2 m_z m_x + n_\theta^2 m_x m_y) \tag{2-10-11}$$

where m_x, m_y, and m_z are components of the effective mass tensor for the directions coinciding with the principal axes of the ellipsoids of equal energy surfaces, and l_θ, m_θ, and n_θ are cosines of angles formed by the normal to the metal-semiconductor interface and the three principal axes of the ellipsoid. This should be done for each valley (i.e., each equivalent minimum of the conduction band) and the resulting Richardson constants should be added up. In silicon there are six equivalent valleys and such a calculation yields

$$m^* = 2m_t + 4(m_l m_t)^{1/2} = 2.05\ m_e \qquad \text{for } \langle 100 \rangle \text{ directions} \tag{2-10-12}$$

$$m^* = 6(m_t^2/3 + 2m_l m_t/3)^{1/2} = 2.15\ m_e \quad \text{for } \langle 111 \rangle \text{ directions} \tag{2-10-13}$$

(see Rhoderick (1977). For germanium we find that

$$m^* = 4(m_t^2/3 + 2m_l m_t/3)^{1/2} = 1.19\ m_e \qquad \text{for } \langle 100 \rangle \text{ directions} \tag{2-10-14}$$

$$m^* = m_t + (m_t^2 + 8m_l m_t/3)^{1/2} = 1.07\ m_e \quad \text{for } \langle 100 \rangle \text{ directions} \tag{2-10-15}$$

Here m_t and m_l are transverse and longitudinal effective masses, respectively. For (111) surfaces of Si and GaAs, A^* is equal to 96 A/(cm^{2}K^2) and 4.4 A/(cm^{2}K^2), respectively.

In practical devices the current-voltage characteristic is more accurately described by the following equation:

$$j = j_{ss}\left[\exp\left(\frac{qV}{m_i k_B T}\right) - 1\right] \tag{2-10-16}$$

Factor m_i in eq. (2-10-16) is called the *ideality factor*. The review of different mechanisms, including the dependence of the barrier height on bias voltage, leading to this dependence was given by Rhoderick (1978).

It is instructive to compare the value of the saturation current density for the Schottky barrier, j_{ss}, with the saturation current density for a p$^+$-n junction, j_s:

$$\frac{j_{ss}}{j_s} \approx (\pi/2)^{1/2}\,\alpha\,\frac{N_D}{N_c}\left(\frac{\tau_{pl}}{\tau_p}\right)^{1/2}\frac{m_n}{m_p}\exp\left[\frac{q(E_g - \phi_b)}{k_B T}\right] \tag{2-10-17}$$

where τ_{pl} is the hole lifetime in the n-type region, τ_p is the hole momentum relaxation time, E_g is the energy in the p-n junction, ϕ_b is the Schottky barrier height, N_D is the donor density, and N_c is the effective density of states in the conduction band (see Problem 2-10-1). Typically, ϕ_b is of the order of $0.6E_g$. The ratio τ_{pl}/τ_p can be of the order of 10^4 in GaAs and even higher in Si. Hence, the saturation current in a Schottky diode is bigger than in a p$^+$-n junction by many orders of magnitude, and a turn-on voltage is smaller by several tenths of a volt.

2-10-2. Current-Voltage Characteristics: Thermionic-Field Emission and Field Emission

In Schottky barriers on highly doped semiconductors the depletion region becomes so narrow that electrons can tunnel through the barrier near the top, where the barrier is thin (see Fig. 2-10-1). This process is called *thermionic-field emission*. The number of electrons with a given energy, E, exponentially decreases with energy as $\exp[-E/(k_B T)]$. On the other hand, the barrier transparency exponentially increases with the decrease in the barrier width (see Section 2-7). Hence, the dominant electron tunneling path occurs at lower energies as doping increases and the barrier becomes thinner. In degenerate semiconductors, especially in semiconductors with small electron effective mass, such as GaAs, electrons can tunnel through the barrier near the Fermi level, and the tunneling current is dominant. Such a mechanism is called *field emission*. The current-voltage characteristic of a Schottky diode in the case of thermionic-field emission or field emission can be calculated by evaluating the product of the tunneling transmission coefficient and the number of electrons as a function of energy and integrating over the states in the conduction band. Such a calculation yields

$$J = J_{stf} \exp\left[\frac{qV}{E_o}\right] \tag{2-10-18}$$

where

$$E_o = E_{oo} \coth\left(\frac{E_{oo}}{k_B T}\right) \tag{2-10-19}$$

and

$$E_{oo} = \frac{qh}{4\pi}\left(\frac{N_D}{m^* \varepsilon_s}\right)^{1/2} = 1.85 \times 10^{-11}\left[\frac{N_D\,(\text{cm}^{-3})}{(m_n/m_e)(\varepsilon_s/\varepsilon_o)}\right]^{1/2} (eV) \tag{2-10-20}$$

(see Padovani and Stratton 1966). The preexponential term was calculated by Crowell and Rideout (1969):

$$J_{stf} = \frac{A^*\, T\,[\pi E_{oo} q(\phi_b - V - \xi)]^{1/2}}{k_B \cosh(E_{oo}/k_B T)} \exp\left[-\frac{q\xi}{k_B T} - \frac{q}{E_o}(\phi_b - \xi)\right] \tag{2-10-21}$$

Here $\xi = (E_c - E_{Fn})/q$ (see Fig. 2-10-1), so that ξ is negative for a degenerate semiconductor. In GaAs the thermionic field emission occurs roughly for $N_D > 10^{17}$ cm^{-3} at 300 K and and for $N_D > 10^{16}$ cm^{-3} at 77 K. In silicon the corresponding values of N_D are several times bigger.

If field emission takes place at very high doping levels, the width of the depletion region becomes so narrow that direct tunneling from the semiconductor to the metal may take place, as shown in Fig. 2-10-1. This happens when E_{oo} becomes much greater than $k_B T$. The current-voltage characteristics in this regime are given by

$$J \approx J_{sf} \exp\left[\frac{qV}{E_{oo}}\right] \tag{2-10-22}$$

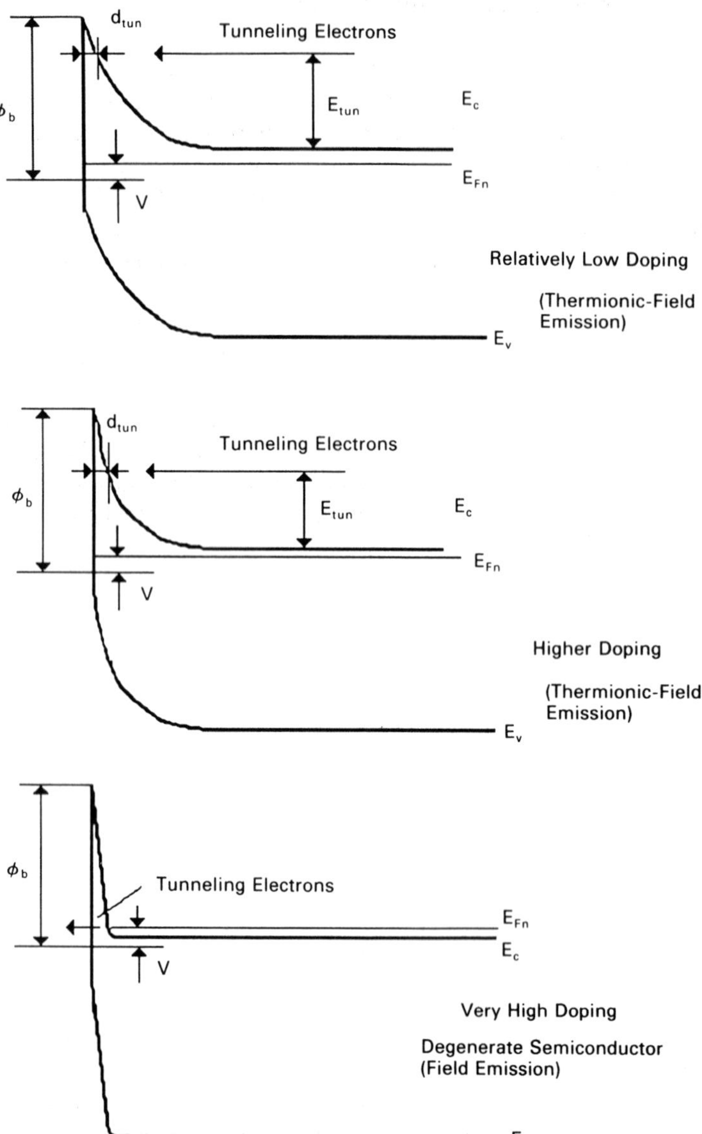

Fig. 2-10-1. Thermionic-field and field emission under forward bias. d_{tun} is the characteristic tunneling length. At low doping levels electrons tunnel across the barrier closer to the top of the barrier. With increase in doping, the characteristic tunneling energy, E_{tun}, decreases. In highly doped degenerate semiconductors electrons near the Fermi level tunnel across a very thin depletion region.

where

$$J_{sf} = \frac{\pi A^* T}{k_B C_1 \sin(\pi k_B T C_1)} \exp\left[-\frac{q\phi_b}{E_{oo}}\right] \qquad (2\text{-}10\text{-}23)$$

and

$$C_1 = (2E_{oo})^{-1} \ln[-4(\phi_b - V)/\xi] \qquad (2\text{-}10\text{-}24)$$

The effective resistance of the Schottky barrier in the field-emission regime is quite low. Therefore, the metal-n^{++} Schottky barriers are used for ohmic contacts (see Section 2-11).

2-10-3. Small-Signal Circuit of a Schottky Diode

The small-signal equivalent circuit of a Schottky barrier is shown in Fig. 2-10-2. It includes the parallel combination of differential resistance of the Schottky barrier,

$$R_d = dV/dI \qquad (2\text{-}10\text{-}25)$$

and the differential capacitance of the space charge region (which may be estimated using the depletion approximation, just as we did for a p^+-n junction; see Section 2-5):

$$C_d = S\left[\frac{qN_D\varepsilon_s}{2(V_{bi} - V)}\right]^{1/2} \qquad (2\text{-}10\text{-}26)$$

Here S is the device cross section. As shown in Fig. 2-10-3 (from Parker 1987), the capacitance of a practical Schottky barrier device is indeed proportional to $(V_{bi} - V)^{-1/2}$.

These circuit elements are in series with series resistance (consisting of the contact resistance and the resistance of the neutral semiconductor region between the ohmic contact and the depletion region and the equivalent inductance) and parasitic inductance. The device geometric capacitance, C_{geom}, is given by

$$C_{geom} = \varepsilon_s S/L \qquad (2\text{-}10\text{-}27)$$

where L is the device length.

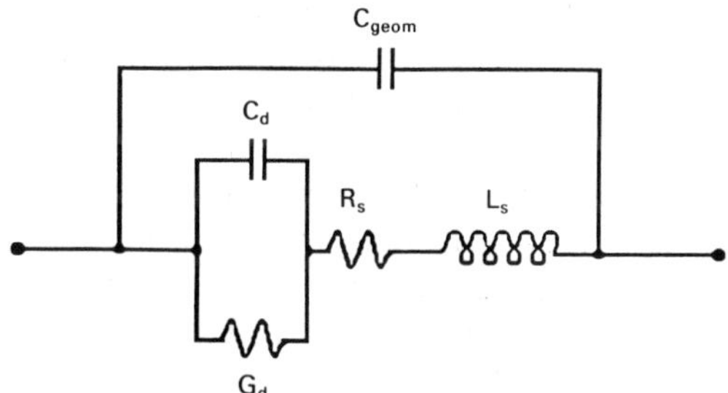

Fig. 2-10-2. Equivalent circuit of Schottky diode.

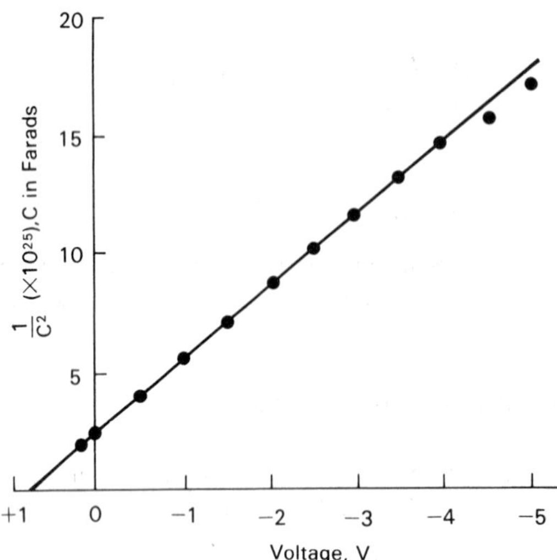

Fig. 2-10-3. $1/C^2$ vs. voltage for Indium-Tin-Oxide-GaAs Schottky barrier diode (from D. G. Parker, General Electric Company Journal of Research, 5, no. 3, pp. 116–123 (1987)).

The major difference between the equivalent circuit of Fig. 2-10-2 and an equivalent circuit for a p-n junction is the absence of the diffusion capacitance in the equivalent circuit of a Schottky diode. This leads to a much faster response under forward bias conditions and allows the use of Schottky diodes as microwave mixers, detectors, etc. (see, for example, Maas 1986).

In most cases $R_s << R_d$ and $C_d >> C$, so that the characteristic time constant limiting the frequency response of a Schottky diode is given by

$$\tau_{\text{Schottky}} = R_s C_d \tag{2-10-28}$$

A good review of early work on Schottky contacts was given by Rhoderick (1978). A more recent bibliography was given by Sharma and Gupta (1980). A thorough review of metal-semiconductor contacts to III-V compounds was published by Robinson (1983).

2-11. OHMIC CONTACTS

The current-voltage characteristics of a Schottky barrier diode and of an "ohmic contact are compared in Fig. 2-11-1a. Ideally, an ohmic contact has a linear current-voltage characteristic and a very small resistance that is negligible compared with the resistance of the active region of a semiconductor device. An ohmic contact to an n-type semiconductor should also ideally be made using a metal with a lower work function than a semiconductor. Unfortunately, very few practical material systems satisfy this condition, and metals usually form Schottky barriers at semiconductor interfaces. Therefore, a practical way to obtain a low resistance ohmic contact is to increase the doping near the metal-semiconductor

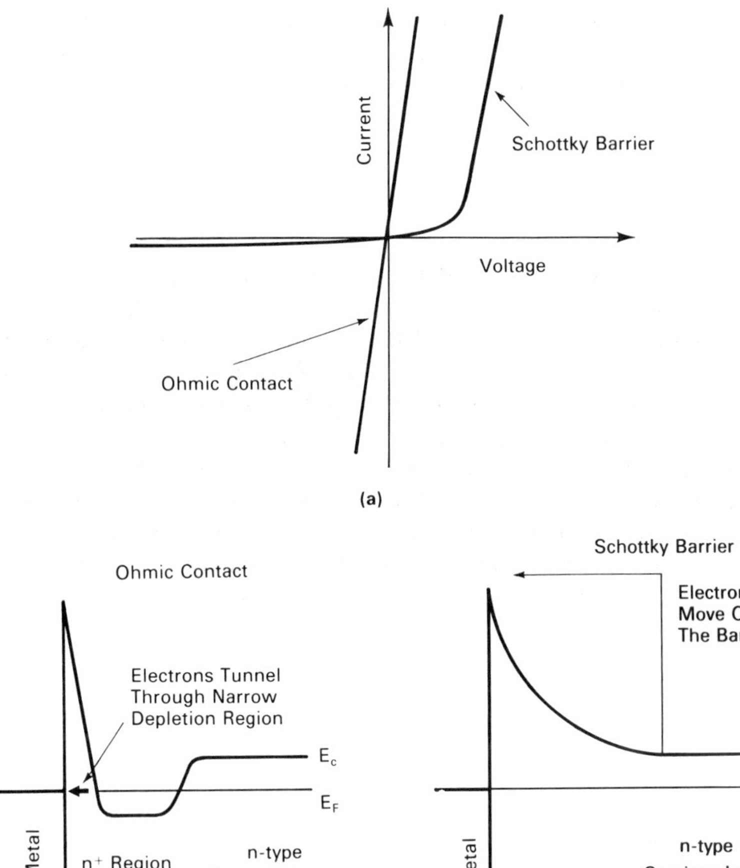

Fig. 2-11-1. (a) Current-voltage characteristics of Schottky barrier diode and of ohmic contact (b) Band diagrams of metal $-n^+$-n ohmic contact and a Schottky contact.

interface to a very high value so that the depletion layer caused by the Schottky barrier becomes very thin and the current transport through the barrier is enhanced by tunneling (field emission regime; see Section 2-10). Band diagrams of a metal $-n^+$-n ohmic contact and a Schottky contact are compared in Fig. 2-11-1b. This comparison clearly illustrates the role played by the n^+ layer. As was discussed in Section 2-10, the current density, j, when the tunneling mechanism is dominant is given by eqs. (2-10-18) to (2-10-22). From these equations we find that a specific contact resistance, r_c, defined as

$$r_c = dV/dj \,|_{V \to 0} \tag{2-11-1}$$

is proportional to

$$r_c \sim \exp[4\pi(\varepsilon_s m_n)^{1/2} q\phi_b/(N_D)^{1/2}] \tag{2-11-2}$$

At relatively low doping levels, when thermionic-field emission and thermionic emission play a dominant role, contact resistance should be much less dependent on the doping concentration. This conclusion is in good agreement with experimental data (see Fig. 2-11-2 from Sze 1981, where we compare experimental data and theoretical calculations for PtSi and Al ohmic contacts to Si).

A conventional approach to reducing the specific contact resistance is to form very high doped regions near the surface by using alloyed ohmic contacts. Ion implantation or diffusion have also been used to create a highly doped region near the surface to facilitate the formation of ohmic contacts. The doping concentration is limited by the impurity solubility and may reach 5×10^{19} cm^{-3} for n-type

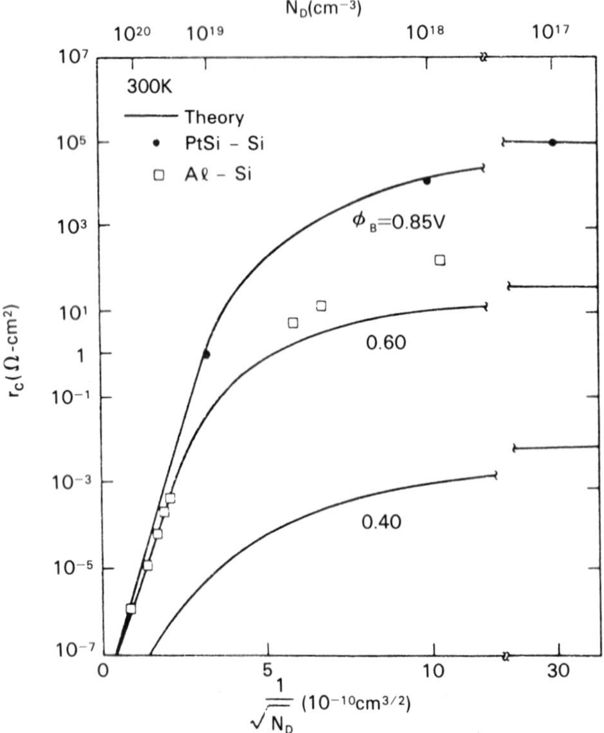

Fig. 2-11-2. Experimental and theoretical values of specific contact resistance for PtSi and Al ohmic contacts to Si (From S. M. Sze, *Physics of Semiconductor Devices*, John Wiley & Sons, New York 1981).

GaAs and 10^{20} cm^{-3} for p-type GaAs (see Chang and Pearson 1964). Using ion implantation, doping concentrations in access of the solubility limit may be reached near the surface (Robinson 1983). Good alloyed ohmic contacts to n-type GaAs have a low contact resistance (less than 10^{-6} Ωcm^2) and high reliability.

Au-Zn (see Gohen and Yu 1971) , Au-Zn-Au (see Sanada and Wada 1980), and Ag-Zn (see Matino and Tokunaga 1979) alloyed contacts are used for p-type GaAs.

As was pointed out by Braslau (1981) the numerous experimental data indicate that the specific contact resistance of alloyed ohmic contacts to GaAs is roughly proportional to $1/N_D$, where N_D is the doping concentration in the semiconductor layer. According to the previous discussion, we could have expected that r_c is proportional to $\exp[4\pi(\varepsilon_s m_n)^{1/2} q\phi_b/(N_{D+})^{1/2}]$, where N_{D+} is the effective doping concentration in the alloyed region (see eq. (2-11-2)). Braslau (1981) explained this dependence by introducing a model that postulates that the alloyed region contacts the semiconductor active layer not in uniform fashion but only at certain points. Then the current "spreads out" as shown in Fig. 2-11-3, and the specific contact resistance is determined by the spreading resistance of the active layer which is proportional to the resistivity

$$r = 1/(q\mu_n N_D)$$

(see Robinson 1983 for further discussion of this model).

A discussion of ohmic contacts to InP and GaP was given by Robinson (1983). Ohmic contacts to III-V alloys were considered by Rideout (1975). Nakato et al. (1980) studied ohmic contacts to p-type InGaAs alloys.

A high contact resistance is often a factor limiting the performance of high-speed semiconductor devices. More detailed experimental and theoretical studies are necessary to learn how to produce low-resistance, reliable, and reproducible contacts.

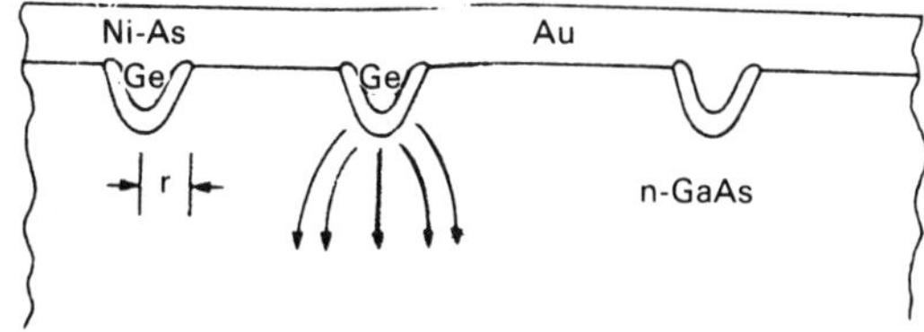

Fig. 2-11-3. Model of alloyed Au-Ge-Ni ohmic contact to GaAs (from N. Braslau, (1981)). Arrows show the lines of current. Conduction takes place through Ge-rich protrusions of negligible resistance. Then the current spreads out into the active layer. The spreading resistance is proportional to the resistivity of the active layer.

The resistance of ideal planar contacts, shown in Figs. 2-11-4a and 2-11-4c, may be found using a *transmission-line model* (TLM), first introduced for diffused ohmic contacts in silicon (see Shockley 1964, Murrmann and Widman 1969, Berger 1969, Berger 1972a, and Berger 1972b).

The current distribution under such a contact (see Fig. 2-11-5) is described by the following equation:

$$\frac{dI}{dx} = -J(x)W \tag{2-11-3}$$

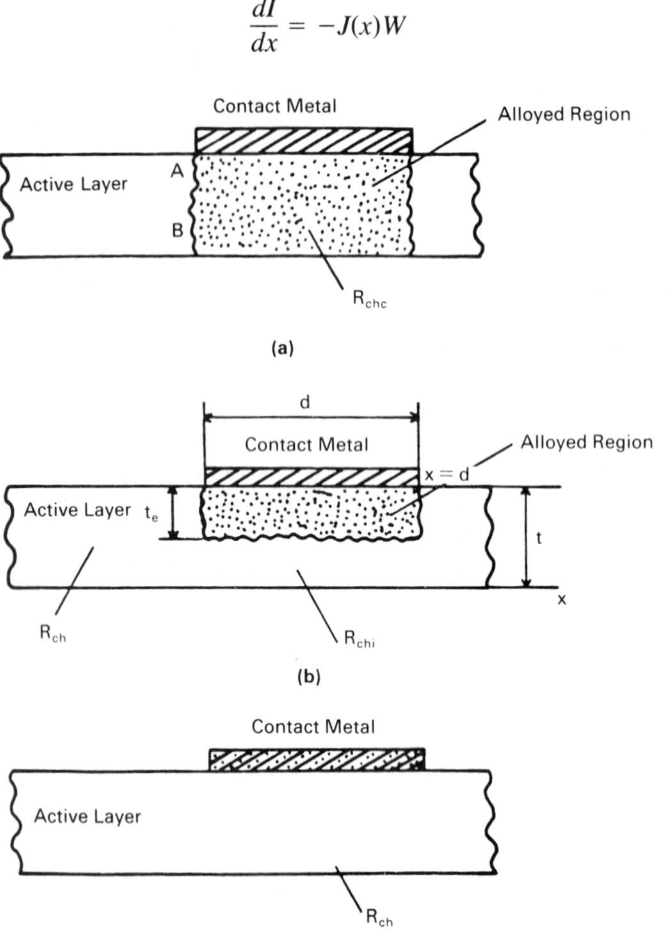

Fig. 2-11-4. Schematic diagram of alloyed ohmic contacts (from M. S. Shur, *GaAs Devices and Circuits,* Plenum, New York (1987). (a) alloyed region goes through the entire active layer, (b) alloyed region does not go through the entire active layer, and (c) nonalloyed contact. R_{ch} is the sheet resistance of the active layer outside the contact, R_{chi} is the sheet resistance of the unalloyed portion of the active layer under the contact, and R_{chc} is the sheet resistance of the alloyed region under the contact when it goes through the entire active region.

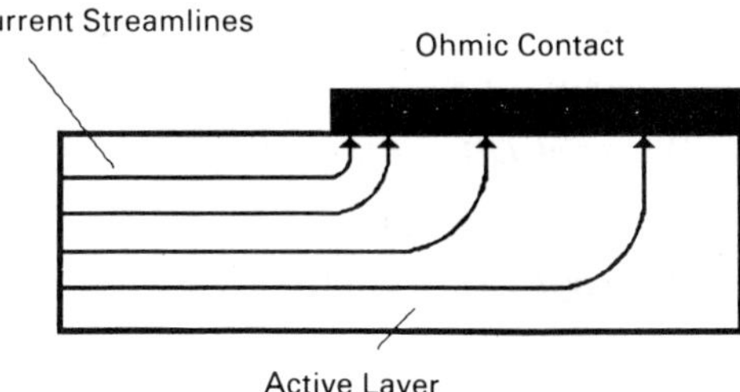

Fig. 2-11-5. Current distribution under the planar contact.

where I is the current, x is the coordinate (in the direction parallel to the current flow in the active layer), and J is the current density,

$$J(x) = V(x)/r_c \tag{2-11-4}$$

Here r_c is the specific contact resistance and $V(x)$ is the channel potential with respect to the potential of the contact metal:

$$\frac{dV}{dx} = -IR_{chc}/W \tag{2-11-5}$$

R_{chc} is the sheet resistance of the semiconductor film under the contact in ohms per square, and W is the contact width. (For the contact shown in Fig. 2-11-4c, $R_{chc} = R_{ch}$, where R_{ch} is the sheet resistance of the semiconductor layer.) Equations (2-11-3) to (2-11-5) may be reduced to an equation similar to the equation describing a transmission line:

$$\frac{d^2V}{dx^2} = \frac{V}{(L_T)^2} \tag{2-11-6}$$

where

$$L_T = (r_c/R_{chc})^{1/2} \tag{2-11-7}$$

is called the *transfer length*. Equation (2-11-6) is only valid when the active-layer thickness is much smaller than L_T.

The boundary conditions for eq. (2-11-6) are

$$\frac{dV}{dx}\bigg|_{(x=0)} = \frac{R_{chc}}{W} I_o \tag{2-11-8}$$

$$\frac{dV}{dx}\bigg|_{(x=d)} = 0 \tag{2-11-9}$$

Here d is the contact length and I_o is the total channel current.

The solution of eq. (2-11-6) with boundary conditions (2-11-8) and (2-11-9) is given by

$$V = A \exp(x/L_T) + B \exp(-x/L_T) \tag{2-11-10}$$

where

$$A = \frac{I_o R_{chc} L_T \exp(d/L_T)}{W[\exp(d/L_T) - \exp(-d/L_T)]} \quad (2\text{-}11\text{-}11)$$

$$B = \frac{I_o R_{chc} L_T \exp(-d/L_T)}{W[\exp(d/L_T) - \exp(-d/L_T)]} \quad (2\text{-}11\text{-}12)$$

From eq. (2-11-10) we find that

$$V(0) = I_o R_c \quad (2\text{-}11\text{-}13)$$

where

$$R_c = R_{chc}(d/W)F_{tlm} \quad (2\text{-}11\text{-}14)$$

is the contact resistance (see Shur 1987). Here

$$F_{tlm} = (L_T/d)\coth(d/L_T) \quad (2\text{-}11\text{-}15)$$

For $d/L_T << 1$, $F_{tlm} \approx (L_T/d)^2$. For $d/L_T >> 1$, $F_{tlm} \approx L_T/d$ (see Fig. 2-11-6).

The value of the potential of the channel at the end of the contact $V(d)$ is also found from eqs. (2-11-10) to (2-11-12):

$$V(d) = \frac{I_o R_{chc} L_T}{W \sinh(d/L_T)} \quad (2\text{-}11\text{-}16)$$

Hence, the *end resistance,* defined as

$$R_{end} = V(d)/I_o \quad (2\text{-}11\text{-}17)$$

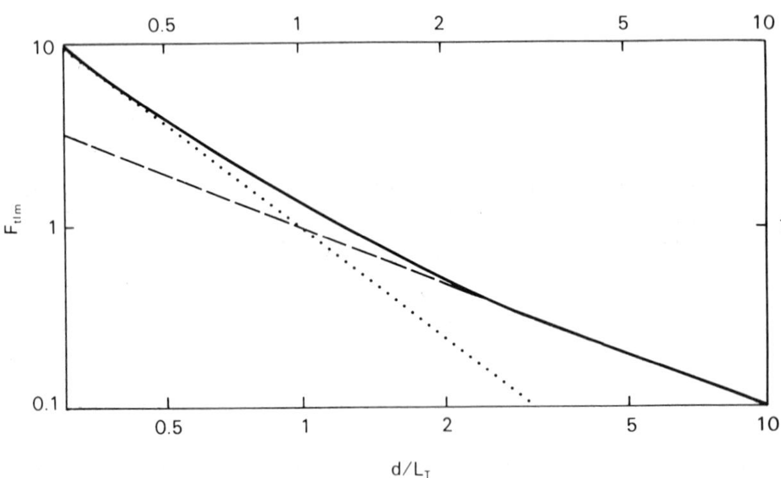

Fig. 2-11-6. F_{tlm} vs. d/L_T (from M. S. Shur, *GaAs Devices and Circuits,* Plenum, New York (1987). Dashed and dotted lines show asymptotic approximations for large and small values of d/L_T respectively.

is given by

$$R_{\text{end}} = R_{\text{chc}}L_T/[\sinh(d/L_T)W] \tag{2-11-18}$$

From eqs. (2-11-14), (2-11-15), and (2-11-18) we find that

$$R_c/R_{\text{end}} = \cosh(d/L_T) \tag{2-11-19}$$

Equations (2-11-17), (2-11-18), and (2-11-19) are used for the contact characterization based on TLM measurements (see Fig. 2-11-7).

A more conventional TLM technique is based on the assumption that r_c is determined by the metal-semiconductor interface and that the semiconductor resistivity under the contact is uniform (though it may be different from the resistivity of the semiconductor layer outside the contact) (see, for example, Keller 1975, Chang 1970, and Reeves and Harrison 1982). In this case the contact resistance R_c and R_{end} are given by eqs. (2-11-14) and (2-11-17), respectively.

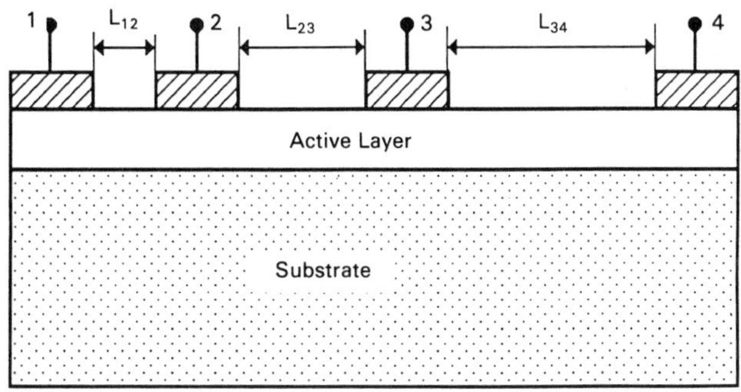

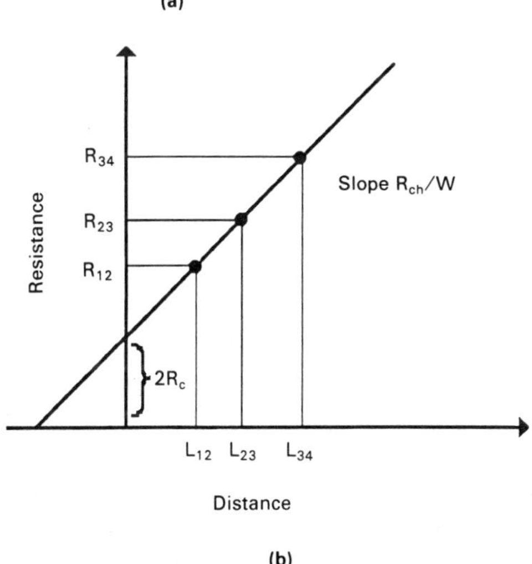

Fig. 2-11-7. Determination of contact and sheet resistance using the Transmission Line Model (TLM) measurements: (a) TLM pattern and (b) resistance between the contacts vs. distance between the contacts.

The TLM pattern used for the contact resistance measurements is shown in Fig. 2-11-7a. The resistance between two adjacent pads of width W separated by distance L is given by

$$R = R_{ch}L/W + 2R_c + R_p \qquad (2\text{-}11\text{-}20)$$

where R_{ch} is the sheet resistance of the active layer between the contacts (i.e., the film resistance per square), R_c is the contact resistance, and R_p is the resistance of the interconnect wires. The value of resistance R_p of the interconnect wires may be estimated by bonding two interconnect wires to the same contact. The value of R_c is determined from the intercept of the R vs. L curve. The sheet resistance of the channel outside the contact R_{ch} is found from the slope of this line (see Fig. 2-11-7b).

When the sheet resistance of the active layer under the contact, R_{chc}, is different from R_{ch}, an additional "end" resistance measurement is needed to deduce R_{chc} and r_c. The end resistance, R_{end}, is defined as

$$R_{end} = V_{2,3}/I_{1,2} \qquad (2\text{-}11\text{-}21)$$

where I_{12} is the current flowing through contact 1 and 2 and $V_{2,3}$ is the voltage difference between contact 2 and floating contact 3 (see Fig. 2-11-7a). An alternative but equivalent technique of measuring of R_{end} was proposed by Reeves and Harrison (1982).

As was discussed previously, in the simple case when the specific contact resistance is determined by either the contact metal-semiconductor interface or by the interface between the alloyed and nonalloyed portions of the active layer, the specific contact resistance, r_c, and the sheet resistance of the channel under the contact, R_{chc}, may be determined from R_c and R_{end} using equations of the transmission-line model:

$$R_c = R_{chc}\,(L_T/W)\coth(d/L_T) \qquad (2\text{-}11\text{-}22a)$$

$$R_{end} = R_{chc}(L_T/W)/\sinh(d/L_T) \qquad (2\text{-}11\text{-}22b)$$

From these equations we find that

$$L_T = d/\cosh^{-1}(R_c/R_{end})$$

Then R_{chc} and r_c are determined by solving eq. (2-11-7):

$$L_T = (r_c/R_{chc})^{1/2}$$

together with eq. (2-11-22a) or eq. (2-11-22b). Reeves and Harrison (1982) measured the values of R_{ch} and R_{chc} using the transmission-line technique (based on eqs. (2-11-14) and (2-11-18)) for contacts to n-type GaAs and to p-type Si. For contacts to silicon they deduced $R_{ch} \approx 2100$ Ω/square and $R_{chc} \approx 430$ Ω/square. For contacts to n-type GaAs they obtained $R_{ch} \approx 430$ Ω/square and $R_{chc} \approx 22$ Ω/square. Hence, the sheet resistance of the channel under the alloyed contact is much smaller than the sheet resistance of the active layer. As a consequence, the transfer length is much larger than the value that is deduced when R_{chc} is assumed to be equal to R_{ch}.

If both the interface between the contact metal and semiconductor and the interface between the alloyed and nonalloyed portions of the active layer (see Fig. 2-11-4b) contribute to the contact resistance, a more sophisticated model will have to be developed to deduce the specific contact resistance from the TLM measurements.

In addition to the TLM technique, the contact resistance may also be measured by other methods. The Cox-Strack method (see Cox and Strack 1967) is used for the experimental determination of the contact resistance for thick (bulk) samples when the contacts may be made to both sides of the wafer. For the configuration shown in Fig. 2-11-8 the total resistance R between the top and the bottom contacts is given by

$$R = R_c + R_b \tag{2-11-23}$$

where

$$R_c = r_c/(\pi a^2) \tag{2-11-24}$$

is the contact resistance and

$$R_b = \rho F_{cs}(a/t)/a \tag{2-11-25}$$

is the bulk resistance of the layer (see Cox and Strack 1967). Here t is the film thickness, a is the radius of the top contact, ρ is the resistivity of the active layer, and the function F takes into account the spreading of the current streamlines of the active layer.

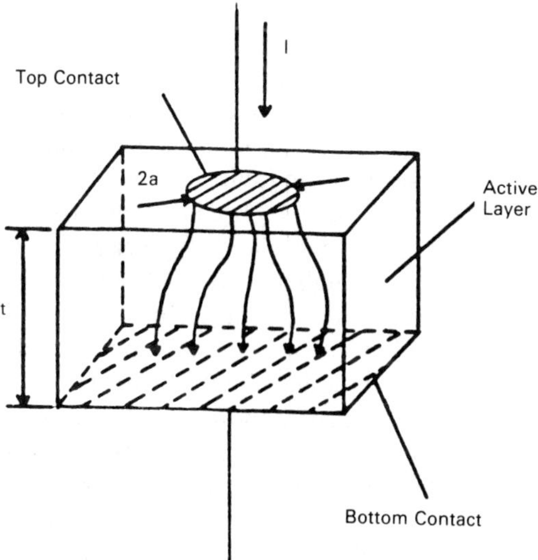

Fig. 2-11-8. Cox-Strack characterization technique.

According to Cox and Strack (1967) function F_{cs} may be approximated as

$$F_{cs}(a/t) \approx \tan^{-1}(2t/a)/\pi \tag{2-11-26}$$

A more accurate numerical calculation of F_{cs} was reported by Brooks and Mathes (1971).

In practical measurements, the values of R are determined for top contacts of different areas and the values of $R - R_b$ are plotted as a function of $1/a^2$ to determine r_c and to find the correction introduced by the interconnect wires (see Robinson 1983).

Other techniques of contact-resistance measurement are reviewed, for example, by Robinson (1983).

2-12. HETEROJUNCTIONS

The first *heterojunction device*—the device using the contact between two different semiconductor materials—was proposed by Shockley (1951). Gubanov (1951) developed the first theory of heterojunctions. Kroemer (1957) published a pioneering paper on a heterojunction bipolar junction transistor. Anderson (1960) presented the results of an experimental study of Ge-GaAs heterojunctions and proposed a simple model that became a starting point for most discussions of heterojunction behavior. He also pointed out that an *accumulation layer*—a layer with high carrier concentration—may form at the heterojunction interface, as shown in Fig. 2-12-1. This accumulation layer is very thin, so that the carrier motion in the direction perpendicular to the heterointerface may be quantized, just as the motion of electrons in a potential well considered in Section 1-6 is quantized (see Fig. 1-6-7). The accumulation layer is therefore frequently called a *two-dimensional electron gas*. This emphasizes that electrons are free to move only in the direction parallel to a heterointerface.

Esaki and Tsu (1969) pointed out that the mobility of carriers in the two-dimensional gas at the heterointerface may be higher than the mobility of carriers in the bulk material. Indeed, in bulk material, electrons are usually supplied by shallow ionized donors and, hence, suffer from ionized impurity scattering. In an accumulation layer, such as is shown in Fig. 2-12-1, a large electron concentration may be created in an undoped material. Hence, impurity scattering is diminished and the low-field mobility can be much higher, especially at low temperatures, when the ionized impurity scattering is dominant. A prerequisite for such an enhancement of the electron low-field mobility is the quantization of the electron motion in the direction perpendicular to the heterointerface. This means that the separation between the energy levels corresponding to this motion should be greater than the thermal energy, $k_B T$. Otherwise, the electronic motion is essentially three-dimensional, and the electron mobility in the accumulation layer may actually be smaller because electrons may experience an additional scattering by the potential walls limiting the accumulation layer. This is usually the case in

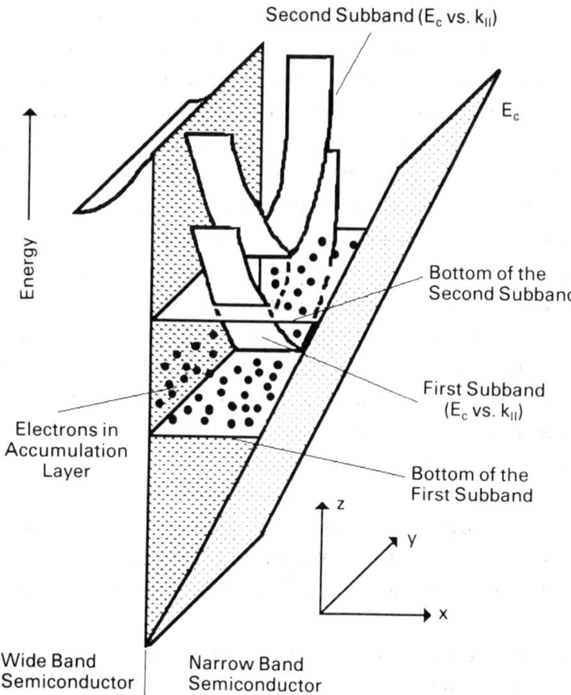

Fig. 2-12-1. Bottom of conduction band near heterointerface with accumulation layer in narrow gap semiconductor (to the right of the conduction band discontinuity). Also shown are two energy subbands (compare with Fig. 1-6-7) that correspond to two quantized levels due to the restricted electron motion in the direction perpendicular to the heterointerface.

silicon, where the electron effective mass is large and 2-d levels are close to each other (As can be seen from eq. (1-2-20) the energy difference between the levels is inversely proportional to the electron effective mass). In gallium arsenide, the electron effective mass is very small ($0.067m_e$), and the mobility enhancement can be easily observed.

Dingle et al. (1978) were first to observe the enhanced electron mobility of the two-dimensional electron gas in GaAs. Mimura et al. (1980) developed the first heterojunction AlGaAs-GaAs transistor that utilized the mobility enhancement of the two-dimensional electron gas. They called this device the HEMT (High Electron Mobility Transistor). (Other terms, such as Modulation Doped Field Effect Transistor (MODFET), Selectively Doped Heterojunction Transistor (SDHT), Two-dimensional Electron Gas Field Effect Transistor (TEGFET), and Heterojunction Field Effect Transistor (HFET) are also used for the same device.) Based on this proliferation of different names, we can assume that this device has been quite successful and has attracted a lot of interest. Since 1980 considerable progress has been achieved in heterostructure field-effect transistors of different types (see Chapter 4). Other important heterostructure devices include Heterojunction Bipolar Transistors (HBTs) (see Chapter 3), heterojunction solar cells, heterojunction photodetectors, heterojunction Light Emitting Diodes (LEDs), heterojunction lasers (see Chapter 5), and novel heterojunction devices (see Chapter 7).

When two different semiconductors are joined together the atoms at the heterointerface have to form chemical bonds. As the lattice constants of these two semiconductor materials are different, atoms at the heterointerface have to adjust by developing strain. If this strain exceeds some critical value, it results in crystal dislocations that are crystal imperfections propagating across many crystalline layers. These dislocations act as scattering centers for electrons and holes, limiting electron and hole mobilities, and as recombination centers, limiting the electron and hole lifetimes. The result may be very poor device properties. One possible way to avoid this problem is to use materials with nearly equal lattice constants, such as GaAs and AlAs (or a ternary compound $Al_xGa_{1-x}As$). Another approach (used, for example, in growing Si-Ge heterostructures) is to use thin alternating layers of the two semiconductor materials, producing a structure that is called a *superlattice* (see Section 7-5). This reduces strain and reduces the number of dislocations. Still another technique is to choose a substrate crystal plane that is slightly offset from a major crystal plane so that the distance between the atoms on the substrate surface approximates the distance between the atoms in the deposited film of another semiconductor material. This may also lead to a deflection of the dislocations, so that they are primarily located near the heterointerface. Such an approach is used for growing GaAs films on silicon substrates (see, for example, Morkoç et al. 1985).

More recently an intriguing idea of growing heterostructure films on porous silicon was proposed by Luryi and Suhir (1986). The surface of the porous silicon consists of small (100–300 Å) random islands separated by 30–50 Å grooves. These grooves may relieve the strain and prevent the formation of dislocations. This idea, however, is still to be checked experimentally.

Another consequence of the lattice mismatch in heterostructures is the appearance of the surface states similar to the surface states at the metal-semiconductor interface in metal-semiconductor contacts (see Section 2-9). These surface states also act as scattering centers for electrons and holes, limiting electron and hole mobilities, and as recombination centers, limiting the electron and hole lifetimes. In short, heterojunction interfaces may be far from ideal, especially when the lattice mismatch is noticeable. Nevertheless, a model for an ideal heterojunction first developed by Anderson provides some useful insight into the device behavior.

According to this model the energy bands in both materials constituting a heterostructure are not affected. Then the problem reduces to a proper alignment of the band edges at the heterointerface. Anderson assumed that the vacuum energy level is continuous and, hence, that the conduction band discontinuity is determined by the difference of the electron affinities (see Fig. 2-12-2):

$$\Delta E_c = X_1 - X_2 \tag{2-12-1}$$

The valence band discontinuity is then given by

$$\Delta E_v = \Delta E_g - \Delta E_c \tag{2-12-2}$$

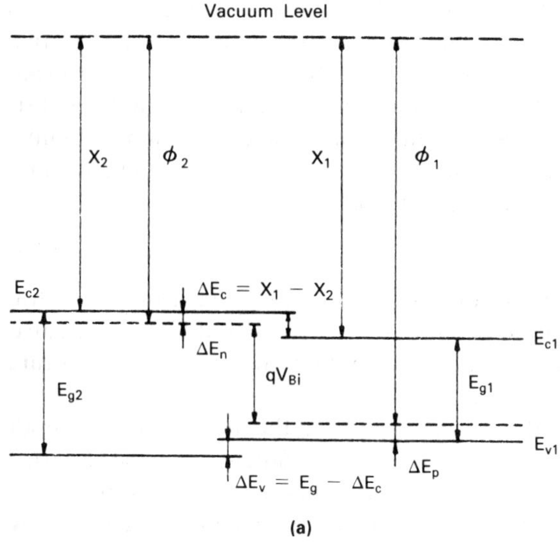

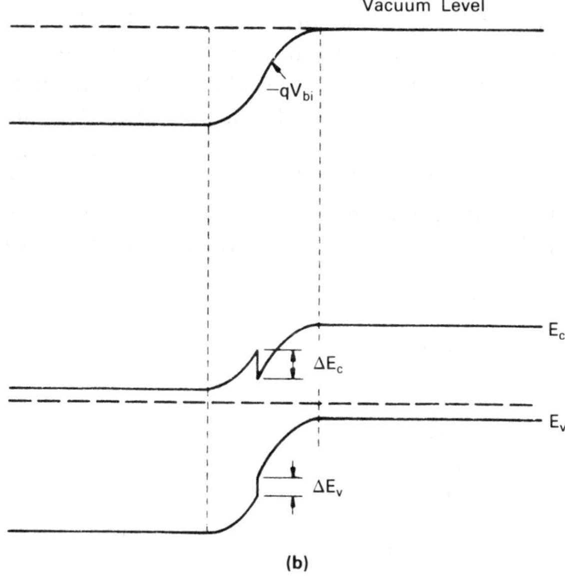

Fig. 2-12-2. (a) Band diagram for two different semiconductor materials (n-type $Al_{0.35}Ga_{0.65}As$ and p-type GaAs. (b) Band diagram of a p-n heterojunction under zero bias according to the Anderson model.

or

$$\Delta E_v = \Delta E_g - X_1 + X_2 \tag{2-12-3}$$

where ΔE_g is the energy gap discontinuity. However, this model is in disagreement with experimental data (see, for example, Bauer and Margaritonto 1987), and alternative models have been proposed (see, for example, the discussion by Kroemer [1983] and Terzoff [1984]).

Terzoff (1984) postulated that the band discontinuities in the heterojunctions are controlled by the same mechanism as the barrier height in the Schottky barrier diodes. This mechanism is the electron tunneling from one material into the energy gap of the other material at the heterointerface, leading to the formation of the interfacial dipole layer. According to this model, the conduction band discontinuity can be related to the difference in the Schottky barrier heights for the two semiconductor materials forming the heterojunction:

$$\Delta E_c = \phi_{b1} - \phi_{b2} \tag{2-12-4}$$

Eizenberg et al. (1986) tested this model experimentally, measuring the Schottky barrier heights and band discontinuities for $Al_xGa_{1-x}As$ (for different values of x) and GaAs using the internal photoemission technique (mentioned in Section 2-8) and found a good correlation (see Fig. 2-12-3).

In most models, however, the only change from the ideal Anderson model is a different choice of $\Delta E_c/\Delta E_g$. In principle, this parameter may be determined from the experiment; however, different experiments have yielded different values, even for the most widely studied heterojunction system, which is a GaAs-$Al_xGa_{1-x}As$ heterostructure.

When two semiconductors form a heterojunction, the Fermi level must be continuous throughout. Just as in a conventional p-n junction, this requirement leads to band bending, as shown in Fig. 2-12-2b. If the conduction band discontinuity, ΔE_c, is known, the built-in voltage, V_{bi}, may be found from Fig. 2-12-2b:

$$qV_{bi} = E_{g1} - \Delta E_n - \Delta E_p + \Delta E_c \tag{2-12-5}$$

Here E_{g1} is the energy gap of the narrower gap material, which we assumed to be doped p-type (see Fig. 2-12-2b), ΔE_n is the difference between the bottom of the

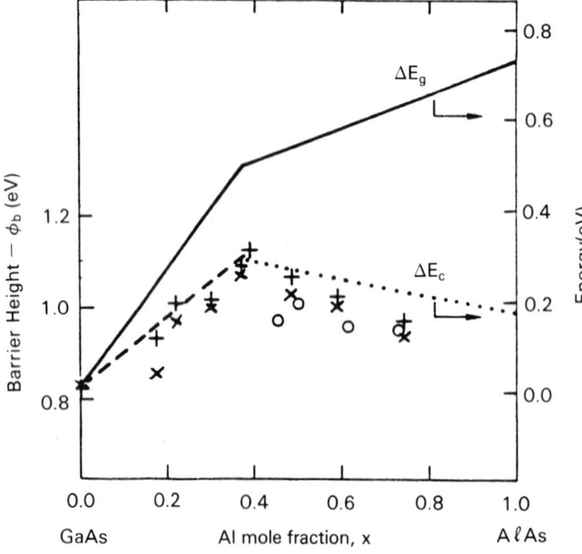

Fig. 2-12-3. Schottky barrier heights and conduction band discontinuities for $Al_xGa_{1-x}As$-GaAs heterojunctions vs. composition, x, of $Al_xGa_{1-x}As$ (from M. Eizenberg, M. Heiblum, M. I. Nathan, N. Braslau, and P. M. Mooney, "Barrier-Heights and Electrical Properties of Intimate Metal-Al-GaAs Junctions," *J. Appl. Physics.*, pp 1516–1522 (1987).

conduction band in the wide-gap n-type material and the Fermi level far from the heterointerface, and ΔE_p is the difference between the Fermi level and the top of the valence band in the narrow-gap p-type material.

Using the depletion approximation, we find how V_{bi} is divided between the p region and n region:

$$V_{bi1} = \frac{\varepsilon_2 N_D}{\varepsilon_2 N_D + \varepsilon_1 N_A} V_{bi} \tag{2-12-6}$$

$$V_{bi2} = \frac{\varepsilon_1 N_A}{\varepsilon_2 N_D + \varepsilon_1 N_A} V_{bi} \tag{2-12-7}$$

The depletion widths x_{dn} and x_{dp} are given by

$$x_{dn} = \left[\frac{2\varepsilon_1\varepsilon_2 N_A V_{bi}}{qN_D(\varepsilon_2 N_D + \varepsilon_1 N_A)}\right]^{1/2} \tag{2-12-8}$$

$$x_{dp} = \left[\frac{2\varepsilon_1\varepsilon_2 N_D V_{bi}}{qN_A(\varepsilon_2 N_D + \varepsilon_1 N_A)}\right]^{1/2} \tag{2-12-9}$$

and the depletion capacitance is given by

$$C_d = \left[\frac{\varepsilon_1\varepsilon_2 N_A N_D}{2(\varepsilon_2 N_D + \varepsilon_1 N_A)V_{bi}}\right]^{1/2} \tag{2-12-10}$$

When a reverse or a small forward bias is applied to a heterojunction device, we have to substitute V_{bi} by $V_{bi} - V$ in the foregoing equations for x_{dn}, x_{dp}, and C_d.

The simplest model for the current-voltage characteristics of a heterojunction is based on the assumption of a thermionic emission mechanism across the heterointerface (similar to that considered for the Schottky diodes in Section 2-10). This model (see Chang 1965) leads to the following expression for the current-voltage characteristics:

$$j = (A^* T q V_{bi}/k_B)\exp(-qV_{bi}/k_B T)[\exp(qV/k_B T) - 1] \tag{2-12-11}$$

Here j is the current density and A^* is the effective Richardson constant (see Problem 2-12-1). A more accurate model that takes into account diffusion-drift processes in the depletion regions and the thermionic emission across the heterointerface was developed by Perlman and Feucht (1964).

REFERENCES

R. A. ANDERSON, *IBM J. Res. Dev.*, 4, p. 283 (1960).

C. L. ANDERSON and C. R. CROWELL, *Phys. Rev.*, B8, p. 2267 (1972).

W. T. ANDERSON, JR., A. CHRISTOU, and J. F. GIULIANI, *IEEE Electron Device Letters*, EDL-2, p. 115 (1981).

S. ASAI et al., *Proc. 5th Conf. on Solid State Devices* (Tokyo), p. 442 (1973).

G. A. BARAFF, *Phys.*, 133, p. A26 (1964).

J. BARDEEN, *Phys. Rev.*, 71, p. 717 (1947).

R. S. BAUER and G. MARGARITONTO, *Physics Today*, p. 3, Jan. (1987).

H. H. BERGER, "Contact Resistance on Diffused Resistors," *IEEE ISSCC Digest of Tech. Papers*, pp. 160–161 (1969).

H. H. BERGER, "Contact Resistance and Contact Resistivity," *J. Electrochem. Soc.*, 119, p. 509 (1972a).

H. H. BERGER, "Models for Contacts to Planar Devices," *Solid State Electronics*, 15, p. 145 (1972b).

D. I. BLOKHINTSEV, *Principles of Quantum Mechanics*, Allyn and Bacon, Boston (1964).

G. BRADERTSCHER, R. P. SALATHE, and W. LUTHY, *Elec. Lett.*, 16, p. 113 (1980).

N. BRASLAU, "Alloyed Ohmic Contacts to GaAs," *J. Vac. Sci. Tech.*, Vol 19, No. 3, p. 803, Sep/Oct. (1981).

R. D. BROOKS and H. G. MATHES, "Spreading Resistance between Constant Potential Surfaces," *Bell System Tech. J.*, 50. pp. 775–784 (1971).

I. F. CHANG, "Contact Resistance in Diffused Resistors," *J. Electrochem. Soc.*, 117, p. 368 (1970).

L. L. CHANG and G. L. PEARSON, "The Solubilities and Distribution Coefficients of Zn in GaAs and GaP," *Physics and Chem. Solids*, 25, pp. 23–30 (1964).

A. CHRISTOU, *Sol. State Electr.*, 22, p. 141 (1979).

R. H. COX and H. STRACK, *Sol. State Electr.*, 10, p. 1213 (1967).

C. R. CROWELL, *Sol. State Electr.*, 8, p. 395 (1965).

C. R. CROWELL and V. L. RIDEOUT, *Sol. State Electr.*, 12, p. 89 (1969).

C. R. CROWELL and S. M. SZE, *Sol. State Electr.*, 9, p. 695 (1966).

R. DINGLE, H. L. STORMER, A. C. GOSSARD, and W. WIEGMANN, *Appl. Phys. Lett.*, 37, p. 805 (1978).

G. ECKHARDT, *Laser and Electron Beam Processing of Materials*, Academic Press, ed. C. W. White and P. S. Peercy, p. 467 (1980).

W. D. EDWARDS, W. A. HARTMAN, and A. B. TORRENS, *Sol. State Electr.*, 15, p. 387 (1972).

M. EIZENBERG, M. HEIBLUM, M. I. NATHAN, N. BRASLAU, and P. M. MOONEY, "Barrier-Heights and Electrical Properties of Intimate Metal-AlGaAs Junctions," *J. Appl. Phys.*, 61, pp. 1516–1522 (1987).

L. ESAKI and R. TSU, "Superlattice and Negative Conductivity in Semiconductors," IBM Res. Note, RC-2418, March (1969).

Y. K. FANG, C. Y. CHANG, and Y. K. SU, "Contact Resistance in Metal-Semiconductor Systems," *Sol. State Electr.*, 22, pp. 933–938 (1979).

H. J. GOHEN and A. Y. C. YU, "Ohmic Contacts to Epitaxial p-GaAs," *Sol. State Electr.*, 14, pp. 515–517 (1971).

R. B. GOLD, R. A. POWELL, and J. F. GIBBONS, "Laser-Solid Intersections and Laser Processes," *AIP Conf. Proc.*, no. 50, p. 635 (1978).

Y. A. GOLDBERG and B. V. TSARENKEV, *Sov. Phys. Semi.*, 3, p. 551 (1970).

H. R. GRINOLDS and G. Y. ROBINSON, *Sol. State Electr.*, 23, p. 973 (1980).

A. S. GROVE, *Physics and Technology of Semiconductor Devices,* Wiley, London (1967).

A. I. GUBANOV, *Zh. Tekh. Fiz.*, 21, p. 304 (1951); *Zh. Eksp. Teor. Fiz.*, 21, p. 721 (1951).

R. P. GUPTA and J. FREYER, *Int. J. Elec.*, 47, p. 459 (1979).

K. HEIME, U. KONIG, E. KOHN, and A. WORTMANN, *Sol. State Electr.*, 17, p. 835 (1974).

T. INADA, S. KATO, T. HARA, and N. TOYADA, *J. Appl. Phys.*, 50, p. 4466 (1979).

W. KELLNER, "Planar Ohmic Contacts to n-type GaAs: Determination of Contact Parameters Using the Transmission Line Model," *Siemens Forsch.-u.Entwickl-Ber.*, 4, p. 137 (1975).

K. KLOHN and Z. WANDINGER, *J. Electrochem. Soc.*, 116, p. 709 (1969).

H. KROEMER, *Proc. IRE,* 45, p. 1535 (1957).

H. KROEMER, *IEEE Electron Dev. Lett.*, EDL-4, no. 2, pp. 25–27 (1983).

E. KUPHAL, "Low Resistance Ohmic Contacts to n- and p-InP," *Sol. State Electr.*, 24, pp. 69–78 (1981).

C. A. LEE, R. A. LOGAN, R. L. BATDORF, J. J. KLEINACK, and W. WIEGMAN, "Ionization Rates of Holes and Electrons in Silicon," *Phys. Rev.*, 134, p. A761 (1964).

C. P. LEE, B. M. WELCH, and W. P. FLEMING, "Reliability of AuGe/Pt and AuGe/Ni Ohmic Contacts on GaAs," *Electronics Letters,* no. 12, pp. 406–407 (1981).

M. H. LEE and S. M. SZE, "Orientation Dependence of Breakdown and Voltage in GaAs." Solid State Electronics, 23, p. 1007 (1980).

S. LURYE and E. SUHIR, *Appl. Phys. Lett.*, 49, no. 3, pp. 140–142 (1986).

S. A. MAAS, *Microwave Mixers,* Artech House, Dedham, Mass. (1986).

S. MARGALIT, D. FEBETE, D. M. PEPPER, G. P. LEED, and A. YARIV, *Appl. Phys. Lett.*, 33, p. 346 (1978).

H. MATINO and M. TOKUNAGA, "Contact Resistance of Several Metals and Alloys to GaAs," *J. Electrochem. Soc.*, 116, pp. 709–711 (1979).

D. C. MILLER, *J. Electrochem. Soc.*, 127, p. 467 (1980).

T. MIMURA, S. HIYAMIZU, T. FUJII, and K. NAMBU, "A New Field Effect Transistor with Selectively Doped GaAs/n-Al_xGa_{1-x} as Heterostructures," *Jpn. J. Appl. Phys.*, 19, pp. L225–227 (1980).

J. L. MOLL, *Physics of Semiconductors,* McGraw-Hill, New York (1964).

H. MORKOÇ, C. K. PENG, T. HENDERSON, W. KOPP, R. FISHER, L. P. ERICKSON, M. LONGERBONE, and R. C. YOUNGMAN, *IEEE Elec. Dev. Lett.*, EDL-6, pp. 381–383 (1985).

R. L. MOZZI, W. FABIAN, and I. J. PIEKARSKI, *Appl. Phys. Lett.*, 35, p. 337 (1979).

H. MURRMANN and D. WIDMAN, "Current Crowding on Metal Contacts to Planar Devices," *IEEE Trans. Elec. Dev.*, ED-16, pp. 1022–1024 (1969).

Y. NAKANO, S. TAKAHASHI, and Y. TOYOSHIMA, "Contact Resistance Dependence on InGaAsP Layers Lattice Matched to InP," *Jap. J. Appl. Phys.*, 19, pp. L495–L497 (1980).

Y. I. NISSIM, J. F. GIBBONS, and R. B. GOLD, *IEEE Trans. Elec. Dev.*, ED-28, p. 607 (1981).

D. C. NORTHROP and E. H. RHODERICK. "The Physics of Schottky Barriers in Variable Impedance Devices," ed. M. J. Howes and D. V. Morgan, John Wiley & Sons (1978).

M. OGAWA, *J. Appl. Phys.*, 51, p. 406 (1980).

M. OGAWA, K. OHATA, T. FURUTSUKA, and N. KAWAMURA, *IEEE Trans. on Micr. Theory and Tech.*, MTT-24, p. 300 (1976).

K. OHATA, T. NOZAKI, and N. KAWAMURA, *IEEE Trans. Elec. Dev.*, ED-24, p. 1129 (1978).

K. OHATA and M. OGAWA, *Proc. 12th Annual Reliability Physics Symposium*, IEEE, New York, p. 278 (1974).

A. H. ORABY, K. MURAKAMI, Y. YUBA, K. GAMO, S. NAMBA, and Y. MASUDA, *Appl. Phys. Lett.*, 38, p. 562 (1981).

A. PADOVANI and R. STRATTON, *Sol. State Electr.*, 9, p. 695 (1966).

D. G. PARKER, *GEC J. of Research*, 5, no. 3, pp. 116–123 (1987).

T. P. PEARSALL, F. CAPASSO, R. E. NAHORY, M. A. POLLACK, and J. R. CHELIKOWSKY, "The Band Structure Dependence of Impact Ionization by Hot Carriers in Semiconductors: GaAs," *Sol. State Electr.*, 21, pp. 297–302 (1978).

S. S. PERLMAN and D. L. FEUCHT, *Sol. State Electr.*, 7, pp. 911–923 (1964).

G. K. REEVES and H. B. HARRISON, "Obtaining the Specific Contact Resistance from Transmission Line Model Measurements," *IEEE Elec. Dev. Lett.*, EDL-3, no. 5, pp. 111–113, May (1982).

E. H. RHODERICK, *Metal-Semiconductor Contacts*, Clarendon Press, Oxford, 1978.

V. L. RIDEOUT, "A Review of the Theory and Technology for Ohmic Contacts to Group III-V Compound Semiconductors," *Sol. State Electr.*, 18, pp. 541–550 (1975).

G. Y. ROBINSON, *Solid State Electr.* 18, p. 331 (1975).

G. Y. ROBINSON, "Schottky Diodes and Ohmic Contacts for the III-V Semiconductors, in *Physics and Chemistry of III-V Semiconductor Interfaces*, ed. C. W. Wilmsen, Plenum New York (1983).

T. SANADA AND O. WADA, "Ohmic Contacts to p-GaAs with Au/Zn/Au Structure," *Jap. J. Appl. Phys.*, 19, pp. L491–L494 (1980).

B. L. SHARMA and S. C. GUPTA, "Metal-Semiconductor Barrier Junctions, *Sol. State Tech.*, 23, pp. 90–95 (1980).

W. SHOCKLEY, U.S. Patent 2,569,347 (1951).

W. SHOCKLEY, *Research and Investigation of Inverse Epitaxial UHF Power Transistors*, Report No. Al-TOR-64-207, Air Force Atomic Laboratory, Wright-Patterson Air Force Base, Ohio, September (1964).

M. S. SHUR, *GaAs Devices and Circuits*, Plenum, New York (1987).

M. S. SHUR, *IEEE Trans. Electron Dev.*, ED-35, Aug. (1988).

B. L. SMITH and E. H. RHODERICK, *Sol. State Electr.*, 14, p. 71 (1971).

W. E. SPICER, P. W. CHYE, P. R. SKEATH, C. Y. SU, and I. LINDAU, *J. Vac. Sci. Tech.*, 16, p. 1422 (1979).

R. STALL, C. E. C. WOOD, K. BOARD, N. DANDEKAR, L. F. EASTMAN, and J. DEVLIN, "A Study of Ge/GaAs Interface Grown by Molecular Beam Epitaxy," *J. Appl. Phys.*, 52, pp. 4062–4069 (1981).

M. J. O. STRUTT, *Semiconductor Devices,* Vol. 1, *Semiconductor and Semiconductor Diodes,* Academic, New York, Chap. 2, (1966).

S. M. SZE, *Physics of Semiconductor Devices,* John Wiley & Sons, New York (1981).

S. M. SZE and G. GIBBONS, *Appl. Phys. Lect.,* 8, p. 111 (1966).

H. TEMKIN, R. J. MCCOY, V. G. KERAMIDAS, W. A. BONNER, *Appl. Phys. Letts.,* 36, p. 444 (1980).

L. E. TERRY and R. W. WILSON, "Metallization Systems for Si Integrated Circuits," *Proc. IEEE,* 57, pp. 1580–1586 (1969).

J. TERSOFF, *Phys. Rev.,* B30, p. 4879 (1984).

W. TSENG, A. CHRISTOU, H. DAY, J. DAVEY, and B. WILKINS, *J. Vac. Sci. Tech,* 19, p. 623 (1981).

H. UNLU and A. NUSSBAUM, *IEEE Trans. Elec. Dev.,* ED-33, pp. 616–619 (1986).

F. VIDIMARI, *Elec. Lett.* 15, p. 675 (1979).

L. S. WEINMAN, S. A. JAMISON, and M. J. HELIX, "Sputtered TiW/Au Schottky Barriers on GaAs, *J. Vac. Sci. Tech.,* 18 no. 3, pp. 838–840, April (1981).

S. H. WEMPLE and W. C. NIEHAUS, *Inst. Phys. Conf. Ser.,* 33b, p. 262 (1977).

J. M. WOODALL, J. L. FREEOUF, G. D. PETTIT, T. JACKSON, and P. KIRSHNER, "Ohmic Contacts to n-type GaAs Using Graded Band Gap Layers of $Ga_xIn_{1-x}As$ Grown by Molecular Beam Epitaxy." T. Vac. Sci. Technol., vol. 19, pp. 626–627 (1981).

J. G. WERTHEN and D. R. SCIFRES, *J. Appl. Phys.* 52, p. 1127 (1981).

M. H. WOODS, W. C. JOHNSON, and M. A. LAMPERT, *Sol. State Electr.,* 16, p. 387 (1973).

M. YODER, *Sol. State Electr.,* 23, p. 117 (1980).

N. YOKOYAMA, S. OHKAWA, and H. ISHIKAWA, *Jap. J. Appl. Phys.,* 14, p. 1071 (1975).

A. VAN DER ZIEL, *Solid State Physical Electronics,* Prentice-Hall, Englewood Cliffs, New Jersey (1976).

YA. B. ZELDOVICH and A. D. MYSHKIS, *Elements of Applied Mathematics,* Nauka (in Russian) (1967).

PROBLEMS

2-2-1. Sketch qualitative distributions of electric field, electric potential, and carrier concentration in thermal equilibrium for a p-i-n structure.

2-2-2. Sketch a qualitative energy band diagram in the thermal equilibrium for a n-i-p-i-n structure where the p-layer in the middle is very thin (see Fig. P2-2-2) and the concentration of acceptors in the p-type layer per unit area, N_A, is relatively small. The dielectric permittivity of the semiconductor is ε. What is the value of the electric field in the i-regions for the dimensions shown in Fig. P2-2-2 ?

Hint: When the concentration of donors in the p-type layer per unit area is relatively small, the p-type layer is totally depleted. Consider the doping level of n-type regions to be very large (i.e., the depletion regions in the n-type regions to be very thin).

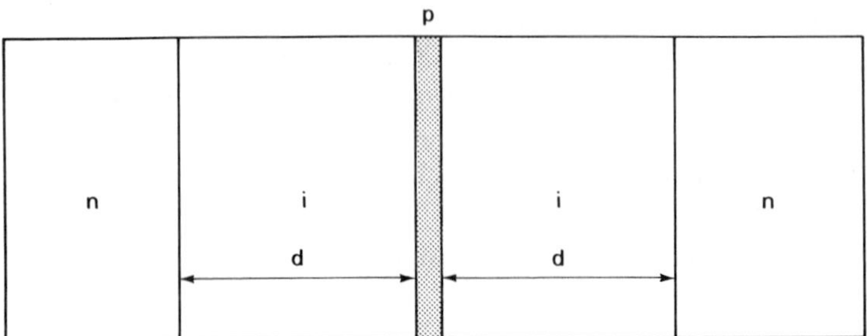

Fig. P2-2-2. n-i-p-i-n structure.

2-3-1. Find the total charge of electrons injected into the p region of a n^+-p silicon diode as a function of the bias voltage, V. Doping density of the p region is N_a, intrinsic carrier concentration is n_i, and the diode temperature is T. The length of the p region is L, and the diffusion length of electrons in the p region is L_n. Consider three cases: (a) arbitrary relation between L and L_n, (b) $L >> L_n$, and (c) $L << L_n$. Assume that at the contact, $n = n_{po}$, where n_{po} is the equilibrium concentration of electrons.

2-3-2. Calculate the Auger recombination current, I_{rA}, in the n region of n^+-p diode as a function of the bias voltage, V. The Auger recombination time is given by

$$\tau_A = \frac{1}{G_n N_D^2}$$

where $G_n = 10^{-31}$ cm^6/s. The donor density $N_D = 10^{19}$ cm^{-3}. The device area $S = 10^{-4}$ m^2, and the hole mobility $\mu_p = 0.04$ m^2/Vs. Assume that the Auger recombination current is given by

$$I_{rA} = Q_p/\tau_A$$

where Q_p is the total charge of holes injected into the n^+ region. The width of the n^+ region $L = 1$ μm.

Hint: First prove that the hole diffusion length in the n^+ region is much larger than L, assuming that the Auger is the dominant recombination process. Neglect the depletion region width in the n^+ region compared with L.

2-3-3. Find the maximum electric field in a p^+-n-n^+ structure as a function of the applied voltage, V, for -3 V $< V <$ 0.6 V. Assume that the voltage drop across p^+ and n^+ regions may be neglected. Assume further that the applied voltage drops only across the depletion region between the p^+ and n regions. The donor and acceptor densities in the n^+ and p^+ regions are $N_{D^+} = 10^{17}$ cm^{-3} and $N_{A^+} = 10^{17}$ cm^{-3}, respectively. The donor density in the n region is $N_D = 10^{15}$ cm^{-3}. The device temperature $T = 300$ K. The length of the n region is 2μm. Energy gap $E_g = 1.12$ eV. Effective density of states in the conduction band $N_c = 3.22 \times 10^{19}$ cm^{-3}. Effective density of states in the valence band $N_v = 1.83 \times 10^{19}$ cm^{-3}. Dielectric permittivity $\varepsilon = 1.05 \times 10^{-10}$ F/m.

2-3-4. Consider a p-n junction diode. The concentration of holes in the n section of the device is described by the continuity eq. (2-3-9)

$$D_p \, \partial^2 p_n/\partial x^2 - (p_n - p_{no})/\tau_{pl} = 0$$

The concentration of shallow ionized donors in the n section is equal to 10^{15} cm^{-3}. The intrinsic carrier concentration is 10^{10} cm^{-3}. The forward voltage applied to the diode is 0.5 V. Assuming that the length of the n section, L, is much smaller than the diffusion length, L_p, calculate and sketch the hole distribution in the n section of the device. Also, assuming that $D_p = 12$ cm^2/s and the lifetime $\tau_{pl} = 1$ μs, how short does the n section have to be to satisfy the condition that $L << L_p$ (use $L = L_p/10$ as the criterion)?

2-3-5. A silicon p-n junction has a resistivity of 0.1 Ω cm and 2 Ω cm for the uniformly doped p and n sections, respectively. If $\mu_n = 1500$ cm^2/Vs, $\mu_p = 450$ cm^2/Vs, and $n_i = 1.45 \times 10^{10}$cm^{-3} at room temperature,

(a) calculate the built-in voltage of the junction

(b) calculate the diode saturation current at room temperature (the minority carrier lifetimes for carriers in the p- and n-type sections are 15 μs and 50 μs, respectively, the sample cross section is 0.05 cm^2)

(c) calculate the temperature dependence (and plot) of the saturation current, neglecting, for simplicity, the temperature dependence of mobility ($E_g = 1.12$ eV)

2-4-1. Calculate the forward current voltage characteristics of a p$^+$-n silicon diode at T = 300 K, taking into account the recombination current in the depletion region. The acceptor and donor densities are $N_A = 10^{17}$ cm^{-3} and $N_D = 10^{15}$ cm^{-3}. The recombination time $\tau = 10^{-8}$ s. The device area $S = 10^{-4}$ m^2. The electron mobility $\mu_n = 0.1$ m^2/Vs. The hole mobility $\mu_p = 0.04$ m^2/Vs. Energy gap $E_g = 1.12$ eV. Effective density of states in the conduction band $N_c = 3.22 \times 10^{19}$ cm^{-3}. Effective density of states in the valence band $N_v = 1.83 \times 10^{19}$ cm^{-3}. Temperature $T = 300$ K. Dielectric permittivity $\varepsilon = 1.05 \times 10^{-10}$ F/m.

2-4-2. Calculate the temperature dependence of the current through a p$^+$-n silicon diode for the forward bias voltage $V = 0.5$ V, in the temperature range from 280 K to 450 K, taking into account the recombination current in the depletion region. The acceptor and donor densities are $N_A = 10^{17}$ cm^{-3} and $N_D = 10^{15}$ cm^{-3}. The recombination time $\tau = 10^{-8}$ s. The device area $S = 10^{-4}$ m^2. The electron mobility $\mu_n = 0.1$ m^2/Vs. The hole mobility $\mu_p = 0.04$ m^2/Vs. Energy gap $E_g = 1.12$ eV. Effective density of states in the conduction band $N_c = 3.22 \times 10^{19}$ (T/300)$^{3/2}$ cm^{-3}. Effective density of states in the valence band $N_v = 1.83 \times 10^{19}$ (T/300)$^{3/2}$ cm^{-3}. Dielectric permittivity $\varepsilon = 1.05 \times 10^{-10}$ F/m. Assume that $L >> L_p$.

2-4-3. Calculate the temperature dependence of the current through a p$^+$-n silicon diode for the reverse bias voltage, V = 3 V, in the temperature range from 280 K to 450 K taking into account the generation current in the depletion region. The acceptor and donor densities are $N_A = 10^{17}$ cm^{-3} and $N_D = 10^{15}$ cm^{-3}. The effective generation time $\gamma_{gen} = 10^{-8}$ s. The device area $S = 10^{-4}$ m^2. The electron mobility $\mu_n = 0.1$ m^2/Vs. The hole mobility $\mu_p = 0.04$ m^2/Vs. Energy gap $E_g = 1.12$ eV. Effective density of states in the conduction band $N_c = 3.22 \times 10^{19}$ (T/300)$^{3/2}$ cm^{-3}. Effective density of states in the valence band $N_v = 1.83 \times 10^{19}$ (T/300)$^{3/2}$ cm^{-3}. Dielectric permittivity $\varepsilon = 1.05 \times 10^{-10}$ F/m.

2-4-4. Repeat the derivation of eq. (2-4-39) for the case in which $N_D > N_A$.

2-5-1. Design a varactor (a device with a voltage-dependent capacitance used as a tuning element in microwave circuits) using a uniformly doped, p^+-n silicon diode. Assume that the maximum electric field in the junction should be less than or equal to 300 kV/cm. The required capacitance variation is from 1 to 2.5 pF. The reverse bias voltage applied to the diode should be less than or equal to 5V. The built-in voltage is 0.8 V. Dielectric permittivity $\varepsilon = 1.05 \times 10^{-10}$ F/m. The design should specify the doping level in the n region and the diode area.

Hint: Use the voltage dependence of the capacitance of a p-n junction under reverse bias.

2-5-2. Derive eq. (2-5-8)

$$C_d = \varepsilon S/x_d = S\left[\frac{qa\varepsilon^2}{12(V_{bi} - V)}\right]^{1/3} \tag{2-5-8}$$

2-6-1. Consider a p^+-n diode (see Fig. 2-6-1). Assume that the width of the n region, W, is much smaller than the hole diffusion length, L_p. The hole lifetime in the n region is τ_p, and the hole diffusion constant is D_p. Find the small-signal admittance of the diode as a function of the forward bias voltage, V, and frequency, ω, if the device temperature is T. Neglect the series resistance and the series inductance. Assume that $\omega\tau_p << 1$.

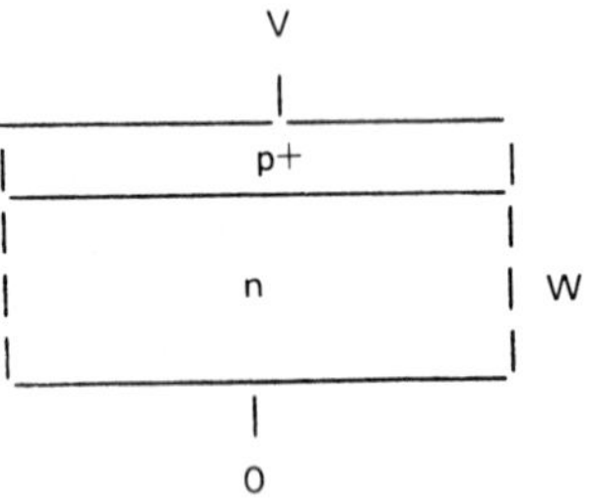

Fig. P2-6-1

2-6-2. The measured capacitance of an abrupt p^+-n junction as a function of reverse bias is shown in Fig. P2-6-2 (see also Table P2-6-2). Determine and sketch the doping profile in the n section, as a function of x_d, for 100 μm $< x_d <$ 500 μm. The device cross section $S = 10^{-2}$ cm^{-2}. Dielectric permittivity $\varepsilon = 1 \times 10^{-10}$ F/m.

TABLE P2-6-2

Rev. bias (V)	Depl. cap. (F)	Rev. bias (V)	Depl. cap. (F)	Rev. bias (V)	Depl. cap. (F)
0 V	1.000×10^{-12}	−0.3 V	8.771×10^{-13}	−0.7 V	7.700×10^{-13}
−1.0 V	7.071×10^{-13}	−1.3 V	6.594×10^{-13}	−1.7 V	6.086×10^{-13}
−2.0 V	5.774×10^{-13}	−2.3 V	5.505×10^{-13}	−2.7 V	5.199×10^{-13}
−3.0 V	5.000×10^{-13}	−3.3 V	3.740×10^{-13}	−3.7 V	2.968×10^{-13}
−4.0 V	2.626×10^{-13}	−4.3 V	2.380×10^{-13}	−4.7 V	2.139×10^{-13}
−5.0 V	2.000×10^{-13}				

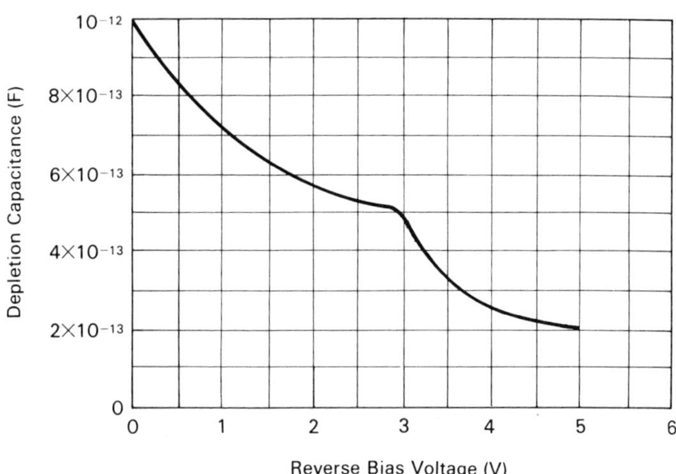

Fig. P2-6-2

2-6-3. The current waveform through a p^+-n junction is shown in Fig. P2-6-3. The effective hole lifetime, τ_p, in the n region is 1 ns. Temperature $T = 300$ K. The diode saturation current $I_s = 0.1$ nA. $I_1 = 1$ mA, $I_2 = 2$ mA. Diffusion coefficient $D_p = 12\ cm^2/s$. The length of the n section is much longer than the hole diffusion length.

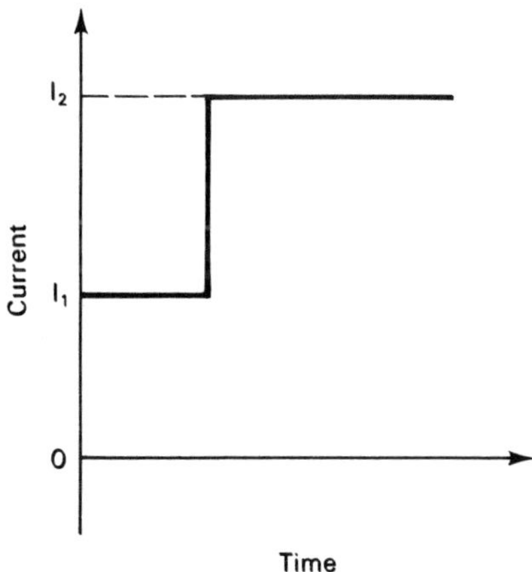

Fig. P2-6-3

(a) Find the voltage across the diode and the charge of minority carriers in the n section before and long after the transient.

(b) Sketch the qualitative time dependence of the voltage across the diode.

(c) Sketch the qualitative dependencies of minority carrier distributions in the n section vs. distance for $t = 0$ and for $t \to \infty$.

2-7-1. Derive eqs. (2-7-16) and (2-7-17):

$$A_r = \frac{4n_b \exp(-ikd_b)}{(1 + n_b)^2 \exp(-ikn_b d_b) - (1 - n_b^2)\exp(ikn_b d_b)} \quad (2\text{-}7\text{-}16)$$

$$B_1 = \frac{[\exp(-ikn_b d_b) - \exp(ikn_b d_b)](1 - n_b^2)}{(1 + n_b)^2 \exp(-ikn_b d_b) - (1 - n_b^2)\exp(ikn_b d_b)} \quad (2\text{-}7\text{-}17)$$

2-7-2. What is the transmission coefficient for the fish in the fishbowl in Fig. 2-7-2. Assume the mass of the fish to be 1 g, the height of the barrier to be 10 cm (from the position of the fish to the top of the fishbowl), and the thickness of the fishbowl wall to be 3 mm. Assume $D_o \sim 1$.

2-7-3. What is the transmission coefficient for an electron for a potential barrier in GaAs? Assume $m_n \approx 0.067 m_e = 6.1 \times 10^{-32}$ kg, and the barrier with the barrier height of $U_b - E = 1$ eV and the barrier width of 20 Å. Assume $D_o \sim 1$.

2-8-1. Using the values of breakdown fields given in Appendixes for Si, Ge, and GaAs (dielectric constants: Si = 11.7, Ge = 16.2, GaAs = 12.9 and $\varepsilon_o = 8.854 \times 10^{-12}$ F/m), calculate and plot (on a double log scale) the breakdown voltages if Si, Ge, and GaAs p^+-n diodes vs. doping density, N_D, in the n section for values of N_D between 10^{14} and 10^{17} cm^{-3}.

2-8-2. The situation when the depletion of region of a p^+-n junciton extends all the way from the metallurgical junction through the neutral region and touches the ohmic contact is called punch-through. Assume that the impact ionization in the depletion region occurs when the maximum electric field in the depletion region exceeds the breakdown field, F_{br}. The concentration of shallow donors in the n region is N_D. The length of the n-type region is L_n. The dielectric permittivity of the semiconductor material is $\varepsilon(1 \times 10^{-10}$ F/m). The built-in voltage of the junction is V_{bi} (0.65 V). The electronic charge is q. Find the ratio of the critical voltage of impact ionization, V_i, over the critical punch-through voltage, V_{pt}. Calculate this ratio for $L_n = 1\ \mu$m, $N_D = 10^{16}$ cm^{-3}, $F_{br} = 500$ kV/cm.

2-9-1. Calculate the surface potential of a free GaAs surface as a function of the GaAs doping density (for both n- and p-type doping in the range between 10^{14} cm^{-3} and 10^{17} cm^{-3}). Assume the density of the surface states to be 2×10^{12} cm^{-2}eV^{-1} and the position of the neutral level, ϕ_o, to be 0.65 eV below the bottom of the conduction band, E_c. The definition of the neutral level is such that the surface is charged negatively when the surface states are filled up to a level above the neutral level, and charged positively when filled up to a level below the neutral level. The effective density of states in the conduction band, $N_c = 4.21 \times 10^{17}$ cm^{-3}, the energy gap $E_g = 1.42$ eV, and the dielectric permittivity $\varepsilon = 1.05 \times 10^{-10}$ F/m.

2-9-2. Derive equation (2-9-18).

2-10-1. Compare the value of the saturation current density for the Schottky barrier, j_{ss}, with the saturation current density for a p^+-n junction, j_s. Show that

$$\frac{j_{ss}}{j_s} \approx (2\pi)^{-1/2} \frac{N_D}{N_c} \left(\frac{\tau_{pl}}{\tau_p}\right)^{1/2} \frac{m_n}{m_p} \exp\left[\frac{q(E_g - \phi_b)}{k_B T}\right]$$

where τ_{pl} is the hole lifetime in the n-type region, τ_p is the hole momentum relaxation time, E_g is the energy in the p-n junction, ϕ_b is the Schottky barrier height, N_D is the donor density, and N_c is the effective density of states in the conduction band.

2-10-2. Calculate (and plot) the real and imaginary parts of impedance for a Schottky barrier detector versus the log of frequency (from 1 MHz to 100 GHz) at doping densities, N_D, 10^{16} and 10^{17} cm^{-3} and temperature of 300 K. The detector is biased with 1 mA of forward bias current (I_{dc}). The Richardson constant $A^* = 4$ A/cm²/K², the barrier height $\phi_b = 0.8$ V, the parasitic series resistance is 25 Ω, and the device cross section $S = 10^{-4}$ cm^{-2}. Dielectric permittivity $\varepsilon = 1.14 \times 10^{-10}$ F/m, and the effective density of states in the conduction band $= 4.7 \times 10^{17}$ cm^{-3} (GaAs). $m_i = 1$. You can neglect the geometric capacitance and series inductance for this problem ($C_{geom} \Rightarrow 0$, $L_s \Rightarrow 0$; see Section 2-10-2).

Hint: $V_{bi} < \phi_b$.

2-10-3. The measured capacitance of the reversed biased silicon Schottky barrier diode is given by

$$\frac{1}{C^2} = \frac{1}{C_o^2} - kV$$

where $C_o = 1$ pF and $k = 2$ $pF^{-2}V^{-1}$. The diode is uniformly doped. Dielectric permittivity $\varepsilon_s \approx 1.05 \times 10^{-10}$ F/m. Device area is 10^{-4} cm^{-2}. The effective density of states in the conduction band $N_c = 2.8 \times 10^{19}$ cm^{-3}. Find the Schottky barrier height.

2-10-4. A Schottky diode and a p^+-n diode are connected as shown in Figs. P2-10-4a and b. The values of the device parameters are

Temperature	$T = 300$ K (thermal voltage 26 meV)
Richardson constant	$A = 5$ A/cm²/(degree K)²
Schottky barrier height	$\phi_b = 0.8$ eV
Ideality factors for both diodes	1
Cross sections of both diodes	$S = 10^{-2}$ cm^{-2}
Diffusion coefficient of holes	$D_p = 9$ cm²/s
Length of the n section	$X_n = 3$ μm
Doping of the n section	$N_a = 10^{15}$ cm^{-3}
Intrinsic carrier concentration	$n_i = 10^{10}$ cm^{-3}
Hole lifetime in the n section	$\tau = 10^{-6}$ s

Assume ideal diode equations. The applied voltage is 0.8 V. Find the electric current and voltage drops across the diodes.

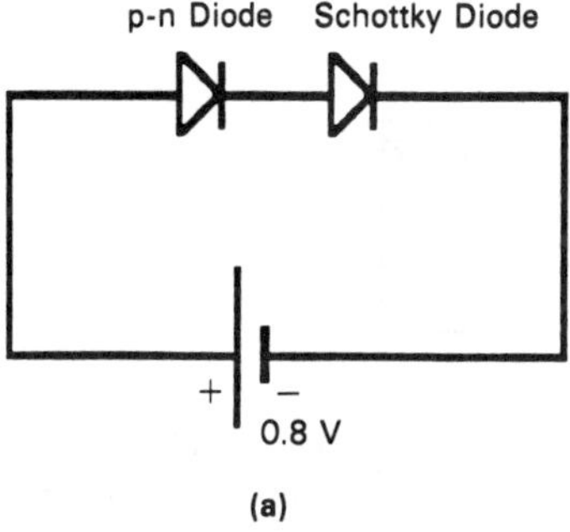

Fig. P2-10-4a

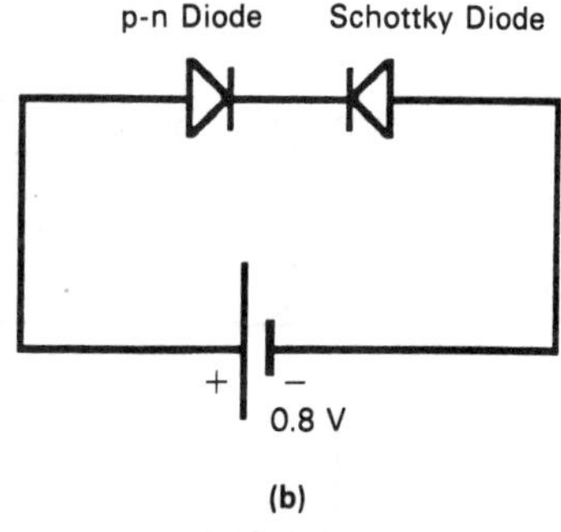

Fig. P2-10-4b

2-11-1. The planar design of a Schottky barrier diode is shown in the following figure. All dimensions shown in the figure are in micrometers.

|<-L = 1--->|<-L_{so} = 1.5----->|<--------L_o = 3 ---------------------->|

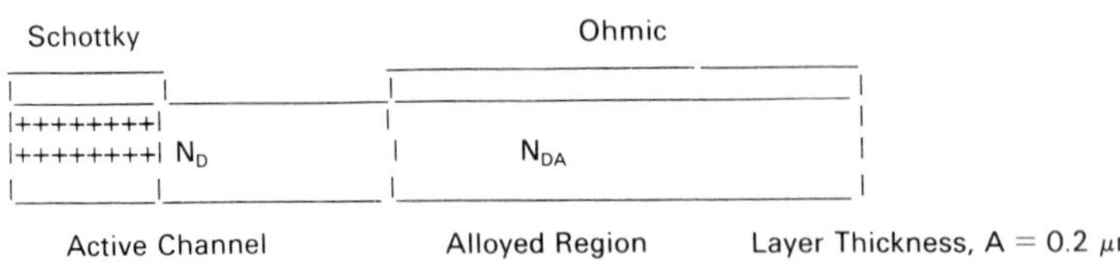

Fig. P2-11-1

The doping of the active layer is 1×10^{23} m^{-3}. The doping of the alloyed region under the ohmic contact is 5×10^{23} m^{-3}. The device width is 5 micrometers. The specific contact resistance $r_c = 5 \times 10^{-6}$ ohm cm^2. The electron mobility is 3000 cm^2/Vs. The Schottky barrier height is 0.8 V. The Richardson constant A = 4.4×10^4 A/(m^{2}K^2). The current flowing through the diode is 1 mA.

(a) What is the contact resistance of the ohmic contact?

(b) What is the series resistance of the Schottky diode? Assume that the built-in voltage of the channel is 0.5 V. Also assume that the cross-section of the channel between the Schottky contact and the ohmic contact is uniform.

Hint: estimate the resistance of the undepleted section of the active channel under the Schottky contact as

$$R_{\text{undepl}} = \frac{L}{2qW\mu N_d A_{\text{ud}}}$$

where μ is the low field mobility, W is width, and A_{ud} is the thickness of the undepleted region.

(c) What is the voltage drop across the diode?

2-11-2. The measured dependence of the resistance between ohmic contacts for the transmission-line-model pattern shown in Fig. P2-11-2a is shown in Fig. P2-11-2b as a function of the distance between the contacts. The thickness of the doped n-type layer is 0.1 μm. The device width is 5 μm. Assume electron mobility of 1000 cm^2/Vs.

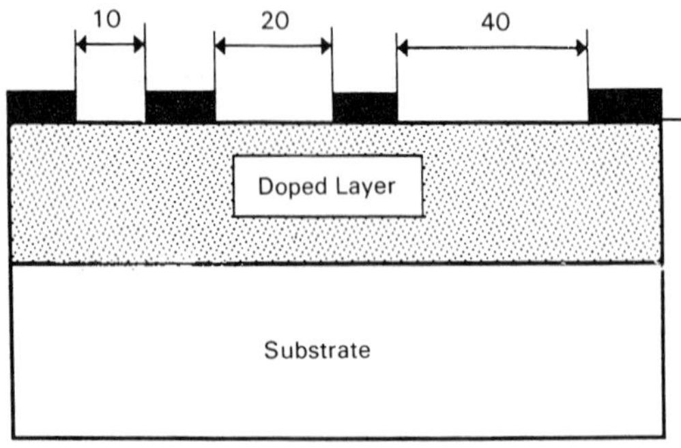

Fig. P2-11-2a Transmission-line-model pattern. Distances shown are in μm. Substrate is undoped.

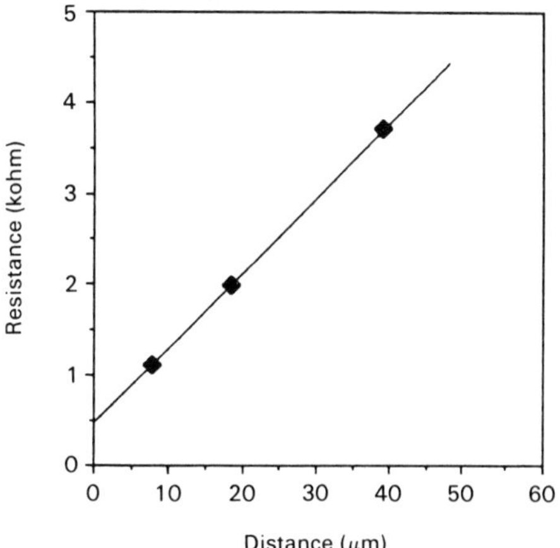

Fig. P2-11-2b

Find the electron concentration in the channel and the specific contact resistance, assuming the same doping level everywhere in the channel. Also assume a long contact length. The measured value of the end resistance for the shortest device is 200 Ω.

2-12-1. Consider an AlGaAs-GaAs p-n junction. Calculate the current-voltage characteristic of the junction using the model proposed by Chang (1965), which leads to the following expression for the current-voltage characteristics:

$$j = (A^*TqV_{bi}/k_B)\exp(-qV_{bi}/k_BT)[\exp(qV/k_BT) - 1]$$

Here j is the current density, $A^* = 4$ (A/cm^2/K^2) is the effective Richardson constant, built-in voltage

$$V_{bi} = E_{gl} - \Delta E_n - \Delta E_p + \Delta E_c$$

$E_{gl} = 1.42$ eV is the energy gap of GaAs (which is doped p-type with shallow acceptor density $N_A = 10^{16}$ cm^{-3}), ΔE_n is the difference between the bottom of the conduction band in AlGaAs (doped with shallow donor density $N_D = 10^{16}$ cm^{-3}) and the Fermi level far from the heterointerface, ΔE_p is the difference between the Fermi level and the top of the valence band in the narrow-gap p-type material, and $\Delta E_c = 0.3$ eV is the conduction band offset. Assume room temperature $T = 300$ K, density of states effective mass of holes in GaAs $m_h = 0.4\ m_e$, density of states effective mass of electrons in AlGaAs $m_h = 0.07m_e$.

3

Bipolar Junction Transistors

3-1. PRINCIPLE OF OPERATION

In 1948 John Bardeen and Walter Brattain published the first paper on a Bipolar Junction Transistor (BJT). W. Shockley (1949) developed the first theoretical model for bipolar junction transistors, which still forms the foundation for the understanding of these devices.

The discovery of the bipolar junction transistor started the revolutionary development of semiconductor electronics. Today the bipolar junction transistor remains one of the most important discrete devices as well as an important component of integrated circuits.

Bipolar junction transistors are three-terminal devices that can be used as amplifiers or logic elements. The schematic device structures of p-n-p and n-p-n transistors along with their circuit symbols are shown in Fig. 3-1-1. As can be seen from the figure, a bipolar junction transistor comprises two back-to-back p-n junctions. An emitter region and a base region form the first p-n junction. The second p-n junction is formed between the base and collector region. If the base region were long compared with the diffusion length of the minority carriers in the

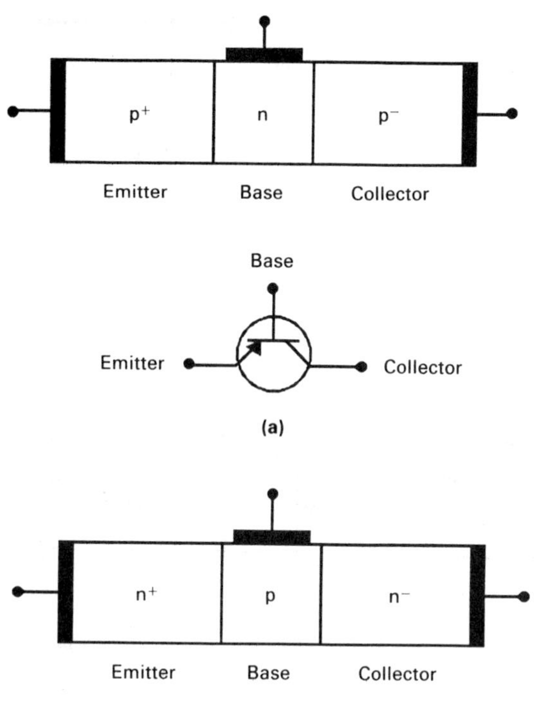

Fig. 3-1-1. Schematic device structures and circuit symbols for p^+-n-p^- (a) and n^+-p-n^- (b) bipolar junction transistors. Superscripts + and − mean higher and lower doping levels, respectively.

base (holes for a p-n-p transistor and electrons for an n-p-n transistor), the equivalent circuit of the device would have included two diodes connected as shown in Figs. 3-1-2a and b. In fact, the opposite is true: the base region is made to be much shorter than the diffusion length of the minority carriers. Therefore, the emitter-base and collector-base junctions affect each other. In the equivalent circuit this may be represented by controlled current sources as shown in Figs. 3-1-2c and 3-1-2d. This interaction between the emitter-base and collector-base junctions (described by controlled current sources in Fig. 3-1-2) is crucial for BJT operation.

As shown in Fig. 3-1-1, the emitter region has the highest doping. The collector region is doped the lowest. The doping level in the base region is smaller than that in the emitter region but higher than that in the collector region. A typical doping profile of an n-p-n bipolar junction transistor is shown in Fig. 3-1-3a. The reasons for such a choice of doping levels as well as for nonuniformity of the doping profile in the emitter and base will be explained shortly. We will first

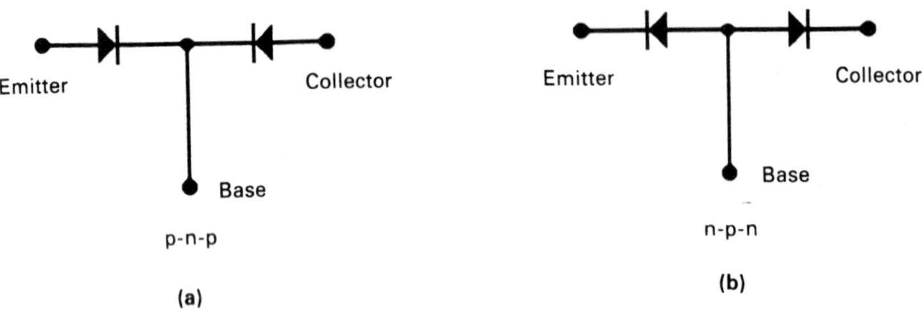

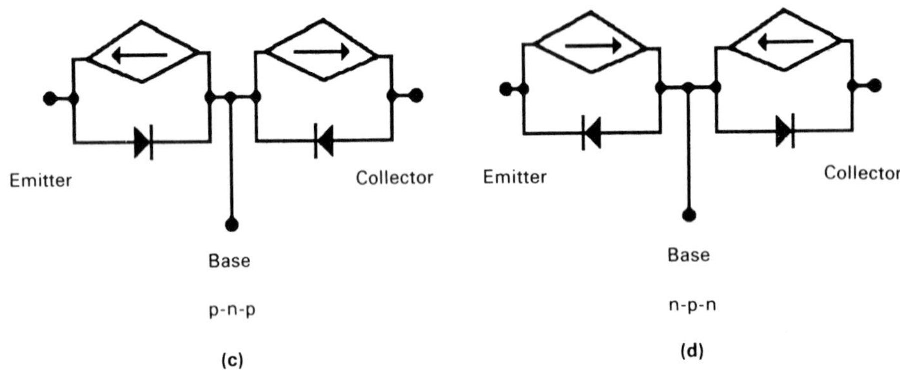

Fig. 3-1-2. Simplified equivalent circuits of *p-n-p* transistor with a very long base (a), *n-p-n* transistor with a very long base (b), more realistic equivalent circuits of *p-n-p* (c) and *n-p-n* (d) transistors with base lengths smaller than diffusion lengths of minority carriers.

consider a transistor with uniform and abruptly changing impurity distributions, as shown in Fig. 3-1-3b.

The qualitative band diagram of an n-p-n bipolar junction transistor in equilibrium is shown in Fig. 3-1-4a. Under typical operating conditions the emitter-base junction is forward biased, whereas the collector-base junction is reverse biased (see Fig. 3-1-4b). Hence, the electron concentration in the base at the emitter-base junction is much larger than the equilibrium concentration of minority carriers (electrons) in the base, $n_{bo} = n_{ib}^2/N_{ab}$, where n_{ib} is the intrinsic carrier concentration in the base region. As shown in Chapter 2, the concentration of minority carriers at the edge of the depletion region is related to the voltage drop across the junction and to the equilibrium concentration of the minority carriers:

$$n_{be} \equiv n_b(x_e) = n_{bo} \exp(V_{be}/V_{th}) >> n_{bo} \tag{3-1-1}$$

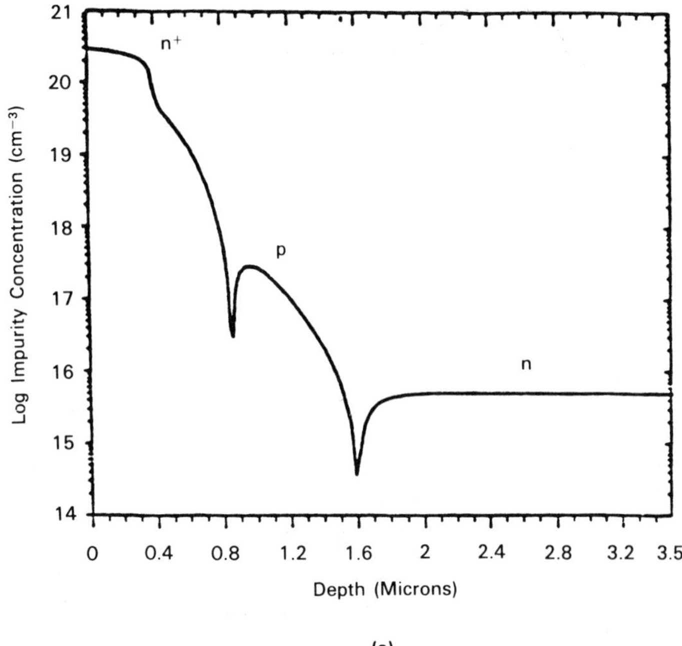

(a)

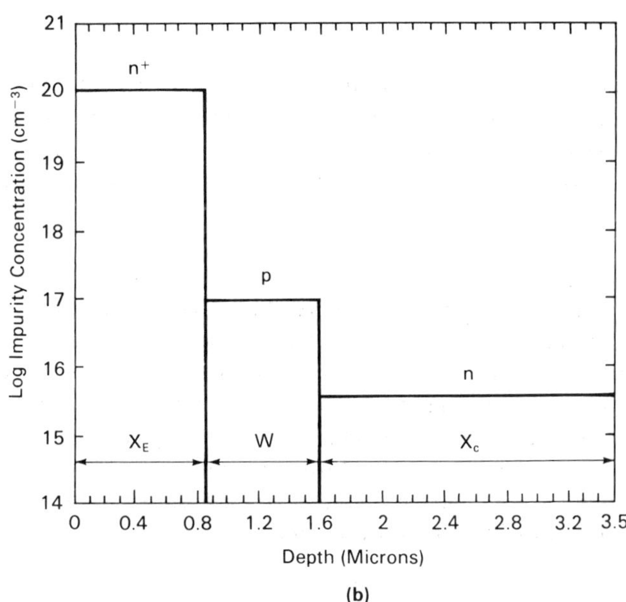

(b)

Fig. 3-1-3. Typical doping profile of a silicon bipolar junction transistor (a) (after Z. Yu and R. W. Dutton, *Sedan III-A Generalized Electronic Material Device Analysis Program,* Program Manual, Stanford University, p. 81, July (1985)) and idealized uniform abruptly changing doping profile (b).

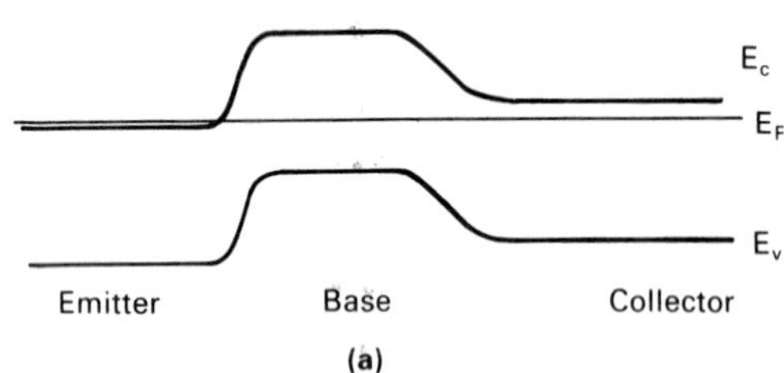

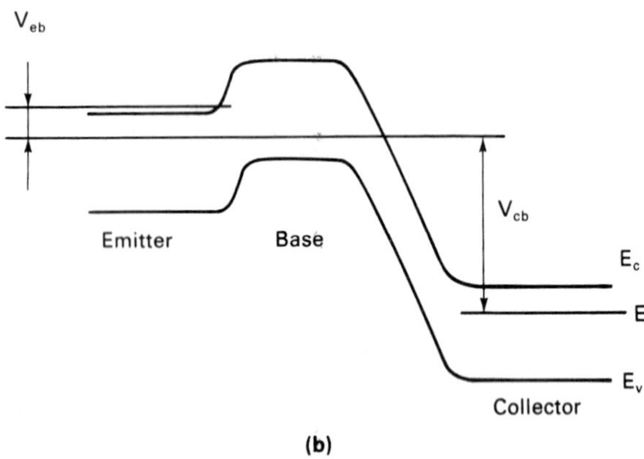

Fig. 3-1-4. Band diagrams of a bipolar junction transistor (a) under equilibrium conditions (no external bias voltage) (b) under typical operating conditions when the emitter-base junction is forward biased and the collector-base junction is reverse-biased. E_c is the bottom of the conduction band, E_v is the ceiling of the valence band, E_F is the Fermi level in Fig. 3-1-4a and the quasi-Fermi level in Fig. 3-1-4b.

Here $V_{th} = k_B T/q$ is the thermal voltage, n_b is the electron concentration in the base, and V_{be} is the emitter-base voltage ($V_{be} >> V_{th}$). The electron concentration in the base at the collector-base junction is much smaller than the equilibrium concentration of minority carriers (electrons) in the base, n_{bo}:

$$n_{bc} \equiv n_b(x_e + W) = n_{bo} \exp(V_{bc}/V_{th}) << n_{bo} \tag{3-1-2}$$

Here V_{bc} is the collector-base voltage ($V_{bc} < 0$), x_e is the width of the emitter region, and W is the base width (see Fig. 3-1-3b). Here we assume that the widths of the depletion regions at the emitter-base and collector-base interface are sufficiently small and can be neglected, compared with the base width.

Just as in a regular n$^+$-p junction (see Chapter 2), the diffusion component of the electron current is dominant in the p region. If the base region is narrow ($W << L_{nb}$, where L_{nb} is the diffusion length of electrons in the base), the carrier recombination in the base is small and, to first order, the electron concentration in the base varies linearly with distance:

$$n_b \approx n_{be}(W - x + x_e)/W \tag{3-1-3}$$

This result is based on the solution of the continuity equation for electrons in the base and is obtained in the same way as the solution of the continuity equation for minority carriers in a short p-n diode (see Section 2-3), i.e., by expanding

$\exp(x/L_{nb})$ and $\exp(-x/L_{nb})$ into a Taylor series. The profile of the electron carrier concentration in the base is shown in Fig. 3-1-5.

The electron diffusion current in the base,

$$I_n = qD_n \frac{\partial n_b}{\partial x} S \tag{3-1-4}$$

remains practically constant and equal to both emitter and collector current. (Here D_n is the electron diffusion coefficient, and S is the device cross section.) The reason is that practically all electrons injected into the base region from the emitter region diffuse toward the collector region and reach the large negative slope of the bottom of the conduction band shown in Fig. 3-1-4b. From that point they are carried away toward the collector contact by the strong electric field of the reverse-biased collector-base p-n junction (down the negative slope of the bottom of the conduction band). Of course, some electrons do recombine with the holes in the base region. However, the number of such electrons is small because the base region is much thinner than the electron diffusion length. If the base region were made much wider than L_{nb}, practically all electrons would have recombined, and we would have had a very poor transistor indeed.

Substituting eq. (3-1-1) into eq. (3-1-3) and the resulting equation into eq. (3-1-4), we obtain

$$I_e \cong I_c \cong |I_n| \cong \frac{qD_n n_{bo} \exp(qV_{be}/k_B T)\, S}{W} \tag{3-1-5}$$

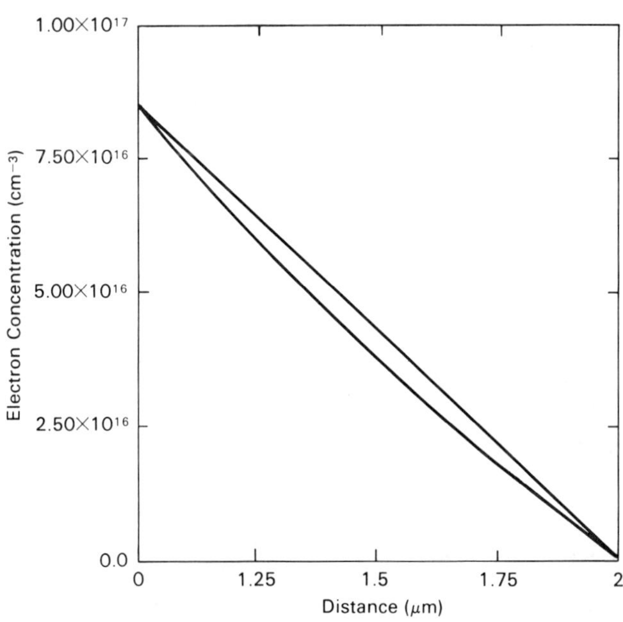

Fig. 3-1-5. Distributions of electrons injected into the base region of a silicon *n-p-n* bipolar junction for different diffusion lengths of electrons in the base: The straight line corresponds to $L_{nb} = 20$ μm and the concave curve corresponds to $L_{ne} = 1$ μm. The curves were computed using subroutine PB-JTM.BAS (see Program Description Section). Parameters used in the calculation are emitter-base voltage $V_{be} = 0.85$ V, collector-base voltage $V_{bc} = -4$ V, emitter doping $N_{de} = 1 \times 10^{19}$ cm^{-3}, base doping $N_{ab} = 2 \times 10^{17}$ cm^{-3}, and Temperature $T = 300$ K.

Hence, a change in the emitter-base voltage leads to an exponential change of both emitter and collector current. Let us now consider how we can operate a bipolar junction transistor as an amplifier. The input signal may be represented by an ac voltage source that we can connect across the emitter-base junction. The load resistance we can connect between the collector terminal and power-supply terminal (this corresponds to the *common-base configuration* [see Fig. 3-1-6a]). A small variation of the emitter-base voltage, ΔV_{be}, caused by the signal leads to nearly equal variation of the emitter and collector currents ($\Delta I_c \approx \Delta I_e$). However, the collector current flows in the loop containing the power supply (bias voltage, $V_{cc} >> V_{be}$), and the variation of voltage drop, $\Delta V_c = \Delta I_c R_c$, across an external load resistance, R_c, can be much greater than ΔV_{be}. (With an appropriate value of R_c, the maximum value of ΔV_c is limited by the collector voltage supply, V_{cc}.) Hence, we can have a voltage gain $\Delta V_c/\Delta V_{be}$. The ac power, supplied by the input signal, is proportional to $\Delta I_e \Delta V_{be}$. The ac power generated in the output loop is proportional to $\Delta I_c \Delta V_c >> \Delta I_e \Delta V_{be}$. Hence, this circuit also provides power gain. This gain occurs because electrons are injected from the emitter

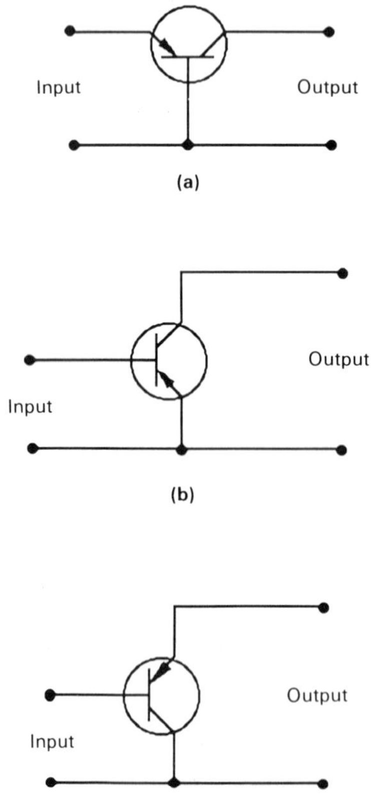

Fig. 3-1-6. (a) Common-base, (b) common-emitter, and (c) common-collector circuit configurations for bipolar junction transistor.

region, through the base, into the collector region, where the voltage drop is large compared with a small forward emitter-base voltage.

Two other possible circuit configurations of a bipolar junction transistor (*common-emitter* and *common-collector* configurations) are illustrated by Figs. 3-1-6b and 3-1-6c, respectively.

The principle of the transistor operation in the common-emitter configuration (see Fig. 3-1-6b) may be explained by considering what happens when we increase the base current by increasing the emitter-base voltage. The base current, I_b, is primarily the hole current, I_{pe}, that flows from the base region into the emitter region. The hole current in the n-type doped emitter is a diffusion current, just as is the electron (minority carrier) current in the p-type doped base. Hence, we have (similar to eq. (3-1-5))

$$I_b \approx I_{pe} \cong qD_pS \left.\frac{\partial p_e}{\partial x}\right|_{x=x_e} = \frac{qD_pp_{eo}\exp(qV_{be}/k_BT)S}{x_e} \tag{3-1-6}$$

where $p_{eo} = n_{ie}^2/N_{de}$ and we assume that $x_e << L_{pe}$, where L_{pe} is the diffusion length of holes in the emitter region (this is always true for practical BJTs). Let us compare eqs. (3-1-5) and (3-1-6). Usually, x_e is of the same order of magnitude as W. However, the emitter region is much more heavily doped than the base region, and hence, $p_{eo} << n_{bo}$. This means that the base current is much smaller than the emitter current, and the transistor in the common-emitter configuration has a current gain. From eqs. (3-1-5) and (3-1-6) we can estimate the common-emitter current gain, $\beta = I_c/I_b \approx I_e/I_b$:

$$\beta \approx \frac{D_nN_{de}x_e}{D_pN_{ab}W} \tag{3-1-7}$$

β is much greater than 1 because $N_{de} >> N_{ab}$ (see Fig. 3-1-3). As $D_n/D_p = \mu_n/\mu_p$ is approximately 2.5 in silicon, it is also clear that n-p-n Si BJTs should have, in general, higher β than p-n-p Si BJTs.

In modern bipolar junction transistors, I_{pe} is the dominant component of the base current. The base current also includes other components, such as the recombination current in the depletion region of the emitter-base junction, the generation current in the depletion region of the collector-base junction, and the recombination current in the neutral base region. However, for a crude estimate of the transistor current gain, all components of the base current, except I_{pe}, may be neglected (see Section 3-3 for a more detailed discussion).

As was shown previously, the operation of a bipolar transistor is based on the exponential variation of the injected carrier density, with the height of the potential barrier between the emitter and base regions controlled by the emitter-base voltage. This leads to the exponential dependence of the emitter and collector currents on the emitter-base voltage (see eq. (3-1-5)) and to a very high transconductance

$$g_m = \partial I_c/\partial V_{be} \tag{3-1-8}$$

compared with a similar parameter for a field-effect transistor, which utilizes a capacitive modulation of charge in a conducting channel (see Chapter 4). High transconductance and current swing make a bipolar junction transistor a device of choice for many high-speed and high-power applications in both discrete and integrated circuits.

On the other hand, the exponential dependence of the emitter and collector currents on temperature (see eq. (3-1-5)) may lead to thermal instability. For example, if one transistor, among many BJTs in an integrated circuit, has a higher emitter current, this device will have a higher temperature caused by the Joule heating. This, in turn, will cause a further increase in the emitter and collector currents, raising the temperature of this particular device even further and creating a weak spot in the circuit that may fail. Such an instability may represent a serious problem for bipolar-junction-transistor integrated-circuit technology. However, this problem may be addressed by choosing an appropriate circuit configuration.

3-2. MINORITY CARRIER PROFILES IN A BIPOLAR JUNCTION TRANSISTOR

As was shown in Section 3-1, a minority carrier profile in a narrow base region ($W << L_{nb}$) is approximately linear. For a steady-state case ($\partial n_b/\partial t = 0$)) this profile is described by the solution of the continuity equation for the minority carriers in the base:

$$D_n \frac{\partial^2 n_b}{\partial x^2} - \frac{n_b - n_{bo}}{\tau_{nb}} = 0 \qquad (3\text{-}2\text{-}1)$$

(compare with eq. (2-3-9)). Here n_b is the electron concentration in the base, τ_{nb} is the electron lifetime in the base region, and $n_{bo} = n_i^2/N_{ab}$. Equation (3-2-1) should be solved for the boundary conditions $n_b(x_e) = n_{bo}\exp(V_{be}/V_{th})$ and $n_b(x_e + W) = n_{bo}\exp(V_{bc}/V_{th})$ (discussed in Section 3-1). Here we assume that the space charge (depletion) regions in the base at the boundaries with the emitter and collector regions are much smaller than the base width, W. This assumption may not be justified for narrow-base devices or for large reverse voltages across the collector-base junctions. Some implications of a possible deviation from this assumption will be considered in Section 3-6.

The solution of eq. (3-2-1) with these boundary conditions is given by

$$n_b(x) = n_{bo} + A_{1b}\exp\left(\frac{x - x_e}{L_{nb}}\right) - A_{2b}\exp\left(-\frac{x - x_e}{L_{nb}}\right) \qquad (3\text{-}2\text{-}2)$$

where

$$A_{1b} = \frac{\Delta n_b(x_e + W) - \Delta n_b(x_e)\exp\left(-\dfrac{W}{L_{nb}}\right)}{2\sinh(W/L_{nb})} \qquad (3\text{-}2\text{-}3)$$

$$A_{2b} = \frac{\Delta n_b(x_e + W) - \Delta n_b(x_e) \exp\left(\frac{W}{L_{nb}}\right)}{2 \sinh(W/L_{nb})} \tag{3-2-4}$$

Here $\Delta n_b(x_e + W) = n_b(x_e + W) - n_{bo}$ and $\Delta n_b(x_e) = n_b(x_e) - n_{bo}$.

When the emitter-base junction is forward biased and the collector-base junction is reverse biased,

$$n_b(x_e) = n_{bo} \exp\left(\frac{qV_{be}}{k_B T}\right) \tag{3-2-5}$$

$$n_b(x_e + W) \cong 0 \tag{3-2-6}$$

and for a typical case when $W/L_{nb} << 1$, eq. (3-2-2) reduces to

$$n_b(x) = n_b(x_e) \frac{W + x_e - x}{W} \tag{3-2-7}$$

The minority carrier profiles calculated for different values of W/L_{nb} are shown in Fig. 3-1-5. As can be seen from the figure, when $W/L_{nb} << 1$ the electron concentration in the base varies linearly, as predicted by eq. (3-2-9). It is also instructive to compare this distribution with the distribution of holes in the n-type region of a short p^+-n diode (see Fig. 2-3-4 and related the discussion in Section 2-3).

The hole concentrations, p_e and p_c, in the emitter and collector regions are found by solving the continuity equations for holes in these regions:

$$D_p \frac{\partial^2 p_e}{\partial x^2} - \frac{p_e - p_{eo}}{\tau_{pe}} = 0 \tag{3-2-8}$$

and

$$D_p \frac{\partial^2 p_c}{\partial x^2} - \frac{p_c - p_{co}}{\tau_{pc}} = 0 \tag{3-2-9}$$

Here τ_{pe} and τ_{pc} are hole lifetimes in the emitter and collector regions, and $p_{eo} = n_i^2/N_{de}$ and $p_{co} = n_i^2/N_{dc}$ are the equilibrium hole concentrations in the emitter and collector regions, respectively. The boundary conditions for eq. (3-2-8) are $p_e(0) = p_{eo}$ and $p_e(x_e) = p_{eo} \exp(V_{be}/V_{th})$. The boundary conditions for eq. (3-2-9) are $p_c(x_e + W) = p_{co} \exp(V_{bc}/V_{th})$ and $p_c(W + x_e + x_c) = p_{co}$. (We again neglect the widths of the depletion regions, even though the depletion regions in the base may strongly affect the device characteristics [see Section 3-6].)

Solving eqs. (3-2-8) and (3-2-9), we find the hole concentrations in the emitter and collector regions of an n-p-n BJT:

$$p_e(x) = p_{eo} + p_{eo}\left[\exp\left(\frac{qV_{be}}{k_B T}\right) - 1\right] \sinh(x/L_{pe})/\sinh(x_e/L_{pe}) \tag{3-2-10}$$

$$p_c(x) = p_{co} + p_{co}\left[\exp\left(\frac{qV_{bc}}{k_B T}\right) - 1\right] \sinh[(x_c + x_e + W - x)/L_{pc}]/\sinh(x_c/L_{pc}) \tag{3-2-11}$$

TABLE 3-2-1. BIAS CONDITIONS FOR FOUR DIFFERENT MODES OF OPERATION OF A BIPOLAR JUNCTION TRANSISTOR

Mode	Emitter-base bias	Collector-base bias
forward active mode	forward	reverse
saturation mode	forward	forward
cutoff mode	reverse	reverse
reverse active mode	reverse	forward

The hole concentration in the neutral base region, p_b, is given by

$$p_b \approx N_{ab} + n_b \tag{3-2-12}$$

This equation follows from the condition of neutrality that means that majority carriers (holes) enter the base region to compensate the charge of minority carriers (electrons) injected into the base. In a low-injection regime $n_b << N_{ab}$ and $p_b \sim N_{ab}$. However, when a relatively high forward bias emitter-base voltage is applied, n_b and p_b may become comparable to or even larger than N_{ab}.

In Fig. 3-2-1 we show the minority carrier profiles in an n-p-n BJT corresponding to four possible modes of the transistor operation: forward active mode, saturation mode, cutoff mode, and reverse active mode. The bias conditions for different modes of operation are summarized in Table 3-2-1.

In the forward active mode (see Fig. 3-2-1a) the emitter-base junction is forward biased and the collector-base junction is reverse biased. The comparison between the hole concentration profile in the emitter region and electron concentration profile shown in Fig. 3-2-1a clearly illustrates how the higher doping in the emitter region leads to a higher injection into the base and, hence, to a larger emitter efficiency:

$$\gamma = \frac{\partial I_{ne}}{\partial \mathrm{I_e}} \tag{3-2-13}$$

where I_{ne} is the electron component of the emitter current, I_e.

In the saturation regime both emitter-base and collector-base junctions are forward biased (see Fig. 3-2-1b). In this regime the injection of electrons from the collector region is in the opposite direction compared with the injection of the electrons from the emitter region. Hence, the current that is proportional to the slope of the electron profile in the base region is smaller than in the active forward mode of operation. This is illustrated by Fig. 3-2-2, which shows a qualitative distribution of the electrons injected into the base from the emitter and collector regions, respectively, and the resulting electron distribution in the base (on a linear scale).

In the cutoff region (see Fig. 3-2-1c) there are very few minority carriers in the base and, hence, transistor currents are very small.

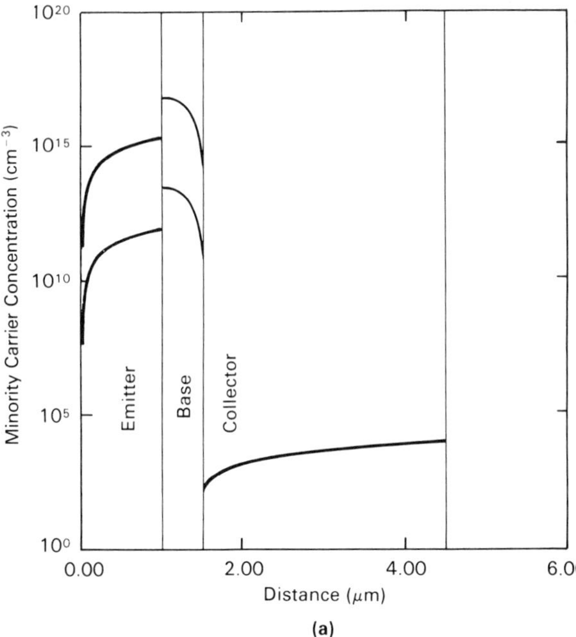

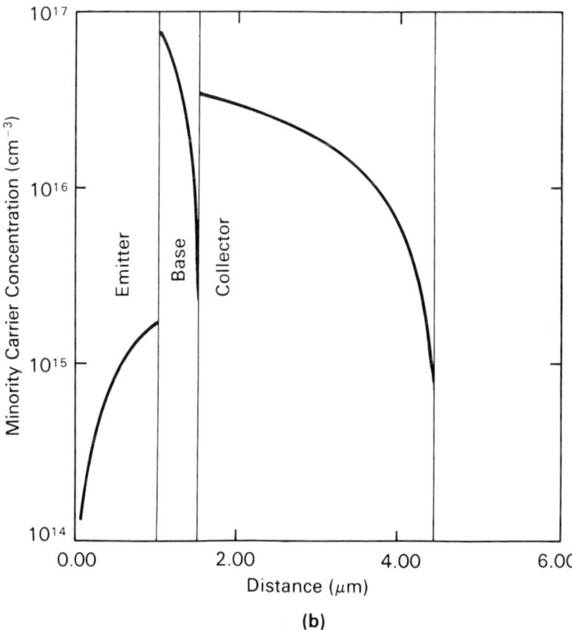

Fig. 3-2-1. Distributions of minority carriers in a silicon *n-p-n* bipolar junction for four different regimes of operation. (a) Forward active mode: Emitter-base voltage $V_{be} = 0.85$ V (top curves) and 0.65 V (bottom curves) and collector-base voltage $V_{bc} = -4$ V. (b) Saturation mode: Emitter-base voltage $V_{be} = 0.85$ V and collector-base voltage $V_{bc} = 0.75$ V. (c) Cutoff mode: Emitter-base voltage $V_{be} = -3$ V and collector-base voltage $V_{bc} = -4$ V. Please notice that there are practically no minority carriers in the transistor in this mode. (d) Reverse active mode: Emitter-base voltage $V_{be} = -3$ V and collector-base voltage $V_{bc} = 0.75$ V. The curves were computed using subroutine PBJTM-.BAS (see Program Description Section). Parameters used in the calculation are emitter doping $N_{de} = 1 \times 10^{19}$ cm^{-3}, base doping $N_{ab} = 2 \times 10^{17}$ cm^{-3}, collector doping $N_{dc} = 1 \times 10^{16}$ cm^{-3}, hole diffusion length in the emitter region $L_{pe} = 10$ μm, electron diffusion length in the base region $L_{nb} = 40$ μm, hole diffusion length in the collector region $L_{pc} = 20$ μm, emitter region width $x_e = 1$ μm, base region width $W = 0.5$ μm, collector region width $x_c = 3$ μm, and temperature $T = 300$ K.

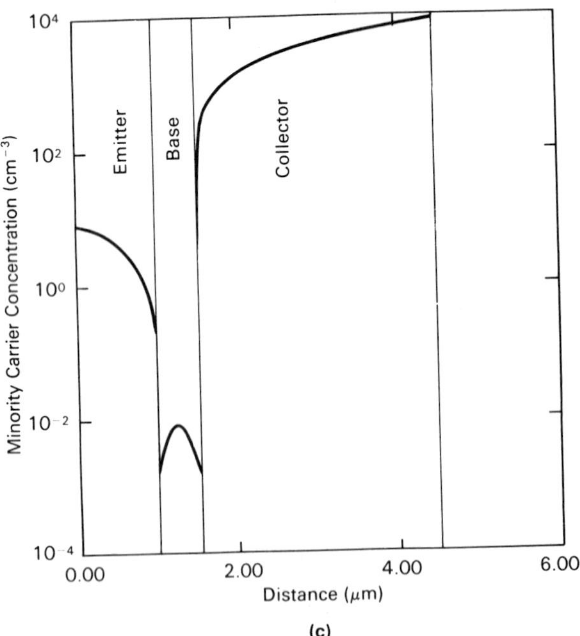

(c)

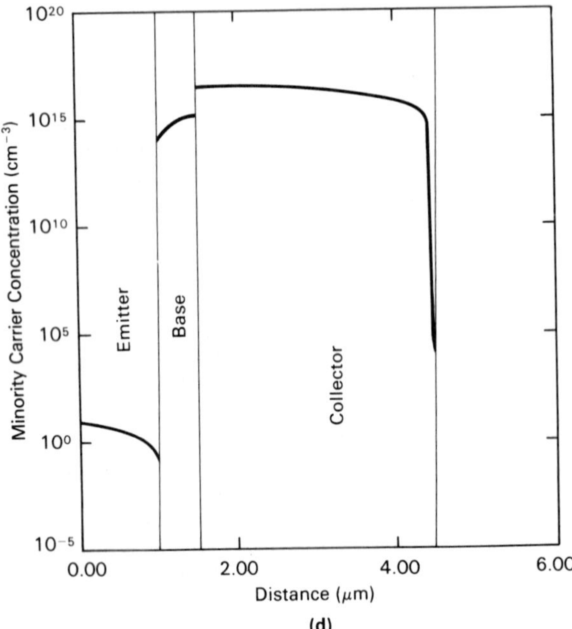

(d)

Fig. 3-2-1. ***Cont.***

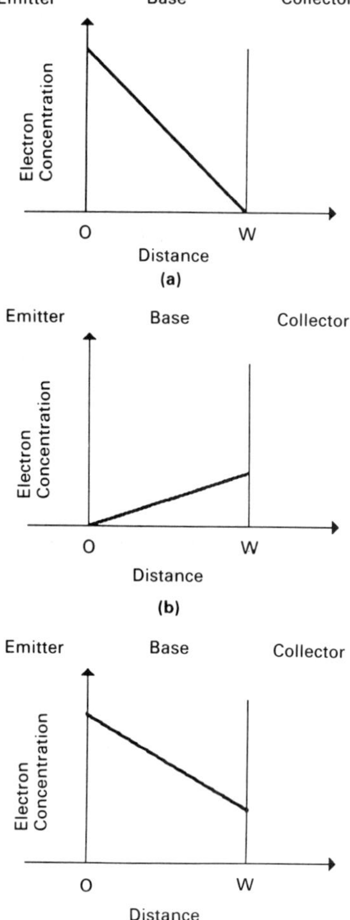

Fig. 3-2-2. Qualitative distributions of electrons injected into base for the saturation regime from (a) emitter, (b) collector, and (c) resulting electron distribution in the base, on a linear scale.

In the reverse active mode (see Fig. 3-2-1d) more holes are injected into the collector region than electrons into the base because of the higher doping in the base. This leads to a very low transistor gain in this regime, which is of little practical importance. This mode of operation may be used for the extraction of transistor parameters (transistor characterization).

3-3. CURRENT COMPONENTS AND CURRENT GAIN

In the forward active mode, the emitter, collector, and base currents, I_e, I_c, and I_b, in an n-p-n bipolar junction transistor may be expressed as follows:

$$I_e = I_{ne} + I_{pe} + I_{re} \tag{3-3-1}$$

$$I_c = I_{nc} + I_{gc} \tag{3-3-2}$$

$$I_b = I_{pe} + I_{re} - I_{gc} + I_{br} \tag{3-3-3}$$

where I_{ne} and I_{nc} are the electron components of the emitter and the collector current, I_{pe} is the hole component of the emitter current, I_{re} is the recombination current in the depletion region of the emitter-base junction, I_{br} is the recombination current in the neutral base region, and I_{gc} is the generation current in the depletion region of the collector-base junction. Different current components in a bipolar junction transistor are shown in Fig. 3-3-1.

The largest current components for the forward active mode are I_{ne} and I_{nc}. They are determined by the electron diffusion current in the base at the emitter-base junction and the collector-base junction, respectively:

$$I_{ne} = qD_nS(\partial n_b/\partial x)|_{x=x_e} \tag{3-3-4}$$

$$I_{nc} = qD_nS(\partial n_b/\partial x)|_{x=x_e+W} \tag{3-3-5}$$

(see eq. (3-1-4)). The electron distribution in the base is given by eq. (3-2-2). Using eqs. (3-3-5) and (3-2-2), and taking into account that for the forward active mode

$$n_b(x_e + W) = 0 \tag{3-3-6}$$

because $V_{bc} << -V_{th} < 0$ (see eq. (3-2-6)), we find that

$$I_{ne} = -qD_nS\Delta n_b(x_e)\coth(W/L_{nb})/L_{nb} \tag{3-3-7}$$

$$I_{nc} = -qD_nS\Delta n_b(x_e)\operatorname{csch}(W/L_{nb})/L_{nb} \tag{3-3-8}$$

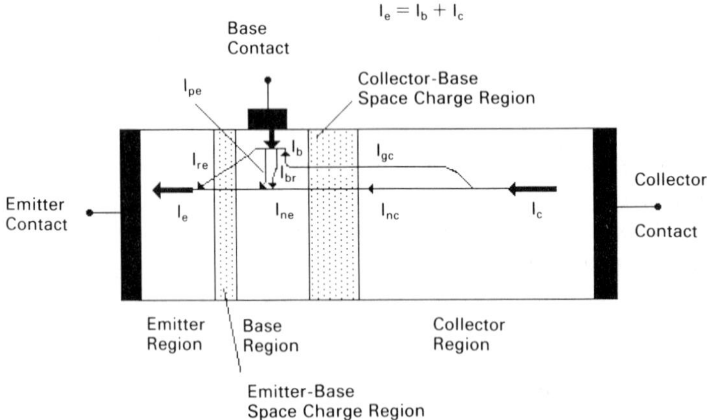

Fig. 3-3-1. Different current components in bipolar junction transistor.

(Definitions and derivatives of hyperbolic functions are given in Appendix 7.) The base recombination current, I_{br}, is given by

$$I_{br} = I_{ne} - I_{nc} \tag{3-3-9}$$

(Using this equation and eqs. (3-3-1) to (3-3-3), we can check that $I_e = I_c + I_b$.) Equation (3-3-9) can be rewritten as

$$I_{br} = -qD_n S \Delta n_b(x_e)[\cosh(W/L_{nb}) - 1]/[\sinh(W/L_{nb})L_{nb}] \tag{3-3-10}$$

The hole component of the emitter current, $I_{pe} = qD_p S\ (\partial p_e/\partial x)$, can be found using eq. (3-2-10):

$$I_{pe} = qD_p \Delta p_e(x_e)\cosh(x_e/L_{pe})S/[\sinh(x_e/L_{pe})L_{pe}] \tag{3-3-11}$$

As was discussed in Section 3-1, for modern BJTs, I_{pe} is larger than I_{br}.

The recombination current in the emitter-base depletion region, I_{re}, may be crudely estimated using the equations given in Section 2-4 (see eq. (2-4-38a)):

$$I_{re} = I_{res} \exp[qV/(m_{re}k_B T)] \tag{3-3-12}$$

where

$$I_{res} = (\pi/2)^{1/2}(k_B T/F_{np})n_i S/\tau_{rec} \tag{3-3-13}$$

Here F_{np} is the electric field at the point in the emitter-base space charge region where $n = p$, m_{re} is an empirical constant determining the voltage dependence of the recombination current (in the simplest case $m_{re} = 2$; see eq. (2-4-38)), and τ_{rec} is the effective recombination time in the emitter-base depletion region. For an n-p-n transistor the electric field F_{np} is given by

$$F_{np} = [qN_{ab}(2V_{bip} - V_{be})/\varepsilon]^{1/2} \tag{3-3-14}$$

where

$$V_{bip} = (k_B T/q)\ln(N_{ab}/n_i) \tag{3-3-15}$$

(see eqs. (2-4-41) and (2-4-42)). Note that for the forward active mode $V_{be} > 0$.

The collector generation current, I_{gc}, may be estimated using eq. (2-4-15):

$$I_{gc} = qn_i x_{dcb} S/\tau_{gen} \tag{3-3-16}$$

where τ_{gen} is the effective generation time and

$$x_{dcb} = [2\varepsilon(V_{bicb} - V_{bc})/(qN_{dc})]^{1/2} \tag{3-3-17}$$

is the width of the depletion region for the collector-base junction. (Note that for the forward active mode $V_{bc} < 0$.) Here

$$V_{bicb} = V_{th} \ln(N_{dc}N_{ab}/n_i^2) \tag{3-3-18}$$

is the built-in voltage of the collector-base junction.

The current gain of a bipolar junction transistor in a common-base configuration, α, is defined as

$$\alpha = \frac{\partial I_c}{\partial I_e} \tag{3-3-19}$$

It may be presented as a product of two factors,

$$\alpha = \alpha_T \gamma \tag{3-3-20}$$

where

$$\gamma = \frac{\partial I_{ne}}{\partial I_e} \tag{3-3-21}$$

is the emitter injection efficiency, and

$$\alpha_T = \frac{\partial I_{nc}}{\partial I_{ne}} \tag{3-3-22}$$

is the base transport factor (here we assume that $I_c \approx I_{nc}$). Some collector current, however, may flow when the emitter terminal is open, so that the relationship between the collector and emitter currents may be presented as

$$I_c = I_{cbo} + \alpha I_e \tag{3-3-23}$$

where I_{cbo} is called a *common-base collector saturation current*. Here we assumed $I_c \approx I_{nc}$ and neglected the dependence of α on I_e. When such a dependence is important, eq. (3-3-23) may still be used. However, in this case the common-base current gain, α, should be redefined as

$$\alpha = (I_c - I_{cbo})/I_e \tag{3-3-23a}$$

The value of α given by eq. (3-3-23a) is called a *common-base dc current gain* (or *current ratio*). Usually the same symbol, α, is used for somewhat different values of α defined by eqs. (3-3-19) and (3-3-23a).

Equation (3-3-23) is valid when the reverse collector-base bias is not too large, so that there is no avalanche breakdown in the collector-base junction (see Section 2-8). Otherwise it should be modified to include the multiplication coefficient, M, for the collector current:

$$I_c = M(I_{cbo} + \alpha I_e) \tag{3-3-23b}$$

where M is the collector multiplication factor. The effect of an avalanche breakdown in the collector-base junction is considered in Section 3-9.

Using eq. (3-1-7), in which we neglected the recombination in the base, and taking into account that $I_e = I_c + I_b$, we can estimate the emitter injection efficiency as

$$\gamma = \frac{(D_n N_{de} x_e / D_p N_{ab} W)}{1 + (D_n N_{de} x_e / D_p N_{ab} W)} \tag{3-3-24}$$

The base transport factor, α_T, may be found from eqs. (3-3-22), (3-3-7), and (3-3-8):

$$\alpha_T = \frac{1}{\cosh(W/L_{nb})} \tag{3-3-25}$$

In practical transistors, $W/L_{nb} << 1$ and

$$\alpha_T \cong 1 - \frac{W^2}{2L_{nb}^2} \tag{3-3-26}$$

Both α_T and the collector multiplication factor are very close to 1. Hence,

$$\alpha \cong \gamma \tag{3-3-27}$$

in most modern bipolar junction transistors.

The common-emitter current gain, β, is defined as

$$\beta = \frac{\partial I_c}{\partial I_b} \tag{3-3-28}$$

Substituting

$$I_b = I_e - I_c \tag{3-3-29}$$

into eq. (3-3-28) we obtain

$$\beta = \frac{\partial I_c}{\partial (I_e - I_c)} = \frac{\partial I_c/\partial I_e}{1 - \partial I_c/\partial I_e} = \frac{\alpha}{1 - \alpha} \tag{3-3-30}$$

As was shown in Section 3-1,

$$\beta \cong \frac{D_n N_{de} x_e}{D_p N_{ab} W} \tag{3-3-31}$$

(see eq. (3-1-7)). Hence, β is inversely proportional to the total number of acceptors in the base region ($N_{ab}W$), which is called the *Gummel number, Q_G*. For a more general case of a nonuniform doping in the base, the Gummel number is defined as

$$Q_G = \int_{x_e}^{w+x_e} N_a(x)\, dx \tag{3-3-32}$$

The estimates for α given by eqs. (3-3-27) and (3-3-24) and for β given by eq. (3-3-31) are valid at moderate injection levels, i.e., when the emitter-base forward bias, V_{be}, is neither too small nor too large. At low injection levels (small V_{be}) the recombination current in the emitter-base depletion region, I_{re}, becomes dominant and reduces the gain. This happens because the pre-exponential factor in eq. (3-3-12) for the recombination current is usually much larger then the diffusion saturation current, $I_{nes} = qD_n Sn_i/L_{nb}$. However, the diffusion component of the current increases proportionally to $\exp(V_{be}/V_{th})$ with an increase in V_{be}, faster than the recombination current (which is proportional to $\exp[V_{be}/(m_{re}V_{th})]$, where m_{re} is greater than 1). Hence, the diffusion component of the current is dominant at larger values of the forward bias. This can clearly be seen from Fig. 2-3-3, which shows the current-voltage characteristics of Si and Ge p-n diodes. Using eq. (3-3-12) we can obtain the following estimate for β at low injection levels (i.e., relatively small emitter-base voltages):

$$\beta \sim \frac{I_{ne}}{I_{re}} \cong \frac{\mu_n F_{np} n_{bo} \tau_{rec} \exp[(V_{be}(1 - 1/m_{re})/V_{th}]}{W n_i} \tag{3-3-33}$$

We should notice that this estimate is based on the definition of the dc common-emitter short-circuit current gain, also called the *dc common-emitter short-circuit current ratio*. Usually, the same symbol, β, is used for somewhat different values of β defined by eq. (3-3-28). The collector current

$$I_c \approx I_{ne} \approx q D_n n_{bo} \exp(V_{be}/V_{th}) S/W \tag{3-3-34}$$

is proportional to $\exp(V_{be}/V_{th})$. Hence, in the low injection regime

$$\beta \approx \frac{\mu_n F_{np} n_{bo} \tau_{rec} [W/(q D_n n_{bo} S)]^{(1-1/m_{re})}}{W n_i} I_c^{(1-1/m_{re})} \tag{3-3-35}$$

Equation (3-3-35) shows that β increases with I_c in the low-injection regime. With the increase in the emitter-base voltage and, hence, with the increase in the collector current, the hole component of the emitter current, I_{pe}, increases proportionally to $\exp(V_{be}/V_{th})$. At the same time, the recombination current, I_{re}, increases at a slower rate, proportionally to $\exp(V_{be}/m_{re}V_{th})$, where m_{re} is greater than 1. Hence, when V_{be} and I_c become large enough, I_{pe} becomes greater than I_{re} and β reaches the value given by eq. (3-3-31). At this value of I_c the emitter injection efficiency becomes the most important factor limiting the common-emitter current gain.

At high emitter and collector currents, *high-injection effects* become important. In this case the number of electrons injected into the base may actually become greater than the Gummel number. As we mentioned before (see eq. (3-2-12)), the number of additional holes in the base should be approximately equal to the number of electrons injected into the base (to maintain the quasi neutrality of the base). Hence, under high-injection conditions when $n_b >> N_{ab}$, where N_{ab} is the concentration of acceptors in the base region,

$$p_b \cong n_b >> N_{ab} \tag{3-3-36}$$

The product

$$p_b(x_e) n_b(x_e) = n_i^2 \exp\left(\frac{V_{be}}{V_{th}}\right) \tag{3-3-37}$$

increases proportionally to $\exp(V_{be}/V_{th})$. For low injection levels $p_b(x_e) \cong N_{ab}$ and

$$\Delta n_b(x_e) \approx n_b(x_e) = \frac{n_i^2}{N_{ab}} \exp\left(\frac{V_{be}}{V_{th}}\right) = n_{bo} \exp(V_{be}/V_{th}) \tag{3-3-38}$$

For high injection levels when $\Delta n_b(x_e) >> N_{ab}$,

$$\Delta n_b(x_e) \cong p_b(x_e) \cong n_i \exp\left(\frac{V_{be}}{2V_{th}}\right) \tag{3-3-39}$$

Therefore, the emitter and collector currents are approximately given by

$$I_c \cong \frac{q\Delta n_b(x_e) D_n S}{W} \cong \frac{q n_i D_n S}{W} \exp\left(\frac{V_{be}}{2V_{th}}\right) \tag{3-3-40}$$

On the other hand, the concentration of holes injected from the base into the emitter region is still low compared with the emitter doping, N_{de}. Hence,

$$I_b \cong \frac{q D_p n_i^2 S}{x_e N_{de}} \exp\left(\frac{V_{be}}{V_{th}}\right) \tag{3-3-41}$$

and

$$\beta = \frac{\partial I_c}{\partial I_b} \cong \frac{D_n N_{de} x_e}{D_p n_i W} \exp\left(-\frac{V_{be}}{2V_{th}}\right) \tag{3-3-42}$$

or

$$\beta = \frac{q D_n^2 N_{de} x_e S}{D_p W^2 I_c} \tag{3-3-43}$$

The overall dependence of the common-emitter current gain on the collector current, I_c, may be obtained by adding the expressions for the inverse common-emitter current gain, $1/\beta$, obtained from equations (3-3-31), (3-3-35), and (3-3-43), as well as the expression for the inverse common-emitter current gain, $1/\beta_T$, limited by the base transport factor, α_T,

$$\beta_T = \frac{\alpha_T}{1 - \alpha_T} \approx 2(L_{nb}/W)^2 \tag{3-3-44}$$

(see eq. (3-3-26)). This leads to the following equation:

$$\beta^{-1} = \frac{D_p W^2 I_c}{q D_n^2 N_{de} x_e S} + \frac{D_p N_{ab} W}{D_n N_{de} x_e} + \frac{W n_i}{\mu_n F_{np} n_{bo} \tau_{rec} [W I_c/(q D_n n_{bo} S)]^{(1-1/m_{re})}} + \frac{W^2}{2L_{nb}^2} \tag{3-3-45}$$

The dependence of β on I_c calculated using eq. (3-3-45) is shown in Fig. 3-3-2. As can be clearly seen from the figure, β increases with the increase of I_c at low injection levels, reaches the maximum value at moderate injection levels, and decreases with further increase in I_c owing to high-injection effects.

In addition to the aforementioned high-injection effects, at high current levels the value of the common-emitter current gain may decrease owing to the carrier injection into the collector region and the related shift of the high-field region from the emitter-base junction into the collector region. This effect (called

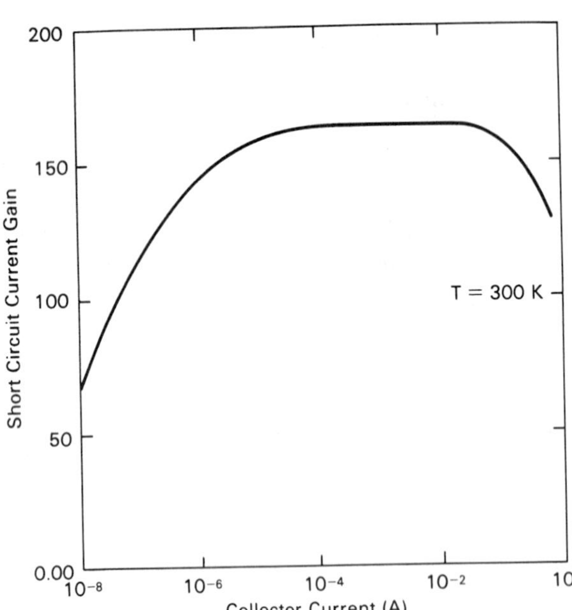

Fig. 3-3-2. Dependence of the maximum common-emitter current gain, β, on collector current for silicon *n-p-n* bipolar junction transistor. The curves were computed using subroutine PBJTGI.BAS (see Program Description Section). Parameters used in the calculation are emitter region width $x_e = 1 \mu m$, base region width $W = 0.5\ \mu m$, device area $S = 1 \times 10^{-4}\ cm^2$, emitter doping $N_{de} = 5 \times 10^{18}\ cm^{-3}$, base doping $N_{ab} = 2 \times 10^{17}\ cm^{-3}$, energy gap $E_g = 1.12$ eV, effective density of states in the conduction band $N_c = 3.22 \times 10^{19}\ cm^{-3}$, effective density of states in the valence band $N_v = 1.83 \times 10^{19}\ cm^{-3}$, temperature $T = 300$ K, electron mobility in the base region 1500 cm²/Vs, hole mobility in the emitter region 450 cm²/Vs, dielectric permittivity $\varepsilon = 1.05 \times 10^{-10}$ F/m, effective recombination time in the emitter-base depletion region $\tau_{rec} = 1 \times 10^{-7}$ s, and diffusion length of electrons in the base $L_{nb} = 40\ \mu m$.

the *Kirk effect*) occurs when the collector current density exceeds the saturation current density:

$$I_c/S > j_{sc} = qN_{dc}v_s \tag{3-3-46}$$

The current density j_{sc} is the highest current density that the relatively lightly doped collector region can support without the injection of extra carriers. As can be seen from eq. (3-3-46), it is especially important in transistors with low doping levels in the collector region. Typically, such transistors are made by utilizing the epitaxial growth of silicon films; they therefore are called *epitaxial transistors*. According to Kirk (1962) the distribution of carriers and electric field resulting from the carrier injection into the collector region may be interpreted as resulting from an effective increase in the base width. A more accurate analysis of this effect is given by Warner and Grung (1983).

Still another factor limiting the common-emitter current gain at high current levels is related to the base resistance between the central section of the transistor and the base contact, i.e., the *base spreading resistance* and the resulting nonuniform distribution of the emitter current called *emitter current crowding*. This effect is considered in the next section.

The maximum value of β, reached at moderate injection levels, may be estimated using eq. (3-3-31). This equation seems to indicate that the common-emitter current gain may be increased simply by raising the emitter doping level.

Unfortunately, in practical devices very high doping levels lead to changes in the material parameters, in particular, to a decrease in the energy gap. According to Lanyon and Tuft (1978) the energy gap narrowing, ΔE_g, in silicon is given by

$$\Delta E_g = \frac{3q^2}{16\pi\varepsilon_s}\left(\frac{q^2 N_d}{\varepsilon_s k_B T}\right)^{1/2} \cong 22.5\left(\frac{N_d}{10^{18}}\right)^{1/2}\left(\frac{300K}{T}\right)^{1/2} \text{ (meV)} \tag{3-3-47}$$

where N_d is in cm^{-3}. This reduction in the band gap leads to an exponential increase of the intrinsic carrier density n_{ie} in the emitter region:

$$n_{ie} = n_{ieo} \exp\left(\frac{\Delta E_g}{2k_B T}\right) \tag{3-3-48}$$

where n_{ieo} is the value of n_{ie} for $N_{ie} = 0$. The concentration of holes injected into the emitter region is proportional to n_{ie}^2/N_{de}, and hence, the band-gap narrowing leads to the exponential reduction of the maximum common-emitter current gain:

$$\beta = \frac{D_n N_{de} x_e}{D_p N_{ab} W} \exp(-\Delta E_g/k_B T) \tag{3-3-49}$$

The dependence of β on N_{de} for different values of N_{ab} calculated using eqs. (3-3-47) and (3-3-49) is shown in Fig. 3-3-3. As can be seen from the figure, the reduction of β, caused by the band-gap narrowing, is quite dramatic.

Additional reduction of β may be caused by the direct recombination of the electrons and holes in the heavily doped emitter region. The important recombi-

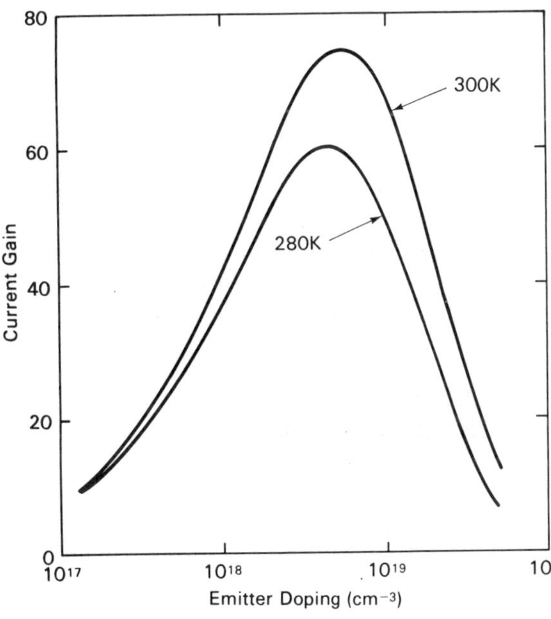

Fig. 3-3-3. Dependence of the maximum common emitter current gain, β, on the emitter doping for a silicon n-p-n bipolar junction transistor. The curves were computed using subroutine PBJTG.BAS (see Program Description Section). Parameters used in the calculation are emitter region width $x_e = 1$ μm, base region width $W = 0.5$ μm, base doping $N_{ab} = 2 \times 10^{17}$ cm^{-3}, temperature $T = 300$ K and 280 K electron mobility in the base region is 1000 cm^2/Vs, hole mobility in the emitter region is 400 cm^2/Vs, and dielectric permittivity $\varepsilon = 1.05 \times 10^{-10}$ F/m.

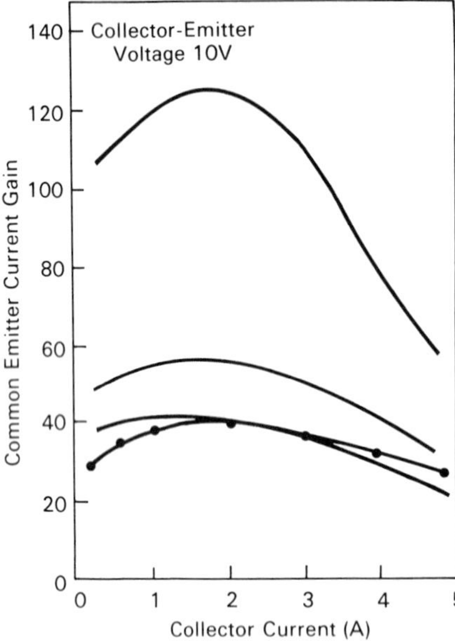

Fig. 3-3-4. Measured and calculated dependences of the maximum common emitter current gain, β, on the collector current for a silicon *n-p-n* bipolar junction transistor (from E. J. McGrath and D. H. Navon, "Factors Limiting Current Gain in Power Transistors," *IEEE Trans. Electron Devices,* ED-24, p. 1255 (1977). © 1977 IEEE). Solid curves are calculated; curve with dark circles is measured; top curve is calculated neglecting the band narrowing and Auger recombination; and the middle curve is calculated neglecting the Auger recombination.

nation process is the Auger recombination, which is the inverse process of avalanche multiplication (see Section 2-8). The Auger recombination time may be estimated as

$$\tau_A \cong \frac{1}{G_n N_{de}^2} \tag{3-3-50}$$

(see Section 1-12) for n^+ emitters in n-p-n transistors and

$$\tau_A \cong \frac{1}{G_p N_{ae}^2} \tag{3-3-51}$$

for p^+ emitters in p-n-p transistors. For silicon coefficient $G_p = 9.9 \times 10^{-32}$ cm^6/s and $G_n = 2.28 \times 10^{-31}$ cm^6/s at room temperature (see Section 1-12). The effect of the Auger recombination on the common-emitter current gain is illustrated by Fig. 3-3-4 (from McGrath and Navon 1977) in which are compared the measured and computed dependencies of β on the collector current.

3-4. BASE SPREADING RESISTANCE AND EMITTER CURRENT CROWDING IN BIPOLAR JUNCTION TRANSISTORS

As can be seen from the schematic structure of a bipolar junction transistor shown in Fig. 3-4-1, a base current flows in the direction parallel to the emitter-base junction. This distributed current flow leads to a voltage drop, $V_{bb'}$, between the

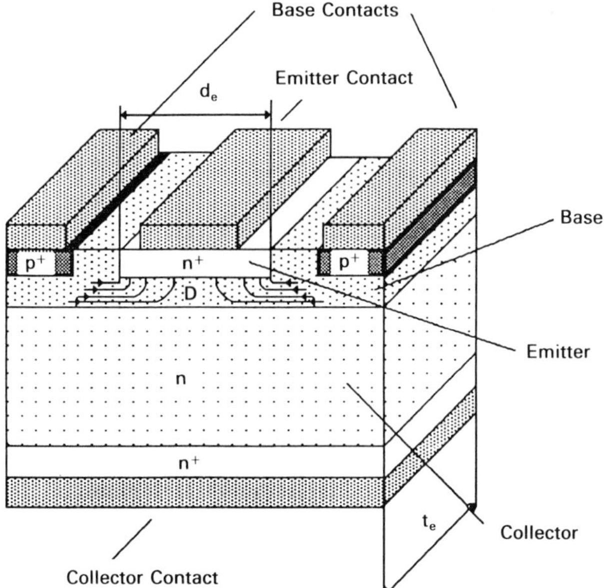

Fig. 3-4-1. Schematic structure of a bipolar junction transistor. Also shown are the streamlines of the base current.

central portion of the emitter-base junction and the periphery of the junction. This voltage drop, $V_{bb'}$, is proportional to the base current:

$$V_{bb'} = r_{bb'} I_b \tag{3-4-1}$$

Resistance $r_{bb'}$ is called the base spreading resistance. An estimate of the base spreading resistance may be obtained by calculating the ohmic resistance of the base region between the emitter-base junction and the base contact. The results of such a calculation are presented in Fig. 3-4-2 (after Warner and Fordemwalt 1965). More accurate calculations should take into account the nonuniformity of the emitter current distribution along the emitter-base junction. This nonuniformity is directly related to the voltage drop along the base caused by the base current. Indeed, if the voltage drop across the central portion of the emitter-base junction is V_{eb}, the voltage drop across the periphery of the emitter-base junction is $V_{eb} + I_b r_{bb'}$. As the emitter current density is exponentially dependent on the voltage across the emitter-base junction, the emitter current density at the periphery of the junction may be much larger than in the center of the junction. This effect is called emitter current crowding.

Emitter current crowding is schematically illustrated in Fig. 3-4-1, which shows the streamlines of the base current. In the extreme case in which the voltage drop $V_{bb'}$ becomes much larger than the thermal voltage $k_B T/q$, most of the emitter current is concentrated near the edges of the emitter-base junction. The

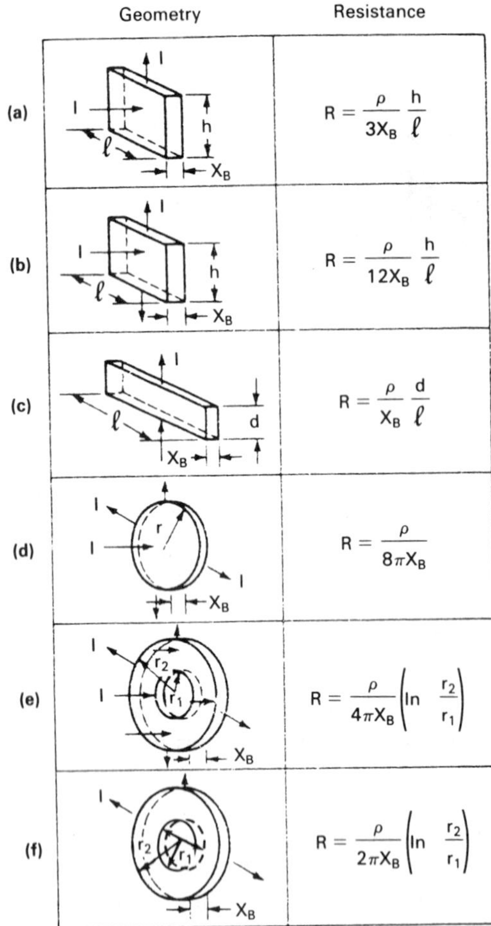

Fig. 3-4-2. Base spreading resistance for different transistor geometries (after R. M. Warner and J. N. Fordemwalt, *Integrated Circuits*, McGraw-Hill, New York 1965).

length of the edge region, Δd_e, where most of the emitter current flows may be crudely estimated as

$$\Delta d_e \approx V_{th}/F_{base} \tag{3-4-2}$$

where

$$F_{base} \approx I_b/(qN_{ab}\mu_p t_e W) = I_c/(qN_{ab}\mu_p t_e W\beta) \tag{3-4-3}$$

Here t_e is the width of the emitter stripe (see Fig. 3-4-1), and μ_p is the hole mobility (we consider here an n-p-n transistor). Assuming $I_c = 2$ mA, $\beta = 50$, $W = 0.5\ \mu$m, $t_e = 5\ \mu$m, $\mu_p = 400$ cm^2/Vs, $N_{ab} = 10^{17}$ cm^{-3}, and $T = 300$ K ($V_{th} = 0.0258$ V), we obtain $\Delta d_e \approx 1\ \mu$m. Hence, the emitter current in this case is primarily flowing in two 1-μm-wide regions near the edges of the emitter stripe. To make the emitter-current distribution at high injection levels more uniform, the emitter stripes

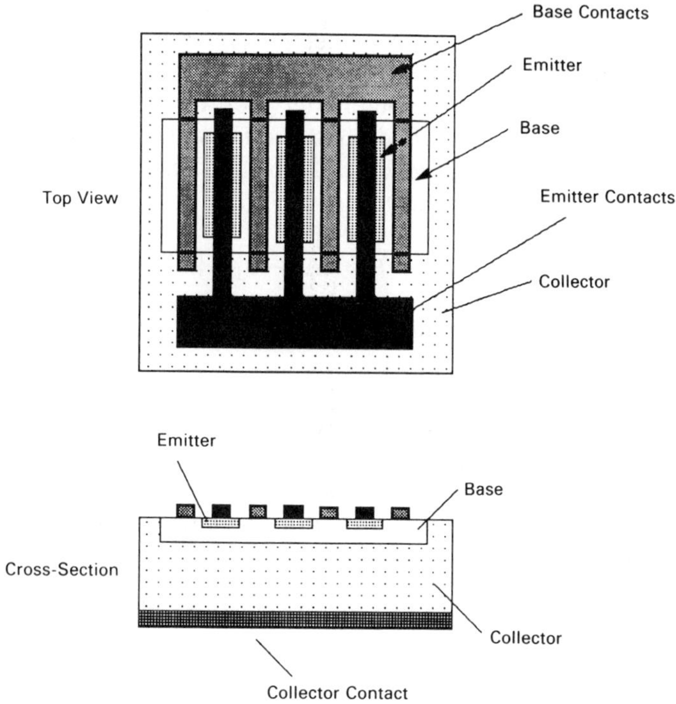

Fig. 3-4-3. Transistor design with interdigitated emitter and base fingers.

should be made narrow, i.e., we should have $d_e < 2\Delta d_e$. This leads to the transistor design with interdigitated emitter and base fingers (see Fig. 3-4-3). Such a design considerably improves the uniformity of the emitter-current distribution.

3-5. EFFECTS OF NONUNIFORM DOPING IN THE BASE REGION: GRADED BASE TRANSISTORS

As can be seen from Fig. 3-1-3a the doping in the base region of a typical bipolar junction transistor is not uniform. The acceptor concentration in the base decreases in the direction from the emitter to the collector region. A transistor with such a doping profile is called a *graded base transistor*. In such a transistor, the concentration of free holes in the base near the emitter-base junction is larger than near the collector-base junction. Hence, the holes diffuse toward the collector, creating the built-in electric field directed from the collector-base junction toward the emitter-base junction. This built-in field leads to the drift of holes counteracting their diffusion flux. Under equilibrium conditions, the drift and diffusion fluxes compensate each other. When minority carriers (electrons) are injected into the base from the emitter region, the built-in field pushes electrons toward the collector, creating a drift flux of electrons towards the collector, in addition to the electron diffusion flux. This increases the electron component of the emitter current and, hence, the common-emitter short-circuit current gain.

Under equilibrium conditions, the Fermi level in the base region is uniform and the variable doping level leads to the slope of the valence and conduction bands that can be found using eq. (1-7-31):

$$E_v = E_F - k_B T \ln\left(\frac{N_v}{N_{ab}}\right) \tag{3-5-1}$$

From this equation we find the built-in electric field in the base, F_b, related to the nonuniform doping:

$$qF_b = dE_v/dx = dE_c/dx = k_B T(dN_{ab}/dx)/N_{ab} \tag{3-5-2}$$

The direction of this field is from the collector-base junction toward the emitter-base junction (see Fig. 3-5-1), and hence, it aids the transport of minority carriers (electrons) across the base. Substituting eq. (3-5-2) into the equation for the electron current in the base region,

$$I_n = qS(\mu_n n_b F_b + D_n\, dn_b/dx) \tag{3-5-3}$$

we obtain the following differential equation for n:

$$dn_b/dx + (n_b/N_{ab})(dN_{ab}/dx) = I_n/(qSD_n) \tag{3-5-4}$$

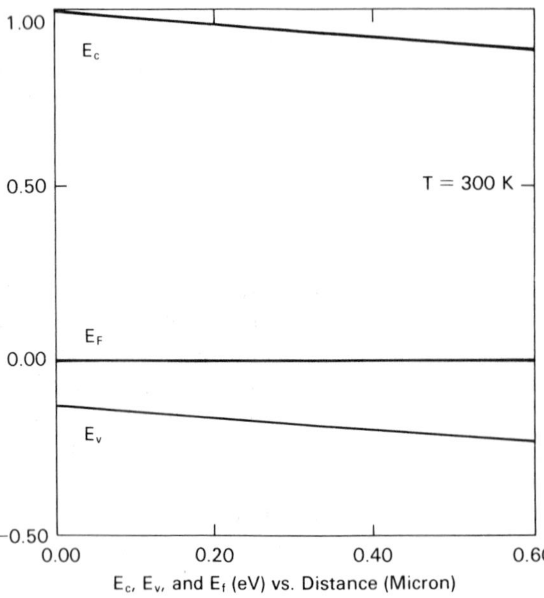

Fig. 3-5-1. Band diagram of nonuniformly doped silicon sample with an exponentially varying acceptor density $N_a = N_o \exp(-x/\lambda_b)$ at room temperature. The band diagram was computed using subroutine PBNU.BAS (see Program Description Section). Parameters used in the calculation are $N_o = 2 \times 10^{17}$ cm^{-3}, $\lambda_b = 0.3$ μm, Energy gap $E_g = 1.12$ eV, Effective density of states in the conduction band $N_c = 3.22 \times 10^{19}$ cm^{-3}, Effective density of states in the valence band $N_v = 1.83 \times 10^{19}$ cm^{-3}, and Temperature $T = 300$ K.

Here I_n is the principal component of the emitter and collector currents (to the first order, $I_{ne} \approx I_{nc} \approx -I_n$). By multiplying eq. (3-5-4) by N_{ab}, we obtain

$$d(N_{ab}n_b)/dx = I_n N_{ab}/(qSD_n) \tag{3-5-6}$$

The integration of eq. (3-5-6) yields

$$n_b(x) = -\left[I_n \int_x^{x_e+W} N_{ab}(x')\,dx'\right]/(qSD_n\, N_{ab}) \tag{3-5-7}$$

(Here W is the base width, and we have taken into account that $n_b(x_e+W) = 0$.) Substituting $x = x_e$ into eq. (3-5-7) and taking into account that $n_b(x_e) \approx (n_i^2/N_{ab})\exp(V_{be}/V_{th})$, we find that

$$I_n = -qSD_n n_i^2 exp(V_{be}/V_{th})/Q_b \tag{3-5-8}$$

where

$$Q_b = \int_{x_e}^{x_e+W} N_{ab}(x)\,dx \tag{3-5-9}$$

is the Gummel number. This parameter (a total doping in the base per unit area) plays a very important role in determining the characteristics of a bipolar junction transistor.

We can now estimate the common-emitter current gain, β_T, limited by the recombination of minority carriers in the base:

$$\beta_T = I_n/I_{br} \tag{3-5-10}$$

The recombination current in the neutral base region, I_{br}, can be estimated as

$$I_{br} = (qS/\tau_{bn}) \int_{x_e}^{x_e+W} n_b(x')\,dx' \tag{3-5-11}$$

where $\tau_{bn} = L_{nb}^2/D_n$ is the electron lifetime in the base region. Substituting eq. (3-5-7) into eq. (3-5-11) and the resulting equation into eq. (3-5-10), we finally obtain

$$\beta_T = L_{nb}^2 \Big/ \int_{x_e}^{x_e+W} N_{ab}^{-1}(x) \left[\int_x^{x_e+W} N_{ab}(x')\,dx'\right] dx \tag{3-5-12}$$

For a uniform doping profile eq. (3-5-14) reduces to $\beta_T = 2L_{nb}^2/W^2$. Let us now consider a doping profile in the base exponentially decreasing toward the base-collector junction ($N_{ab}(x) \sim \exp[-(x - x_e)/\lambda_b]$). Then the evaluation of the integrals in eq. (3-5-12) yields

$$\beta_T = L_{nb}^2/\{\lambda_b W - \lambda_b^2[1 - \exp(-W/\lambda_b)]\} \tag{3-5-13}$$

For $\lambda_b >> W$ (which corresponds to a very small variation of $N_{ab}(x)$ within the base region), eq. (3-5-13) reduces to $\beta_T = 2(L_{nb}/W)^2$, as expected. However, for $\lambda_b << W$, eq. (3-5-13) yields

$$\beta_T = L_{nb}^2/(\lambda_b W) >> 2(L_{nb}/W)^2 \tag{3-5-14}$$

clearly illustrating the improvement in the base transport factor. For example, if the base width $W = 0.2$ μm and $\lambda = 0.05$ μm, the common-emitter current gain limited by the recombination increases by a factor of two.

3-6. OUTPUT CHARACTERISTICS OF BIPOLAR JUNCTION TRANSISTORS AND EARLY EFFECT

The relationship between collector, base, and emitter currents in a bipolar junction transistor, in an forward active mode (when $V_{eb} < 0$ and $V_{cb} > 0$ for an n-p-n transistor), is given by

$$I_e \cong I_c \qquad (3\text{-}6\text{-}1)$$

$$I_b \equiv I_e - I_c << I_e \qquad (3\text{-}6\text{-}2)$$

(see Section 3-1). This is clearly illustrated in Fig. 3-6-1, which shows the current-voltage characteristics of an n-p-n Si BJT in a common-base and common-emitter configuration.

When the emitter is open ($I_e = 0$), some collector current (equal, of course, to the base current in this case) still flows across the collector-base junction. This component of the collector current is the common-base collector saturation current, I_{cbo}. Hence, for the forward active mode the collector current is related to the emitter current as follows:

$$I_c = I_{cbo} + \alpha I_e \qquad (3\text{-}6\text{-}3)$$

The collector current that flows in the collector-emitter circuit when the base is open is called the *common-emitter collector saturation current*, I_{ceo}:

$$I_c = I_{ceo} + \beta I_b \qquad (3\text{-}6\text{-}4)$$

(I_{ceo} and I_{cbo} are sometimes called the collector-emitter leakage current and the collector-base leakage current, respectively). Substituting the base current, I_b, from eq. (3-6-2) we find that

$$I_c = \frac{I_{ceo}}{1 + \beta} + \alpha I_e \qquad (3\text{-}6\text{-}5)$$

Hence, by comparing eqs. (3-6-3) and (3-6-5) we obtain

$$I_{ceo} = (1 + \beta)I_{cbo} \qquad (3\text{-}6\text{-}6)$$

or

$$I_{ceo} = I_{cbo}/(1 - \alpha) \qquad (3\text{-}6\text{-}7)$$

Thus, the collector-emitter leakage current is much larger than the collector-base leakage current. This result can be interpreted as follows. The collector-base leakage current flows in the input circuit (input loop) of a bipolar junction transis-

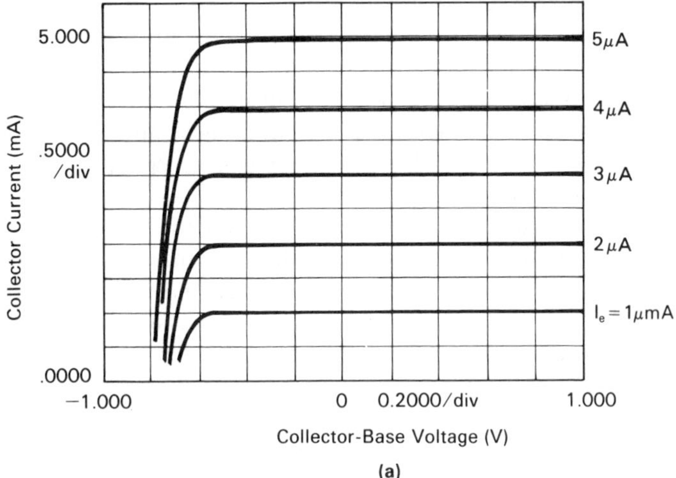

(a)

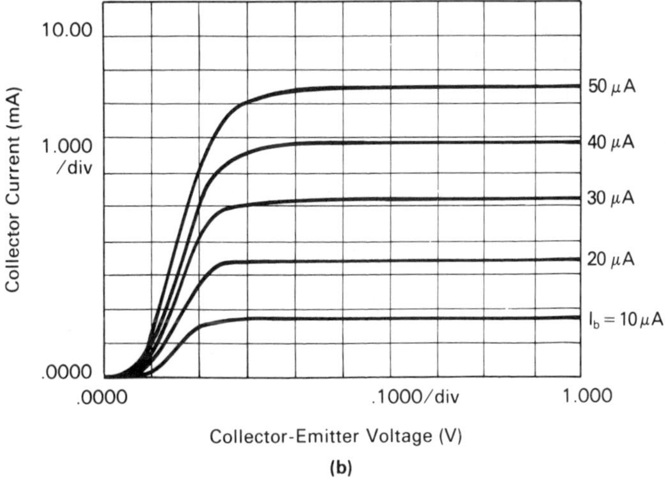

(b)

Fig. 3-6-1. Current-voltage characteristics of *n-p-n* Si BJT for common-base (a) and common-emitter configuration (b).

tor. The collector-emitter leakage current flows in the output circuit (output loop) of a bipolar junction transistor (see Fig. 3-1-6b). The ratio of the current in the output loop over the current in the input loop for the common-emitter configuration is approximately equal to the ratio of the emitter current over the base current; hence, it should be of the order of the short-circuit common-emitter current gain, β, in agreement with eq. (3-6-6).

As can be seen from Fig. 3-6-1a, $I_c \cong I_e$, independently of the collector-emitter voltage, as long as $V_{cb} > 0$ (for an n-p-n transistor). A common-base current gain, α, remains very close to 1 until the collector-base breakdown voltage, BV_{cbo}, is reached. At voltages higher than BV_{cbo}, the common-base current gain is equal to $M\alpha$, where M is the multiplication rate (see eq. (3-3-23b)), and it may actually exceed 1. The breakdown may occur either because of the avalanche multiplication in the collector-base junction (as described by eq. (3-3-23b)) or because of the "punch-through," when the collector-base depletion region increases so much that it merges with the emitter-base depletion region.

Fig. 3-6-1a clearly shows that a change in α with an increase in the collector-base voltage in the forward active mode is very small (not really noticeable on the graph). However, even small changes in α correspond to large changes in the common-emitter current gain:

$$\beta = \frac{\alpha}{1 - \alpha} \tag{3-6-8}$$

(see eq. (3-3-30)). For instance, the change in α from 0.98 to 0.99 corresponds to the change in β from 49 to 99. Hence, we may expect a much larger variation of the collector current with the collector-emitter voltage in the common-emitter configuration when the base current is kept constant. Indeed, an increase in the collector current with the collector-emitter voltage in the common-emitter configuration for the same transistor is noticeable in Fig. 3-6-1b. When the collector-emitter voltage is increased, practically all the increase (in the forward active mode) is equal to the increase in the reverse collector-base voltage. The main reason for the increase in the collector current with the collector-base voltage (at the fixed base current)—i.e., for the increase in β—is the dependence of an effective base width, W_{eff}, on the collector-base voltage,

$$W_{eff} = W - x_{deb} - x_{dcb} \tag{3-6-9}$$

where the depletion widths of the emitter-base and collector-base junctions in the base region, x_{deb} and x_{dcb}, are given by

$$x_{deb} = \left[\frac{2\varepsilon(V_{bieb} - V_{be})}{qN_{ab}}\right]^{1/2} \tag{3-6-10}$$

$$x_{dcb} = \left[\frac{2\varepsilon(V_{bibc} + V_{cb})N_{dc}}{qN_{ab}^2}\right]^{1/2} \tag{3-6-11}$$

Equations (3-6-10) and (3-6-11) are obtained using an elementary theory of p-n junctions based on the depletion approximation (see Sections 2-2 and 2-3).

In the forward active mode, the emitter and collector currents have an exponential dependence on the emitter-base voltage, V_{eb}. Hence, this voltage and the depletion width of the emitter-base p-n junction, x_{deb}, vary relatively little over a wide range of the emitter and collector currents. Most of the variation in W_{eff} comes from the changes in x_{dcb} (see eq. (3-6-11)).

The common emitter current gain, β, may be crudely estimated as

$$\beta = D_n N_{de} x_e / (D_p N_{ab} W_{eff}) \tag{3-6-12}$$

(compare with eq. (3-1-7)). As can be seen from eqs. (3-6-9) and (3-6-11), W_{eff} decreases with an increase in V_{cb}. This leads to the increase in gain and, hence, to the finite output conductance for common-emitter current-voltage characteristics (see Fig. 3-6-1b). This effect is called the *Early effect* (after Early [1952]).

The foregoing equations assume a constant doping in the base region. In fact, as we discussed in Section 3-4, a doping gradient is maintained throughout the base in order to improve the base transport factor, α_T. Therefore, the dependence of W_{eff} on V_{cb} or V_{ce} may be quite different from that predicted by eqs. (3-6-9) to (3-6-11). Empirically, it is found that a linear interpolation of the collector current dependence on the collector-emitter voltage in the common-emitter configuration is adequate in most cases:

$$I_c = (I_{ceo} + \beta I_b)(1 + V_{ce}/V_A) \tag{3-6-13}$$

where V_A is the Early voltage. The positive sign of V_{ce} in eq. (3-6-13) corresponds to the reverse bias for an n-p-n transistor, which is the case that we consider here.

An empirical relation

$$V_A = k_A \frac{q Q_G W}{\varepsilon} \tag{3-6-14}$$

shows that the Early voltage is roughly proportional to the Gummel number, Q_G, and to the base width, W. Here k_A is a numerical constant of the order of unity, q

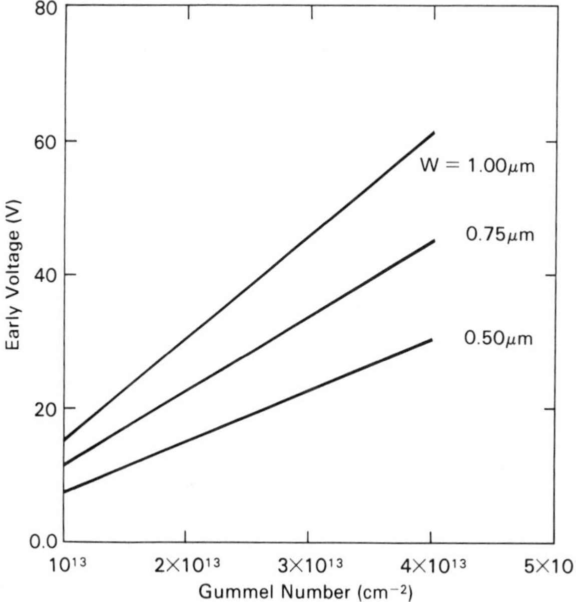

Fig. 3-6-2. Early voltages vs. Gummel number for different base thicknesses. The curves are calculated for Si transistors using program PLOTF with subroutine PEBV.BAS (see Program Description Section).

is the electronic charge, and ε is the dielectric permittivity. The dependence of the Early voltage on the Gummel number for different base thicknesses is shown in Fig. 3-6-2.

3-7. EBERS–MOLL MODEL

The simplest large-signal equivalent circuit of an ideal (intrinsic) bipolar junction transistor consists of two diodes and two current-controlled current sources describing the interaction between the emitter-base and collector-base junctions. (Current sources describing the recombination and generation currents have to be added to this circuit to represent a real BJT, as shown in Fig. 3-7-1.) This simple model for an "intrinsic" transistor (called the *Ebers-Moll model* after Ebers and Moll [1954]) is described by the following equations:

$$I_e' = a_{11}[\exp(qV_{be}/k_BT) - 1] + a_{12}[\exp(qV_{bc}/k_BT) - 1] \qquad (3\text{-}7\text{-}1)$$

$$I_c' = a_{21}[\exp(qV_{be}/k_BT) - 1] + a_{22}[\exp(qV_{bc}/k_BT) - 1] \qquad (3\text{-}7\text{-}2)$$

Coefficients a_{11}, a_{12}, a_{21}, and a_{22} may be related to the material parameters, transistor dimensions, and doping levels using equations given in Section 3-3. For n-p-n transistors and for the sign convention for currents I_e' and I_c' used in Fig. 3-7-1, we find that

$$a_{11} = -qS[D_n n_{bo} \coth(W_{eff}/L_{nb})/L_{nb} + D_p p_{eo}/x_e] \approx -qS[D_n n_i^2/(N_{ab}W_{eff}) + D_p n_i^2/(N_{de}x_e)] \qquad (3\text{-}7\text{-}3)$$

$$a_{12} = qS\{D_n n_{bo}/[L_{nb} \sinh(W_{eff}/L_{nb})]\} \approx qSD_n n_i^2/(N_{ab}W_{eff}) = a_{21} \qquad (3\text{-}7\text{-}4)$$

$$a_{22} = -qS[D_n n_{bo} \coth(W_{eff}/L_{nb})/L_{nb} + D_n p_{co}/x_c] \approx -qSn_i^2[D_n/(N_{ab}W_{eff}) + D_p/(N_{dc}x_c)] \qquad (3\text{-}7\text{-}5)$$

The effective base width, W_{eff}, is given by eq. (3-6-9).

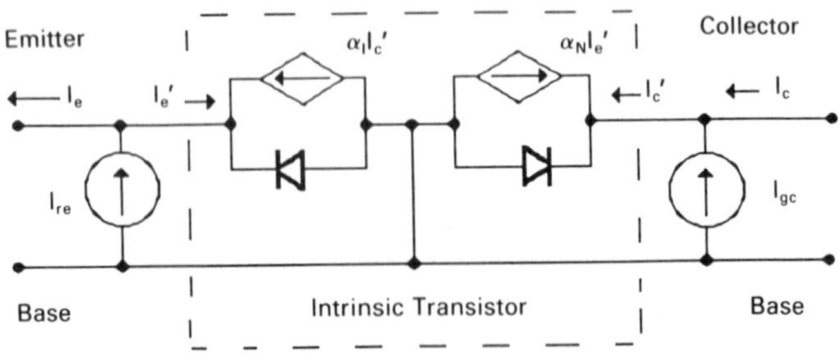

Fig. 3-7-1. Simple large signal equivalent circuit of bipolar junction transistor. The circuit includes current sources accounting for the emitter-base recombination and collector-base generation currents. For a more realistic modeling, additional circuit elements accounting for the Early effect and for parasitic resistances and capacitances should be added (not shown in the figure).

The emitter-base voltage may be expressed in terms of the emitter current of the intrinsic transistor, I'_e, using eq. (3-7-1):

$$\exp(qV_{be}/k_B T)-1 = I'_e/a_{11} - a_{12}[\exp(V_{bc}/V_{th}) - 1]/a_{11} \tag{3-7-6}$$

Substituting eq. (3-7-6) into eq. (3-7-2) we find that

$$I'_c = \left(\frac{a_{12}}{a_{11}}\right) I'_e + (a_{22} - a_{12}a_{21}/a_{11})[\exp(V_{bc}/V_{th}) - 1] \tag{3-7-7}$$

This equation may be rewritten as

$$I'_c = -\alpha_N I'_e - I_{co}[\exp(V_{bc}/V_{th}) - 1] \tag{3-7-8}$$

Here

$$\alpha_N = -a_{12}/a_{11} \tag{3-7-9}$$

is the normal (or forward) common-emitter current gain, and

$$I_{co} = -a_{22} + a_{12}a_{21}/a_{11} \tag{3-7-10}$$

is the common-base collector reverse saturation current. For the active forward mode, V_{bc} is negative, $|V_{bc}|$ is much larger than V_{th}, and eq. (3-7-8) reduces to eq. (3-3-23) if we neglect the recombination current I_{re} and assume that $I_e \approx I'_e$ (see Fig. 3-7-1), $\alpha_N \approx \alpha$, and $I_{co} \approx I_{cbo}$:

$$I_c = I_{cbo} + \alpha I_e \tag{3-3-23}$$

However, eq. (3-3-23) may account for the recombination current, I_{re}, (through the dependence of α on the collector current), and hence, current gains α_N and α and saturation currents I_{co} and I_{cbo} do not have to be exactly equal.

In a similar way, we can derive from eqs. (3-7-1) and (3-7-2) an equation relating the emitter current in the intrinsic transistor to the collector current:

$$I'_e = \left(\frac{a_{21}}{a_{22}}\right) I'_c + (a_{11} - a_{21}a_{12}/a_{22})[\exp(V_{be}/V_{th}) - 1] \tag{3-7-11}$$

This equation may be rewritten as

$$I'_e = -\alpha_I I'_c - I_{eo}[\exp(V_{be}/V_{th}) - 1] \tag{3-7-12}$$

Here α_I is the inverse current gain (sometimes also called inverse alpha) and I_{eo} is the emitter reverse saturation current. Equations (3-7-11) and (3-7-12) describe the inverse mode of operation when the collector-base junction is forward biased and the emitter-base junction is reverse biased.

The Ebers-Moll model for the intrinsic transistor has four parameters: a_{11}, a_{12}, a_{21}, and a_{22} or α_N, α_I, I_{eo}, and I_{co}. The reciprocity relationship for an ideal two-port device requires that

$$a_{12} = a_{21} \tag{3-7-13}$$

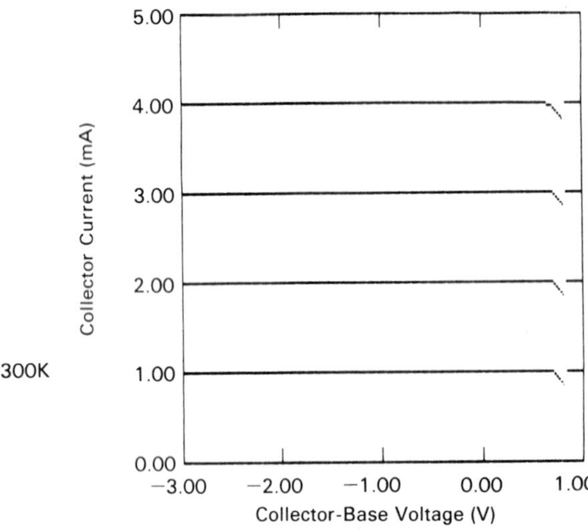

Fig. 3-7-2. Current-voltage characteristics of Si *n-p-n* bipolar junction transistor in common-emitter configuration. The curves are calculated for the values of the emitter current of 1 mA, 2 mA, 3 mA, and 4 mA in the frame of the Ebers-Moll model using program PLOTF with subroutine PEBML.BAS (see Program Description Section).

and hence, only three parameters (for example, α_N, I_{eo}, and I_{co}) are required for this basic transistor model. In practice, however, the reciprocity is not precisely satisfied for real transistors.

The family of the current-voltage characteristics of a Si n-p-n bipolar junction transistor in the common-emitter configuration calculated in the frame of the Ebers-Moll model are shown in Fig. 3-7-2.

This simple version of the Ebers-Moll model does not take into account the dependence of the common-emitter current gain on the injection level (see Fig. 3-3-2). Hence, it may only be applied for moderate emitter and collector currents.

This model may be somewhat improved by adding current sources to account for the emitter-base and collector-base recombination and generation currents (see Fig. 3-7-1). The expressions for these current components are given in Section 3-3 (see eqs.(3-3-12) to (3-3-18)). However, even this improved model does not account for the high-injection effects and for effects related to the band-gap narrowing, Auger recombination, and emitter current crowding. All these effects are very important for practical regimes of transistor operation, and the simplified Ebers-Moll model predicts values of the common-emitter current gain that are too high. A more accurate and realistic model proposed by Gummel and Poon (1970)—the *Gummel-Poon model*—is described in the next section.

3-8. GUMMEL–POON MODEL

The Ebers-Moll model is based on the assumption that BJT currents can be found using simple superposition of the emitter and collector currents for the forward and reverse active modes. However, this assumption becomes invalid in the high-injection regime when the charge of minority carriers in the base becomes comparable with the doping in the base. Another reason why the simple low-injection approach becomes invalid is the dependence of the effective base width on the injection level (base narrowing due to the Early effect or base widening in a high-injection regime (Kirk effect).)

Gummel and Poon (1970) developed an improved model that is applicable in the high-injection regime. They started from rewriting Ebers-Moll equations in an equivalent but more symmetrical form. Following their approach and using equations of the Ebers-Moll model given in Section 3-7, we can express coefficients a_{11}, $a_{12} = a_{21}$, and a_{22} through the emitter reverse saturation current I_{eo}, normal common-base current gain α_N, and inverse common-base current gain α_I:

$$a_{11} = -a_{12}/\alpha_N \tag{3-8-1}$$

$$a_{22} = -a_{12}/\alpha_I \tag{3-8-2}$$

$$-a_{11} + a_{12}^2/a_{22} = I_{eo} \tag{3-8-3}$$

(see eqs. (3-7-9) to (3-7-12)). Solving these three equations with respect to a_{11}, a_{12}, and a_{22}, we obtain

$$a_{12} = \frac{\alpha_N I_{eo}}{1 - \alpha_N \alpha_I} \tag{3-8-4}$$

$$a_{11} = -\frac{I_{eo}}{1 - \alpha_N \alpha_I} \tag{3-8-5}$$

$$a_{22} = -\frac{\alpha_N I_{eo}}{(1 - \alpha_N \alpha_I)\alpha_I} \tag{3-8-6}$$

We now introduce a parameter called an *intercept current,* I_i:

$$I_i = -a_{12} = -\frac{\alpha_N I_{eo}}{1 - \alpha_N \alpha_I} \tag{3-8-7}$$

The logarithm of I_i can be determined from the intercept of the dependence of the logarithm of the emitter current on the collector-base voltage with the base and emitter shorted. Using this notation and the relationship between the common-emitter current gain and common-base current gain,

$$\beta_N = \frac{\alpha_N}{1 - \alpha_N} \tag{3-8-8}$$

$$\beta_I = \frac{\alpha_I}{1 - \alpha_I} \tag{3-8-9}$$

we can rewrite the Ebers-Moll equations (3-7-1) and (3-7-2) as follows:

$$I_e' = (1 + 1/\beta_N)\, I_i\, [\exp(qV_{be}/k_BT) - 1] + I_i\, [\exp(qV_{bc}/k_BT) - 1] \tag{3-8-10}$$

$$I_c' = I_i\, [\exp(qV_{be}/k_BT) - 1] + (1 + 1/\beta_I)\, I_i\, [\exp(qV_{bc}/k_BT) - 1] \tag{3-8-11}$$

Following Gummel and Poon we may now represent the emitter and collector currents for the intrinsic transistor as

$$I'_e = I_{cc} + I_{be} \tag{3-8-12}$$

$$I'_c = -I_{cc} + I_{bc} \tag{3-8-13}$$

where

$$I_{cc} = I_i\,[\exp(qV_{be}/k_BT) - \exp(qV_{bc}/k_BT)] \tag{3-8-14}$$

is the principal component of the emitter and collector currents,

$$I_{be} = \frac{I_i}{\beta_N}\,[\exp(qV_{be}/k_BT) - 1] \tag{3-8-15}$$

and

$$I_{bc} = \frac{I_i}{\beta_I}\,[\exp(qV_{bc}/k_BT) - 1] \tag{3-8-16}$$

Current component I_{bc} is small compared with I'_e in the important forward active mode of operation because β_N is much larger than unity. The idea of the Gummel-Poon model is to simplify the mathematics of the problem by accounting for high-injection effects only for the most important current component, I_{cc}. In this model, eq. (3-8-14) is replaced by

$$I_{cc} = \frac{qQ_G I_i}{Q_b}\,(\exp[qV_{eb}/k_BT] - \exp[qV_{cb}/k_BT]) \tag{3-8-17}$$

where Q_b is the total charge of the majority carriers in the base per unit area and Q_G is the Gummel number (see eq. (3-3-32)). This equation can be derived assuming the constant electron quasi-Fermi level in the base and neglecting the recombination in the base (see, for example, Sze 1981, p. 154). When the injection level is low and the Early effect may be neglected,

$$Q_b = qQ_G \tag{3-8-18}$$

and eq. (3-8-17) becomes identical with eq. (3-8-14).

In a general case, however, the base hole charge, Q_b, is affected by the emitter-base and collector-base space charge regions and by the charge of the hole injected into the base region:

$$Q_b = qQ_G + Q_{be} + Q_{bc} + Q_{dife} + Q_{difc} \tag{3-8-19}$$

where Q_{dife} and Q_{difc} are the charges of holes per unit area, injected into the base and associated with the emitter-base and collector-base diffusion capacitances,

$$Q_{bc} = -qx_{dcb}N_{ab} = -[2q\varepsilon N_{ab}N_{dc}(V_{bibc} + V_{cb})/(N_{ab} + N_{dc})]^{1/2} \tag{3-8-20}$$

is the charge in the depletion region in the base at the collector-base junction, and

$$Q_{be} = -qx_{deb}N_{ab} = -[2q\varepsilon N_{ab}N_{de}(V_{bieb} + V_{eb})/(N_{ab} + N_{de})]^{1/2} \tag{3-8-21}$$

is the charge in the depletion region in the base at the emitter-base junction. Here the extensions of the collector-base and emitter-base depletion regions into the base region, x_{dcb} and x_{deb}, are found using the depletion approximation:

$$x_{dcb} = \left[\frac{2\varepsilon(V_{bicb} + V_{cb})}{qN_{ab}(1 + N_{ab}/N_{dc})}\right]^{1/2} \approx \left[\frac{2\varepsilon N_{dc}(V_{bicb} + V_{cb})}{qN_{ab}^2}\right]^{1/2} \tag{3-8-22}$$

$$x_{deb} = \left[\frac{2\varepsilon(V_{bieb} + V_{eb})}{qN_{ab}(1 + N_{ab}/N_{de})}\right]^{1/2} \approx \left[\frac{2\varepsilon(V_{bieb} + V_{eb})}{qN_{ab}}\right]^{1/2} \tag{3-8-23}$$

V_{bicb} and V_{bieb} are the built-in voltages of the collector-base and emitter-base junctions, respectively. Here we assume that

$$N_{dc} << N_{ab} \tag{3-8-24}$$

and

$$N_{de} >> N_{ab} \tag{3-8-25}$$

Typically, for the forward active mode $Q_{bc} >> Q_{be}$ because V_{bc} is negative and $|V_{bc}| >> |V_{be}|$. Equations (3-8-20) through (3-8-23) are valid for a uniform doping profile. For practical transistors with nonuniform doping in the base, an empirical linearized expression for Q_{bc} and Q_{be} is used:

$$Q_{be} + Q_{bc} = qQ_G V_{ce}/|V_A| \tag{3-8-26}$$

where V_A is the Early voltage. (For n-p-n transistors in the active forward mode V_{ce} is positive). As was mentioned in Section 3-6, the Early voltage is roughly proportional to the Gummel number and to the base width:

$$|V_A| = k_A q Q_G W/\varepsilon \tag{3-8-27}$$

where k_A is a numerical constant of the order of unity (see eq. (3-6-14) and Fig. 3-6-2). Using eqs. (3-8-26) and (3-8-27), we find from eq. (3-8-17) that

$$I_{cc} = \frac{I_i q Q_G[\exp(qV_{be}/k_B T) - \exp(qV_{bc}/k_B T)]}{qQ_G(1 + V_{ce}/|V_A|) + Q_{dife} + Q_{difc}} \tag{3-8-28}$$

The high-injection effects caused by the majority carriers (holes) injected into the base are accounted for by charges Q_{dife} and Q_{difc} in eq. (3-8-19). The changes in Q_{dife} and Q_{difc} describe the dependence of transistor parameters on the injection level. These charges may be presented as

$$Q_{dife} = B\tau_F I_F/S \tag{3-8-29}$$

$$Q_{difc} = \tau_R I_R/S \tag{3-8-30}$$

where B is an empirical factor that is equal to 1 at relatively low injection levels but may be greater than 1 at high injection levels (see, for example, Sze 1981). Parameter B describes the changes in Q_{dife} caused by the Kirk effect. Here τ_F and

τ_R are effective minority carrier lifetimes associated with the forward and reverse currents, respectively, and

$$I_F = I_i q Q_G[\exp(qV_{be}/k_B T) - 1]/Q_b \tag{3-8-31}$$

$$I_R = I_i q Q_G[\exp(qV_{bc}/k_B T) - 1]/Q_b \tag{3-8-32}$$

Substituting eqs. (3-8-31) and (3-8-32) into eq. (3-8-19), we obtain a quadratic equation for Q_b that yields the following expression for Q_b:

$$\begin{aligned} Q_b = \frac{qQ_G + Q_{be} + Q_{cb}}{2} &+ \left\{\frac{(qQ_G + Q_{be} + Q_{cb})^2}{4}\right. \\ &+ qI_i Q_G B\tau_F[\exp(qV_{be}/k_B T) - 1]/S \\ &\left. + qI_i Q_G \tau_R[\exp(qV_{bc}/k_B T) - 1]/S\right\}^{1/2} \end{aligned} \tag{3-8-33}$$

The base current may be now presented as

$$I_b = S\frac{dQ_b}{dt} + I_r + I_F/\beta_N + I_R/\beta_I \tag{3-8-34}$$

where

$$I_r = I_{br} + I_{re} + I_{rc} \tag{3-8-35}$$

and

$$I_{br} = I_1[\exp(qV_{be}/k_B T) - 1] \tag{3-8-36}$$

is the recombination current in the base region (outside of space charge regions) (see eq. (3-3-10)).

$$I_{re} = I_2[\exp(qV_{be}/m_{re}k_B T) - 1] \tag{3-8-37}$$

is the recombination current in the emitter-base space charge region (see eq. (3-3-12)), and

$$I_{rc} = I_3[\exp(qV_{bc}/m_{rc}k_B T) - 1] \tag{3-8-38}$$

is the recombination current in the collector-base space charge region. Ideality factors m_{re} and m_{rc} typically range from 1 to 2.

The emitter and collector currents are given by

$$I_e = I_{cc} + I_{re} + \tau_F\frac{dI_F}{dt} + C_{de}\frac{dV_{be}}{dt} + I_{be} \tag{3-8-39}$$

$$I_c = I_{cc} - I_{rc} - \tau_R\frac{dI_R}{dt} + C_{dc}\frac{dV_{bc}}{dt} + I_{bc} \tag{3-8-40}$$

where

$$C_{de} = \varepsilon S/x_{deb} \tag{3-8-41}$$

$$C_{dc} = \varepsilon S/x_{dcb} \tag{3-8-42}$$

are depletion capacitances for the emitter-base and collector-base junctions, respectively.

Equations (3-8-28) and (3-8-31) to (3-8-42) form a set of equations of the Gummel-Poon model. Some important effects, such as the Early effect or high-injection effects, such as Kirk effect, are directly included in this model. Other effects, such as Auger recombination and band-gap narrowing (see Section 3-3), may be accounted for indirectly by an appropriate choice of the model parameters, such as I_i, m_{re}, m_{rc}, I_1, I_2, and I_3. In addition, parasitic emitter and collector series resistances and a base spreading resistance (see Section 3-4) have to be included for a realistic modeling of a bipolar junction transistor. To take into account emitter and collector series resistances, r_{es} and r_{cs}, we have to substitute "extrinsic" emitter-base and collector-base voltages, V_{eb} and V_{cb}, for "intrinsic" emitter-base and collector-base voltages, V'_{eb} and V'_{cb} in the equations of the Gummel-Poon model.

$$V'_{be} = V_{be} - |I_e|r_{es} \tag{3-8-43}$$

$$|V'_{cb}| = |V_{cb}| - |I_c|r_{cs}$$

The base spreading resistance is more difficult to account for, as it leads to a nonuniform distribution of the emitter and collector current densities and to the effect of emitter current crowding (see Section 3-4). However, in the frame of the Gummel-Poon model, the fallout of the common-emitter current gain, β, caused by the emitter current crowding at high emitter currents may be reproduced, to some extent, by choosing an appropriate effective value of τ_F.

Some other parameters of the Gummel-Poon model may be estimated as follows. The intercept current may be estimated using eqs. (3-8-7) and (3-7-4). The saturation current I_1 is of the order of I_{eo}/α_N. The saturation current I_2 can be estimated using eq. (3-3-13):

$$I_2 = (\pi/2)^{1/2}(k_B T/F_{np})n_i S/\tau_{rec} \tag{3-8-44}$$

Here F_{np} is the electric field at the point in the emitter-base space charge region where $n = p$, and τ_{rec} is the effective recombination time in the emitter-base depletion region. The electric field F_{np} could be crudely estimated as

$$F_{np} = [qN_{ab}(2V_{bip} - V_{be})/\varepsilon]^{1/2} \tag{3-8-45}$$

where

$$V_{bip} = (k_B T/q)\ln(N_{ab}/n_i) \tag{3-8-46}$$

is the built-in voltage of the emitter-base junction and x_{deb} is the depletion width of the emitter-base junction in the base region (see Section 3-3).

The dependencies of the common-emitter current gain, β, on the collector current for an n-p-n Si bipolar junction transistor calculated using the Gummel-Poon model are shown in Fig. 3-8-1. As can be seen from the figure, the model describes the increase of the common-emitter current gain at low collector currents and the drop of β at high currents (compare with Fig. 3-3-2). In practice, the Gummel-Poon model can reproduce measured device characteristics quite accu-

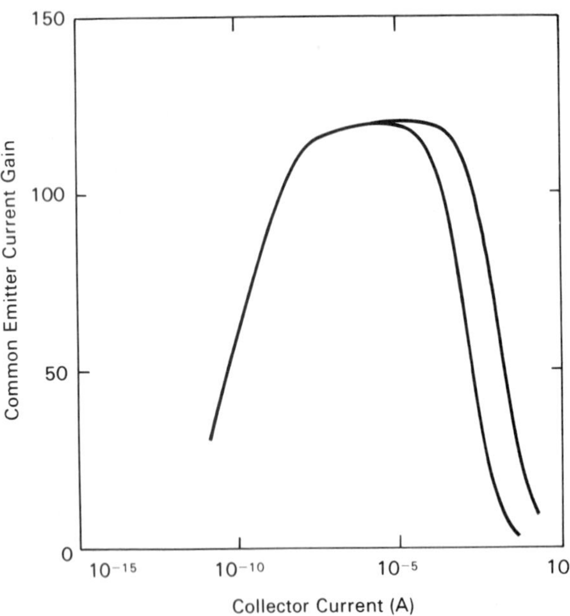

Fig. 3-8-1. Dependence of common-emitter current gain, β, on collector current for *n-p-n* Si bipolar junction transistor calculated using Gummel-Poon model. The transistor operates in the active forward mode (V_{ce} = 4 V, $V_{be} > 0$). The curve decreasing at higher values of the collector current is for $\tau_F = 10^{-8}$ s. The curve decreasing at smaller values of the collector current is for $\tau_F = 10^{-7}$ s. The curves are calculated using program PLOTF with subroutine PGPOON.BAS (see Program Description Section).

rately, but at the price of supplying and adjusting very many different device and material parameters. A version of this model has been implemented in a popular circuit simulator—SPICE—developed at the University of California at Berkeley (see Appendix 24).

3-9. BREAKDOWN IN BIPOLAR JUNCTION TRANSISTORS

There are two important mechanisms of breakdown in bipolar junction transistors: the *avalanche* (or *Zener*) *breakdown* of the collector-base junction and a *punch-through breakdown*. The punch-through breakdown occurs when the reverse collector-base voltage becomes so large that the collector-base depletion region merges with the emitter-base depletion region. The mechanism of avalanche breakdown in bipolar junction transistors is similar to that in p-n diodes (see Section 2-8). However, the critical voltages of the avalanche breakdown in bipolar junction transistors depend on the transistor circuit configuration (i.e., whether it operates in a common-emitter or common-base configuration) and on the external circuit (for example, on the external resistances connected to the transistor terminals).

When the collector-base voltage exceeds some critical value, avalanche breakdown occurs in the collector-base junction, leading to an increase in the collector current. A crude estimate for BV_{cb} may be obtained assuming that the avalanche breakdown occurs when the maximum electric field at the collector-base interface exceeds the breakdown field, F_{br}, (approximately 3×10^7 V/m for Si and 4×10^7 V/m for GaAs transistors):

$$BV_{cb} \approx \frac{\varepsilon F_{br}^2}{2q}(1/N_{ab} + 1/N_{dc}) \approx \frac{\varepsilon F_{br}^2}{2qN_{dc}} \tag{3-9-1}$$

The increase in current for voltages higher than BV_{cb} (see Fig. 3-9-1) is reflected by the multiplication factor M in the expression for the current (see eq. (3-3-23b)). This factor is equal to unity under normal operating conditions and exceeds unity when the avalanche breakdown occurs. When the emitter is open, the multiplication factor due to avalanche breakdown in the collector-base junction, M_{cb}, may be approximated by the following empirical expression:

$$M_{cb} = \frac{1}{1 - \left(\dfrac{V_{cb}}{BV_{cb}}\right)^{m_b}} \tag{3-9-2}$$

where V_{cb} is the collector-base voltage. Constant m_b depends on the doping profile in the collector region and on temperature. Typically, m_b is between 2 and 5 for silicon transistors.

The collector-emitter breakdown voltage in the transistor configuration when the base is open, BV_{ce} (see Fig. 3-9-1), may be related to the collector-base breakdown voltage, BV_{cb}, as follows.

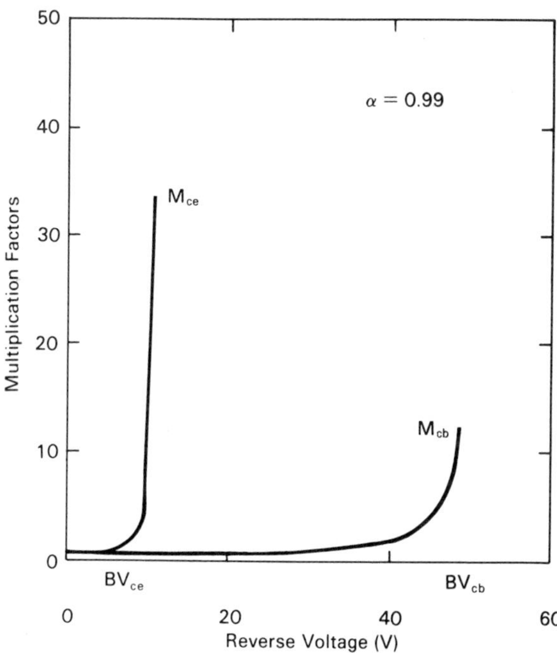

Fig. 3-9-1. Multiplication factors, M_{ce} and M_{cb}, for common-emitter and common-base configurations vs. reverse bias voltage. The curves are calculated using program PLOTF.BAS with subroutines PBJTBR.BAS (see Program Description Section).

When the base is open

$$I_e = I_c \tag{3-9-3}$$

Under the avalanche multiplication conditions, the collector current is given by

$$I_c = M_{cb}(I_{cbo} + \alpha I_e) \tag{3-9-4}$$

(see eq. (3-3-23b)) where

$$\alpha = \gamma\alpha_T \tag{3-9-5}$$

is the common-base current gain (see eq. (3-3-20). Substituting $I_e = I_c$ into eq. (3-9-4) we find that

$$I_c = \frac{M_{cb}I_{cbo}}{1 - \alpha M_{cb}} = M_{ce}I_{ceo} \tag{3-9-6}$$

where

$$M_{ce} = \frac{M_{cb}(1 - \alpha)}{1 - \alpha M_{cb}} \tag{3-9-6a}$$

is the multiplication factor for the common-emitter configuration. From eq. (3-9-6) we obtain the following condition of the breakdown for the common-emitter configuration:

$$\alpha M_{cb} = 1 \tag{3-9-7}$$

Usually, when the breakdown is reached, the reverse collector-base voltage is much greater than the forward voltage bias across the emitter-base junction. Hence, at breakdown, $V_{ce} \approx V_{cb}$. Substituting V_{cb} in eq. (3-9-2) by BV_{ce} and substituting the resulting equation into eq. (3-9-7), we find that

$$\frac{1}{1 - \left(\dfrac{BV_{ce}}{BV_{cb}}\right)^{m_b}} = \alpha^{-1} \tag{3-9-8}$$

Hence,

$$BV_{ce} = BV_{cb}(1 - \alpha)^{1/m_b} \tag{3-9-9}$$

As α is very close to unity and $(1 - \alpha) << 1$, BV_{ce} is much smaller than BV_{cb}. The smaller value of BV_{ce} is related to the amplification of the avalanche base current in the common-emitter configuration.

Under normal operating conditions (when $V_{ce} < BV_{ce}$)

$$I_c = I_{cbo} + \alpha I_e \tag{3-9-10}$$

When the base current is equal to zero,

$$I_e = I_{ceo} \tag{3-9-11}$$

(see eq. (3-6-4)) and $I_e = I_c$. Substituting $I_e = I_{ceo}$ and $I_c = I_{ceo}$ into eq. (3-9-10), we obtain

$$I_{ceo} = I_{cbo}/(1 - \alpha) \tag{3-9-12}$$

(compare with eq. (3-6-7)). Hence, in the common-emitter configuration with an open base, the current is much larger than in the common-base configuration with an open emitter. Again, this is related to a large current gain in the common-emitter configuration. The dependencies of the collector current on the reverse collector bias voltage for common-emitter and common-base configurations are compared in Fig. 3-9-2.

Let us now discuss the effect of the finite base current, I_b, in the common-emitter current-voltage characteristics. In this case

$$I_e = I_c + I_b \tag{3-9-13}$$

Substituting eq. (3-9-13) into eq. (3-9-4) we obtain

$$I_c = \frac{M_{cb}(I_{cbo} + \alpha I_b)}{1 - \alpha M_{cb}} \tag{3-9-14}$$

The dependencies of the collector current on the reverse collector bias voltage for different positive base currents are shown in Fig. 3-9-3. As can be seen from this figure, the common-emitter breakdown voltage is practically independent of the base current. A more interesting situation occurs when the base current is negative, so that the emitter current is less than the collector current. Such a situation

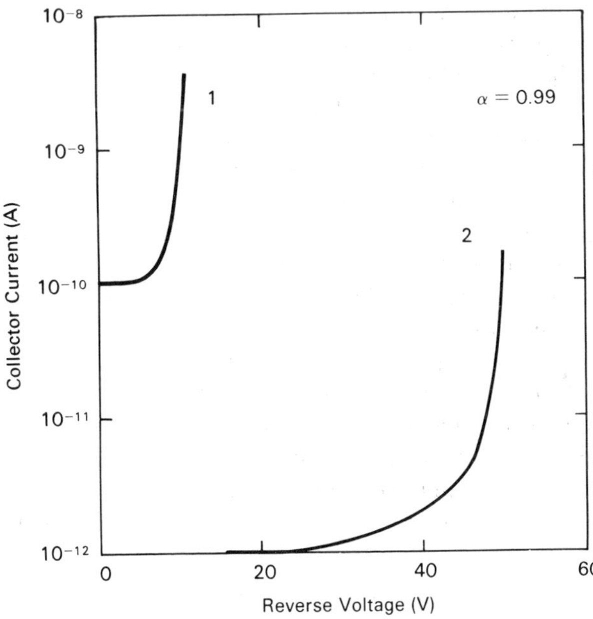

Fig. 3-9-2. The dependences of collector current on reverse collector bias voltage. Curve 1 corresponds to an open base (voltage applied between collector and emitter) and curve 2 corresponds to an open emitter (voltage applied between collector and base). Notice that the collector saturation current for curve 1 (I_{ceo}) is $1/(1 - \alpha)$ times higher (i.e., approximately two orders of magnitudes higher) than the collector saturation current for curve 2 (I_{cbo}). The curves are calculated using program PLOTF.BAS with subroutines PBJTBR.BAS (see Program Description Section).

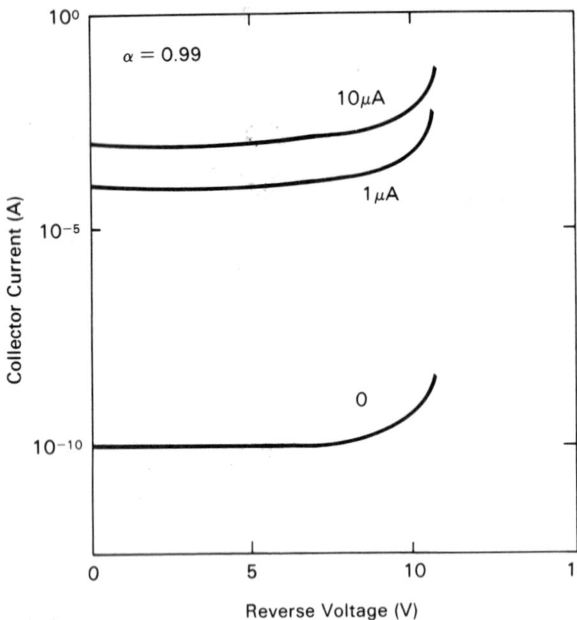

Fig. 3-9-3. The dependences of collector current on reverse collector bias voltage for different positive base currents. The curves are calculated using program PLOTF.BAS with subroutines PBJTBR.BAS (see Program Description Section). Numbers near the curves correspond to the base current.

occurs, for example, when the emitter and base terminals are connected by an external resistor, R_b (see Fig. 3-9-4). In the limiting case of $R_b \to \infty$ this situation corresponds to the common-emitter configuration with an open base. In this limiting case, the breakdown voltage is equal to the common-emitter breakdown voltage, BV_{ce}. However, when R_b is equal to zero the situation is similar to that of the common-base configuration with an open emitter. For finite values of R_b, the breakdown voltage depends on the collector current. When the collector and emitter currents are much larger than the base current, the situation is similar to the situation with the open base because the base current does not really matter. At very low emitter and collector currents the breakdown voltage may approach the common-base breakdown voltage, BV_{cb}. As a consequence, one may observe a negative differential resistance, as schematically shown in Fig. 3-9-4.

Let us now consider a punch-through breakdown. Such a breakdown occurs when the reverse collector-base voltage becomes so large that the collector-base depletion region merges with the emitter-base depletion region, and the effective base width, W_{eff}, given by eq. (3-6-9), becomes equal to zero. Assuming that most of the voltage drop at punch-through is across the collector-base junction and using the depletion approximation for uniform doping profiles in the collector and base regions, we obtain the following expression for the punch-through voltage, V_{pth}:

$$V_{pth} = (1 + N_{ab}/N_{dc})qN_{ab}W^2/(2\varepsilon) \approx qN_{ab}^2W^2/(2\varepsilon N_{dc}) \qquad (3\text{-}9\text{-}15)$$

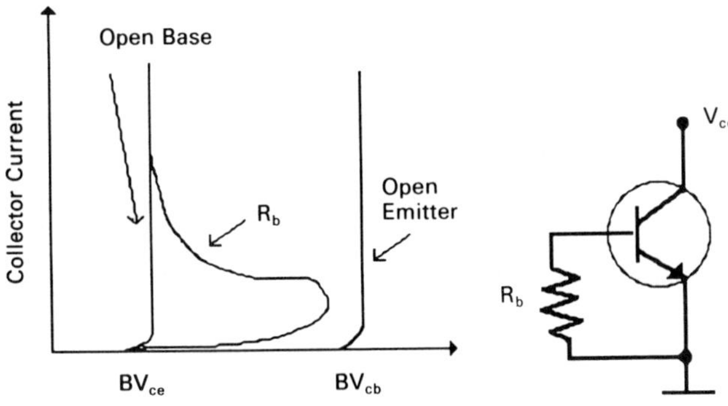

Fig. 3-9-4. Dependences of collector current on reverse collector-base voltage (emitter and base terminals are connected by external resistor, R_b). At very low emitter and collector currents the breakdown voltage may approach the common base breakdown voltage, BV_{cb}. At high emitter and collector currents the breakdown voltage approaches the common collector breakdown voltage, BV_{ce}. This results in a negative differential resistance.

Also shown is the circuit configuration with the resistor R_b in the emitter-base circuit.

The dependencies of the punch-through breakdown voltage on the doping level in the base for two different doping levels in the collector region are shown in Fig. 3-9-5. The ratio of the avalanche breakdown voltage for the common-base configuration, BV_{cb} (see eq. (3-9-1)), over the punch-through voltage, V_{pth}, is given by

$$BV_{\text{cb}}/V_{\text{pth}} \approx \frac{\varepsilon^2 F_{\text{br}}^2}{q^2 Q_{\text{G}}^2} \tag{3-9-16}$$

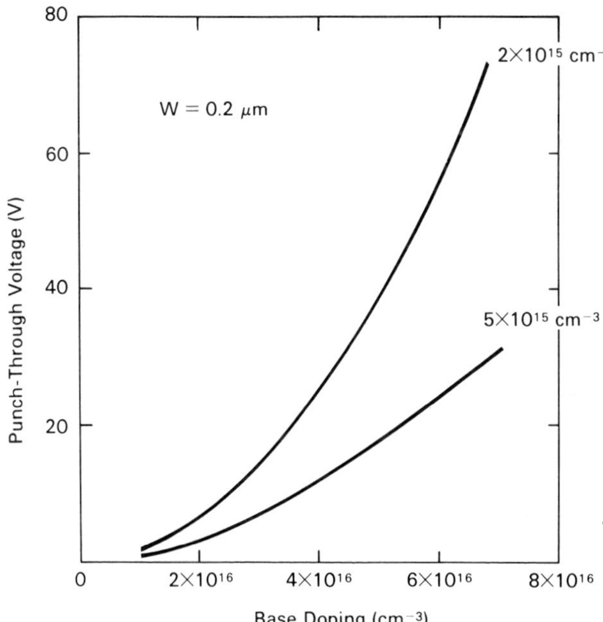

Fig. 3-9-5. Dependence of punch-through voltage, V_{pth}, on doping concentration in base region for different doping levels in collector region. The curves are calculated using program PLOTF.BAS with subroutines PBJTBR.BAS (see Program Description Section). Numbers near the curves correspond to the collector doping densities (in cm^{-3}).

where $Q_G = N_{ab}W$ is the Gummel number. For silicon, $F_{br} \approx 3 \times 10^7$ V/m, $\varepsilon \approx 1.05 \times 10^{-10}$ F/m, and eq. (3-9-16) yields

$$BV_{cb}/V_{pth} \approx \left[\frac{2 \times 10^{12}}{Q_G(\text{cm}^{-2})}\right]^2 \tag{3-9-17}$$

If we assume that the base doping level is $N_{ab} = 10^{17}$ cm^{-3}, for base widths larger than 0.2 μm V_{pth} is larger than BV_{cb}, and hence, the avalanche breakdown is dominant. For thinner bases the punch-through breakdown occurs at voltages smaller than BV_{cb}.

According to the results obtained in the foregoing, the breakdown voltages of both avalanche and punch-through breakdown increase when the collector region doping is lowered. In practical power transistor there is, however, another important effect called the *second breakdown.* This effect (illustrated by Fig. 3-9-6) was first reported by Thorton and Simmons (1958). It occurs at relatively large collector currents. The origin of this effect seems to be a thermal instability. Indeed, the intrinsic carrier concentration in silicon exponentially increases with temperature. If the collector current is kept constant, the temperature increase leads to a decrease of the emitter-base voltage with a negative temperature coefficient of approximately 2 mV/°C for a typical silicon transistor. If the emitter-base voltage is kept constant and the collector current is allowed to change, it increases with temperature, because it is proportional to n_i. A local increase in the collector current density leads to an increase of temperature caused by the Joule heating. This, in turn, leads to a higher collector current density, until a very rapid increase in the current leads to the device destruction. Typically the short circuits between the emitter and collector are produced by local melting of the Al-Si system in the vicinity of the contacts at approximately 577 °C (see Villa 1986). To guarantee safe transistor operation, the collector current and collector-emitter voltage should remain within the *safe operating area* (SOA) on the I_c-V_{ce} plane. This area is limited by the lines corresponding to different failure modes (see Fig. 3-9-7). SOA is an important characteristic of a power bipolar junction transistor.

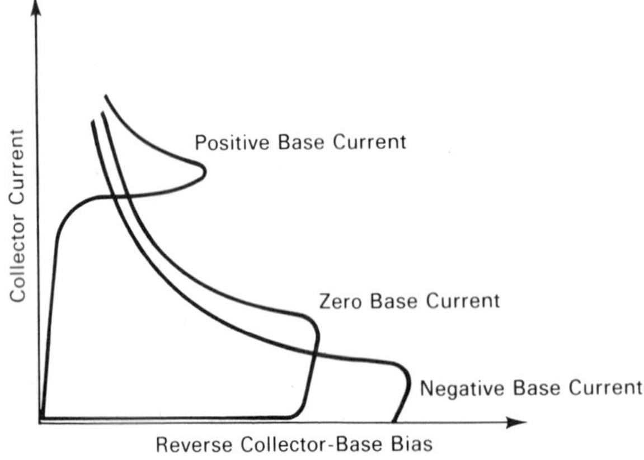

Fig. 3-9-6. Dependencies of the collector current on reverse collector-base voltage for different base currents illustrating the second breakdown. The second breakdown corresponds to smaller collector-base voltage at large collector currents.

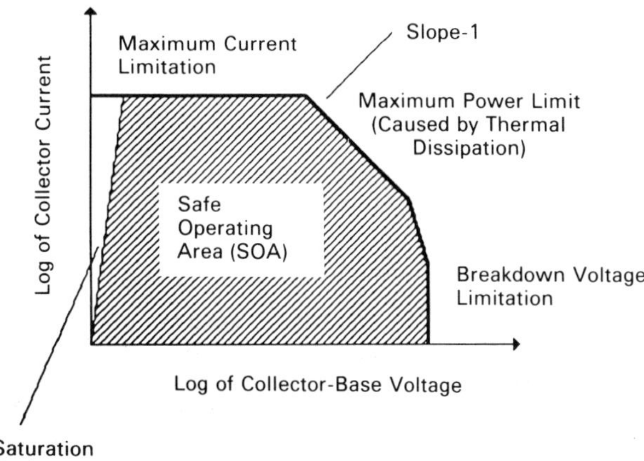

Fig. 3-9-7. Safe Operating Area (SOA) on log (I_c)-$log(V_{ce})$ plane. The Safe Operating Area is limited by the lines corresponding to different failure modes.

3-10. OPERATING POINT AND SMALL-SIGNAL EQUIVALENT CIRCUITS

As was discussed in Section 3-1, a bipolar junction transistor can operate in three different circuit configurations: common-emitter, common-base, and common-collector (see Fig. 3-1-6). Input and output transistor characteristics are different for each configuration. For example, output current-voltage characteristics of an n-p-n Si BJT for common-base and common-emitter configurations are shown in Figs. 3-6-1a and b, respectively. These characteristics are plots of the output current vs. output voltage for different values of the input current. They are usually provided by transistor manufacturers on data sheets. Input characteristics for a common-emitter configuration, for example, are the dependencies of the emitter current on the emitter-base voltage for different collector-base voltages. In a typical circuit, a transistor is connected to a power supply via a resistor, as shown in Fig. 3-10-1, and the bias voltage, V_{cc}, is divided between the voltage

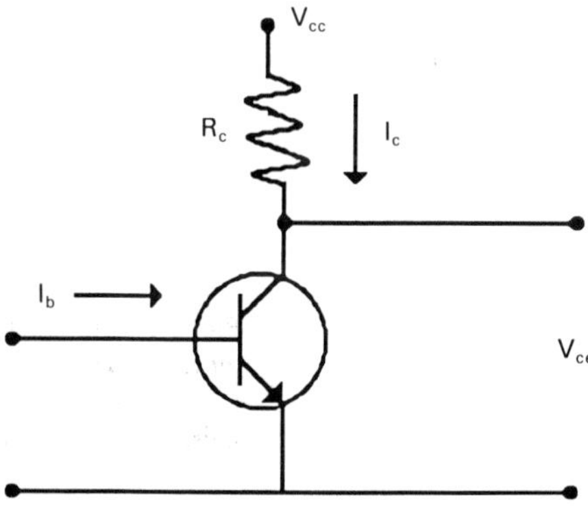

Fig. 3-10-1. Transistor connected to power supply via resistor. The bias voltage, V_{cc}, is divided between the voltage drop across the transistor, V_{ce}, and the voltage drop, I_cR_c, across the resistor.

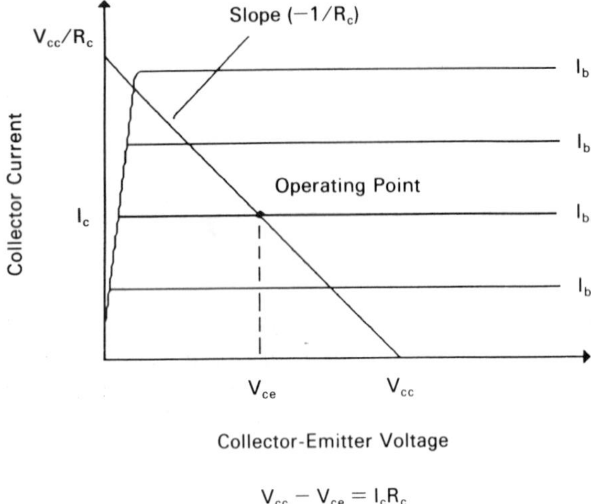

Fig. 3-10-2. Load line superimposed on output current-voltage characteristics of transistor in common-emitter configuration. The intersection of the load line with the output current-voltage curve for a given base current ($I_b = I_{b3}$ in the figure) determines an operating point.

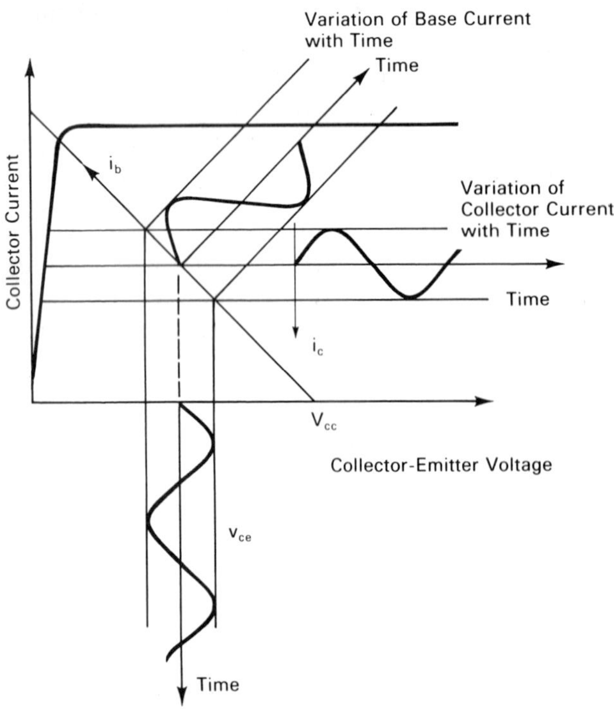

Fig. 3-10-3. Small signal (linear) regime of transistor operation. Notice that the scales for the base current and collector currents are quite different.

drop across the transistor (V_{ce} for the common-emitter configuration) and the voltage drop, I_cR_c, across the resistor:

$$V_{cc} = V_{ce} + I_cR_c \tag{3-10-1}$$

Graphically, this relation may be represented by using a load line, as shown in Fig. 3-10-2. The intersection of the load line with the output current-voltage curve for a given base current determines the *operating point*—also called the *quiescent point*—at which the transistor and resistor currents are equal. At this point, the voltage coordinate in Fig. 3-10-2 is equal to V_{ce} and the current coordinate is equal to I_c. The difference between V_{cc} and V_{ce} is equal to I_cR_c.

One of the important regimes of the transistor operation is a small-signal (or linear) regime when the ac input signal is relatively small, so that the input current, output current, and output voltage vary in the vicinity of the operating point (see Fig. 3-10-3.). In this case, an ac transistor response may be simulated by using a linear two-port network, such as is shown in Fig. 3-10-4. A transistor may be represented by several equivalent circuits, depending on the choice of dependent and independent current and voltage variables. We denote the input ac current and voltage as i_1 and v_1, the output ac current and voltage as i_2 and v_2, the input total current and voltage as I_1 and V_1, and the output total current and voltage as I_2 and V_2. Four frequently used choices for the independent and dependent variables are given in Table 3-10-1. All these choices are, in a sense, equivalent, and the elements of two-by-two matrices of the z, y, h, and g parameters introduced in this table can be expressed through each other. The h parameters described in this table are most frequently used for transistor characterization. For the h parameters, equations describing the relationship between the dependent and independent variables are given by

$$v_1 = h_{11}i_1 + h_{12}v_2 \tag{3-10-2}$$

$$i_2 = h_{21}i_1 + h_{22}v_2 \tag{3-10-3}$$

Similar equations can be written for z, y, and g parameters by substituting v_1 and i_2 in eqs. (3-10-2) and (3-10-3) by the corresponding dependent variables from Table 3-10-1 and by substituting i_1 and v_2 in eqs. (3-10-2) and (3-10-3) by corresponding independent variables from Table 3-10-1.

All these small-signal parameters may be determined from different short-circuit or open-circuit measurements at the input and output ports. For example,

$$h_{11} = v_1/i_1|_{v_2=0} \tag{3-10-4}$$

$$h_{12} = v_1/v_2|_{i_1=0} \tag{3-10-5}$$

$$h_{21} = i_2/i_1|_{v_2=0} \tag{3-10-6}$$

$$h_{22} = i_2/v_2|_{i_1=0} \tag{3-10-7}$$

The h parameter equivalent circuit is frequently used for bipolar junction transistors at relatively low frequencies (below 100 MHz or so). The parameter h_{11} is

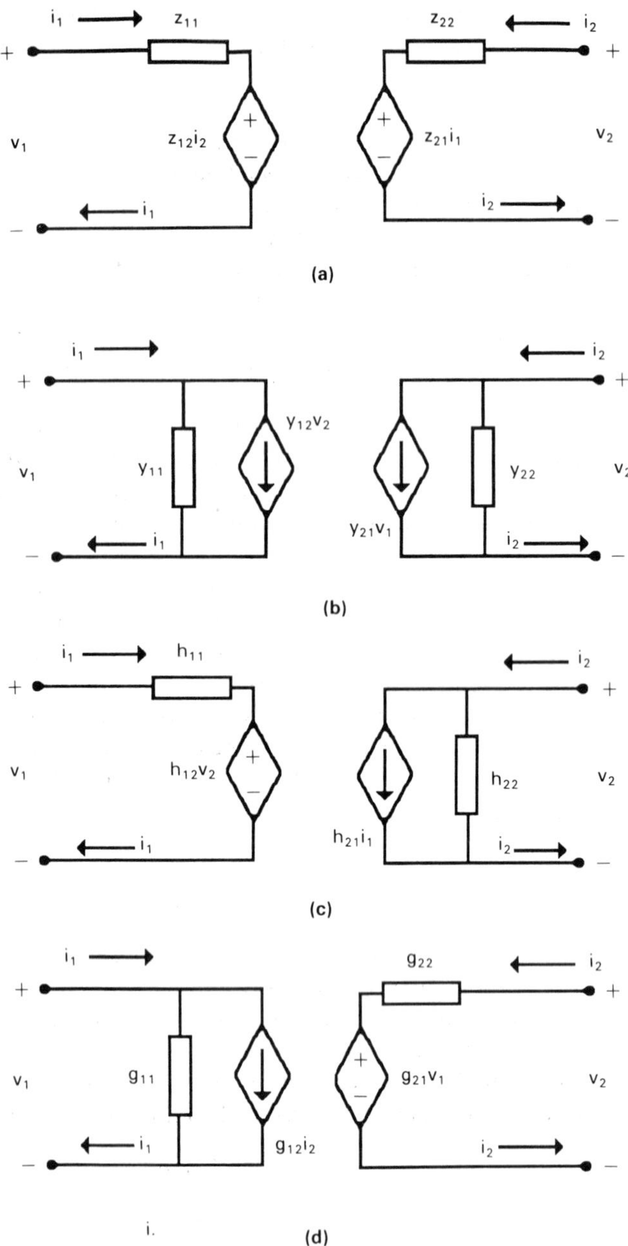

Fig. 3-10-4. Schematic representation of a linear two-port network by different small-signal equivalent circuits. The choice of an equivalent circuit depends on the choice of dependent and independent current and voltage variables. We denote the input ac current and input ac voltage as i_1 and v_1 and the output ac current and output ac voltage is i_2 and v_2: (a) z parameters, (b) y parameters, (c) h parameters, and (d) g parameters.

TABLE 3-10-1. FOUR FREQUENTLY USED CHOICES FOR THE INDEPENDENT AND DEPENDENT VARIABLES AND CORRESPONDING SMALL-SIGNAL PARAMETERS FOR A LINEAR TWO-PORT NETWORK

Independent variables	Dependent variables	Small-signal parameters	Comment
i_1, i_2	v_1, v_2	z parameters	see Fig. 3-10-4a
v_1, v_2	i_1, i_2	y parameters	see Fig. 3-10-4b
i_1, v_2	v_1, i_2	h parameters	see Fig. 3-10-4c
v_1, i_2	i_1, v_2	g parameters	see Fig. 3-10-4d

called a *short-circuit input impedance* (h_i), h_{12} is called an *open-circuit reverse voltage ratio* (h_r), h_{21} is called a *short-circuit forward current ratio* (h_f), and h_{22} is called an *open-circuit output admittance* (h_o). The second subscript is used to denote a transistor configuration; for example, h parameters for the common-emitter transistor circuit configuration are denoted as h_{ie}, h_{re}, h_{fe}, and h_{oe}. The h parameter equivalent circuit for a common-emitter transistor circuit configuration is shown in Fig. 3-10-5. Out of a total of twelve h parameters for three transistor configurations, four parameters, h_{ie}, h_{re}, h_{oe}, and h_{fe}, are usually provided by transistor manufacturers on transistor data sheets. Approximate relationships allowing us to express other h parameters in terms of h_{ie}, h_{re}, h_{oe}, and h_{fe} are given in Table 3-10-2. The h parameters and other characteristics of a general purpose Si n-p-n transistor (Motorola 2N2219A) are given in Appendix 26, which contains information found on a typical data sheet.

The parameters of small-signal equivalent circuits vary with temperature, measuring frequency, and operating point. The variation of the h parameters with collector current for Motorola 2N2219A general-purpose silicon n-p-n transistors is shown in Fig. 3-10-6 (from *Small-Signal Transistor Data,* published by Motorola Semiconductor Products, Inc., Phoenix, Ariz., 1983). As can be seen from this figure the h parameters may vary quite substantially for different transistors of the same type (compare curves marked 1 and 2 in Fig. 3-10-6).

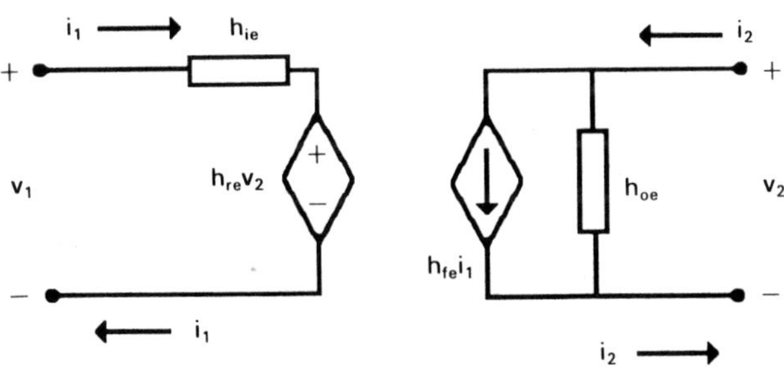

Fig. 3-10-5. *H* parameter equivalent circuit for common-emitter configuration.

TABLE 3-10-2. APPROXIMATE RELATIONSHIPS EXPRESSING DIFFERENT h PARAMETERS IN TERMS OF h_{ie}, h_{re}, h_{oe}, AND h_{fe}

Common-emitter h parameters	Common-base h parameters	Common-collector h parameters
h_{ie}	$h_{ib} = h_{ie}/(h_{fe} + 1)$	$h_{ic} = h_{ie}$
h_{re}	$h_{rb} = h_{ie}h_{oe}/(h_{fe} + 1) - h_{re}$	$h_{rc} = 1$
$h_{fe} = -h_{fb}/(h_{fb} + 1)$	$h_{fb} = -h_{fe}/(h_{fe} + 1)$	$h_{fc} = -h_{fe} - 1$
h_{oe}	$h_{ob} = h_{oe}/(h_{fe} + 1)$	$h_{oc} = h_{oe}$

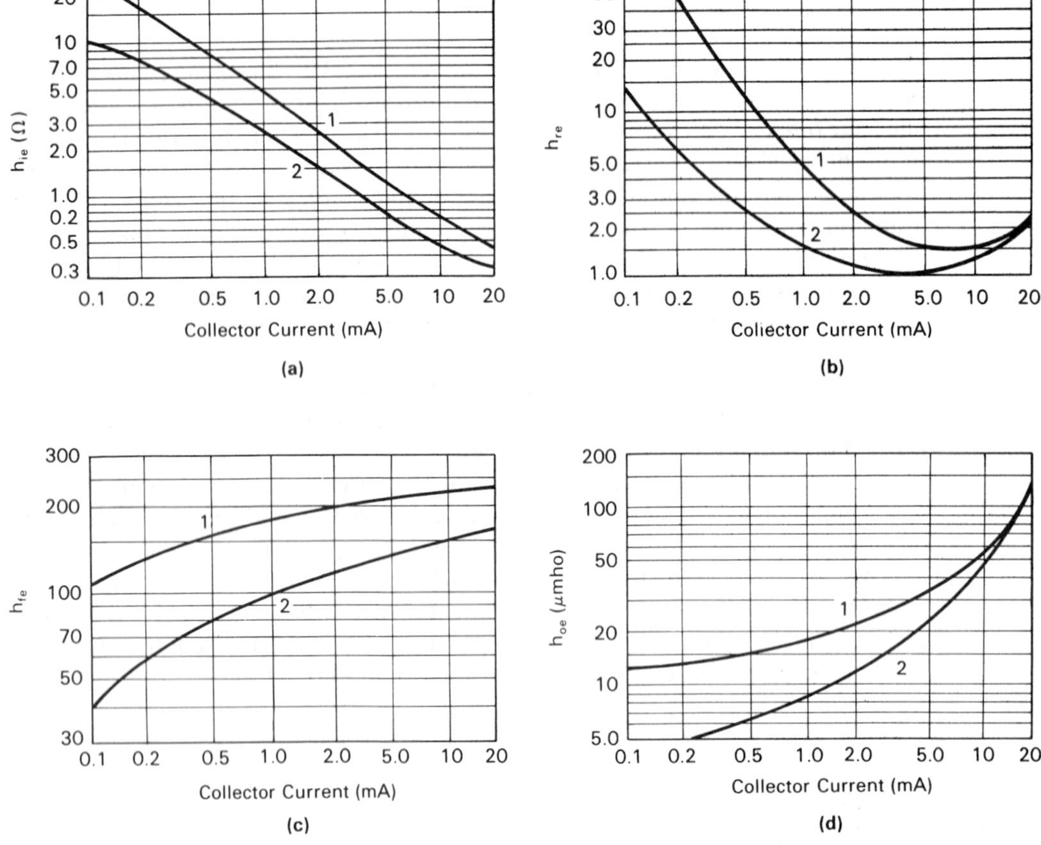

Fig. 3-10-6. Variation of h parameters with collector current for Motorola 2N2219A general purpose silicon n-p-n transistors (from *Small-Signal Transistor Data,* Motorola Semiconductor Products, Inc., Phoenix AR [1983]). $V_{ce} = 10$ V (dc), $f = 1$ kHz, ambient temperature $T_A = 25°\text{C}$: (a) input impedance h_{ie}, (b) voltage feedback ratio h_{re}, (c) current gain h_{fe}, and (d) output admittance h_{oe}. Curves marked 1 are for high gain transistors, curves marked 2 are for low gain transistors of the same type.

Some of the h parameters can be directly related to device parameters. For example, at low frequencies h_{fe} is equal to β and h_{fb} is equal to $-\alpha$. For other h parameters such a relationship is less straightforward. For a common-emitter configuration a *hybrid-π equivalent circuit* (see Fig. 3-10-7) is frequently used. The parameters of this circuit may be easier to relate to the device parameters.

The transconductance, g_m, in Fig. 3-10-7 may be related to the dynamic (differential) resistance, r_e, of the forward-biased emitter-base junction:

$$g_m = \partial I_c/\partial V_{b'e} = \alpha \partial I_e/\partial V_{b'e} \approx \alpha/r_e \approx I_c/V_{th} \tag{3-10-8}$$

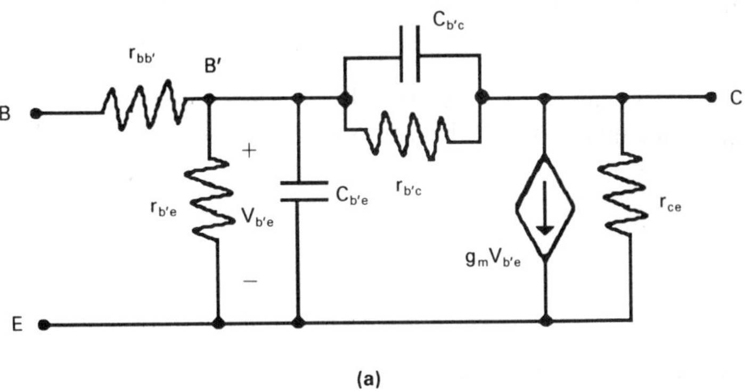

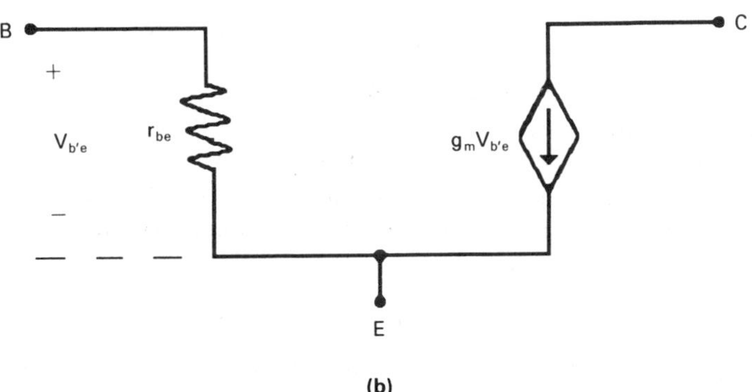

Fig. 3-10-7. Hybrid-π equivalent circuits: (a) complete hybrid-π equivalent circuit and (b) simplified equivalent circuit.

where $V_{th} = k_B T/q$ is the thermal voltage, and

$$r_e \approx V_{th}/I_e \tag{3-10-9}$$

Resistance $r_{b'e}$ in Fig. 3-10-7a may also be related to r_e. Indeed, assuming that resistance r_{ce} in Fig. 3-10-7 is very large, we find that

$$i_c \approx g_m V_{b'e} \tag{3-10-10}$$

At low frequencies, when the effect of capacitance $C_{b'c}$ in Fig. 3-10-7a is negligible,

$$V_{b'e} = i_b r_{b'e} \tag{3-10-11}$$

Substituting eq. (3-10-11) into eq. (3-10-10) we find that

$$r_{b'e} \approx (i_c/i_b)/g_m \approx h_{fe}/g_m \tag{3-10-12}$$

The emitter capacitance, C_e (see Fig. 3-10-8), may be estimated as the sum of the diffusion capacitance of the emitter-base junction, C_{edif}, and the depletion capacitance, C_{ed}:

$$C_e = C_{edif} + C_{ed} \tag{3-10-13}$$

where

$$C_{edif} = dQ_b/dV_{eb'} \tag{3-10-14}$$

The charge of minority carriers, Q_b (electrons in the case of an n-p-n transistor), injected into the base is given by

$$Q_b \approx I_e W^2/(2D_n) \tag{3-10-15}$$

Hence, from eqs. (3-10-14) and (3-10-8),

$$C_{edif} \approx W^2/(2D_n r_e) \tag{3-10-16}$$

leading to the base charging time constant

$$\tau_B = C_{edif} r_e = W^2/(2D_n) \tag{3-10-17}$$

τ_B is the characteristic diffusion time of the minority carriers across the base width, i.e., the effective transit time. In drift transistors with nonuniform doping decreasing from the emitter-base junction toward the collector-base junction, this transit time constant may be considerably reduced by drift caused by the built-in electric field. Such a field is caused by the nonuniformity of the doping profile (see Section 3-5). In this case

$$\tau_B = C_{edif} r_e = W^2/(f D_n) \tag{3-10-18}$$

$f = 2$ for the uniform profile and may be much larger than 2 if the built-in electric field is sufficiently large. However, τ_B cannot be smaller than the minimum transit time of carriers across the base:

$$\tau_{Bmin} = W/v_{sn} \tag{3-10-19}$$

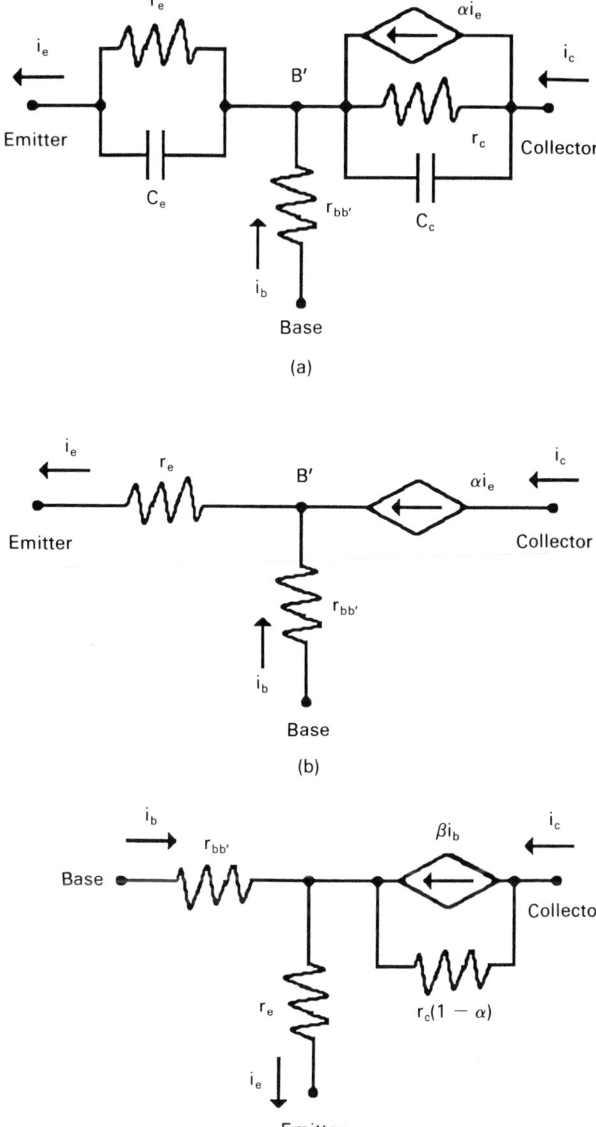

Fig. 3-10-8. *T*-equivalent circuits: (a) complete equivalent circuit for common-base configuration, (b) simplified equivalent circuit for common-base configuration, and (c) low frequency *T*-equivalent circuits for common-emitter configuration.

where v_{sn} is the effective saturation velocity of electrons in the base region. In very short bases this velocity may be substantially larger than the electron saturation velocity in long samples because of ballistic or overshoot effects.

The depletion capacitance of the emitter-base junction may be estimated as

$$C_{ed} \approx S\left[\frac{qN_{ab}\varepsilon}{2(V_{bieb} - V_{be})}\right]^{1/2} \tag{3-10-20}$$

(see Section 2-5). Here N_{ab} is the effective base doping concentration, and V_{bieb} is the built-in voltage of the emitter-base junction.

Resistance $r_{b'c}$ and capacitance $C_{b'c}$ in Fig. 3-10-7 represent the dynamic (differential) resistance and capacitance of the reverse-biased collector-base junction. (The collector-base capacitance $C_{b'c}$ is usually denoted as C_{ob} in manufacturer data sheets; see Appendix 26.) Resistance r_{ce} describes the Early effect considered in Section 3-6. Resistance $r_{bb'}$ is the base spreading resistance (see Section 3-4).

Still another equivalent circuit used in the transistor analysis is a T-equivalent circuit. T-equivalent circuits for the common-base and common-emitter configurations are shown in Fig. 3-10-8. Equations relating the parameters of T- and π-equivalent circuits to h parameters are given in Appendix 25.

3-11. BIPOLAR JUNCTION TRANSISTOR AS A SMALL–SIGNAL AMPLIFIER: CUTOFF FREQUENCIES

In a typical circuit, a transistor operating point is chosen using biasing resistors (resistors R_{B1} and R_{B2} in Fig. 3-11-1). Coupling and bypass capacitances (C_1, C_2, and C_E in Fig. 3-11-1) are used to connect the transistor stage to the rest of the circuit and to isolate the dc bias and ac signal. The rest of the circuit connected to the input may be represented at ac by its Thevenin's equivalent (the series connection of an equivalent source resistance, R_s, and an equivalent voltage source, V_s, as shown in Fig. 3-11-1) or by its Norton's equivalent (the parallel connection of an equivalent current source, $i_s = V_s/R_s$, and an equivalent source conductance, $G_s = 1/R_s$). At low frequencies, the voltage drop across the coupling capacitances becomes significant and the bypass capacitance does not completely shunt resistance R_E. At high frequencies, internal transistor capacitances become important, leading to the deterioration of the transistor performance, as discussed below. However, at mid-frequencies the coupling and bypass capacitances are not important and internal transistor capacitances still do not appreciably affect the transistor operation. In this mid-frequency range the small-signal transistor operation may be analyzed using one of the equivalent circuits discussed in Section 3-10 with an equivalent voltage source with an equivalent series resistance connected to the input and an equivalent load resistance connected to the output. We can now define the voltage, current, and power gains and input and output

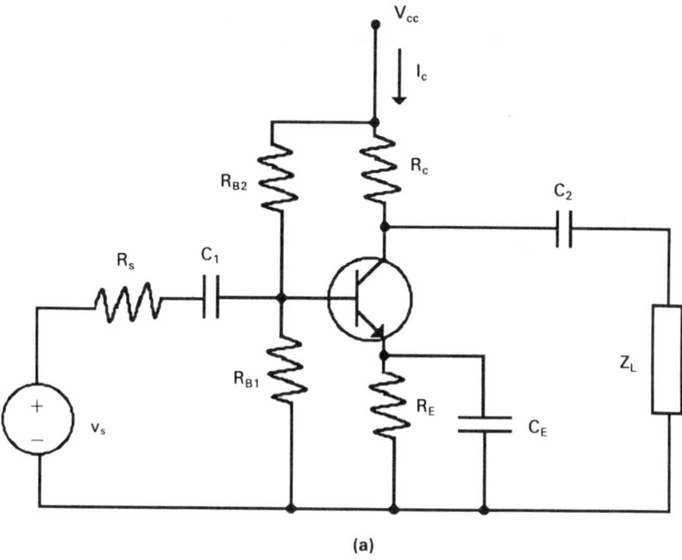

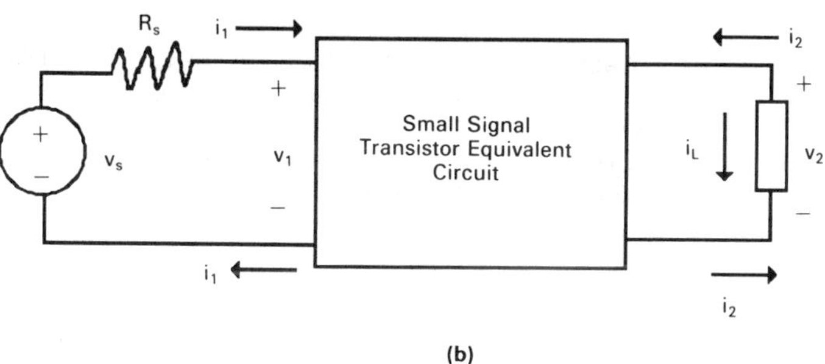

Fig. 3-11-1. a. Typical circuit for a transistor amplifier in a common-emitter configuration. Operating point is chosen using biasing resistances, R_{B1} and R_{B2}. Resistance R_E stabilizes the operating point. Coupling capacitances C_1 and C_2 provide dc isolation from the rest of the circuit represented by the Thevenin's equivalent. The bypass capacitance C_E shunts resistance R_E at ac. b. Small-signal equivalent circuit at midband.

impedances for different transistor configurations and express these parameters in terms of h parameters for these configurations using standard circuit theory (see Table 3-11-1). In Table 3-11-2 we give the values of h parameters for a Motorola 2N2219A general-purpose transistor and the values of A_i, Z_i, A_v, Y_o, A_{vs}, and A_{is} for the common-base, common-emitter, and common-collector configurations. This table may be used for a qualitative comparison of different transistor configurations.

TABLE 3-11-1. VOLTAGE, CURRENT, AND POWER GAINS AND INPUT AND OUTPUT IMPEDANCES FOR DIFFERENT TRANSISTOR CONFIGURATIONS

	Definition	Relation to h parameters, load impedance, Z_L, and source series resistance R_s
Input impedance	$Z_i = v_1/i_1$	$Z_i = h_i - h_f h_r/(h_o + Y_L)$
Output admittance	$Y_o = i_2/v_2$	$Y_o = h_o - h_f h_r/(h_i + R_s)$
Voltage gain	$A_v = v_2/v_1$	$A_v = A_i Z_L/Z_i$
Voltage gain	$A_{vs} = v_2/v_s$	$A_{vs} = A_v Z_i/(Z_i + R_s)$
Current gain	$A_i = i_L/i_1 = -i_2/i_1$	$A_i = -h_f/(1 + h_o Z_L)$
Current gain	$A_{is} = -i_2/i_s$	$A_{is} = A_i R_s/(Z_i + R_s)$
Power gain	$A_p = A_v A_i$	
Power gain	$A_{ps} = A_{vs} A_{is}$	

v_s is ac source voltage
$i_s = v_s/R_s$

TABLE 3-11-2. THE VALUES OF h PARAMETERS FOR A MOTOROLA 2N2219A GENERAL-PURPOSE TRANSISTOR AND THE VALUES OF A_i, Z_i, A_v, Y_o, A_{vs}, AND A_{is} FOR THE COMMON-BASE, COMMON-EMITTER, AND COMMON-COLLECTOR CONFIGURATIONS

Equivalent source resistance $R_s = 2\ \text{k}\Omega$

Equivalent load resistance $R_L = 2\ \text{k}\Omega$

$IC = 10$ mA (dc), $VCE = 10$ V (dc), $f = 1$ kHz

	CE	CB	CC
h_i (Ω)	1250	16.4	1250
h_r	4×10^{-4}	1.1×10^{-5}	1
h_f	75	−0.987	−76
h_o(μmho)	25	0.329	25
Z_i (kΩ)	1.19	0.0164	146.01
Y_o(μmho)	15.77	0.334	24350
A_i	−71.43	0.986	72.38
A_{is}	−44.72	0.977	0.978
A_v	−119.76	120.13	0.991
A_{vs}	−44.72	0.977	0.978

These data are taken from *Small-Signal Transistor Data*, Motorola Semiconductor Products, Inc., Phoenix, Ariz. (1983)

As was mentioned previously, at high frequencies of operation internal transistor capacitances have to be taken into account. Using a T-equivalent circuit for the common-base configuration shown in Fig. 3-10-8a we can evaluate a common base current gain α_ω at frequency ω. The ac emitter current, i_e, is given by

$$i_e = V_{b'e}(1 + j\omega C_e r_e)/r_e = i_{eo}(1 + j\omega C_e r_e) \tag{3-11-1}$$

where i_{eo} is the ac current through resistance r_e. Hence,

$$\alpha_\omega = i_c/i_e = i_c/[i_{eo}(1 + j\omega C_e r_e)] = \alpha/(1 + j\omega C_e r_e) \tag{3-11-2}$$

where α is a common-base current gain at zero frequency. Equation (3-11-1) may be rewritten as

$$\alpha_\omega = \alpha/(1 + j\omega/\omega_\alpha) \tag{3-11-3}$$

where

$$\omega_\alpha = 2\pi f_\alpha = 1/(C_e r_e) \tag{3-11-4}$$

is called the *alpha cutoff frequency*. (Sometimes f_α is denoted as $f_{\alpha b}$). At $f = f_\alpha$ the common-base current gain drops to 0.707 ($1/\sqrt{2}$) of its zero frequency value.

Let us now consider a common-emitter current gain. A simplified hybrid-π equivalent circuit used in the calculation of the short-circuit emitter current gain, β_ω, is shown in Fig. 3-11-2 (compare with Fig. 3-10-7a). The voltage drop $V_{b'e}$ is given by

$$V_{b'e} = \frac{i_i}{g_{b'e} + j\omega(C_{b'e} + C_{b'c})} \tag{3-11-5}$$

Hence,

$$\beta_\omega = -\frac{i_L}{i_i} = \frac{g_m}{g_{b'e} + j\omega(C_{b'e} + C_{b'c})} \tag{3-11-6}$$

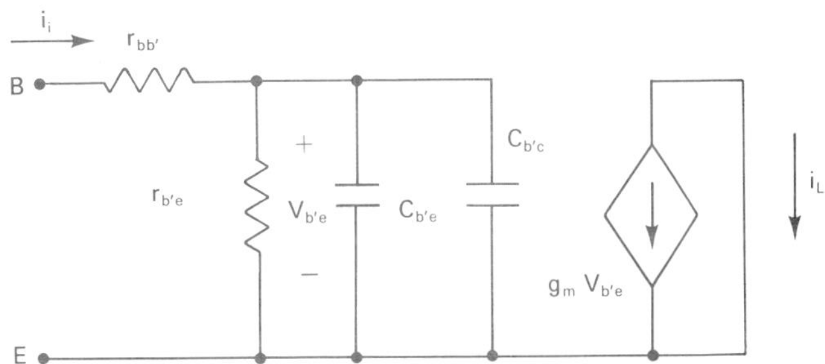

Fig. 3-11-2. A simplified hybrid-π equivalent circuit used in the calculation of the short circuit emitter current gain, β_ω. (Resistance $r_{b'c}$ is neglected.)

Using eq. (3-10-12), eq. (3-11-6) can be rewritten as

$$\beta_\omega = \beta/(1 + j\omega/\omega_\beta) \tag{3-11-7}$$

where

$$\omega_\beta = 2\pi f_\beta = g_{b'e}/(C_{b'e} + C_{b'c}) \tag{3-11-8}$$

is called the *beta cutoff frequency*. At $f = f_\beta$ the common-emitter current gain drops to 0.707 ($1/\sqrt{2}$) of its zero frequency value. Equation (3-11-7) may be rewritten as

$$f_\beta = g_m/[2\pi h_{fe}(C_{b'e} + C_{b'c})] \approx 1/[2\pi h_{fe}(C_{b'e} + C_{b'c})r_e] \tag{3-11-9}$$

In most cases $C_{b'e} >> C_{b'c}$, and hence,

$$f_\beta \approx f_\alpha/h_{fe} \tag{3-11-10}$$

Finally, we introduce the cutoff frequency, f_T, which is defined as the frequency at which the short-circuit common-emitter current gain drops to unity:

$$f_T \approx f_\beta h_{fe} \approx g_m/[2\pi(C_{b'e} + C_{b'c})] \tag{3-11-11}$$

In many cases this frequency is estimated from the measured current gain at a frequency much larger than f_β but much smaller than f_T. At such a frequency

$$\beta_\omega \approx \beta/(j\omega/\omega_\beta) \tag{3-11-12}$$

and hence,

$$|f\beta_\omega| \approx h_{fe} f_\beta = f_T \tag{3-11-13}$$

Here h_{fe} is the low-frequency value of short-circuit common-emitter gain, β.

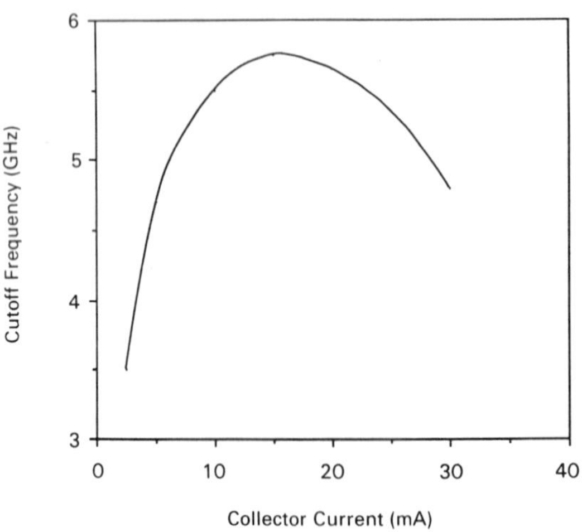

Fig. 3-11-3. Dependence of f_T on the collector current for a Motorola high-frequency silicon *n-p-n* transistor 2N6603 (from *Small-Signal Transistor Data,* Motorola Semiconductor Products, Inc., Phoenix AR, [1983]).

In Fig. 3-11-3 we show the dependence of f_T on the collector current for a high-frequency silicon n-p-n transistor (from *Small-Signal Transistor Data,* published by Motorola Semiconductor Products, Inc., Phoenix, Ariz., 1983).

A more accurate expression for f_T may be obtained by including in eq. (3-11-11) additional time delays associated with the collector depletion layer transit time, τ_{cT}, the collector charging time, τ_c, and the parasitic capacitance, C_p.

The collector transit time may be estimated as

$$\tau_{cT} \approx x_{dc}/(v_{sn}) \qquad (3\text{-}11\text{-}14)$$

where x_{dc} is the width of the collector-base depletion region and v_{sn} is the electron saturation velocity (we are considering here n-p-n transistors).

The collector charging time is given by

$$\tau_c = r_{cs}C_{b'c} \qquad (3\text{-}11\text{-}15)$$

where r_{cs} is the collector series resistance and $C_{b'c}$ is the collector capacitance.

The equation for the cutoff frequency, f_T, may now be rewritten as

$$f_T \approx 1/(2\pi\tau_{eff}) \qquad (3\text{-}11\text{-}16)$$

where the effective delay time

$$\tau_{eff} = \tau_e + \tau_c + \tau_{cT} \qquad (3\text{-}11\text{-}17)$$

where

$$\tau_e = (C_{b'e} + C_{b'c} + C_p)/g_m \approx (k_BT/qI_e)(C_{b'e} + C_{b'c} + C_p) \qquad (3\text{-}11\text{-}18)$$

In most practical transistors, τ_e is the largest contribution to the total delay time. As can be seen from eq. (3-11-18), this time can be reduced by increasing the collector current. However, at very large collector currents, the displacement of the effective base-collector boundary into the collector region caused by the Kirk effect (see Section 3-3) leads to an increase in the effective base width and to an increase in the emitter diffusion capacitance, C_{edif} (see eqs. (3-10-16) and (3-10-17)). This may explain the decrease of f_T at large collector currents (see Fig. 3-11-3).

The analysis of eqs. (3-11-14) to (3-11-18) shows that to achieve a large cutoff frequency, narrow emitter stripes (small area), large emitter currents, very thin base regions, high base doping, and low parasitic capacitances are required. By using very narrow (submicron) emitter stripes and base thicknesses below 0.1 μm, cutoff frequencies of the order of 15 GHz have been obtained (see Kikushi et al. 1986). Much higher cutoff frequencies (as high as 165 GHz; see Y. K. Chen et al. 1989) have been achieved in heterojunction bipolar transistors where high doping in the base region may be combined with high emitter injection efficiency (see Section 3-14).

3-12. BIPOLAR JUNCTION TRANSISTOR AS A SWITCH

A bipolar junction transistor can function as a switch between a high-voltage, low-current state and a low-voltage, high-current state. This mode of operation is illustrated by Fig. 3-12-1. As the base current is increased, the operating point shifts from a high-voltage, low-current state (V_{off}, I_{off}) to the low-voltage, high-current state (V_{on}, I_{on}). The off state corresponds to the cutoff mode of transistor operation, whereas the on state corresponds to the saturation mode. Fig. 3-12-2 shows qualitative distributions of the minority carriers in the base in the cutoff mode, active mode, and saturation mode. The drawing is not to scale, as the density of the minority carrier in the cutoff regime is many orders of magnitude smaller than the density of the minority carriers in the active and saturation modes.

A typical circuit for measuring transistor switching characteristics is shown in Fig. 3-12-3 (from *Small-Signal Transistor Data,* published by Motorola Semiconductor Products, Inc., Phoenix, Ariz., 1983). Fig. 3-12-3 also shows the shape of the input voltage pulse, base current pulse, and output current pulse. Characteristic time constants—the delay time τ_d, the rise time τ_r, the turn-on time, $\tau_{on} = \tau_d + \tau_r$, the storage time τ_s, the fall time τ_f, and the turn-off time, $\tau_{off} = \tau_s + \tau_f$, are defined on that figure. When the input pulse, V_1, is applied, the base current, I_{b1}, is given by

$$I_{b1} \approx (V_1 - V_{besat})/R_s \tag{3-12-1}$$

where V_{besat} is the emitter-base saturation voltage, and R_s is the input source resistance (see Fig. 3-12-3). At the moment when the input pulse is turned off and the input voltage drops to a negative value, V_2, the base current becomes equal to

$$I_{b2} \approx (V_2 - V_{besat})/R_s \tag{3-12-2}$$

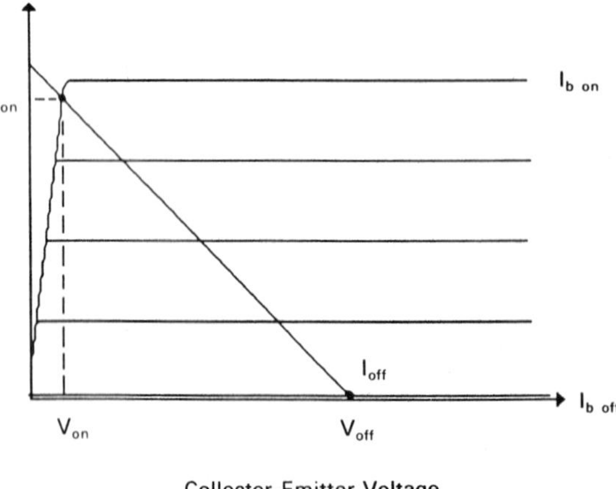

Fig. 3-12-1. Switching of bipolar junction transistor between high-voltage low-current state (V_{off}) and low-voltage high-current state (V_{on}) with the increase in the base current. The off-state corresponds to the cutoff mode of the transistor operation and the on-state corresponds to the saturation mode.

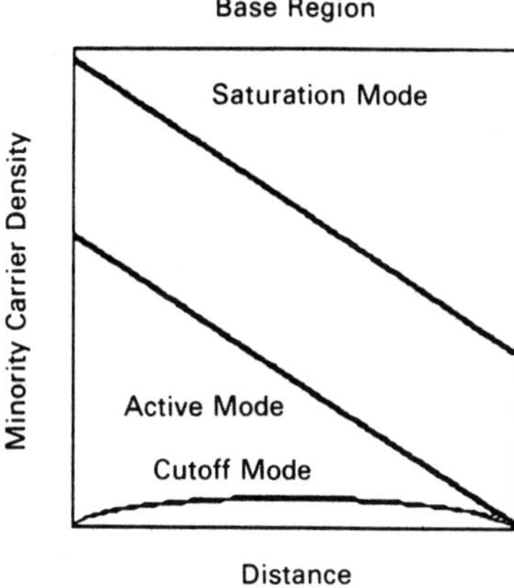

Fig. 3-12-2. Qualitative distributions of the minority carriers in the base in the cutoff, active, and saturation modes. The drawing is not to scale as the density of the minority carrier in the cutoff regime is many orders of magnitude smaller than the density of the minority carriers in the active and saturation modes.

It remains approximately equal to this value during the *storage time,* τ_s, i.e., as long as the minority carrier distribution in the base region corresponds to the saturation mode (see Fig. 3-12-2 and Fig. 3-12-4). When the excess minority carriers in the base recombine, the carrier distribution in the base changes to the distribution corresponding to the active mode (see the figure corresponding to $\tau > \tau_s$ in Fig. 3-12-3). The collector current is proportional to the slope of the minority carrier distribution in the base region. Hence, once the transistor is driven into the active mode and the slope of the minority carrier distribution in the base region starts decreasing, the collector starts dropping as well. At the same time, the emitter-base voltage decreases, leading to an increase of the base current from the negative value, I_{b2}, to zero after the switching process is over. The base current is given by

$$I_b = (V_2 - V_{be})/R_s \tag{3-12-3}$$

at the end of the switching process, and $V_2 = V_{be}$ and $I_b = 0$.

The storage time, τ_s, is one of the most important characteristic times limiting the transistor switching speed. To estimate this time, we first notice that the transistor is driven into the saturation regime when the collector current is given by

$$I_c \approx (V_{cc} - V_{cesat})/R_L \approx V_{cc}/R_L \tag{3-12-4}$$

where R_L is the load resistance (see Fig. 3-12-3)). Hence, the transistor is driven into the saturation regime when the base current

$$I_b \geq I_{ba} \approx V_{cc}/(h_{fe}R_L) \tag{3-12-5}$$

Once the transistor is in saturation, the collector current is given by eq. (3-12-4) independently of by how much the base current exceeds this critical value, I_{ba}. During the storage time, the collector current remains approximately constant (and approximately equal to V_{cc}/R_L) until the transistor reaches the active mode.

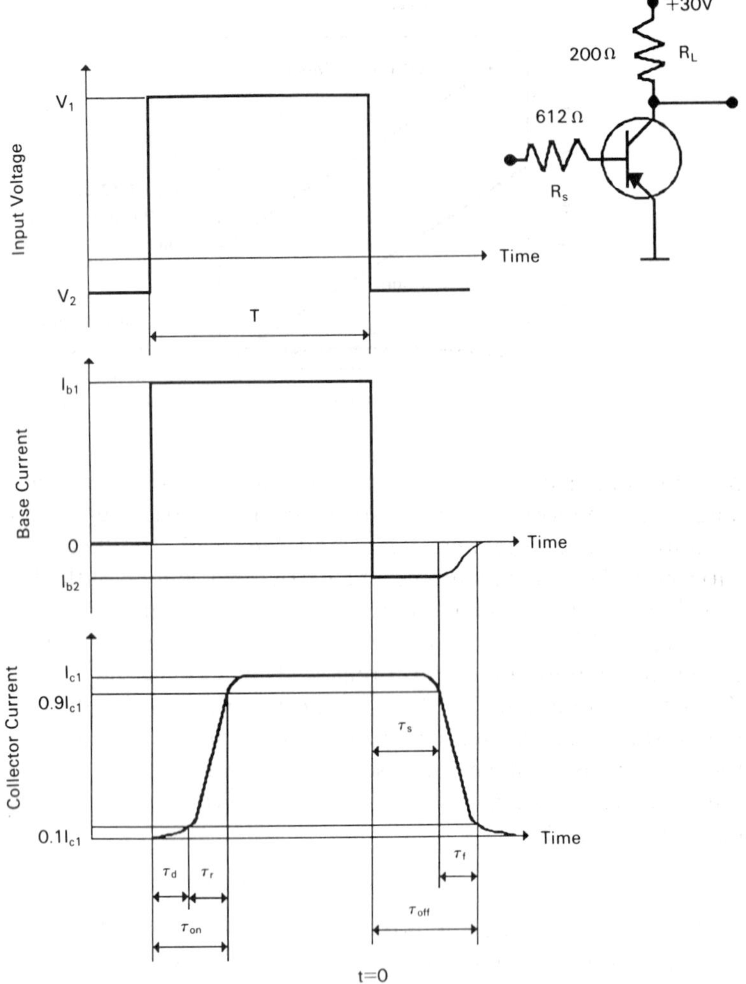

Fig. 3-12-3. Typical circuit for measuring transistor switching characteristics (from *Small-Signal Transistor Data,* Motorola Semiconductor Products, Inc., Phoenix AR [1983]) and shape of the input voltage pulse, base current pulse, and output current pulse. Characteristic time constants (the delay time τ_d, the rise time τ_r, the turn-on time, $\tau_{on} = \tau_d + \tau_r$, the storage time τ_s, the fall time τ_f, and the turn-off time, $\tau_{off} = \tau_s + \tau_f$) are defined on the figure. For the test circuit shown in the figure, $V_1 = 1.9$ V, $V_2 = -0.5$ V.

As was explained previously, the storage time, τ_s, is the time required for the charge in the base in the saturation regime, Q_{bs}, to drop to the value of the charge, Q_{ba}, corresponding to the active normal mode. As is shown in Fig. 3-12-4, the slope of the minority carrier distribution in the base during this time remains

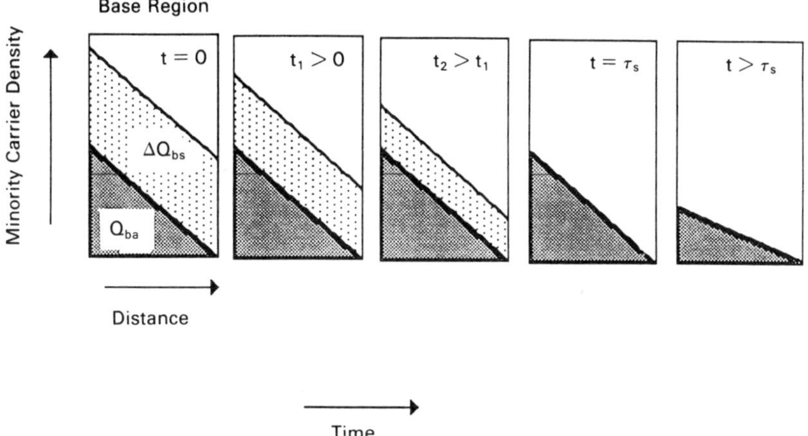

Fig. 3-12-4. Time evolution of minority carrier profiles in the base region when the transistor is being turned off.

approximately constant. Once the transistor is in the active mode, the collector current decreases with time. The difference, ΔI_{bs}, between the base current in the saturation regime, I_{bs}, and the base current at the onset of the active mode, I_{ba},

$$\Delta I_{bs} = I_{bs} - I_{ba} = I_{bs} - V_{cc}/(h_{fe}R_L) \tag{3-12-6}$$

comprises the excess recombination current (caused by the recombination of the excess charge $\Delta Q_{bs} = Q_{bs} - Q_{ba}$ (see Fig. 3-12-4) and the displacement current,

$$\Delta I_{bs} = \Delta Q_{bs}/\tau_{sr} + d\Delta Q_{bs}/dt \tag{3-12-7}$$

where τ_{sr} is the characteristic recombination time related to the minority carrier lifetime in the base, and

$$Q_{ba} = I_{ba}\tau_{sr} \tag{3-12-8}$$

is the charge in the base corresponding to the onset of the saturation regime. Equation (3-12-7) is called the *charge control equation.* For the base current pulse shown in Fig. 3-12-3 for $t = 0$,

$$\Delta I_{bs} = I_{b2} - I_{bs} \tag{3-12-9}$$

The initial condition for ΔQ_{bs} is given by

$$\Delta Q_{bs}(0) = (I_{b1} - I_{ba})\tau_{sr} \tag{3-12-10}$$

Solving eq. (3-12-7) with ΔI_{bs} given by eq. (3-12-9) and with the initial condition given by eq. (3-12-10) yields

$$\Delta Q_{bs} = (I_{b1} - I_{b2})\tau_{sr} \exp(-t/\tau_{sr}) + (I_{b2} - I_{ba})\tau_{sr} \tag{3-12-11}$$

At $t = \tau_s$, $\Delta Q_{bs} = 0$. From this condition we find that

$$\tau_s = \tau_{sr} \ln[(I_{b1} - I_{b2})/(I_{ba} - I_{b2})] \tag{3-12-12}$$

When $t > \tau_s$ the charge control equation becomes

$$I_b = Q_b/\tau_{nl} + dQ_b/dt \tag{3-12-13}$$

where Q_b is the charge of the minority carriers in the base in the active mode and τ_{nl} is the lifetime of the minority carriers in the base region. (τ_{nl} may be different from τ_{sr} because the lifetime may depend on the injection level.) Equation (3-12-13) should be solved together with eq. (3-12-3) and the equation relating Q_b to V_{be}:

$$Q_b = qSn_{po}W \exp(qV_{be}/k_BT)/2 \tag{3-12-14}$$

Such a solution describes the transistor transient during the fall time, τ_f. However, we should emphasize that eq. (3-12-13) does not account for the emitter hole current, I_{pe}, which may be an important or even dominant component of the base current (see Section 3-3). More accurate and realistic analysis of the switching process can be done numerically using the Gummel-Poon model (see Section 3-8). This model is implemented in a popular circuit simulation program, SPICE (see Appendix 24). Characteristic times τ_d, τ_r, τ_s, and τ_f are usually specified by transistor manufacturers in the data sheets (see, for example, Appendix 26).

3-13. BIPOLAR JUNCTION TRANSISTORS IN INTEGRATED CIRCUITS

Bipolar Junction Transistors are used as elements of integrated circuits of different integration scales: Small Scale Integration (SSI), Medium Scale Integration (MSI), Large Scale Integration (LSI), and Very Large Scale Integration (VLSI). The principle of operation and device physics of an integrated Bipolar Junction Transistor is similar to that of a discrete device. However, the integrated device has all contacts on top of a semiconductor wafer. Also, isolation between different devices has to be provided. In addition, multiple transistor structures are used in integrated circuits to provide several logic inputs.

Simple integrated transistor structures are shown in Figs. 3-13-1a and b. Transistors are isolated by the depletion region at the boundary between the n-type collector and the p-type substrate. The substrate should be connected to the most negative potential in the circuit to provide such an isolation. The collector region completely surrounds the base. This leads to a fairly large common-base current gain, as most electrons injected from the emitter region are collected.

One drawback of the structure shown in Fig. 3-13-1a is a large collector series resistance caused by a relatively low doping level in the collector region. This resistance may be considerably reduced by utilizing a buried n^+ collector layer, as shown in Fig. 3-13-1b. Another improvement in the integrated transistor

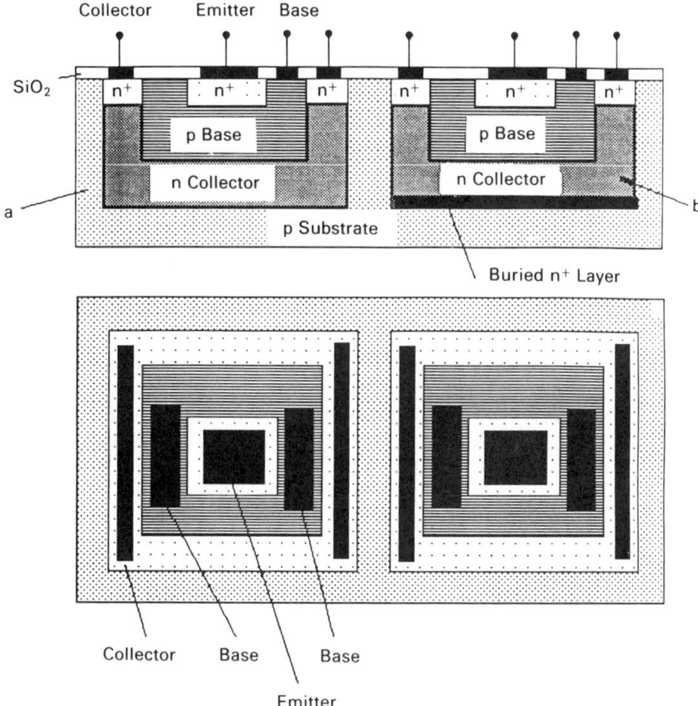

Fig. 3-13-1. Simple integrated transistor structure without (a) and with (b) buried n^+ layer.

structures is the dielectric isolation between different devices. This leads to a considerable reduction of parasitic capacitances caused in the structure of Fig. 3-13-1 by the isolation depletion regions at the p-n junctions between the collector region and the substrate. Integrated n-p-n transistors with buried layer and dielectric isolation may exhibit common-emitter current gain of up to several hundreds, collector breakdown voltages of several tens of volts, and cutoff frequencies up to 15 GHz or higher.

The fabrication process for integrated n-p-n bipolar transistors allows one to fabricate p-n-p transistors on the same wafer. Such a p-n-p transistor (called a *lateral p-n-p transistor*) is shown in Fig. 3-13-2a. For common-collector configurations when the collector contact is attached to the ground the *substrate p-n-p bipolar transistor* (see Fig. 3-13-2b) is sometimes used instead of a lateral p-n-p transistor. Because of a low mobility of holes and relatively long effective base region, lateral transistors have a low common-emitter current gain (typically 50 or less at low collector currents and much less at collector currents of several milliamps) and a low cutoff frequency (usually 100 MHz). However, circuit applications requiring a combination of n-p-n and p-n-p transistors make a lateral p-n-p transistor a popular choice in integrated circuits.

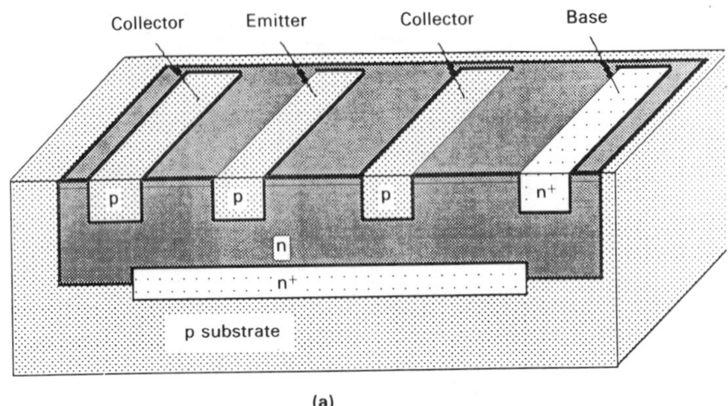

(a)

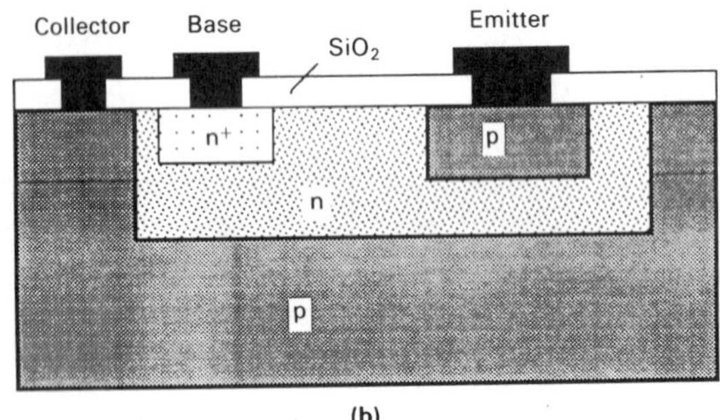

(b)

Fig. 3-13-2. (a) Lateral *p-n-p* transistor. (b) Substrate *p-n-p* bipolar junction transistor.

Recently a new lateral p-n-p device structure that uses sidewall contacts was developed (see Nakazato et al. 1986). This new structure is shown in Fig. 3-13-3. It has polysilicon contacts on silicon dioxide, which greatly reduces parasitic capacitances. Collector and emitter regions were created by boron diffusion from the polysilicon contacts. The diffusion width was 0.5 μm, leading to an effective base thickness of only 0.5 μm for a nominal 1.5 μm base mask. This transistor exhibited a common-emitter current gain of over a hundred and a cutoff frequency of 3 GHz.

To achieve fast switching operation in a bipolar junction transistor it is desirable to reduce or eliminate the storage time (see Section 3-12). This may be achieved by connecting a diode between the base and the collector with a rela-

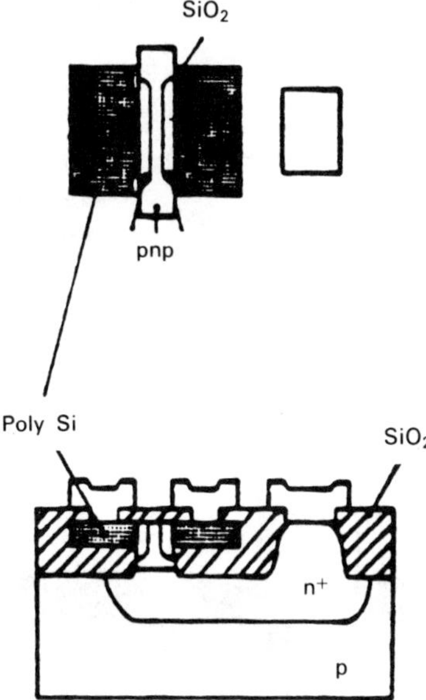

Fig. 3-13-3. Lateral *p-n-p* device structure with sidewall contacts (from K. Nakazato, T. Nakamura, and M. Kato, "A 3 GHz Lateral pnp Transition," *IEEE Tech. Digest,* p. 266, Los Angeles, published by IEEE (1986). © 1986 by IEEE).

tively small cut-in voltage. A Schottky diode used for this purpose in silicon bipolar transistors (see Fig. 3-13-4) starts to conduct when the forward voltage across the diode is approximately 0.4 V, therefore preventing the transistor from being driven into saturation. A Schottky diode may be fabricated using an Al contact to n-type silicon (at the same time, Al forms an ohmic contact with n^+ silicon or p-type silicon). Hence, the transistor with a Schottky "clamp" may be easily fabricated as shown in Fig. 3-13-4. Such a transistor is sometimes called a Schottky transistor. Platinum (which forms platinum silicide after alloying) is frequently used to make a Schottky barrier clamp in Schottky bipolar junction transistors.

Integrated circuit technology allows us to fabricate multiple transistor structure with several collectors or several emitters (see Fig. 3-13-5). Such structures are used in logic gates with multiple inputs or outputs. Figure 3-13-5 shows schematic diagrams of a Transistor-Transistor Logic (TTL) gate and an Integrated Injection Logic (I^2L) gate. Transistor-transistor logic uses multiple-emitter integrated transistors. Integrated injection logic (which is also called merged transistor logic) uses an n-p-n transistor with multiple collectors integrated with a lateral p-n-p transistor. The p-n-p transistor acts as a current source substituting for a resistance in a TTL gate.

One of the fastest silicon integrated-circuit bipolar technologies is *Emitter Coupled Logic* (ECL) (see Fig. 3-13-6). Using polysilicon self-aligned technology utilizing a submicron epitaxial layer and Rapid Thermal Annealing (RTA), Take-

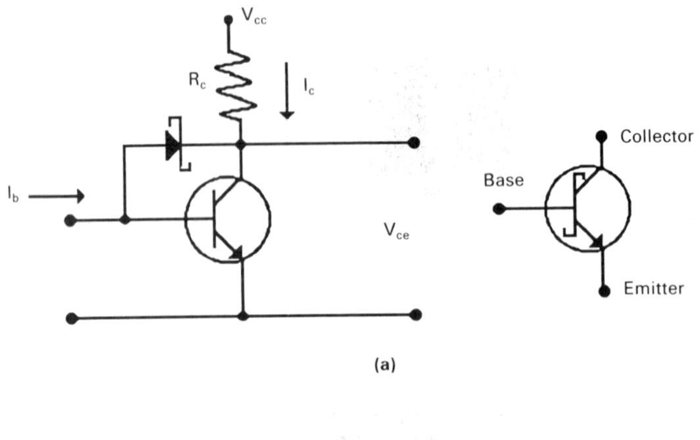

(a)

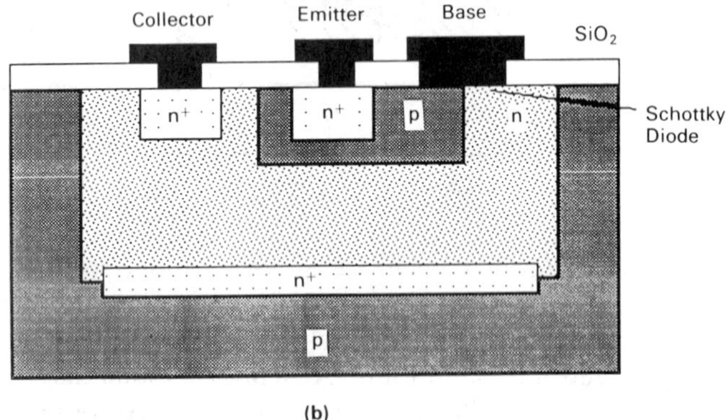

(b)

Fig. 3-13-4. (a) "Clamping" Schottky diode connected between the collector and base of a bipolar junction transistor. (b) Circuit symbol for the integrated transistor with the Schottky diode (the Schottky transistor). (c) Schematic structure of integrated bipolar junction transistor with Schottky clamp (Schottky transistor).

mura et al. (1986) obtained the extremely short switching time of only 52 ps for the collector current of about 2 mA in a silicon ECL circuit.

In general, integrated bipolar technology has a very high speed but at a price of high power consumption. Cray-II supercomputer uses small-scale silicon bipolar integrated circuits (of only several hundred logic gates on a chip) for an ultimate speed. However, in such applications silicon bipolar technology faces stiff competition from new emerging technologies, such as heterojunction bipolar tech-

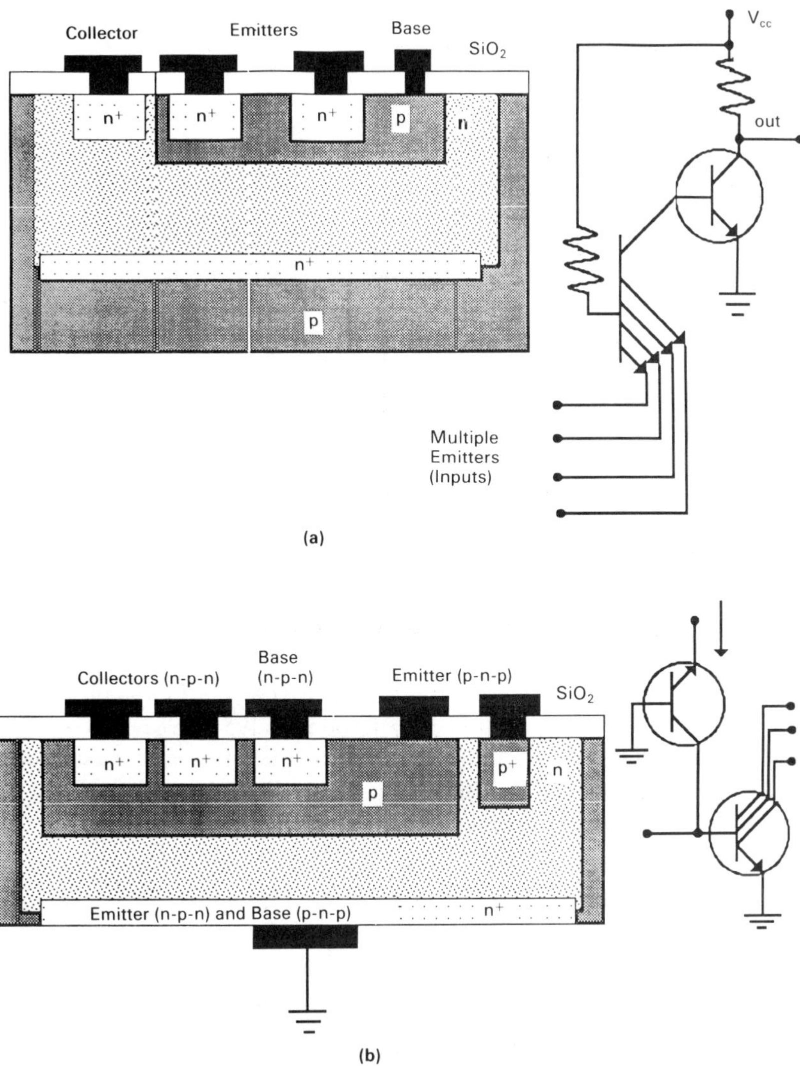

Fig. 3-13-5. Multiple transistor structures with several collectors and emitters: (a) Transistor-Transistor Logic (TTL) gate using a multiple emitter integrated transistor, and (b) Integrated Injection Logic (I^2L) gate.

nology (considered in the next section) and compound semiconductor field-effect transistors (discussed in Chapter 4). The latter technology will be utilized in a new Cray-III supercomputer for reaching computational speeds that have never been possible before.

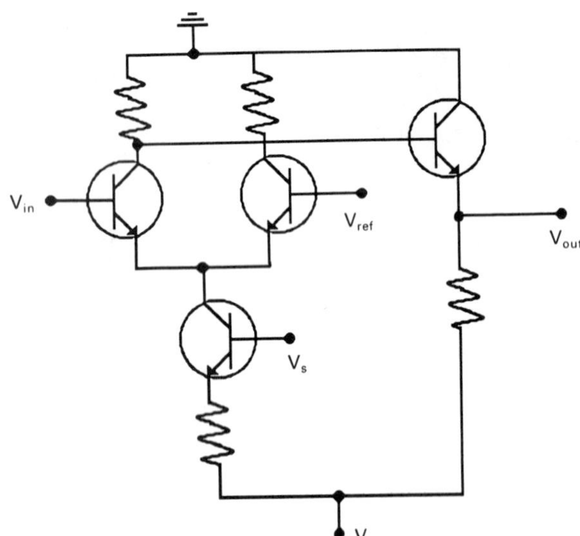

Fig. 3-13-6. Circuit diagram of basic ECL gate.

3-14. HETEROJUNCTION BIPOLAR TRANSISTORS

The idea of a *Heterojunction Bipolar Transistor* (HBT) was proposed by Shockley (1951) and was later developed by Kroemer (1957, 1982, 1983). The principle of operation may be understood by analyzing the different current components flowing in the device, as shown in Fig. 3-3-1. In the forward active mode emitter, collector, and base currents, I_e, I_c, and I_b, in an n-p-n bipolar junction transistor may be expressed as follows (see eqs. (3-3-1) to (3-3-3)):

$$I_e = I_{ne} + I_{pe} + I_{re} \tag{3-14-1}$$

$$I_c = I_{nc} + I_{gc} \tag{3-14-2}$$

$$I_b = I_{pe} + I_{re} - I_{gc} + I_{br} \tag{3-14-3}$$

where I_{ne} and I_{nc} are the electron components of the emitter and the collector current, I_{pe} is the hole component of the emitter current, I_{re} is the recombination current in the depletion region of the emitter-base junction, I_{br} is the recombination current in the neutral base region, and I_{gc} is the generation current in the depletion region of the collector-base junction.

The largest current components for the forward active mode are I_{ne} and I_{nc}. They are determined by the electron diffusion current in the base at the emitter-base junction and the collector-base junction, respectively. The common emitter current gain, β, is given by

$$\beta = \frac{I_c}{I_b} < \beta_{max} = \frac{D_n N_{de} x_e}{D_p N_{ab} W} \exp(-\Delta E_g / k_B T) \tag{3-14-4}$$

(see eq. (3-3-49)). The maximum gain is limited by the hole component of the emitter current (caused by the hole injection from the base into the emitter region). The exponential factor $\exp(-\Delta E_g/k_B T)$ that reduces gain is caused by the band-gap narrowing in the highly doped emitter region (see Section 3-3). As pointed out by Kroemer (1982), a more realistic estimate of β_{max} is given by

$$\beta_{max} = \frac{v_{nb} N_{de} x_e}{v_{pe} N_{ab} W} \exp(-\Delta E_g/k_B T) \tag{3-14-5}$$

where v_{nb} and v_{pe} are effective velocities of electrons in the base and holes in the emitter, which include contributions from the carrier diffusion and drift.

In a heterojunction bipolar junction transistor, the emitter region has a wider band gap than the base. Many HBTs are made using an AlGaAs/GaAs heterostructure because the lattice constants of these two materials are very close and, as a consequence, the heterointerface quality is excellent, with very few dislocations and a low density of interface states. If we assume for illustrative purposes that the discontinuity of the band gaps is entirely related to the valence band discontinuity ΔE_v, the band diagram of an HBT will look as shown in Fig. 3-14-1a. In this case, the expression for β_{max} for a wide-gap emitter device is given by

$$\beta_{max} = \frac{v_{nb} N_{de} x_e}{v_{pe} N_{ab} W} \exp(\Delta E_g/k_B T) \tag{3-14-6}$$

Hence, a very high value of β_{max} may be achieved even when N_{de} is smaller than N_{ab}.

In fact, the band diagram of heterojunction looks like that shown in Fig. 3-14-1b, and as a consequence, eq. (3-14-6) should be rewritten as

$$\beta_{max} = \frac{v_{nb} N_{de} x_e}{v_{pe} N_{ab} W} \exp(\Delta E_v/k_B T) \tag{3-14-7}$$

where $\Delta E_v = \Delta E_g - \Delta E_c$. This leads to a less dramatic increase in β_{max}, but the basic principle remains intact. Moreover, the decrease in the maximum gain related to the decrease of the potential barrier for holes may be partially compensated by the increase in the electron velocity in the base caused by the conduction band spike. As pointed out by Kroemer (1982), electrons in this "spike-notch" structure enter the base with very large energies (close to ΔE_c) and, as a consequence, may have very high velocities (of the order of several times 10^5 m/s). Because of the directional nature of the dominant polar optical scattering in gallium arsenide (see Section 1-9)—a typical material of choice for a narrow-gap semiconductor in a heterostructure—electrons may traverse the base region maintaining a very high velocity, even when they experience scattering. Kroemer described the conduction band spike as "a launching pad" for ballistic electrons. Also, the magnitude of the spike can be reduced by grading the composition of the wide band gap emitter near the heterointerface (see Kroemer 1982) as shown by a dashed line in Fig. 3-14-1b. Such a grading plays an important role in optimizing the emitter injection efficiency of HBTs (see, for example, Grinberg et al. 1984).

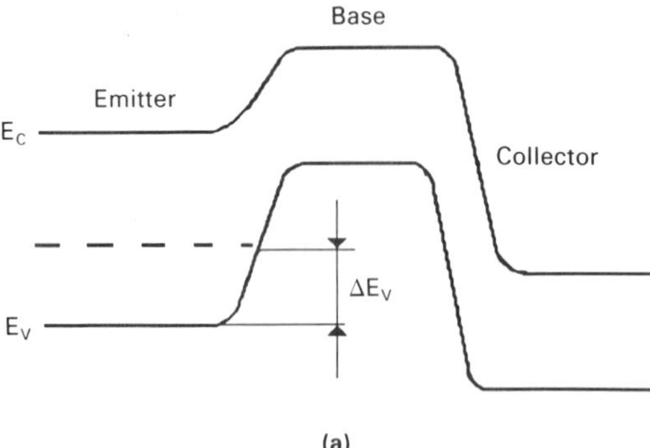

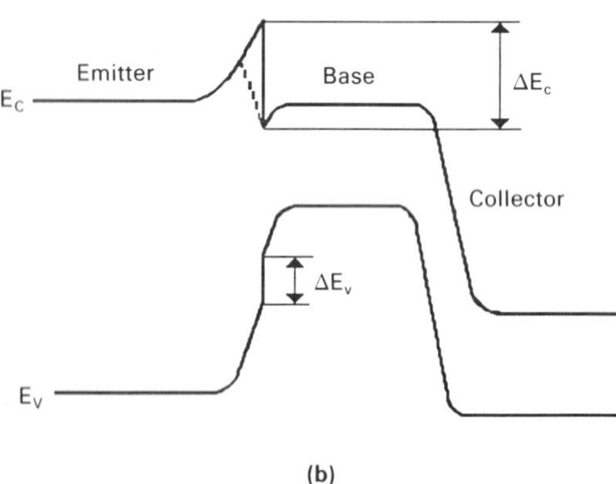

Fig. 3-14-1. Simplified energy band diagrams of a heterojunction bipolar transistor. (a) For simplicity all the difference in the energy gaps is related to the valence band. Dashed line corresponds to a homojunction transistor and is shown for comparison (after H. Kroemer, "Heterostructure Bipolar Transistors and Integrated Circuits," *Proc. IEEE*, 70, pp. 13–25 (1982). © 1982 IEEE). (b) This is a more realistic band diagram of the emitter-base heterojunction. Dashed line shows the bottom of the conduction band for a heterojunction with graded composition, x, in $Al_xGa_{1-x}As$ for $Al_xGa_{1-x}As$/GaAs heterostructure near the heterointerface.

This discussion shows that the emitter injection efficiency of an HBT may be made very high. In this case the transistor gain is limited primarily by the recombination current,

$$\beta = \frac{I_{ne}}{I_{re} + I_{br}} \tag{3-14-8}$$

and could be as high as a few thousands or more if the heterojunction interface is relatively defect free so that I_{re} is not excessively high.

The recombination current, I_{br}, caused by the recombination, in the base, can be estimated as

$$I_{br} \approx qcSn_p(0)W/\tau \tag{3-14-9}$$

where c is a numerical constant of the order of unity, $n_p(0)$ is the concentration of minority carriers (electrons) at the emitter end of the base, and τ is the lifetime. The electron component of the emitter current can be estimated as follows

$$I_{ne} \approx qn_p(0)v_{nb}S \tag{3-14-10}$$

Thus, if I_{re} can be neglected (which may be possible at a relatively high forward bias; see Section 2-4 and Section 3-3), we obtain

$$\beta \approx \frac{I_{ne}}{I_{br}} \approx \tau / t_{TR} \tag{3-14-11}$$

where t_{TR} is the electron transit time across the base. For a sufficiently short base (say, W ~ 0.1 μm), $\beta > 1000$ can be obtained even if the lifetime is only of the order of a nanosecond. Common emitter gains in excess of 1600 have been achieved for AlGaAs/GaAs heterojunction bipolar junction transistors. For comparison, in a conventional homojunction silicon transistor, the dependence of the energy gap on the doping level leads to a shrinkage of the energy gap in the emitter region. One possible mechanism of the energy-gap narrowing is the formation of the impurity band when wave functions of the donor electrons overlap (similar to the band formation from individual atomic levels illustrated by Fig. 1-5-3). This impurity band effectively decreases the energy gap. According to Lanyon and Tuft (1978), the energy-gap narrowing, ΔE_g, in silicon is given by

$$\Delta E_g = \frac{3q^2}{16\pi\varepsilon_s}\left(\frac{q^2 N_{de}}{\varepsilon_s k_B T}\right)^{1/2} \cong 22.5\left(\frac{N_{de}}{10^{18}}\right)^{1/2}\left(\frac{300\text{ K}}{T}\right)^{1/2} \text{meV} \tag{3-14-12}$$

where N_{de} is in cm^{-3}. This reduction in the bandgap leads to an exponential increase of the intrinsic carrier density, n_{ie}, in the emitter region:

$$n_{ie} = n_{ie}(N_{de} = 0) \cdot \exp\left(\frac{\Delta E_g}{2k_B T}\right) \tag{3-14-13}$$

(see Section 3-3.) The concentration of holes injected into the emitter region is proportional to n_{ie}^2/N_{de}, and hence, the band narrowing leads to the exponential reduction of the maximum common-emitter current gain (see eq. (3-14-5)), up to a factor of 20 for $N_{de} = 10^{19}$ cm^{-3}. This energy-gap shrinkage represents one of the dominant performance limitations for conventional Si BJTs (see, for example, Sze 1981).

In addition to high injection efficiencies and, as a consequence, high common-emitter current gains, heterojunction bipolar junction transistors have a number of other advantages over conventional bipolar transistors. As a consequence of higher base doping, the base spreading resistance is smaller. Because of a relatively low doping of the emitter region, the emitter-base capacitance can be made quite small. All these factors result in a higher speed of operation.

The foregoing equations merely illustrate the principle of operation of an HBT. More detailed and accurate analysis of an HBT operation should be based on more detailed models (see, for example, Tiwari 1986). Such an analysis shows that both bulk recombination in the depletion regions and surface recombination, especially at the emitter-base junction periphery, limit the common-emitter short-circuit current gain. As a consequence, in heterojunction bipolar transistors optimized for high speed the common-emitter short-circuit current gain can be as low as 10 or 20. The propagation delay of an HBT logic gate is primarily limited by the time constant equal to the product of the collector capacitance and the load resistance. The propagation delay can be analyzed using the equation proposed by Dumke et al. (1972) for a typical HBT logic gate:

$$\tau_s = 2.5 R_B C_{BC} + R_B \tau_B / R_L + (3C_{BC} + C_L) R_L \qquad (3\text{-}14\text{-}14)$$

where τ_B is the base transit time, R_B is the base resistance, R_L and C_L are the load resistance and load capacitance, and C_{BC} is the collector-base capacitance.

A schematic structure of a high-speed AlGaAs/GaAs HBT are shown in Fig. 3-14-2 (after Chang et al. 1987). This structure has a proton-implanted area that reduces the collector-base capacitance by as much as 60%, leading to an increase in cutoff frequency, f_T, and maximum oscillation frequency, f_{max}, compared with unimplanted devices.

Recently the NTT group (see Nakajima et al. 1986 and Nagata et al. 1987) reported a 5.5 ps/gate propagation delay for AlGaAs/GaAs HBTs with nonalloyed compositionally graded InGaAs/GaAs emitter ohmic contacts. They reduced the specific emitter contact resistance to as low as $1.4 \times 10^{-7}\ \Omega\text{cm}^2$. Emitter and collector dimensions were 2×5 and $4 \times 7\ \mu\text{m}^2$, respectively. The structure also utilized proton-implantion to reduce the collector-base capacitance. This increased the cutoff frequency, f_T, and maximum oscillation frequency, f_{max}, com-

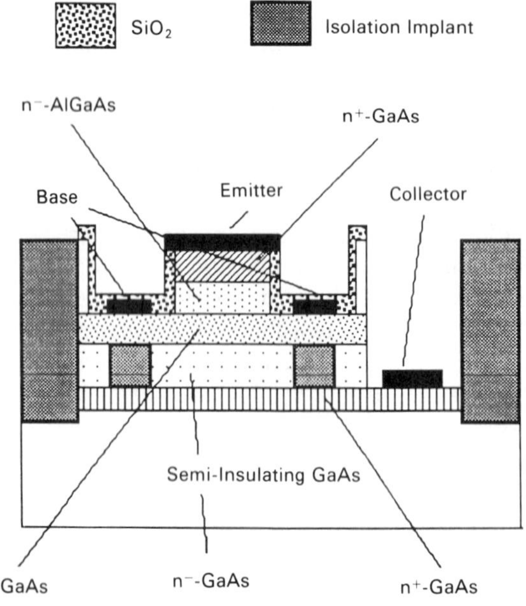

Fig. 3-14-2. A schematic structure of a high speed AlGaAs/GaAs HBT (from M. F. Chang, P. M. Asbeck, K. C. Wang, G. J. Sullivan, N. H. Sheng, J. A. Higgins, and D. L. Miller, "AlGaAs/GaAs Heterojunction Bipolar Transistors Fabricated Using a Self-Aligned Dual-Lift-Off Process," *IEEE Elec. Dev. Lett.*, EDL-8, pp. 303–305 (1987). © 1987 IEEE).

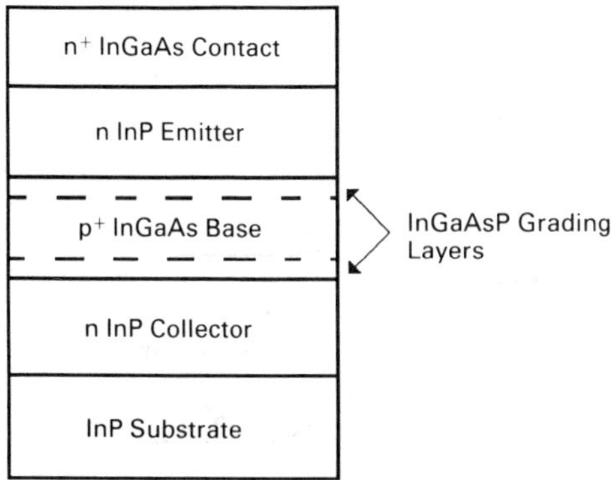

	Thickness (μm)	Doping Level (cm^{-3})
Contact	0.2	$>10^{19}$
Emitter	0.3	2×10^{17}
Grading Layers	0.02	2×10^{17}
Base	0.12	2×10^{18}
Collector	0.3	2×10^{17}

Fig. 3-14-3. The schematic structure of a InGaAs/InP double heterostructure bipolar transistor (from R. N. Nottenburg, J. C. Bischoff, J. H. Abeles, M. B. Panish, and H. Temkin, "Base Doping Effects in InGaAs/InP Double Heterostructure Bipolar Transistors," *IEEE Tech. Digest,* p. 278, Los Angeles, published by IEEE (1986). © 1986 IEEE).

pared with unimplanted devices. For the collector current of 12 mA, Nagata et al. reported $f_{\max} = 60$ GHz and $f_T = 80$ GHz.

Another heterostructure system suitable for HBTs is the InGaAs/InP system (see Nottenburg et al. 1986 and Nottenburg et al. 1987). The schematic structure of a InGaAs/InP double heterostructure bipolar transistor is shown in Fig. 3-14-3. The potential advantages of such a device are related to a smaller built-in emitter junction potential, a higher mobility in InGaAs, a high electron drift velocity in InP, and low surface recombination velocity (see Nottenburg et al. 1986). Nottenburg et al. (1987) reported a high current gain (over 500) for over eight decades of the collector current for InGaAs/InP double heterostructure bipolar transistors. Recently Y. K. Chen et al. (1989) reported cutoff frequency of 165 GHz for such a transistor.

The low-field mobility of ternary compound InGaAs is higher than the mobility of GaAs. Using a high mobility InGaAs as a base material for HBTs, we can improve performance, compared with AlGaAs/GaAs HBTs. A schematic band structure and layer structure for such a device are shown in Fig. 3-14-4 (after Asbeck et al. 1986). A cutoff frequency as high as 40 GHz (see Asbeck et al. 1986)

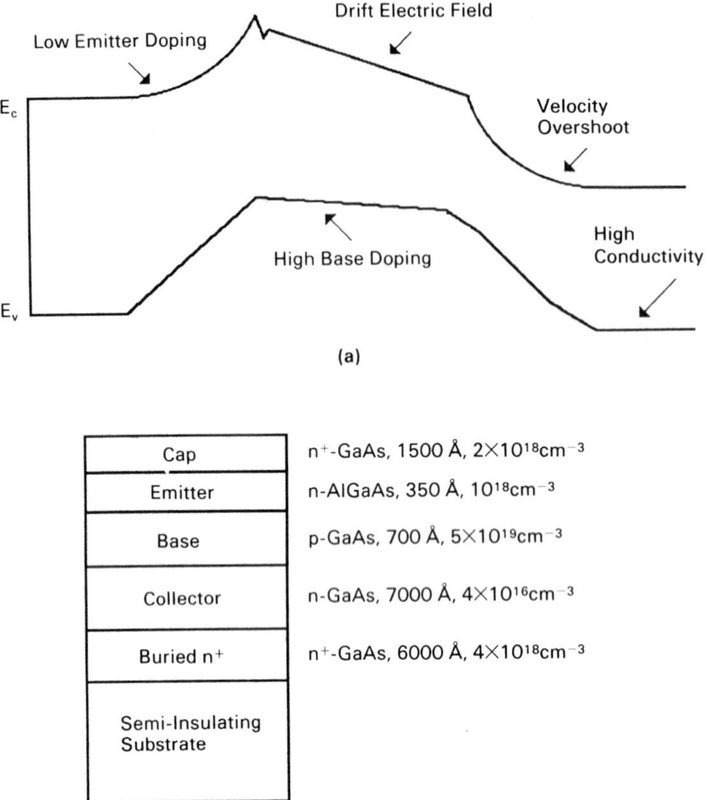

Fig. 3-14-4. (a) Schematic band structure of AlGaAs/InGaAs/GaAs HBT. (b) Schematic layer structure of AlGaAs/InGaAs/GaAs HBT (parts a and b after P. M. Asbeck, M. F. Chang, K. C. Wang, G. J. Sullivan, and D. L. Miller, "GaAlAs/GaInAs/GaAs Heterojunction Bipolar Technology for sub-35 ps Current-Mode Logic Circuits," *Proceedings of the 1986 Bipolar Circuits and Technology Meeting*, IEEE. Minneapolis, MN, pp. 25–26 (1986). © 1986 IEEE).

and a propagation delay as short as 19.7 ps (for 19-stage current mode logic (CML) ring oscillators, see Asbeck et al. 1986 and Sullivan et al. 1986) were obtained.

Even though the base width in HBTs determines transit time across the base and, hence, device speed for this vertical device structure, in practical transistors the lateral dimensions of the emitter stripes are very important. These dimensions determine parasitic capacitances, resistances, and power dissipation. This was clearly demonstrated by Chang et al. (1987). They obtained a 14.5 ps per gate propagation delay in HBT common mode logic ring oscillators using devices with an emitter width of only 1.2 μ. These devices exhibited a cutoff frequency f_T = 67 GHz and a maximum oscillation frequency f_{max} = 105 GHz. Another important

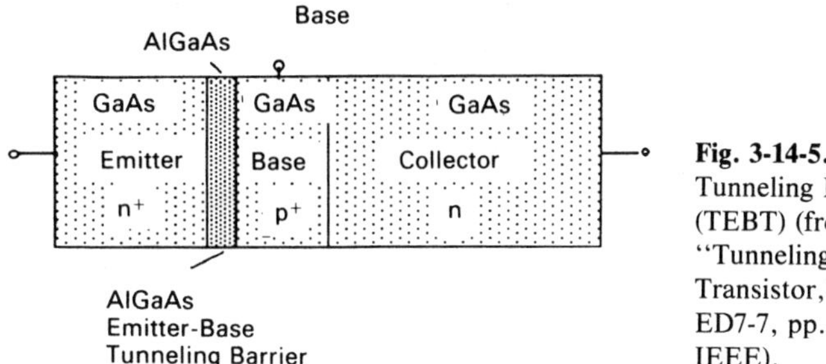

Fig. 3-14-5. Schematic diagram of Tunneling Emitter Bipolar Transistor (TEBT) (from J. Xu and M. Shur, "Tunneling Emitter Bipolar Junction Transistor," *IEEE Elec. Dev. Lett.* ED7-7, pp. 416–418 (1986). © 1986 IEEE).

feature of these devices was a very high base-region doping (up to 10^{20} cm^{-3}), leading to a small base spreading resistance.

Ishibashi and Yamauchi (1987) reported an HBT with a new device structure. They replaced an n-type GaAs collector layer by a double layer that included a relatively thick i-GaAs layer (2000 Å in their devices) and a thin p^+ GaAs layer (200Å thick doped at 2×10^{18} cm^{-3}). The p^+ layer is totally depleted and introduces a potential drop and electric field in the i-layer, resulting in a near ballistic collection of electrons in a certain voltage range. They obtained a very high cutoff frequency of 105 GHz.

Most HBTs are emitter-on-top structures (see Fig. 3-14-2). However, as was pointed out by Kroemer (1982 and 1983), the collector capacitance may be substantially smaller for the collector-on-top configuration. This was recently confirmed by the circuit simulation of HBTs reported by Akagi et al. (1986), who predicted a higher speed for collector-on-top configurations when comparing non-self-aligned structures.

A new type of heterojunction bipolar transistor—a *tunneling emitter bipolar transistor* (TEBT)—was recently proposed by Xu and Shur (1986) and fabricated by Najar et al. (1986) (see Fig. 3-14-5). In this device a wide band gap AlGaAs emitter is replaced by a conventional n^+ GaAs emitter, but a thin compositionally graded AlGaAs layer is inserted between the emitter and base regions. This layer has vastly different tunneling rates for electrons and holes, so that electrons can easily go through but holes are prevented from being injected into the emitter region. This is reminiscent of a "mass filtering" idea proposed by Capasso et al. (1985) for superlattice devices. The TEBT should have smaller emitter contact resistance, higher gain, fewer traps, and higher speed of operation than a conventional HBT.

The progress achieved in developing HBTs is illustrated in Fig. 3-14-6 (from Shur 1988), which shows cutoff frequencies and gate propagation delays as functions of time. These results show that HBTs present a serious challenge to GaAs metal-semiconductor field-effect transistors (MESFETs), AlGaAs/GaAs heterostructure field-effect transistors (HFETs), and other high speed devices (considered in Chapter 4).

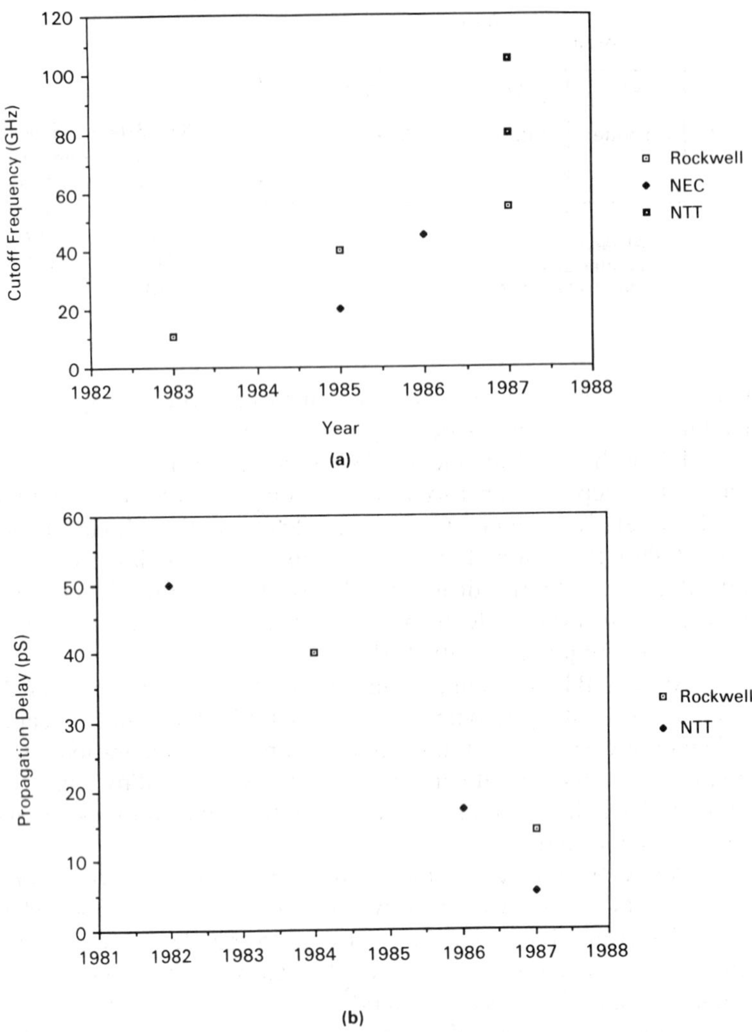

Fig. 3-14-6. HBT cutoff frequencies and gate propagation delays (from Shur [1988]).

REFERENCES

J. AKAGI, J. YOSHIDA, and M. KURATA, "A Model-Based Comparison of Switching Characteristics Between Collector-Top and Emitter-Top HBT's," *IEEE Trans. Elec. Dev., ED-34,* pp. 1413–1418 (1987).

P. M. ASBECK, M. F. CHANG, K. C. WANG, G. J. SULLIVAN, and D. L. MILLER, "GaAlAs/GaInAs/GaAs Heterojunction Bipolar Technology for sub-35 ps Current-Mode Logic

Circuits,'' *Proceedings of the 1986 Bipolar Circuits and Technology Meeting,* IEEE, Minneapolis, Minn, pp. 25–26 (1986).

J. BARDEEN and W. H. BRATTAIN, ''The Transistor: A Semiconductor Triode,'' *Phys. Rev., 74,* p. 230 (1948).

F. CAPASSO, K. MOHAMMED, A. Y. CHO, R. HULL, and A. L. HUTCHINSON, ''Effective Mass Filtering: Giant Quantum Amplification of the Photocurrent in a Semiconductor Superlattice,'' *Appl. Phys. Lett., 47,* no. 4, pp. 420–422 (1985).

D. CASASENT, *Electronic Circuits,* Quantum, New York (1973).

M. F. CHANG, P. M. ASBECK, K. C. WANG, G. J. SULLIVAN, N. H. SHENG, J. A. HIGGINS, and D. L. MILLER, ''AlGaAs/GaAs Heterojunction Bipolar Transistors Fabricated Using a Self-Aligned Dual-Lift-Off Process,'' *IEEE Elec. Dev. Lett., EDL-8,* pp. 303–305 (1987).

Y. K. CHEN, R. N. NOTTENBURG, M. B. PANISH, R. A. HAMM, and D. A. HUMPHREY, Subpicosecond InP/InGaAs Heterostructure Bipolar Transistors, IEEE EDL, vol. 10, No. 6, pp. 267–269, June (1989).

A. N. DAW, R. N. MITRA, and N. K. D. CHOUDHURY, ''Cutoff Frequency of a Drift Transistor,'' *Sol. State Elec., 10,* p. 359 (1967).

W. P. DUMKE, J. M. WOODALL, and V. L. RIDEOUT, ''GaAs-AlGaAs Heterojunction Transistor for High Frequency Operation,'' *Sol. State Elec., 15,* pp. 1339–1343 (1972).

J. M. EARLY, ''Effects of Space-Charge Layer Widening in Junction Transistors,'' *Proc. IRE, 40,* p. 1401 (1952).

J. J. EBERS and J. L. MOLL, ''Large-Signal Behavior of Junction Transistors,'' *Proc. IRE, 42,* p. 1761 (1954).

A. A. GRINBERG, M. SHUR, R. J. FISHER, and H. MORKOÇ, ''Investigation of the Effect of Graded Layers and Tunneling on the Performance of AlGaAs/GaAs Heterojunction Bipolar Transistors,'' *IEEE Trans. Elec. Dev.,* ED-31, no. 12, pp. 1758–1765 (1984).

H. K. GUMMEL and H. C. POON, ''An Integral Charge Control Model of Bipolar Transistors,'' *Bell Syst. Tech. J., 49,* p. 827 (1970).

T. ISHIBASHI and Y. YAMAUCHI, ''A Novel AlGaAs/GaAs HBT Structure for Near Ballistic Collection,'' in *Program of 45th Annual Device Research Conference,* June 22–24, Santa Barbara, p. IV-A6 (1987).

K. KIKUSHI, S. KAMEYAMA, M. KAJIYAMA, M. NISHIO, and T. KOMEDA, ''A High Speed Bipolar LSI Process using Self-aligned Double Diffusion Polysilicon Technology,'' *IEDM Tech. Digest,* p. 420, Los Angeles, published by IEEE, Dec. (1986).

C. T. KIRK, ''A Theory of Transistor Cutoff Frequency (f_T) Fall-Off at High Current Density,'' *IEEE Trans. Elec. Dev., ED-9,* p. 164 (1962).

H. KROEMER, ''Theory of a Wide-Gap Emitter for Transistors,'' *Proc. IRE, 45,* pp. 1535–1537 (1957).

H. KROEMER, ''Heterostructure Bipolar Transistors and Integrated Circuits,'' *Proc. IEEE, 70,* pp. 13–25 (1982).

H. KROEMER, ''Heterostructure Bipolar Transistors: What Should We Build,'' *J. Vac./Sci. Technol., B1,* no. 2, pp. 112–130 (1983).

H. P. D. LANYON and R. A. TUFT, ''Bandgap Narrowing in Heavily Doped Silicon,'' in *Technical Digest of Internation Electron Device Meeting,* IEEE, p. 316 (1978).

M. MADIHIAN, K. HONJO, H. TOYOSHIMA, and S. KUMASHIRO, ''Fabrication and Modeling of a Novel Self-aligned AlGaAs/GaAs Heterojunction Bipolar Transistor with a

Cutoff Frequency of 45 GHz," *IEEE Tech. Digest,* p. 270, Los Angeles, published by IEEE, Dec. (1986).

E. J. McGrath and D. H. Navon, "Factors Limiting Current Gain in Power Transistors," *IEEE Trans. Elec. Dev., ED-24,* p. 1255 (1977).

K. Nagata, O. Nakajima, Y. Yamauchi, H. Ito, T. Nittono, and T. Ishibashi, "High Speed Performance of AlGaAs/GaAs Heterojunction Bipolar Transistor with Non-Alloyed Emitter Contacts," in *Program of 45th Annual Device Research Conference,* June 22–24, Santa Barbara, p. IV-A2 (1987).

F. E. Najar, D. C. Radulescu, Y. K. Chen, G. W. Wicks, P. J. Tasker, and L. F. Eastman, "DC Characterization of the AlGaAs/GaAs Tunneling Emitter Bipolar Transistor," *Appl. Phys. Lett.,* 50, no. 26, p. 1915 (1987).

O. Nakajima, K. Nagata, Y. Yamauchi, H. Ito, and T. Ishibashi, "High Speed AlGaAs/GaAs HBTs with Proton-implanted Buried Layers," *IEEE Tech. Digest,* p. 416, Los Angeles, published by IEEE, Dec. (1986).

K. Nakazato, T. Nakamura, and M. Kato, "A 3 GHz Lateral pnp Transistor," *IEEE Tech. Digest,* p. 266, Los Angeles, published by IEEE, Dec. (1986).

R. N. Nottenburg, J. C. Bischoff, J. H. Abeles, M. B. Panish, and H. Temkin, "Base Doping Effects in InGaAs/InP Double Heterostructure Bipolar Transistors," *IEEE Tech. Digest,* p. 278, Los Angeles, published by IEEE, Dec. (1986).

R. N. Nottenburg, H. Temkin, M. B. Panish, R. Bhat, and J. C. Bischoff, "InGaAs/InP Double Heterostructure Bipolar Transistors with Near Ideal β versus I_c," *IEEE Elec. Dev. Lett.,* EDL-8, p. 282 (1987).

W. Shockley, "The Theory of p-n Junctions in Semiconductors and p-n Junction Transistors," *Bell Syst. Tech. J., 28,* p. 435 (1949).

W. Shockley, U. S. Patent No. 2,569,347, issued 1951.

"Small-Signal Transistor Data," published by Motorola Semiconductor Products, Inc., Phoenix, Ariz. (1983).

M. Shur, "Submicron AlGaAs, AlGaAs/GaAs, and AlGaAs/InGaAs Transistors," in *Integrated Circuits with 0.05 to 0.5 μm Feature Size,* ed. K. Watts, John Wiley & Sons, New York (1988).

G. J. Sullivan, P. M. Asbeck, M. F. Chang, D. L. Miller, and K. C. Wang, "High Frequency Performance of AlGaAs/InGaAs/GaAs Strained Layer Heterojunction Bipolar Transistors," *IEEE Trans. Elec. Dev.,* ED-33, no. 11, p. 1845, Nov. (1986).

S. M. Sze, *Physics of Semiconductor Devices,* 2d ed., John Wiley & Sons, New York (1981).

H. Takemura, T. Kamiya, S. Ohi, M. Sugiyama, T. Tashiro, and M. Nakamae, "Submicron Epitaxial Layer and RTA Technology for Extremely High Speed Bipolar Transistors," *IEEE Tech. Digest,* p. 424, Los Angeles, published by IEEE, Dec. (1986).

C. G. Thorton and C. D. Simmons, "A New High Current Mode of Transistor Operation," *IRE Trans. Elec. Dev., ED-5,* pp. 6–10 (1958).

S. Tiwari, "GaAlAs/GaAs Heterostructure Bipolar Transistors: Experiment and Theory," *IEEE Tech. Digest,* p. 262, Los Angeles, published by IEEE, Dec. (1986).

B. D. Urke and R. S. L. Lutze, "A Comparison of CML and ECL for VLSI Applications," *Proceedings of the 1986 Bipolar Circuits and Technology Meeting,* IEEE, Minneapolis, Minn., pp. 49–50 (1986).

F. F. VILLA, "Improved Second Breakdown of Integrated Bipolar Power Transistors," *IEEE Trans. Elec. Dev., ED-33,* pp. 1971–1976, Dec. (1986).

R. M. WARNER and J. N. FORDEMWALT, *Integrated Circuits,* McGraw-Hill, New York (1965).

R. M. WARNER and B. L. GRUNG, *Transistors: Fundamentals for the Integrated-Circuit Engineer,* John Wiley & Sons, New York (1983).

J. XU and M. SHUR, "Tunneling Emitter Bipolar Junction Transistor," *IEEE Elec. Dev. Lett., EDL-7,* pp. 416–418 (1986).

Z. YU and R. W. DUTTON, *Sedan III: A General Electronic Material Device Analysis Program,* program manual, Stanford University, July (1985).

PROBLEMS

Use the following values of parameters in the problems that follow (unless specified differently in the problem itself).

Auger recombination coefficient $G_p = G_n = 2 \times 10^{-31}$ cm^6/s
Emitter region width $x_e = 1\ \mu$m
Base region width $W = 0.5\ \mu$m
Device area $S = 1 \times 10^{-8}$ m^2
Emitter doping $N_{de} = 5 \times 10^{18}$ cm^{-3}
Base doping $N_{ab} = 2 \times 10^{17}$ cm^{-3}
Collector doping $N_{dc} = 5 \times 10^{15}$ cm^{-3}
Energy gap $E_g = 1.12$ eV
Effective density of states in the conduction band $N_c = 3.22 \times 10^{19}$ cm^{-3}
Effective density of states in the valence band $N_v = 1.83 \times 10^{19}$ cm^{-3}
Temperature $T = 300$ K
Dielectric permittivity $\varepsilon = 1.05 \times 10^{-10}$ F/m
Effective recombination time in the emitter-base depletion region $\tau_{rec} = 1 \times 10^{-7}$ s
Diffusion length of electrons in the base $L_{nb} = 40\ \mu$m.
Diffusion length of holes in the emitter $L_{pe} = 25\ \mu$m.
Electron mobility in the base region $\mu_n = 1000$ cm^2/Vs
Hole mobility in the emitter region $\mu_p = 400$ cm^2/Vs
Device width $t_e = 10\ \mu$m

3-1-1. Calculate the total charge of electrons injected into the base of a n-p-n silicon transistor as a function of the emitter-base voltage for $V_{bc} = 0.6$ V (forward collector-base voltage) and $V_{bc} = -3$ V (reverse collector-base voltage). The base thickness $W = 0.5\ \mu$m, the transistor temperature $T = 300$ K.

3-1-2. Express the transconductance of a bipolar junction transistor in terms of the collector current. Compare this expression with the expression for a differential resistance of a p-n diode.

3-1-3. Consider an n-p-n bipolar junction transistor with a long base region in which the recombination current in the neutral base region, I_{br}, is greater than the emitter hole current, I_{pe}, the recombination current in the emitter-base depletion region,

I_{re}, and the generation current in the collector-base depletion region, I_{gc}. Derive an expression for a common-emitter current gain, β, for such a transistor.

3-1-4. Consider a bipolar junction transistor connected as shown in Fig. P3-1-4. Assume that the current-voltage characteristics of the emitter-base and base collector-junctions can be approximated by

$$I_{be} = I_{bes}[\exp(V_{be}/V_{th}) - 1]$$

$$I_{bc} = I_{bcs}[\exp(V_{bc}/V_{th}) - 1]$$

where $V_{th} = k_B T/q$ is the thermal voltage. Assuming that $V >> V_{th}$, calculate the voltage across the emitter-base junction. Sketch a qualitative dependence of current, I, on voltage, V, for $V > 0$, showing its asymptotic value at high voltages and its slope at the origin.

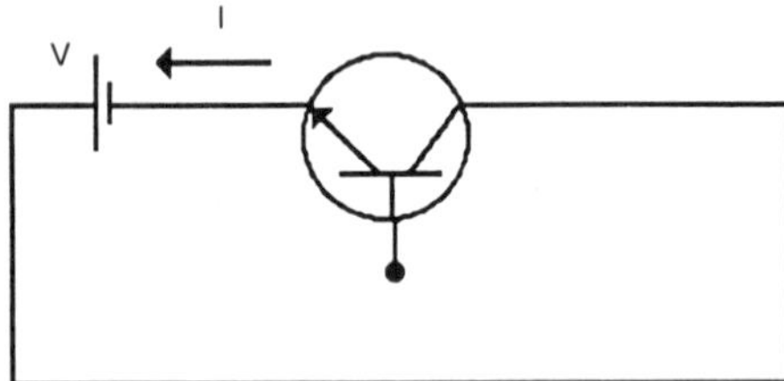

Fig. P3-1-4

3-1-5. Consider a silicon p-n-p bipolar junction transistor connected as shown in Fig. P3-1-5. Voltage $V = 10$ V. Sketch the qualitative band diagram and show a crude voltage scale. Show the position of the hole quasi-Fermi levels in the emitter and collector regions and the position of the electron quasi-Fermi level in the base region. Show the emitter-base and collector-base voltages, V_{eb} and V_{bc}.

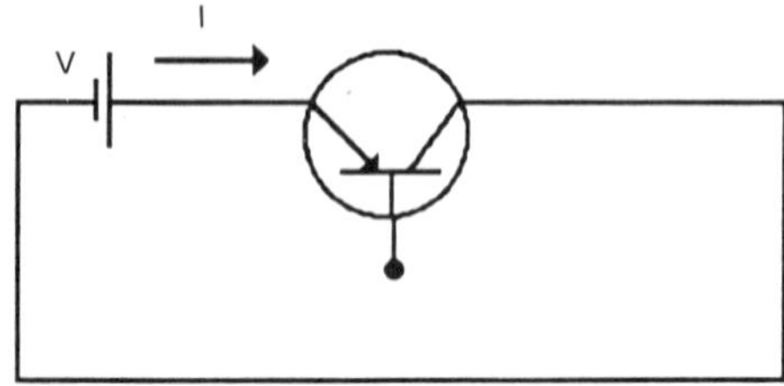

Fig. P3-1-5

3-2-1. A small sinusoidal voltage with the amplitude $V_{be'}$ and frequency ω is applied between the emitter and base of a bipolar junction transistor in addition to the forward dc bias V_{be}:

$$V_{be} = V_{beo} + V_{be'} \sin \omega t$$

The collector-base p-n junction is reverse biased. Find the amplitude of the induced sinusoidal electron charge in the base of an n-p-n bipolar junction transistor, assuming that the frequency is low enough so that the time derivative term in the continuity equation for electrons in the base can be neglected (see eq. (3-2-1)).

Use the continuity equation to estimate how low the frequency should be in order to neglect the time derivative term in the continuity equation and reduce it to eq. (3-2-1).

Assume that the equilibrium value of the electron concentration in the base is n_{bo}, the electron lifetime in the base is τ_{nb}, the base width is W, and the transistor temperature is T.

3-3-1. Calculate common-emitter current gain, β_A, limited by the Auger recombination, as a function of the collector current. Assume that $\beta_A = I_c/I_A$ where the Auger recombination current

$$I_A = qS \int_0^{x_e} p_e(x)\, dx/\tau_A$$

where the Auger recombination time τ_A is given by eq.(3-3-50).

3-4-1. Calculate the dependence of the collector current density, j_c, in the center of a silicon bipolar junction transistor on the base spreading resistance $r_{bb'}$. Neglect the recombination currents.

Hint: assume that $j_c = j_n \exp[(V_{eb} - I_b r_{bb'})/V_{th}]$, where I_b is the base current, $j_n = qD_n n_{bo}/W$. Also assume that $x_e << L_{pe}$. Emitter-base voltage $V_{be} = 0.65$ V.

3-4-2. The effect of the emitter current crowding discussed in Section 3-4 may be analyzed as follows. The emitter current density, j_e, under forward emitter-base bias is given by

$$j_e = (qD_n n_{bo}/W)\exp\{qV_{be}(y)/(k_B T)\} \qquad \text{(P3-4-2-1)}$$

where $V_b(y)$ is the potential along the base (see Fig. 3-4-1).

The base current, $i_b(y)$, is given by

$$i_b = q\mu_p N_{ab} W t_e\, dV_{be}(y)/dy \qquad \text{(P3-4-2-2)}$$

It is also related to the emitter current density as follows:

$$di_b/dy = j_e t_e/(1 + \beta_o) \qquad \text{(P3-4-2-3)}$$

where the common-emitter current gain β_o may be considered approximately constant under moderate injection conditions (see Fig. 3-3-2).

(a) Using eqs. (P3-4-2-1) to (P3-4-2-3), derive a differential equation for $V_{be}(y)$ for the transistor geometry shown in Fig. 3-4-1.

(b) Solve this equation, assuming the following boundary conditions:

$$dV_{be}/dy|_{y=o} = 0$$

$$V_{be}(d_e/2) = V_{be}$$

where V_{be} is the emitter-base voltage.

(c) Plot j_e vs. y for V_{be}= 0.5 V and V_{be} = 0.75 V.

(d) For the foregoing transistor parameters calculate the ratio of the common-emitter current gain (affected by the emitter current crowding) over the ideal current gain, β_o, vs. the emitter current I_e. (Assume that β_o is independent of the injection level.)

3-4-3. Find the value of the emitter current I_e at which the emitter crowding becomes important for a transistor shown in Fig. P3-4-3. $W_1 = 0.1\ \mu m$, $W = 0.1\ \mu m$, $d_e = 2\ \mu m$, $d = 4\ \mu m$, $t_e = 50\ \mu m$, hole mobility $\mu_p = 300\ cm^2/Vs$, base doping $N_{ab} = 5 \times 10^{16}\ cm^{-3}$, $\beta = 50$.

Neglect the widths of the depletion regions. Assume that the contact resistance for the base contacts is $R_c = 1\ \Omega mm$ (that includes the resistance of the material under the contact).

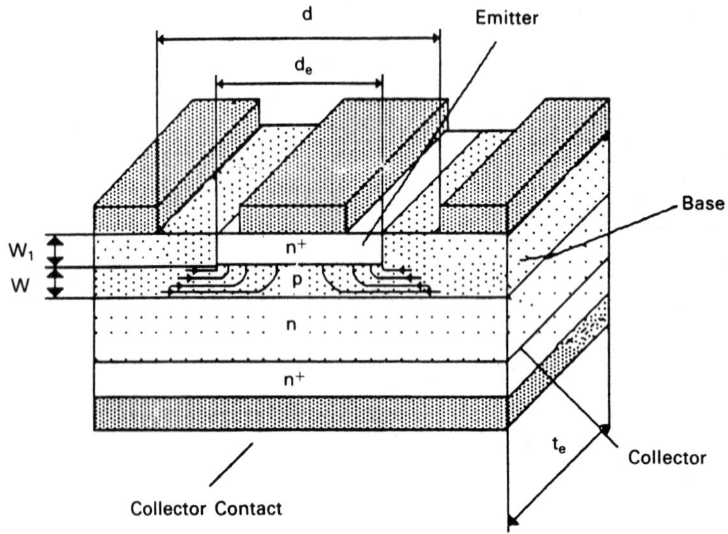

Fig. P3-4-3

Hint: Use the following result: where R is the contribution to the base spreading resistance from the transistor section under the emitter contact, arrows represent the directions of the current flow.

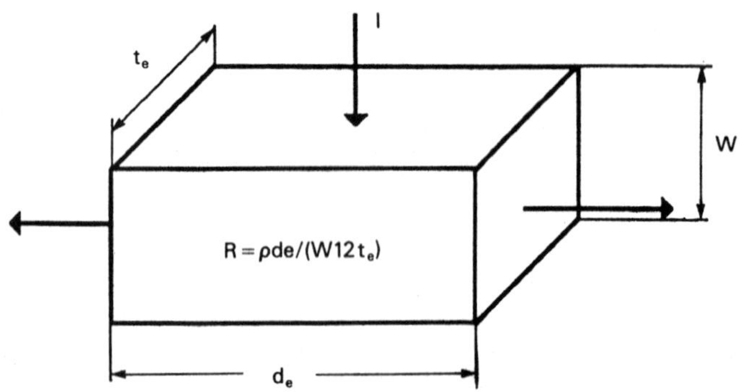

3-5-1. Calculate the base transport factor for a linear distribution of dopants in the base. Assume $N_{ab}(x_e) = 4N_{ab}(x_e + W)$, where W is the base width and x_e is the coordinate corresponding to the emitter-base junction. The diffusion length of electrons in the base is L_{nb}.

3-5-2. Assume an exponential distribution of dopants in the base

$$N_{ab}(x) = N_{ab}(x_e)\exp[-(x - x_e)/\lambda_b]$$

where W is the base width and x_e is the coordinate corresponding to the emitter-base junction. Assume that the common-emitter current gain, β, is limited by the base transport factor. Calculate the dependence of the common-emitter current gain, β, on the ratio of the base width, W, over the characteristic length of the exponential variation of dopants, λ_b. The diffusion length of electrons in the base is $L_{nb} = 10\ \mu$m, the base width is $W = 1\ \mu$m.

3-6-1. **(a)** Derive an expression for the output conductance $\partial I_c/\partial V_{bc}$ for the forward active mode in terms of basic material and transistor parameters such as doping levels, dimensions, and carrier mobilities.

(b) Assume that the output conductance has to be equal to or less than 10% of the transconductance $g_m = \partial I_c/\partial V_{be}$ for the collector current $I_c = 1$ mA. Plot the minimum value of the doping density in the base, N_{ab}, that will satisfy this requirement vs. the base width, W, for the values of W between 0.1 μm and 1 μm.

3-7-1. Using the Ebers-Moll model, express the common-base collector saturation current, I_{cbo}, and the common-emitter saturation current, I_{ceo}, in terms of the basic transistor and material parameters, such as mobilities, base width, device temperature, and doping levels.

Hint: These currents are defined in eqs.(3-6-3) and (3-6-4):

$$I_c = I_{cbo} + \alpha I_e \tag{3-6-3}$$

$$I_c = I_{ceo} + \beta I_b \tag{3-6-4}$$

Assume that the recombination current is negligible.

3-7-2. The nonlinear dc equivalent circuit of a BJT corresponding to the Ebers-Moll model is shown in Fig. 3-7-1. What other nonlinear circuit elements should be added to this circuit to obtain an ac equivalent circuit (a) neglecting parasitic elements and (b) including parasitic elements?

3-7-3. Using the Ebers-Moll model, calculate the output characteristics of a silicon n-p-n bipolar junction transistor (I_c vs. V_{ce}). Plot it with I_b as a parameter, for I_b = 10 μA, 30 μA, and 100 μA. Use the following transistor parameters:

Emitter region width $x_e = 1\ \mu$m
Base region width $W = 0.5\ \mu$m
Emitter region horizontal width $d_e = 5\ \mu$m
Collector region width $x_c = 5\ \mu$m
Emitter doping $N_{de} = 5 \times 10^{18}$ cm^{-3}
Base doping $N_{ab} = 2 \times 10^{17}$ cm^{-3}
Collector doping $N_{dc} = 5 \times 10^{15}$ cm^{-3}
Energy gap $E_g = 1.12$ eV

Effective density of states in the conduction band $N_c = 3.22 \times 10^{19}$ cm^{-3}
Effective density of states in the valence band $N_v = 1.83 \times 10^{19}$ cm^{-3}
Temperature $T = 300$ K
Dielectric permittivity $\varepsilon = 1.05 \times 10^{-10}$ F/m
Effective recombination time in the emitter-base depletion region $\tau_{rec} = 1 \times 10^{-8}$ s
Diffusion length of electrons in the base $L_{nb} = 40$ μm.
Diffusion length of holes in the emitter $L_{pe} = 25$ μm.
Electron mobility in the base region $\mu_n = 0.1$ m^2/Vs
Hole mobility in the emitter region $\mu_p = 0.04$ m^2/Vs
Width of the device $t_e = 10$ μm
Neglect Auger recombination.
In addition, for the first problem, neglect the base spreading resistance and base narrowing (use $W_{eff} \cong W$).

3-9-1. Sketch a qualitative band diagram of a bipolar junction transistor for the reverse collector-base voltage equal to the critical voltage of the punch-through breakdown.

3-10-1. Calculate (plot) h parameters for a silicon transistor for a common-emitter configuration as a function of the emitter current, I_e, for 0.1 mA $< I_e <$ 2 mA. You can assume midband conditions for this problem (i.e, you can neglect the capacitances while calculating the h parameters here). The transistor design is shown in the figure:

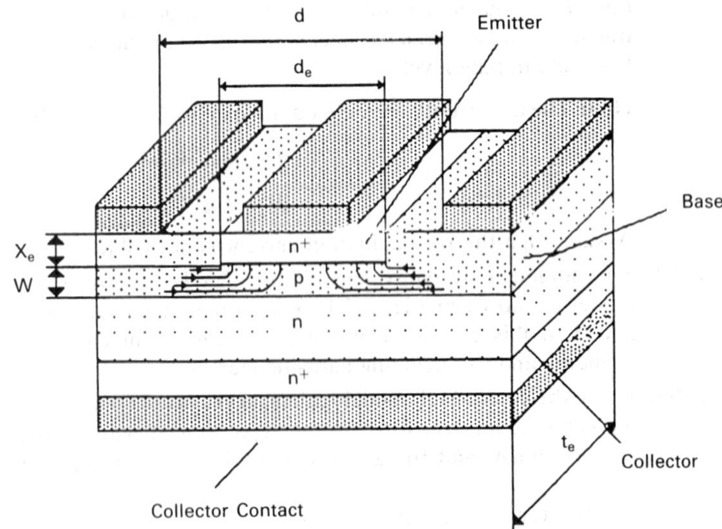

Use the transistor parameters given in Problem 3-7-3. Separation between the base contacts $d = 10$ μm (emitter is centered within d). Assume that the resistance of each base contact is $R_c = 10$ Ω (that includes the resistance of the metal interface and the material directly under the contact only). For the base spreading resistance portion of $r_{bb'}$ use the formula given in the hint for Problem 3-4-3. Early voltage is 100 V ($r_c \cong V_A/I_e$).

Hint: β can be found with the aid of Section 3-3 and in particular eq. (3-3-45) (m_{re} = 2; use $I_c \cong I_e$ in this equation).

3-11-1. Calculate (plot) the cutoff frequency, f_T, as a function of the emitter current, I_e, for 0.1 mA < I_e < 2 mA for the silicon transistor described in Problem 3-10-1. Neglect parasitic interconnect capacitance (C_p). The collector-base capacitance can be assumed to be entirely a depletion capacitance. The depletion capacitance between the base and emitter can be neglected. Assume a bias voltage, collector to base, of 9.4 V. The collector series resistance, including all components, is 5 Ω. The electron saturation velocity, v_{sn}, is 10^5 m/s.

3-12-1. A typical circuit for measuring transistor switching characteristics is shown in Fig. P3-12-1 along with the input voltage waveform. The short-circuit common-emitter gain β = 100. The characteristic recombination time in the base τ_{sr} = 50 ns. The emitter-base saturation voltage V_{besat} = 0.7 V. Calculate the storage time, τ_s, for V_1 = 5 V and V_1 = 10 V.

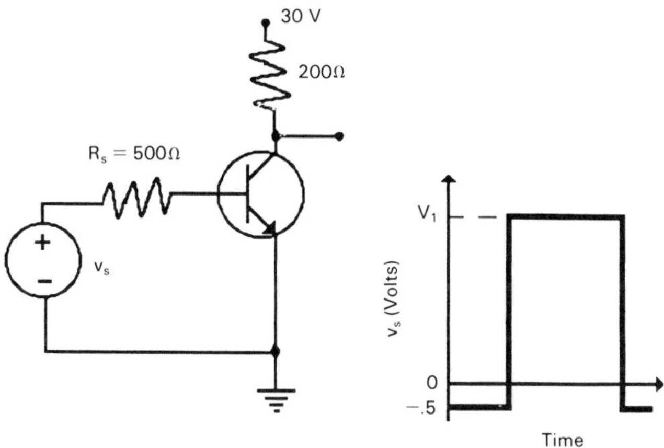

Fig. P3-12-1

3-14-1. In heterojunction bipolar junction transistors the emitter is made from a semiconductor with a larger energy gap. This increases the common-emitter current gain by a factor equal to $\exp(\Delta E_v/k_B T)$, where ΔE_v is the valence band discontinuity, k_B is the Boltzmann constant, and T is temperature. On the other hand, the recombination current in the base region may increase because of the decrease in the effective lifetime, τ_{eff}, which may be crudely estimated as

$$\frac{1}{\tau_{eff}} = \frac{1}{\tau} + R_t v_{th} N_{st}$$

where τ_{eff} is the effective lifetime, τ is the lifetime, v_{th} is the thermal velocity, R_t is the effective radius of a surface state, $v_{th} = \sqrt{(3k_B T/m_n)}$ is the electron thermal velocity, m_n is the electron effective mass, and N_{st} is the density of the surface states. Calculate the common-emitter current gain, β, as a function of N_{st} for an n-p-n AlGaAs/GaAs bipolar junction transistor, assuming that β is limited by the recombination current in the base and by the hole injection current. Hole and electron diffusion coefficients are D_p and D_n, emitter and base doping levels are N_{de} and N_{ab}, and thicknesses of the emitter and base regions are x_e and W.

4

Field-Effect Transistors

4-1. INTRODUCTION

The concept of a Field-Effect Transistor (FET) was first proposed by Lilienfeld (1930). The basic idea of this device is illustrated in Fig. 4-1-1. The device operates as a capacitor with one plate serving as a conducting channel between two ohmic contacts—source and drain contacts. The other plate—the gate—controls the charge induced into the channel. The carriers in the channel come from the source and move across the channel into the drain.

This basic principle has been implemented in a variety of different devices, such as the Metal Oxide Semiconductor Field-Effect Transistor (MOSFET) (or Metal Insulator Semiconductor Field-Effect Transistor—MISFET), the MEtal Semiconductor Field-Effect Transistor (MESFET), the Junction Field-Effect Transistor (JFET), and many others. These devices differ in the gate material, in the location of the gate with respect to the channel, in how the gate is isolated from the channel, and what type of carriers are induced by the gate voltage into the channel (electrons in n-channel devices, holes in p-channel devices, and both electrons and holes in the Double Injection Field-Effect Transistor.)

In a silicon MOSFET (see Fig. 4-1-2a), the gate contact is separated from the channel by a silicon dioxide layer. The excellent quality of the silicon–silicon

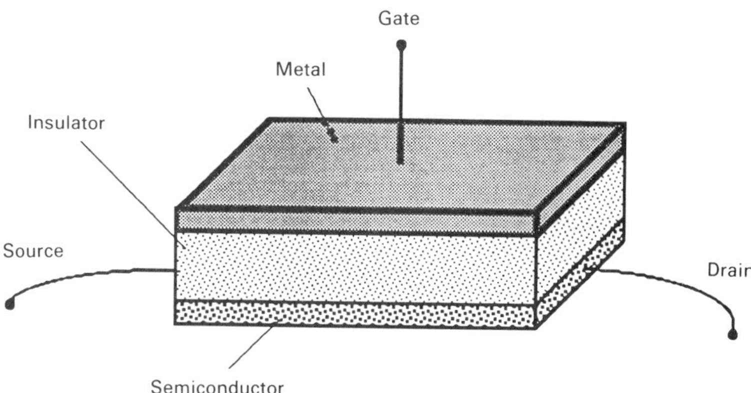

Fig. 4-1-1. The schematic illustration of the concept of a Field-Effect Transistor (FET). The device operates as a capacitor with one plate serving as a conducting channel between the source and drain contacts. The other plate, the gate, controls the charge induced into the channel.

dioxide interface, which has a very low density of interface states, helped the silicon MOSFET become the most important device in modern electronics. The channel in this transistor is created by the electron gas (in n-channel devices) or by the hole gas (in p-channel devices) induced by the gate voltage into the semiconductor at the silicon–insulator interface. In a MESFET (see Fig. 4-1-2b) the channel can be partially or totally depleted by the gate voltage and by the built-in voltage of the Schottky contact forming a gate. The isolation between the gate and channel (formed by the doped semiconductor material) is provided by the Schottky gate depletion layer that is controlled by the gate voltage. The gate voltage changes the width of the depletion region and, hence, the cross section of the conducting channel, modulating the electric current in the channel. In a JFET a p-n junction plays a role similar to the Schottky gate in a MESFET (see Fig. 4-1-2c). Usually, MESFETs or JFETs are fabricated from materials, such as gallium arsenide, that may offer some advantages over silicon but do not possess a stable oxide layer, like the one in silicon.

Most of this chapter is devoted to the silicon MOSFET, which is a workhorse of modern electronic industry. Silicon MOSFETs are used both as discrete devices and as components in monolithic Integrated Circuits (ICs). The minimum device size for such circuits has shrunk to 0.7 micron in commercial IC chips (see Fig. 4-1-3), with a commensurate increase in speed and in the integration scale (to many millions of transistors on a single chip for memory ICs). According to Sze (1988), the cost of a bit in a computer memory chip has dropped to about 6×10^{-3} cent/bit. Sales of MOSFET digital integrated circuits in the United States alone have reached six billion dollars per year in the 1980s and are projected to rise. (The total IC sales in the the United States were 11 billion dollars in 1986 with

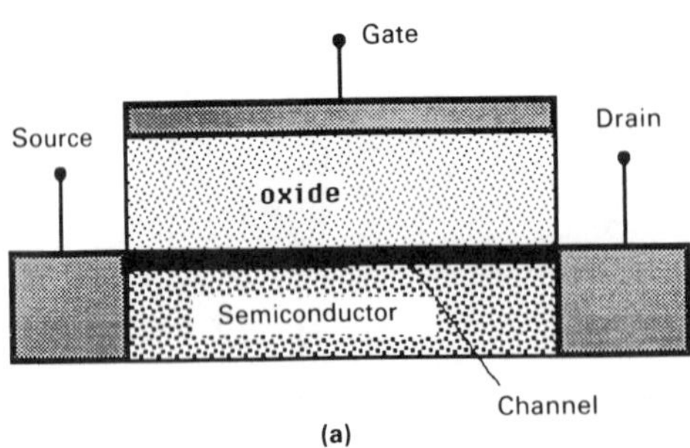

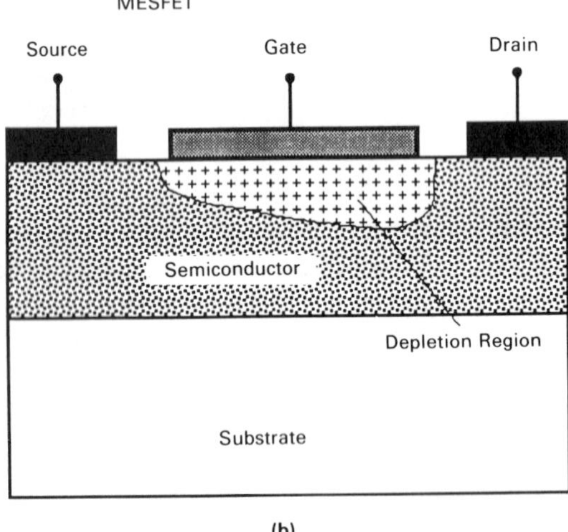

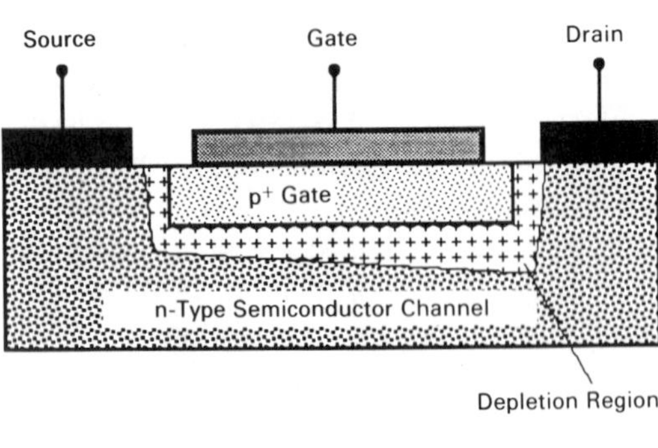

Fig. 4-1-2. Schematic structures of different Field-Effect Transistors (FETs): (a) Metal Oxide Semiconductor Field-Effect Transistor (MOSFET) or Metal Insulator Semiconductor Field-Effect Transistor (MISFET), (b) MEtal Semiconductor Field-Effect Transistor (MESFET), and (c) Junction Field-Effect Transistor (JFET).

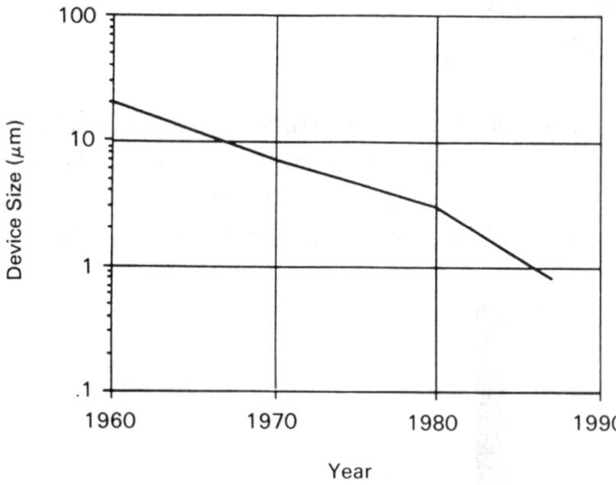

Fig. 4-1-3. Decreasing trend for minimum device size in commercial silicon MOSFET integrated circuits.

approximately 3.5 billion in sales of digital bipolar ICs.) It is expected that IC chips will contain up to 100 million devices per chip by the year 2000, with the cost per bit of memory below 10^{-4} cent.

Complementary MOSFET technology that combines both n-channel and p-channel MOSFETs provides very low power consumption along with high speed. New silicon-on-insulator technology may help achieve three-dimensional integration, packing devices into many layers with a dramatic increase in integration densities. New improved device structures and the combination of bipolar and field-effect technologies may lead to further advances, even beyond what is projected now.

Many books, such as those by Nicolian and Brews (1982), Milnes (1983), Pierret (1983), Schroder (1987), and Tsividis (1987), have been written to deal with MOSFETs because of the great importance of these devices in modern electronics. These books may be used as references for further reading.

In addition to silicon, new materials have emerged. These materials offer advantages for certain applications. In particular, we will consider compound semiconductor field-effect transistors, such as gallium arsenide MESFETs and gallium arsenide–aluminum gallium arsenide heterostructure field-effect transistors. These devices have the highest speed of operation and can be integrated with silicon transistors or with optoelectronic devices based on compound semiconductors. We will also consider amorphous silicon Thin Film Transistors (TFTs) that may find important applications as drivers in large-area low-cost color flat panel displays and high-quality printers, imagers, and copiers, as well as field effect transistors made of wide band semiconductors (silicon carbide and diamond) that offer high power operation at elevated temperatures. However, we do not expect that these new materials will replace silicon. Rather, they will add a new dimension to field effect transistor technology.

4-2. SURFACE CHARGE IN METAL OXIDE SEMICONDUCTOR CAPACITOR

The silicon MOSFET is probably the most important solid-state device for modern electronics. To understand its operation we first consider a Metal Insulator Semiconductor (MIS) capacitor. The schematic MIS capacitor structure and the corresponding band diagram at zero bias voltage are shown in Fig. 4-2-1. Here we

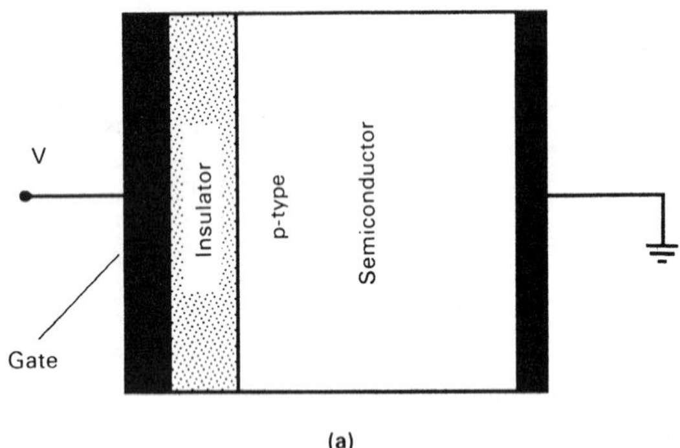

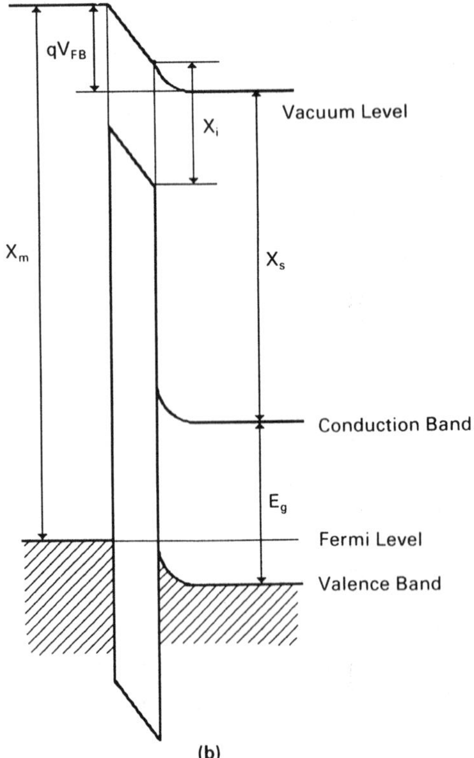

Fig. 4-2-1. (a) Schematic MIS capacitor structure and (b) corresponding band diagram at zero bias voltage. For the chosen value of the flat-band voltage, zero gate bias corresponds to the accumulation of holes at the semiconductor-insulator interface: V_{FB} is a flat band voltage; X_m is the metal work function; X_s is the electron affinity of the semiconductor; X_i is the electron affinity of the insulator; and E_g is the energy gap of the semiconductor.

consider an ideal MIS structure neglecting any charges that can exist in the insulator layer and any possible surface states at the semiconductor–insulator interface. We further assume that the insulator layer has infinite resistivity, so that there is no current across the dielectric layer when the bias voltage is applied.

As can be seen from Fig. 4-2-1, at zero bias voltage the band bending in the semiconductor layer is determined by the difference of the work functions of the metal and semiconductor. This band bending may be compensated by applying the voltage equal to this difference:

$$V_{FB} = X_m - X_s - E_c + E_F \tag{4-2-1}$$

where X_m is the work function of the metal, X_s is electron affinity for the semiconductor, E_c is the conduction band edge, and E_F is the Fermi level. The voltage V_{FB} is called the *flat-band voltage* (see Fig. 4-2-2). Equation (4-2-1) is applicable to an ideal MIS structure. The presence of surface states at the interfaces and/or the presence of fixed charges in the insulator layer will affect the gate voltage required to achieve the flat-band condition.

The situation depicted in Fig. 4-2-1 corresponds to the band bending, such that the Fermi level in the p-type semiconductor at the insulator–semiconductor interface is closer to the top of the valence band. Hence, the concentration of holes near the interface is larger than in the bulk, far from the interface. This corresponds to the *accumulation regime*. When a positive voltage equal to the flat-band voltage, V_{FB}, is applied to the gate, the bands become flat (see Fig. 4-2-2) and the concentration of holes at the insulator–semiconductor interface becomes equal to the equilibrium concentration of holes in this p-type semiconductor.

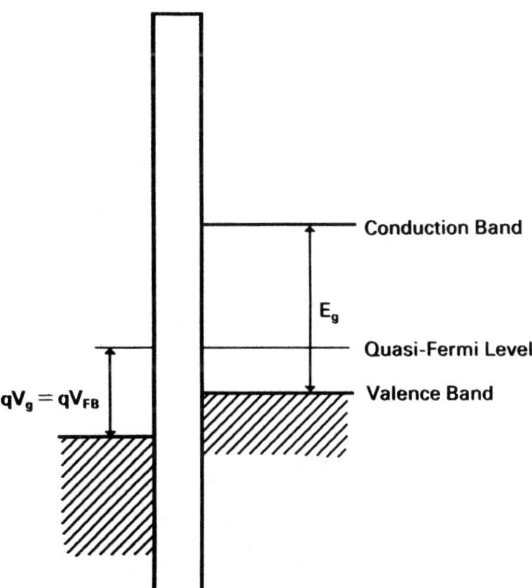

Fig. 4-2-2. Schematic band diagram of MIS capacitor under flat-band conditions. V_g is the gate voltage.

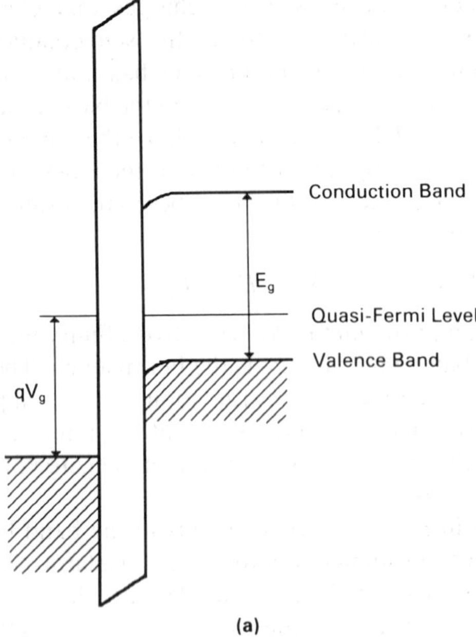

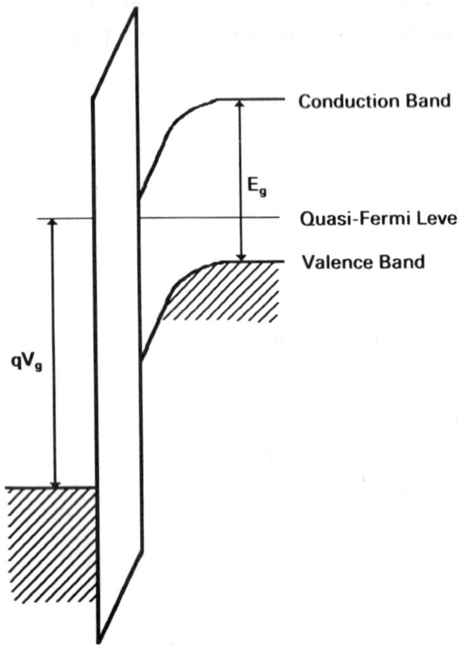

Fig. 4-2-3. Schematic band diagrams for (a) depletion and (b) inversion regimes of the operation of MIS capacitor structure.

When still larger positive bias is applied to the gate, the region close to the insulator–semiconductor interface becomes depleted of holes (the *depletion regime*; see Fig. 4-2-3a). At even larger positive gate voltages the band bending becomes so large that the Fermi level at the insulator–semiconductor interface becomes closer to the bottom of the conduction band than to the top of the valence band (see Fig. 4-2-3b). In this case the concentration of carriers near the interface actually corresponds to that of an n-type semiconductor. This is called *inversion*. The electron concentration in the semiconductor at the semiconductor–insulator interface increases with an increase in the gate voltage. The increase in the band bending leads to an exponential increase in the electron concentration. An increase in the band bending by the thermal voltage, $k_B T/q \approx 25.8$ meV at room temperature, leads to the rise of the volume electron concentration at the surface by a factor of $\exp(1) \approx 2.718$. Hence, even a large increase in the electron concentration—i.e., in the charge induced at the interface—in the inversion regime is accompanied by a relatively small change of the surface potential, V_s. The induced charge is proportional to the gate voltage. Hence, in the inversion regime, the derivative of the surface potential with respect to the gate voltage, dV_s/dV_g, becomes small, compared with the value of this derivative in the depletion regime. In many papers and books the inversion is defined as the *strong inversion* when the difference between the Fermi level and the intrinsic Fermi level at the interface, ϕ_s, becomes equal (and opposite in sign) to this difference in the bulk of the semiconductor, ϕ_b, far from the interface (see Fig. 4-2-4):

$$\varphi_s = -\varphi_b \tag{4-2-2}$$

Here

$$\varphi_b = (k_B T/q)\ln(N_A/n_i) \tag{4-2-2a}$$

where N_A is the shallow acceptor density in the semiconductor, n_i is the intrinsic carrier concentration, and q is the electronic charge. Values of ϕ_s such that $-\phi_b < \phi_s < \phi_b$ correspond to the weak inversion and to the depletion regime, $\phi_s = \phi_b$ corresponds to the flat-band condition, and values of $\varphi_s > \varphi_b$ correspond to the

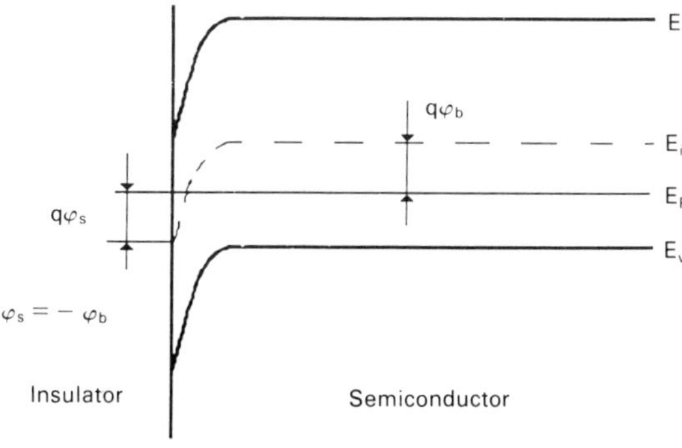

Fig. 4-2-4. Schematic band diagram of MIS capacitor structure at the onset of the strong inversion regime. The inversion is defined as the strong inversion when the difference between the Fermi level at the interface and the intrinsic Fermi level becomes equal (and opposite in sign) to this difference in the bulk of the semiconductor, far from the interface.

accumulation mode. (The surface potential, V_s, counted from the potential of the semiconductor far from the interface is equal to $\varphi_b - \varphi_s$.)

Tsividis (1987) gave a more accurate definition of different inversion regimes. According to his definition, in a weak inversion regime the derivative $|d\phi_s/dV_g| = |dV_s/dV_g|$ is still relatively large, in the strong inversion regime this derivative is relatively small, and eq. (4-2-2) corresponds to the onset of *moderate* inversion. The strong inversion takes place when the band bending is larger than $2\varphi_b$ by several $k_B T/q$. This definition is in agreement with the qualitative discussion of the dependence of dV_s/dV_g on the gate voltage given previously. The range of gate voltages corresponding to the moderate inversion can be as large as 0.5 V or so. Hence, this regime is quite important for modern transistors operating at fairly low power. Nevertheless, for the sake of simplicity, we will not distinguish here between strong and moderate inversion (see Tsividis 1987, p. 55 for a more detailed discussion).

The surface concentrations of electrons and holes are given by

$$n_s = n_{po} \exp[-qV_s/(k_B T)] \tag{4-2-3}$$

$$p_s = p_{po} \exp[qV_s/(k_B T)] \tag{4-2-4}$$

where $p_{po} = N_A$, and N_A is the concentration of shallow acceptors in the p-type semiconductor and $n_{po} = n_i^2/N_A$ is the equilibrium concentration of the minority carriers (electrons) in the bulk. (At the threshold of strong inversion $V_s = -2\varphi_b$). As can be seen from eqs. (4-2-3) and (4-2-4)

$$n_s p_s = n_i^2 \tag{4-2-5}$$

This is the consequence of zero current in the semiconductor (perpendicular to the insulator–semiconductor interface). This corresponds to nonuniform equilibrium and flat (constant as a function of distance) Fermi level in the semiconductor. The electron distribution function in this case is the Fermi-Dirac distribution function with temperature T equal to the device temperature.

The potential distribution in the semiconductor is described by the Poisson equation,

$$d^2V/dx^2 = -\rho(x)/\varepsilon_s \tag{4-2-6}$$

where the space-charge density $\rho(x)$ is given by

$$\rho(x) = q(p - n - N_A) \tag{4-2-7}$$

Here N_A is the concentration of shallow ionized acceptors. Electron and hole concentrations, n and p, may be expressed as

$$n = n_{po} \exp(qV/k_B T) \tag{4-2-8}$$

$$p = p_{po} \exp(-qV/k_B T) \tag{4-2-9}$$

where the potential V is taken to be zero far into the bulk of the semiconductor.

The neutrality condition for the bulk semiconductor region (far from the insulator–semiconductor interface) yields

$$N_A = p_{po} - n_{po} \tag{4-2-10}$$

The substitution of eqs. (4-2-7) to (4-2-10) into eq. (4-2-6) leads to the following equation:

$$d^2V/d^2x = (q/\varepsilon_s)\{n_{po}[\exp(qV/k_BT) - 1] - p_{po}[\exp(-qV/k_BT) - 1]\} \tag{4-2-11}$$

Using the definition of the electric field

$$F = -dV/dx \tag{4-2-12}$$

we can rewrite the left-hand side of eq. (4-2-11) as $(-dF/dx)$. Then dividing this equation by eq. (4-2-12) we find that

$$F(dF/dV) = (q/\varepsilon_s)\{n_{po}[\exp(qV/k_BT) - 1] - p_{po}[\exp(-qV/k_BT) - 1]\} \tag{4-2-13}$$

Integrating this equation with respect to V we obtain

$$F^2 = (2q/\varepsilon_s)\int_0^V \{n_{po}[\exp(qV'/k_BT) - 1] - p_{po}[\exp(-qV'/k_BT) - 1]\}dV' \tag{4-2-14}$$

Here zero potential corresponds to the bulk of the semiconductor far from the insulator–semiconductor interface. Equation (4-2-14) yields

$$F^2 = (2k_BT/\varepsilon_s)\{p_{po}[\exp(-qV/k_BT) + qV/k_BT - 1] + n_{po}[\exp(qV/k_BT) - qV/k_BT - 1]\} \tag{4-2-15}$$

Introducing the Debye radius

$$L_{Dp} = (\varepsilon_s k_BT/q^2N_A)^{1/2} \tag{4-2-16}$$

we can rewrite eq. (4-2-15) as

$$F = (2)^{1/2}\, k_BT/(qL_{Dp})\, f \tag{4-2-17}$$

where

$$f = \pm\{[\exp(-qV/k_BT) + qV/k_BT - 1] + (n_{po}/p_{po})[\exp(qV/k_BT) - qV/k_BT - 1]\}^{1/2} \tag{4-2-18}$$

Here the positive sign should be chosen for a positive potential and the negative sign for a negative potential. In particular, we find for the electric field at the insulator–semiconductor interface that

$$F_s = (2)^{1/2}\, k_BT/(qL_{Dp})\, f(qV_s/k_BT) \tag{4-2-19}$$

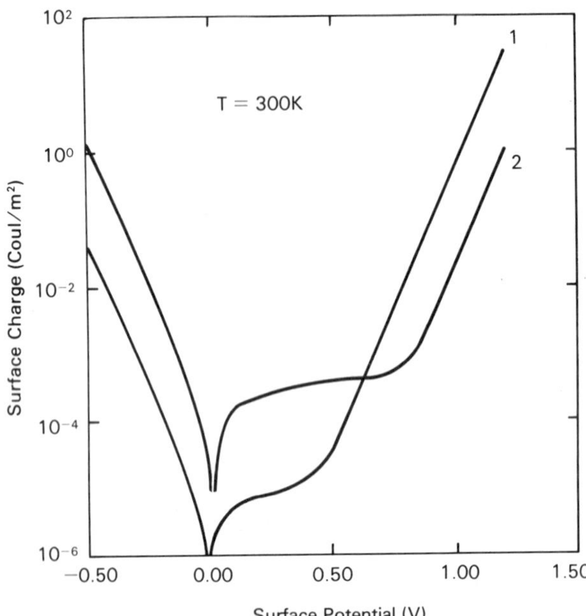

Fig. 4-2-5. Calculated dependence of surface charge on the surface potential for silicon. The curve is calculated using program PLOTF with subroutine PQS.BAS (see Program Description Section). 1–doping level (p-type) $N_A = 10^{13}$ cm^{-3} and 2–the doping level (p-type) $N_A = 10^{16}$ cm^{-3}. Parameters used in the calculation: $T =$ 300 K, flat-band voltage $V_{FB} = 0$ V, energy gap is 1.12 eV, dielectric permittivity $\varepsilon_s = 1.054 \times 10^{-10}$ F/m, effective density of states in the conduction band at room temperature is 3.22×10^{19} cm^{-3}, and effective density of states in the valence band at room temperature is 1.83×10^{19} cm^{-3}.

For a negative surface potential, the surface charge is positive (accumulation mode). The range of surface potential between 0 V and $\approx 2|\varphi_B|$ corresponds to the depletion and weak inversion regime. At the onset of strong inversion the surface potential, $V_s = 2q|\varphi_B| = 2k_BT/q \ln(N_A/n_i)$. At higher surface voltage the absolute value of the surface charge increases proportionate to $\exp(qV_s/2k_BT)$.

where V_s is the value of V at the surface. Using Gauss's law we can relate the total charge in the semiconductor, Q_s, to the surface electric field:

$$|Q_s| = \varepsilon_s|F_s| \tag{4-2-20}$$

The calculated dependence of surface charge on the surface potential for silicon is shown in Fig. 4-2-5. It clearly shows the transition from the accumulation regime through the flat-band condition to the depletion regime, then to the weak inversion regime, to the moderate inversion regime, and, finally, to the strong inversion regime. (As can be shown from the analysis of eqs. (4-2-19) and (4-2-20), in both accumulation and strong inversion regimes, the surface charge is proportional to $\exp[q|V_s|/(2k_BT)]$; see Problem 4-2-9.) The qualitative charge and electric field distributions in a metal insulator semiconductor structure for the accumulation, depletion, flat-band, and inversion regimes are shown in Fig. 4-2-6.

The gate voltage can be related to the surface potential V_s as follows. The electric field at the semiconductor surface is given by eq. (4-2-19). Using the condition of continuity of the electric flux density,

$$\varepsilon_s F_s = \varepsilon_i F_i \tag{4-2-21}$$

(where ε_i is the dielectric permittivity of the insulator layer and F_i is the electric field in the insulator layer), we find the voltage drop $F_i d_i$, where d_i is the insulator thickness, and, finally, the gate voltage:

$$V_g = V_{FB} + V_s + \varepsilon_s F_s / c_i \tag{4-2-22}$$

Here

$$c_i = \varepsilon_i / d_i \tag{4-2-23}$$

is the insulator capacitance per unit area.

In particular, let us calculate the threshold gate voltage, V_T, corresponding to the onset of the strong inversion regime. As can be seen from eq. (4-2-2), at zero substrate potential the onset of the strong inversion occurs when the surface

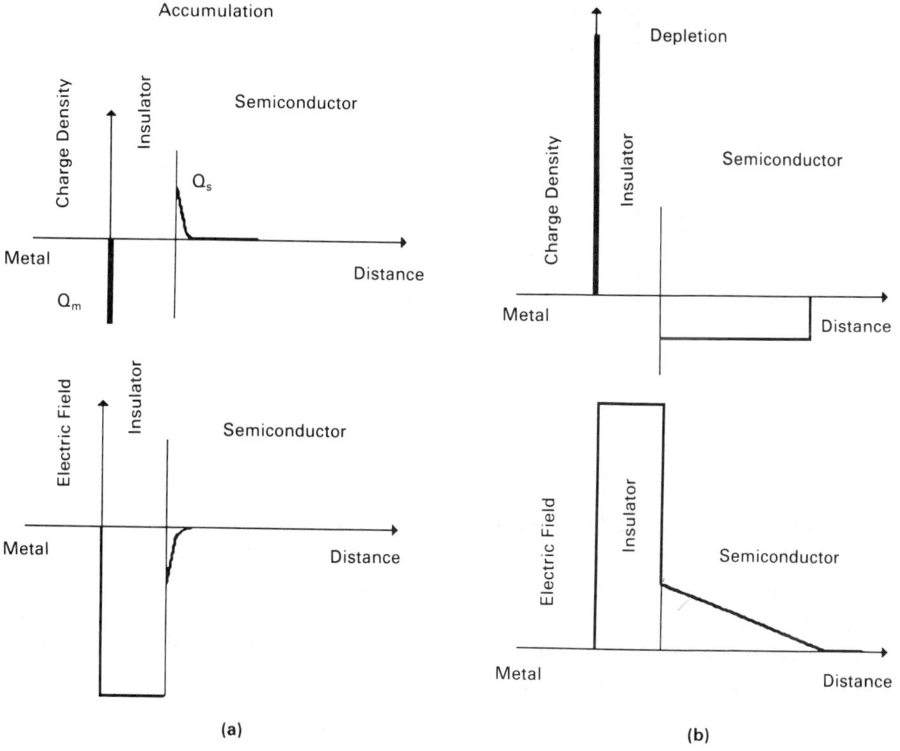

Fig. 4-2-6. Qualitative charge and electric field distributions in a Metal-Insulator-Semiconductor structure for accumulation (a), depletion (b), and inversion (c) regimes. The charge at the metal-insulator interface is practically located at the surface.

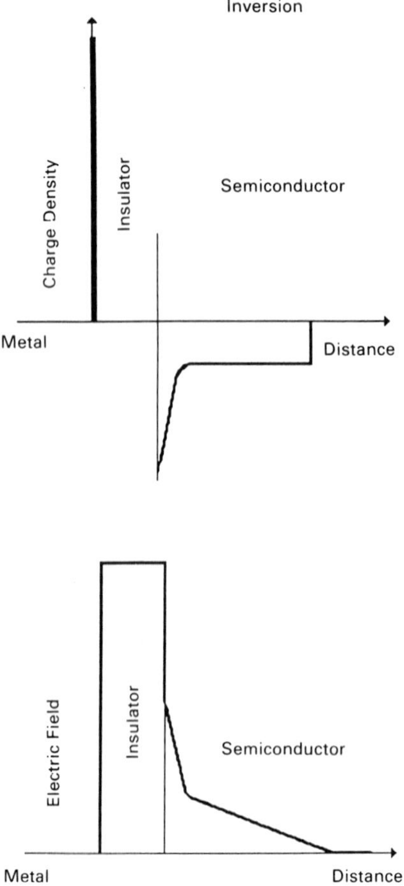

Fig. 4-2-6. ***Cont.***

potential, V_s, measured with respect to the bulk of the semiconductor substrate far from the interface is equal to $2\varphi_b$. (As mentioned previously, this analysis neglects the transitional regime of the moderate inversion; see Tsividis 1987 for a more accurate treatment.) At this surface potential, the charge of the free carriers induced at the insulator–semiconductor interface is still small compared with the charge in the depletion layer, which is given by

$$Q_{dep} = (2\varepsilon_s q N_A V_s)^{1/2} = (4\varepsilon_s q N_A \varphi_b)^{1/2} \tag{4-2-24}$$

This charge causes the electric field at the semiconductor–insulator interface:

$$F_s = (4qN_A\varphi_b/\varepsilon_s)^{1/2} \tag{4-2-25}$$

Using eq. (4-2-22) we find that

$$V_T = V_{FB} + 2\varphi_b + (4qN_A\varphi_b\varepsilon_s)^{1/2}/c_i \tag{4-2-26}$$

where

$$\phi_b = (k_B T/q)\ln(N_A/n_i) \tag{4-2-26a}$$

(see eq. (4-2-2a)). (As was mentioned previously, this threshold voltage corresponds, strictly speaking, to the onset of moderate inversion.)

The threshold voltage, V_T, is one of the most important parameters for metal semiconductor insulator devices. The typical calculated dependence of V_T on temperature, doping level, and dielectric thickness is shown in Fig. 4-2-7.

Let us now discuss the effect of the substrate bias on the threshold voltage. For the MIS structure shown in Fig. 4-2-1a, the application of the substrate bias, V_{sub}, is simply equivalent to changing the gate voltage from V_g to $V_g - V_{sub}$. Hence, at the threshold of the strong inversion V,

$$V_g - V_{sub} = V_T$$

where V_T is given by eq. (4-2-26). However, the situation will be different if the conducting layer of mobile electrons induced at the semiconductor–insulator interface under the conditions of the strong inversion is maintained at some constant potential. This situation can occur in the Metal Oxide Semiconductor Field-

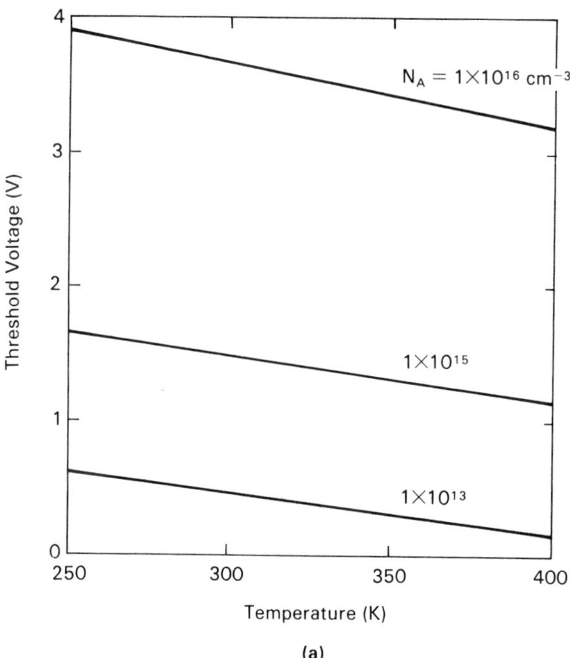

Fig. 4-2-7. Dependences of threshold voltage V_T on (a) temperature, (b) doping level, and (c) insulator thickness. The curves are calculated using program PLOTF with subroutine PVT.BAS (see Program Description Section). Parameters used in the calculation: flat-band voltage $V_{FB} =$ 0 V, energy gap is 1.12 eV, dielectric permittivity of the semiconductor $\varepsilon_s = 1.054 \times 10^{-10}$ F/m, dielectric permittivity of the insulator $\varepsilon_i = 3.454 \times 10^{-11}$ F/m, effective density of states in the conduction band at room temperature is 3.22×10^{19} cm^{-3}, and effective density of states in the valence band at room temperature is 1.83×10^{19} cm^{-3}.

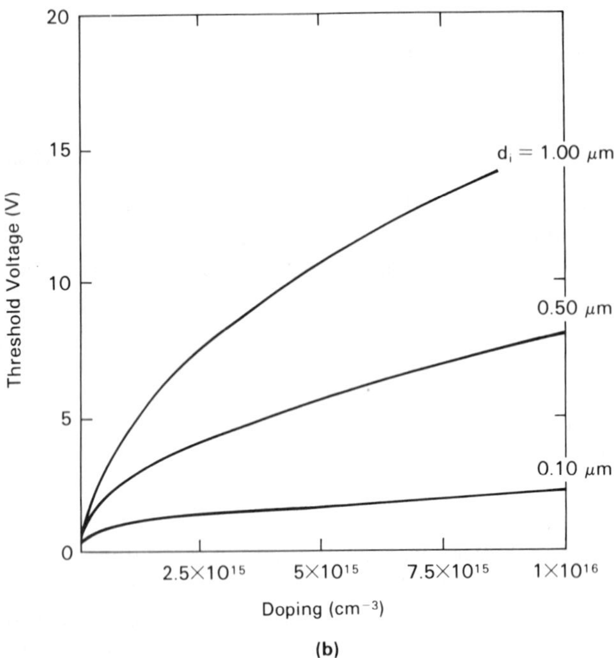

(b)

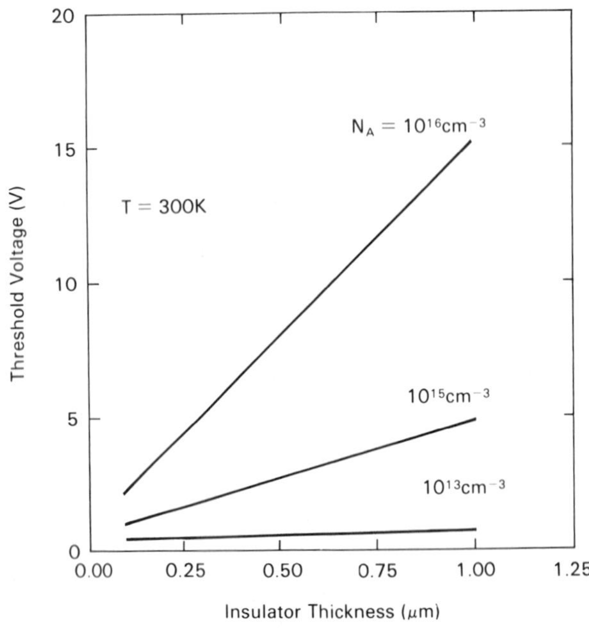

(c)

Fig. 4-2-7. ***Cont.***

Effect Transistor (MOSFET) considered in Section 4-4. Let us assume that the inversion layer is grounded. Then the substrate bias biases the effective inversion n-channel layer–p-type substrate junction, changing the negative charge in the depletion layer. Under such conditions

$$V_T = V_{FB} + 2\varphi_b + [2qN_A(2\varphi_b - V_{sub})\varepsilon_s]^{1/2}/c_i \qquad (4\text{-}2\text{-}26b)$$

This equation is not applicable for two-terminal MIS structures. However, it will be useful in our analysis of MOSFETs.

The threshold voltage may also be affected by *fast surface states* at the insulator–semiconductor interface (see Problem 4-2-2 and Section 4-3) and by fixed charges in the insulator layer. Let us denote the volume density of such charges $N_t(x)$. According to Gauss's law, an incremental charge per unit area $qN_t(x)\,dx$ at distance x from the metal–insulator interface creates an additional electric field $qN_t(x)\,dx/\varepsilon_i$, leading to the additional voltage drop $qN_t(x)x\,dx/\varepsilon_i$. Integrating over the entire insulator layer, we find the threshold voltage shift, ΔV_T, caused by the fixed charges:

$$\Delta V_T = (q/\varepsilon_i)\int_0^{d_i} N_t(x)\,x\,dx \qquad (4\text{-}2\text{-}27)$$

(see Problem 4-2-3). In earlier devices, sodium ions in the silicon dioxide layer created a large positive shift in the threshold voltage, leading to all kinds of problems.

4-3. CAPACITANCE–VOLTAGE CHARACTERISTICS OF MIS STRUCTURE

The capacitance of a semiconductor layer, C_s, for an ideal MIS structure can be calculated as

$$C_s = S\frac{dQ_s}{dV_s} \qquad (4\text{-}3\text{-}1)$$

where Q_s is the total surface charge density in the semiconductor and V_s is the surface potential. Using eqs. (4-2-19) and (4-2-20) and performing this differentiation, we obtain

$$C_s = \frac{C_{so}}{\sqrt{2}}\,\frac{1 - \exp\left(-\dfrac{qV_s}{k_BT}\right) + \dfrac{n_{po}}{p_{po}}\left[\exp\dfrac{qV_s}{k_BT} - 1\right]}{\left|f\left(\dfrac{qV_s}{k_BT}\right)\right|} \qquad (4\text{-}3\text{-}2)$$

Here

$$C_{so} = \varepsilon_s S/L_{Dp} \qquad (4\text{-}3\text{-}3)$$

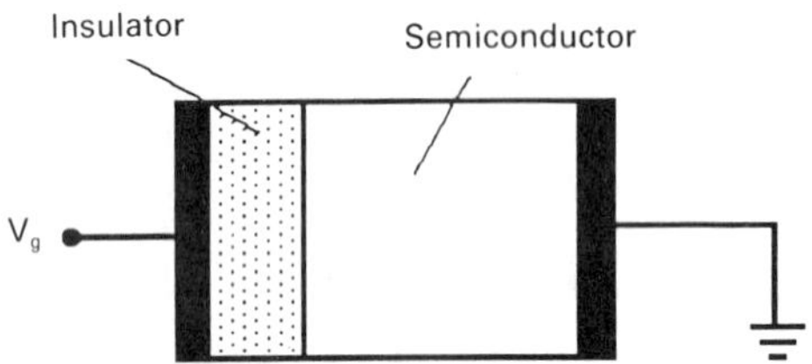

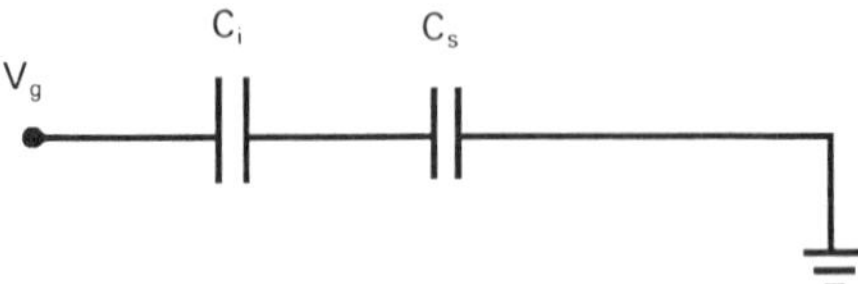

Fig. 4-3-1. Simplified equivalent circuit of MIS structure. This equivalent circuit is only applicable at zero frequency (dc) because the series resistance is neglected.

is the semiconductor capacitance under the flat-band condition (i.e., for $V_s = 0$), and $L_{Dp} = [\varepsilon k_B T/(q^2 N_A)]^{1/2}$ is the Debye radius of the semiconductor.

The total capacitance of an MIS structure, C_{mis}, can now be evaluated as a series combination of C_s and the insulator capacitance, C_i (see Fig. 4-3-1):

$$C_{mis} = \frac{C_i C_s}{C_i + C_s} \tag{4-3-4}$$

where

$$C_i = S\frac{\varepsilon_i}{d_i} \tag{4-3-5}$$

Under equilibrium conditions, the applied gate bias, V_g, is given by

$$V_g = V_i + V_s + V_{FB} \tag{4-3-6}$$

where V_{FB} is the flat-band voltage, and

$$V_i = F_s d_i \frac{\varepsilon_s}{\varepsilon_i}$$

Here F_s is the electric field at the semiconductor interface, ε_i is the dielectric permittivity of the gate insulator, ε_s is the dielectric permittivity of the semiconductor, and V_s is the potential at the semiconductor–insulator interface. The calculated dependence of C_{mis} on applied gate voltage is shown in Fig. 4-3-2 for different values of semiconductor doping.

In the depletion, weak inversion, and strong inversion regimes the semicon-

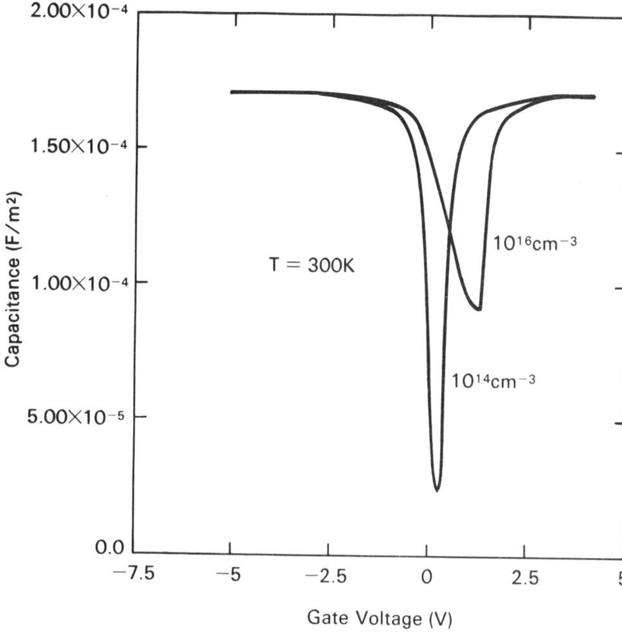

Fig. 4-3-2. *C-V* characteristic for an ideal MIS structure for different values of semiconductor doping. The curves were calculated using program PLOTF with subroutine PCMIS.BAS (see Program Description Section). Parameters used in the calculation are d_i = 300 Å, $\varepsilon_s = 1.05 \times 10^{-12}$ F/m, $\varepsilon_i = 3.45 \times 10^{-12}$ F/m, T = 300 K, $n_i = 1 \times 10^{10}$ cm^{-3}, and $V_{FB} = 0$.

ductor capacitance, C_s, includes two components: the depletion layer capacitance,

$$C_{\text{dep}} = S\frac{\varepsilon_s}{d_{\text{dep}}} \tag{4-3-7}$$

and the capacitance

$$C_{\text{sc}} = C_s - C_{\text{dep}} \tag{4-3-8}$$

related to the free carriers induced in the semiconductor by the gate voltage. Here the depletion width

$$d_{\text{dep}} = [2\varepsilon_s V_s/(qN_A)]^{1/2} \tag{4-3-9}$$

for gate voltages smaller than the threshold voltage, V_T, and

$$d_{\text{dep}} \approx 2[\varepsilon_s \varphi_b/(qN_A)]^{1/2} \tag{4-3-10}$$

for gate voltages larger than the threshold voltage.

The capacitance C_{sc} becomes important in the inversion regime, especially in the regime of strong inversion. In the strong inversion regime, most of the charge is induced at the semiconductor–insulator interface, where the band bending is the largest (see Fig. 4-2-6c). This minority carrier charge comes from electron-hole generation, primarily by the recombination-generation centers in the semi-

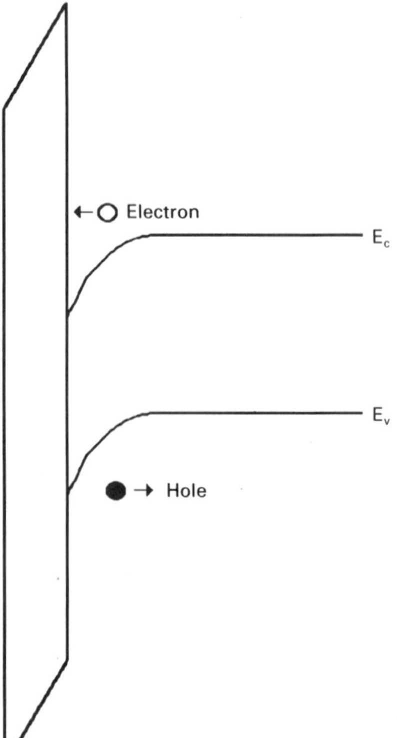

Fig. 4-3-3. Schematic representation of electron-hole separation by the electric field near the semiconductor-insulator interface.

conductor. In principle, the thermal generation of electron-hole pairs across the energy gap can also contribute to the formation of the minority carrier charge in the inversion regime. Once an electron-hole pair is generated, the majority carrier (for example, a hole in p-type material) is removed from the space-charge region by the built-in field that sweeps this carrier out of the depletion region into the substrate. The minority carrier (an electron in p-type material) is swept by the electric field existing in the space-charge region toward the semiconductor–insulator interface. This process is schematically illustrated by Fig. 4-3-3. The build-up of the minority carrier charge at the semiconductor–insulator interface in the inversion regime proceeds at a rate limited by the process of generation of electron-hole pairs. This is reflected by the equivalent circuit of an MIS structure shown in Fig. 4-3-4. This circuit includes resistance R_s, which is the series resistance of the semiconductor layer. For a p-type layer

$$R_s = \frac{L - d_{dep}}{q\mu_p N_A S} \tag{4-3-11}$$

where μ_p is the hole mobility and L is the thickness of the semiconductor layer.

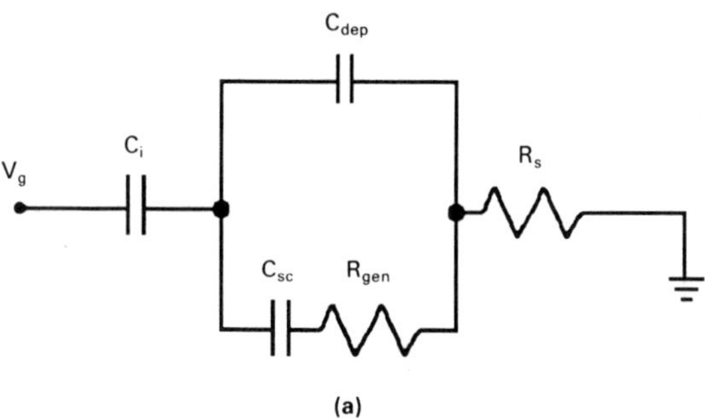

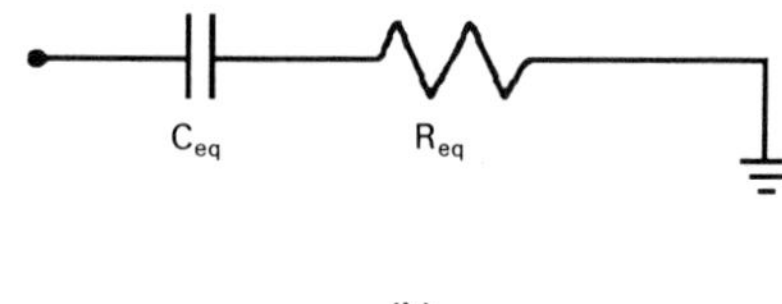

Fig. 4-3-4. Equivalent circuit of MIS structure. R_s is the series resistance of the semiconductor layer. (a) Detailed equivalent circuit. (b) *RC* equivalent circuit.

The differential resistance R_{gen} can be estimated as

$$R_{gen} = \frac{dV_s}{dI_{gen}} \tag{4-3-12}$$

The generation current, I_{gen}, is approximately proportional to the volume of the depletion region:

$$I_{gen} = \frac{qn_i d_{dep} S}{\tau_{gen}} \tag{4-3-13}$$

where τ_{gen} is an effective generation time constant. Using eqs. (4-3-9) and (4-3-13) we find (for gate voltages smaller than the threshold voltage, V_T) that

$$R_{gen} = \sqrt{\frac{2N_A V_s}{\varepsilon_s q}} \frac{\tau_{gen}}{n_i S} \tag{4-3-14}$$

The small-signal impedance of the equivalent circuit, shown in Fig. 4-3-4, can be presented as a series combination of a frequency-dependent equivalent capacitor, C_{eq}, and a frequency-dependent equivalent resistor, R_{eq}, where

$$C_{eq} = \frac{C_{in} C_{eff}}{C_{in} + C_{eff'}} \tag{4-3-15}$$

$$C_{eff} = \frac{(C_{sc} + C_{dep})^2 + \omega^2 C_{dep}^2 C_{sc}^2 R_{gen}^2}{C_{sc} + C_{dep} + \omega^2 C_{dep} C_{sc}^2 R_{gen}^2} \tag{4-3-16}$$

and

$$R_{eq} = R_s + \frac{C_{sc}^2 R_{gen}}{(C_{sc} + C_{dep})^2 + \omega^2 C_{dep}^2 C_{sc}^2 R_{gen}^2} \tag{4-3-17}$$

As can be seen from these equations, in a limiting case of $\omega \Rightarrow 0$,

$$C_{eff} = C_{sc} + C_{dep} \tag{4-3-18}$$

In a limiting case of $\omega \Rightarrow \infty$

$$C_{eff} = C_{dep} \tag{4-3-19}$$

as may be expected from the equivalent circuit shown in Fig. 4-3-4.

The calculated dependence of C_{eff} on the gate voltage V_g is shown in Fig. 4-3-5a. For comparison, we show the experimental curves in Fig. 4-3-5b (after Grove 1961). Measured C-V characteristics allow us to determine important parameters of an MIS structure, including the built-in voltage, the gate dielectric thickness, and the doping of the semiconductor substrate. The maximum measured capacitance, C_{max}, (capacitance C_o in Fig. 4-3-6) yields the dielectric thickness

$$d_i = \varepsilon_i S / C_{max} \tag{4-3-20}$$

The minimum measured capacitance (at high frequency) allows us to find the doping concentration in the semiconductor substrate. First we determine the de-

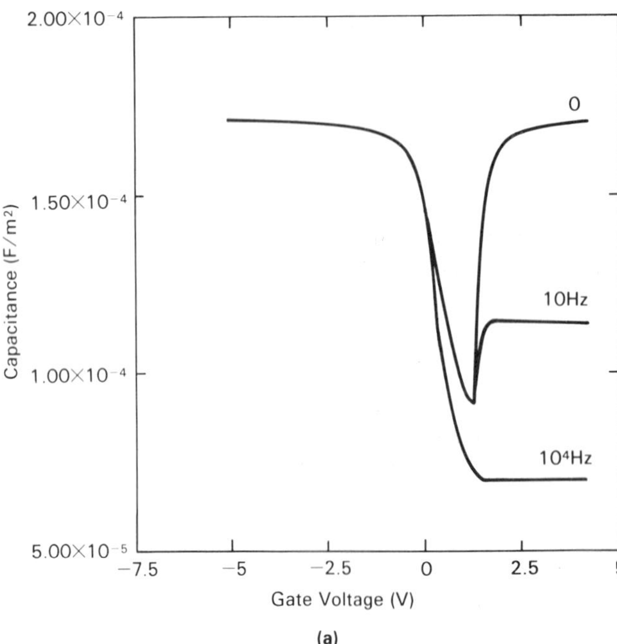

Fig. 4-3-5a. Calculated dependence of gate capacitance on the gate voltage at different frequencies. The curves were calculated using program PLOTF with subroutine PMISF.BAS (see Program Description Section). Parameters used in the calculation are $d_i = 300$ Å, $\varepsilon_s = 1.05 \times 10^{-12}$ F/m, $\varepsilon_i = 3.45 \times 10^{-12}$ F/m, $T = 300$ K, $n_i = 1 \times 10^{10}$ cm^{-3}, $\tau_{gen} = 10^{-8}$ s, and $V_{FB} = 0$.

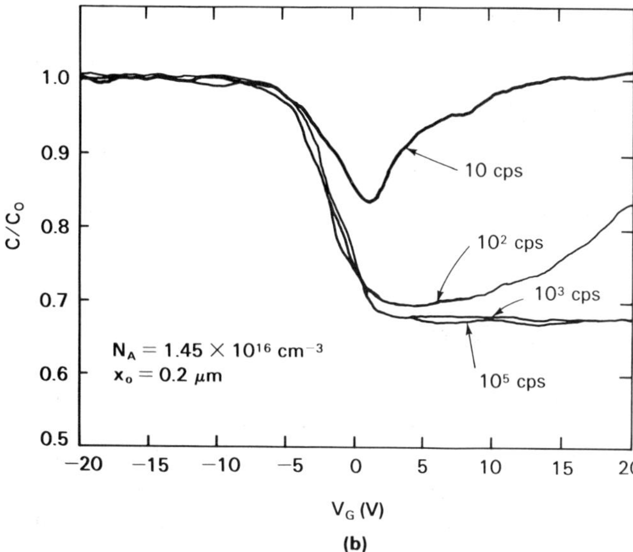

Fig. 4-3-5b. Measured dependences of gate capacitance on the gate voltage at different frequencies (after B. E. Deal, M. Sclar, A. S. Grove, and E. H. Snow, *J. Electrochem. Society*, 114, March (1967)).

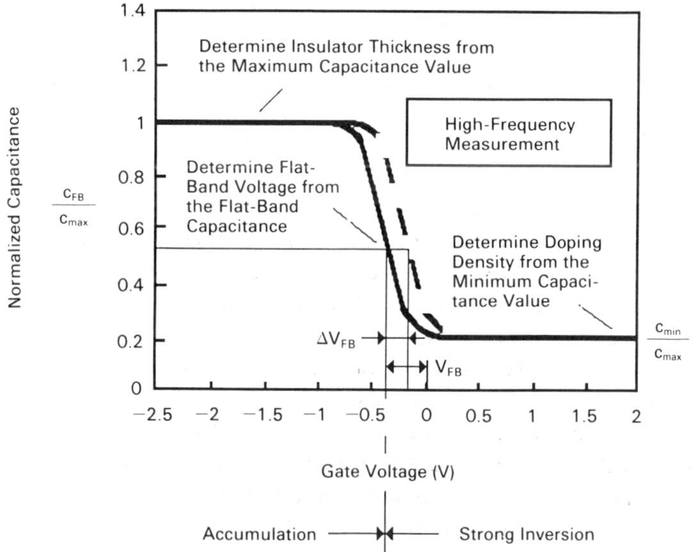

Fig. 4-3-6. Determination of parameters of MIS structure from *C-V* characteristics. Dashed line shows the parallel shift of the *C-V* curve during the high bias-temperature stress test (see text).

pletion capacitance in the strong inversion regime:

$$1/C_{dep} = 1/C_{min} - 1/C_{max} \tag{4-3-21}$$

Then we find the thickness of the depletion region,

$$d_{dep} = \varepsilon_s S/C_{dep} \tag{4-3-22}$$

and finally calculate the doping concentration, N_A, using eqs. (4-3-10) and (4-2-26a):

$$d_{dep} \approx 2[\varepsilon_s \varphi_b/(qN_A)]^{1/2} \tag{4-3-23}$$

$$\phi_b = (k_B T/q)\ln(N_A/n_i) \tag{4-3-24}$$

These two equations can be solved by iteration. For example, we can assume $\varphi_b = 0.2$ eV as our first guess, find N_A from eq. (4-3-23) using this value of ϕ_b, then find a more accurate value of φ_b from eq. (4-3-24), find N_A from eq. (4-3-23) using this new value of ϕ_b, and so on, until a desired accuracy has been reached. Using eqs. (4-3-3) and (4-3-4) we find the device capacitance, C_{FB}, under the flat-band conditions:

$$C_{FB} = \frac{C_i C_{so}}{C_i + C_{so}} = \varepsilon_i/(\varepsilon_s d_i + \varepsilon_i L_{Dp}) \tag{4-3-25}$$

Once d_i and N_A have been determined, eq. (4-3-25) yields the value of $C_{FB}/C_{max} = C_{FB}/C_i$. The value of the gate voltage corresponding to this value of C_{FB}/C_o is equal to the flat-band voltage. This characterization process is illustrated by Fig. 4-3-6.

Strictly speaking, this characterization technique applies to an ideal MIS structure. The capacitance-voltage characteristics of MIS structures may be strongly affected by mobile ions present in the gate dielectric layer and by the surface states at the dielectric–semiconductor interface. Even for the most studied silicon-silicon dioxide interface we do not have a complete understanding of the nature of these states (see Sze 1981 for a detailed discussion). Their density depends very strongly on the oxide-growth procedure and may vary from as high as 10^{15} cm^{-2} (which approximately corresponds to the total number of surface atoms) to as low as 10^{10} cm^{-2}. Just as with surface states introduced in metal-semiconductor structures (see Section 2-9), the surface states at the insulator–semiconductor interface may be acceptorlike or donorlike (depending on their charge states when occupied by an electron and when empty). When the surface potential changes, the charge in the surface states changes as well, leading to the shift of the threshold voltage for the onset of strong inversion (see Problem 4-2-2) and to the change in the C-V characteristics (see Fig. 4-3-7). In an equivalent circuit of an MIS structure, surface states may be represented by an additional

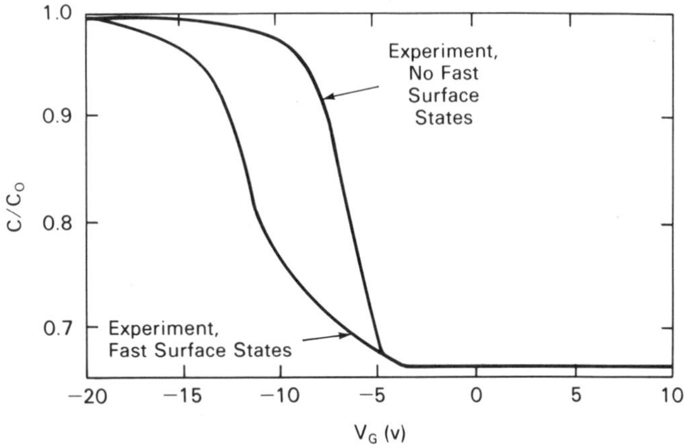

Fig. 4-3-7. Effect of the surface states on *C-V* characteristics of MIS structure (after B. E. Deal, M. Sclar, A. S. Grove, and E. H. Snow, *J. Electrochem. Society,* 114, March (1967)).

series combination of an equivalent capacitance of the surface states, C_{ss}, and additional resistance, R_{ss} (see Fig. 4-3-8). Time constant $C_{ss}R_{ss}$ represents the time response of the surface states. Measurements of the frequency-dependent MIS capacitance and MIS conductance can be interpreted using this equivalent circuit. Such studies yield information about the density of the surface states (see Fig. 4-3-9). As can be seen from this figure, the density of the surface states vs. energy for silicon–silicon dioxide interfaces is strongly dependent on the crystallographic orientation of the interface plane. A smaller density of states for ⟨100⟩ orientation is correlated with the smaller density of the available bonds on the

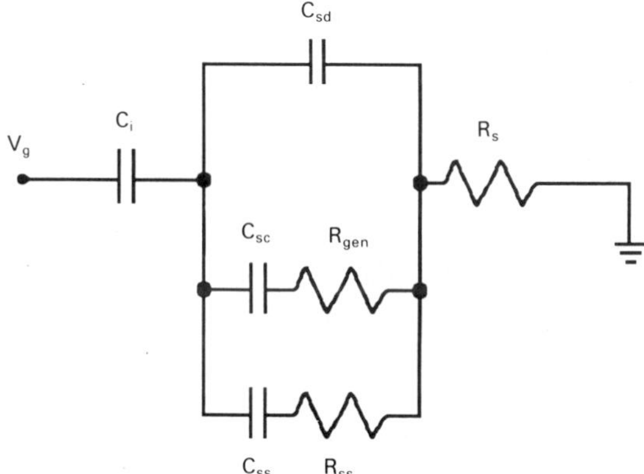

Fig. 4-3-8. Equivalent circuit of an MIS structure accounting for surface states.

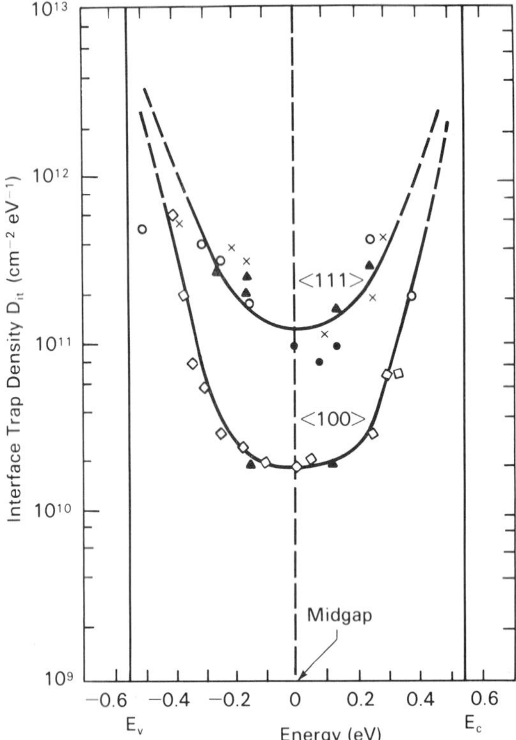

Fig. 4-3-9. Measured density of the surface states for silicon-silicon dioxide interface (after M. H. White and J. R. Cricchi, *IEEE Trans. Electron Devices*, 19, p. 1280 (1972). © 1972 IEEE).

silicon surface. Therefore, silicon MOSFETs are fabricated on ⟨100⟩ substrates (see Sze 1981). Figure 4-3-9 also shows an approximately exponential increase in the density of the surface states in the energy ranges close to the band edges. A more detailed discussion of the interface states for silicon–silicon dioxide interfaces and fixed charges in the silicon dioxide layer is given by Nicolian and Brews (1982).

The effect of the mobile ions in the gate dielectric layer is usually determined using a *high bias-temperature stress test*. During such a test, the applied gate voltage corresponds to the gate electric field on the order of 10^6 V/cm, and the temperature is chosen between 175 °C and 300 °C. The C-V characteristics are measured before and after the test (the application of high gate bias and elevated temperatures usually lasts for approximately 3 minutes). The observed parallel shift of the C-V curves corresponds to the change, ΔV_{FB}, of the flat-band voltage (see Fig. 4-3-6). For a good-quality oxide this change should be less than a few millivolts.

4-4. METAL OXIDE SEMICONDUCTOR FIELD–EFFECT TRANSISTORS (MOSFET): GRADUAL CHANNEL APPROXIMATION AND CHARGE CONTROL MODEL

4-4-1. Principle of Operation

A schematic diagram of a MOSFET is shown in Fig. 4-4-1. Two n^+ regions are diffused or implanted into a p-type silicon substrate in order to form two ohmic contacts called the *source* and the *drain*. A silicon dioxide layer separates the third contact—the gate—from the device channel induced at the interface region in the silicon under the gate.

When a positive voltage is applied to the gate, a negative charge is capacitively induced in the channel, providing a conducting link between the drain and the source. This turns the transistor on. The depletion regions between the p-type substrate and n^+ regions and n channel provide isolation from other devices fabricated on the same substrate. The variation of the gate voltage changes the electron concentration in the channel and, hence, the channel conductance and the device current. When the drain-to-source bias is applied, charge carriers move from the source to the drain across the channel. The channel conductance and, hence, the drain-to-source current, I_D, are modulated by the gate voltage.

When the source and drain contacts are connected, the device is similar to an MIS capacitor with one important difference: highly doped contact regions provide reservoirs from which carriers can enter the channel or to which they can escape from the channel. Hence, up to fairly high frequencies (comparable to the inverse transit time of carriers across the channel), C-V characteristics of the

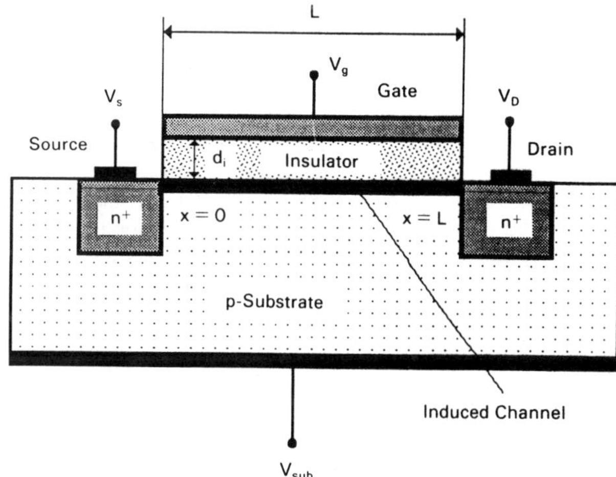

Fig. 4-4-1. Schematic diagram of a MOSFET. Two n^+ regions are diffused or implanted into a p-type silicon substrate in order to form source and drain contracts. The depletion regions between the n^+ regions and n channel and the p-type substrate provides an isolation from the substrate.

device (with the source, drain, and substrate connected) at voltages above threshold voltage look like the C-V characteristics of an MIS structure at very low frequencies (see Fig. 4-3-5).

4-4-2. Gradual Channel Approximation and Constant Mobility Model

To calculate the MOSFET current-voltage characteristics we will use a *gradual channel approximation* proposed by William Shockley. This approximation is based on the assumption that the electric charge density related to the variation of the electric field in the channel in the direction parallel to the semiconductor–insulator interface is much smaller than the electric charge density related to variation of the electric field in the direction perpendicular to the semiconductor–insulator interface, i.e., $|\partial F_x/\partial x| << |\partial F_y/\partial y|$. Hence, the channel potential is assumed to be a "gradually" changing function of position (see Fig. 4-4-2), varying very little over the distances on the order of the insulator thickness, d_i. (This certainly requires that $d_i << L$, where L is the gate length.) According to the gradual channel approximation, we can assume that the charge induced at a given position in the channel may be determined using the formulas derived for an MIS structure if we substitute for the constant surface potential V_s a channel potential

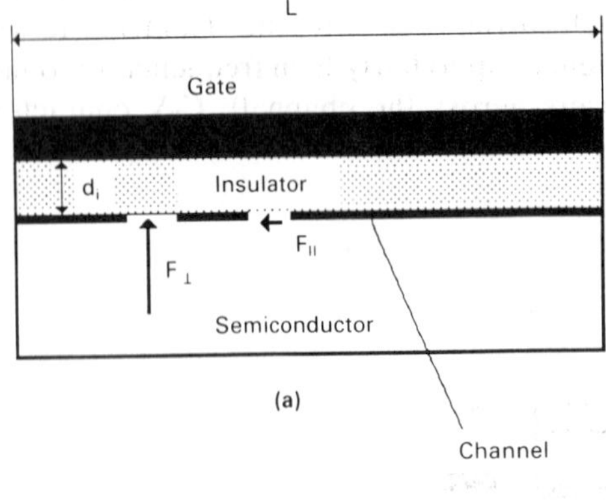

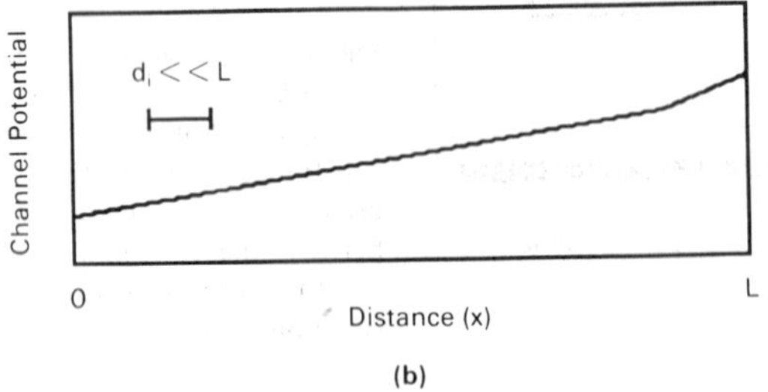

Fig. 4-4-2. Gradual channel approximation: (a) Schematic comparison of the perpendicular and parallel components of the electric field in the channel and (b) Qualitative potential profile in the channel. This approximation is based on the assumption that the electric field in the channel in the direction parallel to the semiconductor-insulator interface, $F_\parallel$, is much smaller than the electric field in the direction perpendicular to the semiconductor-insulator interface, $F_\perp$, and the channel potential is a gradually changing function of position, varying little over the distances of the order of the dioxide thickness, d_i.

$V_c(x)$ in the expression for total surface charge density, Q_s, induced into the semiconductor layer. Here we consider a situation in which the device operates in the above-threshold regime, i.e., the gate voltage is large enough to cause a strong inversion everywhere in the channel. The induced surface charge density, Q_s, is then given by

$$Q_s = -c_i[V_g - 2\varphi_b - V_{FB} - V_c(x)] \tag{4-4-1}$$

where c_i is the insulator capacitance per unit area, and the voltage difference in the brackets is the voltage drop across the insulator layer. (Here we are considering n-channel devices. However, all the results may be applied to p-channel devices with corresponding changes in signs.) We assume that the source is grounded ($V_c(0) = 0$) and that the drain potential is V_D ($V_c(L) = V_D$). For now, we also assume that the semiconductor substrate is connected to the source, i.e., that $V_{sub} = 0$.

The induced density of free electrons, n_{ss}, at each point of the channel can be determined by deducting the surface charge of acceptors in the depletion layer, Q_{dep}, from the total induced charge, Q_m,

$$n_{ss} = c_i[V_g - 2\varphi_b - V_{FB} - V_c(x)]/q - |Q_{dep}|/q \tag{4-4-2}$$

At the source side of the gate where $V_c = 0$ the surface depletion charge density is given by

$$Q_{dep}(x = 0) = -qN_A d_{dep} = -(4\varepsilon_s qN_A\varphi_b)^{1/2} \tag{4-4-3}$$

(see eq. (4-3-10).) However, elsewhere in the channel the total band bending between the bulk of the semiconductor substrate and the surface is $2\varphi_b + V_c$ because the induced n-channel–p-substrate junction is reverse biased by voltage $V_c(x)$. The band diagrams at the source side of the channel and for the drain side of the channel for the direction perpendicular to the semiconductor–insulator interface are compared in Fig. 4-4-3. Figure 4-4-4 shows a qualitative two-dimensional plot of the conduction band edge. These figures clearly illustrate the increase of the band bending in the channel in the direction from the source to the drain. This leads to the increase of the depletion layer and of the depletion charge density,

$$Q_{dep}(x) = -qN_A d_{dep} = -\{2\varepsilon_s qN_A[V_c(x) + 2\varphi_b]\}^{1/2} \tag{4-4-4}$$

as the channel potential provides an additional reverse bias for the induced n-channel–p-substrate junction.

The drain current, I_D, is given by

$$I_D = q\mu_n \frac{dV_c}{dx} n_{ss} W \tag{4-4-5}$$

where μ_n is the low field electron mobility and W is the device width. Here we assume that the electron velocity, v_n, is proportional to the component of the electric field, F, in the channel parallel to the semiconductor–insulator interface,

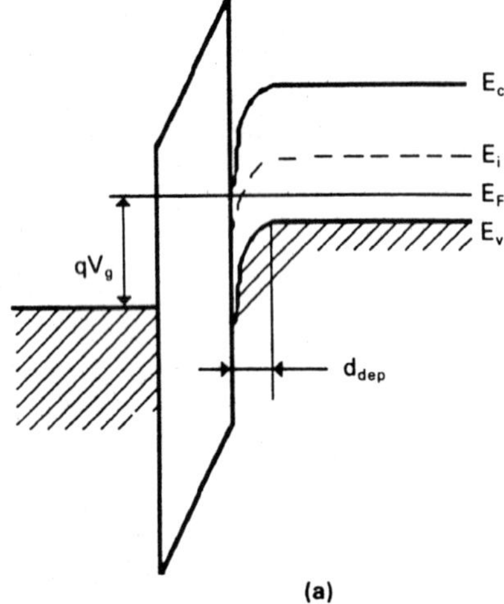

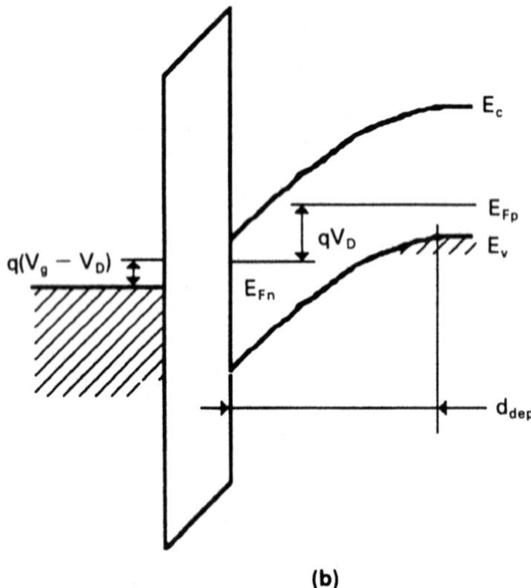

Fig. 4-4-3. Band diagrams at the (a) source and (b) drain sides of the channel for the direction perpendicular to the semiconductor-insulator interface.

$$v_n = \mu_n |F| \tag{4-4-6}$$

In fact, the velocity saturation in a high electric field may play an important role, as will be discussed in Section 4-5.

Equation (4-4-5) can be rewritten as

$$dx = \frac{\mu_n W q n_{ss}(V_c)}{I_D} \, dV_c \tag{4-4-7}$$

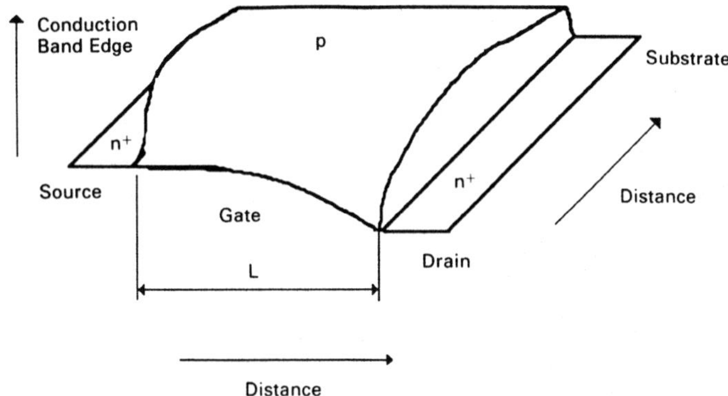

Fig. 4-4-4. Qualitative two dimensional plot of the conduction band edge for an n-channel MOSFET.

Integrating eq. (4-4-7) from zero to the channel length L (this corresponds to the change in V_c from zero at the source side of the gate to V_D at the drain side of the gate), we obtain the following expression for the current-voltage characteristics:

$$I_D = \mu_n(W/L)c_i\left\{\left(V_g - V_{FB} - 2\varphi_b - \frac{V_D}{2}\right)V_D - (2/3)[(2\varepsilon_s q N_A)^{1/2}/c_i][(V_D + 2\varphi_b)^{3/2} - (2\varphi_b)^{3/2}]\right\} \quad (4\text{-}4\text{-}8)$$

This equation is only valid for values of V_D for which the inversion layer still exists even at the drain side of the gate, i.e.,

$$n_{ss}(V_D) \geq 0 \quad (4\text{-}4\text{-}9)$$

The condition $n_{ss} = 0$ is called the *pinch-off condition.* As can be seen from eqs. (4-4-2) and (4-4-3), the pinch-off occurs at the drain side of the gate when

$$V_D = V_{Dsat} = V_g - 2\varphi_b - V_{FB} + (\varepsilon_s q N_A/c_i^2)\{1 - [1 + 2(V_g - V_{FB})c_i^2/(\varepsilon_s q N_A)]^{1/2}\} \quad (4\text{-}4\text{-}10)$$

When $V_g \to V_T$, $V_{Dsat} \to V_g - V_T$, where V_T is the threshold voltage of the onset of the strong inversion:

$$V_T = V_{FB} + 2\varphi_b + \frac{2\sqrt{\varepsilon_s q N_A \varphi_b}}{c_i} \quad (4\text{-}4\text{-}11)$$

(see eq. (4-2-26)). (A more detailed analysis should distinguish between strong and moderate inversion; see comments in Section 4-2 and the book by Tsividis [1987]).

Most MOSFETs have a fourth contact attached to the substrate (see Fig. 4-4-1). Equation (4-4-11) can be modified to account for the substrate bias:

$$V_T = V_{FB} + 2\varphi_b + [2\varepsilon_s q N_A(2\varphi_b - V_{sub})]^{1/2}/c_i \quad (4\text{-}4\text{-}12)$$

Here V_{sub} is the voltage difference between the inversion layer at the semiconductor–insulator interface and the substrate contact. This expression is only valid for a negative or slightly positive substrate bias,

$$V_{sub} < 2\varphi_b$$

when an induced inversion layer–p-substrate junction is either reverse biased or slightly forward biased. A larger positive substrate voltage will forward bias the induced inversion layer–p-substrate junction (as well as n^+ contacts–p-substrate junctions), leading to large leakage currents.

The band diagram of the MOSFET at the source side of the channel for the direction perpendicular to the semiconductor–insulator interface for the case when a substrate bias is applied is shown in Fig. 4-4-5. The calculated dependencies of the threshold voltage on substrate potential are shown in Fig. 4-4-6.

What happens when the drain-to-source voltage approaches V_{Dsat} may be understood by analyzing the electric field distributions under the gate. Integrating eq. (4-4-7) from 0 to x we obtain

$$x = \mu_n W_q \int_0^{V_c(x)} n_{ss}(V')\, dV'/I_D \tag{4-4-13}$$

Using eqs. (4-4-2) and (4-4-4) and performing the integration we find that

$$x = \mu_n W c_i \left\{ \left[V_g - V_{FB} - 2\varphi_b - \frac{V_c(x)}{2} \right] V_c(x) - \frac{2}{3} \frac{\sqrt{2\varepsilon_s q N_A}}{c_i} \left[(V_c(x) + 2\varphi_b)^{3/2} - (2\varphi_b^{3/2}) \right] \right\} \Big/ I_D \tag{4-4-14}$$

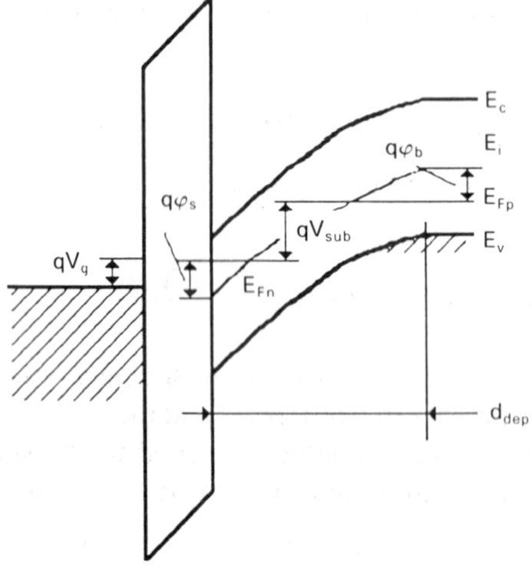

Fig. 4-4-5. Band diagram of an n-channel MOSFET for the direction perpendicular to the semiconductor-insulator interface for negative substrate bias.

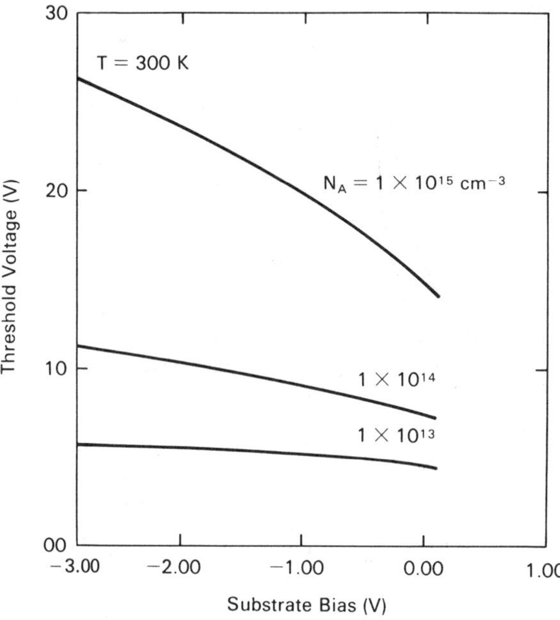

Fig. 4-4-6. Calculated dependences of the threshold voltage on substrate potential for different doping densities of the semiconductor substrate (in cm^{-3}). Negative substrate potential repels electrons in the channel increasing the threshold voltage. The curves were calculated using program PLOTF with subroutine PVTMIS.BAS (see program description section). Parameters used in the calculation are temperature T = 300 K, energy gap E_g = 1.12 eV, effective densities of states in the conduction band and valence band at 300 K are 3.22 × 10^{19} cm^{-3} and 1.83 × 10^{19} cm^{-3}, dielectric permittivity of silicon $\varepsilon_s = 1.05 \times 10^{-10}$ F/m, flat band voltage $V_{FB} = 0$, gate insulator thickness $d_i = 0.2$ μm, and gate insulator dielectric permittivity $\varepsilon_i = 3.45 \times 10^{-11}$ F/m.

The electric field, F, in the channel in the direction parallel to the semiconductor–insulator interface,

$$F = -dV_c/dx \tag{4-4-15}$$

can be found from eq. (4-4-5):

$$|F| = \frac{I_D}{q\mu_n n_{ss}(V_c)W} \tag{4-4-16}$$

Solving eqs. (4-4-14) and (4-4-16) together, we can calculate the field profiles (see Fig. 4-4-7).

As can be seen from this figure and as can be understood from the requirement of the constancy of the current throughout the device, the electric field at the drain side of the gate increases, tending to infinity when $V_D \Rightarrow V_{dsat}$ and $n_{ss}(L) \Rightarrow 0$. The differential drain conductance,

$$g_{ds} = \left.\frac{dI_D}{dV_D}\right|_{V_g = \text{constant}} \tag{4-4-17}$$

tends to zero when $V_D \Rightarrow V_{Dsat}$, and the current-voltage characteristics may be extrapolated in the voltage region $V_D > V_{Dsat}$, assuming a constant (independent of the drain-to-source voltage) drain current in this range of drain-to-source voltages

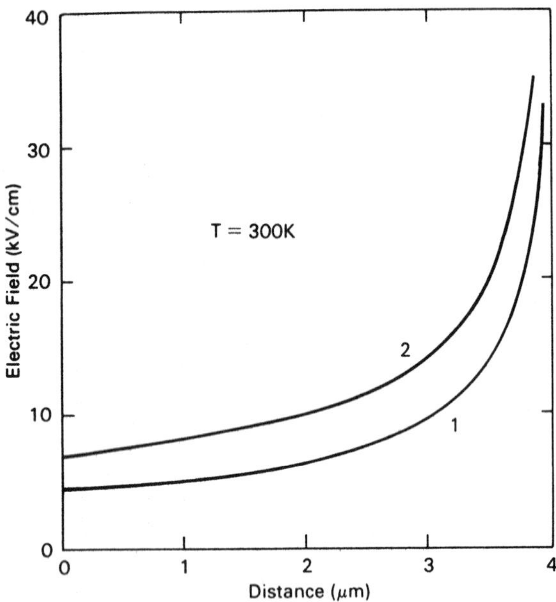

Fig. 4-4-7. Electric field in the channel of a MOSFET vs. the distance at the drain voltage that is nearly equal to the saturation voltage. For curve 1 V_g = 5 V and for curve 2: V_g = 7 V. The curves were calculated using program PLOTF with subroutine PFMIS.BAS (see program description section). Parameters used in the calculation are temperature T = 300 K, substrate doping $N_A = 1 \times 10^{16}$ cm^{-3}, energy gap E_g = 1.12 eV, effective densities of states in the conduction band and valence band at 300 K are 3.22×10^{19} cm^{-3} and 1.83×10^{19} cm^{-3}, dielectric permittivity of silicon $\varepsilon_s = 1.05 \times 10^{-10}$ F/m, flat-band voltage $V_{FB} = 0$, gate insulator thickness $d_i = 0.2$ μm, gate insulator dielectric permittivity $\varepsilon_i = 3.45 \times 10^{-11}$ F/m, gate length $L = 4$ μm, and substrate bias $V_{sub} = 0$.

$$I_D = I_{Dsat} \tag{4-4-18}$$

The drain-to-source saturation current may be found by substituting $V_D = V_{Dsat}$ from eq. (4-4-10) into eq. (4-4-8). This leads to a fairly complicated expression that can be simplified at gate voltages close to the threshold voltage when $V_{Dsat} \approx V_g - V_T$:

$$I_{Dsat} = \mu_n (W/L) c_i \left\{ (V_g/2 - V_{FB} - 2\varphi_b - V_T/2)(V_g - V_T) - \frac{2}{3}\frac{\sqrt{2\varepsilon_s q N_A}}{c_i}\left[(V_g - V_T + 2\varphi_b)^{3/2} - (2\varphi_b)^{3/2}\right] \right\} \tag{4-4-19}$$

Such an approach is only valid when effects related to the saturation of the electron velocity in high electric field are not important. As will be shown in Section 4-5, the velocity saturation effects are, in fact, very important in modern MOSFETs, because of short gate length and high electric fields in the channel (higher than the characteristic electric field of the electron velocity saturation).

Current-voltage characteristics of a MOSFET calculated using this model are shown in Fig. 4-4-8. The calculated dependencies of the drain-to-source saturation current on the gate voltage are shown in Fig. 4-4-9.

For very small drain-to-source voltages, the term in the braces in eq. (4-4-8) may be expanded into the Taylor series, leading to the following simplified expressions for the current-voltage characteristics in the linear region:

$$I_D = \mu_n c_i \frac{W}{L}(V_g - V_T)V_D \tag{4-4-20}$$

This equation can be interpreted as follows. At very small drain voltages the charge induced into the channel does not depend on the channel potential:

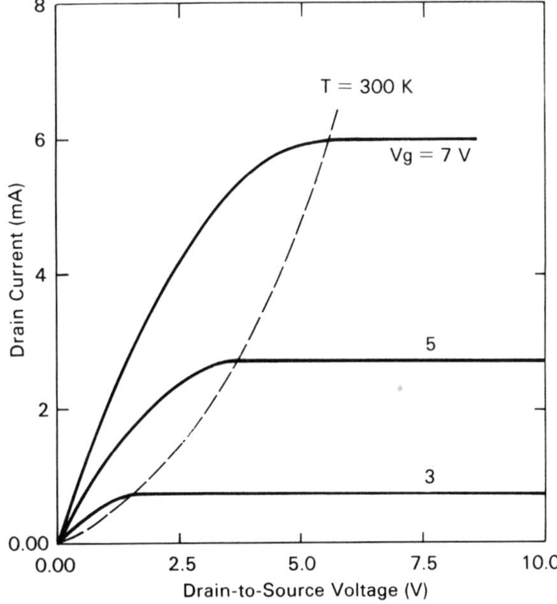

Fig. 4-4-8. Current-voltage characteristics of a MOSFET calculated using the Shockley model (solid curves) and the dependence of the drain saturation current on drain saturation voltage (dashed curve). Numbers near the curves correspond to the gate voltage in volts. The curves were calculated using program PLOTF with subroutine PMOSSH.BAS (see program description section). Parameters used in the calculation are temperature T = 300 K, substrate doping $N_A = 1 \times 10^{14}$ cm^{-3}, energy gap $E_g = 1.12$ eV, effective densities of states in the conduction band and valence band at 300 K are 3.22×10^{19} cm^{-3} and 1.83×10^{19} cm^{-3}, dielectric permittivity of silicon $\varepsilon_s = 1.05 \times 10^{-10}$ F/m, flat-band voltage $V_{FB} = 0$, gate insulator thickness $d_i = 0.2$ μm, gate insulator dielectric permittivity $\varepsilon_i = 3.45 \times 10^{-11}$ F/m, gate length $L = 4$ μm, gate width $W = 100$ μm, substrate bias $V_{sub} = 0$, electron low field mobility $\mu_n = 0.08$ m^2/Vs, and output conductance parameter $\lambda = 0$.

$$qn_{\text{ss}} \approx c_{\text{i}}(V_{\text{g}} - V_{\text{T}}) \tag{4-4-21}$$

In this case the electric field in the channel is nearly constant and given by

$$|F| \approx V_{\text{D}}/L \tag{4-4-22}$$

Multiplying the channel surface charge density qn_{ss} by the electron velocity $v_{\text{n}} = \mu_{\text{n}}|F|$ and by the gate width W yields eq. (4-4-20).

4-4-3. Charge Control Model

A similar approach may be used for a simplified description of the current-voltage characteristics of a MOSFET based on a *charge control model*. In this model we assume that the concentration of free carriers induced into the channel is given by

$$n_{\text{ss}} = c_{\text{i}}(V_{\text{g}} - V_{\text{T}} - V_{\text{c}})/q \tag{4-4-23}$$

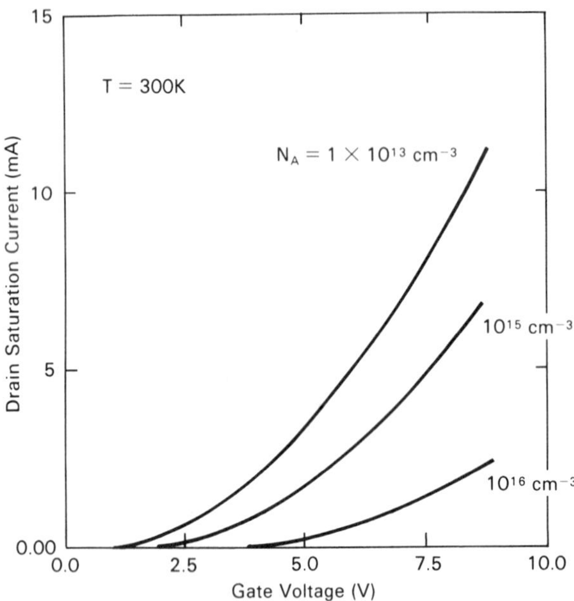

Fig. 4-4-9. Calculated dependence of the drain-to-source saturation current on the gate voltage for different values of doping density (in cm^{-3}). The curves were calculated using program PLOTF with subroutine PMOSSH.BAS (see program description section). Parameters used in the calculation are temperature T = 300 K, energy gap E_g = 1.12 eV, effective densities of states in the conduction band and valence band at 300 K are 3.22×10^{19} cm^{-3} and 1.83×10^{19} cm^{-3}, dielectric permittivity of silicon $\varepsilon_s = 1.05 \times 10^{-10}$ F/m, flat-band voltage $V_{FB} = 0$, gate insulator thickness $d_i = 0.2$ μm, gate insulator dielectric permittivity $\varepsilon_i = 3.45 \times 10^{-11}$ F/m, gate length $L = 4$ μm, gate width $W = 100$ μm, substrate bias $V_{sub} = 0$, electron low field mobility $\mu_n = 0.08$ m^2/Vs, and output conductance parameter $\lambda = 0$.

where V_T is the effective threshold voltage (compare with eq. (4-4-2)). In other words, we neglect the dependence of the charge in the depletion layer, Q_{dep}, on the surface channel potential (see eq. (4-4-4)). The drain current, I_D, can now be presented as

$$I_D = c_i \mu_n \frac{dV}{dx} (V_g - V_T - V_c) W \tag{4-4-24}$$

(compare with eq. (4-4-5)).

Equation (4-4-24) can be rewritten as

$$dx = [c_i \mu_n W (V_g - V_T - V_c)\, dV_c]/I_D \tag{4-4-25}$$

Integrating eq. (4-4-25) from zero to the channel length L (that corresponds to the change in V_c from zero at the source side of the gate to V_D at the drain side of the gate), we obtain the following expressions for the current-voltage characteristics:

$$I_D = \mu_n (W/L) c_i [(V_g - V_T) V_D - V_D^2/2] \quad \text{for } V_D \leq V_{Dsat} = V_g - V_T \tag{4-4-26}$$

$$I_D = (1/2) \mu_n (W/L) c_i (V_g - V_T)^2 \quad \text{for } V_D > V_{Dsat} = V_g - V_T \tag{4-4-27}$$

The current-voltage characteristics calculated using the charge control model (eqs. (4-4-26) and (4-4-27)) are compared with the current-voltage characteristics calculated using the complete Shockley model in Fig. 4-4-10.

An important device characteristic is transconductance, defined as

$$g_m = \left. \frac{\partial I_D}{\partial V_g} \right|_{V_D = \text{constant}} \tag{4-4-28}$$

From eqs. (4-4-26) and (4-4-27) we find that

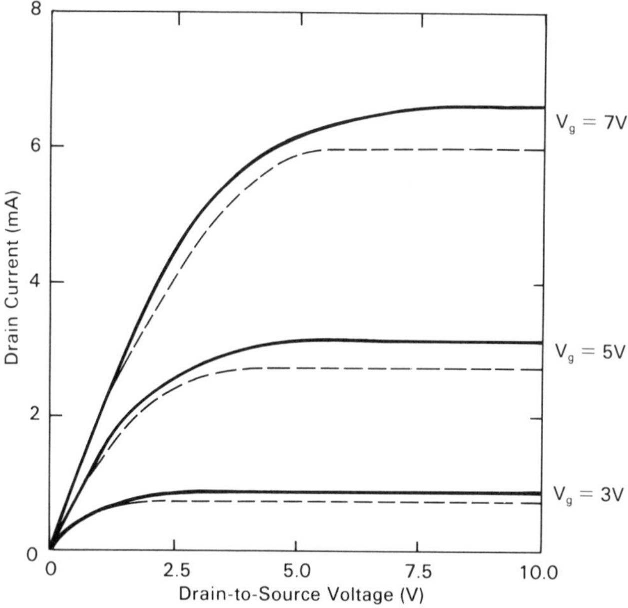

Fig. 4-4-10. Current-voltage characteristics calculated using the charge control model (solid curves) and complete model (dashed curve). The curves were calculated using program PLOTF with subroutine PMOSSH.BAS (see program description section) for the Shockley model and subroutine PMOSCC.BAS for the charge control model. Parameters used in the calculation are temperature T = 300 K, substrate doping $N_A = 1 \times 10^{14}$ cm^{-3}, energy gap $E_g = 1.12$ eV, effective densities of states in the conduction band and valence band at 300 K are 3.22×10^{19} cm^{-3} and 1.83×10^{19} cm^{-3}, dielectric permittivity of silicon $\varepsilon_s = 1.05 \times 10^{-10}$ F/m, flat-band voltage $V_{FB} = 0$, gate insulator thickness $d_i = 0.2$ μm, gate insulator dielectric permittivity $\varepsilon_i = 3.45 \times 10^{-11}$ F/m, gate length $L = 4$ μm, gate width $W = 100$ μm, substrate bias $V_{sub} = 0$, electron low field mobility $\mu_n = 0.08$ m^2/Vs, and output conductance parameter $\lambda = 0$. In the charge control model, the electron saturation velocity was taken to be very high (10^7 m/s) so that velocity saturation effects were not important (see Section 4-5) and direct comparison with the Schottky model was possible.

$$g_m = \beta V_D \qquad \text{for } V_D \le V_{Dsat} = V_g - V_T \tag{4-4-29}$$

$$g_m = \beta(V_g - V_T) \qquad \text{for } VD > V_{Dsat} = V_g - V_T \tag{4-4-30}$$

where the transconductance parameter, β, is given by

$$\beta = \mu_n(W/L)c_i \tag{4-4-31}$$

As can be seen from eqs. (4-4-29) to (4-4-31), higher values of the low field electron mobility, thinner gate insulator layers (i.e., larger gate insulator capacitance $c_i = \varepsilon_i/d_i$), and larger ratios, W/L, lead to a higher transconductance. The dependence of the transconductance on the low field mobility and on the gate length, L, is, however, strongly affected by the velocity saturation effects in short channel devices (see Section 4-5).

4-4-4. Effect of Source and Drain Series Resistances

So far we have considered an ideal device in which the entire voltage drop between the source and drain is across the channel. In fact, both drain and source parasitic series resistances, R_d and R_s, may play an important role limiting the device performance. These resistances may be accounted for by using the following expressions relating the gate and drain voltages to the "extrinsic" (measured) gate-to-source and drain-to-source voltages, V_{gs} and V_{ds}, that include voltage drops across series resistances.

$$V_g = V_{gs} - I_D R_s \tag{4-4-32}$$

$$V_D = V_{ds} - I_D(R_s + R_d) \tag{4-4-33}$$

The calculated dependence of the drain-to-source saturation current on the gate voltage for different source series resistances is shown in Fig. 4-4-11.

The extrinsic transconductance

$$g_m = \frac{\partial I_D}{\partial V_{gs}} \tag{4-4-34}$$

of a field-effect transistor is related to the intrinsic transconductance,

$$g_{mo} = \frac{\partial I_D}{\partial V_g} \tag{4-4-35}$$

of the same device without the source series resistance as follows:

$$g_m = \frac{g_{mo}}{1 + g_{mo}R_s + g_{do}(R_s + R_d)} \tag{4-4-36}$$

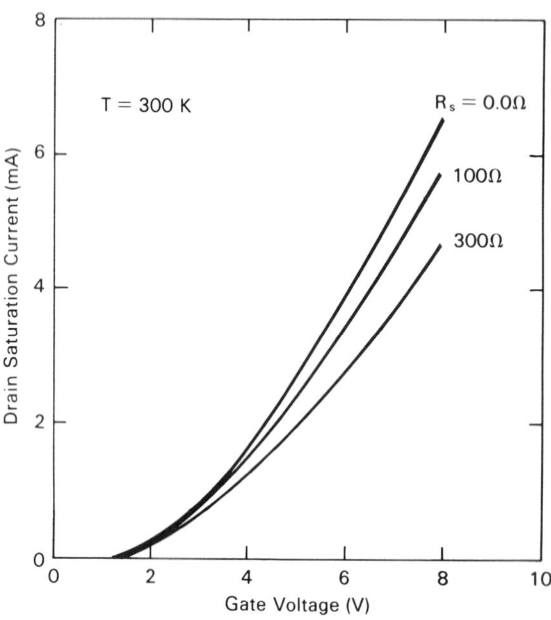

Fig. 4-4-11. Calculated dependence of the drain-to-source saturation current on the gate voltage for different values of source series resistance (in Ω). The curves were calculated using program PLOTF with subroutine PMOSCC.BAS. Parameters used in the calculation are temperature T = 300 K, substrate doping $N_A = 1 \times 10^{14}$ cm^{-3}, energy gap $E_g = 1.12$ eV, effective densities of states in the conduction band and valence band at 300 K are 3.22×10^{19} cm^{-3} and 1.83×10^{19} cm^{-3}, dielectric permittivity of silicon $\varepsilon_s = 1.05 \times 10^{-10}$ F/m, flat-band voltage $V_{FB} = 0$, gate insulator thickness $d_i = 0.2$ μm, gate insulator dielectric permittivity $\varepsilon_i = 3.45 \times 10^{-11}$ F/m, gate length $L = 4$ μm, gate width $W = 100$ μm, substrate bias $V_{sub} = 0$, electron low field mobility $\mu_n = 0.08$ m^2/Vs, output conductance parameter $\lambda = 0$, and the electron saturation velocity $v_s = 10^5$ m/s.

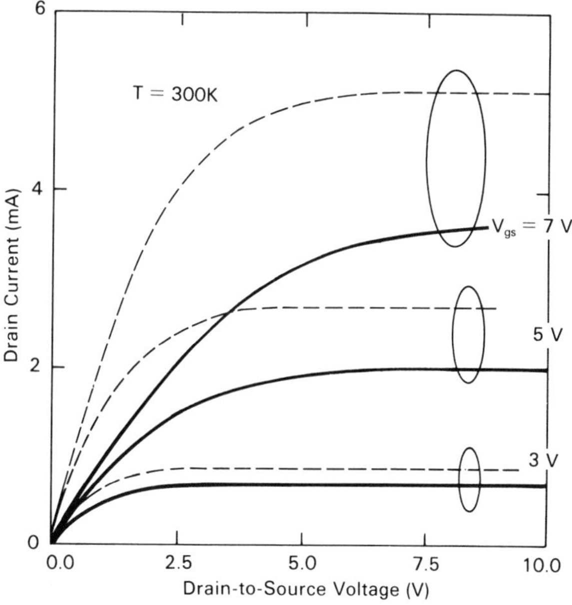

Fig. 4-4-12. Calculated dependence of the drain-to-source saturation current on the drain-to-source voltage for zero source and drain series resistances (dashed curves), and for source series resistance 300 Ω and drain series resistance 300 Ω (solid curves). The source series resistance reduces the drain current, the drain series resistance increases the drain saturation voltage. Both source series resistance and drain series resistance reduce the drain conductance at low drain-to-source voltages. The curves were calculated using program PLOTF with subroutine PMOSCC.BAS. Parameters used in the calculation are temperature T = 300 K, substrate doping $N_A = 1 \times 10^{14}$ cm^{-3}, energy gap $E_g = 1.12$ eV, effective densities of states in the conduction band and valence band at 300 K are 3.22×10^{19} cm^{-3} and 1.83×10^{19} cm^{-3}, dielectric permittivity of silicon $\varepsilon_s = 1.05 \times 10^{-10}$ F/m, flat-band voltage $V_{FB} = 0$, gate insulator thickness $d_i = 0.2$ μm, gate insulator dielectric permittivity $\varepsilon_i = 3.45 \times 10^{-11}$ F/m, gate length $L = 4$ μm, gate width $W = 100$ μm, substrate bias $V_{sub} = 0$, electron low field mobility $\mu_n = 0.08$ m^2/Vs, output conductance parameter $\lambda = 0$, and the electron saturation velocity $v_s = 10^5$ m/s.

where $g_{\mathrm{do}} = \partial I_{\mathrm{D}}/\partial V_{\mathrm{D}}$ is the intrinsic drain conductance (see Problem 4-4-1).

The extrinsic (measured) drain conductance

$$g_{\mathrm{d}} = \frac{\partial I_{\mathrm{D}}}{\partial V_{\mathrm{ds}}} \tag{4-4-37}$$

of a field-effect transistor is related to the intrinsic drain conductance g_{do} of the same device without the source and drain series resistances as follows:

$$g_{\mathrm{d}} = \frac{g_{\mathrm{do}}}{1 + g_{\mathrm{do}}(R_{\mathrm{s}} + R_{\mathrm{d}}) + g_{\mathrm{mo}}R_{\mathrm{s}}} \tag{4-4-38}$$

(see Problem 4-4-3).

The calculated dependence of the drain-to-source saturation current on the drain-to-source voltage for different source and drain series resistances is shown in Fig. 4-4-12 (see Problem 4-4-2).

4-5. VELOCITY SATURATION EFFECTS IN MOSFETs

In modern short-channel MOSFETs the electric fields in the channel may be quite high, and velocity saturation effects are quite important. The measured electron and hole velocity vs. electric field dependencies were discussed in Section 1-9 (see eqs. (1-9-9) to (1-9-21) and Fig. 1-9-2). The electron and hole velocities in inversion layers may be quite different from the values expected for the bulk material. One important dependence ignored in eqs. (1-9-9) to (1-9-21) is the dependence of the electron and hole mobility on the transverse electric field in the channel that determines the effective thickness of the electron or hole gas induced at the insulator–semiconductor interface. As shown in Fig. 4-5-1, carriers experience scattering at the insulator–semiconductor interface, which is more likely in constricted channels, i.e., when the transverse electric field, perpendicular to the interface, is large. The theory of this important effect was considered, for example, by Many et al. (1965). The experimental data of Fang and Fowler (1968) and of Sato et al. (1969) for the dependence of electron and hole mobilities in MOSFETs at room temperature on the gate field, F_g, may be crudely approximated by the following expressions:

$$\mu_n(F_g) = \mu_{no}/(1 + \alpha_{\mu n}F_g)^{1/2} \tag{4-5-1}$$

$$\mu_p(F_g) = \mu_{po}/(1 + \alpha_{\mu p}F_g)^{1/2}$$

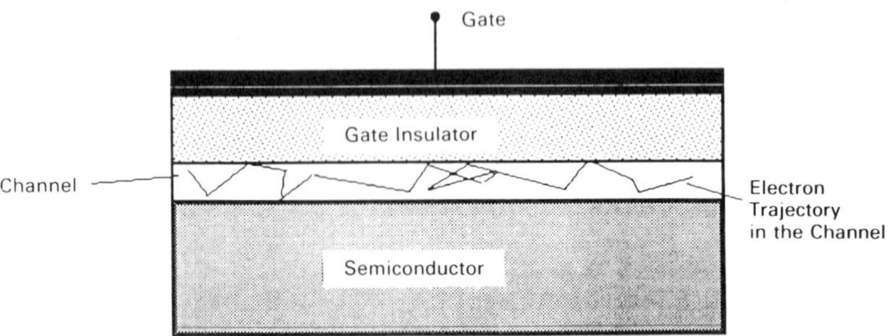

Fig. 4-5-1. Schematic illustration of surface electron scattering in narrow MOSFET channels. In low longitudinal electric fields, electrons in the channel move in random directions with the average velocity determined by the thermal velocity, and with a small drift component of velocity in the direction of the applied electric field. The random electron trajectory leads to surface scattering which is more intense in narrow channels. This explains the reduction of electron mobility with the increase in transverse electric field, F_y, as the effective channel thickness, d_{ch}, is inversely proportional to the transverse electric field $d_{ch} \sim k_B T/(qF_y)$.

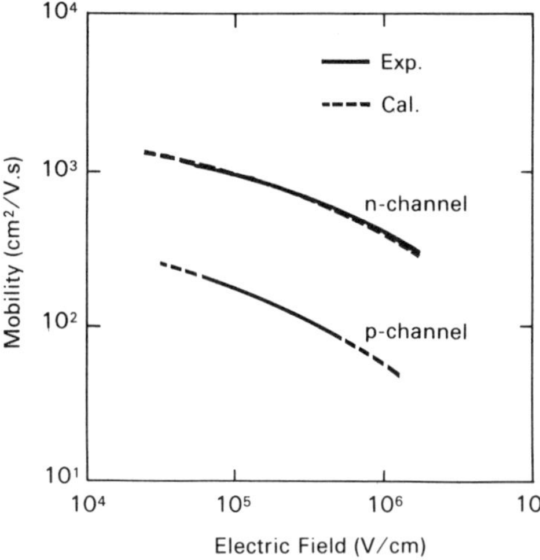

Fig. 4-5-2. Electron and hole mobilities in the inversion layers of *n* and *p* channel MOSFETs as a function of the gate electric field (after Yamaguchi [1979]). Experimental data for *n*-channel devices is from F. F. Fang and A. B. Fowler, *Phys. Rev.*, 169, pp. 619–631 (1968). © 1968 IEEE and experimental data for *p*-channel devices is from T. Sato, Y. Takeishi, and H. Hara, *Jpn. J. Appl. Phys.*, 8, No. 5, pp. 588–598 (1969). © 1969 IEEE.

where μ_{no} and μ_{po} are the electron and hole mobilities for $F_g = 0$, $\alpha_{\mu n} = 1.54 \times 10^{-5}$ cm/V and $\alpha_{\mu n} = 5.35 \times 10^{-5}$ cm/V (see Fig. 4-5-2). Similar experimental data were obtained by Sabnis and Clemens (1979).

However, it is interesting to note that in very constricted channels or at low temperatures, when the electron motion in the direction perpendicular to the semiconductor–insulator interface is quantized, the surface scattering is not important and the electron mobility may actually be enhanced owing to the screening of impurity scattering by a high density of electrons in the channel. Such mobility enhancement, first predicted by Esaki and Tsu (1969) and observed by Dingle et al. (1978) in GaAs, is utilized in high electron mobility transistors or MODFETs (see Section 4-10).

The curves $v_n(F)$ and $v_p(F)$ for silicon calculated using eqs. (1-9-9) to (1-9-14) are shown in Fig. 4-5-3 and Fig. 4-5-4, respectively. In Fig. 4-5-3 we also show a simple two-piece linear approximation for the electron velocity:

$$v_n = \mu_n F \qquad \text{for } F < F_s \tag{4-5-2}$$

and

$$v_n = \mu_n F_s = v_s \qquad \text{for } F \geq F_s \tag{4-5-3}$$

In what follows we use the simple linear approximation given by eqs. (4-5-2) and (4-5-3), as it allows us to derive analytical expressions describing the effects of the velocity saturation in the channel.

The surface carrier concentration of electrons in the channel can be found using the charge control model (see Section 4-4),

$$n_{ss}(x) = \frac{\varepsilon_i}{qd_i}[V_g - V_T - V(x)] \tag{4-5-4}$$

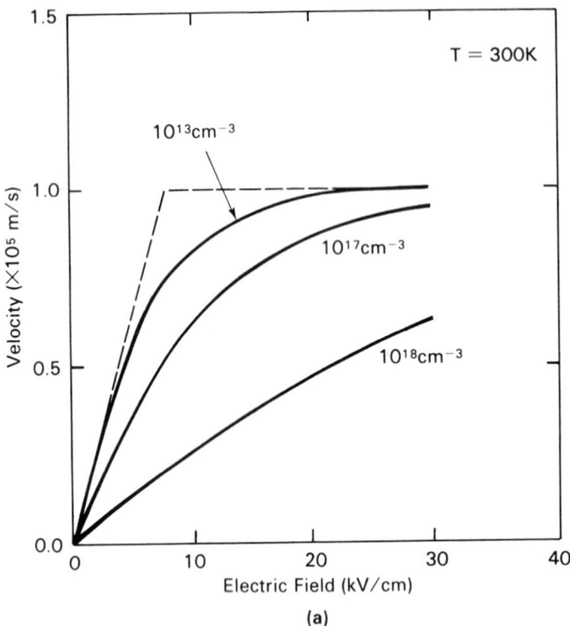

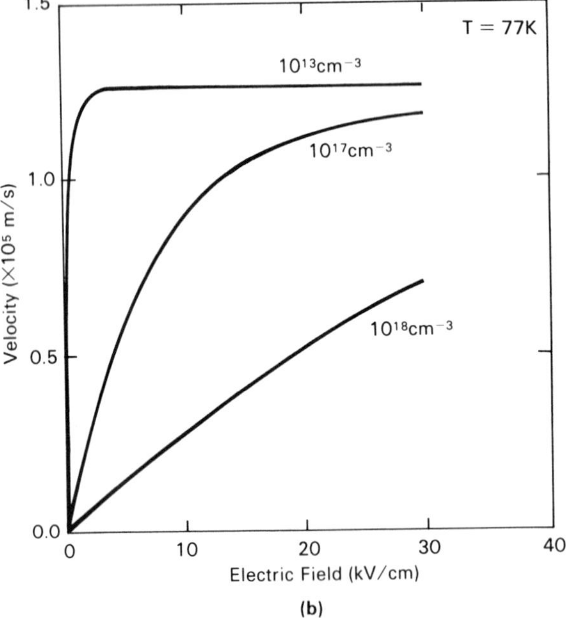

Fig. 4-5-3. Electron velocity and electric field for silicon at 300 K (a) and at 77 K (b) for different values of the doping density. Also shown is a two piece linear approximation for the electron velocity in silicon at room temperature (dashed line in Fig. 4-5-2a). The velocity-field characteristic is calculated using program PLOTF with subroutine PVNSI.BAS (see the Program Description Section).

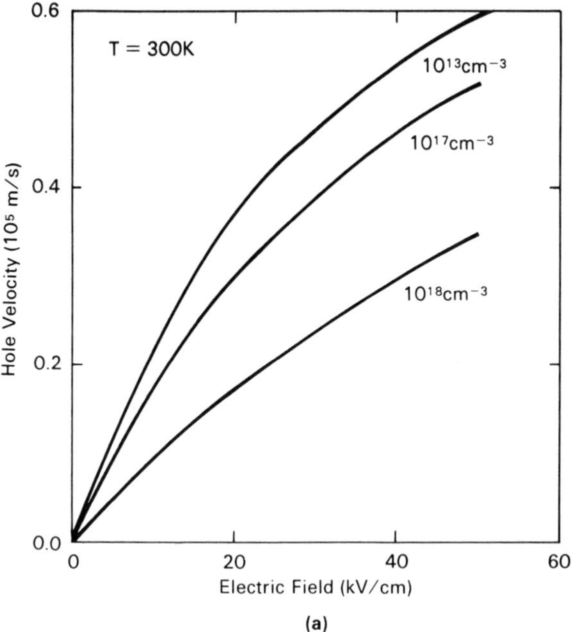

(a)

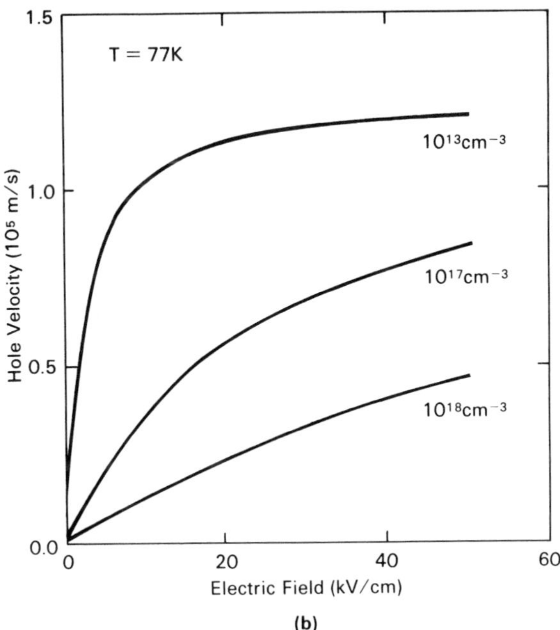

(b)

Fig. 4-5-4. Hole velocity and electric field for silicon at 300 K (a) and at 77 K (b) for different values of the doping density. The velocity-field characteristic is calculated using program PLOTF with subroutine PVPSI.BAS (see the Program Description Section).

where x is the space coordinate along the channel and $V(x)$ is the channel potential. Equation (4-5-4) should be solved together with the equation

$$I_D = qn_{ss}v_n(F)W \qquad (4\text{-}5\text{-}5)$$

which relates the drain-to-source current I_D to the electron velocity $v_n(F)$ in the channel, which is a function of the electric field F in the channel. As in the gradual channel approximation, we assume that the electric field in the channel is parallel to the semiconductor–insulator interface. We also neglect the diffusion current. Substituting eq. (4-5-4) into eq. (4-5-5), replacing F in the resulting equation by $-dV/dx$, and integrating with respect to x from 0 to L, where L is the gate length, we obtain a conventional charge control model equation describing the current-voltage characteristics of a MOSFET at low drain voltages V_D:

$$I_D = \beta(V_{gt}V_D - V_D^2/2) \qquad (4\text{-}5\text{-}6)$$

where

$$V_{gt} = V_g - V_T \qquad (4\text{-}5\text{-}7)$$

and

$$\beta = \mu_n(W/L)c_i \qquad (4\text{-}5\text{-}8)$$

is the transconductance parameter.

The drain-to-source saturation current, I_{Dsat}, is found assuming that the current saturation occurs when the electric field in the channel at the drain side of the gate exceeds the velocity saturation field $F_s = v_s/\mu_n$. This is a more realistic assumption compared with the pinch-off condition at the drain,

$$n_{ss}(L) = 0 \qquad (4\text{-}5\text{-}9)$$

used for determining the current saturation drain voltage in Section 4-4. (Equation (4-5-9) is equivalent to demanding $F(L) \rightarrow \infty$.) We will still use the charge control model assuming the constant electron mobility (see eq. (4-5-2)) in order to describe the longitudinal field distribution in the channel at drain voltages below the saturation voltage.

The absolute value of the electric field in the channel,

$$F = |\mathbf{F}| = dV/dx \qquad (4\text{-}5\text{-}10)$$

at drain voltages below the saturation voltage can be found from eq.(4-4-25):

$$F = I_D/[\beta L(V_{gt} - V)] \qquad (4\text{-}5\text{-}11)$$

Integrating eq. (4-5-11) from 0 to x, we obtain the equation describing the potential, V, in the channel as a function of distance x at drain voltages below the saturation voltage:

$$x = \beta L(V_{gt}V - V^2/2)/I_D \qquad (4\text{-}5\text{-}12)$$

The solution of this equation is given by

$$V = V_{gt} - [V_{gt}^2 - 2I_D x/(\beta L)]^{1/2} \tag{4-5-13}$$

Substituting this equation into eq. (4-5-11), we find the electric field as a function of distance,

$$F = I_D/\{\beta L[V_{gt}^2 - 2I_D x/(\beta L)]^{1/2}\} \tag{4-5-14}$$

and the electric field $F(L)$ at the drain side of the channel (where the electric field is the highest),

$$F(L) = I_D/[\beta L(V_{gt}^2 - 2I_D/\beta)^{1/2}] \tag{4-5-15}$$

From the condition

$$F(L) = F_s \tag{4-5-16}$$

we now find the drain saturation current,

$$I_{Dsat} = \beta V_{sl}^2[(1 + (V_{gt}/V_{sl})^2]^{1/2} - \beta V_{sl}^2 \tag{4-5-17}$$

and drain-saturation voltage (see Problem 4-5-9),

$$V_{Dsat} = V_{gt} + V_{sl} - (V_{gt}^2 + V_{sl}^2)^{1/2} \tag{4-5-18}$$

where

$$V_{sl} = F_s L \tag{4-5-19}$$

At very large values of

$$V_{sl} = v_s L/\mu_n >> V_{gt} \tag{4-5-20}$$

the term in brackets in the right-hand side of eq. (4-5-17) may be expanded into the Taylor series, yielding the following expression for long-channel devices,

$$I_{Dsat} = \beta V_{gt}^2/2 \tag{4-5-21}$$

coinciding with eq. (4-4-27), which does not take into account the velocity saturation effects. For long-channel devices $V_{sl} >> V_{gt}$ and $V_{Dsat} \to V_{gt}$, as predicted by the constant mobility model. Hence, the velocity saturation effects are not very important for long-channel FETs. Assuming that $V_{gt} \approx 3$ V, $\mu_n = 0.08$ m^2/Vs, and $v_s = 1 \times 10^5$ m/s, we find that for

$$L >> 2.4\ \mu\text{m} \tag{4-5-22}$$

velocity saturation effects on the drain saturation current may be neglected. A typical gate length for modern MOSFETs may be smaller than a micron (see Fig. 4-1-3), and hence, the velocity saturation effects are very important for modern devices.

In the limiting case for short-channel devices, when

$$V_{sl} = v_s L/\mu_n << V_{gt} \tag{4-5-23}$$

we obtain from eq. (4-5-17) and (4-5-18)

$$I_{\text{Dsat}} = \beta V_{\text{gt}} V_{\text{sl}} \tag{4-5-24}$$

and $V_{\text{Dsat}} \rightarrow V_{\text{sl}} << V_{\text{gt}} - V_{\text{sl}}$. This means that the drain saturation current is $2V_{\text{sl}}/V_{\text{gt}}$ times smaller than the value predicted by the constant mobility model.

The dependence of the drain saturation current on the gate length is shown in Fig. 4-5-5, where the drain saturation current predicted by the constant mobility model is shown for comparison. As can be seen from the figure, the velocity saturation effects are not important in long-gate devices ($\geq$ 3 μm), as can be expected (see eq. (4-5-22)). However, in short-channel devices the drain satura-

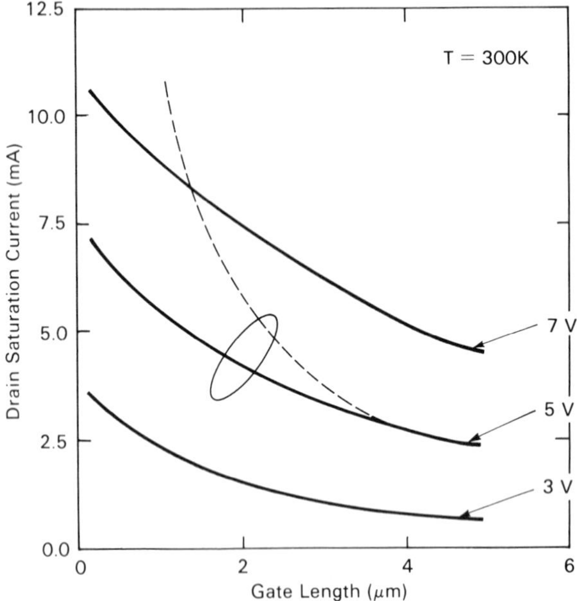

Fig. 4-5-5. The dependence of the drain saturation current on the gate length for gate voltages of 3, 5, and 7 V. The curves were calculated using program PLOTF with subroutine PMOSCC.BAS. Parameters used in the calculation are temperature T = 300 K, substrate doping $N_A = 1 \times 10^{14}$ cm^{-3}, energy gap $E_g = 1.12$ eV, effective densities of states in the conduction band and valence band at 300 K are 3.22×10^{19} cm^{-3} and 1.83×10^{19} cm^{-3}, respectively, dielectric permittivity of silicon $\varepsilon_s = 1.05 \times 10^{-10}$ F/m, flat-band voltage $V_{FB} = 0$, gate insulator thickness $d_i = 0.2$ μm, gate insulator dielectric permittivity $\varepsilon_i = 3.45 \times 10^{-11}$ F/m, gate width $W = 100$ μm, substrate bias $V_{sub} = 0$, electron low field mobility $\mu_n = 0.08$ m^2/Vs, the electron saturation velocity $v_s = 10^5$ m/s and output conductance factor $\lambda = 0$. The drain saturation current predicted by the constant mobility model is shown for comparison (dashed line). This curve was calculated using program PLOTF with subroutine PMOSSH.BAS for the same values of parameters.

tion current is substantially smaller than the saturation current predicted by the constant mobility model.

As was discussed in Section 4-4, drain and source parasitic series resistances, R_d and R_s, may play an important role limiting the device performance. These resistances may be accounted for by using the following expressions relating the gate and drain voltages to the extrinsic (measured) gate-to-source and drain-to-source voltages, V_{gs} and V_{ds}, that include voltage drops across series resistances.

$$V_g = V_{gs} - I_D R_s \tag{4-5-25}$$

$$V_D = V_{ds} - I_D(R_s + R_d) \tag{4-5-26}$$

Substituting $I_D = I_{dsat}$ into eqs. (4-5-25) and (4-5-26) and the resulting equations into eqs. (4-5-17) and (4-5-18) we find that

$$I_{Dsat} = \beta V_{sl}^2 \frac{(1 + 2\beta R_s V_{gst} + V_{gst}^2/V_s^2)^{1/2} - 1 - \beta R_s V_{gst}}{1 - \beta^2 R_s^2 V_{sl}^2} \tag{4-5-27}$$

where $V_{gst} = V_{gs} - V_t$, and

$$(V_{ds})_{sat} = V_{gt} + V_{sl} - (V_{gt}^2 + V_{sl}^2)^{1/2} + I_{Dsat}(R_s + R_d) \tag{4-5-28}$$

Here I_{Dsat} and $(V_{ds})_{sat}$ are the drain-to-source saturation current and drain-to-source saturation voltage in the presence of the source and drain series resistances. The calculated dependence of drain-to-source saturation current on the gate voltage for different source series resistances is shown in Fig. 4-4-11.

In the linear region (i.e., small drain-to-source voltages) the extrinsic drain conductance,

$$g_d = \frac{\partial I_D}{\partial V_{ds}} \tag{4-5-29}$$

of a field-effect transistor is related to the intrinsic drain conductance,

$$g_{do} = \frac{\partial I_D}{\partial V_D} = \beta(V_g - V_T) \tag{4-5-30}$$

of the same device without the source and drain series resistances as follows:

$$g_d = \frac{g_{do}}{1 + g_{do}(R_s + R_d) + g_{mo}R_s} \tag{4-5-31}$$

(see Section 4-4 and Problem 4-4-3). At very small drain voltages the last term in the denominator of eq. (4-5-31) can be neglected.

Equations (4-5-30) and (4-5-44) allow us to propose a convenient interpolation formula for the MOSFET current-voltages characteristics:

$$I_{ds} = I_{Dsat} \tanh[g_d V_{ds}/I_{Dsat})] \tag{4-5-32}$$

This equation describes both limiting cases (small $V_D << V_{Dsat}$ and large $V_D >> V_{Dsat}$) correctly.

In practical devices the current-voltage characteristic does not completely saturate at large drain-to-source voltages. This is related to short-channel (see Section 4-6) and other nonideal effects in MOSFETs. The finite output conductance is typically taken into account empirically by multiplying the calculated drain current by the term $1 + \lambda V_{ds}$, where λ is an empirical parameter leading to the following expression describing MOSFET current-voltage characteristics using the charge control model:

$$I_{ds} = I_{Dsat} \tanh[g_d V_{ds}/I_{Dsat})](1 + \lambda V_{ds}) \tag{4-5-33}$$

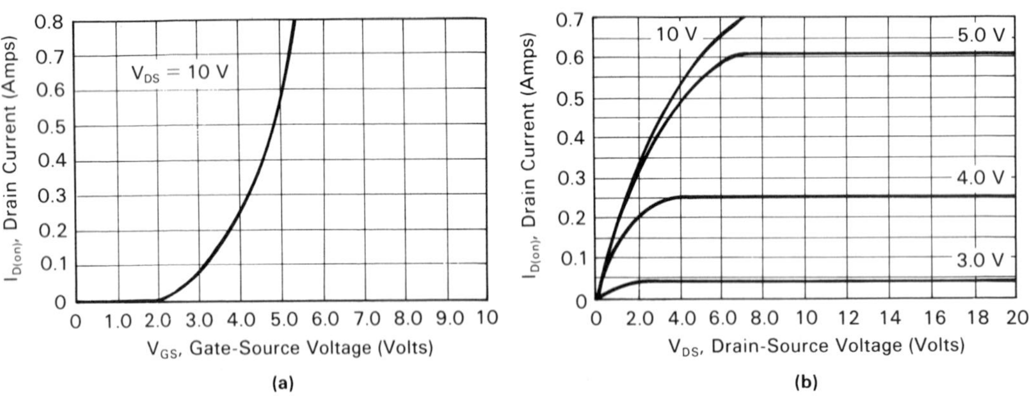

Fig. 4-5-6. Drain-to-source saturation current and gate voltage (a) and current-voltage characteristics for the Motorola MFE9200 MOSFET (from copyright of Motorola, Inc. Used by permission).

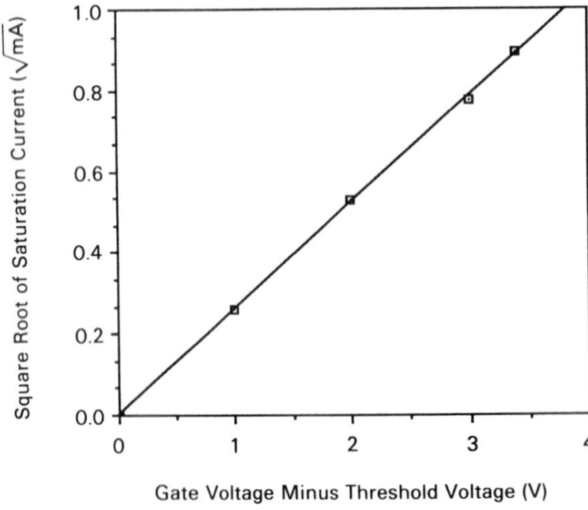

Fig. 4-5-7. Square root of drain-to-source saturation current and gate voltage minus threshold voltage for the Motorola MFE9200 MOSFET.

This model is implemented in the subroutine PMOSCC.BAS (see Program Description section).

In Fig. 4-5-6 we show drain-to-source saturation current vs. gate voltage (a) and current-voltage characteristics (b) for the Motorola MFE9200 MOSFET (from *Small Signal Transistor Data,* published by Motorola Semiconductor Products, Inc., Phoenix, Ariz., 1983). As can be seen from Fig. 4-5-7 (in which the square root of drain-to-source saturation current vs. gate voltage minus threshold voltage is plotted), the transfer curve for this device is very well described by the square-law equation ($I_{Dsat} = (\beta/2)(V_{gs} - V_T)^2$). As can be seen from eq. (4-5-21), this should be expected for long-channel devices. For short-channel devices, how-

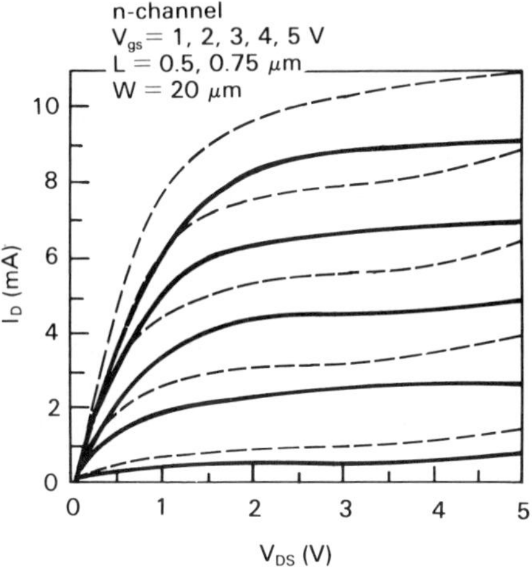

Fig. 4-5-8. Current-voltage characteristics of submicron n-channel MOSFETs (after S. J. Hillenius, R. Liv, G. E. Georgou, R. L. Field, D. S. Williams, S. Kornblit, D. M. Boulin, R. J. F. Jensen, L. G. Salmon, D. S. Deakin, and M. J. Delaney, "Ultrahigh-Speed GaAs Static Frequency Dividers," in *IEDM Technical Digest,* pp. 476–479, IEEE Publication, Los Angles, (1986). © 1986 IEEE). Gate voltages are 1, 2, 3, 4, and 5 V. Dashed lines are for $L = 0.5\ \mu m$, solid line, for $L = 0.75\ \mu m$. Gate oxide thickness is 175 Å; device width is 20 μm; and doping density is 5×10^{16} cm^{-3}.

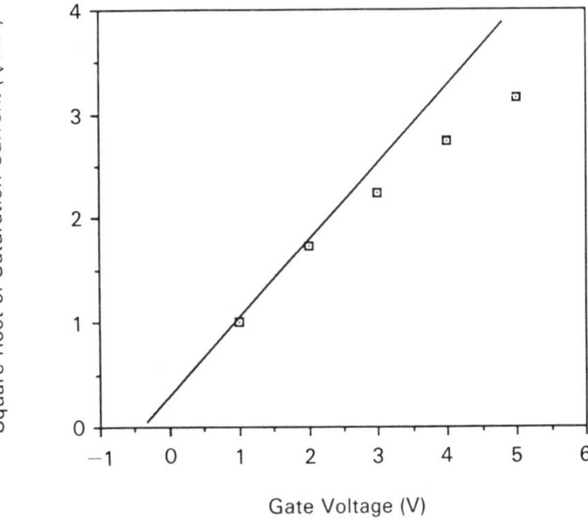

Fig. 4-5-9. The dependence of the square root of the drain-to-source saturation current on the gate voltage for the 0.5 μm-gate device with characteristics shown in Fig. 4-5-8. As can be seen from the figure, the deviation from the square law equation is quite large, especially at large gate voltages.

ever, where the velocity saturation effects become important, the deviations from this simple equation become quite noticeable. This is illustrated by Fig. 4-5-8, in which the current-voltage characteristics of submicron n-channel MOSFETs are shown (after Hillenius et al. 1986). The dependence of the square root of drain-to-source saturation current on the gate voltage for the 0.5-μm-gate device is shown in Fig. 4-5-9. As can be seen from the figure the deviation from the square-law equation is quite large, especially at large gate voltages.

4-6. SHORT–CHANNEL AND NONIDEAL EFFECTS IN MOSFETs

As was shown in Section 4-5 the current-voltage characteristics of long-channel MOSFETs are well described by a constant mobility model. However, in short-channel devices, the effects of velocity saturation play an important role (see Fig. 4-5-5). There are also several other short-channel effects playing an important role in devices with short gate length (on the order of a micron and less). For example, in the analysis of the current-voltage characteristics of MOSFETs given in Sections 4-4 and 4-5, we assumed that the drain-to-source current saturates, i.e., becomes completely independent of the drain-to-source voltage, once this voltage exceeds the drain-to-source saturation voltage, V_{Dsat}. As can be seen from Fig. 4-5-6 this seems to be the case in long-channel devices. However, in short-channel devices, there is a noticeable increase in the drain-to-source current in the saturation region (see, for example, the top [dashed] I_{ds} vs. V_{ds} curve for a 0.5-μm-gate device in Fig. 4-5-8). Also, according to the models of Sections 4-4 and 4-5, the threshold voltage, V_T, is independent of the channel length. In fact, the threshold voltage becomes a strong function of the device length in short-channel devices (see Fig. 4-6-1).

These effects can be qualitatively explained by comparing the distribution of the charges in the depletion layer in long-channel and short-channel devices (see Fig. 4-6-2). In a long-channel device, the depletion regions near the source and

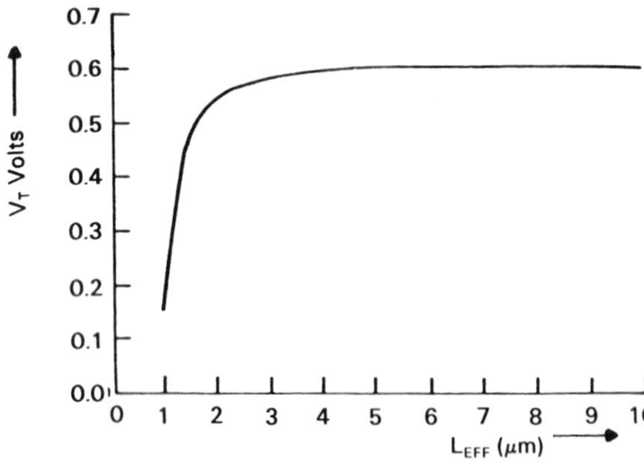

Fig. 4-6-1. Threshold voltage vs. effective channel length (after C. Duvvury, "A Guide to Short Channel Effects in MOSFETs," *TI Engineering Journal*, 1, no. 1, pp. 52–56, July–August (1984). © 1984 IEEE). See also *IEEE Circuit and Devices Magazine*, 2, no. 6, p. 6, November (1986). © 1986 IEEE.

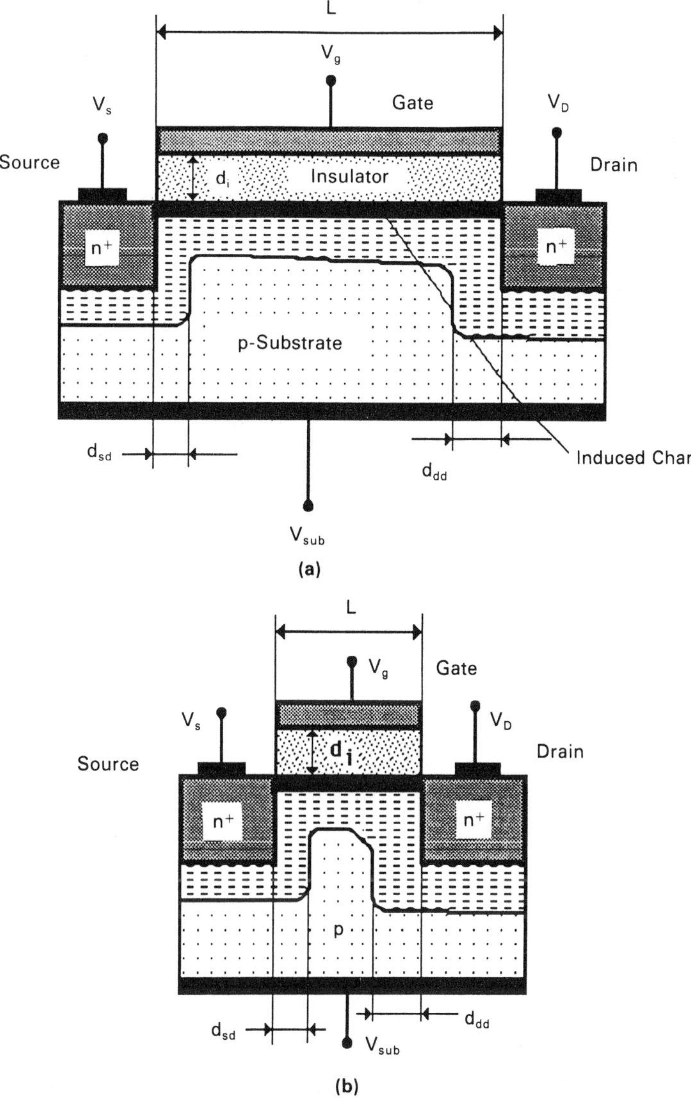

Fig. 4-6-2. Distribution of the charges in the depletion layer in (a) long channel and (b) short channel MOSFETs.

drain are fairly removed from the central section of the channel (see Fig. 4-6-2a). However, in a short-channel device, the widths of these depletion regions, d_{sd} and d_{dd}, become comparable with the gate length, L, and these depletion charges strongly affect the carrier distribution in the channel (see Fig. 4-6-2b). Because the drain voltage provides an additional reverse bias, the depletion width d_{dd} is an increasing function of the drain voltage, and the depletion region at the drain side is especially important. The exact analysis of short-channel effects requires a two-dimensional modeling. However, to the first order, the effect of the increasing depletion width d_{dd} is to reduce the effective channel length in the saturation regime:

$$L_{eff} = L - \Delta L \tag{4-6-1}$$

Here L_{eff} is the effective channel length, such that the voltage drop across the corresponding section of the channel is equal to V_{Dsat}, and ΔL is the depleted portion of the channel related to the depletion region near the drain. This section of the channel supports the remainder of the drain-to-source voltage, $V_D - V_{Dsat}$. With the increase in V_{ds}, the length of the depleted section, ΔL, increases, leading to a shorter effective channel length, L_{eff}, (this effect is called *channel length modulation*) and, hence, to a higher drain saturation current and finite output conductance dI_{Dsat}/dV_{ds} in the saturation region.

A very crude estimate for ΔL can be obtained from the solution of the one-dimensional Poisson's equation,

$$dF/dx = -qN_A/\varepsilon_s \tag{4-6-2}$$

with the following boundary condition

$$F = -F_s \quad \text{at } x = L_{eff} \tag{4-6-3}$$

This leads to

$$F(x) = -qN_A(x - L_{eff})/\varepsilon_s - F_s \tag{4-6-4}$$

and from the condition

$$V_D - V_{Dsat} = -\int_{L_{eff}}^{L} F(x)\, dx \tag{4-6-5}$$

we find that

$$(qN_A/2\varepsilon_s)(\Delta L)^2 + F_s\, \Delta L = V_D - V_{Dsat} \tag{4-6-6}$$

A more accurate and realistic expression for ΔL may be obtained taking into account the density of mobile carriers injected into the depletion region near the drain. Following Hanafi (1986), we assume that these carriers are injected from the inversion layer and spread uniformly into the drain depletion region, leading to the current density

$$J = I_{ds}/(WY) \tag{4-6-7}$$

where

$$Y = (D_d - d_{inv})(x - L_{eff})/\Delta L + d_{inv} \tag{4-6-8}$$

Here D_d is the depth of the drain n^+ region, and d_{inv} is the thickness of the inversion layer (50–100 Å). We also assume that the velocity of the electrons in this region is saturated, so that their volume density is

$$n = J/(qv_s) \tag{4-6-9}$$

Now the one-dimensional Poisson's equation should be rewritten as

$$dF/dx = -q[N_A + J/(qv_s)]/\varepsilon_s \tag{4-6-10}$$

The solution of this equation yields the following expression for ΔL (see Hanafi 1986):

$$(qN_A/2\varepsilon_s)\Delta L^2\{1 + 2I_{ds}D_d[\ln(D_d/d_{inv}) - 1]/(qN_AWv_sD_d)\} + F_s\,\Delta L = V_D - V_{Dsat} \tag{4-6-11}$$

For gate lengths larger than or about 1 μm and drain-to-source voltages smaller than or about 10 V this expression may be simplified to yield

$$\Delta L = L\,\frac{[V_{sl}^2 + 2A(1 + BI_{Dsat})(V_D - V_{Dsat})]^{1/2} - V_{sl}}{A(1 + BI_{Dsat})} \tag{4-6-12}$$

where $V_{sl} = F_sL$,

$$A = qN_AL^2/2\varepsilon_s \tag{4-6-13}$$

and

$$B = [\ln(D_d/d_{inv}) - 1]/(qN_AWv_sD_d) \tag{4-6-14}$$

Another important effect that should be taken into account in short-channel devices is the effect of the charge in the depletion region under the channel that depends on the channel potential. This effect is represented by the second term in the brackets in the right-hand part of eq. (4-4-8), but it is omitted in the simplified charge control model (compare with eq. (4-4-26)). This effect may be taken into account by introducing an additional parameter, a, into the equations of the charge control model (see, for example, Hanafi 1986). The resulting equations are as follows:

Linear Regime ($V_D < V_{Dsat}$)

$$I_D = \beta(V_{gt}V_D - aV_D^2/2) \tag{4-6-15}$$

where

$$V_{gt} = V_g - V_T \tag{4-6-16}$$

and

$$\beta = \mu_n(W/L)c_i \tag{4-6-17}$$

is the transconductance parameter. (In addition, the dependence of electron mobility on the longitudinal and transverse electric field in the channel should be included [see Section 4-5] for a more realistic device modeling.)

Saturation Regime ($V_D > V_{Dsat}$)

$$I_D = \beta_s(V_{gt}V_{Dsat} - aV_{Dsat}^2/2) \tag{4-6-18}$$

Here

$$V_{Dsat} = V_{gt}/a + V_{sl} - [(V_{gt}/a)^2 + V_{sl}^2]^{1/2} \tag{4-6-19}$$

and

$$\beta_s = \mu_n(W/L_{eff})c_i \tag{4-6-20}$$

is the transconductance parameter for the saturation regime. For silicon, parameter a (describing the influence of the bulk substrate depletion layer) can be approximated by the following expression (see Hanafi 1986):

$$a = 1 + \frac{0.5K}{(2\varphi_b - V_{sub})^{1/2}}\left[1 - \frac{1}{1.41 + 0.43(2\varphi_b - V_{sub})}\right] \tag{4-6-21}$$

where

$$K = \partial V_T/\partial(2\varphi_b - V_{sub}) \tag{4-6-22}$$

(see Problem 4-6-1). The threshold voltage, V_T, and parameter K may be determined from the experimental data for a given device.

As can be seen from Fig. 4-6-3, this model is in good agreement with the experimental data.

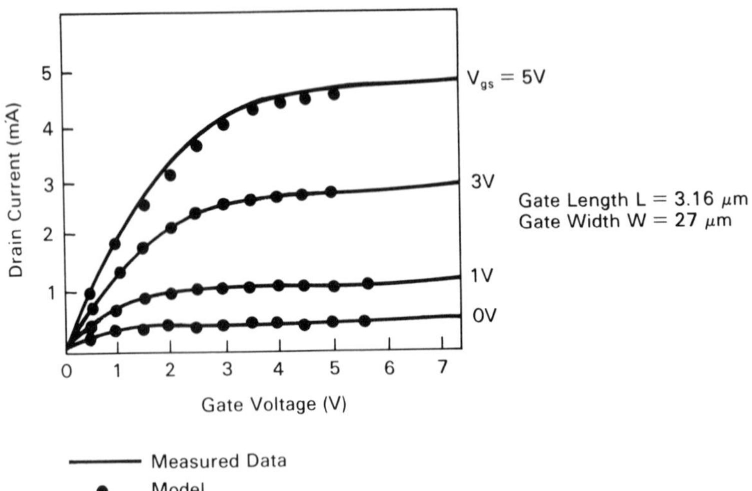

Fig. 4-6-3. Measured and calculated current voltage characteristics of Si MOSFETs (after H. I. Hanafi, "Current Modeling for MOSFET," in *Circuit Analysis, Simulation, and Design*, pp. 71–105, ed. A. E. Ruehli, North Holland, (1986)).

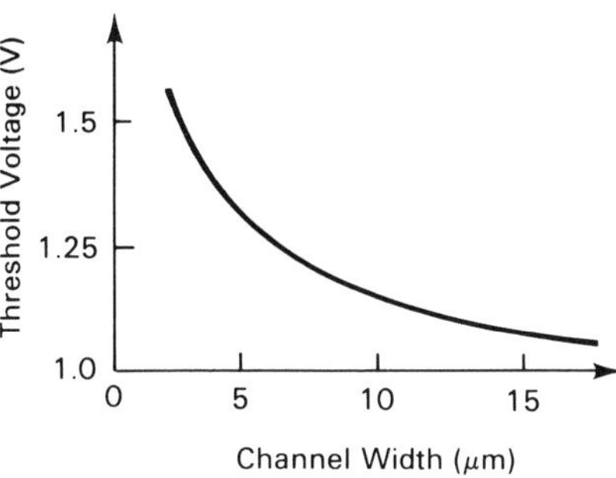

Fig. 4-6-4. Threshold voltage vs. channel width (after J. Compton, "Limitations of and Improvements to MOS Processes," in *MOS Devices: Design and Manufacture,* ed. A. D. Milne, Halsted Press, New York [1983]).

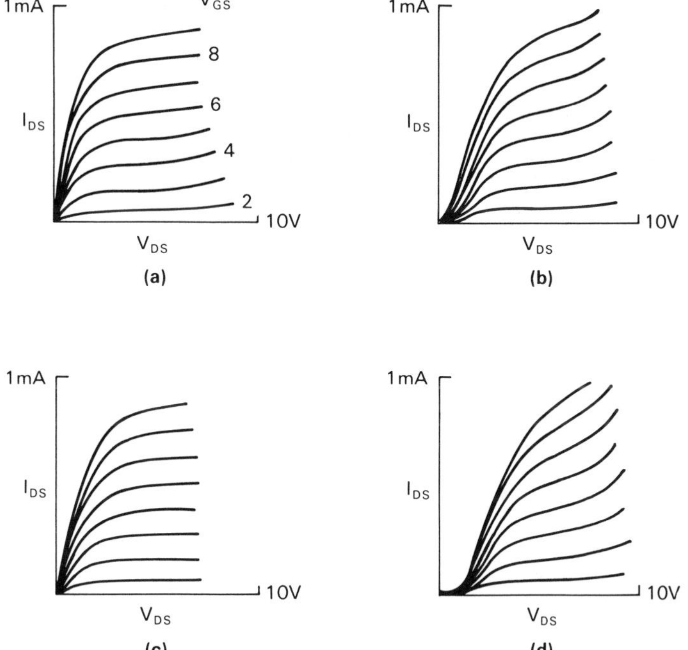

Fig. 4-6-5. Current-voltage characteristics of a silicon MOSFET at 77 K (from L. Forbes, E. Sun, R. Alders, and J. Moll, *IEEE Trans. Electron Devices,* ED-26, p. 1816 (1979). © 1979 IEEE). (a) normal conditions, (b) after stress voltage conditions (after a temporary increase in the drain voltage to the point when the avalanche multiplication takes place leading to the injection of hot electrons near the drain into the silicon dioxide layer), (c) after reset (after applying negative gate bias and positive drain bias the device), and (d) after repeating the stress voltage.

Parameters such as the threshold voltage, V_T, and K depend not only on the device length but also (for narrow devices) on the device width because of the fringing fields that increase the threshold voltage (see Fig. 4-6-4).

Another nonideal effect that may be important, especially in short-channel MOSFETs, is the electron injection into the silicon dioxide gate dielectric layer, where electrons are trapped (see, for example, Ning et al. 1977 and Forbes et al. 1979). This effect is illustrated by Fig. 4-6-5 (from Forbes et al. 1979). Figure 4-6-5a shows the current-voltage characteristics of a silicon MOSFET at 77 K. Figure 4-6-5b shows the current-voltage characteristics of the same device after "stress" voltage conditions, i.e., after a temporary increase in the drain voltage to the point when the avalanche multiplication takes place, leading to the injection of hot electrons near the drain into the silicon dioxide layer. (Hot electrons may also be injected into the oxide even without the avalanche breakdown near the drain; see Sze 1981). The negative charge of these electrons depletes the channel near the drain, leading to the distortion of the current-voltage characteristics. By applying negative gate bias and positive drain bias the device characteristics are "reset" (see Fig. 4-6-5c) because these voltage conditions lead to the electron reemission from the traps. Repeating the stress conditions leads again to the distortion of the current-voltage characteristics (see Fig. 4-6-5d). This effect is used in FAMOS EPROM devices (see Frohman-Bentchkowsky 1971) to alter the device threshold voltage. On the other hand, the trapping effect can be minimized by careful processing of the gate oxide level, yielding devices that exhibit a threshold voltage shift of less than 100 mV in 10 years under accelerated stress test conditions (see Compton 1983).

The avalanche breakdown near the drain may play an important role in determining the current-voltage characteristics and carrier distributions in the device. The effect of the avalanche breakdown on the current-voltage characteristics is illustrated by Fig. 4-6-6 (from Valdya et al. 1986), which demonstrates a sharp increase in the drain current at drain voltages larger than approximately 6.5 V. The effect of avalanche breakdown on the carrier density can be seen from comparison of two-dimensional profiles of electron density (see Fig. 4-6-7; after

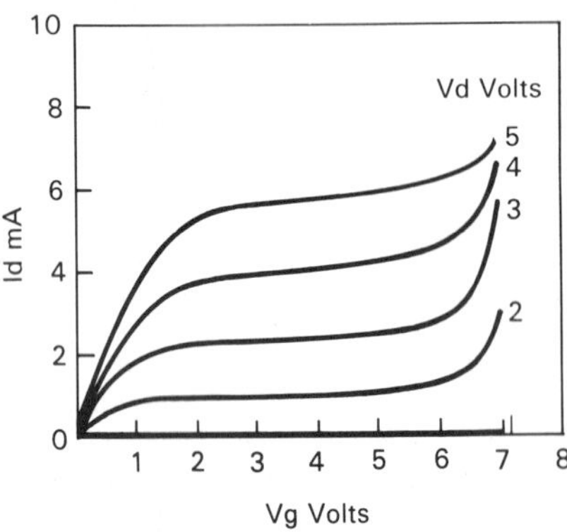

Fig. 4-6-6. The effect of the avalanche breakdown on the current-voltage characteristics (from S. Valdya, E. N. Fuls, and R. L. Johnson, *IEEE Trans. Electron Devices*, ED-33, p. 1321 (1986). © 1986 IEEE).

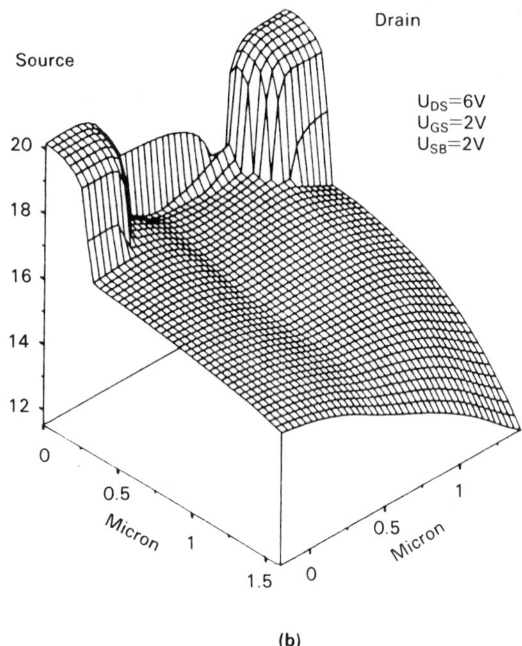

(b)

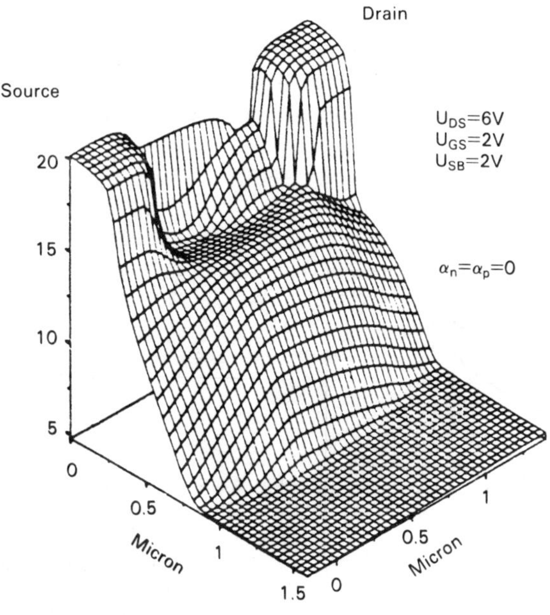

(a)

Fig. 4-6-7. Two dimensional profiles of electron density in a silicon MOSFET (after S. Selberherr, *Analysis and Simulation of Semiconductor Devices,* Springer-Verlag, Wien, New York (1984)). (a) neglecting avalanche breakdown and (b) including avalanche breakdown.

Selberherr 1984). As can be seen from the figure, avalanche breakdown dramatically increases the electron density.

Still another important effect limiting the maximum drain voltage is the punch-through—the effect of merging drain and source depletion regions leading to the space-charge limited current. In the linear region (when the electron velocity, v_n, is proportional to the electric field, i.e., $\mu_n V_D/L_c < v_s$, where L_c is the separation between the contacts, i.e., $L_c \sim L$) this current, I_{sp}, is proportional to V_D^2:

$$I_{sp} = 9\varepsilon_s \mu_n W d_c V_D^2/(8L_c^3) \tag{4-6-23}$$

(This equation is called the Mott-Gurney law after Mott and Gurney [1940]). Here d_c is the depth of the n^+ contacts. In the saturation region when $v_n \approx v_s$,

$$I_{sp} = 2\varepsilon_s v_s W d_c V_D/L^2 \tag{4-6-24}$$

The following interpolation formula provides a good approximation for the space-charge injection current for arbitrary drain voltages:

$$j_{sc} = (2/3)u^2 - (2/27)u^3 \qquad u \leq 3 \tag{4-6-25}$$

and

$$j_{sc} = 2(u - 1) \qquad u > 3 \tag{4-6-26}$$

where $u = \mu_n V_D/(v_s L_c)$ and $j_{sc} = I_{sc}\mu_n L/(\varepsilon_s v_s W d_c)$ (see Chen 1984 and Shur 1987, p. 385).

The effect of the punch-through on the MOSFET current-voltage characteristics is illustrated by Fig. 4-6-8 (from Valdya et al. 1986), which shows a superlinear increase in the drain current with the drain voltage even at gate voltages below an expected threshold voltage.

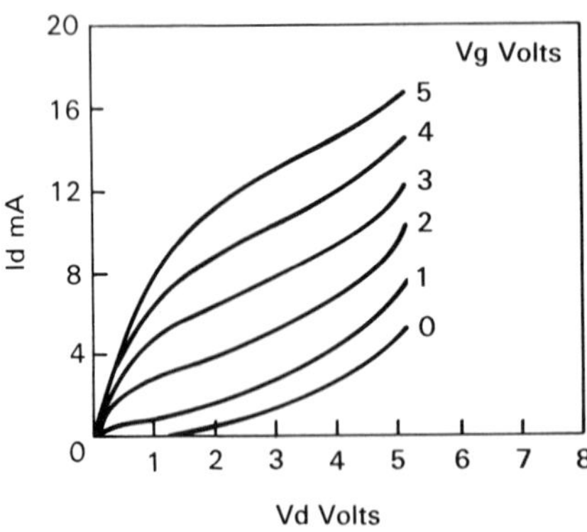

Fig. 4-6-8. The effect of the punch-through on the MOSFET current-voltage characteristics (from S. Valdya, E. N. Fuls, and R. L. Johnson, *IEEE Trans. Electron Devices*, ED-33, p. 1321 (1986). © 1986 IEEE).

4-7. SUBTHRESHOLD CURRENT IN MOSFETs

Ideally, the drain current in a MOSFET flows only when the gate voltage exceeds a well-defined threshold voltage, V_T. The behavior of long-channel devices is often close enough to this ideal situation (see, for example, Fig. 4-5-6a). However, even in long-channel devices there is some small drain current at gate voltages below the threshold. In short-channel devices, this subthreshold current becomes very important.

The subthreshold current flows primarily when the transistor is in the weak inversion regime, i.e., when the surface band bending is in the range between φ_b (which corresponds to the boundary between the depletion regime and weak inversion regime) and $2\varphi_b$ (which corresponds to the boundary between the weak inversion regime and the onset of moderate inversion, occurring when $V_{gs} = V_T$; see Section 4-2.) (For simplicity, we again ignore the distinction between moderate and strong inversion regimes that may be very important for modern devices; see Section 4-2 and Tsividis 1987 for more details.)

The mechanism responsible for the subthreshold current is quite different for long-channel and short-channel devices. In long-channel devices the situation is similar to that in a bipolar junction transistor, where the source plays the role of an emitter, the drain plays the role of a collector, and the region of the p-type substrate in between behaves like a base. The drain voltage drops almost entirely across the drain-substrate depletion region (see Fig. 4-7-1). Hence, the component of the electric field parallel to the semiconductor–insulator interface is small, and the diffusion component of the subthreshold current is dominant, just as for a collector current in a bipolar junction transistor. The subthreshold current can then be evaluated as

$$I_{sub} = -qS_{eff}D_n\, dn/dx \tag{4-7-1}$$

where S_{eff} is the effective cross section for the subthreshold current, D_n is the electron diffusion coefficient ($D_n = \mu_n k_B T/q$), and n is the electron density. If the diffusion length of electrons in the substrate, L_{nd}, is much greater than the channel length, L, the electron density, n, should be a linear function of x, decreasing from the source toward the drain (just like a linear distribution of minority carriers in the base of a bipolar junction transistor; see Fig. 3-1-5).

$$n(x) \approx n_{sd} - (n_{sd} - n_{dd})x/L_1 \tag{4-7-2}$$

where the volume electron concentrations, n_{sd}, at the source side of the channel, and n_{dd}, at the drain side the channel, are given by

$$n_{sd} = n_{po}\exp[qV(y)/k_BT] \tag{4-7-3}$$

$$n_{dd} = n_{po}\exp\{q[V(y) - V_D]/k_BT\} \tag{4-7-4}$$

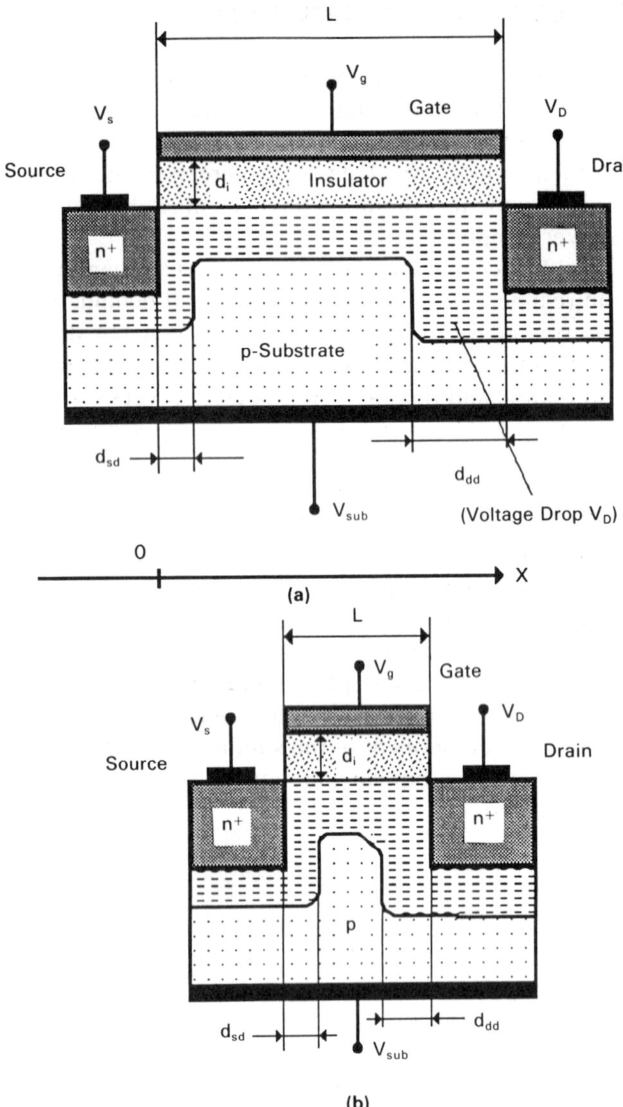

Fig. 4-7-1. Depletion region in (a) long channel and (b) short channel MOSFETs in subthreshold regime.

(again just as expected for the base of a bipolar junction transistor). Here $V(y)$ is the electric potential that is given by

$$V(y) \approx V_s - F_y y \tag{4-7-5}$$

where V_s is the surface potential, y is the coordinate in the direction perpendicular to the semiconductor–insulator interface, and

$$F_y = |Q_{dep}|/\varepsilon_s = (2qV_sN_A/\varepsilon_s)^{1/2} \tag{4-7-6}$$

Since $n_{dd} << n_{sd}$,

$$n(x) \sim n_{sd} - n_{sd}x/L_l \tag{4-7-2a}$$

The distance L_l in eq. (4-7-2) is the length of the undepleted portion of the channel. For long-channel devices, we assume that the depletion widths at the source and drain sides of the channel are small compared with the channel length, L, and hence, $L_l \approx L$.

The effective cross section

$$S_{eff} \approx Wd_c \tag{4-7-7}$$

where d_c is the width of the region where most electrons are located. The electron density, n, at the surface is proportional to $\exp(qV_s/k_BT)$ (see eqs. (4-7-3) and (4-7-4)). The electron density decreases with y proportionally to $\exp(-qF_y y/k_BT)$, and hence, the effective depth, d_c, where the electrons are concentrated, can be estimated as

$$d_c \approx k_BT/[qF_y(0)] \tag{4-7-8}$$

where $y = 0$ corresponds to the insulator–semiconductor interface. Substituting eqs. (4-7-3) and (4-7-4) into eq. (4-7-2) and using eqs. (4-7-7) and (4-7-8), we find that

$$I_{sub} = \varepsilon_s\mu_n(W/L)V_{th}^2(n_i/N_A)^2(V_{th}/V_s)^{1/2}\exp(V_s/V_{th})[(1 - \exp(-V_D/V_{th})]/(2^{1/2}L_{Dp}) \tag{4-7-9}$$

where L_{Dp} is the Debye length, and $V_{th} = k_BT/q$ is the thermal voltage. The surface potential, V_s, at the source can be expressed as a function of the gate voltage using eq. (4-3-6):

$$V_g = V_i + V_s + V_{FB} \tag{4-7-10}$$

where

$$V_i = F_y(0)d_i\varepsilon_s/\varepsilon_i \tag{4-7-11}$$

Using eq. (4-7-7) we find from eqs. (4-7-10) and (4-7-11) that

$$V_s = V_g - V_{FB} + a_s^2 - a_s[a_s^2 + 2(V_g - V_{FB})]^{1/2} \tag{4-7-12}$$

Here

$$a_s = (qN_A\varepsilon_s)^{1/2}d_i/\varepsilon_i \tag{4-7-13}$$

(see Problem 4-7-2). As can be seen from eq. (4-7-9), the subthreshold current is nearly independent of the drain voltage if the drain voltage is larger than several thermal voltages, k_BT/q. This may be expected because in the long-channel devices the drain voltage drops almost entirely across the drain-substrate depletion region (see Fig. 4-7-1), and the subthreshold current is primarily the diffusion current. Moreover, as can be seen from eqs. (4-7-3) and (4-7-4), at large V_D, $n_{dd} << n_{sd}$, and hence, the gradient of n is not affected by V_D (see eq. (4-7-2)). This is similar to the collector current in a bipolar junction transistor being nearly independent of the collector-emitter voltage in the forward active mode (see Chapter 3).

Experimental gate voltage dependence of the subthreshold current in long-channel devices for different drain and substrate voltages is shown in Fig. 4-7-2 (from Troutman 1974). As can be seen from the figure, the subthreshold current is nearly independent of the drain voltage. The substrate bias shifts the threshold voltage (see eq. (4-4-12)) and affects the surface potential, and hence, the subthreshold current. A more detailed analysis of the results presented in Fig. 4-7-2 shows that they are in good agreement with the theory of the subthreshold current for long-channel devices reviewed previously (see also Sze 1981, p. 448).

In short-channel devices the sum of the depletion widths of the source and drain junctions becomes comparable with the channel length (see Fig. 4-7-1b), and their subthreshold current may be evaluated by substituting

$$L_{eff} = L - x_{ds} - x_{dd} \tag{4-7-14}$$

for L in eq. (4-7-9). Here

$$x_{ds} = [2\varepsilon_s(V_{bi} - V_s)/(qN_A)]^{1/2} \tag{4-7-15}$$

$$x_{dd} = [2\varepsilon_s(V_{bi} - V_s + V_D)/(qN_A)]^{1/2} \tag{4-7-16}$$

are effective widths of the depletion regions at the source and the drain, respectively, V_{bi} is the built-in voltage for the n^+ contact p-substrate junction, and the

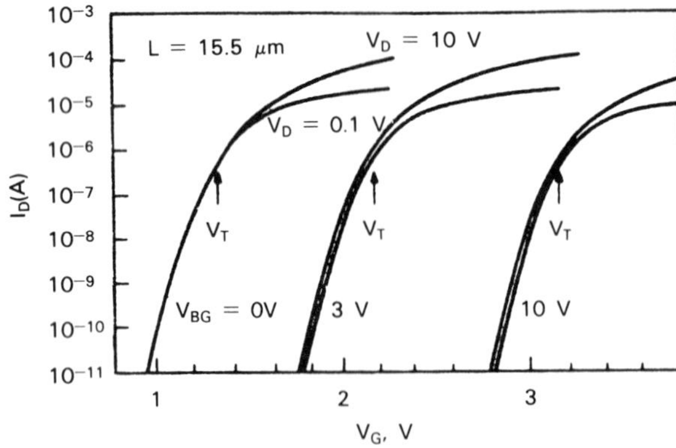

Fig. 4-7-2. Experimental gate voltage dependence of the subthreshold current in long channel devices for different drain and substrate voltages (from R. R. Troutman, *IEEE J. Solid State Circuits* SC-9, p. 55 (1974). © 1974 IEEE).

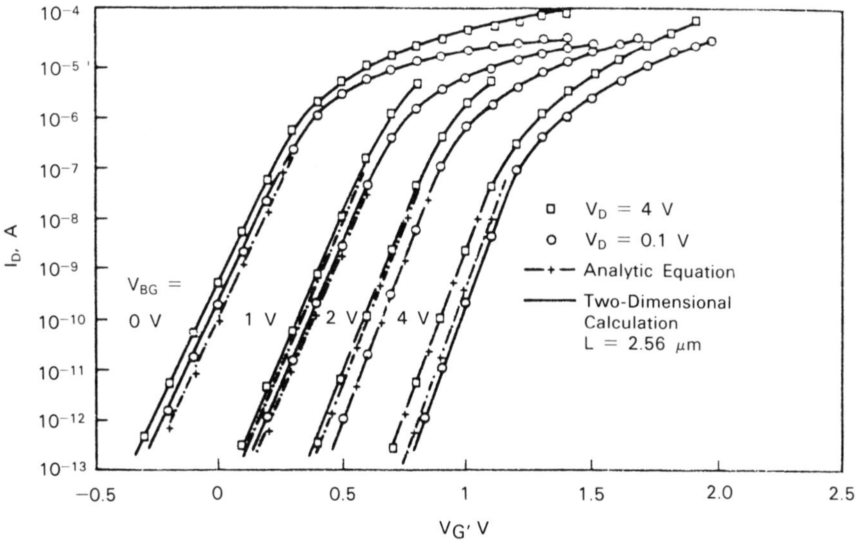

Fig. 4-7-3. Subthreshold current vs. gate voltage in a short channel MOSFET for different drain and substrate voltages (after W. Fichner and H. W. Potzl, *International J. Electronics,* 46, p. 33 (1979)).

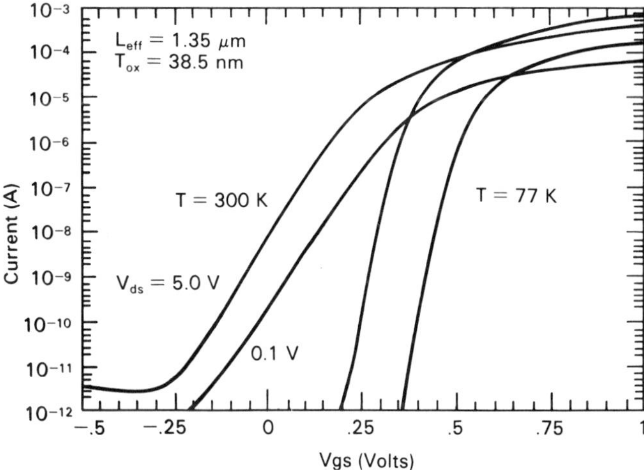

Fig. 4-7-4. Subthreshold current vs. gate voltage at T = 300°K and T = 77°K for an *n*-channel MOSFET (from J. D. Plummer, *IEDM Technical Digest,* p. 378. IEEE Publication, Los Angeles (1986). © 1986 IEEE). T_{ox} is the oxide thickness.

surface potential is now found from the solution of the following equation (see Sze 1981, p. 474):

$$V_s = V_g - V_{FB} - [q\varepsilon_s N_A(V_s - V_{sub})/2]^{1/2}[1 + (L - W_{ds} - W_{dd})/(L - x_{ds} - x_{dd})]/c_i \tag{4-7-17}$$

where V_{sub} is the substrate potential,

$$W_{ds} = [2\varepsilon_s(V_{bi} - V_{sub})/(qN_A)]^{1/2} \tag{4-7-18}$$

$$W_{dd} = [2\varepsilon_s(V_{bi} - V_{sub} + V_D)/(qN_A)]^{1/2} \tag{4-7-19}$$

(see Problem 4-7-1). The comparison of this model with experimental data and two-dimensional simulation is shown in Fig. 4-7-3 (after Fichner and Potzl 1979). As can be seen from the figure, the analytical model is in good agreement with the measured data and a more rigorous numerical simulation.

The temperature dependence of the subthreshold current is illustrated by Fig. 4-7-4 (from Plummer 1986). As can be seen from the figure the slope of the log of the subthreshold current vs. gate voltage curve is indeed proportional to $1/T$, as predicted by the theory. We also notice the dependence of the threshold voltage on the drain voltage typical for short-channel devices.

4-8. MOSFET CAPACITANCES AND EQUIVALENT CIRCUIT

The charge in the MOSFET inversion layer and in the depletion region depends on the gate, source, drain, and substrate (body) potentials. The derivatives of this charge with respect to the terminal voltages may be defined as *MOSFET capacitances*. The equivalent MOSFET circuit that incorporates these capacitances is shown in Fig. 4-8-1 (see, for example, Liu and Nagel 1982). The diodes connected between the substrate and the source and drain contacts represent the leakage

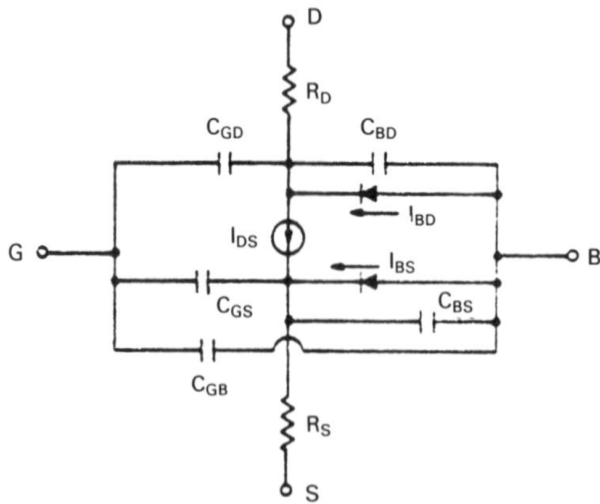

Fig. 4-8-1. MOSFET equivalent circuit (after L. W. Nagel, "SPICE2: A Computer Program to Simulate Semiconductor Circuits," Memorandum No. ERL-M520, Electronics Research Laboratory, College of Engineering, University of California, Berkeley, May 9 (1975)).

currents in the induced channel-to-substrate junctions. This equivalent circuit is used for the transient simulation of MOSFET circuits in the popular circuit simulation program SPICE-II (see Nagel 1975).

The values of these capacitances in SPICE-II are calculated using Meyer's model (see Meyer 1971). In this model the total charge Q_G in the linear regime (when $V_D < V_{Dsat}$) is calculated as

$$Q_G = \int_0^L Q(x)\, dx = W \int_0^L c_i\, V_i(x)\, dx \tag{4-8-1}$$

where $Q(x)$ is the channel charge per unit length, W is the gate width, c_i is the gate insulator capacitance per unit area, and $V_i(x)$ is the voltage drop across the gate insulator. The depletion charge located between the channel and the substrate is attributed in this approach to the gate-to-substrate, gate-to-source, and gate-to-drain capacitances. Using expressions given by the charge control model, we obtain the following expression for the gate charge:

$$Q_G = \frac{2}{3} C_i \frac{(V_{gd} - V_T)^3 - (V_{gs} - V_T)^3}{(V_{gd} - V_T)^2 - (V_{gs} - V_T)^2} \tag{4-8-2}$$

where $C_i = c_i WL$ (see Problem 4-8-1). The gate-to-source, gate-to-drain, substrate-to-source, and substrate-to-drain capacitances are defined as

$$C_{gs} = -\left.\frac{\partial Q_G}{\partial V_s}\right|_{V_g, V_{sub}, V_d = \text{const}} \tag{4-8-3}$$

$$C_{gd} = -\left.\frac{\partial Q_G}{\partial V_d}\right|_{V_g, V_{sub}, V_d = \text{const}} \tag{4-8-3-a}$$

$$C_{bs} = -\left.\frac{\partial Q_{sub}}{\partial V_s}\right|_{V_g, V_{sub}, V_d = \text{const}} \tag{4-8-4}$$

$$C_{bd} = -\left.\frac{\partial Q_{sub}}{\partial V_d}\right|_{V_g, V_{sub}, V_d = \text{const}} \tag{4-8-4-a}$$

where V_g, V_{sub}, V_d are gate, substrate, drain, and source potentials, and Q_{sub} is the charge at the substrate contact (see Tsividis 1987, p. 313).

Using eq. (4-8-2), we obtain

$$C_{gs} = \frac{2}{3} C_i \left[1 - \frac{(V_{gd} - V_T)^2}{(V_{gd} - V_T + V_{gs} - V_T)^2}\right] \tag{4-8-5}$$

and

$$C_{gd} = \frac{2}{3} C_i \left[1 - \frac{(V_{gs} - V_T)^2}{(V_{gd} - V_T + V_{gs} - V_T)^2}\right] \tag{4-8-6}$$

(see Meyer 1971 and Problem 4-8-2). This approach can be extended to describe the saturation regime, subthreshold regime, and accumulation regime, and to include fringing and overlap capacitances, C_{gsf} and C_{gdf}, as well as depletion substrate-to-source, substrate-to-gate, and substrate-to-drain capacitances, C_{bs}, C_{bg}, and C_{bd}. The resulting expressions were summarized by Liu and Nagel (1982) and are given as follows:

Accumulation Region ($V_{gs} \le V_{FB} + V_{sub}$)

$$C_{gs} = C_{gsf}, \; C_{gd} = C_{gdf} \tag{4-8-7}$$

$$C_{gb} = c_i WL \tag{4-8-8}$$

$$C_{bs} = \frac{C_{js}}{\left(1 - \dfrac{V_{sub\text{-}s}}{\phi_b}\right)^{m_B}} \qquad C_{bd} = \frac{C_{jd}}{\left(1 - \dfrac{V_{sub\text{-}d}}{\phi_b}\right)^{m_B}} \tag{4-8-9}$$

Here C_{gsf} and C_{gdf} are fringing and overlap gate-to-source and gate-to-drain capacitances, C_{js} and C_{jd} are substrate-to-source and substrate-to-drain capacitances for zero biases, $V_{sub\text{-}s}$ and $V_{sub\text{-}d}$ are voltages, between the source and the substrate and the drain and the substrate, and ϕ_b is the built-in potential between the substrate and the source and drain regions. The exponent m_B depends on the doping profile ($m_B = 1/2$ for the uniform substrate doping).

Accumulation-Depletion Region ($V_{FB} + V_{sub} \le V_{gs} \le V_T - 0.5$ V)

$$C_{gs} = C_{gsf}, \; C_{gd} = C_{gdf} \tag{4-8-10}$$

$$C_{gb} = \frac{c_i WL}{\sqrt{1 + \dfrac{4(V_{gs} - V_{FB} - V_{sub\text{-}s})}{K_1^2}}} \tag{4-8-11}$$

$$C_{bs} = \frac{C_{js}}{\left(1 - \dfrac{V_{sub\text{-}s}}{\phi_b}\right)^{m_B}} \qquad C_{bd} = \frac{C_{jd}}{\left(1 - \dfrac{V_{sub\text{-}d}}{\phi_b}\right)^{m_B}} \tag{4-8-12}$$

Here $K_1 = (2q\varepsilon_s)^{1/2} N_{sub}/c_i$.

Depletion-Inversion Region ($V_T - 0.5 \text{ V} \le V_{gs} \le V_T$)

$$C_{gs} = C_{gsf} + \frac{2}{3} c_i WL[1 - 4(V_{gs} - V_T)^2] \tag{4-8-13}$$

$$C_{gd} = C_{gdf} \tag{4-8-14}$$

$$C_{gb} = c_i WL \frac{1 + 4(V_{gs} - V_T - 0.5 \text{ V})^2}{\sqrt{1 + \dfrac{4(V_{gs} - V_{FB} - V_{sub\text{-}s})}{K_1^2}}} \tag{4-8-15}$$

$$C_{bs} = \frac{C_{js}}{\left(1 - \dfrac{V_{sub\text{-}s}}{\phi_b}\right)^{m_B}} \left\{1 + \frac{2}{3}\frac{C_{gb}}{C_{js}} WL[1 - 4(V_{gs} - V_T)^2]\right\} \tag{4-8-16}$$

$$C_{bd} = \frac{C_{jd}}{\left(1 - \dfrac{V_{sub\text{-}d}}{\phi_b}\right)^{m_B}} \tag{4-8-17}$$

Equations (4-8-13) to (4-8-16) are interpolation formulas. That is why some coefficients in these equations are given certain numerical values. We should mention that these results have been obtained without taking into account the moderate inversion regime. Tsividis (1987) gives a more accurate analysis of device capacitances.

Saturation Region ($V_{gs} > V_T$ and $V_{ds} \geq V_{Dsat}$)

$$C_{gs} = C_{gsf} + \frac{2}{3} c_i WL \qquad C_{gd} = C_{gdf} \qquad C_{gb} = 0 \tag{4-8-19}$$

$$C_{bs} = \frac{C_{js}}{\left(1 - \dfrac{V_{sub\text{-}s}}{\phi_b}\right)^{m_B}} \left[1 + \frac{2}{3} \frac{C_{gb}}{C_{js}} WL\right] \tag{4-8-20}$$

$$C_{bd} = \frac{C_{jd}}{\left(1 - \dfrac{V_{sub\text{-}d}}{\phi_b}\right)^{m_B}} \tag{4-8-21}$$

Triode Region ($V_{gs} > V_T$ and $0 < V_{ds} < V_{Dsat}$)

$$C_{gs} = C_{gsf} + \frac{2}{3} c_i WL \frac{V_{Dsat}(3V_{Dsat} - 2V_{ds})}{(2V_{Dsat} - V_{ds})^2} \tag{4-8-22}$$

$$C_{gs} = C_{gsf} + \frac{2}{3} c_i WL \frac{(V_{Dsat} - V_{ds})(3V_{Dsat} - V_{ds})}{(2V_{Dsat} - V_{ds})^2} \tag{4-8-23}$$

$$C_{gb} = 0 \tag{4-8-24}$$

$$C_{bs} = \frac{C_{js}}{\left(1 - \dfrac{V_{sub\text{-}s}}{\phi_b}\right)^{m_B}} \left[1 + \frac{2}{3} \frac{C_{gb}}{C_{js}} WL \frac{V_{Dsat}(3V_{Dsat} - 2V_{ds})}{(2V_{Dsat} - V_{ds})^2}\right] \tag{4-8-25}$$

$$C_{bd} = \frac{C_{jd}}{\left(1 - \dfrac{V_{sub\text{-}d}}{\phi_b}\right)^{m_B}} \left[1 + \frac{2}{3} \frac{C_{gb}}{C_{js}} WL \frac{(V_{Dsat} - V_{ds})(3V_{Dsat} - V_{ds})}{(2V_{Dsat} - V_{ds})^2}\right] \tag{4-8-26}$$

A typical dependence of the MOSFET capacitances on the gate-to-source voltage calculated using the foregoing expressions is shown in Fig. 4-8-2.

Meyer's model is simple and straightforward. However, it has serious shortcomings, especially for the description of short-channel devices. In SPICE circuit simulations, the charges related to the capacitances are determined using

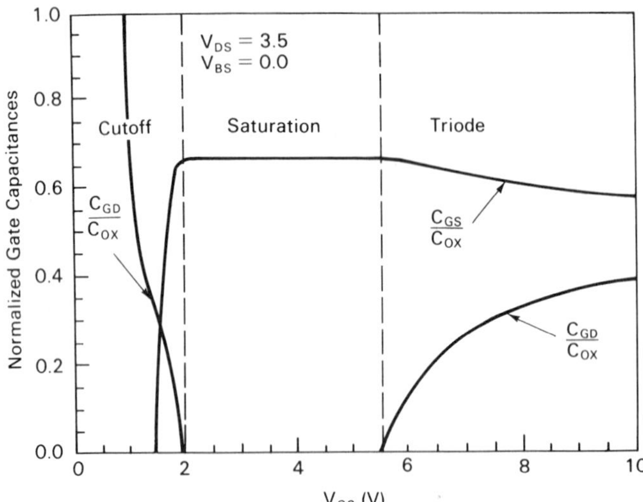

Fig. 4-8-2. Dependence of the MOSFET capacitances on the gate-to-source voltage (from S. Liu and L. W. Nagel, *IEEE J. Solid State Circuits*, SC-17, no. 6, pp. 983–998 (1982). © 1982 IEEE).

the numerical integration. Depending on the integration step, integration errors may lead to the nonconservation of charge (see, for example, Ward and Dutton 1978). These problems may be avoided by partitioning the total charge in the channel into charges related to the gate, source, drain, and substrate charges and using these charges as independent variables in the circuit analysis. Such a model was described by Yang (1986), who showed that it has a better convergence and accuracy than Meyer's model.

A simplified MOSFET equivalent circuit is shown in Fig. 4-8-3a. Using this circuit, we can analyze the MOSFET performance as a small-signal common-source amplifier (see Fig. 4-8-3b). Using the standard circuit analysis, we find the following expression for the voltage gain:

$$A_v = \frac{v_o}{v_i} = \frac{-g_m + j\omega C_{gd}}{g_d + j\omega C_{gd}} \tag{4-8-27}$$

Using the equivalent circuit shown in Fig. 4-8-3c we find the short-circuit current gain

$$A_i = \frac{i_L}{i_i} = \frac{-g_m}{j\omega(C_{gs} + C_{gd})} \tag{4-8-28}$$

From this equation we can find the cutoff frequency, f_T, at which the absolute value of the short-circuit current gain is equal to 1:

$$f_T = \frac{g_m}{2\pi C_g} \tag{4-8-29}$$

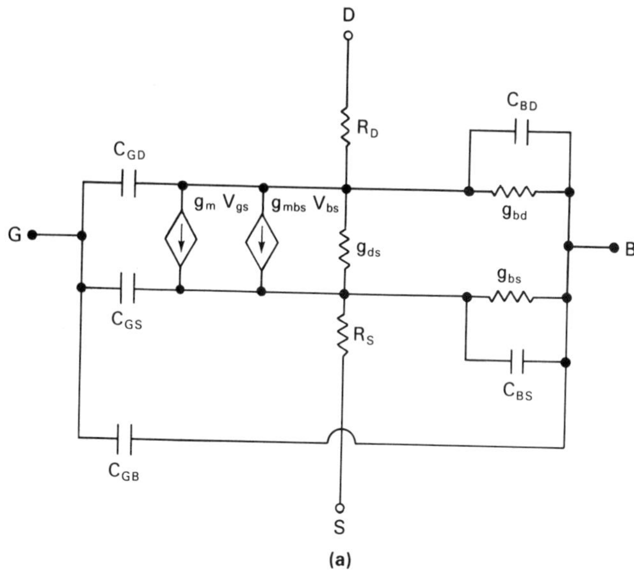

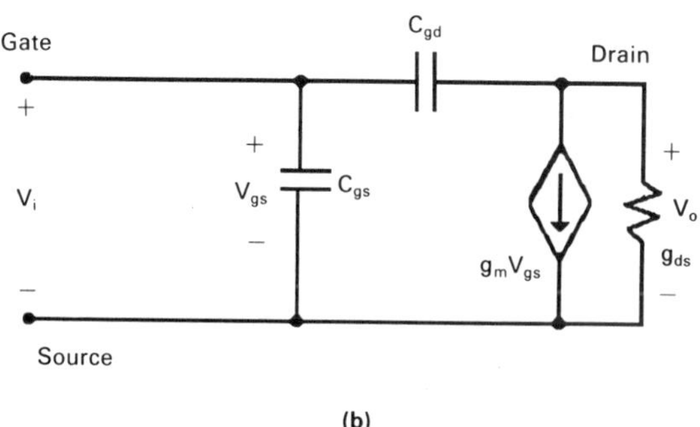

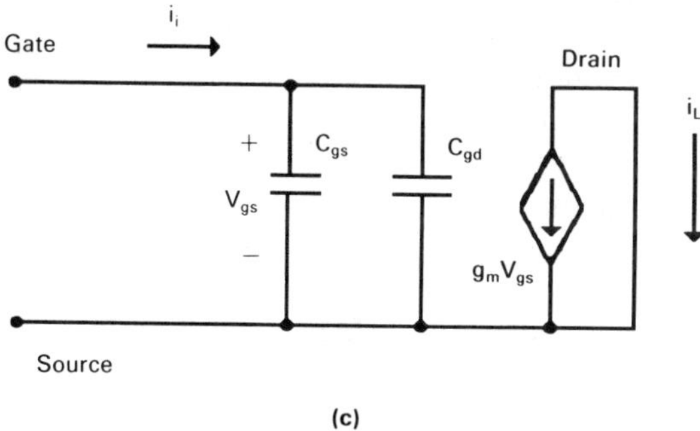

Fig. 4-8-3. MOSFET (a) small signal equivalent circuit (from Liu and Nagel [1982]), (b) simplified equivalent circuit, and (c) simplified equivalent circuit for calculation of the short circuit current gain.

where $C_g = C_{gs} + C_{gd}$. We can assume that $C_g \sim C_i$ and that

$$g_m \sim C_i v_{eff}/L \tag{4-8-30}$$

where v_{eff} is the effective velocity of electrons in the channel. (Equation (4-8-30) can be obtained as follows: The drain current $I_d = qn_s v_{eff} W$, where n_s is the electron concentration in the channel per unit area. Hence, $I_d \approx (Q_G/WL)v_{eff}W$, and $g_m = \partial I_d/\partial V_g \approx (\partial Q_G/\partial V_g)v_{eff}/L \approx C_i v_{eff}/L$.) When $V_D < V_{Dsat}$

$$v_{eff} \sim \mu_n V_D/L \tag{4-8-31}$$

(see Section 4-4). When $V_D > V_{Dsat}$, v_{eff} is less than v_s and can be estimated using the theory developed in Section 4-5. Equation (4-8-30) leads to the following estimate for f_T:

$$f_T = \frac{1}{2\pi\tau_{tr}} \tag{4-8-32}$$

where $\tau_{tr} = L/v_{eff}$ is the transit time of electrons in the channel. Assuming the effective electron velocity to be of the order of 5×10^4 m/s we obtain that the characteristic switching time for a MOSFET, τ_{tr}, is on the order of $t_s \sim \tau_{tr}$ (ps) $\sim$ $20/L$ (μm) and f_T (GHz) $\sim 8/L$ (μm). In fact, the measured switching times are at least several times larger because of the parasitic and fringing capacitances, C_p, that have to be added to the gate capacitance, leading to the following estimate for f_T:

$$f_T = \frac{g_m}{2\pi(C_g + C_p)} = \frac{1}{2\pi\tau_{tr}(1 + C_p/C_g)} \tag{4-8-33}$$

This expression may be used to determine how the device speed scales with the device dimensions (see Section 4-11).

4-9. ENHANCEMENT AND DEPLETION MODE MOSFETS. COMPLEMENTARY MOSFETS (CMOS) AND SILICON ON SAPPHIRE

MOSFET threshold voltages depend on the doping level in the semiconductor substrate (see eq. (4-4-12)). This allows us to control the threshold voltage by varying the doping density near the semiconductor–insulator interface using ion implantation. An additional implant of acceptor-type impurities shifts the threshold voltage toward more positive values, yielding an *enhancement-mode transistor* (compare Figs. 4-9-1a and 4-9-1b). A donor implant creates a conducting channel between the n^+ source and n^+ drain that has to be depleted by the applied gate voltage in order to turn the device off. Such an implant yields a *depletion-mode* device (see Fig. 4-9-1c).

Qualitative transfer characteristics for enhancement- and depletion-mode n-channel and p-channel devices are shown in Fig. 4-9-2. Figures 4-9-3 and 4-9-4

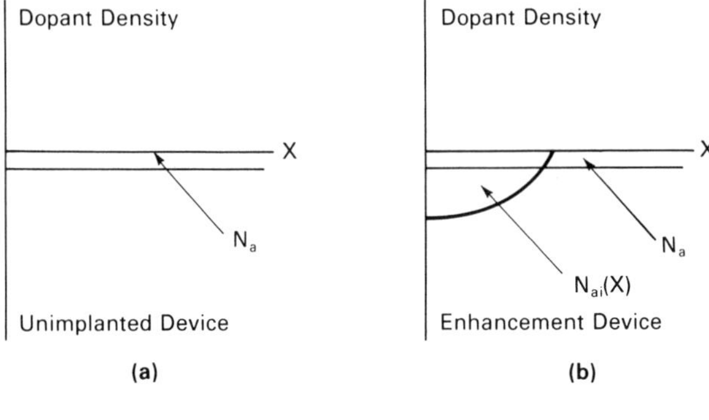

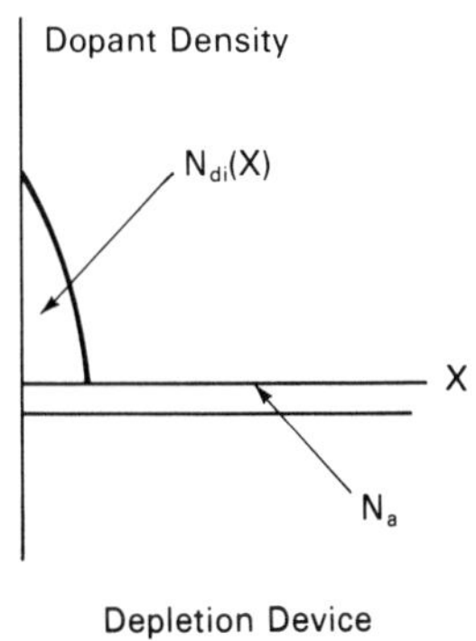

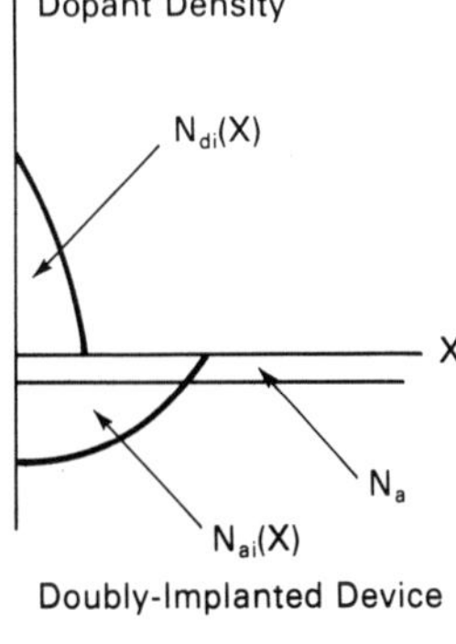

Fig. 4-9-1. Qualitative doping profiles of (a) unimplanted, (b) enhancement, (c) depletion, and (d) double implanted *n*-channel MOSFETs (after B. S. Song and P. R. Gray, *IEEE J. Solid State Circuits*, SC-17, no. 2, pp. 291–298 (1982). © 1982 IEEE).

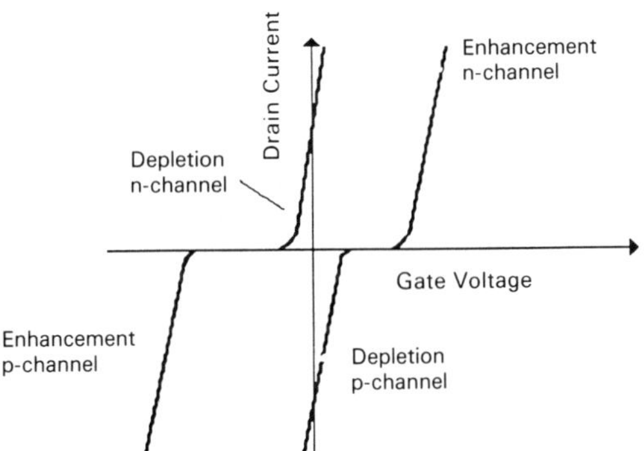

Fig. 4-9-2. Qualitative transfer characteristics for enhancement and depletion mode *n*-channel and *p*-channel MOSFETs.

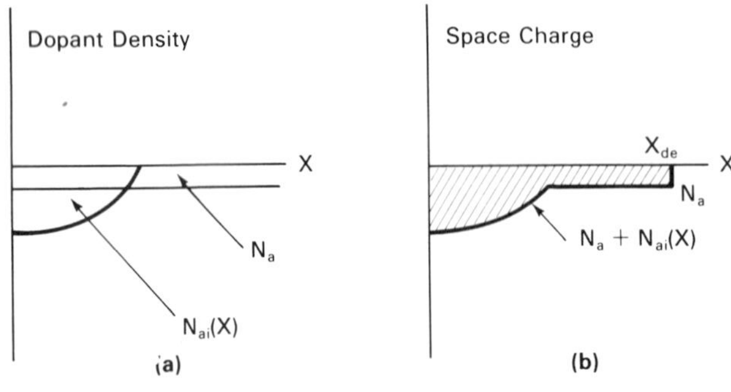

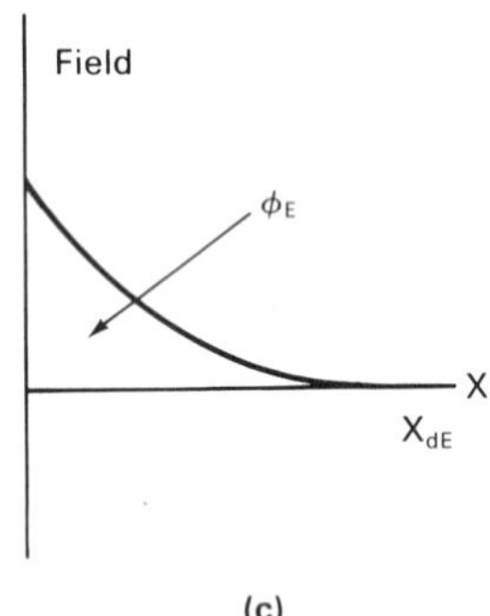

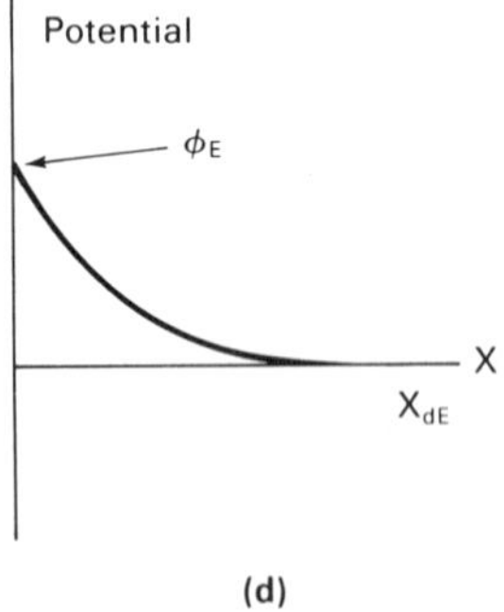

Fig. 4-9-3. Qualitative profiles of space charge, electric field, and electric potential for enhancement *n*-channel MOSFET at the threshold (after B. S. Song and P. R. Gray, *IEEE J. Solid State Circuits*, SC-17, no. 2, pp. 291–298 (1982). © 1982 IEEE).

show qualitative profiles of space charge, electric field, and electric potential for enhancement and depletion n-channel devices at the threshold (after Song and Gray 1982). As can be seen from these profiles, a principal effect of the ion implantation is to introduce an additional charge of ionized impurities, shifting the threshold voltage accordingly. The threshold voltage is also shifted by a change in the built-in potential between the channel and the substrate. Hence, for an enhancement-mode device the threshold voltage increase, ΔV_T, caused by acceptor implantation is given by

$$\Delta V_T = \Delta V_{bi} + q\Delta n_A/c_i \tag{4-9-1}$$

where

$$\Delta V_{bi} = k_B T/q \ln(N_{Aimax}/N_A) \tag{4-9-2}$$

is the shift in the built-in voltage, and

$$\Delta n_A = \int_0^{x_{dE}} [N_{Ai}(x) + N_A(x)]\, dx - \int_0^{x_d} N_A\, dx \tag{4-9-3}$$

is the additional surface density of implanted acceptors in the depletion layer. Here $N_{Ai}(x)$ is the acceptor concentration in the implanted device, N_{Aimax} is the maximum value of the acceptor concentration in the implanted device, N_A is the

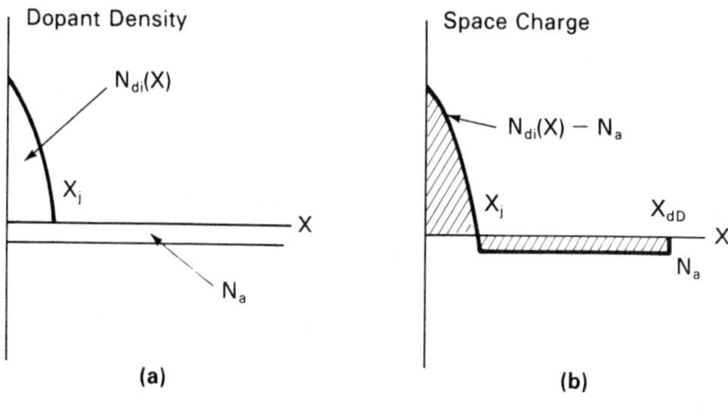

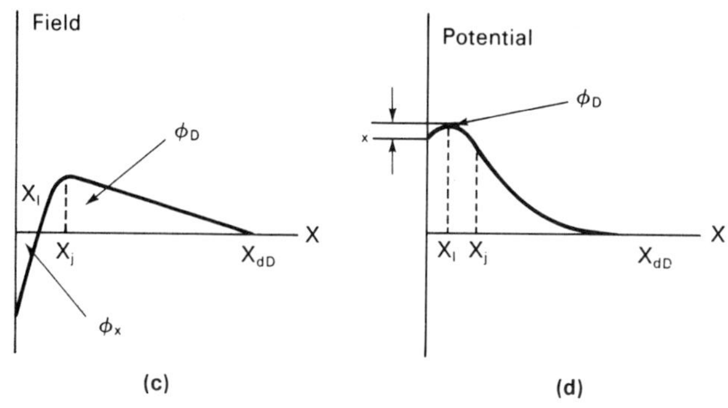

Fig. 4-9-4. Qualitative profiles of space charge, electric field, and electric potential for depletion *n*-channel MOSFET at the threshold (after B. S. Song and P. R. Gray, *IEEE J. Solid State Circuits*, SC-17, no. 2, pp. 291–298 (1982). © 1982 IEEE).

acceptor concentration in the unimplanted device, x_{dE} is the boundary of the depletion region at the threshold in the implanted device, and x_d is the boundary of the depletion region at the threshold in the unimplanted device (see Song and Gray 1982). Similarly, in the depletion-mode MOSFET the threshold voltage shift, ΔV_T, caused by a donor implantation is given by

$$\Delta V_T = \Delta V_{bi} - \Phi_x - q\Delta n_D / c_i \tag{4-9-4}$$

where

$$\Delta V_{bi} = k_B T/q \ln(n_l/N_A) \tag{4-9-5}$$

is the shift in the built-in voltage, and

$$q\Delta n_D = q \int_0^{x_{dD}} [N_{Di}(x) - N_A(x)]\, dx + q \int_0^{x_d} N_A\, dx \tag{4-9-6}$$

is the additional surface charge of implanted impurities in the depletion layer. Here $N_{Di}(x)$ is the donor concentration in the implanted device, n_l is the concentration of electrons at $x = x_l$ in the implanted device (see Fig. 4-9-4d), x_{dD} is the boundary of the depletion region at the threshold in the implanted device, x_d is the boundary of the depletion region at the threshold in the unimplanted device, and

Φ_x is the potential drop from the the maximum potential point, x_1, to the surface (see Fig. 4-9-4):

$$\Phi_x = (q/\varepsilon_s) \int_0^{x_1} x[N_{Di}(x) - N_A(x)]\, dx \tag{4-9-7}$$

The dependence of the threshold voltages for an unimplanted MOSFET, enhancement MOSFET, doubly implanted MOSFET, and depletion MOSFET on temperature is shown in Fig. 4-9-5. As can be seen from this figure, the ion implantation can be used to control the threshold voltage over a wide range of temperatures (see also Problems 4-9-1 to 4-9-4).

Ion-implantation and diffusion technologies can be used to fabricate both n-channel and p-channel devices on the same wafer (Complementary Metal Oxide Semiconductor (CMOS) technology). A typical CMOS structure is shown in Fig. 4-9-6. A p-channel device is fabricated using two p^+ implants into an n-type silicon substrate for the source and drain contacts. Two n^+ regions are implanted in a p well to provide the source and drain contacts for an n-channel MOSFET; n^+ and p^+ guard rings (also called channel stops) serve to isolate n- and p-channel devices.

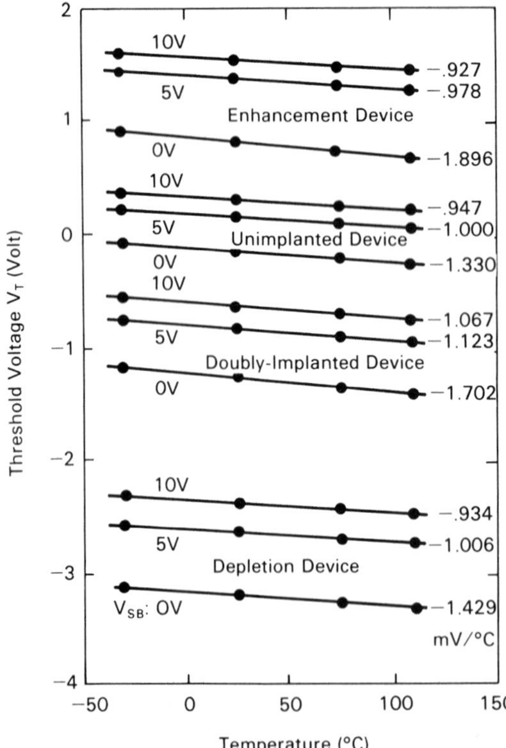

Fig. 4-9-5. The temperature dependence of the threshold voltages for an unimplanted MOSFET, enhancement MOSFET, doubly implanted MOSFET, and depletion MOSFET (after B. S. Song and P. R. Gray, *IEEE J. Solid State Circuits*, SC-17, no. 2, pp. 291–298 (1982). © 1982 IEEE).

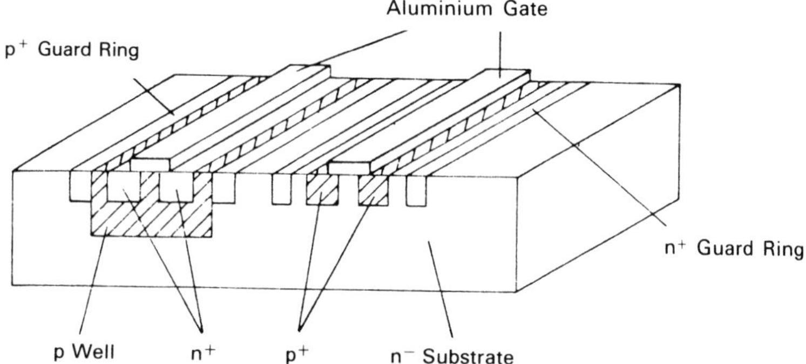

Fig. 4-9-6. A typical CMOS structure (after J. Compton, "Limitations of and Improvements to MOS Processes," in *MOS Devices: Design and Manufacture*, ed. A. D. Milne, Halsted Press, New York (1983)).

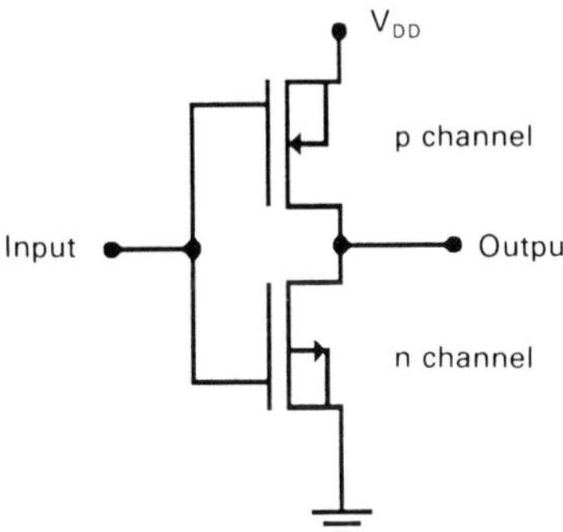

Fig. 4-9-7. A CMOS logic gate

Typically, n- and p-channel CMOS devices are connected as shown in Fig. 4-9-7 to form a CMOS logic gate. The two transistors have the same input, but their threshold voltages are such that either only one transistor or none is turned on for a given gate voltage (see Fig. 4-9-8). Such a CMOS gate has a very low power consumption in either of the two stable states and a large voltage swing. Another advantage of CMOS technology is that it can operate with a wide range of supply voltages (from 15 volts to 3 volts or less) with a large output voltage swing (from zero to the supply voltage for the CMOS inverter shown in Fig. 4-9-7). CMOS devices can be also integrated with bipolar devices. This *BICMOS* technology allows one to combine low-power logic signals processed by CMOSFETs with high power output provided by bipolar transistors. It is widely used in High Voltage Integrated Circuits (HVICs) (see, for example, Baliga 1986).

The drawbacks of the CMOS technology include a more complicated fabrication process (compared with n-channel MOSFET technology) and a possibility of "latch-up" when the CMOS gate nearly short-circuits the power supply. This

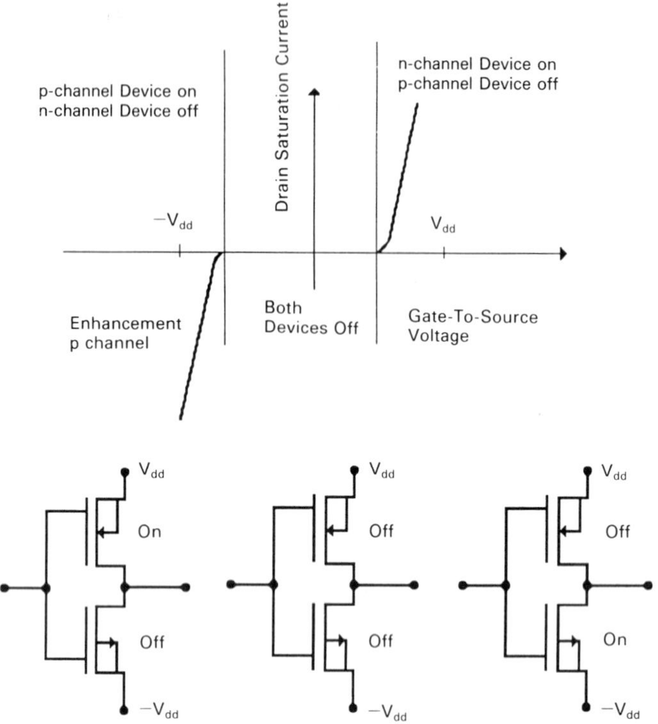

Fig. 4-9-8. Schematic illustration of operation of CMOS gate.

effect could be explained by referring to the CMOS structure and identifying two parasitic bipolar transistor (vertical n-p-n and lateral p-n-p; see Fig. 4-9-9a). The parasitic vertical n-p-n transistor is formed by the n^+ contact (emitter), p-type tub (base), and n-type substrate or epilayer (collector). The parasitic lateral p-n-p transistor is formed by the p^+ contact (emitter), n-type substrate or epilayer (base), and p-type tub (collector). The parasitic vertical transistor may have a high short-circuit common-emitter current gain (above 100). The short-circuit common-emitter current gain of the p-n-p transistor is typically less or even much less than 1. If the product of these two gains is larger than unity, the positive feedback in the loop shown in Fig 4-9-9a leads to a latch-up. This effect can be prevented by using guard rings as shown in Fig. 4-9-6. Another technique is to use a layout with an increased distance between the p-well and p^+ contacts of the p-channel device. A heavily doped buried layer may also be used to increase the doping in the base of the vertical n-p-n transistor and reduce the gain (see Fig. 4-9-9b).

A very promising, but yet to be refined and still expensive, approach is to use insulating substrate—Silicon-On-Insulator (SOI)—technology. This eliminates any substrate current and reduces parasitic capacitances related to the

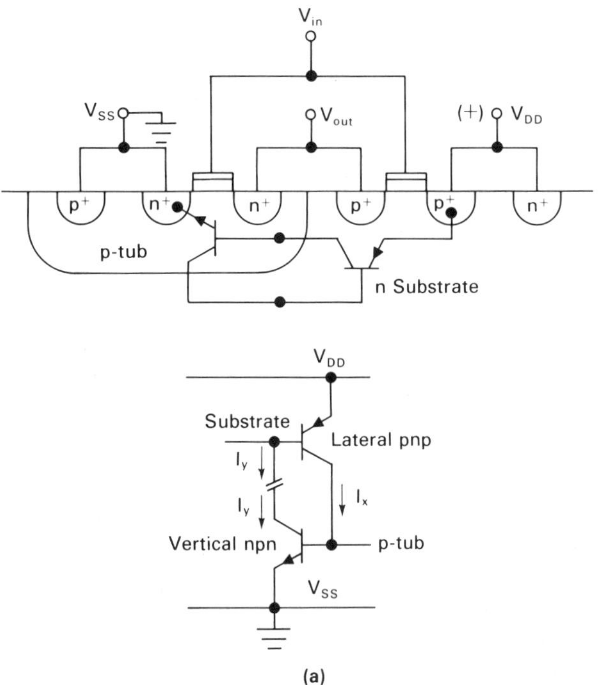

(a)

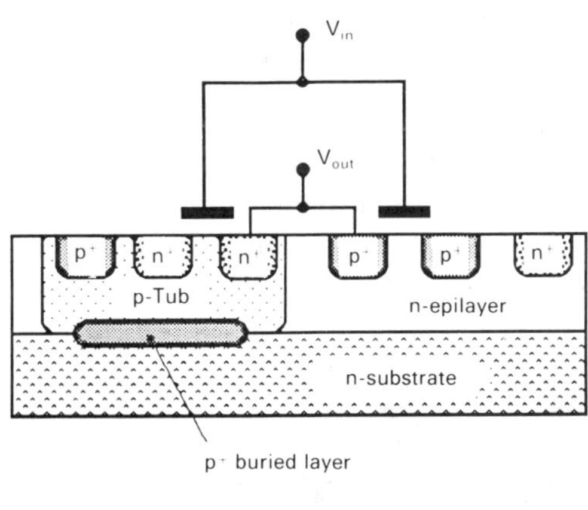

(b)

Fig. 4-9-9. (a) CMOS structure with two parasitic bipolar transistor (*n-p-n* and lateral *p-n-p*) identified and its equivalent circuit (from D. G. Ong, *Modern MOS Technology, Processes, Devices, and Design*, McGraw-Hill, New York (1984)). (b) Use of a buried p^+ layer in CMOS structure in order to minimize latch-up problems.

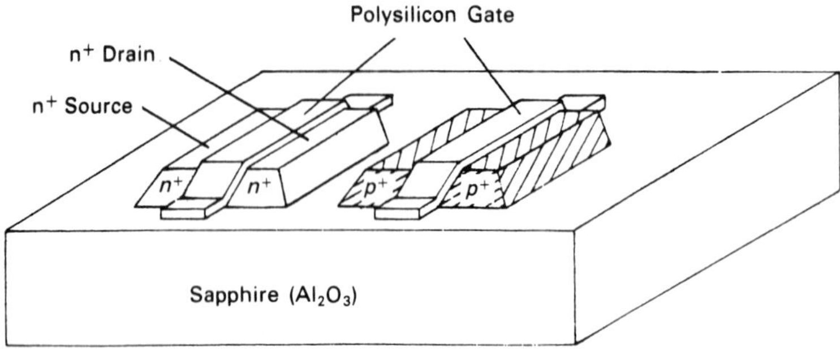

Fig. 4-9-10. Silicon-On-Sapphire, SOS technology (after J. Compton, "Limitations of and Improvements to MOS Processes," in *MOS Devices: Design and Manufacture,* ed. A. D. Milne, Halsted Press, New York (1983)).

depletion region formed in the semiconducting substrate. A schematic diagram of a CMOS gate on a sapphire substrate (Silicon-On-Sapphire (SOS) technology) is shown in Fig. 4-9-10. Silicon islands are epitaxially grown on sapphire and used for n- and p-channel transistors. This technology requires a significantly smaller number of fabrication steps and can result in a more compact layout, as the need for guard rings is eliminated. However, the high cost of sapphire substrates and the difficulties related to the epitaxial growth of silicon on sapphire have slowed down the application of this technology. Still, this is a well-developed technology, and silicon-on-sapphire chips are produced in significant quantities by a number of semiconductor manufacturers. Other examples of silicon-on-insulator technology include using oxygen-ion-implanted silicon as a substrate (called SIMOX technology), using recrystallized polycrystalline silicon utilizing Zone Melting Recrystallization (ZMR) technology (see Partridge 1986), and using oxidized porous silicon (see Zorinsky et al. 1986).

Further refinement of SOI technologies will also open up the possibility of vertical integration, leading to the development of three-dimensional integrated circuits (see Fig. 4-9-11; from Akasaki and Nishimura (1986), who discuss the basic concepts and technologies for three-dimensional ICs). "Stacked" CMOS

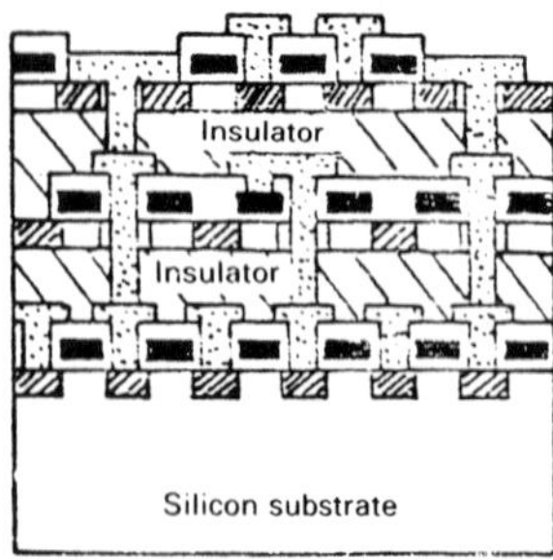

Fig. 4-9-11. Schematic drawing of a 3-*d* integrated circuit (from Y. Akasaki and T. N. Shimura, *IEDM Technical Digest,* p. 488, IEEE Publication, Los Angeles (1986). © 1986 IEEE.)

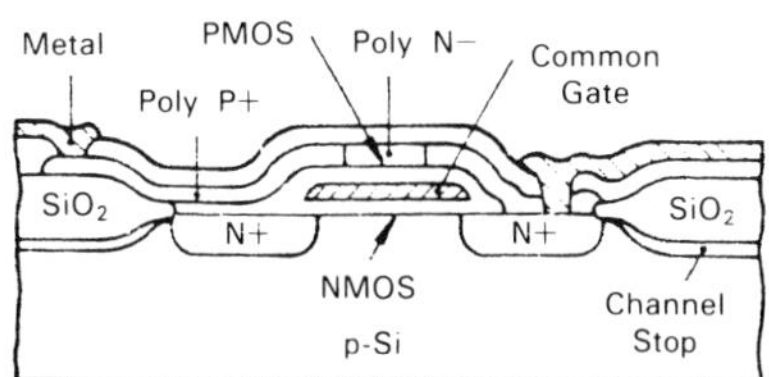

Fig. 4-9-12. Stacked CMOS structure (from C. E. Chen, H. W. Lam, S. D. S. Malhi, and A. R. Pnizzotto, *IEEE Electron Device Lett.*, EDL-4, no. 8, pp. 272–274 (1983). © 1983 IEEE).

(see Gibbons and Lee 1980), in which p- and n-channel devices share a common gate (with the p-channel device on the top and the n-channel device at the bottom of the gate), have also been proposed to increase packing density and alleviate isolation problems (see Fig. 4-9-12; from Chen et al. 1983).

4-10. METAL SEMICONDUCTOR FIELD-EFFECT TRANSISTORS

In MEtal Semiconductor Field-Effect Transistors (MESFETs), the Schottky barrier gate contact is used to modulate the channel conductivity (see Fig. 4-1-2b). This allows one to avoid problems related to traps in the gate insulator in MOSFETs, such as hot electron trapping (see Section 4-6), threshold voltage shift due to charge trapped in the gate insulator, etc. Silicon has a very stable natural oxide that can be grown with a very low density of traps, and the problems related to the gate insulators for silicon MOSFETs have been minimized. However, compound semiconductors, such as GaAs, do not have such a stable oxide; therefore, most of the compound semiconductor devices use Schottky gates. The drawback of MESFET technology is a limited gate voltage swing due to the low turn-on voltage of the Schottky gate. However, this limitation is less important in low-power circuits operating with a low supply voltage.

Compound semiconductor field-effect transistors occupy an important niche in the electronics industry. GaAs FET amplifiers, oscillators, mixers, switches, attenuators, modulators, and current limiters are widely used, and high-speed integrated circuits based on GaAs FETs have been developed. The basic advantages of GaAs devices include a higher electron velocity and mobility, which lead to smaller transit time and faster response, while semi-insulating GaAs substrates reduce parasitic capacitances and simplify the fabrication process. Other material systems, such as AlGaAs/InGaAs or InGaAs/InP, have also exhibited superior device properties for applications in very-high-speed circuits.

The poor quality of oxide on GaAs and a correspondingly high density of surface states at the GaAs–insulator interface make it difficult to fabricate GaAs MOSFETs or MISFETs. Only very recently has the new approach of oxidizing a thin silicon layer grown by molecular beam epitaxy on the GaAs surface offered hope for the development of a viable GaAs MOSFET technology (see Tiwari et al. 1988). Schottky barrier metal semiconductor field-effect transistors (MESFETs), junction field-effect transistors (JFETs), and heterostructure AlGaAs/GaAs transistors are the most commonly used GaAs devices. In many cases GaAs MESFETs and JFETs are fabricated by direct ion implantation into a GaAs semi-insulating substrate.

GaAs technology has a significant advantage over silicon in terms of speed, power dissipation, and radiation hardness. However, compound semiconductor technology is much less developed. In an elemental semiconductor, such as silicon, the device quality depends on the material purity. In a compound semiconductor, such as GaAs, the material composition is of utmost importance. For example, defects that degrade device performance may be caused by a deficiency of arsenic atoms. Another serious problem associated with GaAs MESFETs, JFETs, and heterostructure field-effect transistors is the gate leakage current. This current limits the allowed gate voltage swing, thus reducing the noise margin in circuits. Finally, the fabrication processes for compound semiconductor devices are not as well developed or understood as for silicon.

Compound semiconductor technology has clearly emerged as a leading contender for applications in ultra-high-speed, low-power integrated circuits, primarily because of the higher electron velocity in GaAs. The higher velocity is a consequence of a lighter electron effective mass. This advantage is especially pronounced in short-channel devices, where ballistic and overshoot effects play a dominant role (see Chapter 2). Van Zeghbroek et al. (1986) reported a cutoff frequency of 106 GHz for a 0.25-micron-gate GaAs MESFET. Using 0.2-μm-gate GaAs MESFETs, Johnston et al. (1986) fabricated a divide-by-two static frequency divider operated at frequencies up to 17.9 GHz. These numbers clearly illustrate the high speed potential of GaAs MESFET technology.

In n-channel MESFETs, n^+ drain and source regions are connected by an n-type channel. This channel is partially depleted by voltage applied to the gate. In the normally-off (enhancement-mode) MESFETs, the channel is totally depleted by the gate built-in potential even at zero gate voltage. The threshold voltage of enhancement-mode devices is positive. In normally-on (depletion-mode) MESFETs, the conducting channel has a finite cross section at zero gate voltage (see Fig. 4-10-1).

A schematic diagram of a depletion region under the gate of a MESFET for a finite drain-to-source voltage is shown in Fig. 4-10-2. The depletion region is wider closer to the drain because the positive drain voltage provides an additional reverse bias across the channel-to-gate junction. Also shown in Fig. 4-10-2 is the boundary region between the depletion region and undepleted conduction channel. The thickness of this region is of the order of several Debye lengths and may be comparable to the thickness of the conducting channel. Nevertheless, in the qualitative analysis that follows, we will assume that the boundary between the depletion region and the undepleted conduction channel is abrupt. The shape of the depletion region and the device current-voltage characteristics may be found using the same gradual channel approximation that was used for MOSFETs in Section 4-4.

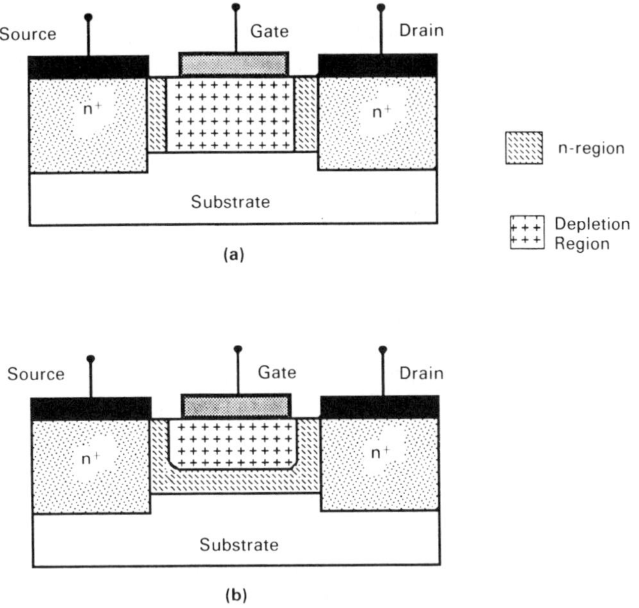

Fig. 4-10-1. Depletion regions in (a) normally-off and (b) normally-on MESFETs at zero gate-to-source and drain-to-source voltages.

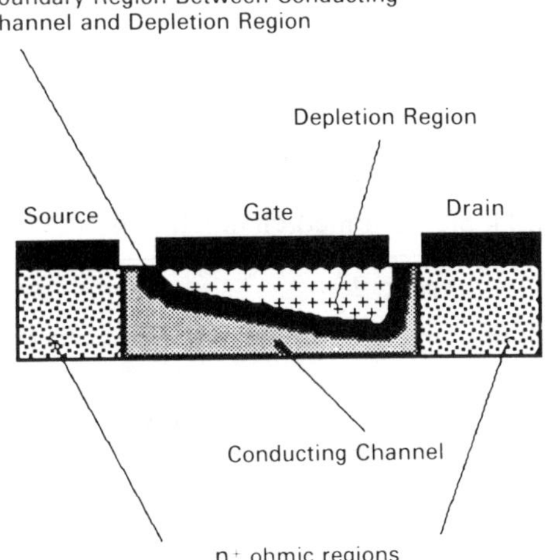

Fig. 4-10-2. Schematic diagram of a depletion region under the gate of a MESFET for finite drain-to-source voltage.

For simplicity we consider here the uniform n-type doping profile in the channel. We will first consider small drain-to-source voltages such that the longitudinal electric field everywhere in the channel is smaller than the velocity saturation field, F_s, so that the electron drift velocity is proportional to the longitudinal electric field F,

$$v_n = \mu_n F \tag{4-10-1}$$

We also assume that the conducting channel is neutral and that the space-charge region between the gate and the channel is totally depleted. The boundary between the neutral conducting channel and the depleted region is assumed to be sharp. According to the gradual channel approximation (similar to that discussed in Section 4-4), the potential across the channel varies slowly with distance, x, so that at each point the width of the depleted region can be found from the solution of the Poisson equation valid for a one-dimensional junction. Under such assumptions we find the incremental change of channel potential to be

$$dV = I_{ds}\, dR = \frac{I_{ds}\, dx}{q\mu_n N_D W[A - A_d(x)]} \tag{4-10-2}$$

where I_{ds} is the channel current, dR is the incremental channel resistance, x is the coordinate along the channel, A is the thickness of the active layer, $A_d(x)$ is the thickness of the depletion layer (see Fig. 4-10-2), and W is the gate width.

The depletion region thickness at a distance x is given by

$$A_d(x) = \left\{\frac{2\varepsilon[V(x) + V_{bi} - V_G]}{qN_D}\right\}^{1/2} \tag{4-10-3}$$

Substituting eq. (4-10-3) into eq. (4-10-2) and integrating with respect to x from $x = 0$ (the source side of the gate) to $x = L$ (the drain side of the gate), we derive the equation called the *fundamental equation of field-effect transistors:*

$$I_{ds} = g_o\left\{V_D - \frac{2\,[(V_D + V_{bi} - V_G)^{3/2} - (V_{bi} - V_G)^{3/2}]}{3V_{po}^{1/2}}\right\} \tag{4-10-4}$$

where

$$g_o = q\mu_n N_D WA/L \tag{4-10-5}$$

is the conductance of the undepleted doped channel, L is the gate length,

$$V_{po} = qN_D A^2/2\varepsilon \tag{4-10-6}$$

is the channel pinch-off voltage, and V_D is the voltage drop in the channel under the gate.

Equation (4-10-4) is applicable only for such values of the gate voltage, V_G, and drain-to-source voltage, V_D, that the neutral channel still exists even in the narrowest spot at the drain side, i.e.,

$$A_d(L) = \left\{\frac{2\varepsilon[V_D + V_{bi} - V_G]}{qN_D}\right\}^{1/2} < A \tag{4-10-7}$$

If we neglect the effects of velocity saturation in the channel, we may asume that when the pinch-off condition

$$A_d(L) = A \tag{4-10-8}$$

is reached, the drain-to-source current saturates. That is why the saturation voltage V_{Dsat} predicted by the constant mobility model (also called the Shockley model) is given by

$$V_{Dsat} = V_{po} - V_{bi} + V_G \tag{4-10-9}$$

Substitution of eq. (4-10-9) into eq. (4-10-4) leads to the following expression for the drain-to-source saturation current:

$$(I_{ds})_{sat} = g_o \left[V_{po}/3 + \frac{2(V_{bi} - V_G)^{3/2}}{3V_{po}^{1/2}} - V_{bi} + V_G \right] \tag{4-10-10}$$

From eq. (4-10-4) we find that in the linear region, the intrinsic device transconductance (i.e., the transconductance calculated without taking into account parasitic source and drain series resistances),

$$g_m = \partial I_{ds}/\partial V_G|_{V_D=\text{const}} \tag{4-10-11}$$

is given by

$$g_m = g_o \frac{(V_D + V_{bi} - V_G)^{1/2} - (V_{bi} - V_G)^{1/2}}{V_{po}^{1/2}} \tag{4-10-12}$$

For small drain-to-source voltages,

$$V_D << V_{bi} - V_G \tag{4-10-13}$$

eqs. (4-10-4) and (4-10-12) can be simplified to

$$I_{ds} = g_o \left[1 - \left(\frac{V_{bi} - V_G}{V_{po}} \right)^{1/2} \right] V_D \tag{4-10-14}$$

$$g_m = g_o V_D / \{2[V_{po}^{1/2}(V_{bi} - V_G)^{1/2}]\} \tag{4-10-15}$$

From eq. (4-10-10) we find the transconductance in the saturation region:

$$(g_m)_s = g_o \left[1 - \left(\frac{V_{bi} - V_G}{V_{po}} \right)^{1/2} \right] \tag{4-10-16}$$

According to the Shockley model, current saturation occurs when a conducting channel is pinched off at the drain side of the gate. At this point, the cross section of the conducting channel, predicted by the Shockley model, is zero, and hence, the electron velocity has to be infinitely high in order to maintain a finite drain-to-source current. In reality, the electron velocity saturates in high electric fields, and this causes the current saturation. In Section 4-5 we discussed the importance of the field dependence of electron mobility to the understanding of current saturation in silicon MOSFETs. A similar approach can also be used for GaAs MESFETs.

Velocity saturation is first reached at the drain side of the gate, where the electric field is the highest, according to the Shockley model. It occurs when

$$F(L) = F_s \tag{4-10-17}$$

where $F(L)$ is the magnitude of the electric field in the conducting channel at the drain side of the gate. Using the dimensionless variables

$$u = V(x)/V_{po} \tag{4-10-18}$$

$$u_G = (V_{bi} - V_G)/V_{po} \tag{4-10-19}$$

$$u_i = V_D/V_{po} \tag{4-10-20}$$

and

$$z = x/L \tag{4-10-21}$$

eq. (4-10-17) may be rewritten as

$$|du/dz|_{z=1} = \alpha \tag{4-10-22}$$

where

$$\alpha = F_s L/V_{po} \tag{4-10-23}$$

At drain-to-source voltages smaller than the saturation voltage, the electric field in the channel may be found from the equations of the Shockley model given in Section 2-1:

$$|du/dz| = \frac{u - (2/3)(u + u_G)^{3/2} + (2/3)u_G^{3/2}}{1 - (u + u_G)^{1/2}} \tag{4-10-24}$$

The dimensionless saturation voltage,

$$u_s = V_{Dsat}/V_{po} \tag{4-10-25}$$

is then determined from eq. (4-10-22), which may be rewritten as

$$\alpha = \frac{u_s - (2/3)(u_s + u_G)^{3/2} + (2/3)u_G^{3/2}}{1 - (u_s + u_G)^{1/2}} \tag{4-10-26}$$

For large $\alpha >> 1$ the solution of eq. (4-10-26) approaches

$$u_s + u_G = 1 \tag{4-10-27}$$

which is identical to the corresponding equation for the Shockley model (see eq. (4-10-9)). The opposite limiting case, $\alpha << 1$, corresponds to the complete velocity saturation model:

$$u_s = \alpha \tag{4-10-28}$$

For this solution to be valid it is also necessary to have

$$\alpha << 2(1 - u_G^{1/2})u_G^{1/2} \tag{4-10-29}$$

The numerical solution of eq. (4-10-26) can be interpolated using a simple interpolation formula (see Shur 1982):

$$u_s \approx \frac{\alpha(1 - u_G)}{\alpha + 1 - u_G} \tag{4-10-30}$$

The saturation current is given by

$$(I_{ds})_{sat} = qN_d v_s W[A - A_d(L)] \tag{4-10-31}$$

where the depletion width $A_d(L)$ at the drain side of the gate is found as

$$A_d(L) = A(u_G + u_s)^{1/2} \tag{4-10-32}$$

In dimensionless units, eq. (4-10-31) may be rewritten as

$$i_s = \alpha[1 - (u_s + u_G)^{1/2}] \tag{4-10-33}$$

where

$$i_s = (I_{ds})_{sat}/(g_o V_{po}) \tag{4-10-34}$$

In the limiting case $\alpha \to \infty$ (which corresponds to a long-gate device with a small pinch-off voltage), eq. (4-10-33) reduces to the corresponding equation of the Shockley model (see eq. (4-10-10)). In the opposite case, ($\alpha << 1$ (short gate and/or large pinch-off voltage),

$$i_s = \alpha(1 - u_G^{1/2}) \tag{4-10-35}$$

This expression corresponds to a simple analytical model of GaAs MESFETs that assumes the complete velocity saturation in the channel (see Shur 1978 and Williams and Shaw 1978).

A simple interpolation formula approximates the result of the numerical solution for all values of α (see Shur 1982):

$$i_s = \frac{\alpha}{1 + 3\alpha}(1 - u_G)^2 \tag{4-10-36}$$

This equation may be rewritten as

$$(I_{ds})_{sat} = \beta(V_G - V_T)^2 \tag{4-10-37}$$

where

$$\beta = \frac{2\varepsilon\mu_n v_s W}{A(\mu_n V_{po} + 3v_s L)} \tag{4-10-38}$$

and

$$V_T = V_{bi} - V_{po} \tag{4-10-39}$$

In practice, the square law (i.e., eq. (4-10-37)) is fairly accurate for devices with relatively low pinch-off voltage ($V_{po} \leq 1.5 \sim 2$ V). For devices with higher pinch-off voltages, an empirical expression proposed by Statz et al. (1987),

$$(I_{ds})_{sat} = \frac{\beta(V_G - V_T)^2}{1 + b(V_G - V_T)} \tag{4-10-40}$$

provides an excellent fit to the experimental data.

As can be seen from eq. (4-10-38), the values of β (and, hence, the values of the transconductance for a given voltage swing) increase with the decrease of the active layer thickness, A. (If the device threshold voltage is kept constant, the decrease in A requires an increase in doping proportional to A^2; see eq. (4-10-6).) The increase of β is accompanied by a similar increase in gate capacitance, C_g,

$$C_g = \varepsilon LW/A \tag{4-10-41}$$

However, parasitic capacitances do not increase with an increase in doping or with a decrease in active layer thickness; therefore, thin and highly doped active layers should lead to a higher speed of operation.

Another important advantage of highly doped devices is the reduction of the active layer thickness for a given value of pinch-off voltage. This tends to minimize short-channel effects, which become quite noticeable when $L/A < 3$ (see Dambkes et al. 1983).

The source and drain series resistances, R_s and R_d, may play an important role in determining the current-voltage characteristics of GaAs MESFETs. These resistances can be taken into account as follows: The gate-to-source voltage, V_{GS}, is given by

$$V_{GS} = V_G + I_{ds}R_s \tag{4-10-42}$$

Substituting $V_G = V_{GS} - (I_{ds})_{sat}R_s$ into eq.(4-10-37) and solving for $(I_{ds})_{sat}$, we obtain

$$(I_{ds})_{sat} = \frac{1 + 2\beta R_s(V_{GS} - V_T) - [1 + 4\beta R_s(V_{GS} - V_T)]^{1/2}}{2\beta R_s^2} \tag{4-10-43}$$

In device modeling suitable for computer-aided design, one has to model the current-voltage characteristics in the entire range of the drain-to-source voltages, not only in the saturation regime. Curtice (1980) proposed the use of a hyperbolic tangent function for the interpolation of MESFET current-voltage characteristics:

$$I_{ds} = (I_{ds})_{sat}(1 + \lambda V_{DS})\tanh[g_{ch}V_{DS}/(I_{ds})_{sat}] \tag{4-10-44}$$

where $(I_{ds})_{sat}$ is given by eq. (4-10-43),

$$g_{ch} = g_{chi}/[1 + g_{chi}(R_s + R_d)] \tag{4-10-45}$$

is the channel conductance at low drain-to-source voltages, and

$$g_{chi} = g_o\{1 - [(V_{bi} - V_{GS})/V_{po}]^{1/2}\} \tag{4-10-46}$$

is the intrinsic channel conductance at low drain-to-source voltages predicted by the Shockley model.

Constant λ in eq. (4-10-44) is an empirical constant that accounts for output conductance. This output conductance may be related to short-channel effects (see Pucel et al. 1975) and also to parasitic currents in the substrate, such as a space-charge limited current. Hence, output conductance may be greatly reduced by using a heterojunction buffer, which prevents carrier injection into the substrate (see, for example, Eastman and Shur 1979).

The drain-to-source voltage is given by

$$V_{DS} = V_D + I_{ds}(R_s + R_d) \tag{4-10-47}$$

As was previously mentioned, gate current also plays an important role in GaAs MESFET circuits. It becomes important when a positive voltage is applied to the Schottky gate junction. An accurate analytical model for the gate current has not yet been developed. A practical approach used to fit the experimental data and to account for the gate current in circuit simulations is to introduce two equivalent diodes connecting the gate contact with the source and drain contacts, respectively. This approach leads to the following expression for the gate current, I_g:

$$I_g = I_{gs} + I_{gd} \tag{4-10-48}$$

where

$$I_{gs} = I_{go} \exp\{q[V_{GS} - I_g R_g - (I_{ds} + I_{gs})R_s]/(n_g k_B T)\} \tag{4-10-49}$$

$$I_{gd} = I_{go} \exp\{q[V_{GD} - I_g R_g - (I_{ds} + I_{gd})R_d]/(n_g k_B T)\} \tag{4-10-50}$$

Equations (4-10-43) through (4-10-50) form the complete set of equations for the analytical MESFET model used in the GaAs circuit simulator UM-SPICE (see, for example, Hyun et al. 1985). The parameters of the model are related to the device geometry, doping, and material parameters such as saturation velocity and low field mobility.

Figure 4-10-3 compares the I_{DS} vs. V_{DS} characteristics calculated using this model with the experimental data. MESFET device parameters used in these calculations were determined from measured I-V characteristics using the following procedure: First, a suitable value of λ is chosen from the I_{DS} vs. V_{DS} curve for one value of V_{GS}. Then saturation currents are extrapolated to $V_{DS} = 0$ for all values of V_{GS}, keeping λ constant. To obtain the source resistance R_s, $(I_{DS})^{1/2}$ vs. $V_{GS} - I_{DS}R_s$ were plotted for different values of R_s, until a best least-square fit is obtained. The slope and intercept of this line give us β and V_T, respectively. The value of built-in voltage, V_{bi}, is determined from the gate I-V characteristic. The channel thickness and doping are determined using eq. (4-10-6) and the implant dose data (i.e., the $N_d A$ product). Assuming that $R_s = R_d$, we obtain the intrinsic channel resistance, R_i, from the slope of the I_{DS} vs. V_{DS} characteristic in the linear region at large V_{GS}:

$$R_i = R_{DS} - R_s - R_d \tag{4-10-51}$$

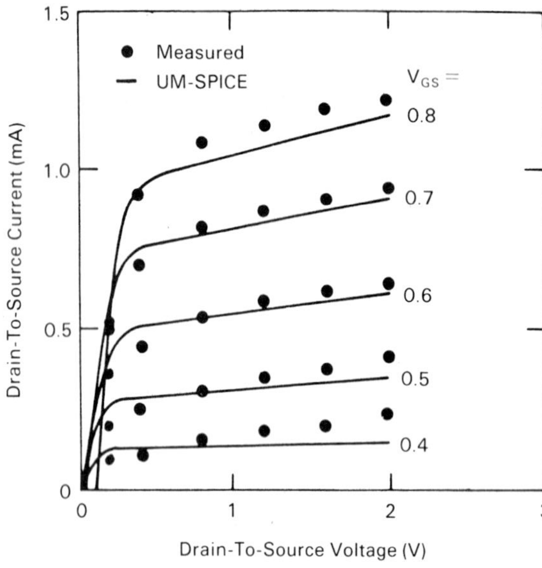

Fig. 4-10-3. Current-voltage characteristics of GaAs MESFET (from C. H. Hyun, M. Shur, and A. Peczalski, *IEEE Trans. Electron Devices*, ED-33, p. 1421–1426 (1986b). © 1986 IEEE). Parameters used in the calculation are gate width $W = 20\ \mu$m, gate length $L = 1\ \mu$m, threshold voltage $V_T = 0.21$ V, $\lambda = 0.15$ 1/V, mobility $\mu = 0.25$ m²/Vs, saturation velocity $v_s = 1.3 x 10^5$ m/s, source series resistance $R_s = 50\ \Omega$, drain series resistance $R_d = 50\ \Omega$, channel doping density $N_d = 7.24 \times 10^{16}$ cm^{-3}, gate saturation current density $J_s = 0.255$ A/m², and Schottky diode ideality factor $n_{id} = 1.44$.

R_i is related to the gate voltage as follows:

$$R_i = \frac{L}{qA\mu_n N_d W\{1 - [(V_{bi} - V_{GS})/V_{po}]^{1/2}\}} \tag{4-10-52}$$

This equation is used to determine μ_n. Once μ_n is known, the saturation velocity is calculated using eq. (4-10-38). For more accurate device characterization more sophisticated characterization techniques may be required, such as end resistance measurements (see, for example, Shur 1987).

In the short-channel GaAs MESFET, effects related to the overshoot transport become especially important. These effects occur because the electron transit time under the gate becomes comparable to the energy relaxation time. Their importance is illustrated by Fig. 4-10-4 (from Cappy et al. 1980), which shows the velocity profiles in short-channel MESFETs fabricated from different semiconductors as well as peak values of velocity achieved in long samples under stationary conditions. As can be seen from the figure, considerably larger velocities are expected in short-channel devices. However, in the simple model described previously, this effect can be taken into account, to a certain extent, by increasing the effective value of the electron saturation velocity.

The foregoing analytical models are suitable for computer-aided design of GaAs MESFETs and GaAs MESFET circuits. However, these models do not explicitly take into account many important and complicated effects, such as deviations from the gradual channel approximation (which may be especially important at the drain side of the channel; see Pucel 1975), possible formation of a high field region (i.e., a dipole layer) at the drain side of the channel (see

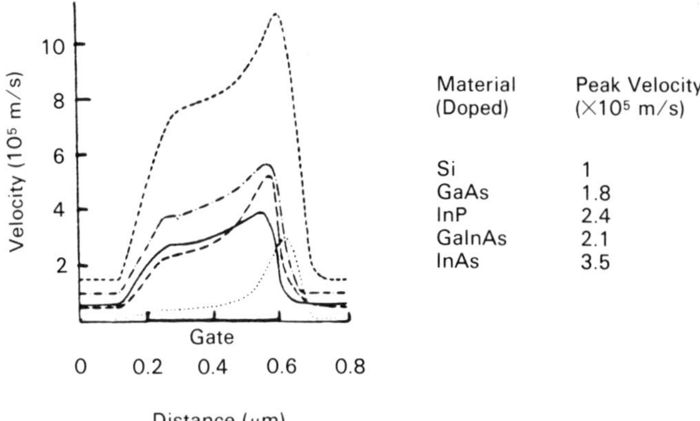

Fig. 4-10-4. Electron velocity in relation to distance in the channel of a field effect transistor for different semiconductor materials (from A. Cappy, B. Carnes, R. Fauquembergues, G. Salmer, and E. Constant, "Comparative Potential Performance of Si, GaAs, GaInAs, InAs Submicrometer-gate FETs," *IEEE Trans. Electron Devices*, ED-27, pp. 2158–2168 (1980). © 1980 IEEE).

.... Si
——— ——— InP
——————— GaAs
–·–·–·–·–· GaInAs
- - - - - - - - - - - - InAs

Peak velocities achieved in long samples under stationary conditions are shown for comparison.

Engelmann and Liehti 1977, Shur and Eastman 1978), and Fjeldly 1986), inclusion of diffusion and incomplete depletion at the boundary between the depletion region and the conducting channel (see Yamaguchi and Kodera 1976), ballistic or overshoot effects (Ruch 1972, Maloney and Frey 1975, Shur 1976, Warriner 1977, Cappy 1980), effects of donor diffusion from the n^+-contact regions into the channel (Chen et al. 1987), effects of the passivating silicon nitride layers (Asbeck et al. 1984 and Chen et al. 1987), effects of traps (Chen et al. 1986). These effects may still be included indirectly by adjusting the model parameters, such as μ, v_s, and N_d. In a rigorous way they can only be treated using numerical solutions. Such solutions provide an insight into the device physics. However, for a practical circuit simulator used in circuit design, analytical or very simple numerical models remain a must. Besides considerations related to computer time involved in the simulation of hundreds of transistors in a circuit, simple models make device characterization easier because of a relatively small number of parameters. Hence, these parameters can be readily measured and compared for different wafers, fabrication processes, etc.

Internal device capacitances play an important role in determining the speed of GaAs MESFETs and GaAs MESFET circuits. The simplest approach is to model two FET capacitances—gate-to-source capacitance, C_{gs}, and gate-to-drain capacitance, C_{ds}—as capacitances of equivalent Schottky barrier diodes connected between the gate and source and drain, respectively:

$$C_{gs} = \frac{C_{go}}{(1 - V_{gs}/V_{bi})^{1/2}} \tag{4-10-53}$$

$$C_{gd} = \frac{C_{go}}{(1 - V_{gd}/V_{bi})^{1/2}} \tag{4-10-54}$$

Here

$$C_{go} = \frac{\varepsilon WL}{2A_o} = \frac{WL}{2}\left(\frac{q\varepsilon N_d}{2\,V_{bi}}\right)^{1/2}$$

$$A_o = \left(\frac{2q\varepsilon V_{bi}}{qN_d}\right)^{1/2}$$

This capacitance model was applied, for example, to junction field-effect transistors (JFETs) in the circuit simulator SPICE. However, this model is com-

Depletion Region at Gate Voltage $V_g = V_{g1}$

Change in the Depletion Region When the Gate Voltage is Increased by a Small Amount

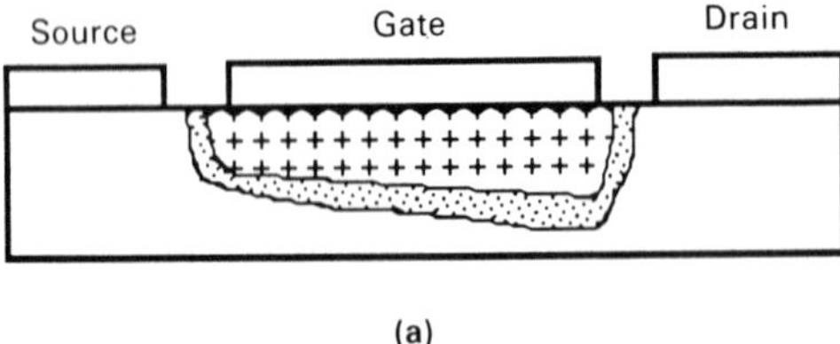

(a)

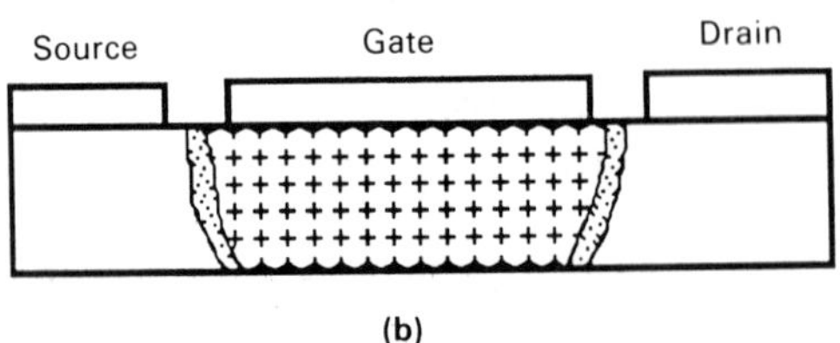

(b)

Fig. 4-10-5. Changes in the depletion region shape in response to a small variation of gate voltage: (a) gate voltage above threshold and (b) gate voltage below threshold. Below the threshold, the channel under the gate is totally depleted and the variation of the charge under the gate is only related to the fringing (sidewall) capacitance.

pletely inadequate for gate voltages smaller than the threshold voltage. Indeed, below the threshold, the channel under the gate is totally depleted and the variation of the charge under the gate is only related to the fringing (sidewall) capacitance (see Fig. 4-10-5).

A better capacitance model was proposed by Takada et al. (1982). This model essentially assumes that at gate voltages above the threshold (at gate voltages V_g higher than $V_{t2} > V_T$) the device capacitances are equal to the Schottky gate capacitances and the fringing sidewall capacitances:

$$C_{gs} = \frac{C_{go}}{(1 - V_{gs}/V_{bi})^{1/2}} + \pi \varepsilon W/2 \tag{4-10-55}$$

$$C_{gd} = \frac{C_{go}}{(1 - V_{gd}/V_{bi})^{1/2}} + \pi \varepsilon W/2 \tag{4-10-56}$$

(The fringing sidewall capacitances are represented by the second terms in the right-hand side of eqs. (4-10-55) and (4-10-56).) Below the threshold ($V_g < V_{t1} < V_t$) the internal capacitances are equal to the fringing sidewall capacitances:

$$C_{gs} = \pi \varepsilon W/2 \tag{4-10-57}$$

$$C_{gd} = \pi \varepsilon W/2 \tag{4-10-58}$$

For the gate voltage range $V_{t1} < V_g < V_{t2}$, capacitances are found by using an interpolation formula. In Fig. 4-10-6 we compare the predictions of the model with experimental data (from Hyun et al. 1986) and with the simple SPICE model given by eqs. (4-10-53) and (4-10-54)).

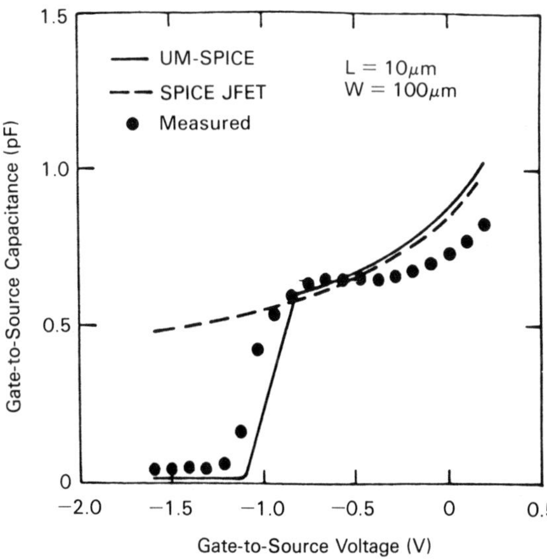

Fig. 4-10-6. Gate-to-source capacitance of a long channel GaAs MESFET at zero drain-to-source voltage (from C. M. Hyun, M. Shur, and A. Peczalski, *IEEE Trans. Electron Devices*, ED-33, pp. 1421–1426 (1986). © 1986 IEEE. The curve marked UM-SPICE is calculated using the model proposed by Takada et al. [1982]).

4-11. HETEROSTRUCTURE FIELD-EFFECT TRANSISTORS

GaAs MESFETs discussed in Section 4-10 have a limited gate voltage swing because a turn-on voltage of a Schottky gate is only 0.8 V or so. In particular, the gate voltage swing is very limited for normally-off devices, i.e., for devices with a positive threshold voltage. An increase in the gate voltage swing along with improvements in speed and other advantages can be achieved by *heterostructure technology*—a new and rapidly emerging field in compound semiconductor electronics.

The idea of using heterostructures—layered structures of different semiconductor materials—belongs to William Shockley, who also invented the first practi-

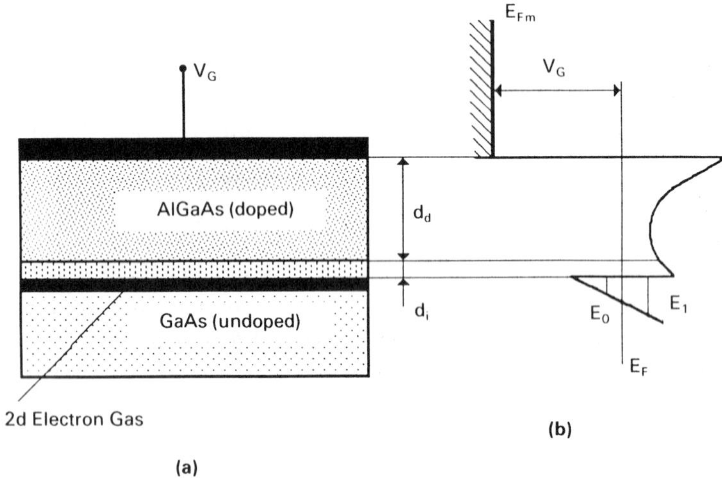

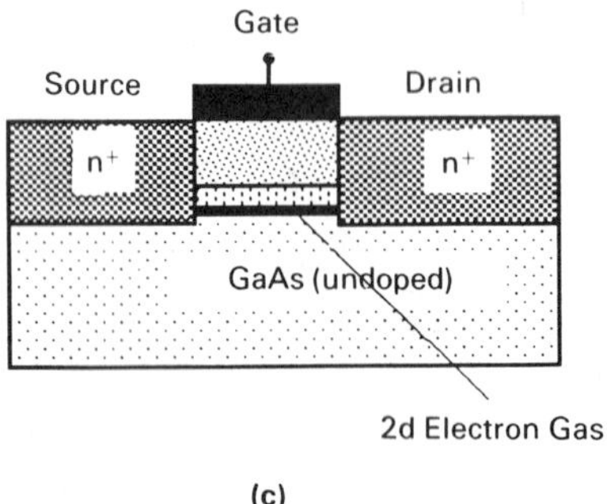

Fig. 4-11-1. Schematic (a) diagram of n-channel modulation doped structure, (b) band diagram of modulation doped structure, and (c) cross-section of self-aligned n-channel MODFET. d_d is the thickness of doped AlGaAs layer; d_i is the thickness of undoped AlGaAs spacer layer; E_o and E_1 are energies of two lowest subbands; E_F is electron quasi-Fermi level; and V_G is applied gate voltage.

cal heterojunction device, the heterojunction bipolar transistor (see Section 3-14). Heterostructure AlGaAs/GaAs and AlGaAs/InGaAs/GaAs Field-Effect Transistors (HFETs), also called Modulation Doped Field-Effect Transistors (MODFETs), High Electron Mobility Transistors (HEMTs), Selectively Doped Heterojunction Transistors (SDHTs), and Two-dimensional Electron Gas Field-Effect Transistors (TEGFETs), offer excellent ultra-high-speed performance. Propagation delays as low as 5.8 ps at 77 K (see Shah et al. 1986) and 6 ps at 300 K (see Mishra et al. 1988) have been obtained in HFET ring oscillator circuits.

Most HFETs are n-channel devices. A schematic cross section of an n-channel HFET and the band diagram of the structure are shown in Fig. 4-11-1. As can be seen from Fig. 4-11-1, a two-dimensional electron gas is formed in the unintentionally doped GaAs buffer layer at the heterointerface. More recently, complementary n- and p-channel Heterostructure Insulated Gate Field-Effect Transistors (HIGFETs) were developed (see Fig. 4-11-2), offering the potential for high-speed, low-power operation (see Cirillo et al. 1985, Daniels et al. 1987, Kiehl et al. 1987, Daniels et al. 1988, and Ruden et al. 1988). The lowest power silicon circuits use Complementary Metal Oxide Semiconductor (CMOS) technology in which n- and p-channel devices are connected in series (see Fig. 4-9-8). In a CMOS inverter there is very little current in both stable states, so that the power

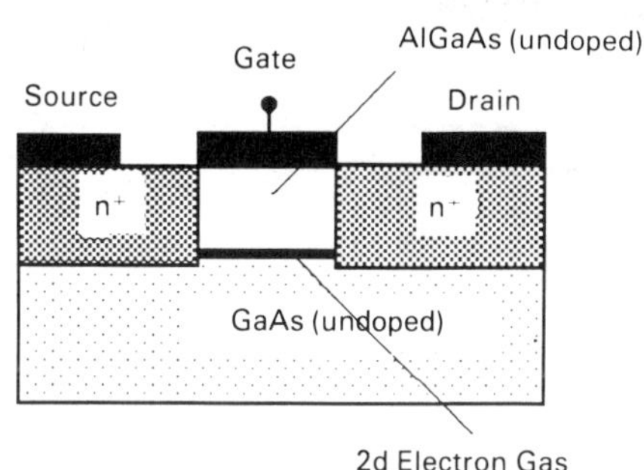

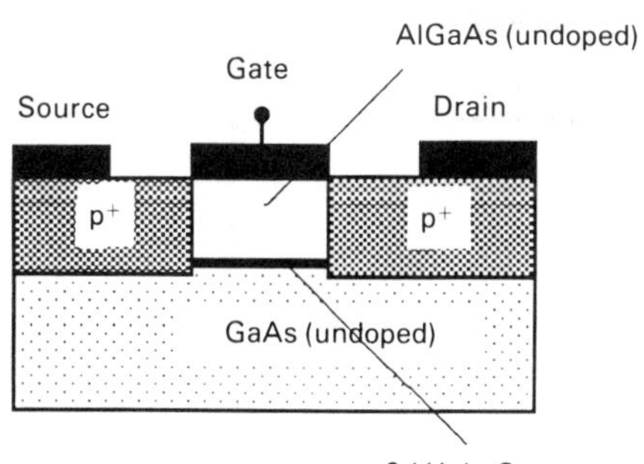

Fig. 4-11-2. Complementary *n*- and *p*-channel Heterostructure Insulated Gate Field-Effect Transistors.

is consumed only during switching. HIGFETs represent a similar technology. As can be seen from Fig. 4-11-2, both n- and p-channel devices are fabricated by using self-aligned n^+ and p^+ implants in the same AlGaAs/GaAs layer that does not have any intentional dopants whatsoever. This helps eliminate detrimental effects related to dopant-induced traps in AlGaAs that lead to several undesirable effects in conventional HFETs, such as a large threshold voltage shift with temperature and persistent photoconductivity (see, for example, Shur 1987, p. 583).

A different type of HFET utilizes a plane of dopants in the AlGaAs layer. The band diagram and schematic cross section of this device (which, to our knowledge, was first proposed by Eastman [1983]) is shown in Fig. 4-11-3. This device has two distinct advantages. First of all, in planar doped devices (sometimes called delta-doped or pulse-doped devices), the effect of traps is diminished. Also, as will be discussed presently, planar doped devices can operate with larger gate-voltage swings. The same goal—trying to diminish the effects of traps in the AlGaAs layer—led to the development of superlattice HFETs (see Baba

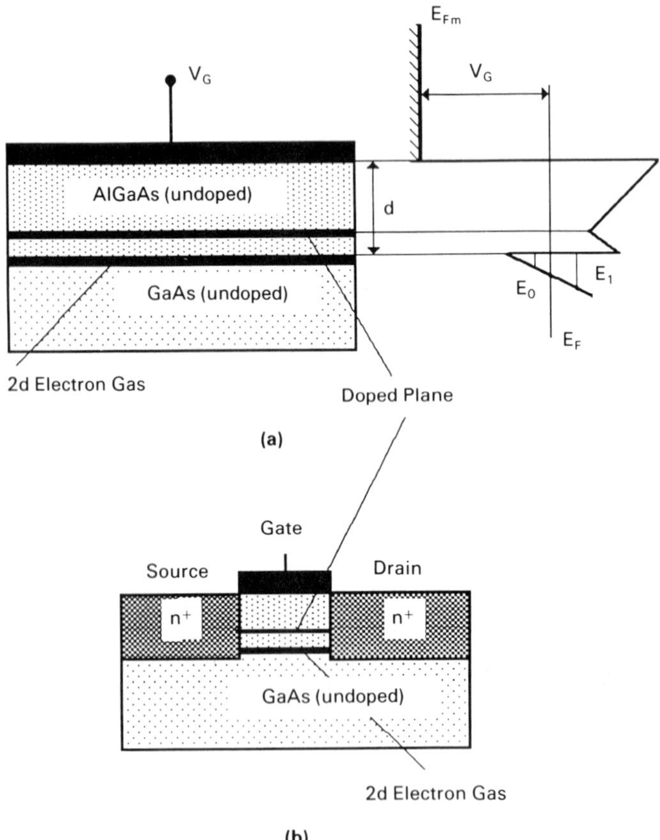

Fig. 4-11-3. Band diagram and schematic cross-section of planar doped HFET.

1983, Shah et al. 1986, Shur et al. 1987). In this device, dopants are incorporated into several thin GaAs quantum wells that form a superlattice structure in the AlGaAs layer.

However, all these different types of HFETs operate in a similar way. By applying a positive voltage to the gate, one may capacitively induce electrons into the narrow channel in gallium arsenide at the heterointerface. The composition and doping profile of the AlGaAs layer (which is sometimes called the charge control layer) determine the device threshold voltage and gate voltage swing. The electron gas induced at the heterointerface (called two-dimensional electron gas) is similar to the electron gas in the inversion layer in silicon MOSFETs. As in a MOSFET, the density of this gas and, hence, the device current is modulated by the gate voltage. However, the small effective mass of electrons in GaAs ($m^* = 0.067m_e$) leads to much more pronounced quantum effects. This can be understood by using a deep, narrow quantum well (considered in Section 1-2) as an example. In this case the separation between the quantum levels is inversely proportional to the electron effective mass (see eq. (1-2-20)). As can be seen from Fig. 4-11-1, the splitting of the energy subbands in the electron gas (called the two-dimensional electron gas) is quite large compared with the thermal voltage V_{th} = 25.84 meV at 300 K. As was proposed by Esaki and Tsu (1969), the high electron concentration and the spatial separation of donors diminishes the effects of impurity scattering in the two-dimensional electron gas, thereby enhancing the electron mobility. This effect of the mobility enhancement was first experimentally observed by Dingle et al. (1978). Mimura et al. (1980) made the first heterostructure transistor in which the increase in the electron mobility improved the device characteristics. They called the new device a High Electron Mobility Transistor (HEMT).

Heterostructure field-effect transistors fully preserve and enhance all the advantages of standard gallium arsenide technology, such as higher electron mobility and velocity than in silicon, overshoot and ballistic effects in very short structures, and low parasitic capacitance related to the semi-insulating properties of semi-insulating GaAs substrates. But in addition to even higher mobility than in GaAs MESFETs there are two important advantages: a very small (and well-controlled) separation between the gate and the channel and a higher turn-on voltage for the gate current.

The importance of a smaller separation between the gate and the channel can be illustrated by considering a FET cutoff frequency, f_t, that may be chosen as the figure-of-merit characterizing device speed:

$$f_t = \frac{g_m}{2\pi(C_{gate} + C_L)} \tag{4-11-1}$$

Here g_m is the device transconductance, C_{gate} is the gate capacitance, and C_L is the effective load capacitance, including fringing and parasitic capacitances. The device transconductance is proportional to the gate capacitance and to the electron velocity in the channel, v, and inversely proportional to the gate length, L,

$$g_m \approx \frac{C_{gate} v}{L} \tag{4-11-2}$$

Equation (4-11-2) may be obtained as follows: the induced electron density in the channel is

$$n_s \approx C_{gate}(V_g - V_t)/(qLW) \tag{4-11-3}$$

where W is the gate width. The drain current

$$I_{ds} \approx q n_s v W \tag{4-11-4}$$

and $g_m = \partial I_{ds}/\partial V_g$, leading to eq. (4-11-2). Using eq. (4-11-2) we can rewrite eq. (4-11-1) as

$$f_t = \frac{1}{2\pi\tau_{tr}} \frac{1}{1 + \dfrac{C_L}{C_{gate}}} \tag{4-11-5}$$

where $\tau_{tr} = L/v$ is the transit time of electrons under the gate. In a typical integrated circuit a substantial fraction of the load capacitance, C_L, may be determined by the gate capacitance of the next stage ($C_L = C_{gate} + C_{int}$, where C_{int} is the interconnect and parasitic capacitance). In this case eq. (4-11-5) becomes

$$f_t = \frac{1}{2\pi\tau_{tr}} \frac{1}{2 + \dfrac{C_{int}}{C_{gate}}} \tag{4-11-6}$$

Equation (4-11-6) reflects the fact that the device transconductance is proportional to the electron velocity and to the gate capacitance. It shows that the higher gate capacitances (and higher transconductances) lead to a higher device speed (as the ratio C_{gate}/C_L increases). The second advantage, a substantially higher gate turn-on voltage, is illustrated by Fig. 4-11-4, where we compare gate current-voltage characteristics of a typical GaAs MESFET and an AlGaAs/GaAs MODFET. These advantages allow heterostructure FETs to occupy an important place among different high-speed technologies (see Fig. 4-11-5). Future applications of heterostructure FET technology are in ultra-high-speed, low-power, radiation-hard signal-processing, and memory ICs.

In what follows we describe a charge control model for HFETs that is in good agreement with numerical one-dimensional and two dimensional simulation and experimental results (see Grinberg and Shur 1989 and Byun et al. [to be published]). The device threshold voltage, V_t, can be found by integrating Poisson's equation for the AlGaAs charge control layer under the conditions when the electric field at the heterointerface is equal to zero. This leads to the following expression (see Problem 4-11-1):

$$V_t = \phi_b - \Delta E_c - (q/\varepsilon_1) \int_0^d N_d(x) x \, dx \tag{4-11-7}$$

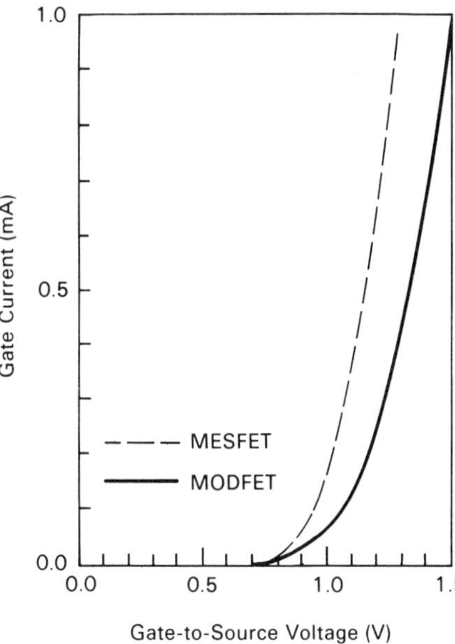

Fig. 4-11-4. Curve 1 shows the gate current-voltage characteristics for typical GaAs MESFET and curve 2, the AlGaAs MODFET.

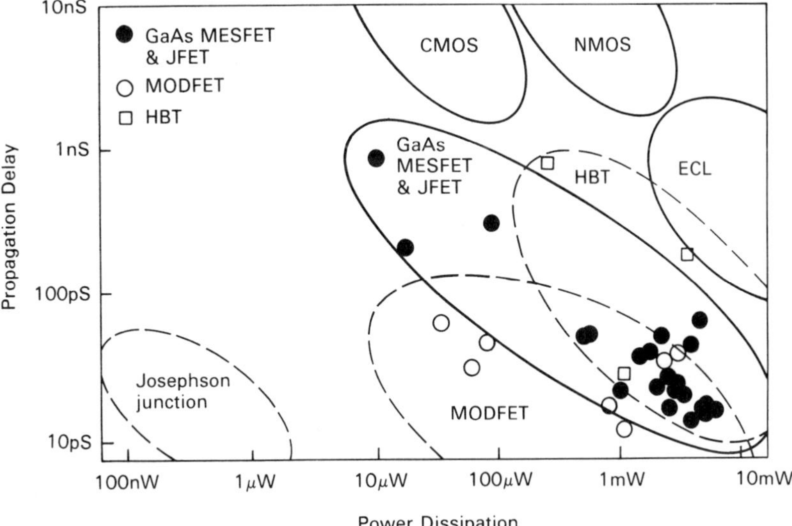

Fig. 4-11-5. Power and delay time for different high speed technologies (from H. Hasegawa, M. Abe, P. M. Asbeck, A. Higashizaka, Y. Koto, and M. Ohmori, "GaAs LSI/VLSI: Advantages and Application," in *Extended Abstracts of the 16th International Conference on Solid State Devices and Materials,* Kobe, Japan, pp. 413–414 (1984)).

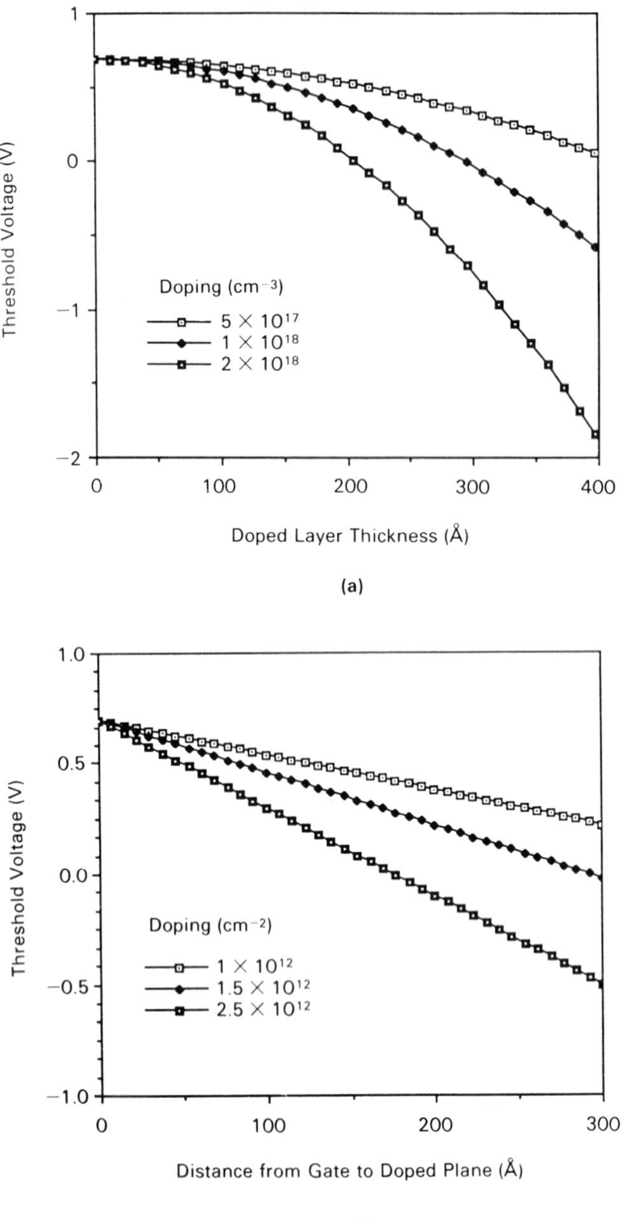

Fig. 4-11-6. Threshold voltage of (a) conventional and (b) planar doped HFETs. (Fig. 4-11-6b is from P. C. Chao, M. Shur, R. C. Tiberio, K. H. G. Duh, P. M. Smith, J. M. Ballingall, P. Ho, and A. A. Jabra, "DC and Microwave Characteristics of 0.1 μm Gate-Length Planar-Doped Pseudomorphic HEMTS," *IEEE Trans. Electron Devices*, 36, no. 3, pp. 461–473 (1989). © 1989 IEEE).

where ε_l is the dielectric permitivity of AlGaAs layer, ϕ_b is the metal barrier height, N_d is the doping density in the $Al_XGa_{1-X}As$ layer, q is the electronic charge, d is the distance between the gate and the heterointerface, Δd is the effective thickness of the two-dimensional gas, and x is the space coordinate ($x = 0$ at the boundary between the gate and the AlGaAs layer). For a uniformly doped AlGaAs layer ($N_d(x)$ = const), the threshold voltage, found from eq. (4-11-7), is given by

$$V_t = \phi_b - \Delta E_c - qN_d d_{dd}^2/(2\varepsilon_l) \tag{4-11-8}$$

where d_{dd} is the thickness of the doped AlGaAs layer. For a planar-doped structure, the evaluation of the integral in eq. (4-11-7) yields

$$V_t = \phi_b - \Delta E_c - qn_d d_d/\varepsilon_l \tag{4-11-9}$$

where d_d is the distance between the metal gate and the doped plane, and n_d is the surface concentration of donors in the doped plane. The calculated dependence of V_t on d_d for uniformly doped and planar-doped devices is shown in Fig. 4-11-6.

The dependence of the concentration of the two-dimensional gas, n_s, on the gate voltage can be approximated by a straight line (as predicted by eq. (4-11-3)) only in a limited range of gate voltages. This deviation becomes especially important at large gate voltages. This can be understood by analyzing the energy band diagram of the modulation-doped structure. As an example, we plot simplified band diagrams for planar-doped HFETs for different values of the gate voltage, V_g, in Fig. 4-11-7 (from Chao et al. 1989). As can be seen from this figure, at large gate voltage swings, the electron quasi-Fermi level in the AlGaAs structure reaches the bottom of the conduction band, so that the carriers are induced into the conduction band minimum in the AlGaAs layer. As most of these carriers

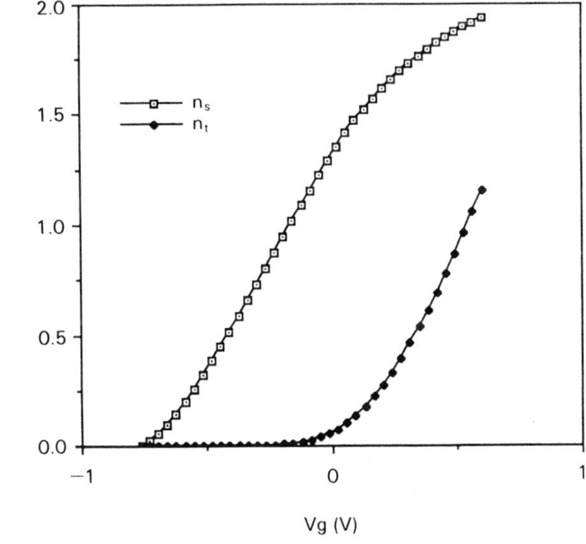

Fig. 4-11-7. Surface carrier densities of electrons in the two-dimensional electron gas, n_s, and in the AlGaAs layer, n_t, in relation to gate voltage. At gate voltages close to zero, real space transfer of electrons into the AlGaAs layer becomes important, leaving fewer electrons induced into the quantum well and limiting the gate voltage swing (from P. C. Chao, M. Shur, R. C. Tiberio, K. H. G. Duh, P. M. Smith, J. M. Ballingall, P. Ho, and A. A. Jabra, "DC and Microwave Characteristics of 0.1 μm Gate-Length Planar-Doped Pseudomorphic HEMTS," *IEEE Trans. Electron Devices,* 36, no. 3, pp. 461–473 (1989). © 1989 IEEE).

(with concentration n_t per unit area) are trapped by traps in AlGaAs, they do not contribute much to the device current. Figure 4-11-8 shows the computed dependencies of n_s and n_t on the gate voltage. This calculation is based on the self-consistent solution of Poisson's equation and Schrödinger's equation, as described by Stern (1974). As can be seen from Fig. 4-11-8, at high gate voltages the concentration of electrons in the AlGaAs layer increases sharply, and the slope of the n_s vs. V_g dependence decreases. At such gate voltages, the dependence of n_s on V_g is quite different from linear.

To derive the expression for the current-voltage characteristics, we start from the equation for the drain current valid at low drain-to-source voltages when the electric field in the channel is small and the electron velocity v is proportional to the electric field, F:

$$I_D \approx q\mu n_{xs} F \tag{4-11-10}$$

where $n_{xs}(x)$ is the surface carrier density at position x of the channel, μ is the electron mobility, q is the electronic charge, and

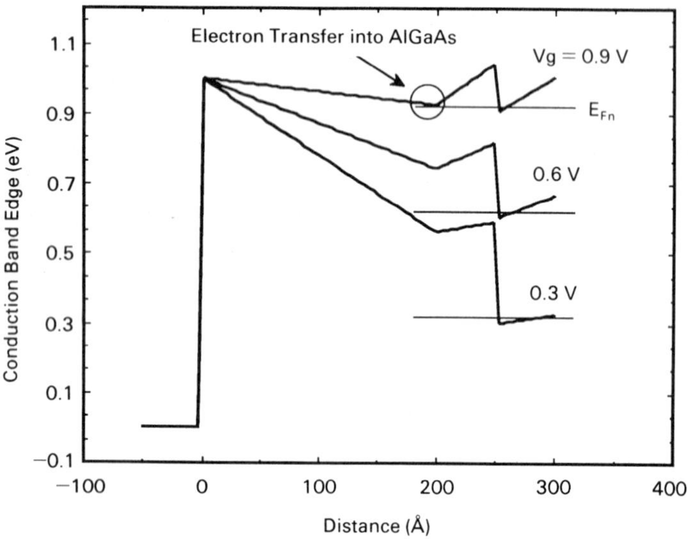

Fig. 4-11-8. Band diagrams for planar doped HEMTs at different values of gate voltage. At small gate voltages, the energy separation between the electron quasi-Fermi level and the conduction band minimum in the AlGaAs charge-control layer is large and the electron concentration in the AlGaAs layer is negligible. At large gate voltages, the electron quasi-Fermi level nearly touches the conduction band minimum in the AlGaAs charge-control layer, and the transfer of electrons into this layer becomes dominant, sharply reducing the device transconductance (after P. C. Chao, M. Shur, R. C. Tiberio, K. H. G. Duh, P. M. Smith, J. M. Ballingall, P. Ho, and A. A. Jabra, "DC and Microwave Characteristics of 0.1 μm Gate-Length Planar-Doped Pseudomorphic HEMTS," *IEEE Trans. Electron Devices*, 36, no. 3, pp. 461–473 (1989). © 1989 IEEE).

$$F = dV/dx \tag{4-11-11}$$

is the absolute value of the electric field. In the frame of the gradual channel approximation,

$$n_{xs} \approx n_s - \varepsilon_1 V(x)/(qd_{eff}) \tag{4-11-12}$$

where ε_1 is the dielectric permittivity of the AlGaAs layer, $d_{eff} = d + \Delta d_o$ is the distance between the gate and the channel, parameter Δd_o is determined from the comparison with numerical calculations (see Byun et al. [to be published]), and $n_s = n_s(V_g)$ is the surface carrier density at the source (independent of V). Substituting eqs. (4-11-11) and (4-11-12) into eq. (4-11-10), we obtain the following expression for the current-voltage characteristic at drain voltages smaller than the saturation voltage, V_D:

$$I_D \approx q\mu(W/L)[n_s V_D - \varepsilon_1 V_D^2/(2qd_{eff})] \tag{4-11-13}$$

If we now use the approximate relation between n_s and V_g, proposed by Drummond et al. (1982),

$$n_s \approx C_o(V_g - V_t)/q \tag{4-11-14}$$

where

$$C_o \approx \varepsilon_1/(d + \Delta d) \tag{4-11-15}$$

and $\Delta d \approx 80$ Å is the effective thickness of the two-dimensional electron gas, we obtain

$$I_D \approx \mu\varepsilon_1(W/L)[(V_g - V_t)V_D/(d + \Delta d) - V_D^2/(2d_{eff})] \tag{4-11-16}$$

As was done in Section 4-10 for GaAs MESFETs, we will assume that the current-voltage characteristics in short-channel devices saturate when the electric field, F, at the drain side of the channel reaches the electric field of the velocity saturation, $F_s = v_s/\mu$, where v_s is the electron saturation velocity. The electric field at the drain, F, can be found as

$$F = I_D/(q\mu W n_d) \tag{4-11-17}$$

or, using eq. (4-11-12),

$$F = I_D/\{q\mu W[n_s - \varepsilon_1 V_D/(qd_{eff})]\} \tag{4-11-18}$$

Substituting $F = F_s$ into eq. (4-11-18) and using eq. (4-11-13), we find the drain saturation voltage, V_D,

$$V_{Dsat} = (qdn_s/\varepsilon_1)[1 + a - (1 + a^2)^{1/2}] \tag{4-11-19}$$

where

$$a = \varepsilon_1 F_s L/(qn_s d_{eff}) \tag{4-11-20}$$

and n_s is a function of V_g (see Section 4-2). From eqs. (4-11-13) and (4-11-19) we find the drain saturation current (see Grinberg and Shur 1989):

$$I_{Dsat} = qn_s\mu F_s W[(1 + a^2)^{1/2} - a] \tag{4-11-21}$$

This model uses a piece-wise linear approximation for the electron velocity:

$$v = \begin{cases} \mu F & F \le F_s \\ v_s & F > F_s \end{cases} \tag{4-11-22}$$

where μ is the low-field mobility, F is the electric field, v_s is the effective saturation velocity, and $F_s = v_s/\mu$. Based on the results of the Monte Carlo simulation (see, for example, Cappy et al. 1980) and transient calculations (see Shur 1976), we know that v_s in short-channel devices is very different from the electron saturation velocity in a long-bulk semiconductor sample: it is considerably higher and is dependent on the effective gate length, L_{eff}. One-dimensional simulations for GaAs n-i-n structures led to the following interpolation formula for v_s (see Shur and Long 1982):

$$v_s \approx (0.22 + 1.39L_{eff}) \times 10^7/L_{eff} \qquad \text{(cm/s)} \tag{4-11-23}$$

where the effective gate length, L_{eff}, is in microns. Also, numerical simulations clearly show that the electron velocity in short structures does not saturate in high electric fields. As a matter of fact, the velocity depends on both electric field and potential. Hence, this simple model relies on an effective value of the electron velocity, in high electric fields, that gives a general idea of the electron velocities in a device channel but does not give any information about the exact shape of the velocity profile in the channel. For devices with recessed gates the effective gate length can be estimated as

$$L_{eff} \approx L + \beta_L(d + \Delta d) \tag{4-11-24}$$

where Δd is the effective thickness of the two-dimensional electron gas, and β_L is a constant. The length $\beta_L(d + \Delta d)$ in the equation represents the total lateral depletion width and is a function of gate recess width in the heterostructure FET. Chao et al. (1988) estimated that $\beta_L \approx 2$.

The source and drain series resistances, R_s and R_d, may play an important role in determining the current-voltage characteristics of heterostructure FETs. These resistances can be taken into account as follows: The extrinsic (applied) gate-to-source voltage V_{GS} is given by

$$V_{gs} = V_g + (I_{ds} + I_g)R_s \tag{4-11-25}$$

where I_g is the gate current. For depletion-mode HFETs, the gate current does not usually play an important role because the electron transfer into the AlGaAs layer (see Fig. 4-11-8) severely limits the device transconductance at lower voltages than the turn-on voltage for the gate leakage current. On the other hand, for enhancement-mode HFETs, the gate current can play a dominant role and, as

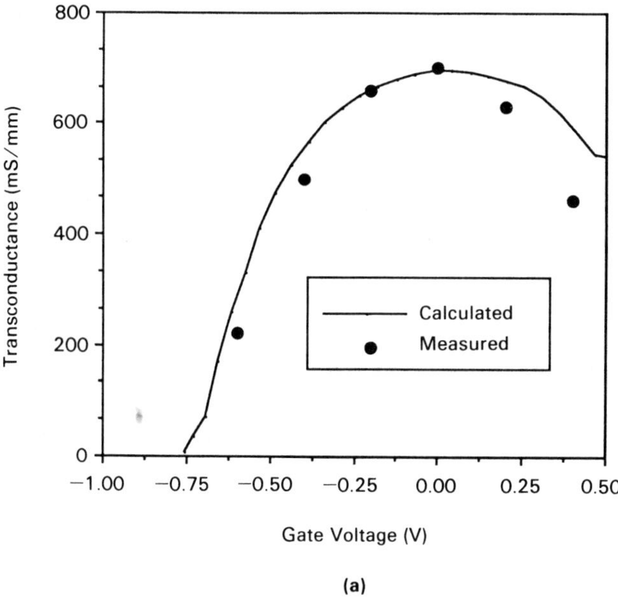

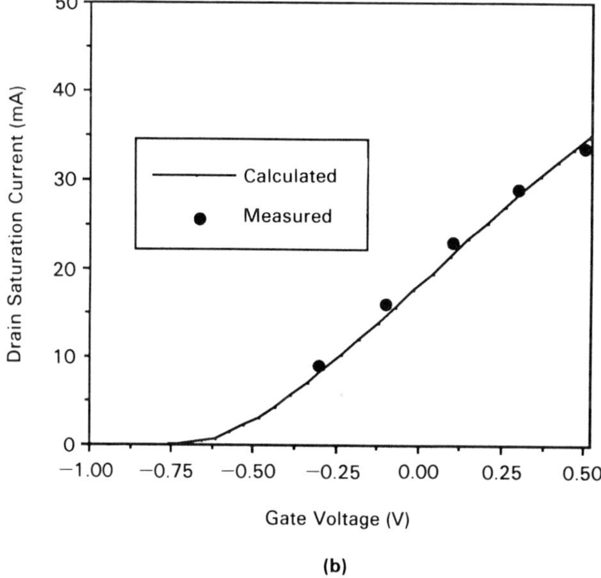

Fig. 4-11-9. Calculated and measured (a) device transconductance and (b) drain saturation current for planar-doped AlGaAs-InGaAs-GaAs HEMTs (from P. C. Chao, M. Shur, R. C. Tiberio, K. H. G. Duh, P. M. Smith, J. M. Ballingall, P. Ho, and A. A. Jabra, "DC and Microwave Characteristics of 0.1 μm Gate-Length Planar-Doped Pseudomorphic HEMTS," *IEEE Trans. Electron Devices*, 36, no. 3, pp. 461–473 (1989). © 1989 IEEE).

discussed, may even affect the value of the intrinsic drain current, I_{ds}. Equation (4-11-25) has to be solved together with eqs. (4-11-19) and (4-11-21) to find the drain saturation current as a function of the gate voltage. Figure 4-11-9 (from Chao et al. 1988) shows a comparison of this model with experimental data for the device transconductance and saturation current. Figure 4-11-10 (from Chao et al. 1989) presents a comparison between a planar-doped barrier and a conventional modulation-doped structure having the same pinch-off voltage. As can be seen from the figure, in a conventional HFET the electrons are induced into the AlGaAs layer at smaller gate voltages. This leads to a decrease in both maximum transconductance and the gate voltage swing.

The effect of gate length on device transconductance is primarily related to the velocity enhancement due to the ballistic and overshoot effects in short device structures. Figure 4-11-11 shows the computed dependencies of the device transconductance on the gate voltage, assuming that the effective saturation velocity is given by eq. (4-11-23). As can be seen from the figure, the transconductance nearly doubles when the device length is scaled down from 1 μm to 80 nm.

The output current-voltage characteristics can be calculated using the same approach as for GaAs MESFETs:

$$I_{ds} = I_{Dsat}(1 + \lambda V_{DS})\tanh[g_{ch}V_{DS}/I_{Dsat}] \tag{4-11-26}$$

where I_{Dsat} is given by eq. (4-10-21),

$$g_{ch} = g_{chi}/[1 + g_{chi}(R_s + R_d)] \tag{4-11-27}$$

is the channel conductance at low drain-to-source voltages, and

$$g_{chi} = q\mu n_s W/L \tag{4-11-28}$$

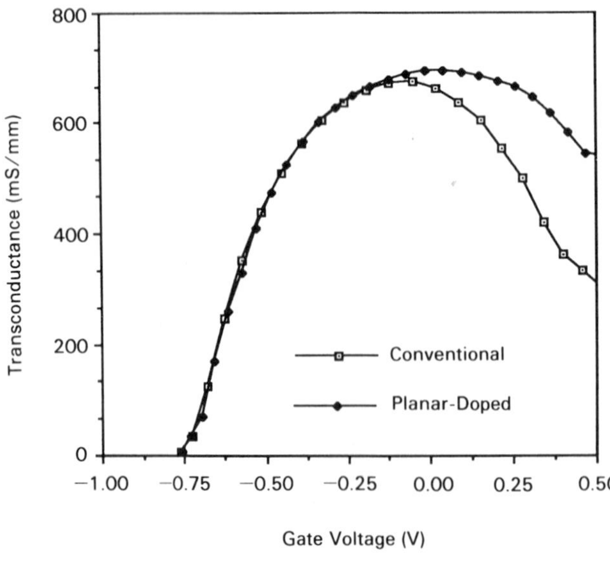

Fig. 4-11-10. Comparison of transconductance between planar-doped and conventional modulation-doped structures with the same threshold voltage. A broader high transconductance region is observed for the planar-doped structure (from P. C. Chao, M. Shur, R. C. Tiberio, K. H. G. Duh, P. M. Smith, J. M. Ballingall, P. Ho, and A. A. Jabra, "DC and Microwave Characteristics of 0.1 μm Gate-Length Planar-Doped Pseudomorphic HEMTS," *IEEE Trans. Electron Devices*, 36, no. 3, pp. 461–473 (1989). © 1989 IEEE).

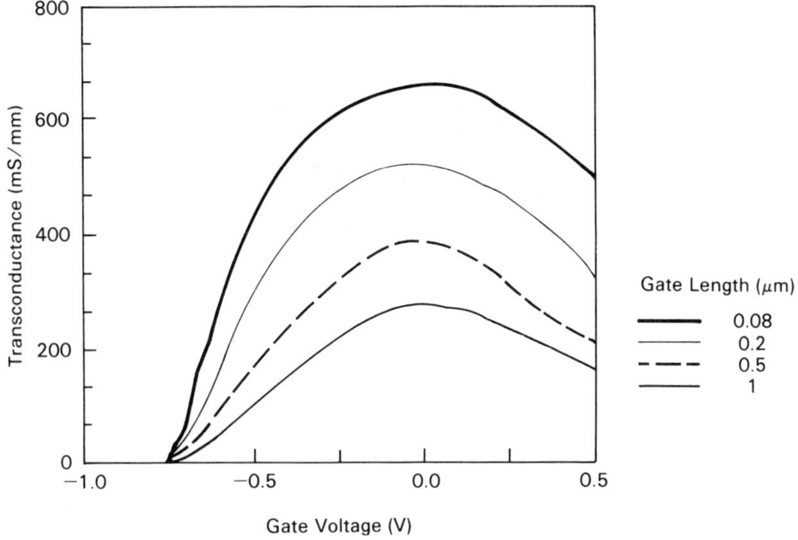

Fig. 4-11-11. Effect of gate-length on device transconductance. The device transconductance increases with the decrease in gate-length due to the increase of the effective electron velocity caused by the ballistic and overshoot effects (from P. C. Chao, M. Shur, R. C. Tiberio, K. H. G. Duh, P. M. Smith, J. M. Ballingall, P. Ho, and A. A. Jabra, "DC and Microwave Characteristics of 0.1 μm Gate-Length Planar-Doped Pseudomorphic HEMTS," *IEEE Trans. Electron Devices*, 36, no. 3, pp. 461–473 (1989). © 1989 IEEE).

is the intrinsic channel conductance at low drain-to-source voltages. The constant, λ, in eq. (4-11-26) is an empirical constant that accounts for output conductance. This output conductance may be related to short-channel effects as well as effects related to the space-charge injection into the buffer layer (see, for example, Han et al. 1988).

This model has been implemented in the GaAs circuit simulator UM-SPICE (see Hyun et al. 1986). The equivalent circuits of HFETs used in UM-SPICE are shown in Fig. 4-11-12. Figure 4-11-12a shows a conventional equivalent circuit similar to that used for other field-effect transistors. This circuit was used in the old version of UM-SPICE (see Hyun et al. 1986). In this circuit, the gate current can be modeled by equivalent Schottky diodes connected from the gate to the source and the drain, respectively. Using the well-known diode equation, we find that for the total gate current,

$$I_g = J_s WL \left[\exp\left(\frac{V_{gs}}{nV_{th}}\right) + \exp\left(\frac{V_{gd}}{nV_{th}}\right) - 2 \right] \tag{4-11-29}$$

where V_{th} is the thermal voltage, n is the diode ideality factor, and J_s is the reverse saturation current density. A more accurate model describing gate current in

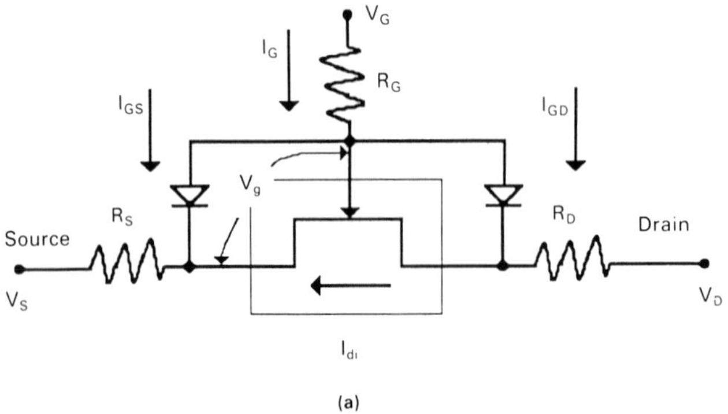

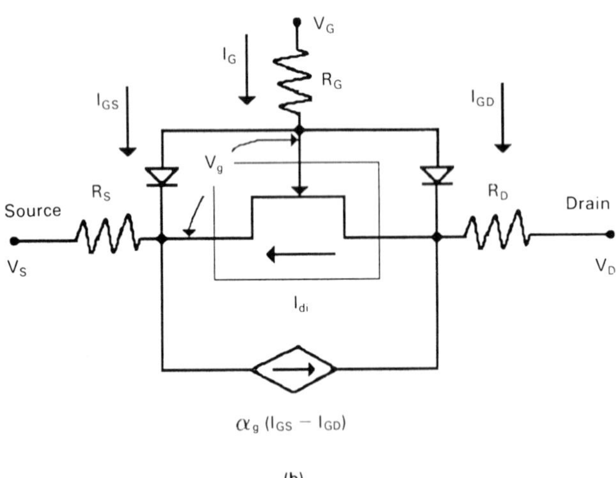

Fig. 4-11-12. HFET equivalent circuits: (a) conventional equivalent circuit similar to that used for other field-effect transistors and (b) equivalent circuit that takes into account the effect of the gate current on the channel current (after P. P. Ruden, M. Shur, A. I. Akinwande, and P. Jenkins, "Distributive Nature of Gate Current and Negative Transconductance in Heterostructure Field-Effect Transistors," *IEEE Trans. on Electron Devices*, ED-36, no. 2, pp. 453–456, February (1989). © 1989 IEEE).

MODFETs was developed by Ponse et al. (1985), Ruden et al. (1988), Chen et al. (1988), and Baek and Shur (1988). This model takes into account that, as shown in Fig. 4-11-13, the gate current mechanism changes at $V_g = \phi_b - \Delta E_c$ where ϕ_b is the Schottky barrier height at the metal–semiconductor interface and ΔE_c is the conduction band discontinuity. At smaller gate voltages the gate current is pri-

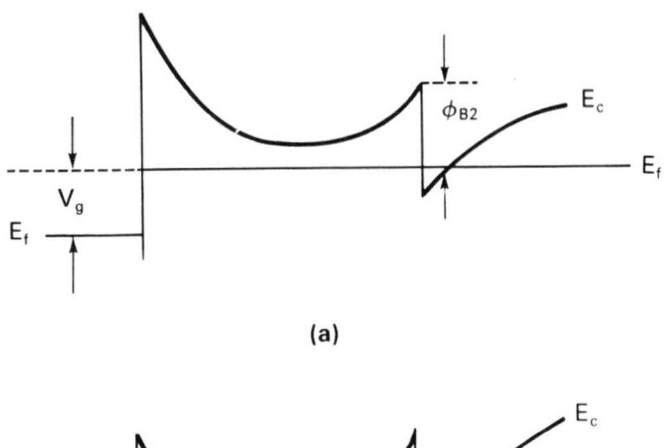

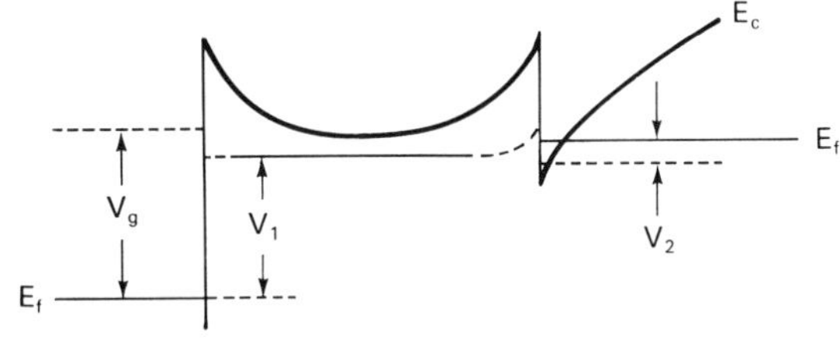

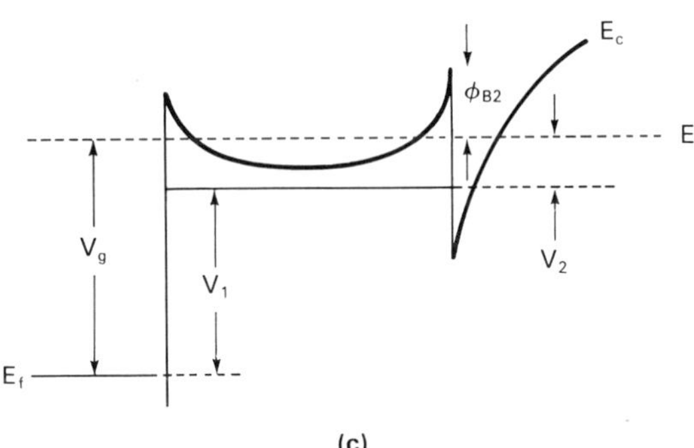

Fig. 4-11-13. HFET band diagram at different gate biases (after C. H. Chen, S. Baier, D. Arch, and M. Shur, "A New and Simple Model GaAs Heterojunction FET Characteristics," *IEEE Trans. Electron Devices*, ED-35, no. 5, pp. 570–577, May (1988). © 1988 IEEE). (a) $V_g < \phi_b - \Delta E_c$. (b) $V_g \approx \phi_b - \Delta E_c$. (c) $V_g > \phi_b - \Delta E_c$. The gate current mechanism changes at $V_g = \phi_b - \Delta E_c$. ϕ_{b2} is the effective barrier height for high gate voltages. V_1 and V_2 are gate voltage drops across the Schottky barrier and the conduction band discontinuity respectively.

marily determined by the Schottky barrier at the metal–semiconductor interface. At larger gate voltages the gate current is limited by conduction band discontinuity or, more precisely, by the effective barrier height equal to the difference between the bottom of the conduction band in the AlGaAs layer at the heterointerface and the electron quasi-Fermi level in the two-dimensional gas. This barrier height changes very little with the gate voltage (and only as a consequence of the dependence of the quasi-Fermi level on electron concentration in the two-dimensional electron gas and, hence, on the gate voltage). Thus, the interpolation of the experimental dependence of the gate current on the gate voltage in this regime, given by the diode equation, has a very large ideality factor (usually from 5 to 20). As can be seen from Fig. 4-11-14, this model agrees quite well with experimental data for both n-channel and p-channel devices.

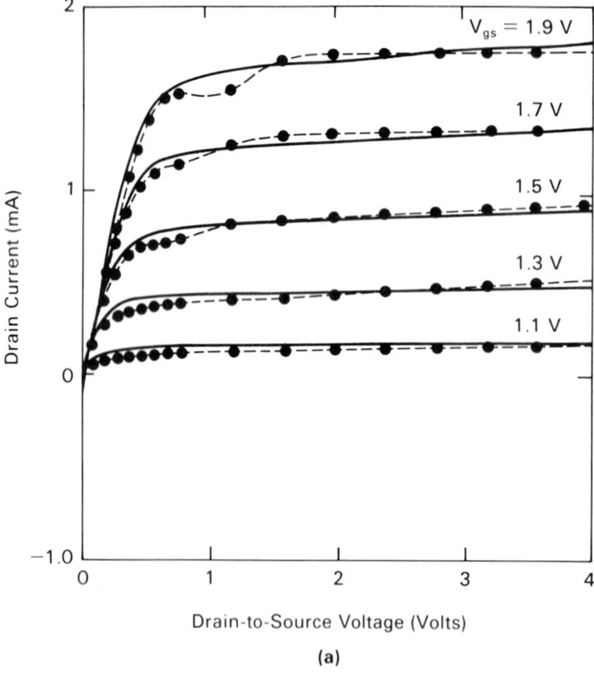

(a)

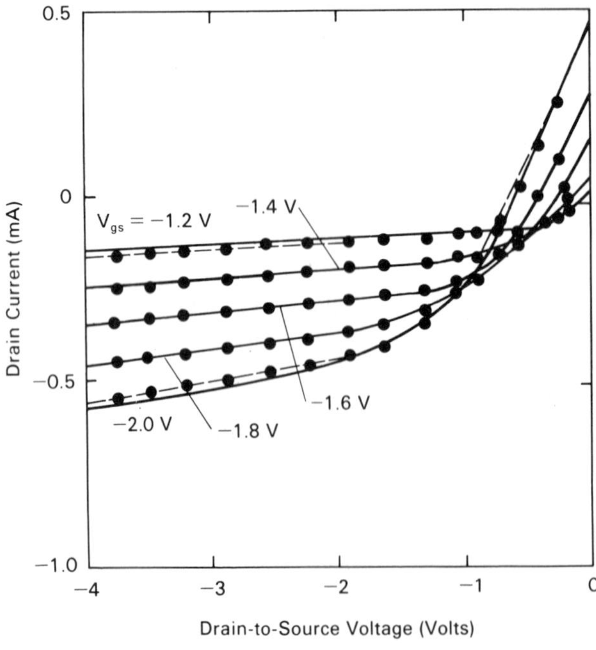

(b)

Fig. 4-11-14. Current-voltage characteristics for *n*-channel (a) and *p*-channel (b) Heterostructure Insulated Gate FETs (after J. H. Baek, M. Shur, R. R. Daniels, D. K. Arch, J. K. Abrokwah, and O. N. Tufte, *IEEE Trans. Electron Devices*, ED-34, August (1987). © 1987 IEEE). Dots show measured data and the solid line represents calculated curves.

An interesting effect that is seen in the measured current-voltage characteristics at high gate voltages is a negative differential resistance (see the curve corresponding to V_{gs} = 1.9 V in Fig. 4-11-14a). This effect may be much more pronounced (see Shur et al. 1986) and is related to the real space transfer mechanism, first proposed by Hess et al. (1979), i.e., to the transfer of hot electrons from the channel over the barrier created by the conduction band discontinuity. A similar effect was observed in HFETs by Y. Chen et al. (1987).

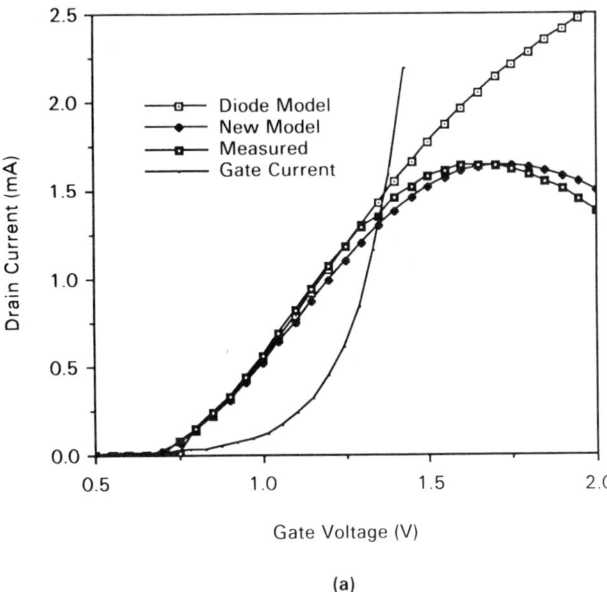

(a)

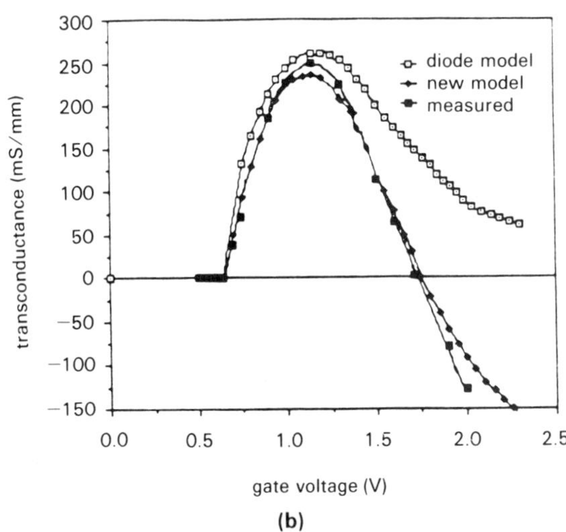

(b)

Fig. 4-11-15. Measured and calculated (a) drain current and (b) transconductance for self-aligned AlGaAs-GaAs MODFET (from P. P. Ruden, M. Shur, A. I. Akinwande, and P. Jenkins, "Distributive Nature of Gate Current and Negative Transconductance in Heterostructure Field-Effect Transistors," *IEEE Trans. on Electron Devices,* ED-36, No. 2, pp. 453–456, February (1989). © 1989 IEEE). The parameters used in the calculation are threshold voltage V_t = 0.582 V, thickness of doped AlGaAs layer d_d = 281 Å, undoped AlGaAs spacer layer thickness d_i = 75 Å, mole fraction of Al in $Al_xGa_{1-x}As$ $x = 0.3$, electron saturation velocity $v_s = 1.6 \times 10^5$ m/s, electron mobility $\mu = 0.35$ m²/Vs, gate saturation current $I_{gs} = 1.14 \times 10^{-7}$ A, effective ideality factor $m = 5.61$, drain and source series resistances $R_d = R_s$ 49 Ω, gate series resistance $R_g = 45$ Ω, gate length L = 1 μm, gate width W = 10 μm, output conductance parameter $\lambda = 0.07$, α_g = 0.26, and V_{DS} = 3 V.

The equivalent circuit shown in Fig. 4-11-12b takes into account the effect of gate current on the channel current. Indeed, the gate current is distributed along the channel, with the largest gate current density near the source side of the channel. This leads to the redistribution of electric field along the channel, with an increase in electric field near the source side of the device and a decrease in channel current. A numerical solution of coupled differential equations describing the gate and channel current distributions along the channel was given by Baek and Shur (unpublished). This calculation provides a justification for the new equivalent circuit shown in Fig. 4-11-12b. As illustrated by Fig. 4-11-15, this new equivalent circuit allows us to obtain much better agreement with the experimental data.

The nonlinear capacitances, $C_{gs}(V_{gs}, V_{ds})$ and $C_{gd}(V_{gd}, V_{ds})$, shown in Fig. 4-11-12, represent channel charge storage effects. These capacitances can be estimated using a modified Meyer's model (see Meyer 1971 and Section 4-8):

$$C_{gs} = -(2C_g/3)\left[1 - \frac{(V_{Dsat} - V_{ds})^2}{(2V_{dss} - V_{ds})^2}\right] \tag{4-11-30}$$

and

$$C_{gd} = -(2C_g/3)\left[1 - \frac{V_{Dsat}^2}{(2V_{Dsat} - V_{ds})^2}\right] \tag{4-11-31}$$

in the linear region. Here V_{Dsat} is the drain saturation voltage. In the saturation

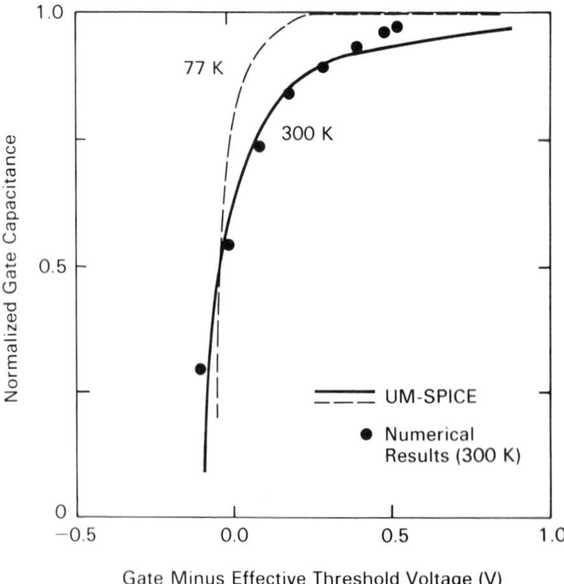

Fig. 4-11-16. HFET gate capacitance vs. gate voltage (after C. H. Hyun, M. Shur, and N. C. Cirillo, Jr., "Simulation and Design Analysis of AlGaAs/GaAs MODFET Integrated Circuits," *IEEE Trans. CAD ICAS*, CAD-5, pp. 284–292 (1986). © 1986 IEEE).

region we have $C_{gs} = 2C_g/3$ and $C_{gd} = 0$. The capacitance C_g is the total gate-to-channel capacitance at $V_{ds} = 0$. An approximate analytical expression for C_g,

$$C_g = C_o \left\{ 1 - \frac{(1 + \sqrt{2})^2 V_{CO}^2}{[-(2 + \sqrt{2})V_{CO} + V_{gst} + \Delta V_T]^2} \right\} \tag{4-11-32}$$

was used to fit the numerically calculated C_g vs. V_{gs} curve, which includes the Fermi level dependence on n_s (see Fig. 4-11-16; from Hyun et al. 1986). Here $\Delta V_T = 0.03V$ and $V_{CO} = 3V_{th}$. As can be seen from Fig. 4-11-16, there is good agreement between numerical calculation and eq. (4-11-32). The constant capacitances C_{gsp}, C_{gdp}, and C_{dsp} are the interelectrode capacitances.

4-12. AMORPHOUS SILICON THIN FILM TRANSISTORS

Conventional electronic devices are fabricated on crystalline wafers of silicon cut from a crystal boule grown from a melt. These wafers are fragile, relatively expensive, and limited in size (to up to approximately 6 inches in diameter). In many applications, amorphous materials that may be deposited in an inexpensive continuous process on a variety of different large-area substrates can compete with crystalline silicon or even open up completely new applications. The most important amorphous material for electronic applications is amorphous silicon. In 1972 Spear and LeComber demonstrated that amorphous silicon films prepared by the glow discharge decomposition of silane gas (SiH_4) have a relatively low density of defect states in the energy gap (see Spear and LeComber 1972). The amorphous silicon material obtained by this process is, in fact, an amorphous silicon-hydrogen alloy with a fairly large concentration of hydrogen. The hydrogen atoms tie up dangling bonds that are present in amorphous silicon in large numbers and decrease the density of localized states in the energy gap. These localized states play a dominant role in determining the transport properties of amorphous Si:H.

Very large area (2 × 4 feet and larger) high quality amorphous silicon (a-Si:H) films may be inexpensively produced in a continuous process, making this material very attractive for applications in electronics, such as large area displays, xerography, imagers, and photovoltaics (see Section 5-3). Amorphous Ge:H, amorphous C:H, amorphous SiC:H, amorphous Si:F, and other amorphous materials have also been produced, opening up opportunities for different amorphous heterostructure devices.

Properties of a-Si:H are quite different from those of crystalline material. a-Si:H is a direct-gap semiconductor with the energy gap close to 1.7 eV and electron and hole band mobilities on the order of 10 cm^2/Vs. It has a much larger absorption coefficient of light than crystalline silicon (which is especially important in photovoltaic and optoelectronic applications). One of the most important differences between a-Si:H and a crystalline semiconductor is the presence of a large number of localized states in the energy gap. An approximate distribution of

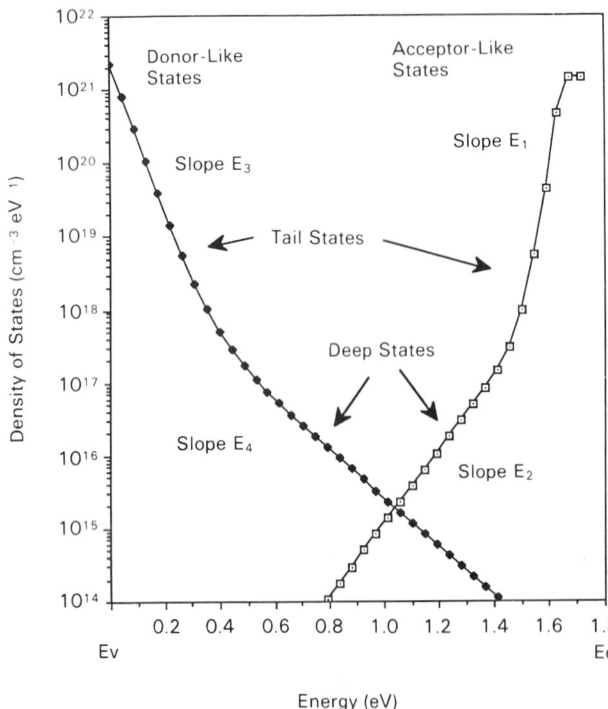

Fig. 4-12-1. An approximate distribution of localized states for intrinsic (undoped) a-Si. E_1, E_2, E_3, E_4 are the characteristic energies of the four exponential distributions: E_1 for tail acceptor like states, E_2 for deep localized acceptor like states, E_3 for tail donor like states, and E_4 for deep localized donor like states.

localized states for intrinsic (undoped) a-Si is shown in Fig. 4-12-1. (An additional peak of the density of the acceptorlike localized states appears in doped amorphous silicon films at approximately 0.4 eV below the edge of the conduction band.) The localized states in the upper half of the energy gap (closer to the bottom of the conduction band) behave like acceptorlike states, whereas the states in the bottom half of the energy gap behave like donorlike states. Donorlike states are positively charged when empty and neutral when filled; acceptorlike states are neutral when empty and negatively charged when filled. As shown in the figure, the localized states can be divided into tail acceptorlike states, deep acceptorlike states, deep donorlike localized states, and tail donorlike states.

The distribution of the localized states is not symmetrical, with more donorlike states than acceptorlike states. Hence, the position of the Fermi level in an undoped uniform a-Si sample in the dark, that is found from the neutrality condition, is shifted closer to the bottom of the conduction band, E_c, (usually in the undoped material $E_c - E_F \approx 0.6$ eV). The position of the Fermi level is determined by the deep localized states and is found from the neutrality condition,

$$Q_d + Q_a = 0 \tag{4-12-1}$$

where Q_d is the charge of the deep donor states and Q_a is the charge of the deep acceptor states. These charges can be found in an analytical form by using a zero temperature approximation, i.e., by assuming that all states below the Fermi level are filled and all states above the Fermi level are empty. As was mentioned previously, donorlike states are positively charged when empty and neutral when filled; acceptorlike states are neutral when empty and negatively charged when filled. Hence, for the most simplistic model of the density of states that represents the densities of the deep donorlike and acceptorlike states by exponential functions of energy (see Fig. 4-12-1) we find that

$$Q_d = q \int_{E_F}^{E_c} g_d(E)\, dE \approx q g_{vd} \int_{E_F}^{\infty} \exp[(E_v - E)/E_d]\, dE = q g_{vd} E_d \exp[(E_v - E_F)/E_d] \tag{4-12-2}$$

$$Q_a = -q \int_{E_v}^{E_F} g_a(E)\, dE \approx q g_{cd} \int_{-\infty}^{E_F} \exp[(E - E_c)/E_a]\, dE = q g_{cd} E_a \exp[(E_F - E_c)/E_a] \tag{4-12-3}$$

Here $g_d(E)$ is the density of deep donorlike states, $g_a(E)$ is the density of deep acceptorlike states, g_{vd} is the density of deep donorlike states extrapolated to the valence band edge, E_v, and g_{cd} is the density of deep acceptorlike states extrapolated to the conduction band edge, E_c. Substituting eqs. (4-12-2) and (4-12-3) into the neutrality condition given by eq. (4-12-1), we find the position of the equilibrium Fermi level, E_{FO}, for an intrinsic material:

$$E_c - E_{FO} = E_g \frac{E_a}{E_a + E_d} - \frac{E_a E_d}{E_a + E_d} \ln \frac{g_{vd} E_d}{g_{cd} E_a} \tag{4-12-4}$$

For intrinsic amorphous Si, $E_a \approx 86$ meV, $E_d \approx 129$ meV, $g_{cd} \approx 8 \times 10^{24}$ m^{-3}eV^{-1}, $g_{vd} \approx 2 \times 10^{25}$ m^{-3}eV^{-1} (see Fig. 4-12-1), and $E_c - E_{FO} \approx 620$ meV. Dopants (typically phosphorus for n-type and boron for p-type material) can shift the position of the Fermi level, and amorphous silicon p-n junctions can be made as was first shown experimentally by Spear et al. (1976) (see Problem 4-12-1).

The position of the Fermi level with respect to the the conduction band edge near the amorphous silicon–insulator interface may be changed by inducing carriers via the field effect, similar to the field effect at the crystalline silicon–insulator interface. This effect has been utilized in amorphous silicon Thin Film Transistors (TFTs) (see Fig. 4-12-2). Amorphous silicon alloy TFTs have the potential to become a viable and important technology for large-area, low-cost integrated circuits. These circuits are currently being used to drive large-area liquid crystal displays. Basic integrated circuits and addressable image-sensing arrays have also been implemented (see, for example, Ito et al. 1985, Thompson and Tuan 1986, Kodama et al. 1982, Hiranaka et al. 1984, and Matsumura et al. 1981).

The low field-effect mobility in a-Si TFTs (typically 0.5 to 1 cm^2/Vs) has long been an obstacle to many practical applications of this technology. However, theoretical and experimental results show that a value for the field-effect mobility close to the electron band mobility (presently estimated at 10 cm^2/Vs or higher)

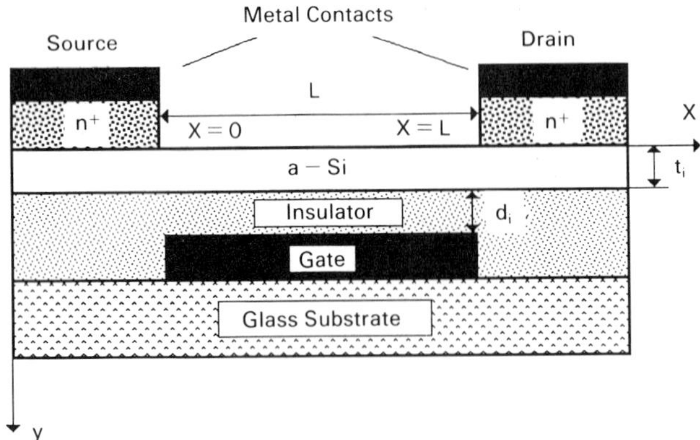

Fig. 4-12-2. Schematic diagram of amorphous silicon Thin Film Transistor or TFT (after M. Shur, M. Hack, and J. G. Shaw, "Capacitance-Voltage Characteristics of Amorphous Silicon Thin Film Transistors," *J. Appl. Phys.*, submitted for publication (1989a). © 1989 IEEE).

may be achieved when the charge induced in the channel is sufficiently large to fill the localized states by electrons induced into the channel (see Shur et al. 1985 and Leroux and Chenevas-Paule 1985). The operation of an a-Si TFT is quite different from that of a crystalline MOSFET. In addition to the below-threshold and above-threshold regimes, there are two new regimes of operation (see Shur et al. 1986). These new regimes occur at high densities of the induced charge in the a-Si TFT channel: a *crystallinelike* regime when the free electron concentration exceeds the localized charge concentration at the a-Si–insulator interface, and a *transitional* regime (between the crystallinelike and above-threshold regimes) at lower densities of induced charge when almost all localized states in the energy gap of the amorphous silicon near the interface are filled.

In the below-threshold regime, nearly all induced charge goes into the deep localized acceptorlike states in the energy gap of a-Si as well as into the surface states at the a-Si–insulator interface. With an increase in the gate voltage, more states are filled, and the Fermi level at the amorphous silicon–insulator interface moves closer to the conduction band. This leads to an increase in the concentration of mobile carriers in the conduction band, which rises superlinearly with gate voltage. With a further increase in the gate voltage, the Fermi level enters the tail states (see Fig. 4-12-1). The characteristic energy of the exponential variation of the tail states is smaller or comparable to $k_B T$ at room temperature. As a consequence, once the Fermi level is in the tail states, most of the charge is actually induced into the states above the Fermi level. Hence, the shift of the Fermi level with gate voltage is considerably smaller than in the below-threshold regime. This regime is an above-threshold regime (see Shur and Hack 1984). There are two important differences between this regime in the a-Si TFTs and the above-threshold regime in crystalline MOSFETs: in a-Si TFTs, most of the induced charge still goes into the tail states with only a small fraction (several percent or so) going into the conduction band, and the Fermi level moves closer to the edge of the conduction band with an increase in gate voltage causing an increase in the field-effect mobility with increasing gate voltage.

When the induced charge is increased even further, we may reach a situation when all tail states at the a-Si–insulator interface are almost completely filled and the Fermi level touches the bottom of the conduction band. An additional increase in the induced charge is divided between the charge going into the conduction band and the charge induced into the tail states farther from the a-Si–insulator interface. The fraction of the mobile charge is at first small, but it increases with the increase in gate voltage. We call this regime a transitional regime because it corresponds to the transition between the above-threshold regime and the crystallinelike regime when the Fermi level at the a-Si–insulator interface has moved high enough into the conduction band so that, finally, most of the induced charge goes into the conduction band. In the latter regime, the field-effect mobility is close to the band mobility and the operation of an a-Si TFT is truly similar to the operation of a crystalline field-effect transistor. We should notice, however, that the gate voltage necessary to achieve the crystallinelike regime is about 50 V to 100 V for an a-Si TFT with an insulator 1000 Å thick and a relative permittivity of approximately 3.9. This is a large voltage; as the material and insulator properties improve, this regime may be achieved at smaller gate voltages.

In what follows we consider the theory of a-Si TFTs following Shur et al. (1989). We start from the equation relating the drain-to-source current, I_d, to the electric field in the channel, F_x, and the concentration per unit area of free conduction band electrons in the channel, n,

$$I_d = -q\mu n_{ind} F_x W \tag{4-12-5}$$

where q is the electronic charge, μ is the band mobility, W is the gate width, and

$$F_x = -\frac{dV}{dx} \tag{4-12-6}$$

Here V is the channel potential. According to the gradual channel approximation,

$$n_{ind} \approx n_{inds} - \frac{\varepsilon_i V}{qd} \tag{4-12-7}$$

where ε_i is the dielectric permittivity of the insulator, d is the dielectric thickness, n_{ind} is the surface charge concentration in the channel per unit area, and n_{inds} is the surface charge concentration in the channel at the source side of the channel per unit area. The concentration of free conduction band electrons in the channel, n_s, per unit area, can be related to the induced charge concentration, introducing the field-effect mobility, μ_{FET}:

$$n_s = (\mu_{FET}/\mu) n_{ind} \tag{4-12-8}$$

Here the field-effect mobility, μ_{FET}, is a function of the induced charge and, hence, the gate voltage. Substituting eqs. (4-12-6), (4-12-7), and (4-12-8) into eq.

(4-12-5) and integrating with respect to y from 0 to L (see Fig. 4-12-1) and with respect to V from 0 to V_D, we obtain

$$I_d = q\frac{W}{L}\int_0^{V_D} \mu_{FET}(n_{ind})n_{ind}\, dV \tag{4-12-9}$$

which can be rewritten as

$$I_d = q^2\frac{W}{L}[n_{inds}(V_g)]^2 (d/\varepsilon_i)\int_z^1 \mu_{FET}(n_{inds}z')z'\, dz' \tag{4-12-10}$$

where

$$z' = 1 - \frac{\varepsilon_i V}{qdn_{inds}} \tag{4-12-11}$$

$$z = 1 - \frac{\varepsilon_i V_D}{qdn_{inds}} \tag{4-12-12}$$

In a conventional crystalline MOSFET

$$n_{inds} \simeq \frac{\varepsilon_i}{qd}(V_g - V_t) \tag{4-12-13}$$

$$\mu_{FET} = \mu \tag{4-12-14}$$

Here V_t is the threshold voltage of a strong inversion in a metal-insulator-semiconductor structure. The substitution of eqs. (4-12-13) and (4-12-14) into eq. (4-12-10) yields the standard equation for a crystalline long-channel metal oxide-field-effect transistor (MOSFET) in the linear regime. However, in the case of an a-Si TFT, n_s is not a simple linear function of the gate voltage. In fact, as was discussed previously, this dependence has several different regions (corresponding to subthreshold, above threshold, and crystallinelike regimes of operation).

The surface induced charge density at the drain is given by

$$n_{indd} \approx n_{inds} - \varepsilon_i V_D/(qd) \tag{4-12-15}$$

In the gradual channel approximation the drain saturation voltage, V_{Dsat}, is reached when the value of nd is equal to zero (the pinch-off condition at the drain). Hence,

$$V_{Dsat} = qn_{inds}\, d/\varepsilon_i \tag{4-12-16}$$

(In fact, as will be discussed presently, current flow in the pinch-off region leads to a positive output conductance in the saturation regime.) For a crystalline transistor eq. (4-12-16) reduces to the standard expression

$$V_{Dsat} = V_g - V_t \tag{4-12-17}$$

Substituting eq. (4-12-16) into eq. (4-12-10) we obtain the expression for the drain saturation current for an a-Si TFT:

$$I_{Dsat} = q^2 \frac{W}{L} [n_{inds}(V_g)]^2 (d/\varepsilon_i) \int_0^1 \mu_{FET}(n_{inds}z)z\, dz \tag{4-12-18}$$

At small drain-to-source voltages $1 - z << 1$ and eq. (4-12-10) reduces to

$$I_d = q\mu_{FET}(n_{inds}) \frac{W}{L} n_{inds}(V_g)\, V_D \tag{4-12-19}$$

Hence, the intrinsic channel conductance, $g_{chi} = \partial I_d/\partial V_D$, at small V_D is given by

$$g_{chi} = q\mu_{FET}(n_{inds}) \frac{W}{L} n_{inds}(V_g) \tag{4-12-20}$$

For a crystalline transistor eqs. (4-12-18) and (4-12-20) reduce to the conventional formulas (see Section 4-4)

$$I_{Dsat} = 0.5 \frac{W}{L} \mu(\varepsilon_i/d)(V_g - V_t)^2 \tag{4-12-21}$$

$$g_{chi} = \mu \frac{W}{L} (\mu_i/d)(V_g - V_t) \tag{4-12-22}$$

In an amorphous silicon TFT the field-effect mobility is much smaller than the band mobility, μ, and depends on the induced charge. This dependence can be evaluated by solving Poisson's equation in the direction perpendicular to the gate and channel and assuming $V_{ds} = 0$. The distribution of the electric field, F_y, perpendicular the TFT channel for zero drain-to-source voltage is found from

$$\frac{dF_y}{dy} = \frac{\rho}{\varepsilon} \tag{4-12-23}$$

where y is the space coordinate (perpendicular to the gate), ε is the dielectric permittivity of amorphous silicon, and ρ is the space-charge density:

$$\rho = -q(N_{loc} + n) \tag{4-12-24}$$

Here q is the electronic charge, N_{loc} is the concentration of localized charge, and n is the concentration of electrons in the conduction band given by

$$n = N_c F_{1/2}(\xi) \tag{4-12-25}$$

where $F_{1/2}$ is the Fermi-Dirac integral (see Appendix 21). Here $\xi = [-(\Delta E_{FO} - qV)/k_B T]$, N_c is the density of extended states in the conduction band, and

$$\Delta E_{FO} = E_c(\text{bulk}) - E_F \tag{4-12-26}$$

where E_F is the equilibrium position of the Fermi level, and V is the electric potential. We choose the potential $V = 0$ to coincide with the bottom of the conduction band far from the gate, where an a-Si is assumed to be in uniform equilibrium.

The density of localized charge for n-channel devices (where the electron quasi-Fermi level in a-Si lies in the upper half of the energy gap) may be found as

$$N_{loc} = \int_{E_{FO}}^{E_c} g_A(E)\, dE/\{1 + \exp[(E - E_F)/k_B T]\} \qquad (4\text{-}12\text{-}27)$$

where g_A is the density of localized states, which may be approximated by the sum of two exponential functions representing the deep and tail states, respectively (see Fig. 4-12-1):

$$g_A(E) = g_{FO} \exp[(E - E_{FO})/E_2] + g_{ct} \exp[(E_c - E)/E_1] \qquad (4\text{-}12\text{-}28)$$

Here E_{FO} is the Fermi level in amorphous silicon at equilibrium, E_c is the bottom of the conduction band, g_{FO} is the density of the deep localized acceptorlike states at $E = E_{FO}$, $E_2 = E_d = k_B T_2$ is the characteristic energy of the exponential variation of the deep localized states, $E_1 = k_B T_1$ is the characteristic energy of the exponential variation of the tail states, and g_{ct} is the density of the tail states at $E = E_c$.

Using the relationship between electric field and electric potential,

$$F_y = -dV/dy \qquad (4\text{-}12\text{-}29)$$

we find from Poisson's equation (see Problem 4-12-2) that

$$F_y^2 = -(2/\varepsilon) \int_0^V \rho(V')\, dV' \qquad (4\text{-}12\text{-}30)$$

The total induced charge, Q_{ind}, the localized charge, Q_{loc}, and the free surface electron charge, qn_s, are given by

$$Q_{ind} = \varepsilon F_y(0)\, [(2\varepsilon) \int_0^{V_o} |\rho(V')|\, dV']^{1/2} \qquad (4\text{-}12\text{-}31)$$

$$Q_{loc} = \int_0^{V_o} \frac{N_{loc}(V')\, dV'}{F_y(V')} \qquad (4\text{-}12\text{-}32)$$

$$qn_s = \int_0^{V_o} \frac{n(V')\, dV'}{F_y(V')} \qquad (4\text{-}12\text{-}33)$$

The potential distribution is found from

$$y = -\int_{V_o}^{V} dV'/F(V') \qquad (4\text{-}12\text{-}34)$$

where V_o is the total band-bending. (V_o is equal to the surface potential of the a-Si–insulator interface.)

Finally, the gate voltage, V_g, is related to the total induced charge, Q_{ind}, as follows:

$$V_g \approx Q_{ind}/C_o + V_{FB} \qquad (4\text{-}12\text{-}35)$$

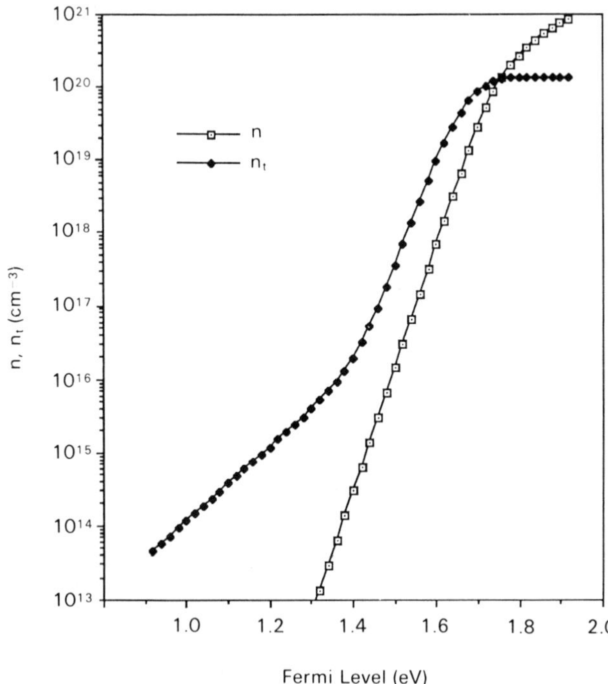

Fig. 4-12-3. Numerically calculated densities of free and trapped charges in an a-Si (after M. Shur, M. Hack, and J. G. Shaw, "Capacitance-Voltage Characteristics of Amorphous Silicon Thin Film Transistors," *J. Appl. Phys.*, submitted for publication (1989a). © 1989 IEEE).

Here

$$C_o = \varepsilon_i / t_i \tag{4-12-36}$$

is the insulator capacitance per unit area, ε_i is the insulator permittivity, t_i is the insulator thickness, and V_{FB} is the flat-band voltage.

Numerically calculated densities of free and trapped charges in an a-Si are shown in Fig. 4-12-3. The surface density of free electrons, n_s, at the a-Si–insulator interface as a function of gate voltage is shown in Fig. 4-12-4. These dependencies allow us to evaluate $\mu_{FET} = n_s/n_{ind}$ as a function of the concentration of electrons, n_{ind}, induced into the TFT channel. Then the current-voltage characteristics can be calculated using eqs. (4-12-10) and (4-12-18). This model can be improved to account for source and drain series resistances and for the space-charge injection effects that are important in the saturation regime (see Shur et al. 1989). Typical measured and calculated current-voltage characteristics of an a-Si TFT are shown in Fig. 4-12-5. This model has also been used to calculate the small-signal capacitances of an a-Si TFT. The complete model has been implemented in a circuit simulator and used for the design of a-Si integrated circuits (see Shur et al. 1989).

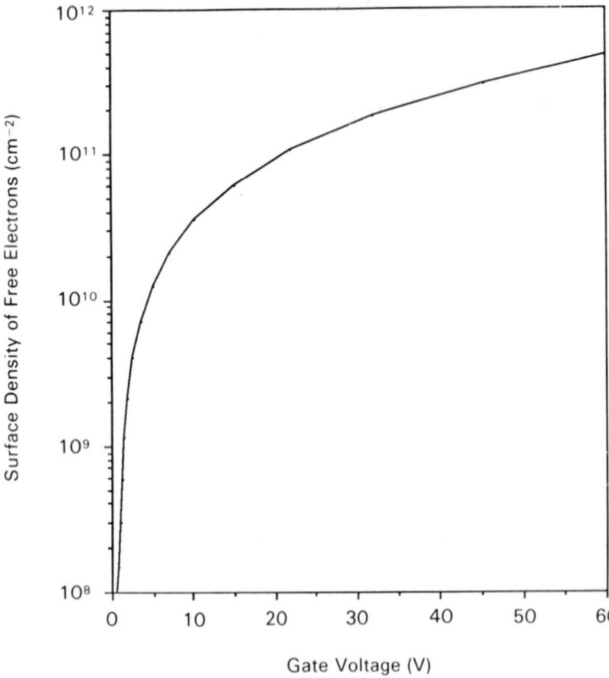

Fig. 4-12-4. Surface density of free electrons at an a-Si-insulator interface as a function of gate voltage (after M. Shur, M. Hack, and J. G. Shaw, "Capacitance-Voltage Characteristics of Amorphous Silicon Thin Film Transistors," *J. Appl. Phys.*, submitted for publication (1989a).

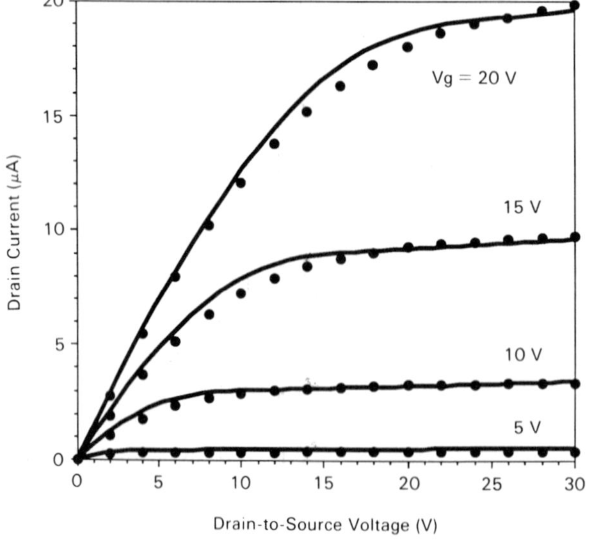

Fig. 4-12-5. Measured (dots) and calculated (solid lines) current-voltage characteristics of an a-Si TFT (after M. Shur, M. Hack, and J. G. Shaw, "Capacitance-Voltage Characteristics of Amorphous Silicon Thin Film Transistors," *J. Appl. Phys.*, Sept. 15, (1989a).

4-13. AMORPHOUS SILICON HIGH–VOLTAGE THIN–FILM TRANSISTORS

In a conventional a-Si thin film transistor (TFT) the electric field in the channel at the drain side of the gate increases rapidly at drain-to-source voltages, V_{ds}, approaching or exceeding the drain-to-source saturation voltage, V_{Dsat}. As a consequence, the device may experience a breakdown at relatively small drain-to-source voltages.

A new amorphous silicon transistor proposed by Tuan (1986) has an insulated gate electrode placed under the source contact but only extending over a portion of the source-drain distance (see Fig. 4-13-1). The source-drain spacing is effectively divided into two regions: a region above the gate electrode, which operates like the conventional accumulation-mode channel in low-voltage amorphous silicon TFTs, and the region between the edge of the gate electrode and the drain contact. This latter region is called the "dead" region. This region sustains the high voltages applied to the drain of the device, and current flow in this region is in most cases space-charge limited. Hence, this device, called an amorphous silicon High-Voltage Thin-Film Transistor (HVTFT), has a much higher drain-to-source breakdown voltage.

In the "on" state, with positive voltage applied to the gate electrode, an accumulation region (of electrons) is formed in a thin channel at the semiconductor–insulator interface in the portion of the active layer above the gate electrode, of length L_1. This region will effectively behave as a conventional low-voltage TFT with a dead region of amorphous silicon, of length L_2, connecting this accumulation channel to the actual drain contact.

In the "off" state there is no accumulation layer above the gate electrode and virtually no current flow.

Following Hack et al. (1987), we use the gradual channel approximation and the constant field-effect mobility model to describe the region under the gate in the on state. For simplicity we neglect the dependence of the field-effect mobility on gate voltage (see Section 4-12). Hence, the saturation current I_{sat} is given by

$$I_{sat} = \mu_{FET} c_i W V_{gt}^2 / (2L_1) \tag{4-13-1}$$

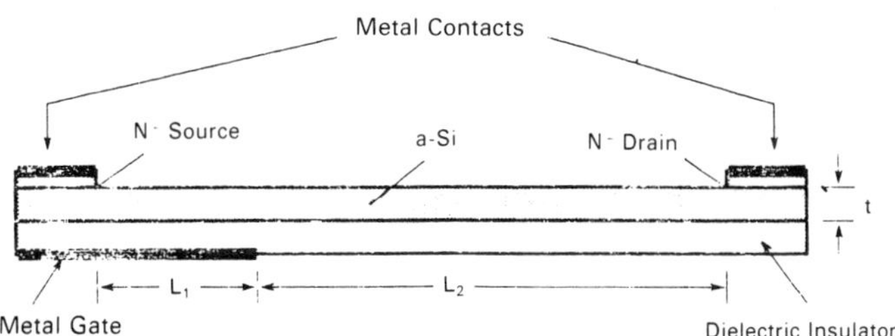

Fig. 4-13-1. High-Voltage Thin-Film Transistor or HVTFT (after H. C. Tuan, *Mat. Res. Soc. Symp. Proc.*, 70, p. 651 (1986). Insulator thickness is 0.3 μm.)

where μ_{FET} is the field-effect mobility, c_i the gate capacitance per unit area, W is the device width, and V_{gt} is the gate voltage minus the threshold voltage (see Section 4-4).

The field F_D in the channel parallel to the semiconductor–insulator interface at the boundary of region 1 (at the drain side of this region) is given by

$$F_D = IV_{gt}/[2L_1 I_{sat}(1 - I/I_{sat})^{1/2}| \quad (4\text{-}13\text{-}2)$$

(see Section 4-4). From eq. (4-13-2) it can be seen that as I approaches I_{sat}, F_D tends to infinity. This limits the injection of charge from this accumulation region 1 into the dead region. Thus, as will now be shown, the overall transistor current cannot exceed I_{sat} because when I approaches I_{sat}, F_D becomes very large, preventing injection of charge from region 1 to region 2. Once region 1 is nearly saturated, any increased drain voltage will be dropped across the dead region 2.

The voltage drop across the accumulation channel, V_1, for a current I is given by

$$V_1 = V_{gt} - [V_{gt}^2 - (2IL_1/c_i W\mu_{FET})]^{1/2} \quad (4\text{-}13\text{-}3)$$

To determine the current-voltage characteristics of region 2 (dead region) for the case of space-charge-limited current flow, Poisson's equation must be solved along the dead region from the edge of the gate electrode to the drain contact. Owing to the relatively large density of states in the energy gap of amorphous silicon, the space charge is primarily determined by the trapped charge, n_t, whereas the conductance of this region is determined by the number of mobile electrons, n.

For a deep localized state, distribution, $g(E)$, varies exponentially with energy, having a characteristic temperature T_o (see Section 4-12):

$$g(E) = g_c \exp[(E - E_c)/k_B T_o] \quad (4\text{-}13\text{-}4)$$

where g_c is the density of deep states extrapolated to the conduction band edge. If we assume that the injected charge density is much greater than the equilibrium charge density, we may interrelate n and n_t by

$$n = N_c[qn_t/g_c k_B T_o]^{\alpha} \quad (4\text{-}13\text{-}5)$$

(see Problem 4-13-1), where N_c is the effective density of states and $\alpha = T_o/T$. Poisson's equation along the dead region 2 is given by

$$dF/dx = qn_t/\varepsilon_s = (k_B T_o/\varepsilon_s)g_c(n/Nc)^{1/\alpha} \quad (4\text{-}13\text{-}6)$$

Integrating eq. (4-13-6) with the boundary condition $F = F_B$ at $x = 0$ and noting that the current density $j = nq\,\mu F$, where μ is the electron band mobility, gives

$$F^{(\alpha+1)/\alpha} = x[(\alpha + 1)/\alpha](k_B T_o/\varepsilon_s)g_c\,[j/q\mu N_c]^{1/\alpha} + F_B^{(\alpha+1)/\alpha} \quad (4\text{-}13\text{-}7)$$

To obtain the potential drop V_2 across region 2, for a current $I = jWt$, where t is the thickness of the active layer, equation (4-13-7) is integrated with respect to x, yielding

$$V_2 = (1/K\beta_2)\,[(KL_2 + H)^{\beta_2} - H^{\beta_2}] \tag{4-13-8}$$

where

$$K = (\alpha + 1)g_c k_B T_o/\alpha\varepsilon_s\,[I/q\mu N_c Wt]^{1/\alpha}$$

$$H = F_B^{(1+\alpha)\alpha}$$

$$\beta_2 = (1 + 2\alpha)/(1 + \alpha)$$

If the length of the dead region, L_2, is too long for current flow to be space-charge limited (i.e., if its ohmic resistance is lower than that determined by space-charge conduction at a given applied bias), V_2 will simply be given by $V_2 = IR_2$, where R_2 is the resistance of region 2. To apply this one-dimensional model to the real two-dimensional situation, we have to include a large fringing capacitance between the source and drain accounting for the electric field stream lines. Hence, we have introduced a correction factor into eq. (4-13-6) such that the permittivity of the silicon, ε_s, has been multiplied by L/t.

The current-voltage characteristics of the overall transistor can now be calculated as follows: First, we choose current I, ($I < I_{sat}$). Hence, the device output current also becomes virtually independent of drain voltage, and any further increase in drain voltage will be dropped across the dead region.

In the off state ($V_g = 0$ V) the gate electrode effectively screens the source electrode from the drain electrode. This is illustrated in Figure 4-13-2, which shows a two-dimensional potential contour for a 20-μm device encapsulated in a material of relative permittivity 6, with the gate electrode overlap L_1 being 5 μm. From Figure 4-13-2, we can see that whereas there is a high electric field in the silicon parallel to the silicon–insulator interface between the gate and drain electrodes, there is virtually no field in this direction between the source and the edge of the gate because the gate and source potentials are constant. It is this screening effect that causes this device to have very low off currents, even with 400 volts applied to the drain electrode. In effect, the gate electrode prevents the source from seeing the drain potential, and hence, the space-charge-limited current that would otherwise flow at high V_{ds} is suppressed at V_{ds} = 400 volts for a 20-μm

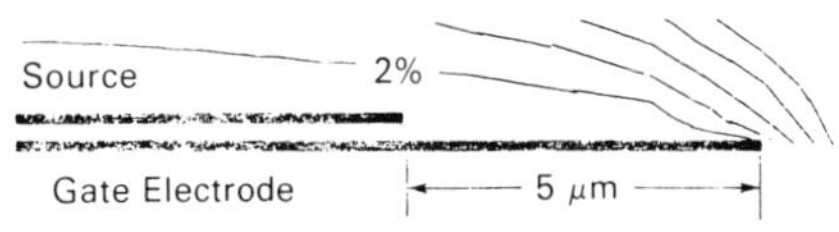

Fig. 4-13-2. Equipotential contours near the source and gate electrodes of the transistor shown in Figure 4-13-1 for $V_s = V_g = 0$ volts (after M. Hack, H. C. Tuan, J. Shaw, M. Shur, and P. Yap, *Mat. Res. Soc. Symp. Proc.*, (1987).)

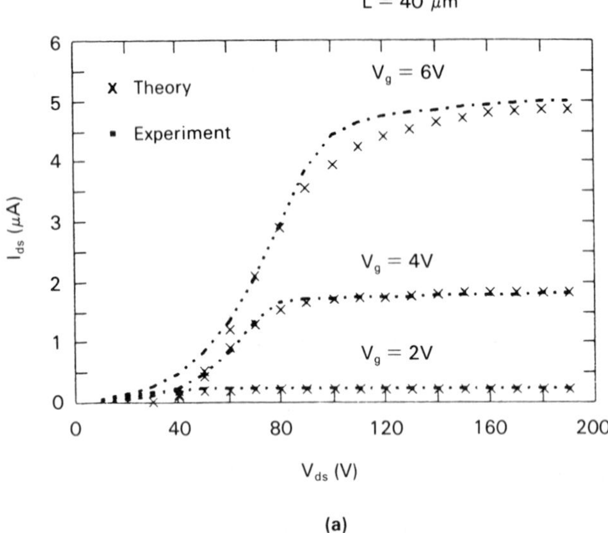

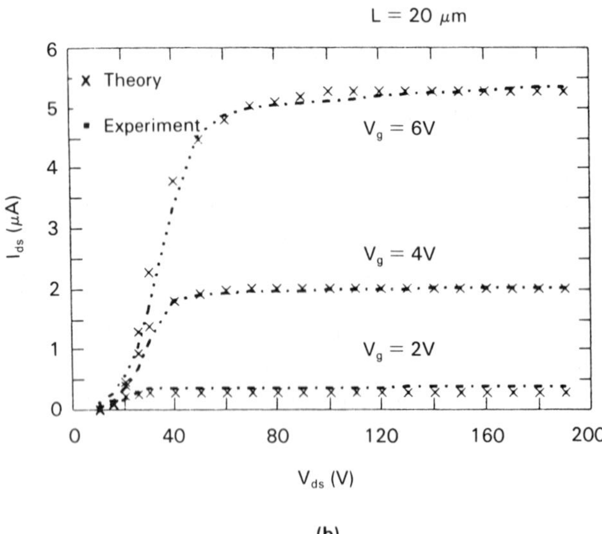

Fig. 4-13-3. Measured and calculated source-drain I-V characteristics for high voltage a-Si transistors (after M. Hack, H. C. Tuan, J. Shaw, M. Shur, and P. Yap, *Mat. Res. Soc. Symp. Proc.*, (1987)). (a) L = 40 μm. (b) L = 20 μm. Parameters use in the calculation: μ = 10 cm²/Vs, gate overlap L_1 = 4 μm, μ_{FE} = 0.5 cm²/Vs, g_c = 1.3 × 10^{18} cm^{-3} eV^{-1}, T_o = 1500 K, c_i = 1.34 × 10^{-8} F/cm², N_c = 10^{19} cm^{-3}, W = 240 μm, and t = 0.1 μm.

device. In fact, the current at $V_g = 0$ was over four orders of magnitude smaller than at V_g = 10 volts (see Tuan 1986).

Figure 4-13-3 shows both measured and calculated I-V characteristics for 20- and 40-μm devices. The agreement between the theory and experiment is very good. As can be seen from these figures, the devices are capable of withstanding hundreds of volts applied to the drain without breakdown. This makes these amorphous silicon transistors appealing for high-voltage, low-current applications.

4-14. AMORPHOUS SILICON DOUBLE–INJECTION FIELD–EFFECT TRANSISTORS

The field-effect transistors described in this chapter are essentially unipolar devices utilizing just one type of carriers, either electrons or holes. There are also devices that combine the principle of operation of a field-effect transistor—i.e, capacitive modulation of charge—with bipolar operation utilizing both electrons and holes. Such devices include a gate-modulated p-n junction (see Richman 1970), an Insulated Gate Thyristor (IGS; see Sze 1981), a Lateral Unidirectional Bipolar-type Insulated gate transiSTOR (LUBISTOR; see Omura 1982), a COn-ductivity-Modulated Field-Effect Transistor (COMFET; see Baliga et al. 1982 and Goodman et al. 1983), a Bipolar Inversion Channel Field-Effect Transistor (BIC-FET; see Taylor and Simmons 1985a and Taylor and Simmons 1985b), and an amorphous silicon Double-Injection Field-Effect Transistor (DIFET; see Hack et al. 1986a, Hack et al. 1986b, and Xu et al. 1987). In this section we will consider the principle of operation of an amorphous silicon Double-Injection Field-Effect Transistor (DIFET).

This device is based on the capacitive modulation of a quasi-neutral plasma of electrons and holes created in a semiconductor by double injection or by an external radiation source. In the double-injection field-effect transistor (DIFET; see Fig. 4-14-1) the double injection current in a forward biased p^+-i-n^+ diode is modulated by an electric field induced by a gate electrode covering the complete path of the channel current. In this device the capacitively induced charge leads to an increase of the electron-hole plasma density, while maintaining the plasma quasi neutrality. Hence, in contrast to a conventional field-effect transistor, the free carrier densities of electrons and holes are much greater than the total induced charge. This leads to a greatly increased transconductance and current swing. Using an amorphous silicon alloy, DIFET currents over twenty times higher than for comparable state-of-the-art amorphous silicon FETs operating under similar conditions have been achieved.

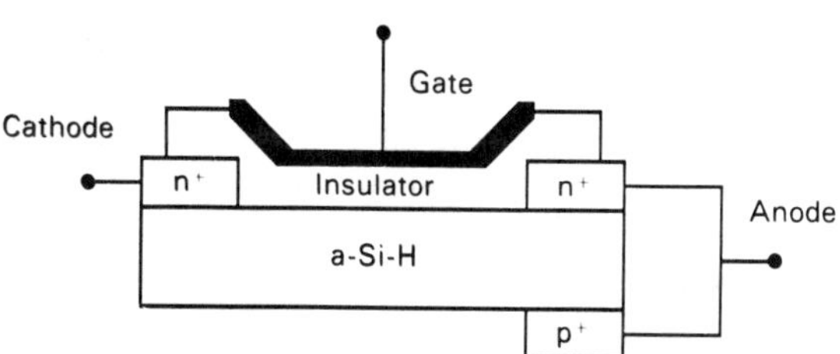

Fig. 4-14-1. Double-Injection Field-Effect Transistor or DIFET (after Hack et al. [1986]).

This principle of operation has some similarity to BJT operation, in which the injection of a carrier from the emitter into the base region causes the injection of the carrier of the opposite polarity from the base contact, which maintains the quasi neutrality in the base. In a similar fashion, in a DIFET the capacitive inducement of charge in the conducting channel leads to the increased plasma density, so that plasma quasi neutrality is maintained.

This principle can be applied to both crystalline and amorphous materials; however, in amorphous materials it leads to important additional advantages. In such materials, the field-effect mobility is limited by localized states in the energy gap, and it can be increased by filling more of these states as a result of higher carrier concentrations in the DIFET.

In a double-injection field-effect transistor (DIFET), the capacitively induced charge in the conducting channel leads to an increased plasma density so as to maintain plasma quasi neutrality. In contrast with a conventional field-effect transistor, the charge densities of both electrons and holes induced in the channel may be much larger than the total net induced charge. This leads to a greatly increased transconductance.

The electron-hole plasma in a DIFET is never neutral. Electrons and holes respond to the gate field. However, their distributions adjust so as to preserve quasi neutrality and so that the change in the net space charge is equal to the charge induced by the gate. This enables us to construct devices in which the double-injection current of the p^+-i-n^+ diode is either considerably enhanced or depleted by the gate field. For enhancement (desirable for high-current, high-speed performance), the gate must act over the complete channel so that the concentration of both carriers can be increased by orders of magnitude, in a controlled fashion, enhancing the conductance of the semiconductor region all the way from anode to cathode. For depletion-mode operation, it is necessary only to reduce both carrier concentrations in one portion of the source-drain region; therefore, a gate need not cover the whole of the channel. This is exactly what is done in the conductivity-modulated field-effect transistor (COMFET; see Baliga et al. 1982 and Goodman et al. 1982), in which the gate acts over only a portion of the current path to control its flow into the base of a bipolar transistor. The DIFET, however, can operate in a linear enhancement mode, where the majority carrier concentration is greater than in a MOSFET, for any given gate field.

The enhancement of a double-injection current by a field applied to the gate electrode is illustrated by Fig. 4-14-2, which shows a one-dimensional computer simulation of the free carrier concentrations in a 5-μm-thick amorphous silicon alloy p^+-i-n^+ diode at 8 V forward bias with the gate floating, as well as with an applied bias of +10 V. It is based on a self-consistent solution of the full set of transport equations incorporating Shockley-Read-Hall recombination kinetics at a continuous distribution of localized states in the mobility gap of these alloys. To model the action of the gate, Poisson's equation along the channel of the device was modified to include the capacitive modulation term (see Hack et al. 1986):

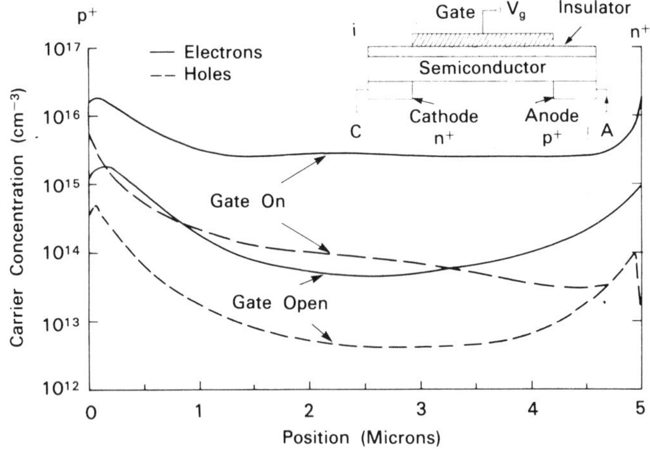

Fig. 4-14-2. One-dimensional computer simulation of the free carrier concentrations along the channel of a DIFET for gate floating and for an applied gate bias of +10V (after Hack et al. [1986]).

$$\rho'(x) = q\{p(x) + p_t(x) - n_t(x) - n(x) + (1 - a)c_g[V_g - V(x)]/(qZ)\} \qquad (4\text{-}14\text{-}1)$$

where $\rho'(x)$ is the fraction of the space-charge density creating the electric field along the channel, $p_t(x)$ and $n_t(x)$ are the densities of trapped positive and negative charge, respectively, $p(x)$ and $n(x)$ are the free carrier densities, c_g is the effective gate capacitance, V_g is the gate potential, $V(x)$ is the channel potential, and Z is the effective two-carrier channel depth (see Hack et al. 1986b). Most of the gate charge, $(1 - a)c_gV_g$, will enter the channel so as to maintain a high electric field perpendicular to the channel, and as a first approximation, we assume that a small component of this charge, ac_gV_g, increases the charge density creating the electric field along the channel (i.e., a is a constant less than unity).

As can be seen from Fig. 4-14-2 the application of a voltage to the gate enhances both the electron and hole concentrations throughout the channel, except for the narrow accumulation regions near the p$^+$ and n$^+$ contacts. As a consequence, in double-injection systems the resultant plasma density can be much larger than the injected or induced charge density (see Lampert and Mark 1970).

Additional insight into this problem can be realized by considering the trap charge densities in the channel. To the first approximation, the double-injection gain factor equals τ_p/t_{ptr}, where τ_p is the hole lifetime and t_{ptr} is the hole transit time. Hence, for a sample length L, we may write

$$qn \sim (\tau_p/t_{ptr})(ac_gV_g/Z + CV/L) + (1 - a)c_gV_g/Z \qquad (4\text{-}14\text{-}2)$$

$$qp_t \sim qn_t - c_gV_g/Z \qquad (4\text{-}14\text{-}3)$$

where C is the effective source-to-drain capacitance per unit area of the channel.

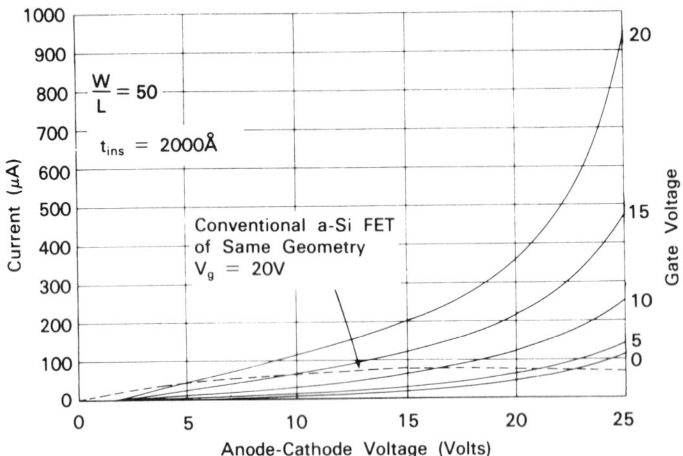

Fig. 4-14-3. Experimental current-voltage characteristics of a DIFET together with state-of-the-art a-Si TFT of similar dimensions (after Hack et al. [1986]).

Short-channel DIFETs have triodic characteristics, as qp_t is larger than c_gV_g/Z and the gain factor τ_p/t_{ptr} increases with increasing anode-cathode voltage V_{AK}. It is this new mode of operation that causes the high current output of the DIFET. The portion of the charge induced in the channel by the gate is amplified by the double-injection gain, resulting in higher carrier concentrations than those possible in MOSFETs. For long-channel devices, the c_gV_g/Z term in eqs. (4-14-2) and (4-14-3) dominates, and for $V_{AK} < V_g - V_T$, where V_T is the threshold voltage, the device current does not saturate, as in a conventional FET. However, even for $V_{AK} > V_g - V_T$ the current in a DIFET does not saturate. This is because the portion of the device near the anode, where the gate potential is less than the channel potential minus the threshold voltage, is a double-injection system whose length will increase considerably with increasing V_{AK}. Thus, increasing V_{AK} effectively reduces the channel length of the accumulation region under the gate, increasing the output current.

The importance of modulating the concentration in the channel of both types of carriers is that the gate field now acts on the space charge arising from the difference in the number of positive and negative charges in the active layer, whereas the conductance of this region depends on the sum of the conductances of mobile carriers. Hence, for any given gate field, much higher channel conductances can be obtained in a two-carrier system than in a conventional FET. This advantage can also be illustrated by comparing the current-voltage characteristics of double-injection p^+-i-n^+ diodes with single-injection n^+-i-n^+ devices (see Hack and den Boer 1985).

Fig. 4-14-3 shows the measured anode–cathode current-voltage characteristics of an amorphous silicon alloy DIFET with a nominal gate length of 20 μm, a

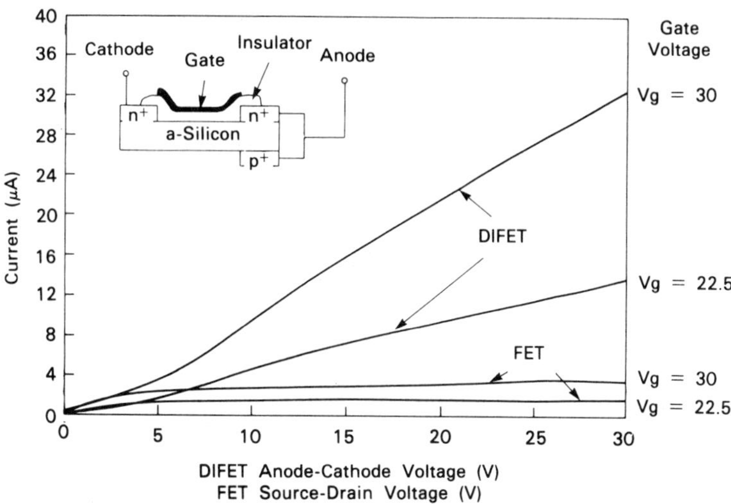

Fig. 4-14-4. Current-voltage characteristics of the device shown in the inset demonstrating the enhancement of the DIFET regime over the MOSFET regime (after Hack et al. [1986]).

width of 1000 μm, and an insulator thickness of 2000 Å. The DIFET does not turn on until the anode–cathode voltage exceeds the built-in potential (~ 1.5 V) of the p^+-i-n^+ diode. At high V_{AK}, the characteristics are triodic, as with increasing V_{AK} the gain in the channel (τ_p/t_{ptr}) increases. Also, superimposed on Fig. 4-14-3 are characteristics of a state-of-the-art amorphous FET of a similar geometry with a field-effect mobility of 0.5 cm^2/Vsec and a threshold voltage of 2 V, operating with V_g = 20 V. For V_{ds} = 25 V the measured DIFET current is approximately fourteen times that of a regular FET. It is this high current capability that may enable transistors operating on these new physical principles to achieve high switching speeds.

Fig. 4-14-4 shows the current-voltage characteristics of the DIFET (see the insert). With the anode grounded and positive voltage applied to the cathode, the device behaves like a conventional FET, whereas with the cathode grounded, the same device operates as a DIFET. These results clearly show the enhancement of the current that is obtained by employing gate-modulated double injection.

REFERENCES

Y. Akasaki and T. Nishimura, *IEDM Technical Digest*, p. 488, IEEE Publication, Los Angeles (1986).

D. K. Arch, M. Shur, J. K. Abrokwah, and R. R. Daniels, *J. Appl. Phys*, 61, pp. 1503–1509 (1987).

N. D. ARORA, J. R. HAUSER, and D. J. RULSON, *IEEE Trans. Electron Devices*, ED-29, p. 292 (1982).

T. BABA, T. MUZUTANI, M. OGAWA, and K. OHATA, *Jpn. J. Appl. Phys.*, 23, p. L654 (1983).

J. H. BAEK and M. SHUR, "Mechanism of Negative Transconductance in Heterostructure Field Effect Transistors," unpublished paper.

J. H. BAEK, M. SHUR, K. LEE, and T. VU, *IEEE Trans. Electron Devices*, ED-32, pp. 2426–2430 (1985).

J. H. BAEK, M. SHUR, R. R. DANIELS, D. K. ARCH, J. K. ABROKWAH, K. LEE, and T. VU, *IEEE Trans. Electron Devices*, ED-32, pp. 2426–2430 (1985).

J. H. BAEK, M. SHUR, R. R. DANIELS, D. K. ARCH, J. K. ABROKWAH, *IEEE Electron Device Lett.*, EDL-7, pp. 519–521 (1986).

J. H. BAEK, M. SHUR, R. R. DANIELS, D. K. ARCH, J. K. ABROKWAH, and O. N. TUFTE, *IEEE Trans. Electron Devices*, ED-34, August (1987).

B. J. BALIGA, M. S. ADLER, P. V. GRAY, R. P. LOVE, and N. ZOMMER, *IEDM Technical Digest*, p. 264, IEEE Publication, Washington D.C. (1982).

B. J. BALIGA, "Power Integrated Circuits—a Brief Overview," *IEEE Trans. Electron Devices*, ED-33, no. 12, pp. 1936–1939 (1986).

M. S. BIRITTELLA, W. C. SEELBACH, and H. GORONKIN, *IEEE Trans. Electron Devices*, ED-29, pp. 1135–1142 (1982).

J. S. BLACKMORE, *Solid State Elec.*, 25, no. 11, pp. 1067–1076 (1982).

Y. BYUN, B. MOON, K. LEE, and M. SHUR, *IEEE Electron Device Letters*, to be published.

A. CAPPY, B. CARNES, R. FAUQUEMBERGUES, G. SALMER, and E. CONSTANT, "Comparative Potential Performance of Si, GaAs, GaInAs, InAs Submicrometer-gate FETs," *IEEE Trans. Electron Devices*, ED-27, pp. 2158–2168 (1980).

D. M. CAUGLEY and R. E. THOMAS, *Proc. IEEE*, 55, pp. 2192–2193 (1967).

P. C. CHAO, M. SHUR, R. C. TIBERIO, K. H. G. DUH, P. M. SMITH, J. M. BALLINGALL, P. Ho, and A. A. JABRA, "DC and Microwave Characteristics of 0.1 μm Gate-Length Planar-Doped Pseudomorphic HEMTS," *IEEE Trans. Electron Devices*, 36, no. 3, pp. 461–473 (1989).

C. E. CHEN, H. W. LAM, S. D. S. MALHI, and A. R. PNIZZOTTO, *IEEE Electron Device Lett.*, EDL-4, no. 8, pp. 272–274 (1983).

T. H. CHEN, "High Speed GaAs Device and Integrated Circuit Modeling and Simulation," Ph.D thesis, University of Minnesota (1984).

C. H. CHEN, M. SHUR, and A. PECZALSKI, "Trap Enhanced Temperature Dependence of the Threshold Voltage of GaAs MESFET's," *IEEE Trans. Electron Devices*, ED-33, pp. 792–798 (1986).

C. H. CHEN, A. PECZALSKI, M. SHUR, and H. K. CHUNG, "Orientation and Ion-implanted Transverse Effects in Self-Aligned GaAs MESFETs," *IEEE Trans. Electron Devices*, ED-34, no. 7, pp. 1470–1481, July (1987).

C. H. CHEN, S. BAIER, D. ARCH, and M. SHUR, "A New and Simple Model for GaAs Heterojunction FET Characteristics," *IEEE Trans. Electron Devices*, ED-35, no. 5, pp. 570–577, May (1988).

Y. K. CHEN, D. C. RADULESCU, G. W. WANG, A. N. LEPORE, P. J. TASKER, L. F. EASTMAN, and E. STRID, "Bias-Dependent Microwave Characteristics of an Atomic

Planar-Doped AlGaAs/InGaAs/GaAs Double Heterojunction MODFET," in *Proceedings of IEEE MTT Symposium,* Las Vegas, June (1987).

N. C. Cirillo, Jr., M. Shur, P. J. Vold, J. K. Abrokwah, R. R. Daniels, and O. N. Tufte, "Realization of n-channel and p-channel High Mobility (Al,Ga)As-GaAs Heterostructure Insulated Gate FETs on a Planar Wafer Surface," *IEEE Electron Device Lett.,* EDL-6, pp. 645-647 (1985).

N. C. Cirillo, Jr., M. Shur, and J. K. Abrokwah, "Inverted GaAs/AlGaS Modulation Doped Transistors with Extremely High Transconductances," *IEEE Electron Device Lett.,* EDL-7, pp. 71–74 (1986).

J. Compton, "Limitations of and Improvements to MOS Processes," in *MOS Devices: Design and Manufacture,* ed. A. D. Milne, Halsted Press, New York (1983).

R. R. Daniels, P. Ruden, M. Shur, D. Grider, and T. Nohava, *IEDM Technical Digest,* Los Angeles, December (1987).

R. R. Daniels, P. P. Ruden, M. Shur, D. E. Grider, T. Nohava, and D. Arch, "Quantum Well p-channel AlGaAs/InGaAs/GaAs Heterostructure Insulated Gate Field Effect Transistors with Very High Transconductance," *IEEE Electron Device Lett.,* EDL-9, July (1988).

B. E. Deal, M. Sclar, A. S. Grove, and E. H. Snow, *J. Electrochem. Society,* 114, March (1967).

R. Dingle, H. L. Stormer, A. C. Gossard, and W. Wiegmann, *Appl. Phys. Lett.,* 37, p. 805 (1978).

T. J. Drummond, H. Morkoç, K. Lee, and M. Shur, *IEEE Electron Device Lett.,* EDL-3, pp. 338–341 (1982).

C. Duvvury, "A Guide to Short Channel Effects in MOSFETs," *TI Engineering Journal,* 1, no. 1, pp. 52–56, July-August (1984); see also *IEEE Circuit and Devices Magazine,* 2, no. 6, p. 6, November (1986).

L. F. Eastman, private communication (1983).

R. W. H. Engelman and C. A. Liehti, *IEEE Trans. Electron Devices,* ED-24, pp. 1288–1296 (1977).

L. Esaki and R. Tsu, Internal Report RC 2418, IBM Research, March 26 (1969).

Frohman-Bentchkowsky, *IEEE J. Solid State Circuits,* SC6 (1971).

F. F. Fang and A. B. Fowler, *Phys. Rev.,* 169, pp. 619–631 (1968).

T. A. Fjeldly, "Analytical Modeling of the Stationary Domain in GaAs MESFET's," *IEEE Trans. Electron Devices,* ED-33, pp. 874–880 (1986).

W. Fichner and H. W. Potzl, *International J. Electronics,* 46, p. 33 (1979).

L. Forbes, E. Sun, R. Alders, and J. Moll, *IEEE Trans. Electron Devices,* ED-26, p. 1816 (1979).

J. F. Gibbons and K. F. Lee, *IEEE Trans. Electron Device Lett.,* EDL-1, p. 117 (1980).

A. M. Goodman, J. P. Russel, L. A. Goodman, C. J. Nuese, and J. M. Neilson, *IEDM Technical Digest,* p. 79, IEEE Publications, Washington, D.C., (1983).

A. A. Grinberg and M. Shur, "New Analytic Model for Heterostructure Field Effect Transistors," *J. Appl. Phys.,* 65, no. 5, p. 2116, March 1 (1989).

M. Hack, S. Guha, and M. Shur, *Phys. Rev.,* B30, no. 12, pp. 6991–6999 (1984).

M. Hack and W. den Boer, *J. Appl. Phys.,* 58, p. 1554 (1985).

M. HACK, M. SHUR, and W. CZUBATYJ, *Mat. Res. Soc. Proc.*, ed. D. Adler, Y. Hamakawa, and A. Madan, 70, p. 643 (1986a).

M. HACK, M. SHUR, and W. CZUBATYJ, *Appl. Phys. Lett.*, 48, no. 20, p. 1386 (1986b).

M. HACK, H. C. TUAN, J. SHAW, M. SHUR, and P. YAP, *Mat. Res. Soc. Symp. Proc.*, (1987).

C. J. HAN , P. P. RUDEN, D. GRIDER, A. FRAASCH, K. NEWSTROM, P. JOSLYN, and M. SHUR, "Short Channel Effects in Submicron Self-Aligned Gate Heterostructure FETs," *IEDM Technical Digest,* IEEE Publications, San Francisco, December (1988).

H. I. HANAFI, "Current Modeling for MOSFET," in *Circuit Analysis, Simulation, and Design,* pp. 71–105, ed. A. E. Ruehli, North Holland (1986).

H. HASEGAWA, M. ABE, P. M. ASBECK, A. HIGASHIZAKA, Y. KATO,, and M. OHMORI, "GaAs LSI/VLSI: Advantages and Applications," in *Extended Abstracts of the 16th International Conference on Solid State Devices and Materials,* Kobe, Japan, pp. 413–414 (1984).

K. HESS, H. MORKOÇ, H. SHICHIJO, and B. G. STREETMAN, "Negative Differential Resistance through Real Space Transfer," *Appl. Phys. Lett.*, 35, p. 459 (1979).

S. J. HILLENIUS, R. LIU, G. E. GEORGOU, R. L. FIELD, D. S. WILLIAMS, A. KORNBLIT, D. M. BOULIN, R. J. F. JENSEN, L. G. SALMON, D. S. DEAKIN, and M. J. DELANEY, "Ultrahigh-Speed GaAs Static Frequency Dividers," in *IEDM Technical Digest,* pp. 476–479, IEEE Publications, Los Angeles (1986).

K. HIRANAKA, T. YAMAGUCHI, S. YANAGISAWA, *IEEE Electron Device Lett.*, EDL-5, p. 224 (1984).

C. H. HYUN, M. SHUR, and N. C. CIRILLO, JR., "Simulation and Design Analysis of AlGaAs/GaAs MODFET Integrated Circuits," *IEEE Trans. CAD ICAS,* CAD-5, pp. 284–292 (1986a).

C. H. HYUN, M. SHUR, and A. PECZALSKI, *IEEE Trans. Electron Devices,* ED-33, pp. 1421–1426 (1986b).

H. ITO, Y. NISHIHARA, M. NOBUE, M. FUSE, T. NAKAMURA, T. OZAWA, S. TOMIYAMA, R. WEISFIELD, H. C. TUAN, and M. J. THOMPSON, *IEDM Technical Digest,* p. 436, IEEE Publication, Washington, D.C., December (1985).

L. JONHSTON and W. T. LYNCH, *IEDM Technical Digest,* p. 252, IEEE Publication, Los Angeles (1986).

R. A. KIEHL, D. A. FRANK, S. L. WRIGHT, and J. H. MAGERLEIN, "Device Physics of Quantum-Well Heterostructure MI^{3}SFET's," *IEDM Technical Digest,* pp. 70–73, IEEE Publication, Washington, D.C. (1987).

T. KODAMA, N. TAKAGI, S. KAWAI, Y. NASU, S. YANAGISAWA, and K. ASAMA, *IEEE Electron Dev. Lett.*, EDL-3, p. 187 (1982).

M. A. LAMPERT and P. MARK, *Current Injection in Solids,* Academic Press, New York (1970).

T. LEROUX and A. CHENEVAS-PAULE, "Anderson Transition in Accumulation Layers of a-Si:H Thin Film Transistors," in *Proceedings of the 11th International Conference on Amorphous and Liquid Semiconductors,* Rome, Italy, ed. F. Evangelisti and J. Stuke, North Holland, Amsterdam, pp. 443–446 (1985).

J. E. LILIENFELD, U.S. Patent 1,745,175 (1930).

S. LIU and L. W. NAGEL, *IEEE J. Solid State Circuits,* SC-17, no. 6, pp. 983–998 (1982).

T. J. MALONEY and J. FREY, *IEEE Trans. Electron Devices,* ED-22, pp. 357–358 (1975).

A. MANY, Y. GOLDSTEIN, and N. B. GROVER, *Semiconductor Surfaces,* North Holland, Amsterdam (1965).

M. MATSUMURA, H. HAYAMA, Y. NARA, and K. ISHIBASHI, *Jap. J. Appl. Phys.,* 20, suppl. 20-1, p. 311 (1981).

T. MIMURA, S. HIYAMIZU, T. FUJII, and K. NAMBU, "A New Field Effect Transistor with Selectively Doped GaAs/n-$Al_xGa_{1-x}As$ Heterostructures," *Jpn. J. Appl. Phys.,* 19, L225–227 (1980).

J. E. MEYER, "MOS Models and Circuit Simulation," *RCA Review,* 32, pp. 42–63, March (1971).

A. D. MILNES, editor, *MOS Devices, Design and Manufacture,* Edinburgh University Press, Edinburgh (1983).

U. K. MISHRA, J. F. JENSEN, A. S. BROWN, M. A. THOMPSON, L. M. JELLOIAN, and R. S. BEAUBIEN, "Ultra-High-Speed Digital Circuit Performance in 0.2 μm Gate-Length AlInAs/GaInAs HEMT Technology," *IEEE Electron Device Lett.,* 9, no. 9, pp. 482–484 (1988).

L. W. NAGEL, "SPICE2: A Computer Program to Simulate Semiconductor Circuits," Memorandum No. ERL-M520, Electronics Research Laboratory, College of Engineering, University of California, Berkeley, May 9 (1975).

E. H. NICOLIAN and J. R. BREWS, *MOS Physics and Technology,* John Wiley & Sons, New York (1982).

T. H. NING, C. M. OSBURN, and H. N. YU, *J. Electr. Materials,* 6, pp. 65–76 (1977).

Y. Y. OMURA, *Proceedings of the 14th Conference (1982 International) on Solid State Devices,* Tokyo, pp. 263–266 (1982).

D. G. ONG, *Modern MOS Technology, Processes, Devices, and Design,* McGraw-Hill, New York (1984).

S. PARTRIDGE, *IEDM Technical Digest,* p. 428, IEEE Publications, Los Angeles (1986).

R. F. PIERRET, *Field Effect Devices,* Addison-Wesley Modular Series on Solid State Devices, vol. 4, Addison-Wesley, Reading, Mass. (1983).

J. D. PLUMMER, *IEDM Technical Digest,* p. 378, IEEE Publications, Los Angeles (1986).

F. PONSE, W. T. MASSELINK, and H. MORKOÇ, *IEEE Trans. Electron Devices,* ED-32, p. 1017 (1985).

P. RICHMAN, U.S. Patent 3,544,864, December 1 (1970).

J. G. RUCH, *IEEE Trans. Electron Devices,* ED-19, pp. 652–654 (1972).

P. P. RUDEN, C. J. HAN, and M. SHUR, "Gate Current Modulation of Modulation-Doped Field-Effect Transistors," *J. Appl. Phys.,* 64 no. 3, pp. 1541–1546, August (1988).

P. P. RUDEN, M. SHUR, D. K. ARCH, R. R. DANIELS, D. E. GRIDER, and T. NOHAVA, "Quantum Well p-Channel AlGaAs/InGaAs/GaAs Heterostructure Insulated Gate Field Effect Transistors," submitted for publication (1989a).

P. P. RUDEN, M. SHUR, A. I. AKINWANDE, and P. JENKINS, "Distributive Nature of Gate Current and Negative Transconductance in Heterostructure Field Effect Transistors," *IEEE Trans. on Electron Devices,* ED-36, no. 2, pp. 453–456, February (1989b).

A. G. SABNIS and J. T. CLEMENS, *IEDM Technical Digest,* p. 18, IEEE Publications (1979).

T. SATO, Y. TAKEISHI, and H. HARA, *Jpn. J. Appl. Phys.*, 8, no. 5, pp. 588–598 (1969).

D. K. SCHRODER, *Advanced MOS Devices,* Addison-Wesley Modular Series on Solid State Devices, Addison-Wesley, Reading, Mass. (1987).

S. SELBERHERR, *Analysis and Simulation of Semiconductor Devices,* Springer-Verlag, Wien, New York (1984).

N. J. SHAH, S. S. PEI, and C. W. TU, *IEEE Trans. Electron Devices,* ED-33, p. 543 (1986).

M. SHUR, *Electronics Lett.*, 12, no. 23, pp. 615–616 (1976).

M. SHUR, *Electronics Lett.*, 18, no. 21, pp. 909–911 (1982).

M. SHUR, C. HYUN, M. HACK, Z. YANIV, M. YANG, and V. CANNELLA, "Localized States Distribution and Characteristics of Amorphous Silicon Alloy FETs," in *Proceedings of the 11th International Conference on Amorphous and Liquid Semiconductors,* Rome, Italy, ed. F. Evangelisti and J. Stuke, North Holland, Amsterdam, pp. 1401–1404 (1985).

M. SHUR, C. H. HYUN, and M. HACK, *J. Appl. Phys.*, 59, no. 7, pp. 2488–2497, April (1986).

M. SHUR and D. LONG, "Performance Prediction for Submicron GaAs SDFL Logic," *IEEE Electron Device Lett.*, EDL-3, p. 124 (1982).

M. SHUR and M. HACK, "Physics of Amorphous Silicon Based Alloy Field Effect Transistors," *J. Appl. Phys.*, 55, no. 10, pp. 3831–3842, May (1984).

M. SHUR, M. HACK, and C. HYUN, "Flat-Band Voltage and Surface States in Amorphous Silicon Alloy Field Effect Transistors," *J. Appl. Phys.*, 56, p. 382 (1984).

M. SHUR, D. K. ARCH, R. R. DANIELS, and J. K. ABROKWAH, "New Negative Resistance Regime of Heterostructure Insulated Gate Transistor (HIGFET) Operation," *IEEE Electron Device Lett.*, EDL-7, no. 2, pp. 78–80, February (1986).

M. SHUR, *GaAs Devices and Circuits,* Plenum, New York (1987).

M. SHUR, J. K. ABROKWAH, R. R. DANIELS, and D. K. ARCH, "Mobility Enhancement in Highly Doped GaAs Quantum Wells," *J. Appl. Phys.*, 61, no. 4, pp. 1643–1645, February 15 (1987).

M. SHUR, M. HACK, and J. G. SHAW, "New Analytic Model for Amorphous Silicon Thin Film Transistors," *J. Appl. Phys.*, vol. 66, no. 7, pp. 3371–3380 (1989).

M. SHUR, M. HACK, and J. G. SHAW, "Capacitance-Voltage Characteristics of Amorphous Silicon Thin Film Transistors," *J. Appl. Phys.*, vol. 66, no. 7, p. 3381 (1989).

Small signal transistor data, published by Motorola Semiconductor Products, Inc., Phoenix, Ariz. (1983).

B. S. SONG and P. R. GRAY, *IEEE J. Solid State Circuits,* SC-17, no. 2, pp. 291–298 (1982).

W. E. SPEAR and P. G. LECOMBER, *J. Non-Crystal. Solids,* 8-10, p. 727 (1972).

W. E. SPEAR, P. G. LECOMBER, S. KINMOND, and M. H. BRODSKY, *Appl. Phys. Lett.*, 28, p. 105 (1976).

F. STERN, *CRC Crit. Rev. Solid State Sci.*, p. 499 (1974).

S. M. SZE, *Physics of Semiconductor Devices,* John Wiley & Sons, New York (1981).

S. M. SZE, editor, *VLSI Technology,* 2nd ed., McGraw-Hill, New York (1988).

T. TAKADA, K. YOKOYAMA, M. IDA, and T. SUDO, *IEEE Trans. Microwave Theory and Technique,* MTT-30, pp. 719–723 (1982).

G. W. TAYLOR and J. G. SIMMONS, *IEEE Trans. Electron Devices,* ED-32, no. 11, pp. 2345–2367, November (1985a).

G. W. TAYLOR and J. G. SIMMONS, *IEEE Trans. Electron Devices,* ED-32, no. 11, pp. 2368–2377, November (1985b).

M. J. THOMPSON and H. C. TUAN, *IEDM Tech. Digest,* IEEE Publications, Los Angeles, p. 192, December (1986).

S. TIWARI, S. L. WRIGHT, and J. BATEY, "Unpinned GaAs MOS Capacitors and Transistors," *IEEE Electron Device Lett.,* 9, no. 9, pp. 488–489 (1988).

R. R. TROUTMAN, *IEEE J. Solid State Circuits,* SC-9, p. 55 (1974).

Y. P. TSIVIDIS, *Operation and Modeling of the MOS Transistor,* McGraw-Hill, New York (1987).

H. C. TUAN, *Mat. Res. Soc. Symp. Proc.,* 70, p. 651 (1986).

S. VALDYA, E. N. FULS, and R. L. JOHNSON, *IEEE Trans. Electron Devices,* ED-33, p. 1321 (1986).

R. A. WARRINER, *Solid State Electron Devices,* 1, p. 105 (1977).

M. H. WHITE and J. R. CRICCHI, *IEEE Trans. Electron Devices,* 19, p. 1280 (1972).

J. XU, M. SHUR, and M. HACK, *Appl. Phys. Lett.* (1987).

K. YAMAGUCHI, *IEEE Trans. Electron Devices,* ED-26, p. 1068 (1979).

K. YAMAGUCHI and H. KODERA, *IEEE Trans. Electron Devices,* ED-23, pp. 545–553 (1976).

Z. YU and R. W. DUTTON, *Sedan III: A General Electronic Material Device Analysis Program,* program manual, Stanford University, July (1985).

B. J. VAN ZEGHBROECK, W. PATRICK, H. MEIER, P. VETTIGER, and P. WOLF, "High Performance GaAs MESFET's," *IEDM Technical Digest,* IEEE Publication, Los Angeles, pp. 832–834, December (1986).

E. ZORINSKY, D. SPRATT, and R. VIRKUS, *IEDM Technical Digest,* p. 488, IEEE Publication, Los Angeles (1986).

PROBLEMS

4-2-1. Calculate the temperature dependence of the surface charge, Q_s, per unit area in the p-type silicon sample doped at 10^{15} cm^{-3} for the surface potential $V_s = V_b$ and $2V_b$ in the temperature range between 150 K and 450 K where the bulk potential $V_b = (k_B T/q)\ln(N_A/n_i)$, the effective densities of states in the conduction and valence band are $N_c = 3.22 \times 10^{19}\ (T/300)^{3/2}$ cm^{-3} and $N_v = 1.83 \times 10^{19}\ (T/300)^{3/2}$ cm^{-3}, respectively, the energy gap $E_g = 1.12$ V, and the dielectric permittivity of the semiconductor $\varepsilon_s = 1.05 \times 10^{-10}$ F/m.

4-2-2. For an MOS structure find the charge per unit area in fast surface states (Q_{st}) as a function of surface potential V_s for the surface states density uniformly distributed in energy throughout the forbidden gap with density $D_{st} = 10^{11}$ cm^{-2} eV^{-1}, at the boundary between the oxide and semiconductor. Assume that the surface states with energies above the neutral level, ϕ_o, are acceptor type (i.e., negative when occupied by an electron, neutral otherwise) and that the surface states with energies below the neutral level, ϕ_o, are donor type (i.e., neutral when occupied by an electron, and positively charged otherwise). Also assume that all surface states with energies below the Fermi level are filled and that all surface states with energies above the Fermi level are empty (i.e., assume zero temperature statistics for the surface states). The energy gap is $E_g = 1.12$ eV, the neutral level ϕ_o is 0.3 V above the bottom of the valence band, the effective densities of states in the conduction band and in the valence band are 3.22×10^{19} cm^{-3}eV^{-1} and 1.83×10^{19} cm^{-3}eV^{-1}, respectively, and semiconductor doping $N_A = 10^{15}$ cm^{-3}. Temperature $T = 300$ K.

4-2-3. Express the capacitance of an MOS structure per unit area in the presence of uniformly distributed fast surface states as described in Problem 4-2-2 in terms of the oxide capacitance, C_{ox}, per unit area. Semiconductor capacitance $C_s = -dQ_s/dV_s$ (where Q_s is the charge is the semiconductor and V_s is the surface potential) and surface states capacitance $C_{ss} = -dQ_{st}/dV_s$, where Q_{st} is the charge in fast surface states per unit area.

4-2-4. Charge density 2.4×10^{-7} coulombs/cm^2 is uniformly distributed in the oxide in a MOSFET structure. The oxide thickness $d_i = 1500$ Å.

(a) Calculate the contribution to the threshold voltage from the charges.

(b) Calculate for the case when all the charges are located at the silicon–silicon dioxide interface.

(c) Calculate for the linear charge distribution within the oxide (zero charge density at the Si–SiO_2 interface, maximum charge density at the metal–SiO_2 interface.)

4-2-5. Plot the dependence of the surface charge, Q_s, per unit area in the p-type silicon sample at 300 K on doping for the doping range 10^{14} cm$^{-3} < N_A < 10^{17}$ cm^{-3} for the surface potential $V_s = V_b$ and $2V_b$. Here $V_b = (k_B T/q) \ln(N_A/n_i)$, the effective densities of states in the conduction and valence band are $N_c = 3.22 \times 10^{19}$ cm^{-3} and $N_v = 1.83 \times 10^{19}$ cm^{-3}, respectively, the energy gap $E_g = 1.12$ V, and the dielectric permittivity of the semiconductor $\varepsilon_s = 1.05 \times 10^{-10}$ F/m.

4-2-6. Sketch the field and voltage distribution in a MOS structure at the threshold gate voltage. Oxide thickness $d_i = 0.1$ μm, doping level $N_A = 10^{15}$ cm^{-3}. The dielectric permittivity of the semiconductor $\varepsilon_s = 1.05 \times 10^{-10}$ F/m, the oxide dielectric permittivity is $\varepsilon_i = 3.45 \times 10^{-11}$ F/m, the substrate bias $V_{sub} = 0$ V, the effective densities of states in the conduction and valence band are $N_c = 3.22 \times 10^{19}$ cm^{-3} and $N_v = 1.83 \times 10^{19}$ cm^{-3}, respectively, the energy gap $E_g = 1.12$ V, and $T = 300$ K. Assume zero flat-band voltage.

4-2-7. Consider two nearly identical silicon MOSFETs. The only difference is that the oxide layer in one of them is clean, and the other one is contaminated with sodium ions that produce a positive charge density. The concentration of sodium ions is 1×10^{16} cm^{-3}. The dielectric thickness is 0.1 μm. The dielectric permittivity of SiO_2 is 3.45×10^{-11} F/m.

(a) What is the difference (including sign) in the device threshold voltages if these devices are n-channel devices?

(b) What is the difference (including sign) in the device threshold voltages if these devices are p-channel devices?

(c) Assume that the threshold voltage of the n-channel clean device is 1 V and that the threshold voltage of the p-channel clean device is −1 V.

Sketch qualitative dependencies of the drain-to-source saturation current on the gate voltage for all four devices (clean and contaminated n-channel and clean and contaminated p-channel). Label the thresholds and shifted thresholds on the V_g axis.

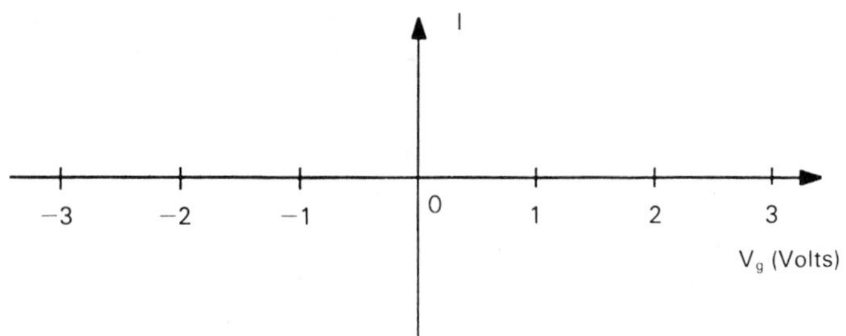

4-2-8. Consider the energy band diagram of a metal-insulator-semiconductor-insulator-metal structure shown in Fig. P4-2-8. E_g (in volts) = 1.12 V (You can assume symmetric bands, $N_c \cong N_v$, for this problem.)

(a) What is the flat-band voltage?

(b) Sketch the band diagram when the left metal plate is at 2 V and the right metal plate is grounded. $\varepsilon_s/\varepsilon_{ox} = 3$. What is the strength of the electric field in silicon? What is the position of the quasi-Fermi level in silicon? In your band diagram, neglect induced charges in the silicon.

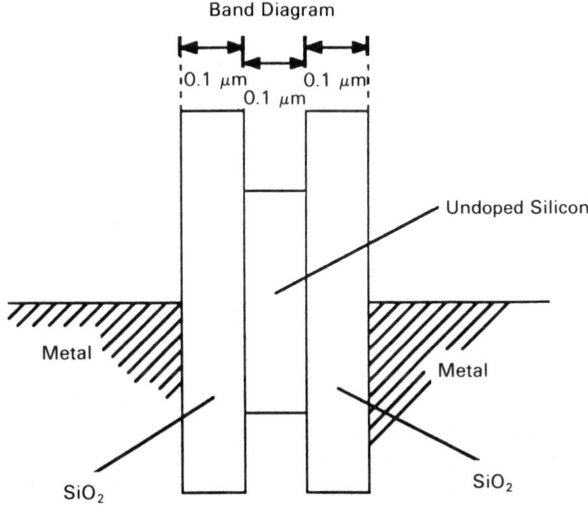

Fig. P4-2-8

4-2-9. Using eqs. (4-2-19) and (4-2-20),

$$F_s = (2)^{1/2}\, k_B T/(qL_{Dp})\, f(qV_s/k_B T) \tag{4-2-19}$$

$$|Q_s| = \varepsilon_s |F_s| \tag{4-2-20}$$

show that in both accumulation and strong inversion regimes the surface charge is proportional to $\exp[q|V_s|/(2k_B T)]$.

4-3-1. Find the maximum width of the depletion region for an ideal MOS capacitor on p-type Si with $N_A = 10^{15}$ cm^{-3} as a function of the substrate bias V_{sub} for -2 V $< V_{sub} < 0.1$ V. Assume the following parameters: the dielectric permittivity of the semiconductor $\varepsilon_s = 1.05 \times 10^{-10}$ F/m, the oxide dielectric permittivity is 3.45×10^{-11} F/m, the intrinsic carrier concentration $n_i = 1.5 \times 10^{10}$ cm^{-3}, and temperature $T = 300$ K.

4-3-2. For the device in Problem 4-3-1, calculate the threshold voltage, V_T, and the capacitance of the structure at low and high frequency for $V >> V_T$. The oxide thickness $d_{ox} = 0.2$ μm, and the oxide dielectric permittivity is 3.45×10^{-11} F/m.

4-3-3. **(a)** The field distribution in a MOS structure is shown in Fig. P4-3-3. Sketch the potential distribution.

Flat band voltage $V_{FB} = 0$

(b) What is the surface potential, V_s?

(c) What is the threshold voltage, V_T?

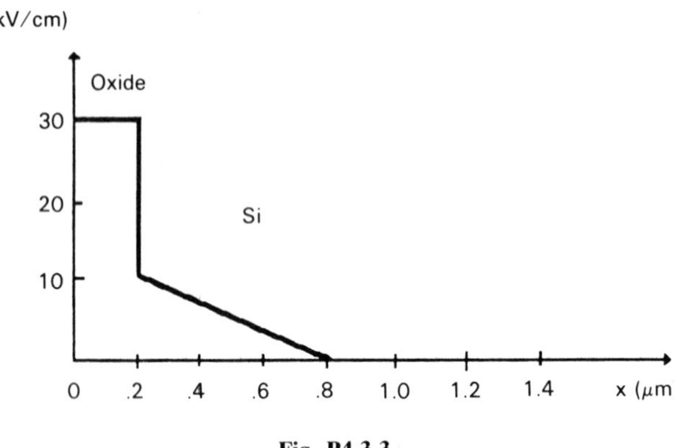

Fig. P4-3-3

4-3-4. Plot the temperature dependence of the maximum width of the depletion region for an ideal MOS capacitor on p-type Si with $N_A = 10^{15}$ cm^{-3} in the temperature range 100 K $< T <$ 350 K. Assume the following parameters: the dielectric permittivity of the semiconductor $\varepsilon_s = 1.05 \times 10^{-10}$ F/m, the effective densities of states in the conduction and valence band (at $T = 300$ K) are $N_{co} = 3.22 \times 10^{19}$ cm^{-3} and $N_{vo} = 1.83 \times 10^{19}$ cm^{-3}, respectively, the energy gap $E_g = 1.12$ V, and the substrate bias $V_{sub} = 0$ V.

4-4-1. Show that the measured (extrinsic) transconductance,

$$g_m = \frac{\partial I_{ds}}{\partial V_{gs}}$$

of a field-effect transistor is related to the intrinsic drain conductance and transconductance of the same device without the source series resistance,

$$g_{do} = \frac{\partial I_D}{\partial V_D} \qquad g_{mo} = \frac{\partial I_D}{\partial V_g}$$

as follows:

$$g_m = \frac{g_{mo}}{1 + g_{mo}R_s + g_{do}(R_s + R_d)}$$

where the gate-to-source voltage, V_{gs}, is given by

$$V_{gs} = V_g + I_{ds}R_s$$

Here the gate voltage, V_g, is the voltage difference between the gate potential and the channel potential at the source side of the channel.

4-4-2. Calculate the dependencies of the drain-to-source current on the drain-to-source voltage for a gate voltage of $V_g = 5$ V for a silicon MOSFET for the following values of the source and drain series resistance, respectively: $R_s = 0\ \Omega$ and $R_s = 100\ \Omega$, and $R_d = 0\ \Omega$ and $R_d = 100\ \Omega$. The device parameters are as follows: gate length $L = 4\ \mu$m, gate width is 100 μm, electron mobility in the channel $\mu_n = 1000$ cm^2/Vs, the dielectric permittivity of the gate insulator $\varepsilon_i = 3.45 \times 10^{-11}$ F/m, the dielectric permittivity of silicon $\varepsilon_s = 1.05 \times 10^{-10}$ F/m, flat-band voltage $V_{FB} = 0$, substrate bias $V_{sub} = 0$, temperature $T = 300$ K, and substrate doping $N_A = 10^{15}$ cm^{-3}. Neglect the velocity saturation effects.

4-4-3. Show that the measured (extrinsic) drain conductance,

$$g_d = \frac{\partial I_D}{\partial V_{ds}}$$

of a transistor is related to the intrinsic drain conductance and intrinsic transconductance of the same device without the source series resistance,

$$g_{do} = \frac{\partial I_D}{\partial V_D} \qquad g_{mo} = \frac{\partial I_D}{\partial V_g}$$

as follows:

$$g_m = \frac{g_{do}}{1 + g_{mo}R_s + g_{do}(R_s + R_d)}$$

Here the gate voltage, V_D, is the voltage difference between the potential at the drain side of the channel and the channel potential at the source side of the channel, and V_{ds} is the drain-to-source voltage that includes voltage drops across the drain and source series resistances R_d and R_s:

$$V_{ds} = V_D + I_D(R_s + R_d)$$

4-4-4. Derive an expression for the drain saturation current of an n-channel field-effect transistor, I_{Dsat}, (using the gradual channel approximation and neglecting velocity saturation, i.e., assuming the constant mobility model) taking into account the source series resistance, R_s. Use the following MOSFET parameters:

$$\text{Gate oxide thickness } d_i = 175 \text{ Å}$$

$$\text{Device width } W = 100\ \mu\text{m}$$

$$\text{Gate length } L = 4\ \mu\text{m}$$

$$\text{Threshold voltage } V_T = -1 \text{ V}$$

$$\text{Gate insulator dielectric permittivity } \varepsilon_i = 3.45 \times 10^{-11} \text{ F/m}$$

$$\text{Substrate bias } V_{sub} = 0$$

$$\mu_n = 800 \text{ cm}^2/\text{Vs}$$

$$V_g = 5 \text{ V}$$

Plot I_{Dsat} vs. R_s for $2\ \Omega < R_s \leq 30\ \Omega$.

4-4-5. Equations for the current voltage characteristics of a long-channel MOSFET are

$$I_D = \mu_n \frac{W}{L} c_i \left[(V_g - V_T) V_D - \frac{V_D^2}{2} \right] \quad \text{for } V_D \leq V_{Dsat} = V_g - V_T \tag{4-4-26}$$

$$I_D = \frac{1}{2} \mu_n \frac{W}{L} c_i (V_g - V_T)^2 \quad \text{for } V_D > V_{Dsat} = V_g - V_T \tag{4-4-27}$$

where μ_n is the low field mobility, W is the gate width, L is the channel length, V_g is the gate voltage (with the source as reference), I_D is the drain current, V_D is the drain voltage (with the source as reference), V_{Dsat} is the drain saturation voltage, c_i is the insulator capacitance per unit area, and V_T is the threshold voltage. When the drain and the gate are connected, consider the following two cases:

(a) $V_T > 0$

(b) $V_T < 0$

Find I_D in terms of V_D in each case.

4-4-6. Show that the current-voltage characteristics of a long-channel MOSFET below saturation can be expressed as

$$I_D = \left[\frac{\mu_n q^2 W}{2Lc_i} \right] (n_s^2 - n_d^2) \tag{P4-4-6-1}$$

where μ_n is the low field mobility, W is the gate width, L is the channel length, q is the electronic charge, I_D is the drain current, c_i is the insulator capacitance per unit area, n_s is the concenration of electrons in the channel at the source (per unit area), and n_d is the concentration of electrons in the channel at the drain (per unit area).

4-5-1. Using the following MOSFET parameters,

$$\text{Gate oxide thickness } d_i = 175 \text{ Å}$$

$$\text{Device width } W = 20\ \mu\text{m}$$

$$\text{Doping density } N_A = 5 \times 10^{16} \text{ cm}^{-3}$$

$$\text{Temperature } T = 300 \text{ K}$$

Threshold voltage $V_T = -1$ V

Gate insulator dielectric permittivity $\varepsilon_i = 3.45 \times 10^{-11}$ F/m

Substrate bias $V_{sub} = 0$

and using equations from Section 1-9 for the electron mobility and saturation velocity, calculate the current-voltage characteristics for gate voltages $V_{gs} = 1$ V, 2 V, 3V, 4 V, and 5 V for n-channel devices with gate length $L = 0.5$ μm and $L = 0.75$ μm. Compare the results of the calculation with the current-voltage characteristics measured by Hillenius et al. and shown in Fig. 4-5-7. Comment on possible reasons for the differences between computed and measured curves.

Hint: Use a linear piece-wise approximation for the velocity-field dependence $v_n(F)$.

4-5-2. Using the following device parameters,

Gate oxide thickness $d_i = 175$ Å

Device width $W = 20$ μm

Doping density $N_D = 5 \times 10^{16}$ cm^{-3}

Temperature $T = 300$ K

Threshold voltage $V_T = 1$ V

Gate insulator dielectric permittivity $\varepsilon_i = 3.45 \times 10^{-11}$ F/m

Substrate bias $V_{sub} = 0$

and using equations from Section 1-9 for the hole mobility and velocity, calculate the current-voltage characteristics for gate voltages $V_{gs} = -1$ V, -2 V, -3 V, -4 V, and -5 V for p-channel devices with gate length $L = 0.5$ μm and $L = 0.75$ μm. Compare the results of the calculation with the current-voltage characteristics measured by Hillenius et al. and shown in Fig. P4-5-2.

Hint: Use a linear piece-wise approximation for the velocity-field dependence $v_p(F)$.

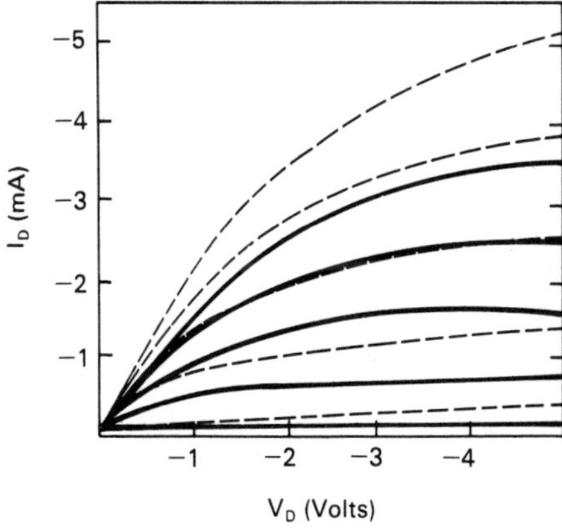

Fig. P4-5-2. I_d vs. V_d for p-channel device. $V_{gs} = -1$ V, -2 V, -3 V, -4 V, and -5 V. $L = 0.75$ μm (solid lines) and 0.5 μm (dashed lines). $W = 20\mu$m.

4-5-3. Using eq. (4-5-27) show (a) that at small values of $V_{gt} = V_{gs} - V_T << V_{sl}$

$$I_{Dsat} \approx \frac{\beta V_{gt}^2}{2(1 - \beta^2 R_s^2 V_{sl}^2)} \tag{P4-5-3-1}$$

or for small values of the source series resistance when $\beta R_s V_{sl} << 1$

$$I_{Dsat} \approx \beta V_{gt}^2/2 \tag{P4-5-3-2}$$

(b) that at large values of $V_{gt} = V_{gs} - V_T << V_{sl}$

$$I_{Dsat} = \frac{\beta V_{gt} V_{sl}}{1 + \beta R_s V_{sl}} \tag{P4-5-3-3}$$

or for small values of the source series resistance when $\beta R_s V_{sl} << 1$

$$I_{Dsat} = \beta V_{gt} V_{sl} \tag{P4-5-3-4}$$

(c) assuming that the source series resistance is small, use the asymptotic expressions for the drain-to-source saturation current given above to deduce β and V_{sl} from the experimental data shown in Figs. 4-5-6 and 4-5-7. Compare the results with the expected values of β and V_{sl}.

4-5-4. Derive an expression for the MOSFET transconductance in the saturation regime, taking into account the velocity saturation effects assuming zero source series resistance (hint: use eq. (4-5-27)).

Plot this transconductance as a function of the gate length using the following parameters: temperature $T = 300$ K, substrate doping $N_A = 1 \times 10^{14}$ cm^{-3}, energy gap $E_g = 1.12$ eV, effective densities of states in the conduction band and valence band at 300 K are 3.22×10^{19} cm^{-3} and 1.83×10^{19} cm^{-3}, dielectric permittivity of silicon $\varepsilon_s = 1.05 \times 10^{-10}$ F/m, flat-band voltage $V_{FB} = 0$, gate insulator thickness $d_i = 0.2$ μm, gate insulator dielectric permittivity $\varepsilon_i = 3.45 \times 10^{-11}$ F/m, gate width $W = 100$ μm, substrate bias $V_{sub} = 0$, electron low field mobility $\mu_n = 0.08$ m^2/Vs, the electron saturation velocity $v_s = 10^5$ m/s, and the output conductance factor $\lambda = 0$.

Compare the results with the transconductance in the saturation regime calculated using the constant mobility model.

4-5-5. Consider the MOSFET structure shown in Fig. P4-5-5.

Assume that the volume electron charge density, n_{sd}, injected into the region of silicon between the drain side of the channel under the gate and the drain (dead region) can be estimated as

$$n_{sd} \sim I_{ds}/(q\mu_n F_d W t_{eff})$$

where the effective channel depth of the dead region, t_{eff}, is roughly equal to the drain contact depth, and the longitudinal electric field, F_d, within this region is approximately

$$F_d \sim F_D + q n_{sd} x/\varepsilon_s$$

where F_d is the electric field at the drain side of the channel under the gate and x is counted from the gate side of the drain as shown in Fig. P4-5-5. Calculate the current-voltage characteristics of such a device for voltages across the gated por-

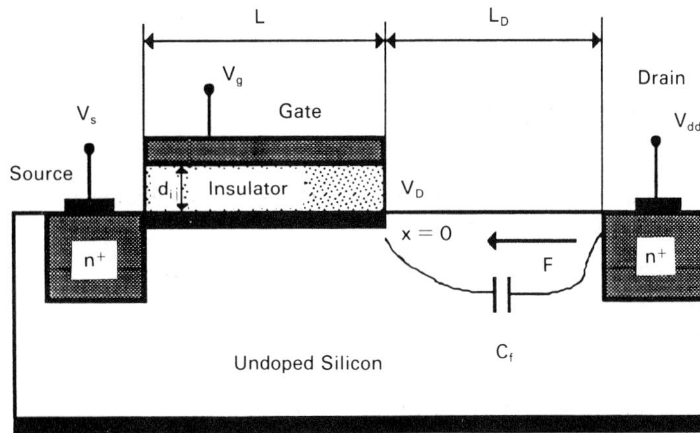

Fig. P4-5-5.

tion of the channel that are smaller than the voltage corresponding to the velocity saturation voltage (i.e., for $V_D < V_{Dsat}$, where V_{Dsat} is given by eq. (4-5-18)). Assume the following parameters:

$$\text{Gate oxide thickness } d_i = 500 \text{ Å}$$
$$\text{Device width } W = 20\ \mu\text{m}$$
$$\text{Gate length } L = 5\ \mu\text{m}$$
$$\text{Length } L_D = 20\ \mu\text{m}$$
$$\text{Doping density } N_A = 0$$
$$\text{Temperature } T = 300 \text{ K}$$
$$\text{Threshold voltage } V_T = 1 \text{ V}$$
$$t_{eff} = 0.7\ \mu\text{m}$$

Use equations of Section 1-9 for the electron mobility and saturation velocity. Calculate the current-voltage characteristics for gate voltages V_{gs} = 1 V, 2 V, 3 V, 4 V, and 5 V.

Hint: Use a constant mobility model for the section of the device under the gate.

4-5-6. The average transverse electric field in the channel can be estimated as

$$F_y \approx [|Q_{dep}| + |Q_c|/2]\varepsilon_s$$

where the depletion charge per unit area is given by

$$Q_{dep} = -qN_A d_{dep} = -\{2\varepsilon q N_A[V_D - V_{sub} + 2\varphi_b]\}^{1/2} \qquad \text{(P4-5-6-1)}$$

for the linear region ($V_D < V_{Dsat}$), and

$$Q_{dep} = -qN_A d_{dep} = -\{2\varepsilon q N_A[V_{Dsat} - V_{sub} + 2\varphi_b]\}^{1/2} \qquad \text{(P4-5-6-2)}$$

for the saturation regime ($V_D \geq V_{Dsat}$) (see Sze 1981). The channel charge

$$|Q_c| = c_i[V_g - 2\varphi_B - V_{FB} - V_D)] - |Q_{dep}| \quad \text{(P4-5-6-3)}$$

for the linear region ($V_D < V_{Dsat}$), and

$$|Q_c| = c_i[V_g - 2\varphi_B - V_{FB} - V_{Dsat})] - |Q_{dep}| \quad \text{(P4-5-6-4)}$$

for the saturation regime ($V_D \geq V_{Dsat}$). The longitudinal electric field in the channel may be estimated as

$$F = V_D/L \quad \text{(P4-5-6-5)}$$

for the linear region ($V_D < V_{Dsat}$) and as

$$F = V_{Dsat}/L \quad \text{(P4-5-6-6)}$$

for the saturation regime ($V_D \geq V_{Dsat}$).

(a) Using these expressions and equations given in Section 1-9, calculate the gate voltage dependence (in the range $V_T < V_g < 10$ V) of the effective mobility in n-channel MOSFETs for the following parameters: threshold voltage $V_T = 1$ V, gate length $L = 2\ \mu m$, substrate doping $N_A = 10^{16}$ cm, gate oxide thickness $d_i = 500$ Å, oxide dielectric permittivity 3.45×10^{-11} F/m, silicon dielectric permittivity 1.05×10^{-10} F/m, temperature $T = 300$ K, and drain voltages $V_D = 0.01$ V, 2 V, and 5 V.

(b) Repeat the same calculations for $T = 77$ K.

4-5-7. Using the following MOSFET parameters,

Gate oxide thickness $d_i = 500$ Å

Device width $W = 100\ \mu m$

Gate length $L = 5\ \mu m$

Electron mobility $\mu_n = 0.8\ m^2/Vs$

Threshold voltage $V_T = 1$ V

Gate insulator dielectric permittivity $\varepsilon_i = 3.45 \times 10^{-11}$ F/m

(a) calculate the drain-to-source saturation current for the gate voltage $V_{gs} = 5$ V and the following values of the source series resistance: $R_s = 0\ \Omega$, $5\ \Omega$, and $10\ \Omega$ (assume that the output transconductance parameter λ is equal to zero and that the drain series resistance $R_d = 5\ \Omega$).

(b) calculate the drain-to-source saturation current for the gate voltage $V_g = 5$ V and the drain-to-source voltage $V_{ds} = 10$ V, assuming the drain series resistance $R_d = 5\ \Omega$, the source series resistance $R_s = 5\ \Omega$, and the output transconductance parameter $\lambda = 0.1$

Hint: Use equations for a long-channel device.

4-5-8. Using the following MOSFET parameters,

Device width $W = 100\ \mu m$

Gate length $L = 1\ \mu m$ (short channel)

Electron mobility $\mu_n = 0.8\ m^2/Vs$

$$\text{Electron saturation velocity } v_s = 10^5 \text{ m/s}$$

$$\text{Threshold voltage } V_T = 1 \text{ V}$$

$$\text{Gate insulator dielectric permittivity } \varepsilon_i = 3.45 \times 10^{-11} \text{ F/m}$$

$$\text{Source series resistance } R_s = 5\ \Omega$$

$$\text{Gate voltage } V_{gs} = 5 \text{ V}$$

plot the drain-to-source saturation current as a function of the insulator thickness, d_i, for 100 Å $\leq d_i \leq$ 1000 Å. Comment on the results.

4-5-9. Derive an expression for the drain-to-source saturation voltage, V_{Dsat}, of a MOSFET using the velocity saturation model considered in Section 4-5. Assume zero source series resistance ($R_s = 0$).

4-5-10. The dependence of the drain-to-source saturation current of a field-effect transistor on the gate voltage is given by

$$I_{Dsat} = k(V_{gs'} - V_T)^2$$

where $V_{gs'}$ is the intrinsic gate voltage (i.e., the voltage difference between the gate and the source side of the channel), and V_T is the threshold voltage. Assume $k = 10^{-3}$ A/V^2, $V_T = 0$.

Calculate the extrinsic transconductance, dI_{ds}/dV_{gs}, for the two circuits shown in Fig. P4-5-10.

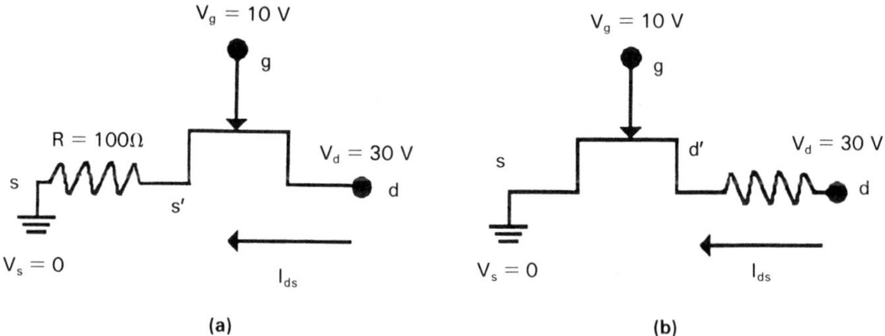

Fig. P4-5-10

4-6-1. Calculate the dependence of parameters K and a introduced in Section 4-6 on the substrate doping in the 10^{14} cm^{-3} $\leq N_A \leq 10^{17}$ cm^{-3} range. Use the following device parameters: threshold voltage $V_T = 1$ V, gate length $L = 2\ \mu$m, gate oxide thickness $d_i = 500$ Å, energy gap $E_g = 1.12$ eV, effective densities of states in the conduction band and in the valence band are 3.22×10^{19} cm^3 and 1.83×10^{19} cm^{-3}, respectively, oxide dielectric permittivity is 3.45×10^{-11} F/m, silicon dielectric permittivity is 1.05×10^{-10} F/m, temperature $T = 300$ K, and drain voltage $V_D = 0$.

4-6-2. Plot the effective channel length as a function of electron mobility, μ_n, for a silicon MOSFET for 600 cm^2/Vs $\leq \mu \leq$ 1200 cm^2/Vs. Use the following device parameters: the threshold voltage $V_T = 1$ V, the gate voltage $V_{gs} = 3$ V, temperature $T =$ 300 K, gate length $L = 2$ μm, device width $W = 20$ μm, gate oxide thickness $d_i =$ 500 Å, energy gap $E_g = 1.12$ eV, effective densities of states in the conduction band and in the valence band are 3.22×10^{19} $(T/300)^{3/2}$ cm^{-3} and 1.83×10^{19} $(T/300)^{3/2}$ cm^{-3}, respectively, oxide dielectric permittivity is 3.45×10^{-11} F/m, silicon dielectric permittivity is 1.05×10^{-10} F/m, drain-to-source voltage $V_{ds} = 10$ V, substrate voltage $V_{sub} = 0$, depth of drain recessed region $D_d = 0.3$ μm, inversion layer thickness $d_{inv} = 100$ Å, electron saturation velocity $v_s = 10^5$ m/s, and drain and substrate doping $N_A = 10^{15}$ cm^{-3}, 10^{16} cm^{-3}, and 10^{17} cm^{-3}.

Hint: Use eq. (4-6-12).

4-6-3. Consider an n-channel MOSFET with p-type substrate.

(a) Calculate the source and drain currents for the substrate bias $V_{sub} = +0.7$ V with the source and drain grounded. The drain and source contacts have the same dimension. The doping density of the p-type substrate is $N_a = 10^{16}$ cm^{-3}. The energy gap is $E_g = 1.12$ eV, the intrinsic density of electrons $n_i =$ 10^{10} cm^{-3}, temperature $T = 300$ K, the diffusion length of electrons in the p-type substrate is 20 μm, and the electron mobility $\mu_n = 1000$ cm^2/Vs.

(b) Estimate the source and drain currents for $V_{sub} = +2$ V. Assume that the drain and source series contact resistances are 20 Ω each.

4-7-1. Calculate and plot the subthreshold current of a short-channel Si MOSFET as a function of the gate voltage in the range of gate voltages between $V_T - 0.5$ V and V_T for the drain voltages $V_D = 0.1$ V and $V_D = 10$ V. Use the following parameters: threshold voltage $V_T = 1$ V, $N_A = 1 \times 10^{15}$ cm^{-3}, electron mobility $\mu_n = 800$ cm^2/Vs, gate length $L = 1$ μm, gate oxide thickness $d_i = 500$ Å, energy gap $E_g = 1.12$ eV, effective densities of states in the conduction band and in the valence band are 3.22×10^{19} cm^{-3} and 1.83×10^{19} cm^{-3}, respectively, oxide dielectric permittivity 3.45×10^{-11} F/m, silicon dielectric permittivity 1.05×10^{-10} F/m, and temperature $T = 300$ K.

4-7-2. Check the derivation of eqs. (4-7-9) and (4-7-12) step by step (show your work). Calculate and plot the subthreshold current of a long-channel channel Si MOSFET as a function of the gate voltage in the range of gate voltages between $V_T - 0.999$ V and V_T for the drain voltages $V_D = 0.01$ V, 0.1 V and 10 V. Use the following parameters: threshold voltage $V_T = 1$ V, $N_A = 1 \times 10^{15}$ cm^{-3}, electron mobility $\mu_n = 800$ cm^2/Vs, width $W = 1$ mm, gate length $L = 20$ μm, gate oxide thickness $d_i = 500$ Å, energy gap $E_g = 1.12$ eV, effective densities of states in the conduction band and in the valence band are $3.22 \times 10^{19}(T/300)^{3/2}$ cm^{-3} and $1.83 \times 10^{19}(T/300)^{3/2}$ cm^{-3}, respectively, oxide dielectric permittivity 3.45×10^{-11} F/m, silicon dielectric permittivity 1.05×10^{-10} F/m, and temperature $T = 300$ K. $V_{FB} = 0$.

4-8-1. Derive eq. (4-8-2).

4-8-2. Derive eqs. (4-8-5) and (4-8-6).

4-9-1. Consider the following idealized ion implantation profile near the semiconductor–insulator interface in a Si MOSFET ($x = 0$ corresponds to the semiconductor–insulator interface; see Fig. P4-9-1). Calculate the threshold voltage shift as a function N_{Ai} for 10^{14} cm$^{-3} \leq N_{Ai} \leq 10^{17}$ cm^{-3} for $d_{imp} = 0.08$ μm. Use the following parameters: $N_A = 1 \times 10^{15}$ cm^{-3}, gate oxide thickness $d_i = 500$ Å, energy gap $E_g =$

1.12 eV, effective densities of states in the conduction band and in the valence band are 3.22×10^{19} cm^{-3} and 1.83×10^{19} cm^{-3}, respectively, oxide dielectric permittivity 3.45×10^{-11} F/m, silicon dielectric permittivity 1.05×10^{-10} F/m, and temperature $T = 300$ K. Assume shallow ionized acceptors. Define the threshold voltage as the voltage the electron concentration at the surface, $n(0)$, is equal to $N_A + N_{Ai}$.

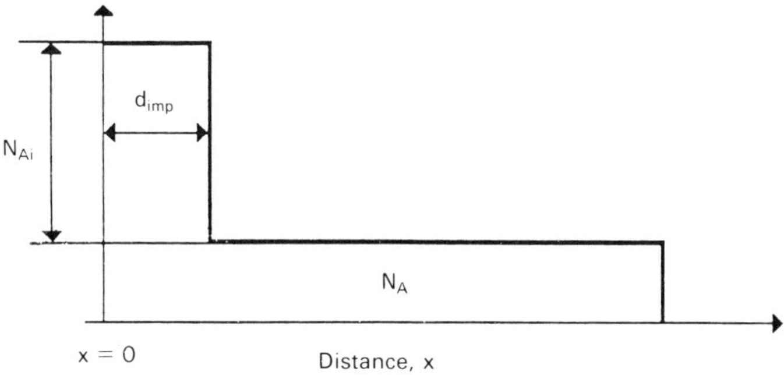

Fig. P4-9-1

4-9-2. Consider the following idealized ion implantation profile near the semiconductor–insulator interface in a Si MOSFET ($x = 0$ corresponds to the semiconductor–insulator interface; see Fig. P4-9-2). Calculate the threshold voltage shift as a function N_{Di} for 10^{14} cm$^{-3} \leq N_{Di} \leq 10^{17}$ cm^{-3} for $d_{imp} = 0.1$ μm. Use the following parameters: threshold voltage $V_T = 1$ V, $N_A = 1 \times 10^{15}$ cm^{-3}, gate oxide thickness $d_i = 500$ Å, energy gap $E_g = 1.12$ eV, effective densities of states in the conduction band and in the valence band are 3.22×10^{19} cm^{-3} and 1.83×10^{19} cm^{-3}, respectively, oxide dielectric permittivity is 3.45×10^{-11} F/m, silicon dielectric permittivity is 1.05×10^{-10} F/m, and temperature $T = 300$ K. Assume shallow ionized donors and acceptors.

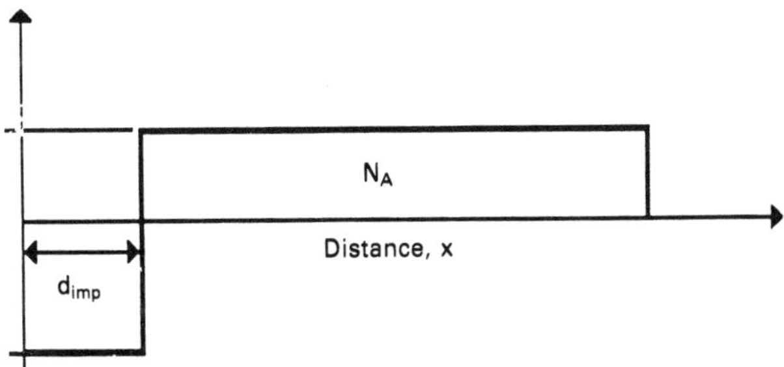

Fig. P4-9-2

4-9-3. Consider the following idealized ion implantation profile near the semiconductor–insulator interface in a Si MOSFET ($x = 0$ corresponds to the semiconductor–insulator interface; see Fig. P4-9-3). Use the following parameters: $N_{Ai} = 10^{17}$

cm^{-3}, $d_{imp} = 0.1\ \mu\text{m}$. $N_A = 1 \times 10^{15}\ \text{cm}^{-3}$, gate oxide thickness $d_i = 500$ Å, energy gap $E_g = 1.12$ eV, effective densities of states in the conduction band and in the valence band are $3.22 \times 10^{19}\ \text{cm}^{-3}$ and $1.83 \times 10^{19}\ \text{cm}^{-3}$, respectively, oxide dielectric permittivity 3.45×10^{-11} F/m, silicon dielectric permittivity 1.05×10^{-10} F/m, temperature $T = 300$ K, and flat-band voltage $V_{FB} = 0$. Calculate and plot charge density, electric field, and potential in the semiconductor as functions of the distance from the semiconductor–insulator interface at the threshold.

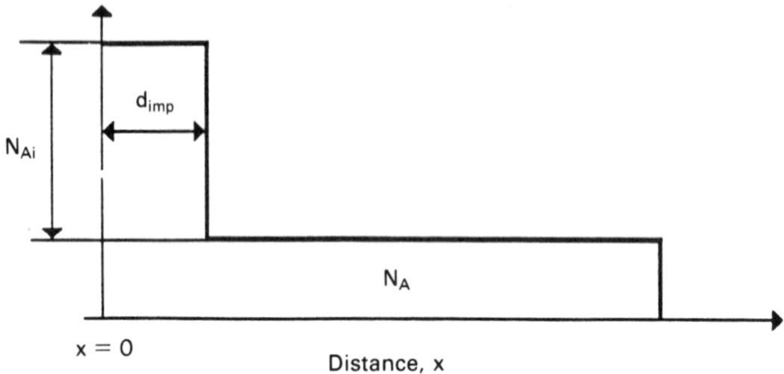

Fig. P4-9-3

4-9-4. Consider the following idealized ion implantation profile near the semiconductor–insulator interface in a Si MOSFET shown in Fig. P4-9-3 ($x = 0$ corresponds to the semiconductor–insulator interface). Use the following parameters: $N_{Ai} = 1 \times 10^{17}$ cm^{-3}, $D_{imp} = 0.1\ \mu\text{m}$. $N_A = 1 \times 10^{15}\ \text{cm}^{-3}$, gate oxide thickness $D_i = 500$ Å, energy gap $E_g = 1.12$ eV, effective densities of states in the conduction band and in the valence band are $3.22 \times 10^{19}\ \text{cm}^{-3}$ and $1.83 \times 10^{19}\ \text{cm}^{-3}$, respectively, oxide dielectric permittivity 3.45×10^{-11} F/m, silicon dielectric permittivity 1.05×10^{-10} F/m, temperature $T = 300$ K, and flat-band voltage $V_{FB} = 0$. At the threshold voltage for this device,

(a) calculate the depletion width in the semiconductor, x_{de}, and sketch the charge density profile from the semiconductor–insulator interface (which is defined as $x = 0$ for this part and for parts b and c)

(b) neglecting any charge within the oxide, calculate the electric field and sketch it from $x = -0.1\ \mu m$, showing numerical values at the interfaces and

(c) sketch the voltage over the same range as in part b, showing numerical values at the interfaces.

4-11-1. Derive eq. (4-11-12):

$$V_T = \phi_b - \Delta E_c - (q/\varepsilon_1)\int_0^d N_d(x)x\,dx \qquad (4\text{-}11\text{-}12)$$

4-12-1. Assume the following parameters for a-Si: $E_a \approx 86$ meV, $E_d \approx 129$ meV, $g_c \approx 8 \times 10^{24}\ m^{-3}eV^{-1}$, $g_v \approx 2 \times 10^{25}\ m^{-3}eV^{-1}$ (see Fig. 4-12-1), and $E_c - E_{FO} \approx 620$ meV. Calculate and plot the position of the electron Fermi level in the uniform a-Si sample in the dark as a function of the concentration of shallow donors, N_D, for $10^{15}\ cm^{-3} < N_d < 10^{17}\ cm^{-3}$.

Hint: Assume that all electrons supplied by the donors fill the deep acceptorlike states.

4-12-2. Derive eq. (4-12-30).

4-13-1. Derive eq. (4-13-5):

$$n = N_c[qn_t/g_c k_B T_o]^\alpha \qquad (4\text{-}13\text{-}5)$$

5

Photonic Devices

5-1. INTRODUCTION

When we shine light or other electromagnetic radiation at the surface of a semiconductor some of it is reflected back, some is absorbed in the material, and some may pass through (see Fig. 5-1-1). The propagation of an electromagnetic wave in a semiconductor may be described by introducing a complex refraction index, n_r^*:

$$n_r^* = n_r(1 - i\chi) \tag{5-1-1}$$

Here n_r is the index of refraction, which is equal to the ratio of the speed of light in vacuum, c, (approximately 3×10^{10} cm/s; see Appendix 1) to the speed of light in the semiconductor, c_s:

$$n_r = c/c_s \tag{5-1-2}$$

The *absorption index*, χ, is related to the absorption coefficient, α:

$$\alpha = 4\pi\chi/\lambda \tag{5-1-3}$$

Here

$$\lambda = 2\pi c/\omega \tag{5-1-4}$$

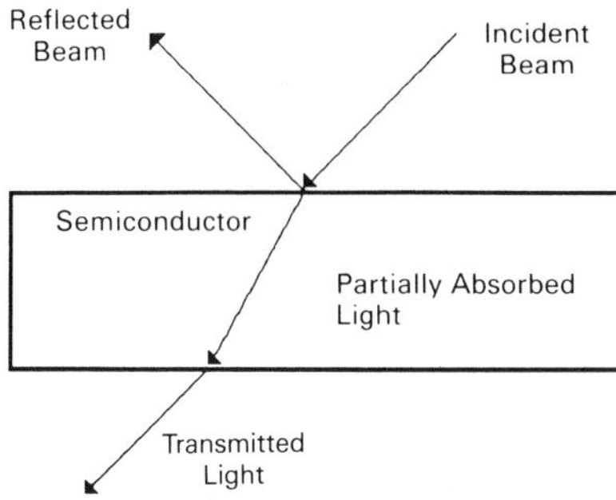

Fig. 5-1-1. Incident, reflected, absorbed, and transmitted light.

is the wavelength. The equation describing the propagation of the electromagnetic wave can be written as follows:

$$F = F_m \exp(-ax/2)\exp[i\omega(t - n_r^* x/c] \qquad (5\text{-}1\text{-}5)$$

where F is the electric field of the electromagnetic wave, F_m is the amplitude, ω is the frequency, and x is the direction of propagation. The complex refraction index can be related to the complex dielectric permittivity, ε^*:

$$n_r^* = (\varepsilon^*/\varepsilon_o)^{1/2} \qquad (5\text{-}1\text{-}6)$$

where $\varepsilon_o = 8.854 \times 10^{-12}$ F/m is the dielectric permittivity in vacuum,

$$\varepsilon^* = \varepsilon - i\sigma/\omega \qquad (5\text{-}1\text{-}7)$$

where σ is the conductivity of the semiconductor. From eqs. (5-1-6) and (5-1-7) we find that

$$\varepsilon = \varepsilon_o n_r^2(1 - \chi^2) \qquad (5\text{-}1\text{-}8)$$

$$\sigma = 2n_r^2\chi\omega\varepsilon_o \qquad (5\text{-}1\text{-}9)$$

$$\alpha = \sigma/(n_r^2 c\varepsilon_o) \qquad (5\text{-}1\text{-}10)$$

(see Problem 5-1-1).

The reflection coefficient, R, which is equal to the ratio of the intensity of the reflected wave to the intensity of the incident wave, is related to n_r^* as follows:

$$R = \left|\frac{n_r^* - 1}{n_r^* + 1}\right| = \frac{(n_r - 1)^2 + n_r^2\chi^2}{(n_r + 1)^2 + n_r^2\chi^2} \qquad (5\text{-}1\text{-}11)$$

The number of absorbed photons is proportional to the total number of photons and, hence, to the light intensity, F_{int}. This leads to the following equation:

$$\frac{dF_{int}(x)}{dx} = -\alpha F_{int}(x) \qquad (5\text{-}1\text{-}12)$$

where α is an absorption coefficient and x is the coordinate in the direction of the light propagation. The distance $1/\alpha$ is called the *light penetration depth* (see Fig. 5-1-2).

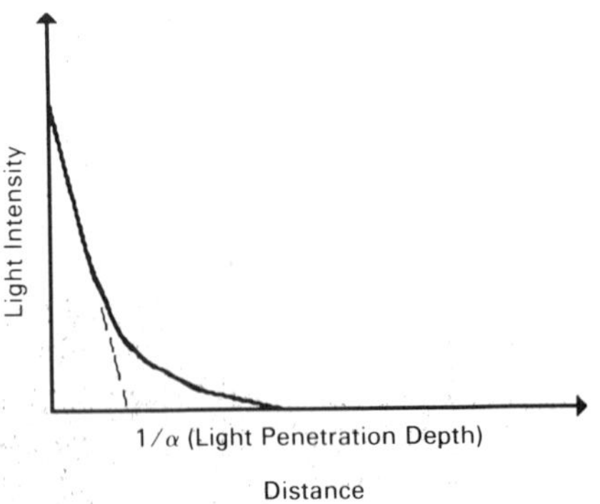

Fig. 5-1-2. Light intensity in semiconductor vs. distance from surface.

The dependencies of absorption coefficients on photon wavelength and energy, λ (μm) and E (eV) = $1.24/\lambda$ (μm), for different semiconductor materials are compared in Fig. 5-1-3. The absorption coefficient drops sharply for photon energies smaller than the energy gap, i.e., below the band edge. Absorbed radiation with the photon energy larger than the energy gap leads to the creation of elec-

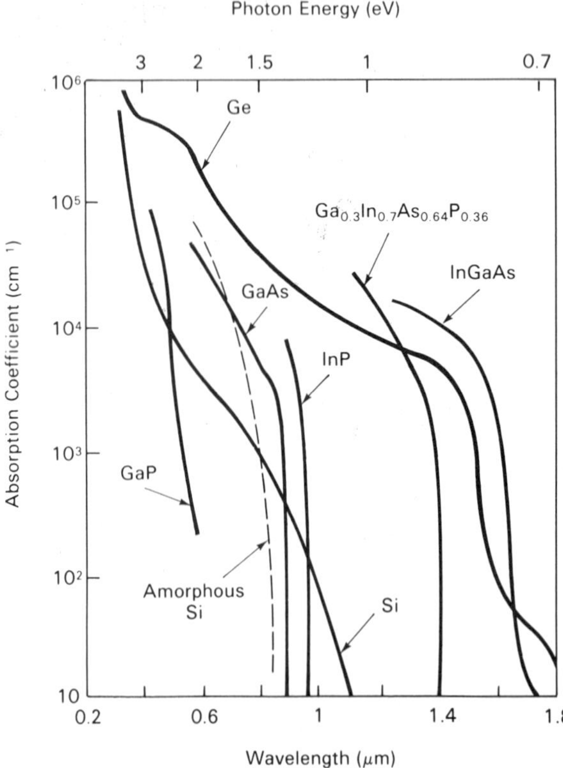

Fig. 5-1-3. Absorption coefficients for different semiconductors.

tron-hole pairs in a semiconductor. In a semiconductor device, such as a p-n diode, these electron-hole pairs may be separated by the built-in electric field, producing either voltage or electric current, or both, depending on the external circuit. This is called the *photovoltaic effect*. The solar (or photovoltaic) cell, used to generate electricity, and the photodetector, used to register electromagnetic radiation, are two important photovoltaic devices. Solar cell technology is fairly mature and may become very important if either the price of oil goes up or if the environmental concerns related to the greenhouse effect will force us to switch from coal and oil to solar energy.

The inverse effect of generating electromagnetic radiation due to the passage of electric current may also take place in a semiconductor device. This effect is called *electroluminescence*. A typical example would be a forward-biased p-n junction where the recombination of the electron-hole pairs produced by injection causes the emission of light. Such a device is called a Light-Emitting Diode (LED). If mirrors are provided (typically formed by cleaved crystallographic surfaces of a semiconductor that such a device is made of) and the injection level exceeds some critical value, this device may function as a semiconductor laser.

In this chapter we will consider crystalline and amorphous solar cells, photodetectors, light-emitting diodes, and semiconductor lasers. We will also briefly discuss integrated optical and electronic devices, i.e., optoelectronic integrated circuits.

5-2. CRYSTALLINE SOLAR CELLS

Let us consider a p^+-n junction uniformly illuminated with light, creating G_L electron-hole pairs per unit volume per second. The steady state continuity equation for holes in the n-type region (see eq. (2-3-9)) may be rewritten as

$$Dp \frac{\partial^2 \Delta p_n}{\partial x^2} - \frac{\Delta p_n}{\tau_{pl}} + G_L = 0 \tag{5-2-1}$$

where $\Delta p_n = p_n - p_{no}$, D_p is the hole diffusion coefficient, and τ_{pl} is the minority carrier (hole) lifetime. The boundary conditions are given by

$$\Delta p_n(x \Rightarrow \infty) = G_L \tau_{pl} \tag{5-2-2}$$

and

$$\Delta p_n(x = 0) = p_{no} \left[\exp\left(\frac{V}{V_{th}}\right) - 1 \right] \tag{5-2-3}$$

Here $V_{th} = k_B T/q$ is the thermal voltage and $x = 0$ corresponds to the p-n junction boundary. (We assume that the width of the depletion region, x_d, is much smaller than the hole diffusion length, $L_p = (D_p \tau_{pl})^{1/2}$, and the width of the n region, W_n, is much larger than L_p.) The solution of eq. (5-2-1), with the boundary conditions given by eqs. (5-2-2) and (5-2-3), yields

$$\Delta p_n = \{p_{no}[\exp(V/V_{th}) - 1] - G_L\tau_{pl}\}\exp(-x/L_p) + G_L\tau_{pl} \tag{5-2-4}$$

For $V = 0$, we obtain

$$\Delta p_n(V = 0) = G_L\tau_{pl}\left[1 - \exp\left(\frac{-x}{L_p}\right)\right] \tag{5-2-5}$$

leading to the hole component of the light-generated current at zero bias given by

$$|I_{pL}| = qD_pS\left.\frac{d\Delta p_n}{dx}\right|_{x=0} = qG_LL_pS \tag{5-2-6}$$

In a similar way we can obtain an expression for the electron component of the light-generated current at zero bias, I_{nL}. Hence, the total light-generated current

$$I_L = qG_L(L_p + L_n)S \tag{5-2-7}$$

This equation is valid when the thicknesses of the p and n regions, W_p and W_n, are much greater than the electron and hole diffusion lengths, respectively (i.e., $W_p >> L_n$ and $W_n >> L_p$). In the opposite limiting case ($W_p << L_n$ and $W_n << L_p$) the light-generated current for the uniform generation rate of the electron-hole pairs is given by

$$I_L = qG_L(W_n + W_p)S/2 \tag{5-2-8}$$

(assuming infinite surface recombination at the ohmic contacts; see Problem 5-2-1).

In a real solar cell the generation rate is not uniform and depends on the incident light spectrum and reflection coefficients of the cell surfaces. The solar spectrum is shown in Fig. 5-2-1, which also shows wavelengths corresponding to the energy gaps of several semiconductor materials used in photovoltaic applications. The spectrum referred to as *AM1* (*air mass 1*) *illumination* is for solar radiation at sea level when the sun is at zenith. The intensity of this radiation is P_{in} = 92.5 mW/cm^2. The solar radiation just outside the earth's atmosphere is called *AM0* (*air mass 0*) *illumination* (see the AM0 spectrum in Fig. 5-2-1). The incident power for the AM0 illumination is approximately 135 mW/cm^2. The losses in the atmosphere are primarily caused by the ultraviolet absorption in ozone and infrared absorption in water vapor.

An upper bound for I_L can be obtained assuming that all photons absorbed in a solar cell produce electron-hole pairs and contribute to the light-generated current:

$$I_L \leq q\int \alpha_c(\lambda)N_{ph}(\lambda)\,d\lambda \tag{5-2-9}$$

Here $\alpha_c(\lambda)$ is the fraction of incident photons absorbed in the cell and $N_{ph}(\lambda)\,d\lambda$ is the number of incident photons with wavelength between λ and $\lambda + \delta\lambda$. Assuming that all photons in the solar spectrum with energies higher than the silicon band gap are absorbed in a cell, one can obtain 54 mA/cm^2 for the AM0 illumination and 44 mA/cm^2 under standard test conditions (called AM1.5 radiation with P_{in} = 100

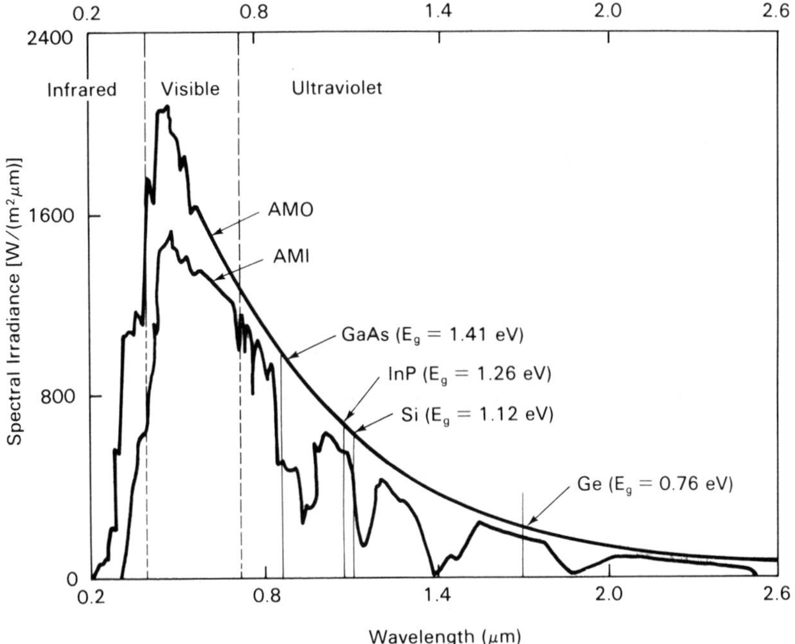

Fig. 5-2-1. AM1 and AM0 solar spectrum.

mW/cm^2; see Green 1982). For comparison, one of the fairly efficient silicon solar cells exhibited light-generated current close to 35.5 mA/cm^2 under these test conditions (AM1.5 radiation with P_{in} = 100 mW/cm^2 and at T = 28° C; see Green et al. 1984).

The light-generated current may be substantially increased by using a special textured front surface (with an antireflection coating). Such a surface traps light because of the total internal reflection (see Fig. 5-2-2). An additional increase of the light-generated current may be achieved by using a back-reflecting surface that effectively increases the length of the material where the light is absorbed. Such a design increases the measured value of the light-generated current to 41.5 mA/cm^2 for AM1.5 illumination bringing it very close to the theoretical limit of about 44 mA/cm^2 for AM1.5 illumination (see Sinton et al. 1986).

The idea of using a special antireflection coating is based on the dependence of the reflection coefficient from the boundary between two media with refraction indexes n_1 and n_2 on the relative refraction index, $n = n_1/n_2$:

$$R = (n - 1)^2/(n + 1)^2 \tag{5-2-10}$$

Covering the surface with an antireflection coating with a refraction index $n_3 = (n_1 n_2)^{1/2}$ minimizes the total reflection (see Problem 5-2-2).

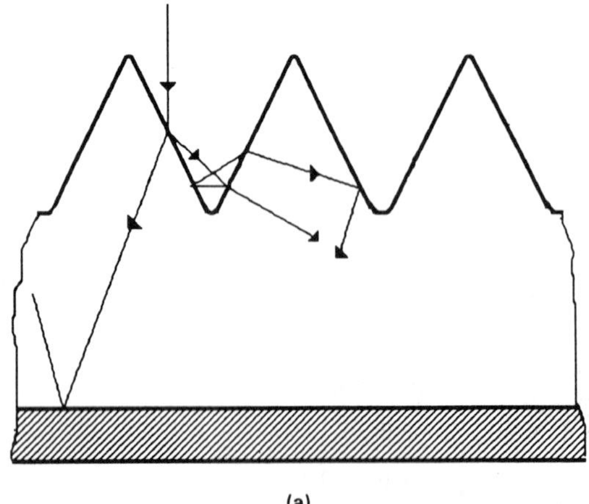

(a)

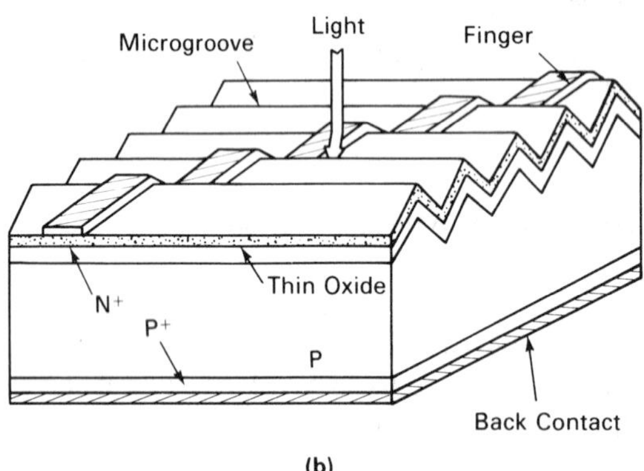

(b)

Fig. 5-2-2. (a) Light trapping in a cell with a front-textured pyramidal surface and back-reflecting surface. (b) Design of efficient silicon solar cell using microgrooves on top surface in order to trap the incident light (from M. A. Green, Z. Jinhua, A. W. Blakers, M. Taouk, and S. Narayanan, "25% Efficient Low-Resistivity Silicon Concentrator Solar Cells," *IEEE Electron Device Lett.*, EDL-7, no. 10, pp. 583–585, October (1986). © 1984 IEEE).

The total diode current, I, under illumination is given by

$$I = I_L + I_s \left[1 - \exp\left(\frac{V + R_s I}{m_{id} V_{th}}\right)\right] \tag{5-2-11}$$

where the second term on the right-hand side of eq. (5-2-11) represents the dark current, I_s is the dark saturation current, V is the applied voltage, m_{id} is the diode ideality factor, and R_s is the diode series resistance.

A typical silicon p-n junction solar cell is made using either a p^+-n junction or an n^+-p junction because p^+-n and n^+-p junctions have higher built-in voltages than p-n junctions. The built-in electric field separating the electrons and holes generated by light can at most provide the built-in potential. Hence, the built-in voltage gives the upper bound of the open circuit voltage, and p^+-n and n^+-p junctions have higher open-circuit voltages than p-n junctions. A p^+ layer in p^+-n cells and an n^+ layer in n^+-p cells are made very thin because the diffusion length of minority carriers in highly doped layers is very small. The dark saturation current, I_s, may be estimated using the solution of the continuity equation for the minority carriers in a lightly doped region. In particular, solving the continuity equation for electrons in the p region of an n^+-p cell, we find that

$$I_s = qS(n_i^2/N_A)(D_n/\tau_{pl})^{1/2}\coth(W_p/L_n) \tag{5-2-12}$$

(see Green 1982). For very pure material the lifetime, τ_{pl}, is limited by the Auger recombination. For p-type material

$$\tau_{pl} = 1/(G_p N_A^2) \tag{5-2-13}$$

For n-type material

$$\tau_{nl} = 1/(G_n N_D^2) \tag{5-2-14}$$

where for silicon, $G_p = 9.9 \times 10^{-32}$ cm^6/s and $G_n = 2.28 \times 10^{-31}$ cm^6/s (see Section 1-12 and Dziewior and Schmid 1982). Other recombination mechanisms may also be important, but eqs. (5-2-13) and (5-2-14) give the upper bound for the recombination time.

As can be seen from eq. (5-2-11), the built-in voltage, V_{oc}, of a solar cell is determined by

$$0 = I_L - I_s\left[\exp\left(\frac{V_{oc}}{m_{id}V_{th}}\right) - 1\right] \tag{5-2-15}$$

Equations (5-2-12) to (5-2-15) lead to the following limit for the open-circuit voltage (see Green 1984):

$$V_{oc} = \frac{m_{id}k_B T}{q}\ln\left[\frac{I_L}{qSn_i^2(D_n G_p)^{1/2}\coth(W_p/L_n)} + 1\right] \tag{5-2-16}$$

It is interesting to note that this limit is independent of doping level. Using typical values of parameters for silicon yields the maximum value of $V_{oc} \approx 716$ mV under AM0 illumination at 25° C. For comparison, an open-circuit voltage of 641 mV

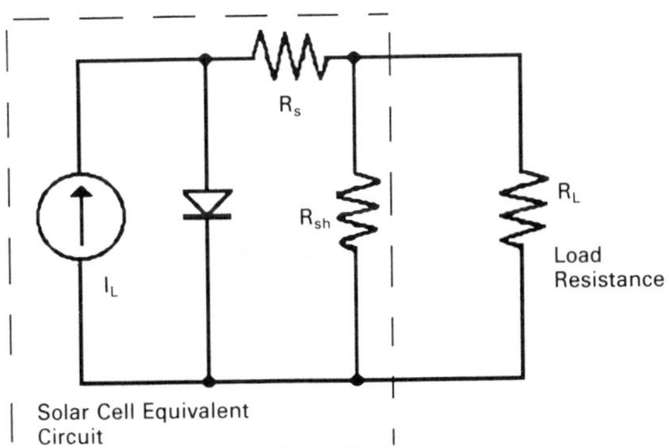

Fig. 5-2-3. Equivalent circuit of a solar cell.

was reported by Green et al. (1984) for a very efficient silicon solar cell under AM1.5 conditions, and even higher values (up to 690 mV) have been reported by Green et al. (1982).

In practical cells there may also be a parasitic shunting resistance, R_{sh}, leading to an additional term in eq. (5-2-11):

$$I = I_L + I_s\left[1 - \exp\left(\frac{V + R_s I}{m_{id} V_{th}}\right)\right] - V/R_{sh} \tag{5-2-17}$$

Equation (5-2-17) may be represented by the equivalent circuit shown in Fig. 5-2-3.

Measured current-voltage characteristics of a high efficiency silicon solar cell are shown in Fig. 5-2-4 (after Green et al. 1984), which shows the operating point corresponding to the maximum power,

$$P_{max} = I_{pm} V_{pm} \tag{5-2-18}$$

that can be obtained from this solar cell. The operating point can be chosen by using an appropriate load resistance,

$$R_L = V_{pm}/I_{pm} \tag{5-2-19}$$

The ratio

$$FF = \frac{P_{max}}{I_{sc} V_{oc}} \tag{5-2-20}$$

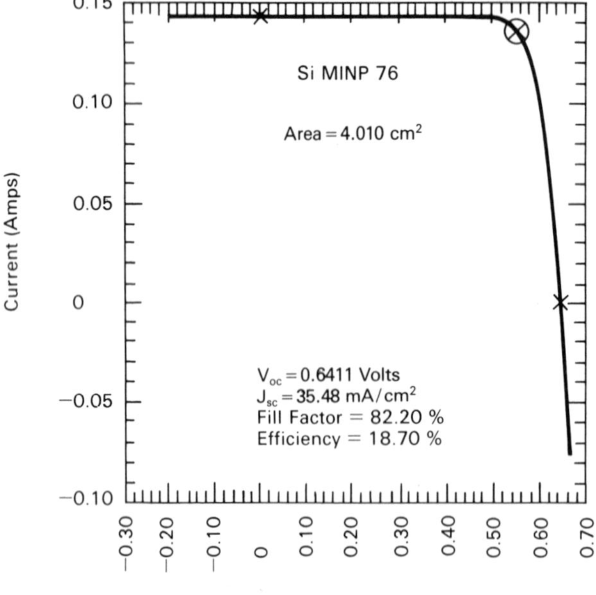

Fig. 5-2-4. Measured current-voltage characteristics of a high efficiency silicon solar cell (after M. A. Green, A. W. Blakers, J. Shi, E. M. Keller, and S. R. Wenham, *IEEE Trans. Electron Devices*, ED-31, no. 5, p. 679 (1984). © 1984 IEEE).

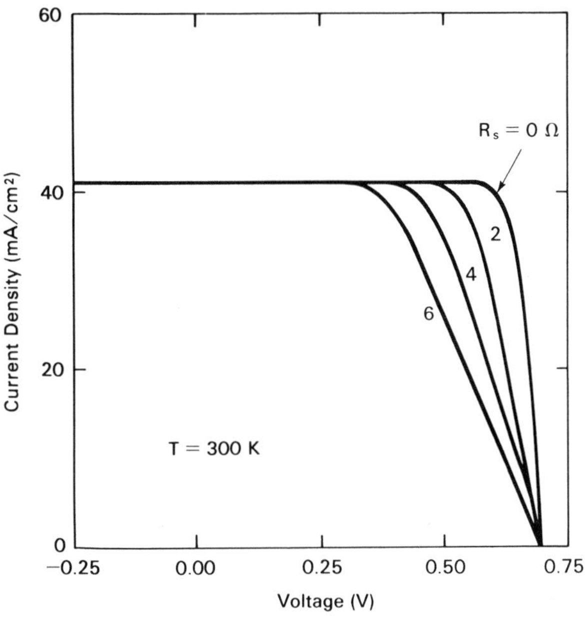

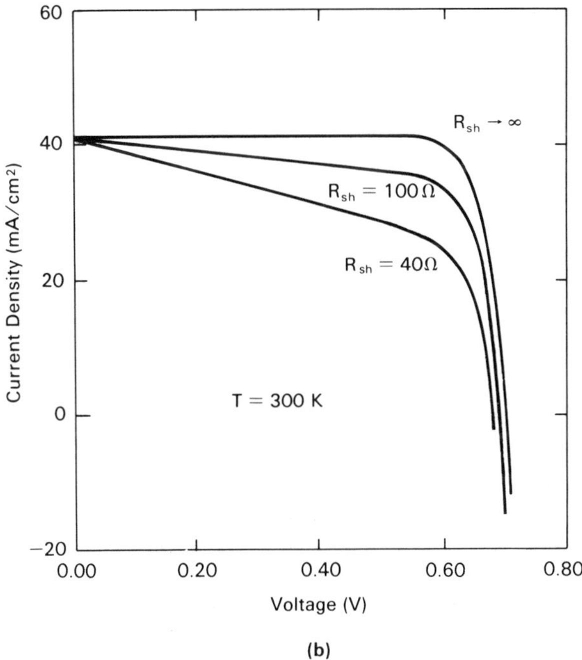

Fig. 5-2-5. Current-voltage characteristics of a solar cell for different values of (a) series and (b) shunt resistances. Curves shown in Fig. 5-2-5a were calculated using program PLOTF with subroutine PSCIV.BAS (see Program Description Section) for the series resistances 0, 2, 4, and 6 Ωcm^2 and an infinitely large shunt resistance. Curves shown in Fig. 5-2-5b were calculated using program PLOTF with subroutine PSCSH.BAS (see Program Description Section) for zero series resistance and for shunt resistances 40, 100 Ωcm^2 and an infinitely large shunt resistance. Parameters used in the calculation are light generated current 41 mA/cm^2, dark saturation current 10^{-9} mA/cm^2, ideality factor 1.1, and T = 300 K.

is called the *fill factor*. Here I_{sc} is the short-circuit current (i.e., $I_{sc} = I(V = 0)$). Using this definition we can introduce the following expression for the maximum solar cell efficiency:

$$\eta = \frac{I_{sc} V_{oc} FF}{P_{in}} \tag{5-2-21}$$

where P_{in} is the power of the incident radiation.

The current-voltage characteristics of a solar cell calculated using eq. (5-2-17) for different values of series and shunt resistances are shown in Fig. 5-2-5. This figure clearly demonstrates how the increase in the series resistance decreases the fill factor and how the decrease in the shunt resistance decreases the fill factor and the open-circuit voltage and, hence, reduces the solar cell efficiency.

As can be seen from eq. (5-2-16) V_{oc} is proportional to $|V_{th} \ln(n_i^2)|$, i.e., to the energy gap, E_g. This can be understood by considering the physical mechanism of charge separation in a p-n junction. When electron-hole pairs are created within the depletion layer, they are separated by the built-in electric field. Hence, the potential difference is limited by the built-in voltage, which, in turn, is determined by the energy gap (see Fig. 2-2-3). On the other hand, only photons with energies larger than the band gap are absorbed in a semiconductor, and hence, the light-generated current decreases with the increase in energy gap. As a consequence, a solar cell efficiency is the largest for the energy gap about 1.4 eV at room temperature (see Fig. 5-2-6).

The design of an efficient silicon solar cell is shown in Fig. 5-2-7 (from Green et al. 1984). The current-voltage characteristics of this cell are shown in Fig. 5-2-4. We notice the use of an ultra thin SiO_2 layer to passivate the silicon sur-

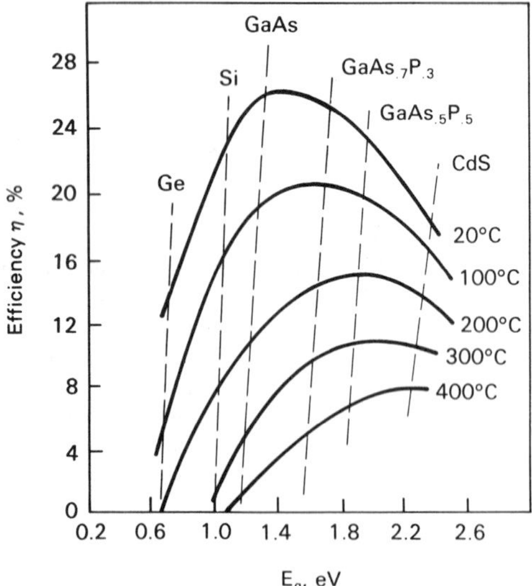

Fig. 5-2-6. Solar cell efficiency as a function of the energy gap (after P. Rappoport and J. J. Wysocki, "The Photovoltaic Effect in GaAs, CdS, and Other Compound Semiconductors," *Acta Electron*, 5, p. 364 (1961). © 1961 IEEE).

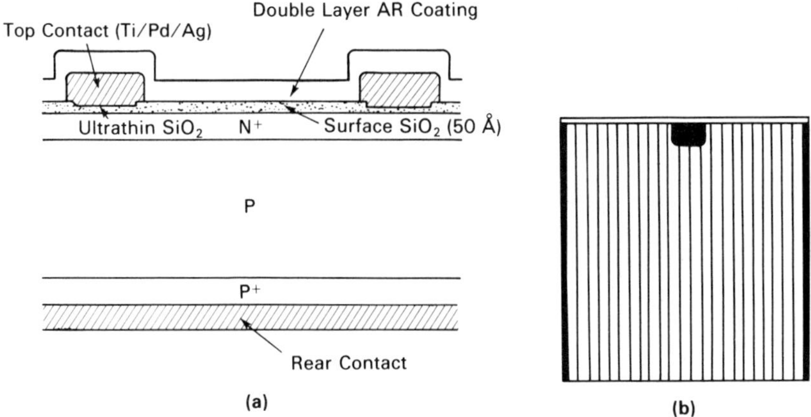

Fig. 5-2-7. Efficient silicon solar cell (from M. A. Green, A. W. Blakers, J. Shi, E. M. Keller, and S. R. Wenham, *IEEE Trans. Electron Devices,* ED-31, no. 5, p. 679 (1984). © 1984 IEEE). (a) Cell design and (b) top contact geometry. The figure shows the top view of the cell.

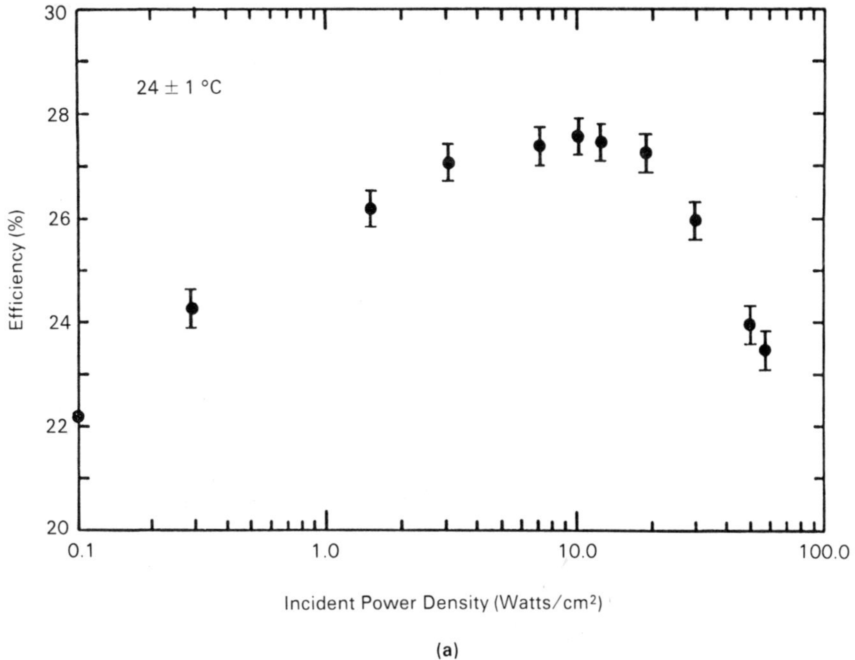

Fig. 5-2-8. Measured dependences of (a) efficiency (b) open circuit voltage and fill factor on the incident power density of light for a silicon cell with efficiency of 27.5% at 10 W/cm^2 (100 suns) c. Solar cell design. (after R. A. Sinton, Y. Kwark, J. Y. Gan, and R. M. Swanson, *IEEE Electron Device Lett.,* EDL-7, no. 7, p. 567 (1986). © 1986 IEEE).

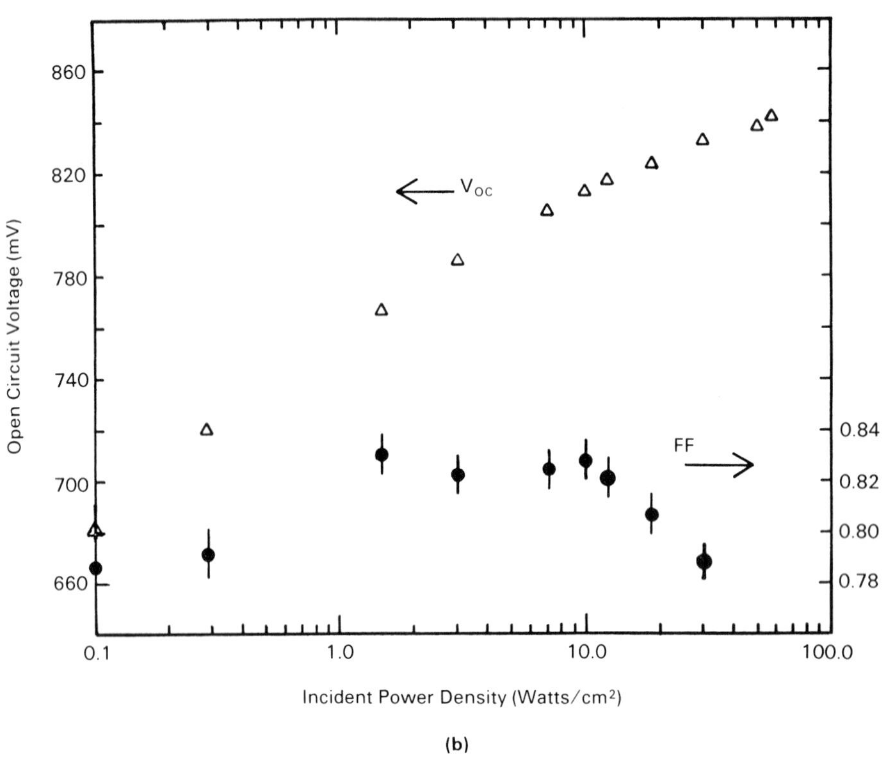

(b)

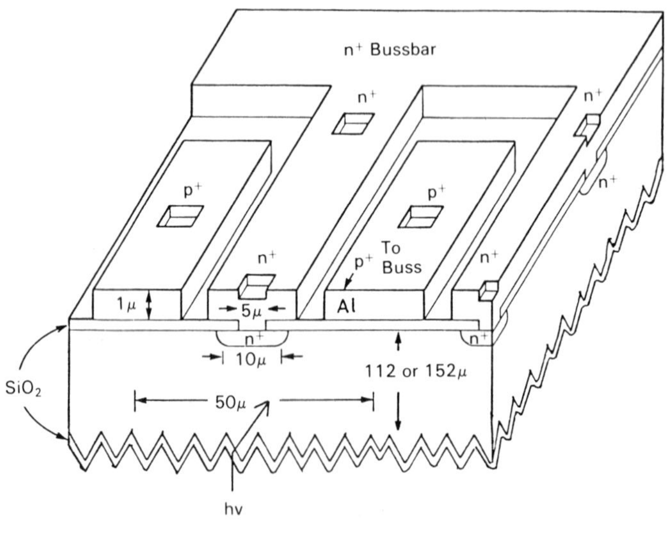

(c)

Fig. 5-2-8. Cont.

face. This passivation decreases the surface recombination rate. The oxide layer must be extremely thin, as the contact between the top metal and the active layer is via tunneling through the oxide layer. Another feature of this cell is the use of a double antireflection coating.

As can be seen from eqs. (5-2-7), (5-2-8), and (5-2-16), both I_L and V_{oc} increase with the increase in the light intensity. The short-circuit current is of the order of I_L. Hence, the efficiency of a solar cell, η, is proportional to $I_L V_{oc}/P_{in}$, where P_{in} is the power of the incident radiation (see eq. (5-2-21)). From eqs. (5-2-7), (5-2-16), and (5-2-21) we find that at large intensities

$$\eta \sim \text{const } FF \ln(P_{in}) \tag{5-2-22}$$

Hence, the solar cell efficiency should increase with intensity until thermal effects at very large intensities cause the efficiency to decrease. The measured dependence of the open-circuit voltage, efficiency, and fill factor on the incident power density of light is shown in Figs. 5-2-8a and 5-2-8b for a silicon cell with a record efficiency of 27.5% at 10 W/cm^2 (100 suns; after Sinton et al. 1986). These cells are called *concentrator cells* because concentrators of the sunlight (lenses or mirrors) are used to increase the intensity. The efficiency reported by Sinton et al. (1986) for the AM1 illumination was 22% at 24° C. The cell design is shown in Fig. 5-2-8c. The cell of this type is called a *back-side point-contact solar cell.* In this design, both n^+ and p^+ contacts are located on the back surface of the cell in a checkerboard pattern. The difference between the Fermi levels in these regions leads to the built-in potential. A silicon material (sandwiched between two thin SiO_2 layers) is very lightly doped (390 Ωcm material). The collection of carriers occurs in a complicated three-dimensional pattern. The advantage of this type of cell is that none of the front surface is covered by the contact (compare with Fig. 5-2-7).

More traditional cells with contacts on both sides have also exhibited very high efficiencies (up to 25%; see Green et al. 1986). The design of such a cell uses microgrooves on the top surface to trap the incident light, as shown in Fig. 5-2-2b.

As can be understood from the foregoing discussion of the principle of a solar cell operation, the photovoltaic effect is based on the separation of the photogenerated carriers caused by a built-in potential. Such a potential may also exist in Schottky barrier diodes, Metal Insulator Semiconductor (MIS) structures, Semiconductor Insulator Semiconductor (SIS) structures, Metal Insulator N/P semiconductor (MINP) structures, and heterostructures. Hence, in addition to solar cells utilizing p-n junctions for the carrier collection, Schottky barrier solar cells, MIS cells, SIS cells, MINP cells, and heterojunction cells have also been fabricated.

In a heterostructure cell a top narrow layer of a wide band-gap semiconductor serves as a "window" for the sunlight that reaches the narrow-gap semiconductor with little loss (see Fig. 5-2-9; from Parekh and Barnett 1984). The top wide-band layer is typically very heavily doped. As was mentioned previously, this leads to a higher built-in voltage, and hence, to a higher open-circuit voltage and a higher cell efficiency. High doping also reduces the parasitic series resis-

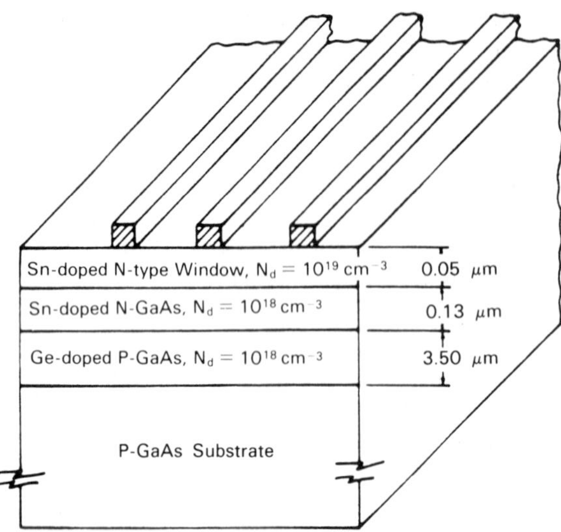

Fig. 5-2-9. Optimized heterostructure GaAs solar cell with top heavily doped wide band GaP layer (from R. H. Parekh and A. M. Barnett, *IEEE Trans. Electron Devices,* ED-31, no. 5, p. 689 (1984). © 1984 IEEE).

tance. A 26% efficient AlGaAs/GaAs concentrator cell (with the efficiency measured at 70 W/cm^2 of incident power) was reported by Hamaker et al. (1985). Gallium phosphide (that has an indirect band gap of 2.2 eV and a direct band gap of 2.75 eV) has also been used as window material for gallium arsenide (see Parekh and Barnett 1984). Tandem heterojunction solar cells can be employed for even higher efficiency (theoretically over 30% at room temperature under AM1 conditions).

In addition to silicon and gallium arsenide many other materials, such as CdTe, $CuInSe_2$, and polycrystalline silicon, have also been used for photovoltaic cells with efficiencies below or slightly above 10% (see the comparison of different technologies in Section 5-3 [Table 5-3-1]).

In 1976 Carlson and Wronski reported a new type of solar cell -- an *amorphous silicon solar cell* (see Carlson and Wronski 1976). Since that time remarkable progress has been achieved in this technology, leading to continuous mass production of large-area cells at a reasonable cost. The physics and applications of amorphous silicon cells are considered in the next section.

5-3. AMORPHOUS SILICON SOLAR CELLS

As was mentioned in Section 4-12, in 1972 Spear and LeComber demonstrated that amorphous silicon films prepared by the glow discharge decomposition of silane have a very low density of defect states in the energy gap (see Spear and LeComber 1972 and Section 4-12). The amorphous silicon material obtained by this process is, in fact, an amorphous silicon-hydrogen alloy with a fairly large concentration of hydrogen. The hydrogen atoms tie up silicon dangling bonds and decrease the density of localized states in the energy gap. These localized states play a dominant role in determining the transport properties of amorphous Si. Amorphous germanium-hydrogen and amorphous silicon carbide–hydrogen alloys have been also obtained by the same process. The energy gaps of these materials range from 1.1 eV for amorphous germanium alloys up to 2.5 eV for amorphous silicon carbide alloys.

An approximate distribution of localized states for intrinsic (undoped) a-Si is shown in Fig. 4-12-1. The localized states in the upper half of the energy gap (closer to the bottom of the conduction band) behave like acceptorlike states, whereas the states in the bottom half of the energy gap behave like donorlike states (see Section 4-12).

The position of the Fermi level in an undoped a-Si sample in the dark is determined by the localized states and is close to the energy corresponding to the minimum of the density of states distribution. Dopants (typically phosphorus for n-type and boron for p-type material) can shift the position of the Fermi level, and amorphous silicon p-n junctions can be made as was first shown by Spear et al. (1976).

The positions of the electron and hole quasi Fermi levels may also be controlled by light. Amorphous silicon is highly photoconductive. Typically the dark conductivity of a-Si is on the order of 10^{-10} to 10^{-9} $(\Omega cm)^{-1}$, and its conductivity under AM1 illumination is on the order of 10^{-5} to 10^{-4} $(\Omega cm)^{-1}$. The possibility of doping amorphous silicon in a controlled fashion, its high photoconductivity, and its high absorption coefficient (see Fig. 5-1-3) make this material very attractive for applications in solar cells, sensors, imagers, and other photonic devices, as well as in temperature and pressure sensors.

The most important advantage of amorphous silicon for photovoltaic applications is a demonstrated ability to produce large-area, low-cost a-Si cells on flexible substrates in a continuous (roll-to-roll) fabrication process (up to 2×4 feet in size).

The physics of a-Si cells is quite different from that of conventional crystalline devices. Most photogenerated carriers in a-Si are trapped in the localized states, with only a small fraction in the delocalized states (higher in energy than the "mobility edge" separating localized and delocalized states) in the conduction and valence bands. The electron and hole band mobilities in the conduction and valence bands are low (about 10 to 20 cm^2/Vs for electrons and 1 to 10 cm^2/Vs for holes). Most efficient a-Si cells are p-i-n devices. The thickness of the active region is chosen to be only 0.5 μm or so because of a very small collection length related to low carrier mobilities. The back surface is texturized to enhance light collection. Photogenerated carriers are collected by drift currents (not by the diffusion currents as in crystalline cells). The recombination rate, electric field distribution, and carrier profiles are highly nonuniform, and a numerical solution of continuity equations together with Poisson's equation is required for a realistic description of such cells (see Hack and Shur 1985).

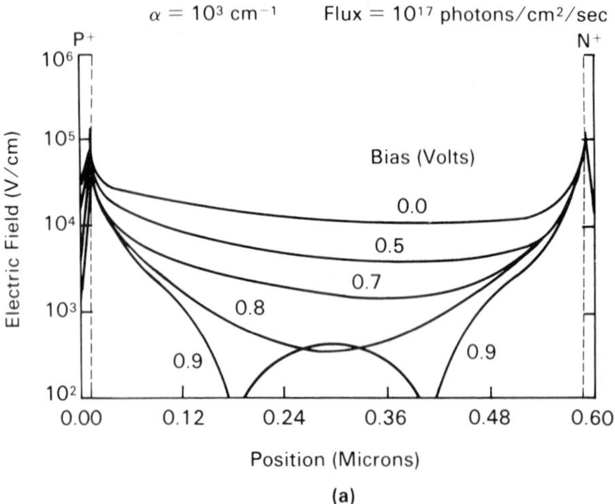

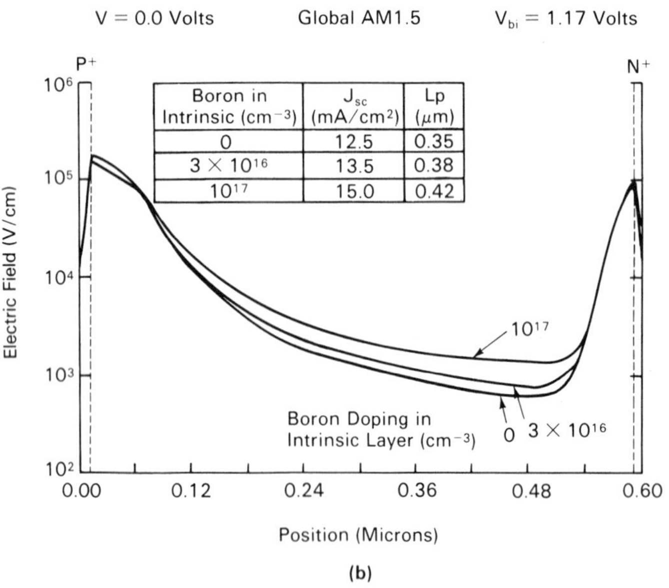

Boron in Intrinsic (cm^{-3})	J_{sc} (mA/cm^2)	Lp (μm)
0	12.5	0.35
3×10^{16}	13.5	0.38
10^{17}	15.0	0.42

Fig. 5-3-1. Typical electric field distributions in amorphous silicon solar cell: (a) for different bias voltages for uniform illumination (after M. Hack and M. Shur, *Technical Digest of the International PVSEC-1,* Kobe, Japan, pp. 645–648 (1984a). © 1984 IEEE). (b) for different boron content in the intrinsic (*i*) layer (after M. Hack and M. Shur *IEEE Trans. Electron Devices,* ED-31, no. 5, pp. 539–542 (1984b). © 1984 IEEE).

Typical electric field distributions in an amorphous silicon solar cell for different bias voltages for uniform illumination are shown in Fig. 5-3-1. As can be seen from the figure, the electric field peaks at the interfaces between p^+ and i and between i and n^+ regions. The electric field is much smaller in the center of the cell where it actually changes sign at voltages close to the open-circuit voltage. The large, nearly constant interface fields are maintained because of the large charge densities of ionized acceptors and donors in the p^+ and n^+ regions, respectively. As is shown in Fig. 5-3-1b, using a boron (i.e., p-type) doping close to the p^+ layer makes the field profile more uniform, leading to better collection, especially at voltages close to V_{oc}. This improves the fill factor and, in a cell with relatively low values of the built-in potential, V_{bi}, also increases the open-circuit voltage.

The efficiency of amorphous silicon cells may be improved using a wide band-gap amorphous semiconductor window, similar to an AlGaAs layer in AlGaAs/GaAs solar cells (see Section 5-2). A schematic diagram of such a cell with an amorphous hydrogenated silicon carbide window (see Tawada et al. 1982) is shown in Fig. 5-3-2a. Measured current-voltage characteristics of amorphous silicon carbide–amorphous silicon cells are shown in Fig. 5-3-2b.

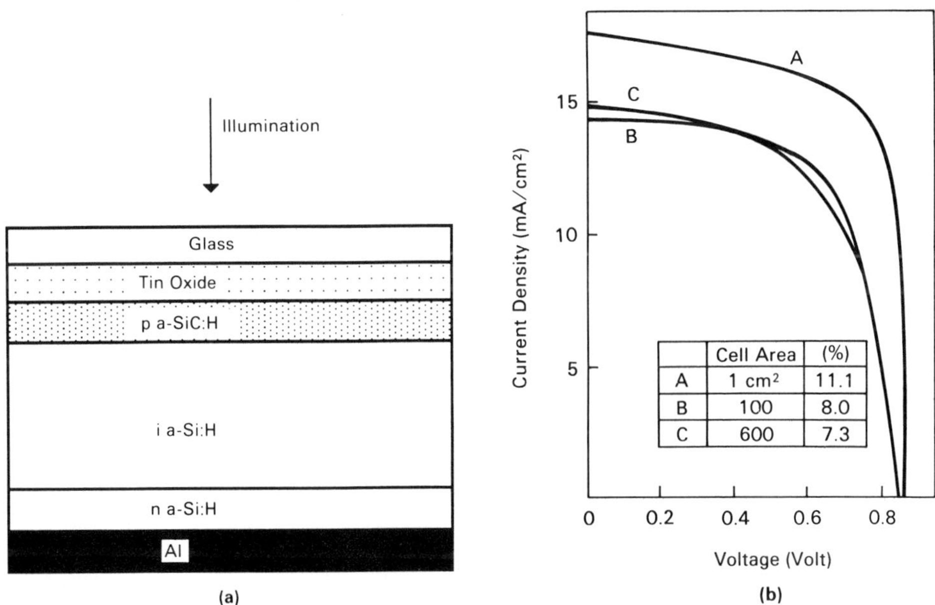

Fig. 5-3-2. (a) Schematic diagram of an a-Si : H solar cell with amorphous hydrogenated silicon carbide window. (b) Measured current-voltage characteristics of amorphous silicon carbide-amorphous silicon cells (after H. Sakai, K. Maruyama, T. Yoshida, Y. Ichikawa, T. Hama, M. Ueno, M. Kamiyama, and Y. Uchida, *Technical Digest of the International PVSEC-1,* Kobe, Japan, p. 591 (1984). © 1984 IEEE).

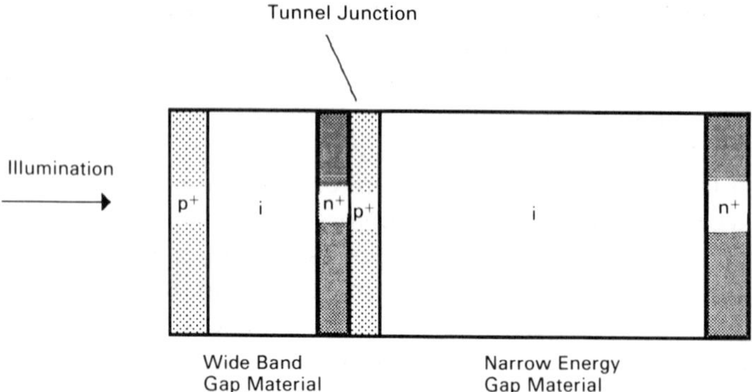

Fig. 5-3-3. Schematic diagram of multiple layer (tandem) a-Si solar cell where wide gap and narrow gap *p-i-n* cells are connected in series. Such a connection is achieved by doping p^+ and n^+ layers very high so that they form a tunnel junction in the middle of the cell.

Further improvement in efficiency may be achieved by utilizing multiple layer (tandem) cells where wide-gap and narrow-gap p^+-i-n^+ cells are connected in series. Such a connection is achieved by doping p^+ and n^+ layers very high so that they form a tunnel junction in the middle of the cell (see Fig. 5-3-3). The open-circuit voltage of such a structure is equal to the sum of the open-circuit voltages

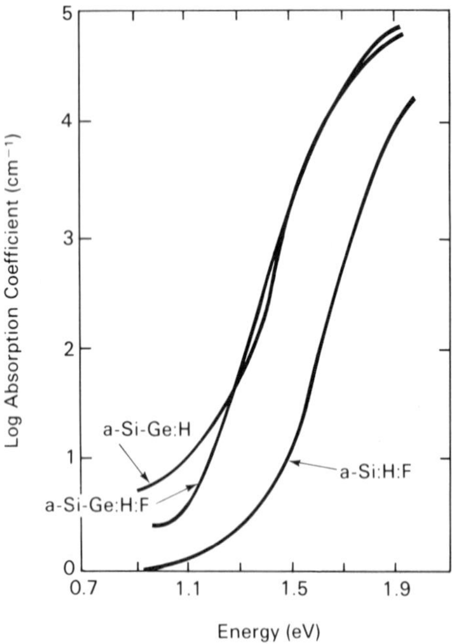

Fig. 5-3-4. Sub-band optical absorption of a-Si:H:F, a-SiGe:H, and a-SiGe:H:F alloys (from S. Guha, J. Non-Crystalline Solids, 77 & 78, pp. 1451–1460 (1985)).

of the individual cells. The light-generated currents of the wide-band-gap and narrow-band-gap cells have to be closely matched in order to achieve high efficiency.

Very high efficiencies have already been achieved for a-Si solar cells, including large-area, mass-produced cells (up to 10% or so for one-square-foot cells). The obtained values of the short-circuit current are on the order of 15 mA/cm^2, and typical open-circuit voltages are close to 900 to 950 mV, with fill factors on the order of 0.7 to 0.75. Maximum efficiencies reach 13% to 14%. Further improvements in this technology may be expected if high-quality a-Si:Ge:H alloys with narrower energy gaps are developed; see Fig. 5-3-4, which shows how the absorption edge in a-Si:Ge:H (corresponding to the energy band gap) is shifted toward smaller energies. This will allow us to fabricate a-Si tandem cells in which better collection of the red portion of the solar spectrum will lead to higher efficiency (see Guha 1985).

The efficiency of a-Si solar cells strongly depends on the density of the localized states in the intrinsic amorphous silicon. As was first shown by Stabler and Wronski (1977), the illumination by light with photon energies larger than the energy gap leads to new light-induced defect states. As a consequence, the cell performance degrades with time, and a typical a-Si cell can lose nearly half its efficiency after light soaking for 24 hours. Computed time dependence of efficiency, fill factor, short-circuit current and open-circuit voltage of an a-Si cell under AM1.5 illumination is shown in Fig. 5-3-5 for different thicknesses of the intrinsic layer. This computation does not take into account the self-annealing process for the light-induced defects. Such a process is inverse with respect to the process of the creation of light-induced defects. It may stabilize the solar cell parameters at a very large time scale.

As can be seen from Fig. 5-3-5, cells illuminated through the p$^+$ layer are more stable. This difference is caused by the asymmetrical distributions of acceptorlike and donorlike localized states (see Fig. 4-12-1) and by the boron doping of the intrinsic layer. Also, thinner cells are more stable. This is because a higher built-in electric field in thinner cells makes the device less sensitive to the effective diffusion length of carriers. This effective diffusion length decreases with an increase in the density of the light-induced defect states. This result explains why tandem cells (in which individual i layers are made thinner than in a conventional cell) are more stable. Also, the stability strongly depends on the light intensity. As shown by Hack and Shur (1986) the degradation time scale is inversely proportional to $1/f^2$, where f is flux. In a multijunction (tandem) cell only part of the overall solar spectrum is collected by each cell (short-wave radiation is primarily collected by the wide-energy-gap material and long-wave radiation is mostly collected by the narrow-gap material). This effectively reduces light-induced defects created by flux and greatly increases the stability of tandem cells. In addition, the stability of a-Si cells may be increased by modifying the absorption properties of the intrinsic layer where the light enters the device (see Hack and Shur 1986). A stable and highly efficient a-Si solar cell with initial efficiency of 9.3% dropping

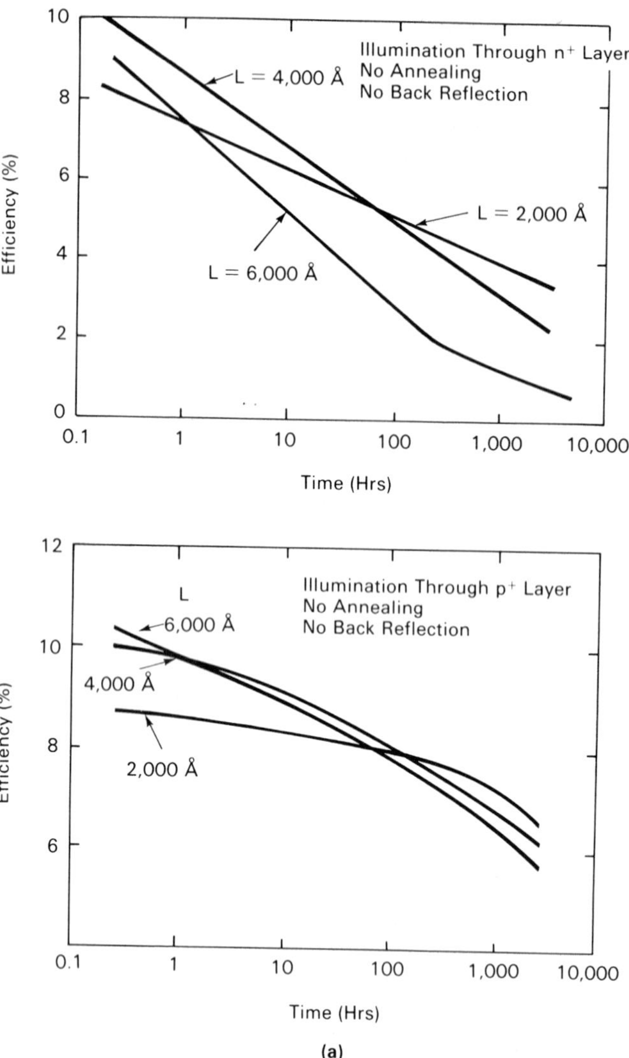

Fig. 5-3-5. Computed time dependences of the (a) efficiency, (b) fill factor, (c) short-circuit current, and (d) open-circuit voltage of an a-Si cell under AM1.5 illumination are shown for different thicknesses of the intrinsic layer (from M. Hack and M. Shur, "Implications of Light-induced Defects on the Performance of Amorphous Silicon Alloy p-i-n Solar Cells," *J. Appl. Phys.*, 59, no. 6, pp. 2222–2228, (1986). The time scale can be adjusted for different illumination intensities as $1/f^2$ where f is flux.

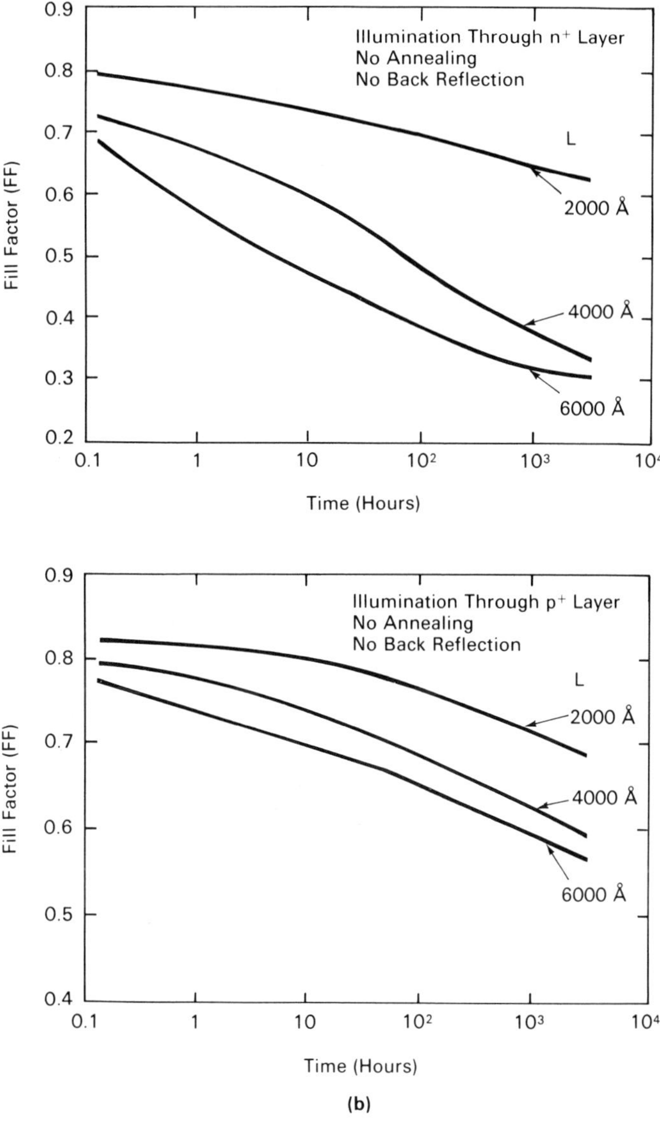

(b)

Fig. 5-3-5. Cont.

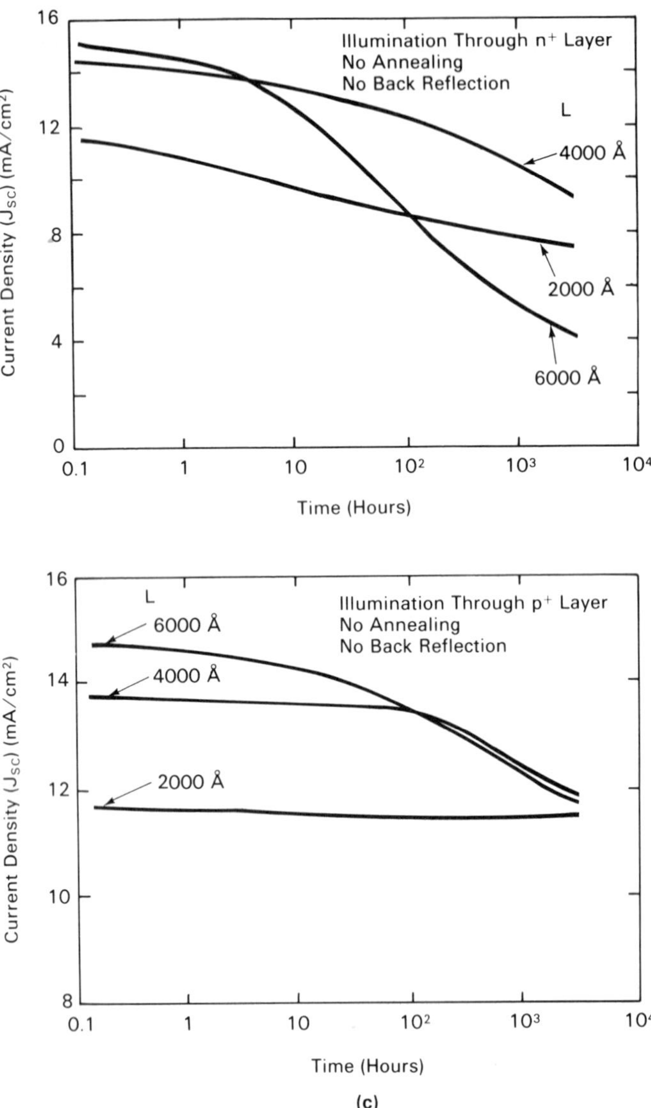

(c)

Fig. 5-3-5. Cont.

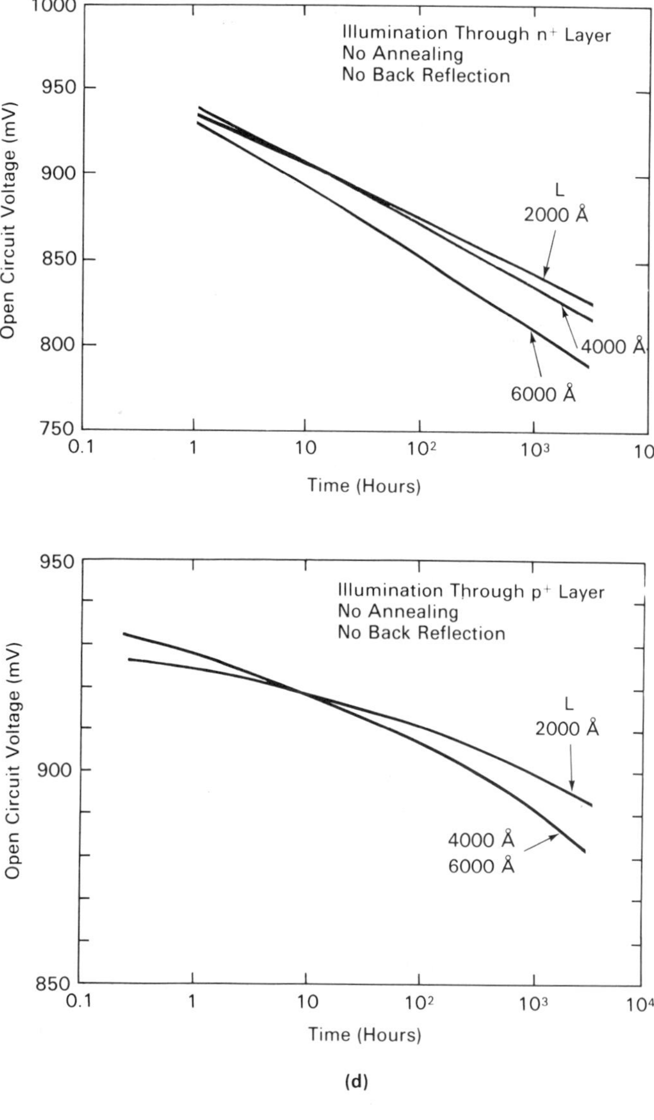

Fig. 5-3-5. Cont.

very little during 1000 hours under AM1 illumination has been reported (see Guha 1985).

Fig. 5-3-6 (from Hamakawa 1987) shows the cost of solar-cell modules per peak watt of produced electricity. It gives a clear idea of the bright prospects for commercial applications of amorphous silicon solar-cell technology. Table 5-3-1 (from Hamakawa 1987) compares this technology with photovoltaic technologies.

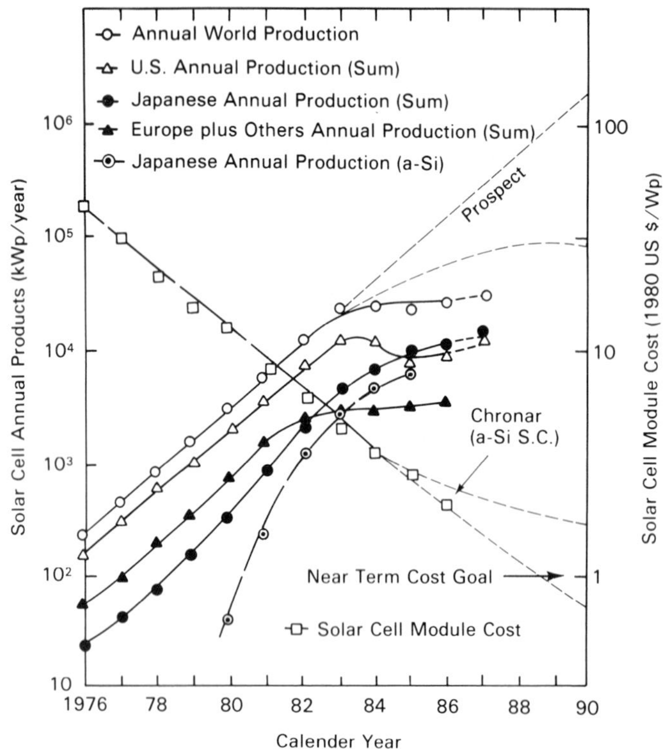

Fig. 5-3-6. The cost of solar cell modules per peak watt of produced electricity (from Y. Hamakawa, "Recent Advances in Solar Cell Technology," *Technical Digest of the International PVSEC-3*, Tokyo, Japan, pp. 147–152 (1987).

TABLE 5-3-1. COMPARISON OF KEY PHOTOVOLTAIC TECHNOLOGIES (from Y. Hamakawa, "Recent Advances in Solar Cell Technology," *Technical Digest of the International PVSEC-3,* Tokyo, Japan, pp. 147–152 (1987)).

Substrate	Key technology	Efficiency	AM	Year	Group
Single cryst. Si	μ-grooved passivated emitter	20.9	1.5	1985	Univ. NSW
	point contact	22.2	1.5	1986	Stanford U.
Compound	GaAs LPE	23.7	1	1987	Spire
crystalline		19.7	0	1985	Mitsubishi
	GaAs on Si (MOCVD)	15.0*	1.5	1987	NTT
	InP	22.0*	1.5	1986	NTT
	GaAsP/GaAs	16.5	1.5	1987	Spire
Polycrystalline	SILSO	14.1			Kyocera
Si	spin coating	10.8	1.5	1985	Hoxan
Amorphous	H.J. with a-SiC	12.0*		1985	SEL/TDK
		11.8*	1	1987	Osaka U.
		11.7*		1987	Sanyo
		11.5*		1987	Fuji

TABLE 5-3-1. Cont.

Substrate	Key technology	Efficiency	AM	Year	Group
	with a-SiC 100 cm^2	10.0*		1987	Sanyo
	with graded gap a-SiC	11.9*		1987	Solarex
	with Superlattic p-	11.2*		1987	Sanyo
	a-Si:F:H	11.8*		1987	ECD
	stacked solar cell				
	with a-SiGe	13.1*		1986	ECD
	with poly Si	13.3*	1	1985	Osaka U.
	4 terminal with $CuInSe_2$	14.2*		1987	ARCO
	a-Si/a-Si/a-SiGe	10.6*	1.5	1987	Mitsubishi
		13.0		1987	ECD
Thin film compound	H.J. CdS/CdTe	1.5	1.5	1987	SMU/SERI
		12.8*	1.5	1983	Matsushita
	$CdS/CuInSe_2$	9.6		1987	Boeing
		12.5*	1.5	1987	Boeing
		12.5*	1.5	1987	ARCO

* active-area efficiency

5-4. PHOTODETECTORS

Photodetectors detect optical signals and convert them into electrical signals. They are used in fiber-optics communication systems, in image processing, in establishing optical links between electrical circuits, etc. A solar cell can operate as a photodetector. However, other types of photodetectors, such as photoconductors, reverse-biased p-i-n diodes (which are basically reverse biased solar cells), Schottky barrier diodes, avalanche photodiodes, and phototransistors may be more suitable for a variety of different applications.

Photoconductivity is caused by the light-generated electrons in the conduction band or light-generated holes in the valence band. Different transitions caused by illumination and leading to photoconductivity are schematically shown in Fig. 5-4-1. These transitions include intrinsic (valence band to conduction band) and extrinsic (donor to conduction band or acceptor to valence band) transitions.

The most important characteristic of an intrinsic photoconductor is its energy gap, which determines the longest wavelength of light, $\lambda_g(\mu m) = 1.24/E_g(eV)$, still absorbed by a photoconductor (see Fig. 5-1-3, in which the wavelength dependence of the absorption coefficient is shown for several important semiconductors).

The properties of extrinsic photoconductors, which rely on extrinsic transitions (see Figs. 5-4-1b and c), depend on doping. In many cases such photoconductors operate at cryogenic temperatures to avoid the conductivity domination by carriers thermally excited from relatively shallow impurity levels.

For a uniform illumination, the generation rate of carriers is given by

$$G_p(\omega) = Q_c P_l/(\hbar\omega L) \tag{5-4-1}$$

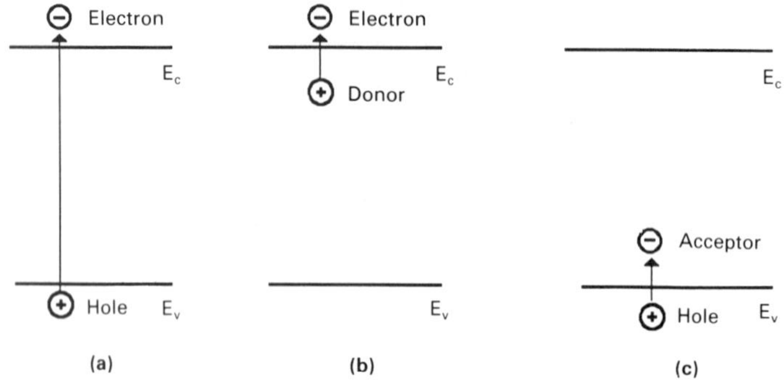

Fig. 5-4-1. Transitions caused by photons in a semiconductors: (a) intrinsic (valence band to conduction band) transition, (b) extrinsic (deep donor to conduction band) transition, and (c) extrinsic (deep acceptor to valence band) transition.

where P_l is the absorbed optical power per unit area (in W/m^2), L is the active thickness of the photoconductive film, and Q_c is the coupling efficiency, i.e., the number of electron-hole pairs created per each absorbed photon. If we assume that the density of the light-generated carriers is much greater than the concentration of carriers in the dark, the recombination rate is given by

$$R = n/\tau_l \qquad (5\text{-}4\text{-}2)$$

where n is the concentration of electron-hole pairs and τ_l is the effective lifetime. In the steady state, $R = G_p(\omega)$. If electron photoconductivity is dominant, the photocurrent is given by

$$I_{ph} = q\mu_n nVS/L \qquad (5\text{-}4\text{-}3)$$

where μ_n is the electron mobility, V is the applied voltage, and S is the area. Let us compare this current with the maximum short-circuit light-generated current, I_{lg}, produced in a p$^+$-n solar cell. The current, I_{lg}, is determined by the number of photons absorbed per unit time:

$$I_{lg} = qQ_cP_lS/(\hbar\omega) \qquad (5\text{-}4\text{-}4)$$

It is called the *primary photocurrent*. The device photocurrent gain, G_{ph}, is defined as

$$G_{ph} = I_{ph}/I_{lg} \qquad (5\text{-}4\text{-}5)$$

Using eqs. (5-4-1) to (5-4-4) and the condition $R = G_p(\omega)$, we find that

$$G_{ph} = \tau_l/t_{tr} \qquad (5\text{-}4\text{-}6)$$

where

$$t_{tr} = L^2/\mu_n V \qquad (5\text{-}4\text{-}7)$$

is the carrier transit time. This gain may be very high (as high as a million) in short devices made from materials with long lifetimes and high values of low-field mobility. Gains as high as 1000 for silicon photoconductors and 50 to 100 for InGaAs photoconductors have been achieved (see Forrest 1986).

The detection time for photogenerated carriers is determined by the transit time of the fastest carrier. However, the photoconductivity persists as long as the lifetime of the photogenerated carriers. Hence, the response time of a photoconductor is inversely proportional to the lifetime of the carriers. Typically, it varies from 10^{-3} s to 10^{-8} s. As the gain is proportional to the effective lifetime, there is a clear trade-off between the gain and speed of a photoconductor.

Another important figure of merit for photodetectors is the Noise Equivalent Power (NEP), defined as the incident optical power required to produce a signal-to-noise ratio of 1 in a bandwidth of 1 Hertz. The related characteristic is detectivity, D, which is inversely proportional to NEP and is typically measured using a blackbody radiation source. Usually photoconductors have a fairly high noise, determined by the thermal noise of the device resistance in the dark.

An $In_{0.53}Ga_{0.47}As$ photoconductor grown on a semi-insulating InP substrate is shown in Fig. 5-4-2a (from Antresyan and Chen 1984). Such a device can be used in long-wavelength communication systems ($\lambda \approx 1.3$–1.55 μm). Its spectral response is shown in Fig. 5-4-2b. The device has a very low dark noise and high gain (see Fig. 5-4-2c). Figure 5-4-2c shows how the gain increases with the bias voltage owing to the decrease of the carrier transit time across the active region of the photoconductor.

The most widely used photodetectors are reverse-biased p-i-n diodes (see Fig. 5-4-3a). The band diagram of such a device is shown in Fig. 5-4-3b. The maximum current that can be collected by a p-i-n diode is given by

$$I_l = q \int_0^L G_p(x)\, dx \tag{5-4-8}$$

where

$$G_p = f_p \alpha \exp(-\alpha x) \tag{5-4-9}$$

is the electron-hole pair generation rate (assuming that each absorbed photon produces an electron-hole pair), and f_p is the incident photon flux (per unit area per second). Substituting eq. (5-4-9) into eq. (5-4-8) and performing the integration we find that

$$I_l = q f_p [1 - \exp(-\alpha L)] \tag{5-4-10}$$

This leads to the following estimate for the collection efficiency:

$$Q_c = (1 - R)[1 - \exp(-\alpha L)] \tag{5-4-11}$$

Here R is the reflection coefficient for incident light, so that $1 - R$ is the fraction of photons entering the semiconductor.

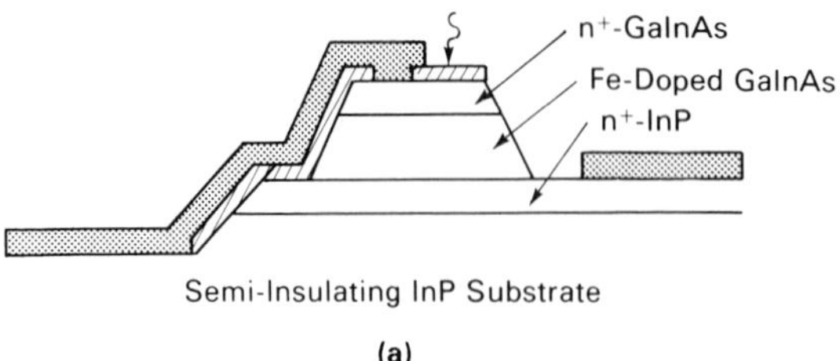

(a)

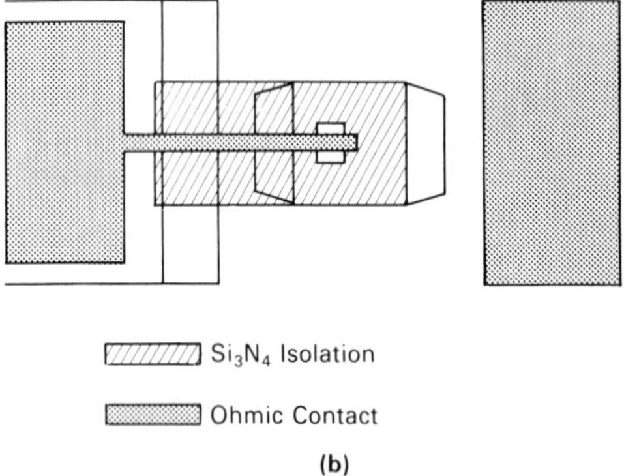

(b)

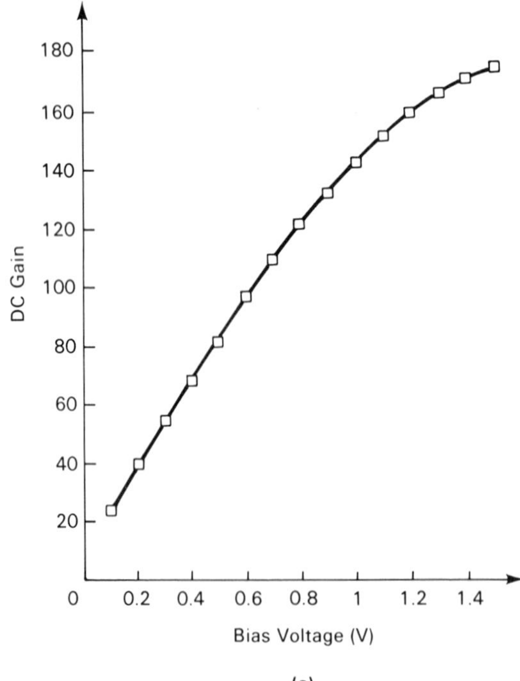

(c)

Fig. 5-4-2. $In_{0.53}Ga_{0.47}As$ photoconductor grown on semi-insulating InP substrate (from A. Antresyan and C. Y. Chen, *IEEE Trans. Electron Devices*, ED-33, no. 2, p. 188 (1984). © 1984 IEEE) (a) schematic diagram, (b) spectral response, and (c) dependence of gain on bias voltage.

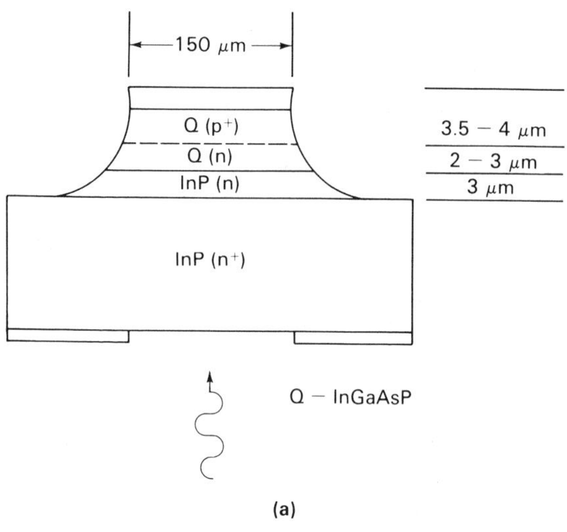

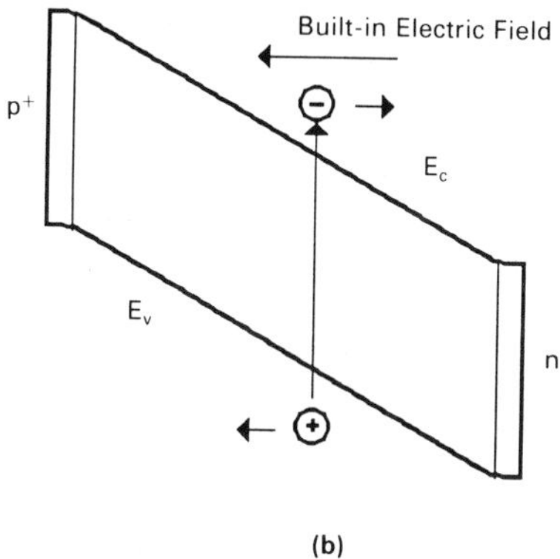

Fig. 5-4-3. (a) *p-i-n* photodiode illuminated from the substrate (after C. A. Burrus, A. G. Dentai, and T. P. Lee, *Electron. Lett.*, 15, no. 20, pp. 655–657 (1979). © 1979 IEEE). (b) Band diagram of a reverse-biased *p-i-n* diode.

As can be seen from eq. (5-4-11) longer intrinsic regions (larger L) have a higher quantum efficiency. On the other hand, the carrier transit time increases proportionally to L. Hence, longer devices have a smaller frequency at which the quantum efficiency starts to drop. However, this limiting frequency of operation may be very high (of the order of 10 GHz for typical InGaAs p-i-n photodiodes). This is much larger than the highest response frequencies of photoconductors, which are close to 100 MHz. Even higher frequencies of operation (up to 100 GHz or higher) may be achieved using metal-semiconductor (Schottky barrier) diodes. The noise in a p-i-n diode is typically several orders of magnitude smaller than in a photoconductor, in which the noise is amplified by the internal photoconductor gain.

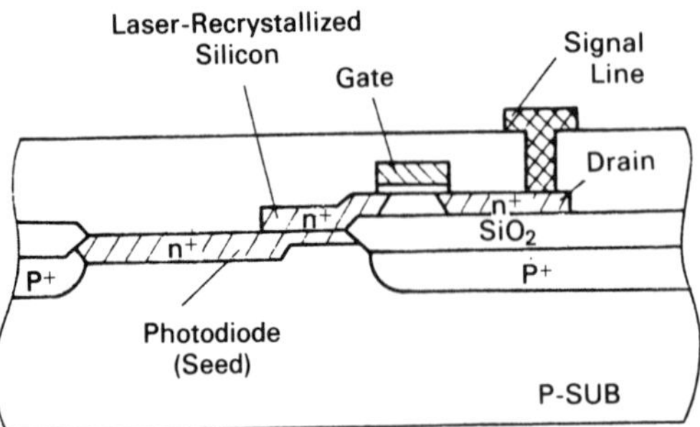

Fig. 5-4-4. Unit cell of an imager (from K. Senda, E. Fujii, Y. Hiroshima, and T. Takamura, *IEDM Technical Digest,* p. 369, Los Angeles, Calif., IEEE Publications (1986). © 1986 IEEE).

Photodiodes may be integrated with field-effect transistors to fabricate image sensors. As an example, Fig. 5-4-4 shows the unit cell of such an imager (see Senda et al. 1986). As can be seen from the figure, the integrated photodiode is connected to the source of the field-effect transistor.

An Avalanche Photodiode (APD) is also either a reverse-biased p-i-n structure or a reverse-biased Schottky barrier diode. However, the reverse voltage is chosen so high that the device experiences an avalanche breakdown. In this regime, carriers generated by light create other carriers as a result of the impact ionization. This leads to internal gain in APDs. However, a regular p-i-n or Schottky barrier diode cannot be used as an APD because in regular structures parasitic effects (such as a breakdown along the edge of the device or the formation of an electron-hole microplasma near a local defect) occur at voltages lower than the critical voltage of bulk impact ionization. This necessitates the use of device structures such as those shown in Fig. 5-4-5 in which special guard rings are used to suppress the edge breakdown.

In the bipolar junction phototransistor, carriers are generated in the base–collector junction. In an n-p-n device, electrons generated in this depletion region directly contribute to the collector current, whereas holes contribute to the base current. The ratio of the electron component of the emitter current over the hole component of the emitter current is large (and proportional to $N_{de}W/(N_{ab}X_e)$, where N_{de} is the doping concentration in the emitter region, N_{ab} is the doping

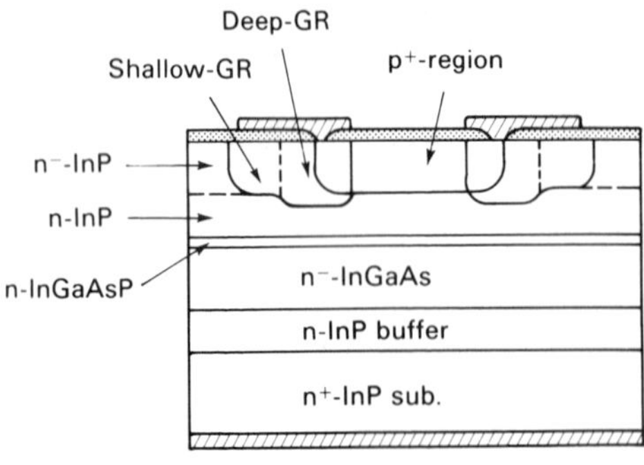

Fig. 5-4-5. An Avalanche PhotoDiode or APD with Preferential Lateral Extended Guard Ring or PLEG (from K. Taguchi, T. Torikai, Y. Sugimoto, K. Makita, H. Ishihara, S. Fujita, and K. Minemura, *IEEE Electron. Device Lett.,* EDL-7, no. 4, p. 257 (1986). © 1986 IEEE).

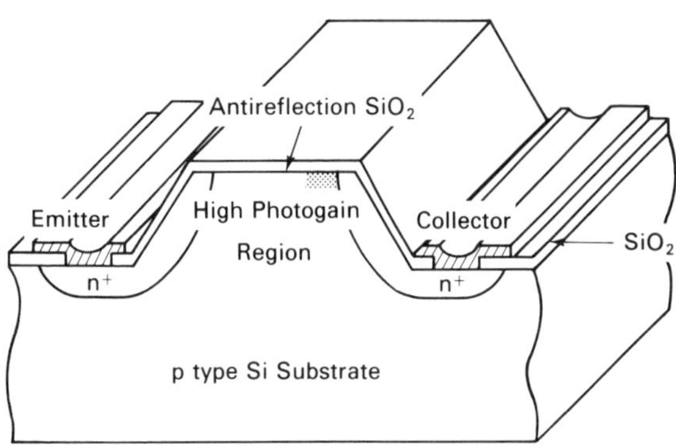

Fig. 5-4-6. Schematic structure of lateral silicon phototransistor (after S. Y. Huang, S. Esener, and S. H. Lee, *IEEE Trans. Electron Devices,* ED-33, no. 4, p. 433 (1986). © 1986 IEEE).

concentration in the base region, X_e is the emitter width, and W is the base width). This is the same current gain as in a conventional bipolar junction transistor when holes are injected into the base region by the base current in a common-emitter active forward mode. The collector current, I_{ceo} (with the base terminal floating), is given by

$$I_{ceo} = I_{ph}(1 + h_{fe}) \tag{5-4-12}$$

Here I_{ph} is the current caused by the photogenerated electrons. A schematic structure of a lateral silicon phototransistor is shown in Fig. 5-4-6. The current gain may be further enhanced using Heterojunction Bipolar Phototransistors (HBTs) (see Brian and Lee 1985).

The metal oxide semiconductor field-effect transistor (MOSFET; see Chapter 4) can also be used as a light sensor. The operation of this device is based on a change in the surface potential under illumination, leading to a modulation of the source current. The schematic diagram of such a device is shown in Fig. 5-4-7a.

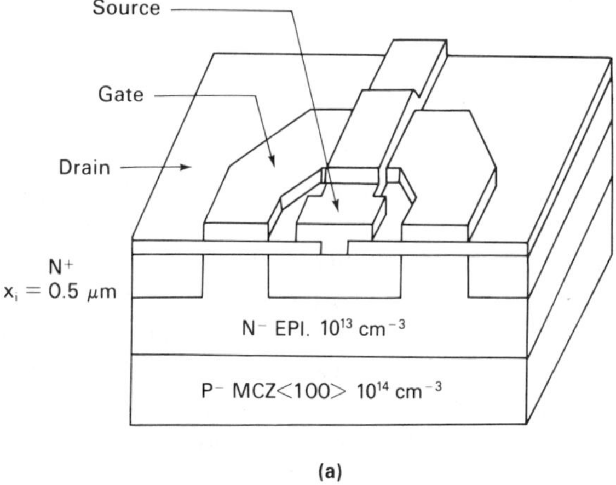

Fig. 5-4-7. Metal Oxide Semiconductor Field Effect Transistor (MOSFET) used as a light sensor (from T. Nakamura, K. Matsumoto, R. Hyuga, and A. Yusa, *IEDM Technical Digest,* p. 353, Los Angeles, Calif., IEEE Publications (1986). © 1986 IEEE). (a) schematic diagram, (b) effect of illumination on the potential energy profiles between the source and drain and between semiconductor-insulator surface and substrate, and (c) light transfer characteristics.

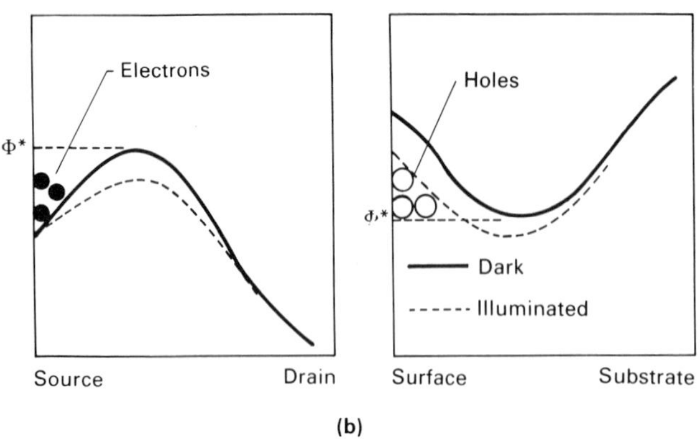

(b)

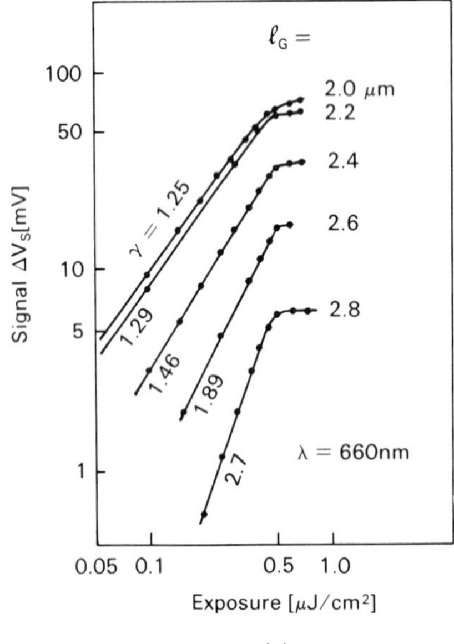

(c)

Fig. 5-4-7. Cont.

The device responds to the illumination when the light spot is focused on the space between the source and gate contacts. Figure 5-4-7b shows the effect of illumination on the potential energy profiles between the source and drain and between the semiconductor–insulator surface and the substrate. The light transfer characteristics (voltage output vs. light exposure) for different gate lengths are shown in Fig. 5-4-7c. Nakamura et al. (1986) describe a new MOS image sensor in which such MOS phototransistors are used in a 210 × 165-pixel array.

Sensitivities of optical receivers using different photodetectors are compared in Fig. 5-4-8 (from Goodfellow et al. 1985).

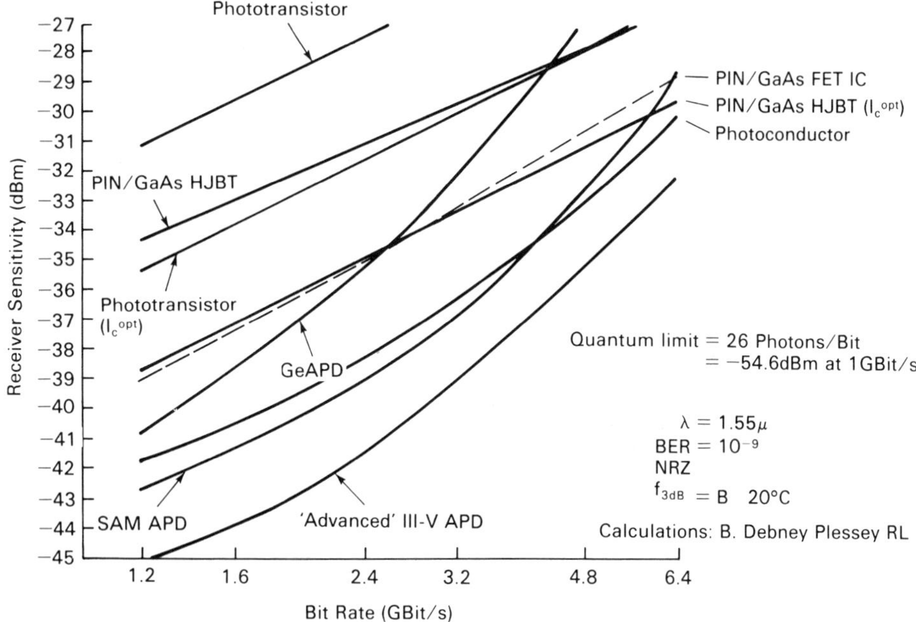

Fig. 5-4-8. Sensitivities of optical receivers as a function of data rate (from R. C. Goodfellow, B. T. Debney, G. J. Rees, and J. Buus, *IEEE Transactions Electron Devices*, ED-32, no. 12, p. 2562 (1985). © 1985 IEEE).

5-5. LIGHT EMISSION IN SEMICONDUCTORS: ELECTROLUMINESCENCE AND LIGHT-EMITTING DIODES

In solar cells and photodetectors illumination by light causes an electric current flow in a device. An opposite effect, light emission caused by electric current flowing through the sample, is called *electroluminescence* (see Round 1907).

There are several mechanisms of light excitation. For example, when a powder of semiconductor material (typically ZnS) is embedded in transparent material with a high dielectric constant (glass or plastic), sandwiched between two conducting electrodes, an *ac electroluminescence* may be caused by applying ac voltage to the contacts. The mechanism of the light emission is believed to be impact ionization at the grain boundaries caused by the acceleration of electrons in the nonuniform electric field. The light output, P_l, depends on the applied voltage amplitude, V:

$$P_l = P_{lo}(\omega) \exp[-(V_o/V)^{1/2}] \tag{5-5-1}$$

where $P_{lo}(\omega)$ is a frequency dependent constant and V_o is the characteristic voltage. This equation is satisfied for light intensities varying over eight orders of magnitude (see Diemer 1956).

The schematic structure of an ac thin film electroluminescent display is shown in Fig. 5-5-1 (from Tannas 1986). The active substance in this display (called phosphor) is manganese doped ZnS. One of the conducting contacts is made transparent (usually from indium tin oxide which is transparent and conducting material).

Electroluminescence may also be caused by an avalanche breakdown in reverse-biased p-n junction or Schottky barrier diodes as well as by tunneling in reverse- or forward-biased p-n junctions. The most widely used electroluminescence excitation technique is *injection electroluminescence,* resulting from the radiative recombination of electrons injected into the p-region and holes injected into the n-region of a p-n junction under a forward bias. This technique is used in light-emitting diodes (LEDs). Radiative recombination is much more probable in direct-gap semiconductors such as GaAs, InP, $GaAs_{1-x}P_x$ (for x less than 0.45), and $Al_xGa_{1-x}As$ (for x less than 0.45). In indirect-band-gap semiconductors, the probability of radiative recombination may be enhanced by adding special impurities to help conserve momentum during indirect transitions. Examples of such impurities are nitrogen, sulphur, and CdO or ZnO incorporated into GaP (see Groves et al. 1971). The energy gaps of different semiconductors that may be used in LEDs are compared with the spectral sensitivity of the human eye in Fig. 5-5-2.

In homojunction LEDs the radiative recombination is primarily between the band and the impurity. The response time of such an LED is determined by

$$\tau_r = 1/(B_R N_A) \tag{5-5-2}$$

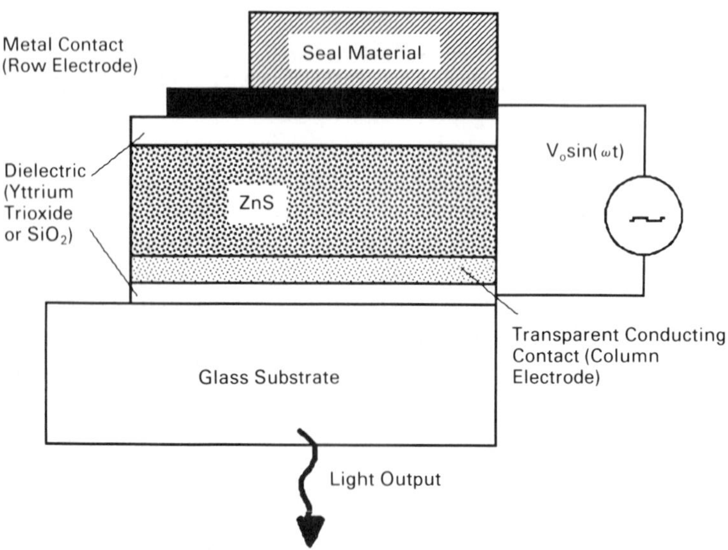

Fig. 5-5-1. Schematic structure of an ac thin film electroluminescent display (after L. E. Tanas, Jr., *Spectrum,* p. 37, October (1986).

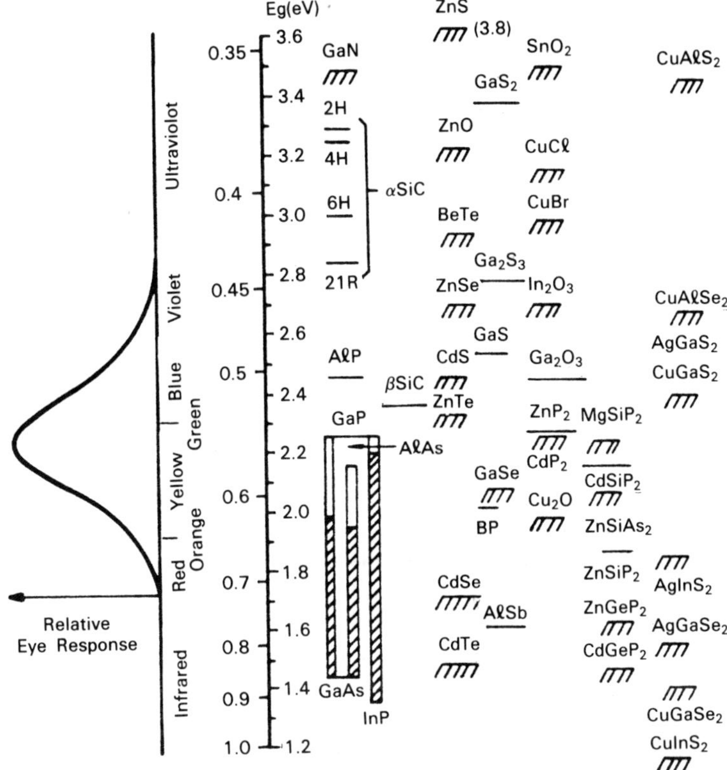

Fig. 5-5-2. Energy gaps of semiconductor materials that may be used in LEDs compared with spectral sensitivity of human eye (after A. A. Bergh and P. J. Dean, *Light-Emitting Diodes*, Clarendon, Oxford (1976); see also S. Sze, *Physics of Semiconductor Devices*, John Wiley & Sons, New York (1981).

where B_R is the radiative recombination coefficient and N_A is the impurity concentration (see Goodfellow et al. 1985). There are also nonradiative recombination processes (such as Auger recombination) that reduce the device efficiency.

GaAs homojunction LEDs operate at the wavelength $\lambda = 0.9\ \mu$m, i.e., they emit infrared light. In Fig. 5-5-3 we compare the response of the human eye and the response of a silicon phototransistor with the output characteristics of GaAs LEDs (from *Optoelectronics Data Book,* TI Instruments, 1983). Materials such as GaP with different impurities or $GaAs_{1-x}P_x$ and $Al_xGa_{1-x}As$ are used to make visible yellow, red, or green LEDs. The light intensity is determined by the current level (see Fig. 5-5-4c).

The loss coefficient of an optical silica fiber widely used in optical communications as a function of a wavelength has local minima at $\lambda = 1.55\ \mu$m (loss of 0.2 db/km) and at $\lambda = 1.3\ \mu$m (loss of 0.6 dB/km). Consequently, LEDs operating at

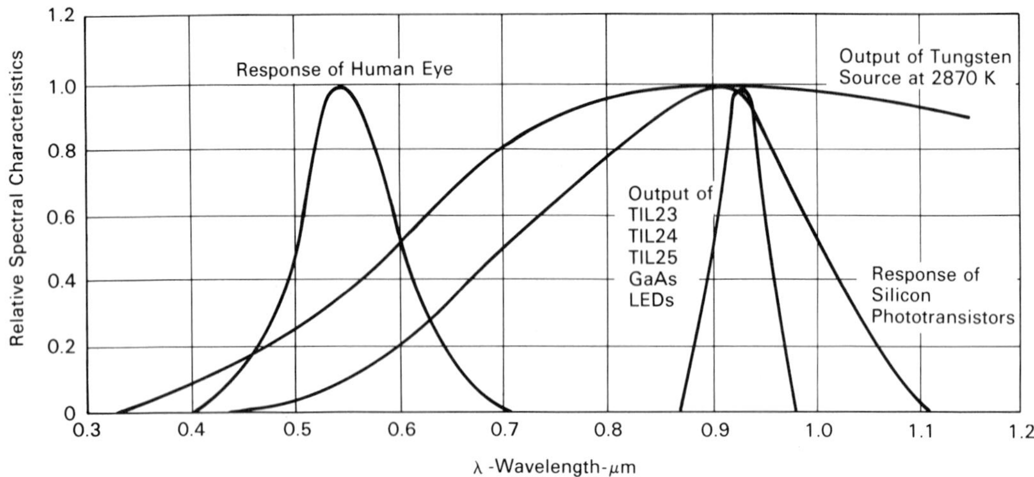

Fig. 5-5-3. Comparison between response of human eye, response of silicon phototransistors, and output characteristics of GaAs LEDs (from Optoelectronics Data Book, TI Instruments [1983]). Courtesy of Texas Instruments Incorporated.

these wavelengths are used in optical communications. Fig. 5-5-5 shows the design of a high-speed 1.3-μm-wavelength InGaAs/InP LED. This device has an optimized doping profile and small dimensions for the active region, resulting in a small parasitic capacitance and higher speed. Optical fiber transmission over a

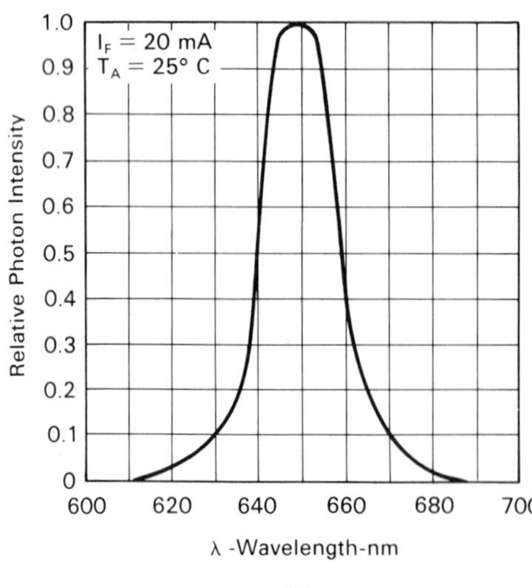

Fig. 5-5-4. $GaAs_{1-x}P_x$ light emitting diodes (from Optoelectronics Data Book, TI Instruments [1983]): (a) output characteristics, (b) LED construction (all dimensions are in mm), and (c) relative light intensity of a $GaAs_{1-x}P_x$ LED and current level (from Optoelectronics Data Book, TI Instruments [1983]). Courtesy of Texas Instruments Incorporated.

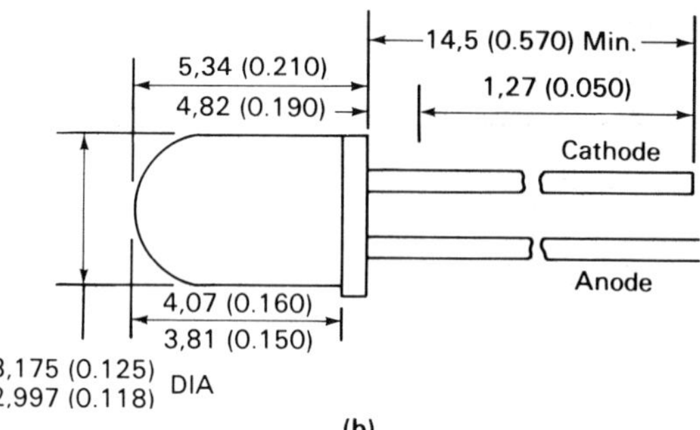

(b)

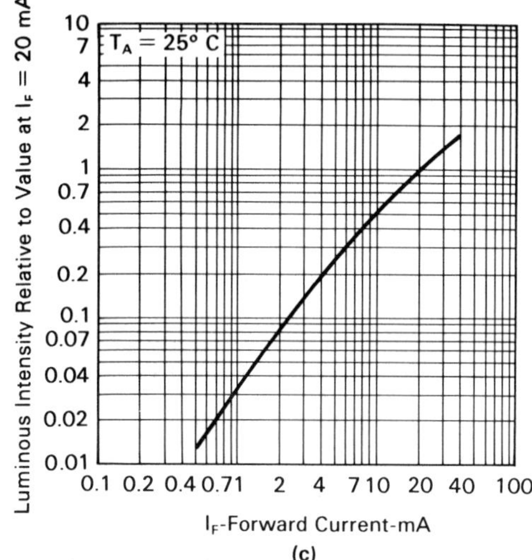

(c)

Fig. 5-5-4. Cont.

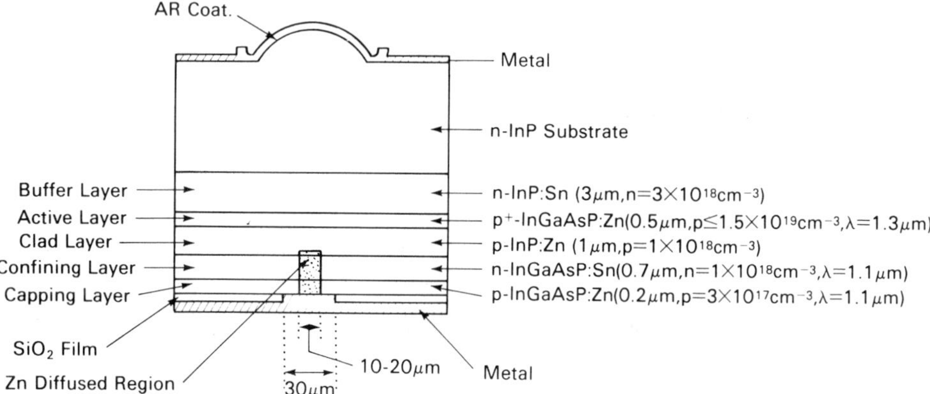

Fig. 5-5-5. Design of high speed 1.3 μm wavelength InGaAs-InP LED (after A. Suzuki, T. Uji, Y. Inomoto, J. Hayashi, Y. Isoda, and H. Nomura, *IEEE Trans. Electron Devices*, ED-32, no. 12, p. 2609 (1985). © 1985 IEEE).

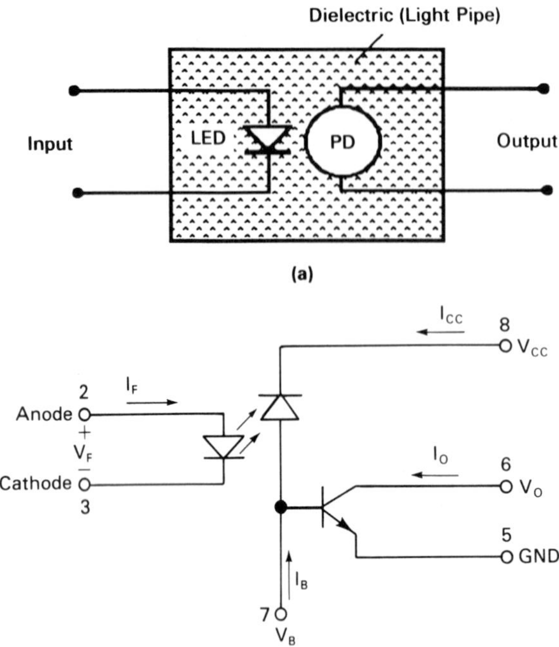

Fig. 5-5-6. (a) A schematic diagram of an optoisolator and (b) schematic of high speed optocoupler (from Optoelectronics Designer's catalog, Hewlett Packard, p. 7-5 [1985]).

500-m span has been demonstrated with this LED with rates up to 2 Gbits/s (see Suzuki et al. 1985).

LEDs are also used in alphanumeric displays and in optoisolators (also called optocouplers) that electrically isolate the input and output signal and prevent noise transfer. A schematic diagram of an optoisolator is shown in Fig. 5-5-6a. Figure 5-5-6b shows the schematic of a high-speed optocoupler that may

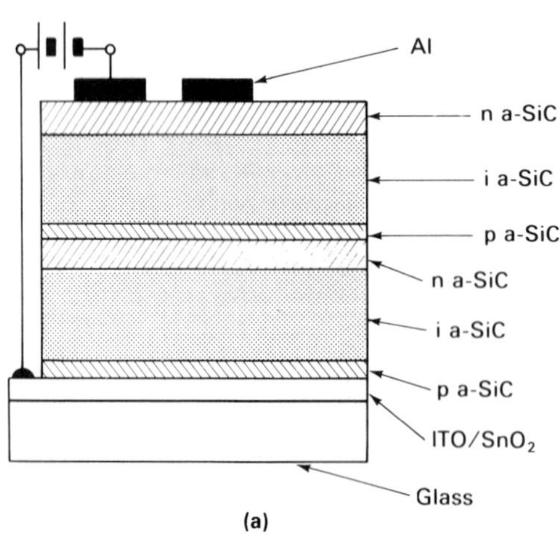

Fig. 5-5-7. Amorphous silicon carbide visible light emitting diodes (after D. Kruangam, M. Deguchi, T. Endo, W. Guang-Pu, H. Okamoto, and Y. Hamakawa, *Extended Abstracts of the 18th (International) Conference on Solid State Devices and Materials,* Tokyo, pp. 683–686 (1986). (a) schematic structure of an a-SiC tandem *p-i-n—p-i-n* LED, (b) band diagram of an a-SiC tandem *p-i-n—p-i-n* LED, (c) emission pattern of an a-SiC tandem *p-i-n—p-i-n* a-SiC LED, (The pattern is orange. The tiger area is 10 × 15 mm. The injection current is 10 mA.) (d) schematic structure of an a-SiC superlattice LED, (e) band diagram of an a-SiC superlattice LED.

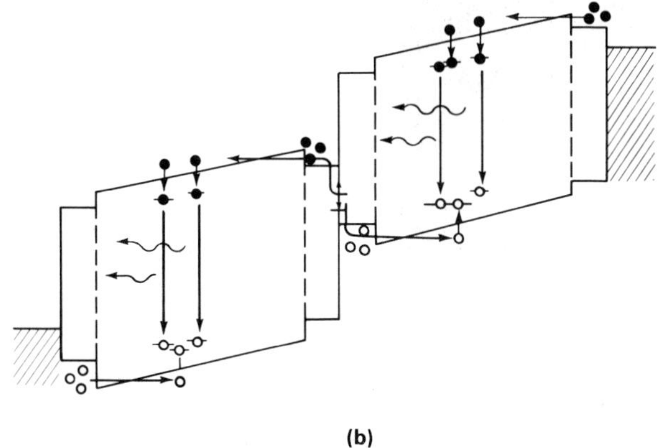

(b)

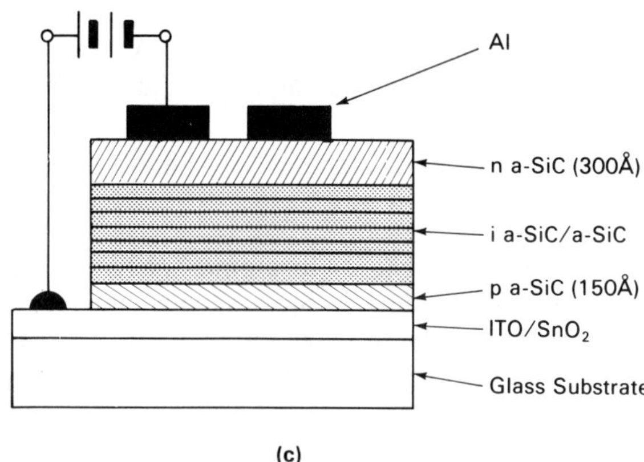

(c)

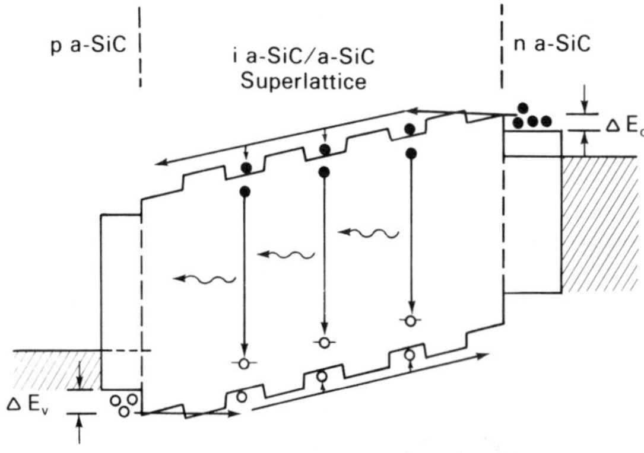

(d)

Fig. 5-5-7. Cont.

(e) **Fig. 5-5-7. Cont.**

be used in TTL or CMOS circuits (see *Optoelectronics Designer's Catalog,* pp. 7–5, Hewlett Packard, 1985).

Amorphous silicon carbide visible LEDs have recently been developed (see Kruangam et al. 1986). These devices use tandem a-SiC p-i-n layers or superlattice a-SiC p-i-n layers (see Fig. 5-5-7). An example of the emission pattern of a tandem p-i-n/p-i-n a-SiC LED is shown in Fig. 5-5-7e. The pattern is orange and the "tiger" area is 10 × 15 mm. The injection current is 10 mA. Such devices have potential application in low-cost, low-voltage, large-area flat panel displays.

5-6. SEMICONDUCTOR LASERS

Radiative transitions between two energy states in atomic systems, including semiconductors, may be spontaneous or induced. Spontaneous emission occurs without an inducement from a radiative electromagnetic field. In the particular case of a system with two states (see Fig. 5-6-1), the spontaneous emission can be described by the following equation:

$$(dN_2/dt)_{\text{spont}} = -a_{12}N_2 \tag{5-6-1}$$

where N_2 is the number of particles in state 2 (which is a higher energy state) and a_{12} is the spontaneous emission rate. The transition rate a_{12} is determined by the square of the matrix element of the particle dipole moment,

$$a_{12} = 2q^2\omega^3(x_{12}^2 + y_{12}^2 + z_{12}^2)/(3hc^3\varepsilon) \tag{5-6-2}$$

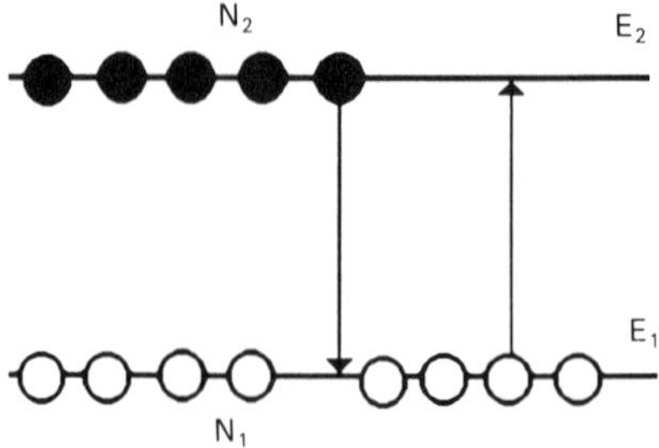

Fig. 5-6-1. Two state system.

where

$$x_{12} = \int \varphi_1^*(\mathbf{r}) x \varphi_2(\mathbf{r})\, d^3\mathbf{r} \tag{5-6-3}$$

$$y_{12} = \int \varphi_1^*(\mathbf{r}) y \varphi_2(\mathbf{r})\, d^3\mathbf{r} \tag{5-6-4}$$

$$z_{12} = \int \varphi_1^*(\mathbf{r}) z \varphi_2(\mathbf{r})\, d^3\mathbf{r} \tag{5-6-5}$$

and $\varphi_1(\mathbf{r})$ and $\varphi_2(\mathbf{r})$ are wave functions corresponding to states 1 and 2, respectively (see, for example, Yariv 1975). As can be seen from these equations, the transition probability of spontaneous emission is independent of the electromagnetic wave intensity (i.e., independent of the number of photons available). Also, for spontaneous emission the transition rate a_{21} from state 1 with low energy to state 2 with higher energy is zero.

The transition probability for induced transitions is proportional to the intensity of the electromagnetic field. The transition rates for transitions from state 2 to state 1 and from state 1 to state 2 for induced transitions are given by

$$(dN_2/dt)_{\text{induced}} = -b_{12} f(\nu) N_2 \tag{5-6-6}$$

$$(dN_1/dt)_{\text{induced}} = -b_{21} f(\nu) N_1 \tag{5-6-7}$$

where $f(\nu)$ is the radiation density at a frequency ν. (As will be shown presently, b_{12} must be equal to b_{21}.)

Under equilibrium conditions, the total transition rates from state 2 to state 1 and from state 1 to state 2 must be equal, i.e.,

$$N_2[b_{12} f(\nu) + a_{12}] = N_1 b_{21} f(\nu) \tag{5-6-8}$$

In thermal equilibrium the ratio N_2/N_1 is equal to the Boltzmann factor:

$$N_2/N_1 = \exp(-\Delta E_{21}/k_B T) \tag{5-6-9}$$

where

$$\Delta E_{21} = E_2 - E_1 = h\nu \tag{5-6-10}$$

For blackbody radiation (i.e., for the case when the electromagnetic radiation is in the thermal equilibrium with the body)

$$f(\nu) = \frac{8\pi n_r^3 h\nu^3}{c^3} \frac{1}{\exp(h\nu/k_B T) - 1} \tag{5-6-11}$$

where c is the speed of light in a vacuum and n_r is the refraction index. Substituting eqs. (5-6-9) and (5-6-11) into eq. (5-6-8), we find that

$$\frac{8\pi n_r^3 h\nu^3}{c^3} \frac{1}{\exp(h\nu/k_B T) - 1} = \frac{a_{12}}{b_{21}\exp(h\nu/k_B T) - b_{12}} \tag{5-6-12}$$

Equation (5-6-11) can be satisfied at all temperatures and frequencies if and only if

$$b_{12} = b_{21} \tag{5-6-13}$$

and

$$\frac{a_{12}}{b_{12}} = \frac{8\pi n_r^3 h\nu^3}{c^3} \tag{5-6-14}$$

(see Einstein 1917). Hence, the induced transition rate, $W_i = b_{12}f(\nu)$, is given by

$$W_i = \frac{c^3}{8\pi n_r^3 h\nu^3 t_s} f(\nu) \tag{5-6-15}$$

where $t_s = 1/a_{12}$ is the characteristic time of spontaneous transitions. This equation is applicable for electromagnetic radiation with an energy density f per unit frequency. For a monochromatic electromagnetic field, with frequency ν and energy density $f(\nu)$, this expression should be replaced by

$$W_i = \frac{c^3}{8\pi n_r^3 h\nu^3 t_s} f(\nu) g_{ls}(\nu) \tag{5-6-16}$$

(see, for example, Yariv 1985), where $g_{ls}(\nu)$ is called the lineshape function. This function describes the strength of the interaction of the monochromatic electromagnetic field with a transition at frequency ν_0. Typically, this function has a *Lorentzian shape,*

$$g_{ls}(\nu) = \frac{\Delta\nu}{2\pi[(\nu - \nu_0)^2 + (\Delta\nu/2)^2]} \tag{5-6-17}$$

where $\Delta\nu$ is called the *line width* and is related to the finite lifetime of a state, τ_l,

$$\Delta\nu = 1/(\pi\tau_l) \tag{5-6-18}$$

The lineshape function is normalized so that

$$\int_{-\infty}^{\infty} g_{ls}(\nu)\, d\nu = 1 \tag{5-6-19}$$

The energy density $f(\nu)$ can be related to the electromagnetic wave intensity I_ν:

$$f = n_r I_\nu / c \tag{5-6-20}$$

Substituting eq. (5-6-20) into eq. (5-6-16), we obtain

$$W_i = \frac{c^2 I_\nu}{8\pi n_r^2 h\nu^3 t_s} g_{ls}(\nu) \tag{5-6-21}$$

Let us now consider the propagation of an electromagnetic wave with a frequency ν in a medium with N_2 particles per unit volume in state 2 and N_1 particles per unit volume in state 1 (see Fig. 5-6-2). The change of the intensity due to the induced transitions is then given by

$$dI_\nu/dx = (N_2 - N_1) W_i h\nu \tag{5-6-22}$$

Using eq. (5-6-20), this equation may be rewritten as

$$dI_\nu/dx = \gamma I_\nu \tag{5-6-23}$$

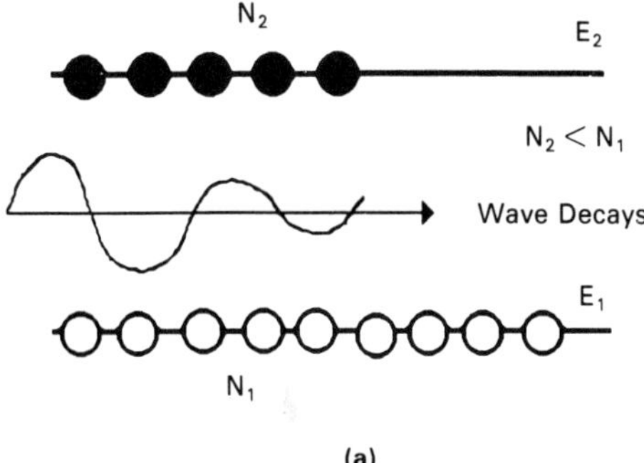

(a)

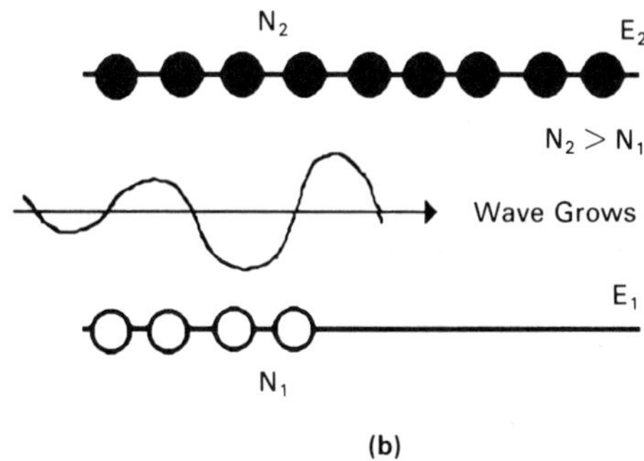

(b)

Fig. 5-6-2. Propagation of electromagnetic wave in a two level system: (a) $N_2 < N_1$ (the electromagnetic wave decays); (b) $N_2 > N_1$ (the electromagnetic wave grows).

where

$$\gamma = (N_2 - N_1)\frac{c^2 g_{ls}(\nu)}{8\pi n_r^2 \nu^2 t_s} \tag{5-6-24}$$

This means that the intensity of the propagating wave either decays or grows exponentially in the medium depending on the sign of γ:

$$I_\nu = I_\nu(x = 0)\exp(\gamma x) \tag{5-6-25}$$

The gain constant, γ, is negative when $N_2 < N_1$ and positive otherwise. Under thermal equilibrium conditions

$$N_2/N_1 = \exp(-h\nu/k_B T) \tag{5-6-26}$$

(see eq. (5-6-9)). However, if the inversion condition, $N_2 > N_1$, is somehow created, the propagating wave will be amplified. Such a nonequilibrium condition can be formally expressed by choosing a negative value for temperature in eq. (5-6-26) and is frequently referred to as a condition of *negative temperature*.

The electromagnetic field intensity, I_ν, is related to the electric field amplitude of the wave as follows:

$$I_\nu = \varepsilon F_o^2(c/n_r)/2 \tag{5-6-27}$$

The power, P_v, absorbed in a unit volume can be related to the imaginary part of the electronic susceptibility, χ'':

$$P_v = \omega\varepsilon\chi'' |F_o|^2/2 \tag{5-6-28}$$

This power is equal to dI_ν/dx, and hence, using eqs. (5-6-22) to (5-6-24) and (5-6-27) and (5-6-28), we obtain

$$\chi'' = \frac{(N_1 - N_2)\lambda^3}{8\pi^3 n_r t_s \Delta\nu[1 + 4(\nu - \nu_o)^2/\Delta\nu^2]} \tag{5-6-29}$$

The gain constant is related to χ'' as follows:

$$\gamma = -(k/n_r^2)\chi'' \tag{5-6-30}$$

where the wave vector

$$k = n_r\omega/c = 2\pi n_r/\lambda \tag{5-6-31}$$

When an amplifying medium is placed between two mirrors (the Fabry-Perot etalon) and the amplification in the medium exceeds the losses caused by an imperfect reflection from the mirrors, the generation of a coherent light beam can take place in the absence of an incident wave. The light emission is coherent when photons generated as a result of induced transitions are in phase. The light emission is monochromatic because the photon frequency, ν, is equal to $(E_2 - E_1)/h$. This idea was first proposed by Schawlow and Townes (1958) and led to the development of lasers (which is an acronym for Light Amplification by Stimulated Emission of Radiation).

Bernard and Duraflourg (1961) and Dumke (1962) showed that the coherent emission of light (lasing) may be achieved in semiconductors. This was first achieved by Hall et al. (1962), Nathan et al. (1962), and Quist et al. (1962), who observed the pulsed coherent emission of infrared light with a wavelength of 0.84 μm from GaAs p-n junctions at cryogenic temperatures. Holonyak and Bevacqua (1962) were first to obtain a coherent emission of visible light from $GaAs_{1-x}P_x$ junctions. Kroemer (1963) and Alferov and Kazarinov (1963) (see also Alferov 1967) invented a double heterostructure semiconductor laser that made it possible to obtain the continuous operation of a semiconductor-junction laser at room temperature (see Hayshi et al. 1970).

In a p-n junction semiconductor laser, the inversion is caused by injection of holes into the n-type region and electrons into the p-type region. The line width parameter, $\Delta\nu$, in semiconductors is determined by the effective relaxation time, τ_p:

$$\Delta\nu \approx 1/(\pi\tau_p) \tag{5-6-32}$$

(Typically, τ_p is of the order of a picosecond.) Hence, eq. (5-6-24) can be rewritten as

$$\gamma(\omega) = \frac{(N_2 - N_1)\lambda^2\tau_p}{4\pi n_r^2 t_s[1 + (\omega - \omega_o)^2\tau_p^2]} \tag{5-6-33}$$

(assuming the Lorentzian shape of the line).

In a semiconductor the transition rate, R_{cv}, from an occupied state, E_2, in the conduction band to an empty state, E_1, in the valence band is proportional to the Fermi-Dirac occupation (distribution) functions,

$$R_{cv} \sim f_c(E_2)[1 - f_v(E_1)] \tag{5-6-34}$$

where the distribution functions

$$f_c(E) = 1/\{\exp[(E - E_{Fn})/k_B T] + 1\} \tag{5-6-35}$$

$$f_v(E) = 1/\{\exp[(E - E_{Fp})/k_B T] + 1\} \tag{5-6-36}$$

and E_{Fn} and E_{Fp} are electron and hole quasi-Fermi levels, respectively. The inversion density, $d(N_2 - N_1)$, for the states with wave vectors in the range between k and $k + dk$ is then given by

$$d(N_2 - N_1) = g(k)\{f_c(E_2)[1 - f_v(E_1)] - f_v(E_1)[1 - f_c(E_2)]\}\, dk/V \tag{5-6-37}$$

The second term in the braces on the right-hand side of eq. (5-6-37) is proportional to the transition rate from state 1 to state 2. The function

$$g(k) = Vk^2/\pi^2 \tag{5-6-38}$$

is the number of states per unit volume in k space for a semiconductor crystal with volume V. This expression for g may be derived as follows: We assume that the crystal dimensions are L_x, L_y, and L_z. The values of the wave vector components k_x, k_y, and k_z are quantized so that L_x, L_y, and L_z are equal to the integer number of wavelengths,

$$L_x = 2\pi n_x/k_x \tag{5-6-39}$$

$$L_y = 2\pi n_y/k_y \tag{5-6-40}$$

$$L_z = 2\pi n_z/k_z \tag{5-6-41}$$

where n_x, n_y, and $n_z = 1, 2, 3, \ldots$ Hence, the distance between two nearest allowed values of k_x in k space is $\Delta k_x = 2\pi/L_x$. Similarly, $\Delta k_y = 2\pi/L_y$, and $\Delta k_z = 2\pi/L_z$. Hence, the volume occupied by one state in k space is equal to

$$\Delta k_x \Delta k_y \Delta k_z = (2\pi/L_x)(2\pi/L_y)(2\pi/L_z) = 8\pi^3/V \tag{5-6-42}$$

where V is the crystal volume (see Section 1-5). The volume of the spherical shell in the k space located between k and $k + \Delta k$ is equal to $4\pi k^2\, dk$. The number of states in this volume is given by

$$g(k)\, dk = 2 \times 4\pi k^2\, dk/(8\pi^3/V) = Vk^2\, dk/\pi^2 \tag{5-6-43}$$

(compare with eq. (5-6-38) and see Problem 5-6-1 and Section 1-6.) (The factor of 2 on the right-hand side of eq. (5-6-43) accounts for two states, with two different values of spin having the same vector **k**.) Equation (5-6-37) may be rewritten as

$$d(N_2 - N_1) = g(k)[f_c(E_2) - f_v(E_1)]\, dk/V \tag{5-6-44}$$

The wave vector k is related to the transition frequency, ω, as follows:

$$\hbar\omega = E_2 - E_1 = E_g + \hbar^2k^2/(2m_n) + \hbar^2k^2/(2m_p) \tag{5-6-45}$$

where E_g is the energy gap and m_n and m_p are electron and hole effective masses, respectively. Here we assume spherical and parabolic bands (see Section 1-6).

Using eq. (5-6-33), we can now obtain the expression for an incremental gain constant related to the transitions associated with the wave vectors k between k and $k + dk$:

$$d\gamma(k) = \frac{k^2\lambda^2\tau_p[f_c(E_2) - f_v(E_1)]\,dk}{4n_r^2 t_s \pi^3[1 + (\omega - \omega_o)^2\tau_p^2]} \tag{5-6-46}$$

The gain constant, γ, may be found by integrating this expression over the first Brillouin zone. Such an integration is easier to carry out by changing the integration variable to ω. This yields

$$\gamma(\omega_o) = \int_0^\infty \frac{g_{\text{eff}}(\omega)\lambda^2\tau_p[f_c(\omega) - f_v(\omega)]\,d\omega}{4n_r^2 t_s \pi[1 + (\omega - \omega_o)^2\tau_p^2]} \tag{5-6-47}$$

Here

$$g_{\text{eff}}(\omega) = \frac{1}{2\pi^2}\left[\frac{2m_n m_p}{\hbar(m_n + m_p)}\right]^{3/2}(\omega - E_g/\hbar)^{1/2} \tag{5-6-48}$$

The spontaneous transition rate, t_s, may be taken equal to the recombination lifetime, τ_l.

The integration limit in eq. (5-6-47) can be taken as infinity because only the values of ω close to ω_o contribute to the integral as a consequence of the lineshape function having a narrow peak. Moreover, it is a good approximation to substitute for this function a delta function, $\delta(\omega - \omega_o)$. This allows us to evaluate the integral in eq. (5-6-47), leading to the following expression:

$$\gamma(\omega_o) = \frac{\lambda_o^2}{8n_r^2 t_s \pi^2}\left[\frac{2m_n m_p}{\hbar(m_n + m_p)}\right]^{3/2}(\omega_o - E_g/\hbar)^{1/2}[f_c(\omega_o) - f_v(\omega_o)] \tag{5-6-49}$$

where $\lambda_o = c/\nu_o$. Hence, the condition $\gamma(\omega_o) > 0$, which is necessary for amplification, requires

$$f_c(\omega_o) > f_v(\omega_o) \tag{5-6-50}$$

or

$$1/\{\exp[(E_c - E_{Fn})/k_BT] + 1\} > 1/\{\exp[(E_v - E_{Fp})/k_BT] + 1\} \tag{5-6-51}$$

which in turn requires

$$\hbar\omega < E_{Fn} - E_{Fp} \tag{5-6-52}$$

This equation was first obtained by Bernard and Durafourg (1961).

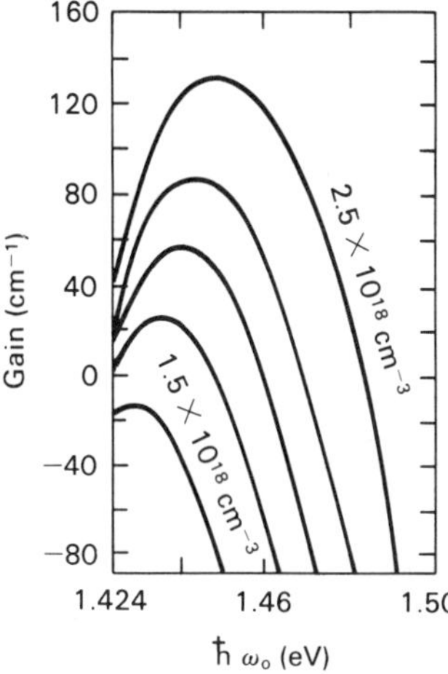

Fig. 5-6-3. Dependence of optical gain in GaAs on frequency and injected carrier density (from K. Vahala, C. Chiu, S. Margalit, and A. Yariv, *Appl. Phys. Lett.*, 42, p. 631 (1983).

Using eq. (5-6-47), Vahala et al. (1983) calculated the dependence of optical gain in a GaAs p-n junction on frequency and injected carrier density (see Fig. 5-6-3). The peak value of the gain for GaAs can be approximated as

$$\gamma_{max} \approx B_g(N_{inj} - N_{th}) \tag{5-6-53}$$

where $B_g \approx 1.5 \times 10^{-16}$ cm^2 at 300 K and $B_g \approx 5 \times 10^{-16}$ cm^3 at 77 K, N_{inj} is the injected carrier density in cm^{-3}, and $N_{th} \approx 1.3 \times 10^{18}$ cm^{-3} (see Yariv 1985, p. 479).

Most semiconductor lasers use Al_xGa_{1-x} As/GaAs (or Al_xGa_{1-x} As/Al_yGa_{1-y} As) and $In_xGa_{1-x}As_yP_{1-y}$/InP heterojunction diodes. In a heterojunction laser, electrons are injected into the narrow active region of the material, with the smaller band gap sandwiched between the p and n regions of the material with the wider band gap. The energy gaps of the active region and confining regions are controlled by controlling the composition (i.e., the atomic percentage of Al in $Al_xGa_{1-x}As$ and the atomic percentages of In and P in $In_xGa_{1-x}As_yP_{1-y}$). The energy gap for GaAs is 1.424 eV and 2.168 eV for AlAs. For $Al_xGa_{1-x}As$

$$E_g = 1.424 + 1.247x \qquad x < 0.45 \tag{5-6-54}$$

and

$$E_g = 1.900 + 0.125x + 0.143x^2 \qquad x \geq 0.45 \tag{5-6-55}$$

(see Adachi 1985 and Appendix 9). For $x < 0.45$, $Al_xGa_{1-x}As$ is a direct-gap material; for $x \geq 0.45$, $Al_xGa_{1-x}As$ is an indirect-gap material. The wavelength corresponding to the energy gap of $Al_xGa_{1-x}As$ ($\lambda = 1.24/E_g$) is shown in Fig. 5-6-4 as a function of x for $0 < x < 0.45$.

A schematic diagram of an AlGaAs/GaAs laser is shown in Fig. 5-6-5. Composition and refraction index profiles for a Double Heterostructure (DH)

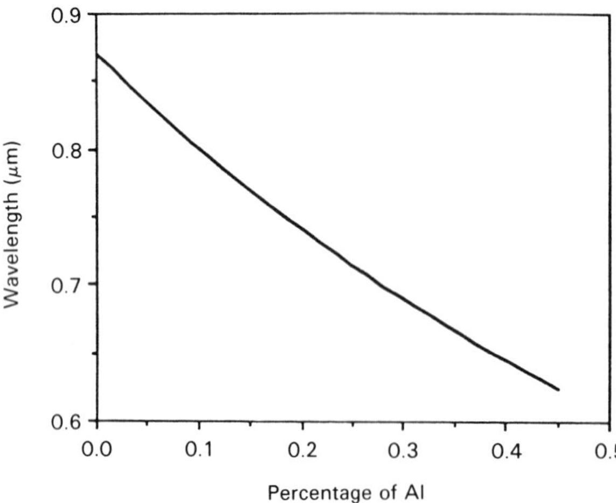

Fig. 5-6-4. The wavelength corresponding to the energy gap of $Al_xGa_{1-x}As$ ($\lambda = 1.24/E_g$) as a function of composition x for $0 < x < 0.45$.

AlGaAs/GaAs laser are shown in Fig. 5-6-6. The variation of the refraction index provides additional confinement for emitted photons.

The gain, G_l, for a given photon mode can be evaluated as follows (see Yariv 1985):

$$G_l = \frac{\text{generated power per unit length}}{\text{power carried by the beam}} - \text{additional losses} \qquad (5\text{-}6\text{-}56)$$

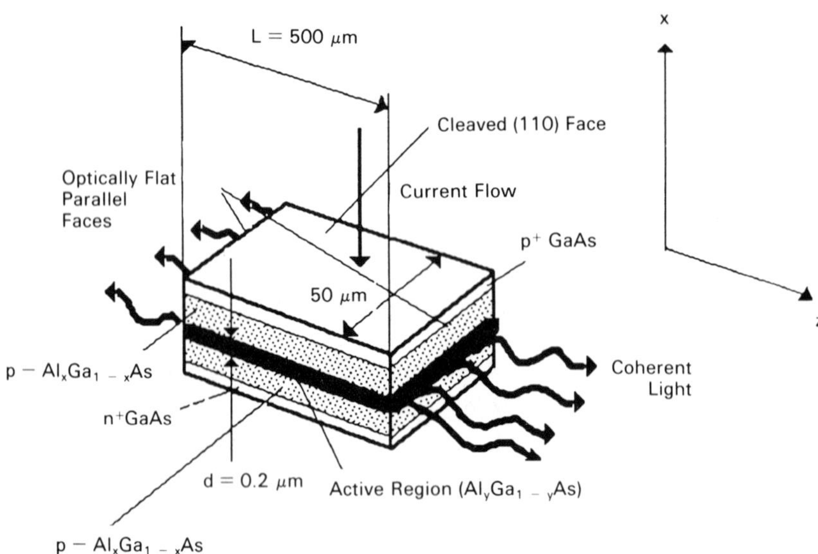

Fig. 5-6-5. Schematic diagram of a double heterostructure AlGaAs-GaAs laser. The dimensions shown are approximate but typical values. Cleaved (110) faces form a Fabry-Perot etalon.

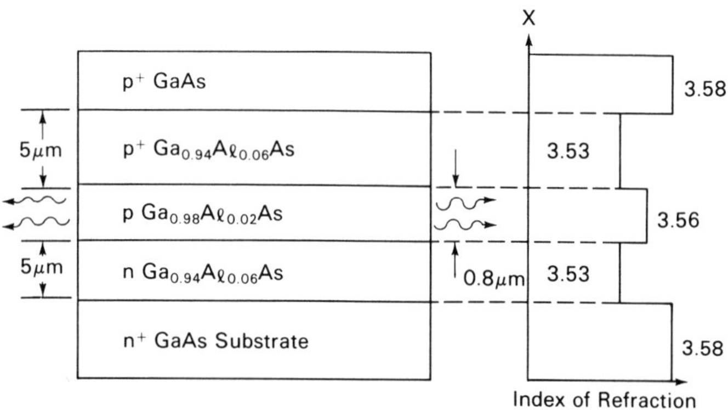

Fig. 5-6-6. Composition and refraction index profiles for a double heterostructure AlGaAs-GaAs laser (from R. G. Hunsperger, *Integrated Optics: Theory and Technology,* 2d ed., Springer-Verlag (1984).

or

$$G_l = \frac{-\alpha_n \int_{-\infty}^{-d/2} |F|^2\, dx + \gamma \int_{-d/2}^{d/2} |F|^2\, dx - \alpha_p \int_{d/2}^{\infty} |F|^2\, dx}{\int_{-\infty}^{\infty} |F|^2\, dx} - \alpha_R - \alpha_s \tag{5-6-57}$$

where d is the thickness of the active region, α_n and α_p are loss constants in the n and p regions, respectively, γ is the gain constant in the active region, F is the amplitude of the electromagnetic field, α_R represents the reflection loss, and α_s represents losses caused by scattering at imperfections (primarily at the heterojunction interfaces); the coordinates x and z are shown in Fig. 5-6-5. This expression can be rewritten as

$$G_l = \gamma\Gamma_a - \alpha_n\Gamma_n - \alpha_p\Gamma_p - \alpha_R - \alpha_s \tag{5-6-58}$$

where Γ_a, Γ_n, and Γ_p are the ratios of the integrals defined using eq. (5-6-57) and

$$\Gamma_a + \Gamma_n + \Gamma_p = 1 \tag{5-6-59}$$

Equation (5-6-58) leads to the following threshold condition for lasing (gain = 0):

$$\gamma\Gamma_a = \alpha_n\Gamma_n + \alpha_p\Gamma_p + \alpha_R + \alpha_s \tag{5-6-60}$$

or

$$\gamma\Gamma_a = \alpha_R + \alpha_{loss} \tag{5-6-61}$$

where

$$\alpha_{loss} = \alpha_n\Gamma_n + \alpha_n\Gamma_n + \alpha_s \tag{5-6-62}$$

The expression for the reflection loss, α_R, can be derived by considering the power flow in a Fabry-Perot etalon during lasing (see Fig. 5-6-7). The power, as a

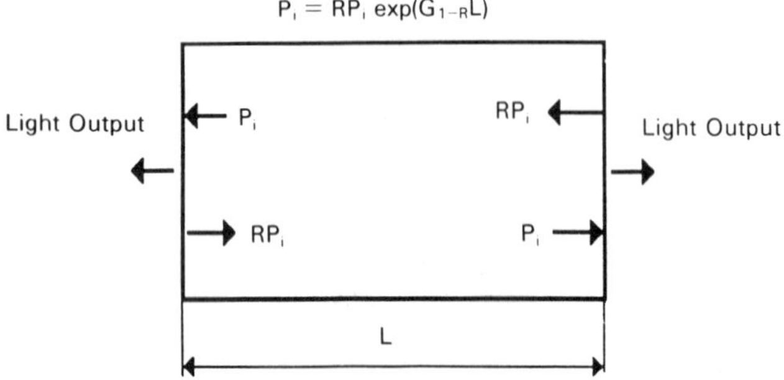

Fig. 5-6-7. Incident and output power in a Fabri-Perot etalon.

function of distance, is given by

$$P = RP_i \exp(G_{l-R}x) \tag{5-6-63}$$

where P_i is the power incident on each of two internal reflecting faces of the laser, R is the reflection coefficient, and G_{l-R} is the gain that does not include the losses due to reflection, i.e.,

$$G_{l-R} = G_l + \alpha_R \tag{5-6-64}$$

In the steady state the total loop gain should be equal to unity, i.e.,

$$P_i = RP_i \exp(G_{l-R}L) \tag{5-6-65}$$

Hence,

$$1 = R \exp(G_{l-R}L) \tag{5-6-66}$$

i.e.,

$$G_{l-R} + (\ln R)/L = 0 \tag{5-6-67}$$

Comparing this equation with the requirement of the total zero loop gain for the steady-state oscillations,

$$G_{l-R} - \alpha_R = 0 \tag{5-6-68}$$

we find that

$$\alpha_R = -(\ln R)/L \tag{5-6-69}$$

Usually, this term represents the largest loss for double heterostructure (DH) lasers. For the GaAs–air interface, the reflection coefficient is

$$R = (n_r - 1)^2/(n_r + 1)^2 = 0.31 \tag{5-6-70}$$

($n_r = 3.5$), and hence, this loss is $1.17/L$ (cm^{-1}). Other losses (such as losses related to nonradiative recombination) may be roughly estimated as α_{loss} = 10 cm^{-1} and $\Gamma_\alpha \sim 1$ (see Yariv 1985). This allows us to estimate the required gain constant, γ, and (using eq. (5-6-53)) the threshold concentration of electrons and holes injected into the active region.

The concentration of electrons and holes injected into the active region, N_{inj}, is related to the diode current by the requirement that in the steady state this carrier density must be equal to the electron-hole recombination rate,

$$j = qN_{inj}d/\tau_l \tag{5-6-71}$$

where τ_l is the effective lifetime. Using eqs. (5-6-61) and (5-6-53), we find the following equation for the threshold current density:

$$j_{th} = q\{N_{th} + [\alpha_{loss} - (\ln R)/L]/B_g\}d/\tau_l \tag{5-6-72}$$

Assuming $\alpha_{loss} = 10\ cm^{-1}$, $\tau_l = 5$ ns, $B_g \approx 1.5 \times 10^{-16}\ cm^2$, $-(\ln R)/L = 1.17/L$ (cm^{-1}), and $L = 500\ \mu m$, we obtain

$$j_{th}/d = 4830\ A/(cm^2 \mu m) \tag{5-6-73}$$

which is in reasonable agreement with the results shown in Fig. 5-6-8.

A typical dependence of light output power for an AlGaAs/GaAs laser diode on the device current is shown in Fig. 5-6-9.

AlGaAs/GaAs lasers emit light with wavelengths of 0.8 to 0.9 μm. As was mentioned in Section 5-5, silica optical fibers have their smallest loss at approxi-

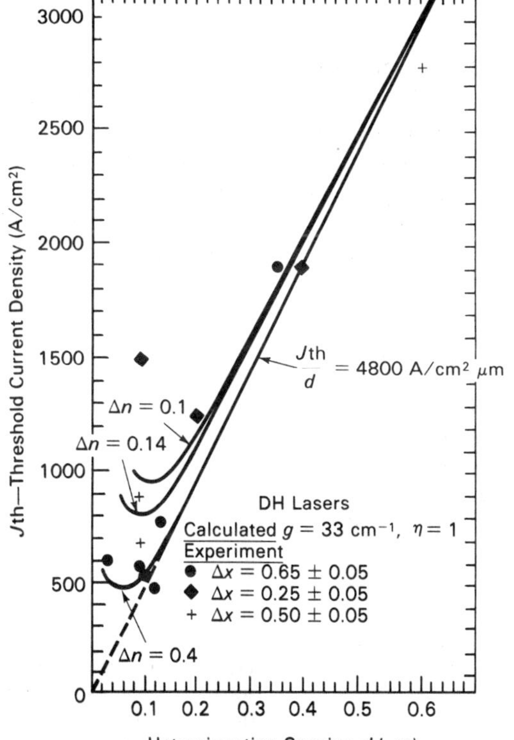

Fig. 5-6-8. Dependence of threshold current density on active layer thickness for double heterostructure lasers (after H. Kressel and J. K. Butler, *Semiconductor and Heterojunction LEDs*, Academic Press, New York (1977). L = 500 μm.

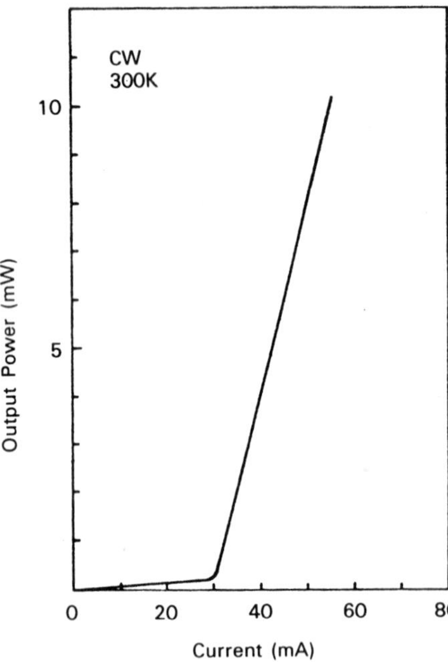

Fig. 5-6-9. Light output power an AlGaAs-GaAs laser diode vs. current (from T. Sugino, A. Yoshikawa, A. Yamamoto, M. Hirose, G. Kano, and I. Teramoto, *IEDM Technical Digest*, p. 618, Los Angeles, Calif., IEEE Publications (1986). © 1986 by IEEE). The threshold current for this particular laser is approximately 30 mA.

mately 1.55 μm (see Fig. 5-6-10a). Another important characteristic of an optical fiber is *chromatic dispersion,* which is a measure of the dependence of the optical signal propagation velocity in a fiber on the light wavelength (see Fig. 5-6-10b).

Fig. 5-6-11 shows a qualitative optical emission spectrum of a laser diode. As can be seen from the figure, several modes can be present (with one mode having the largest intensity). Because of chromatic dispersion, the optical signals carried by two modes with a wavelength difference of just 1 nm (10 Å) will travel

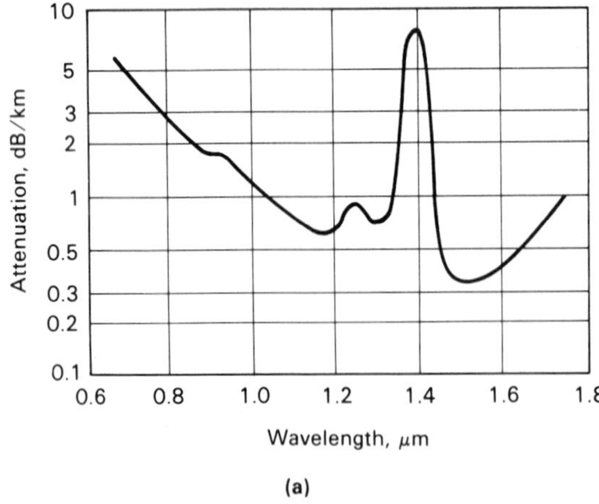

Fig. 5-6-10a. Attenuation loss vs. wavelength for a silica optical fiber (from T. Bell, *Spectrum*, pp. 38–45, December (1983). © 1983 by IEEE).

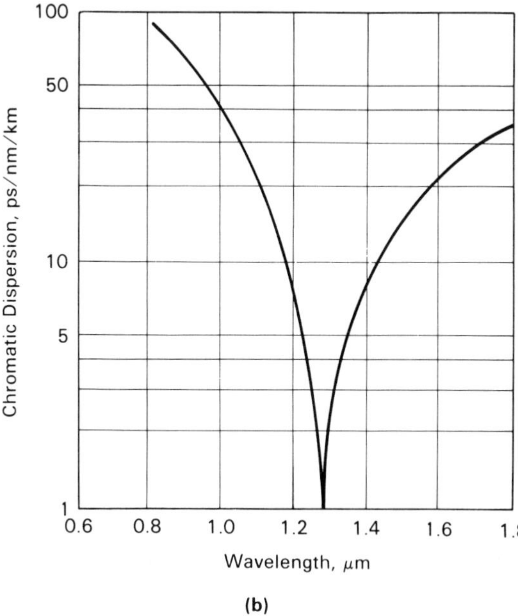

Fig. 5-6-10b. Chromatic dispersion vs. wavelength for a silica optical fiber (from T. Bell, *Spectrum*, pp. 38–45, December (1983). © 1983 by IEEE).

through a 1-km fiber with a time difference of about 100 ps for $\lambda \approx 0.8$ μm (see Fig. 5-6-10b). In addition, the two signals will have a time difference related to the *modal dispersion,* which is related to the different ray trajectories for light rays with different wavelengths reflected from the fiber core. Hence, first, single-mode operation is highly desirable for optical communications using optical fibers, and second, wavelengths of $\lambda = 1.3$ μm (where the chromatic dispersion is nearly zero) or 1.55 μm (where the loss in a silica fiber is the smallest) are more suitable for optical communications.

InGaAsP/InP lasers operate in the range between 1.3 and 1.6 μm, depending on the composition of the active layer, and have become the lasers of choice for long-distance fiber-optic communication links. In addition, different approaches

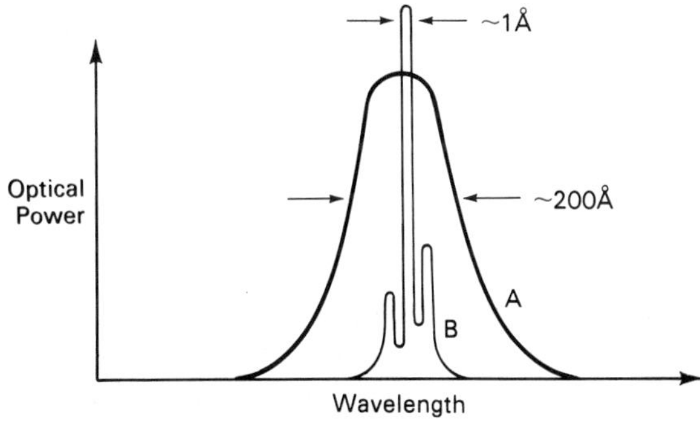

Fig. 5-6-11. Qualitative optical emission spectrum of laser diode for two different current levels (from Hunsperger [1984]).

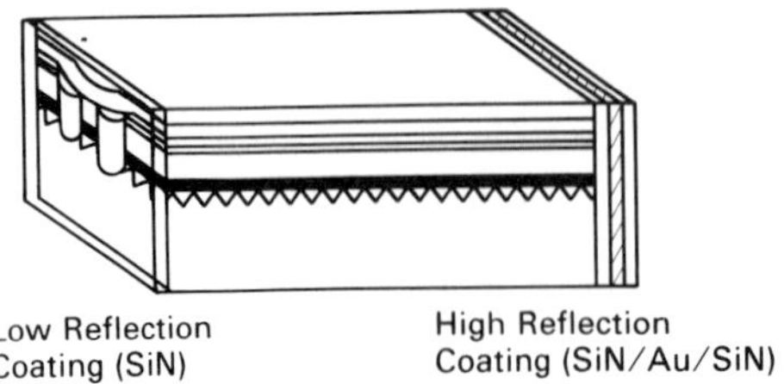

Fig. 5-6-12. Distributive FeedBack (DFB) InGaAs-InP laser diode (from K. Kobayashi and I. Mito, *IEEE Trans. Electron Devices*, ED-32, no. 12, p. 2594 (1985). © 1985 by IEEE).

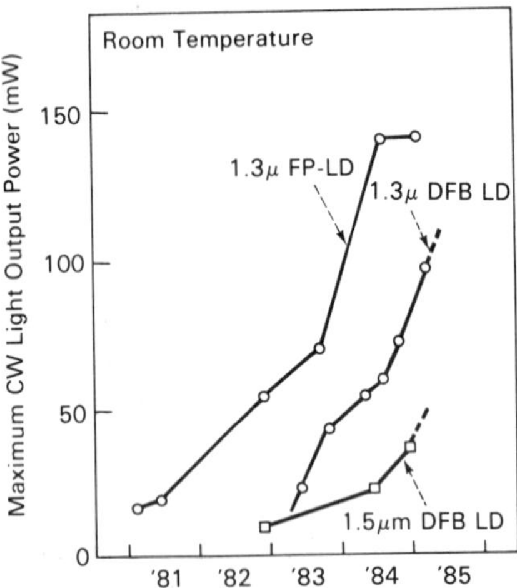

Fig. 5-6-13. The progress in the development of long wave lasers (from K. Kobayashi and I. Mito, *IEEE Trans. Electron Devices*, ED-32, no. 12, p. 2594 (1985). © 1985 by IEEE).

have been proposed to realize single-mode operation (see Bell 1983). One widely used approach is based on using Distributive FeedBack (DFB) laser diodes, in which distributive Bragg reflectors are used instead of a Fabry-Perot etalon to provide reflections. Such reflectors act as frequency-selective mirrors because the constructive and destructive diffraction interference patterns are extremely sensitive to the wavelength of light. A corrugated interface is used to implement such diffraction gratings (see Fig. 5-6-12). The corrugation period of 2000 Å for 1.3 μm devices and 2300 Å for 1.55 μm devices with corrugation depth of approximately 1000 Å was used by Kobayashi and Mito (1985) to fabricate DFB InGaAsP/InP laser diodes. The progress in the development of such devices is illustrated by Fig. 5-6-13.

5-7. INTEGRATED OPTOELECTRONICS

Monolithic integration of lasers, photodetectors, transistors, and other elements should allow us to achieve for optoelectronic systems what conventional integrated circuits achieved for electronics: higher speed, better performance, higher reliability, and much lower cost compared with circuits consisting of discrete components.

In conventional very-large-scale integrated circuits (VLSI), in which delays caused by interconnect wires become very significant, the replacement of these connections by optical links may improve performance. Optical links have the advantages of ultra high speed, low power dissipation, and immunity to outside interference. Optical fibers have a very large bandwidth (of many gigabits). Hundreds of signals can be transmitted through the same optical fiber. On the other hand, it is more difficult to localize light beams than electrical signals and prevent light leakage from an optical waveguide on a chip.

In fiber-optic communication systems, the overall speed is often limited not by the laser response but rather by the electronics available for the modulation of the laser light. Here again, integration of optical and electronic components may help to increase the overall system speed, achieving multigigabit data links. Very simple circuits (such as time-division multiplexers) may be implemented with 20 to 30 transistors. However, more complex functions require as many as 2000 transistors. Hence, both applications—optical links in VLSI and multigigabit fiber-optic communication systems—will require the integration of optical elements and VLSI circuits. In the longer run, integration of electronic and optical elements may make optical computing a reality, leading to new generations of supercomputers (see Bell 1986).

Compound semiconductors seem to be a natural material choice for the integration of electronic and optical elements. GaAs/AlGaAs or InGaAsP/InP structures can be used in lasers, photodetectors, and high-speed transistors. Compound semiconductor integrated circuits can be used by themselves or for the implementation of small transceiver chips, connecting silicon VLSI chips or circuit boards with silicon chips (see Fig. 5-7-1a). A schematic diagram of a GaAs transceiver chip is shown in Fig. 5-7-1b.

New opportunities for the integration of compound semiconductor transceivers and silicon ICs are provided by recently emerged Molecular Beam Epi-

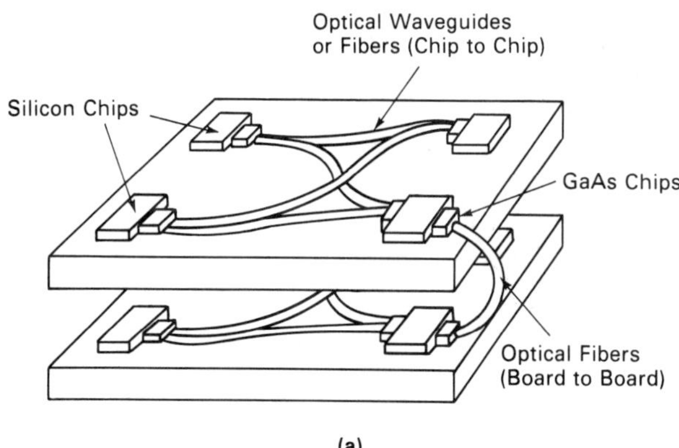

Fig. 5-7-1. Optical interconnect between VLSI silicon chips and circuit boards using GaAs transceiver chips (from L. D. Hutchenson, P. Haugen, and A. Husain, *Spectrum,* pp. 30–35, March (1987). © 1987 by IEEE. (a) Schematic diagram of optical interconnects. (b) Schematic diagram of a GaAs transceiver chip. (c) Optical output couplers that may be used in such a chip, etched-mirror reflector output coupler and grating output coupler.

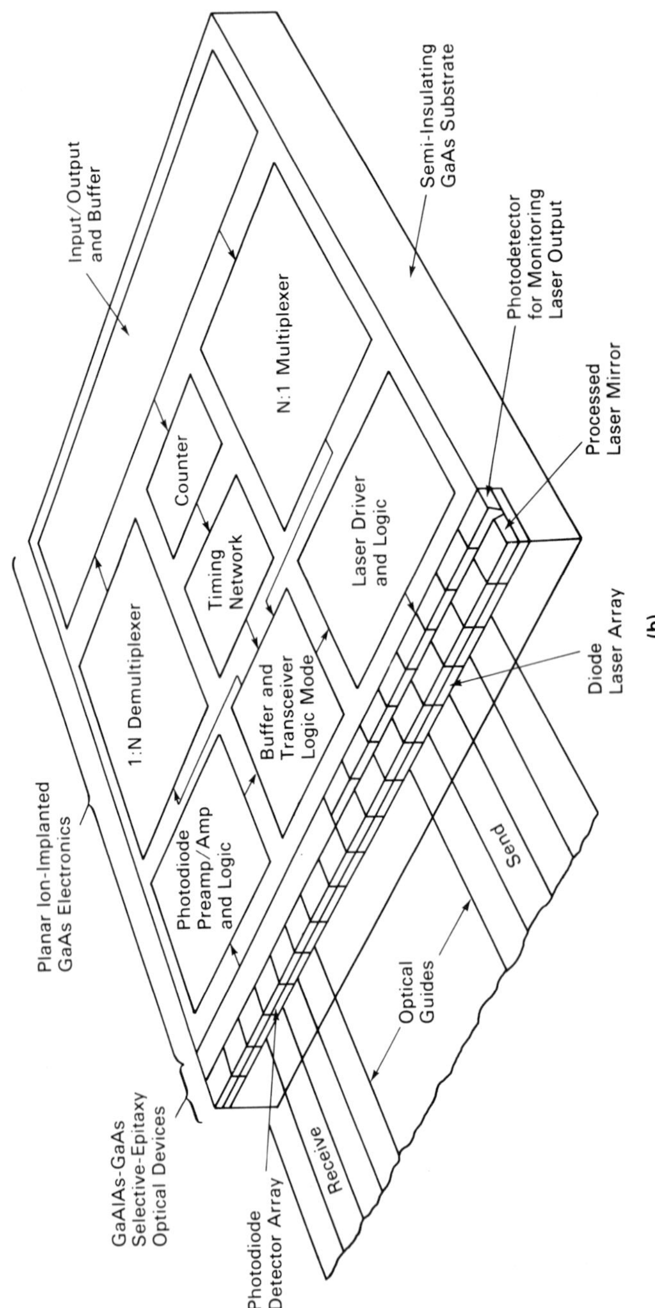

(b)

Fig. 5-7-1. Cont.

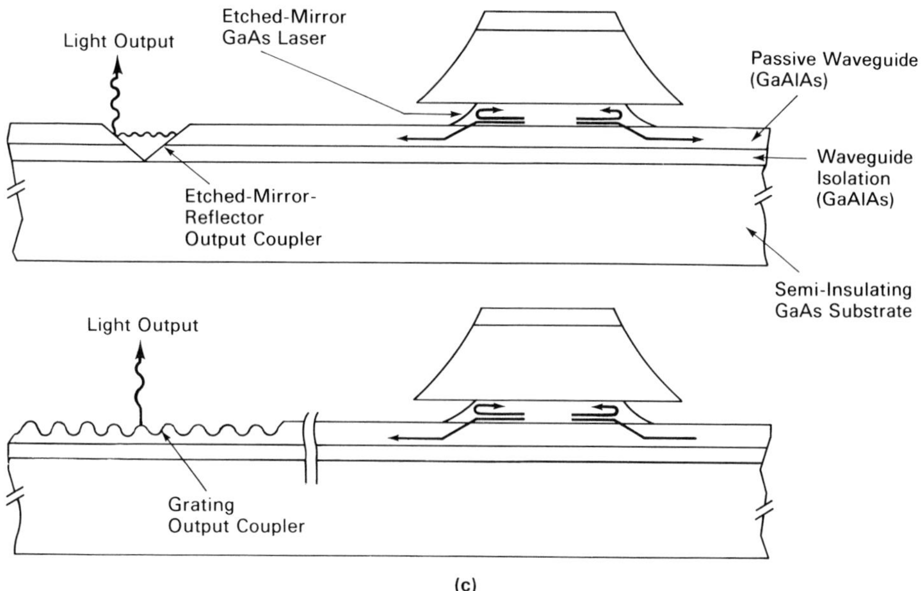

Fig. 5-7-1. Cont.

taxy (MBE) and Metal Organic Chemical Vapor Deposition (MOCVD) technologies that allow one to grow gallium arsenide and related compounds on a silicon substrate (see, for example, Fisher et al. 1986, Aksun and Morkoç 1986, and Hashimoto et al. 1985). Hashimoto et al. (1985) described light-emitting diodes (LEDs) grown on silicon substrates using MOCVD technology. Choi et al. (1986) reported on the monolithic integration of GaAs/AlGaAs LEDs and silicon MOSFETs.

Fig. 5-7-2 shows an OptoElectronic Integrated Circuit (OEIC) in which a high-power GaAs/AlGaAs laser is integrated with a GaAs metal semiconductor field-effect transistor (MESFET) high-frequency modulator integrated circuit. This modulator includes a five-stage ring oscillator with an output buffer MESFET. The output waveform was nearly sinusoidal with a period of 1250 ps and light output power of up to 30 mW.

An OEIC transmitter chip including a GaAs/AlGaAs laser, a power monitor photodiode, three GaAs FETs, and an impedance-matching resistance is shown in Fig. 5-7-3. The threshold current was only 10 mA. Fig. 5-7-4 also shows the transmitter light output obtained when 2 Gb/s random pulses were fed into the input FET gate V_{G2}. The operating speed of the driving circuit (determined in a separate experiment) exceeded 2 Gb/s (see Nobuhara et al. 1986).

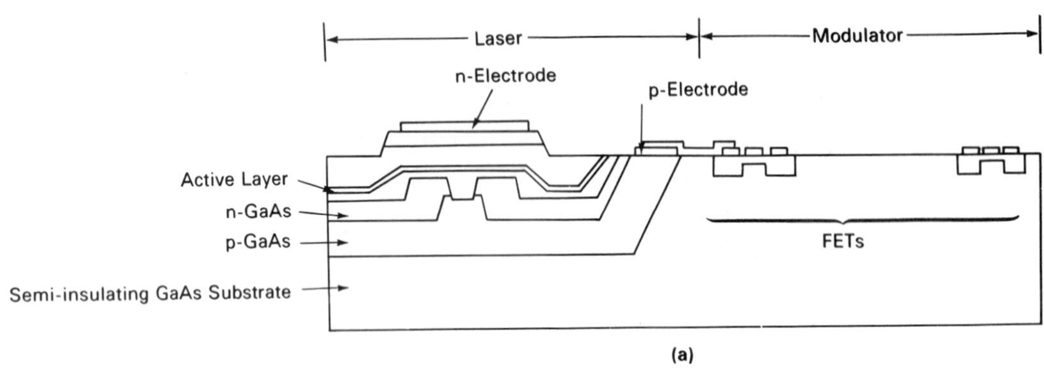

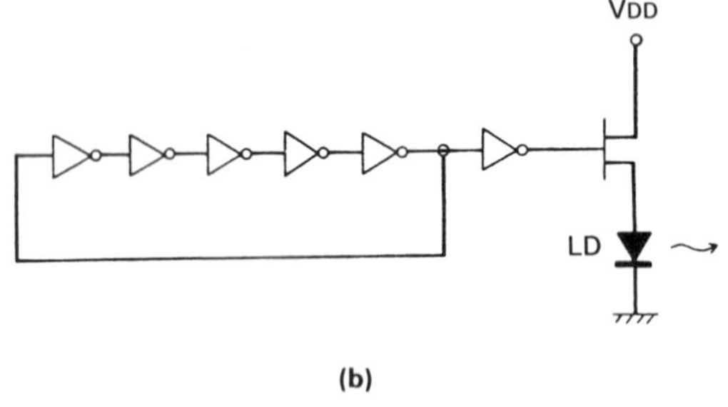

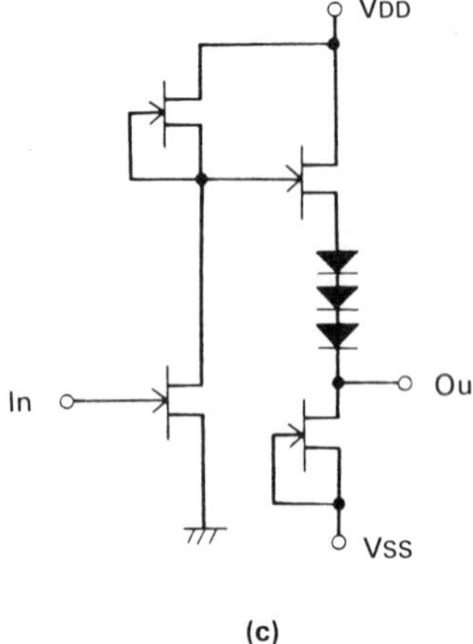

Fig. 5-7-2. OptoElectronic Integrated Circuit (OEIC) where high power GaAs-AlGaAs laser is integrated with GaAs Metal Semiconductor Field-Effect Transistor (MESFET) high frequency modulator (from K. Hamada, N. Yoshikawa, H. Shimizu, T. Otsuki, A. Shimano, K. Itoh, G. Kano, and Y. Teramoto, *Extended Abstracts of the 18th (1986 International) Conference on Solid State Materials and Devices*, Tokyo, pp. 181–184 (1986). (a) Schematic diagram. (b) Circuit diagram. (c) Circuit diagram of the buffer stage.

Fig. 5-7-4 shows a low-capacitance p-i-n GaAs FET integrated-circuit receiver. Fig. 5-7-4a shows a scanning electron microphotograph of a low capacitance GaInAs/InP p-i-n diode that is designed to be "flip-chip" bonded onto the GaAs IC as shown in Fig. 5-7-4b. An integrated GaAs preamplifier is shown in Fig. 5-7-4c.

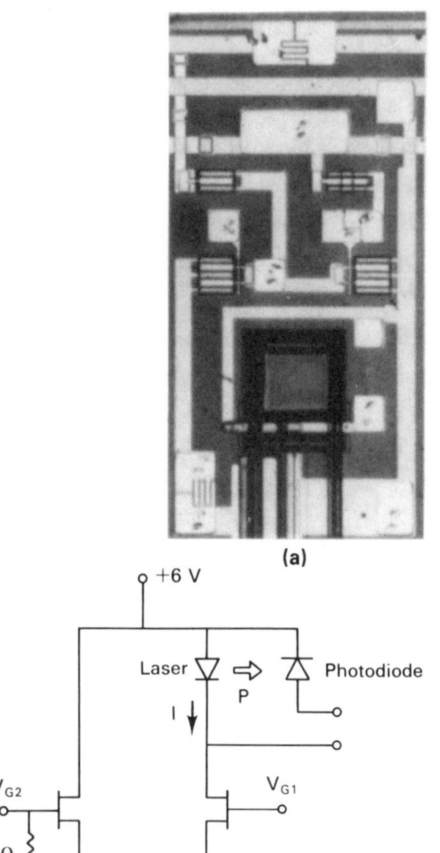

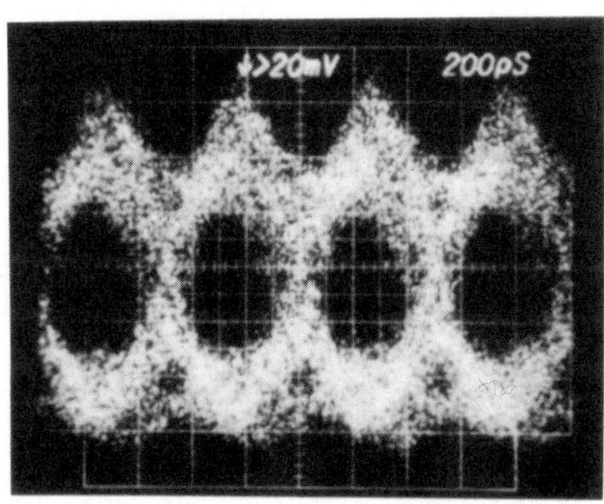

Fig. 5-7-3. OptoElectronic Integrated Circuit (OEIC) transmitter chip including GaAs-AlGaAs laser, power monitor photodiode, three GaAs FETs, and impedance-matching resistance (from H. Nobuhara, T. Sanada, M. Kuno, M. Makiuchi, T. Fujii, and O. Wada, *Extended Abstracts of the 18th (1986 International) Conference on Solid State Materials and Devices,* Tokyo, pp. 185–188 (1986). (a) Photomicrograph. (b) Circuit diagram. (c) Transmitter light output obtained when 2 Gb/s random pulses were fed into the input FET gate V_{G2}.

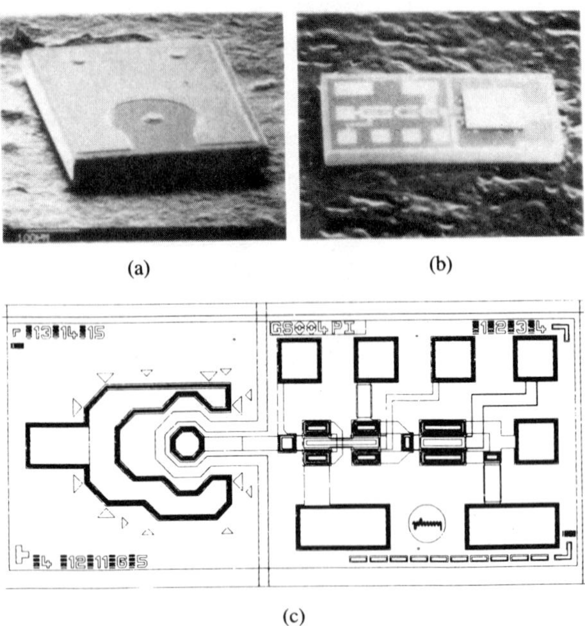

Fig. 5-7-4. Low capacitance *p-i-n* GaAs FET integrated circuit receiver (from R. C. Goodfellow, B. T. Debney, G. J. Rees, and J. Buus, *IEEE Transactions Electron Devices*, ED-32, no. 12, p. 2562 (1985). © by IEEE. (a) Scanning electron microphotograph of a low capacitance GaInAs-InP *p-i-n* diode that is designed to be flip-chip bonded onto the GaAs IC. (b) Inverted GaInAs-InP *p-i-n* diode mounted on the GaAs IC. (c) An Integrated GaAs preamplifier.

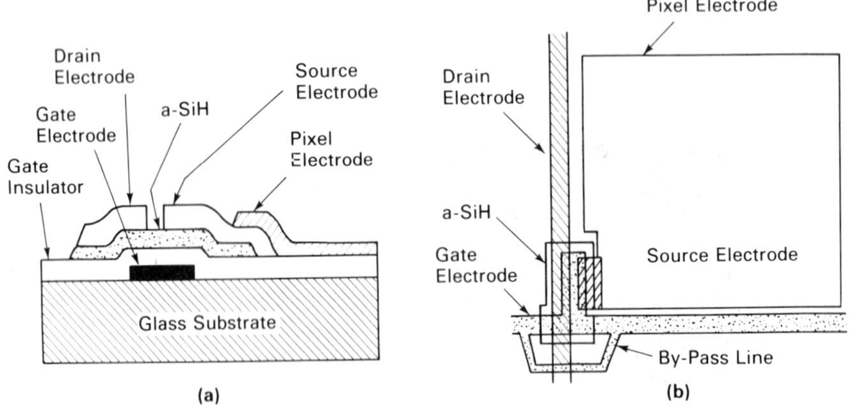

Fig. 5-7-5. Amorphous silicon LCD display (after M. Yamano and H. Takesada, *Proceedings of the Eleventh International Conference on Amorphous and Liquid Semiconductors*, Rome, pp. 1383–1388 (1986). (a) Cross-section of one pixel. (b) Plane view of one pixel. (c) Schematic diagram.

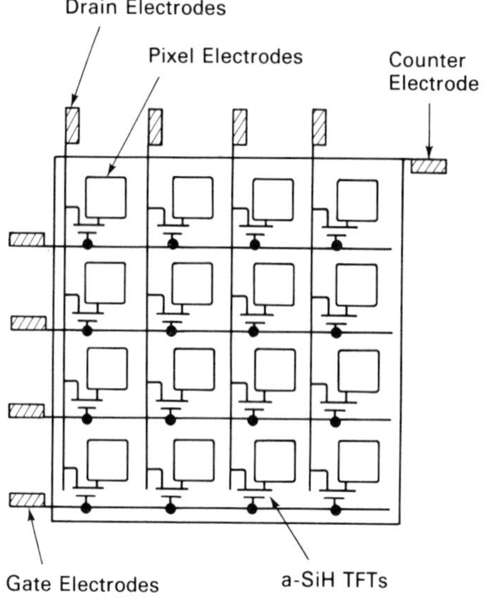

(c)

Fig. 5-7-5. Cont.

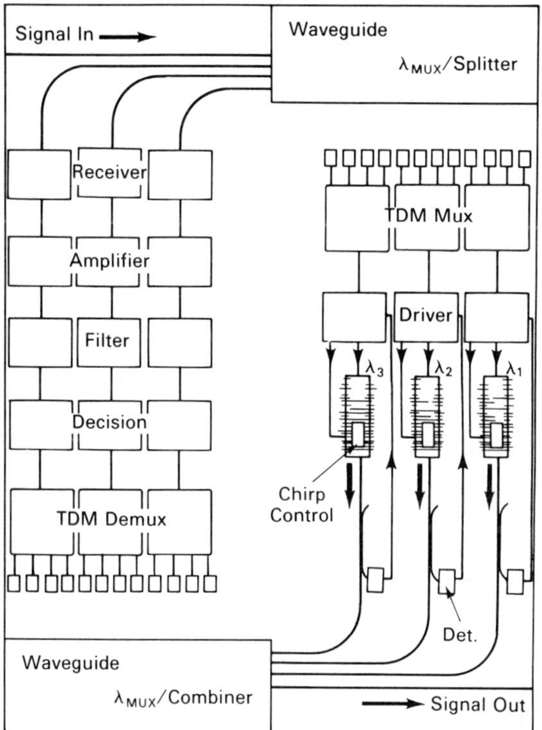

Fig. 5-7-6. Schematic diagram of a future optoelectronic chip with Time Division Multiplexing (TDM) and demultiplexing (after R. C. Goodfellow, B. T. Debney, G. J. Rees, and J. Buus, *IEEE Transactions Electron Devices,* ED-32, no. 12, p. 2562 (1985). © 1985 by IEEE).

Amorphous silicon has also been used for optoelectronic applications. In 1983 H. Yajima et al. at the Electrotechnical Laboratory of MITI, Japan, fabricated an optoelectronic circuit utilizing amorphous silicon (see Bell 1986). The device was fabricated on a lithium niobate substrate (10×30 mm). The four micron wide optical waveguides were made by diffusing titanium into the substrate. The device used amorphous silicon photodetectors. Amorphous silicon photoreceptors are now used in vidicons (see Imamura et al. 1979), in solid-state images (see Kusano et al. 1983), in integrated line sensors (see Morozumi et al. 1984), in copying machines (see Shimizu 1986), and in other devices. Perhaps one of the most dramatic demonstrations of the capabilities of a-Si technology is a 5-inch full-color flat-panel television set with a Liquid Crystal Display (LCD) described by Yamano and Takesada (1986). The cross section and plane view of one pixel of such a display is shown in Fig. 5-7-5. Figure 5-7-5c shows a schematic diagram of the LCD panel.

As optoelectronic integrated-circuit technology matures, more sophisticated and larger-scale integrated circuits will become possible. A schematic diagram of such a future optoelectronic chip is shown in Fig. 5-7-6 (after Goodfellow et al. 1985). A good general introduction to integrated optoelectronics is given by Hunsperger (1984).

REFERENCES

M. I. Aksun and H. Morkoç, *IEDM Technical Digest,* p. 752, Los Angeles, Calif., IEEE Publications (1986).

Zh. I. Alferov, *Fiz. Tekh. Poluprovodn.,* 1, p. 436 (1967).

Zh. I. Alferov and R. F. Kazarinov, Patent 181737 (1963).

A. Antresyan and C. Y. Chen, *IEEE Trans. Electron Devices,* ED-33, no. 2, p. 188 (1984).

T. Bell, *Spectrum,* pp. 38–45, December (1983).

T. Bell, *Spectrum,* pp. 34–57, August (1986).

A. A. Bergh and P. J. Dean, *Light-Emitting Diodes,* Clarendon, Oxford (1976).

M. G. A. Bernard and G. Duraflourg, "Laser Conditions in Semiconductors," *Physica Status Solidi,* 1, p. 699 (1961).

M. Brian and T. P. Lee, *IEEE Trans. Electron Devices,* ED-32, no. 12, p. 2673 (1985).

C. A. Burrus, A. G. Dentai, and T. P. Lee, *Electron. Lett.,* 15, no. 20, pp. 655–657 (1979).

D. E. Carlson and C. R. Wronski, *Appl. Phys. Lett.,* 28, p. 671 (1976).

H. K. Choi, G. W. Turner, T. H. Windhorn, and B. Y. Tsaur, *IEEE Electron Device Lett.,* EDL-7, pp. 500–502 (1986).

G. Diemer, Phillips Research Reports, 11, pp. 352–399 (1956).

W. P. Dumke, "Interband Transitions and Maser Action," *Phys. Rev.,* 127, p. 1559 (1962).

J. Dziewior and W. Schmid, *Appl. Phys. Lett.,* 31, p. 346 (1982).

A. Einstein, "Zur Quantentheorie der Strahlung," *Phys. Z.,* 18, pp. 121–128 (1917).

R. Fisher, N. Chand, W. Kopp, C. K. Peng, H. Morkoç, K. R. Gleason, and D. Scheitlin, *IEEE Trans. Electron Devices,* ED-33, pp. 206–213 (1986).

S. R. Forrest, *Spectrum,* p. 76, May (1986).

R. C. Goodfellow, B. T. Debney, G. J. Rees, and J. Buus, *IEEE Trans. Electron Devices,* ED-32, no. 12, p. 2562 (1985).

M. A. Green, *Solar Cells: Operating Principles, Technology, and Device Applications,* Prentice-Hall, Englewood Cliffs, N.J. (1982).

M. A. Green, A. W. Blakers, J. Shi, E. M. Keller, and S. R. Wenham, *IEEE Trans. Electron. Devices,* ED-31, no. 5, p. 679 (1984).

M. A. Green, A. W. Blakers, M. R. Willison, E. Gauja, and T. Szpitalak, "Towards a 700 mV Silicon Cell," in *Conf. Record,* 16th IEEE Photovoltaic Specialists Conference, San Diego, Calif., September 1982, IEEE Pub. 82CG 1821-8, pp. 1219–1222 (1982).

M. A. Green, Z. Jinhua, A. W. Blakers, M. Taouk, and S. Narayanan, "25% Efficient Low-Resistivity Silicon Concentrator Solar Cells," *IEEE Electron Device Lett.,* EDL-7, no. 10, pp. 583–585, October (1986).

W. O. Groves, A. H. Herzog, and M. G. Craford, *Appl. Phys. Lett.,* 19, p. 184 (1971).

S. Guha, *J. Non-Crystalline Solids,* 77 and 78, pp. 1451–1460 (1985).

M. Hack and M. Shur, *Technical Digest of the International PVSEC-1,* Kobe, Japan, pp. 645–648, (1984a).

M. Hack and M. Shur, *IEEE Trans. Electron Devices,* ED-31, no. 5, pp. 539–542 (1984b).

M. Hack, S. Guha, and M. Shur, *Phys. Rev.,* B30, no. 12, pp. 6991–6999 (1984).

M. Hack and M. Shur, *J. Appl. Phys.,* 58, no. 2, pp. 997–1020 (1985a).

M. Hack and M. Shur, *J. Appl. Phys.,* 59, no. 6, pp. 2222–2228 (1985b).

R. N. Hall, G. E. Genner, J. D. Kingsley, T. J. Soltys, and R. O. Carlson, "Coherent Light Emission from GaAs Junctions," *Phys. Rev. Lett.,* 9, p. 366 (1962).

K. Hamada, N. Yoshikawa, H. Shimizu, T. Otsuki, A. Shimano, K. Itoh, G. Kano, and Y. Teramoto, *Extended Abstracts of the 18th (1986 International) Conference on Solid State Materials and Devices,* Tokyo, pp. 181–184 (1986).

Y. Hamakawa, "Recent Advances in Solar Cell Technology," *Technical Digest of the International PVSEC-3,* Tokyo, Japan, pp. 147–152 (1987).

H. C. Hamaker et al., *Appl. Phys. Lett.,* 47, no. 7, pp. 762–764 (1985).

A. Hashimoto, Y. Kawarada, T. Kamijoh, M. Akiyama, D. Watanabe, and M. Sakuta, *IEDM Technical Digest,* pp. 658–661, IEEE Publications (1985).

I. Hayshi, M. B. Panish, P. W. Foy, and S. Sumski, "Junction Lasers Which Operate Continuously at Room Temperature," *Appl. Phys. Lett.,* 17, p. 109 (1970).

N. Holonyak, Jr., and S. F. Bevacqua, "Coherent (Visible) Light Emission from $Ga(As_{1-x}P_x)$ Junction," *Appl. Phys. Lett.,* 1, p. 82 (1962).

S. Y. Huang, S. Esener, and S. H. Lee, *IEEE Trans. Electron Devices,* ED-33, no. 4, p. 433 (1986).

R. G. Hunsperger, *Integrated Optics: Theory and Technology,* 2d ed., Springer-Verlag (1984).

L. D. HUTCHESON, P. HAUGEN, and A. HUSAIN, *Spectrum,* pp. 30–35, March (1987).

Y. IMAMURA, S. ATAKA, Y. TAKASAKI, C. KUSANO, T. HIRAI, and E. MARUYAMA, *J. Appl. Phys.,* 35, p. 349 (1979).

K. KOBAYASHI and I. MITO, *IEEE Trans. Electron Devices,* ED-32, no. 12, p. 2594 (1985).

H. KRESSEL and J. K. BUTLER, *Semiconductor Lasers and Heterojunction LEDs,* Academic Press, New York (1977).

H. KROEMER, "A Proposed Class of Heterojunction Injection Lasers," *Proc. IEEE,* 51, p. 1782 (1963).

D. KRUANGAM, M. DEGUCHI, T. ENDO, W. GUANG-PU, H. OKAMOTO, and Y. HAMAKAWA, *Extended Abstracts of the 18th (International) Conference on Solid State Devices and Materials,* Tokyo, pp. 683–686 (1986).

C. KUSANO, S. ISHIOKA, Y. IMAMURA, Y. TAKASAKI, Y. SHIMAMURA, T. HIRAI, and E. MARUYMA, *IEDM Technical Digest,* p. 509, IEEE Publications (1983).

S. MOROZUMI, H. KURIHARA, H. OSHIMA, T. TAKESHITA, and K. HASEGAWA, *Extended Abstracts of the 16th Conference on Solid State Devices and Materials,* Kobe, p. 559 (1984).

T. NAKAMURA, K. MATSUMOTO, R. HYUGA, and A. YUSA, *IEDM Technical Digest,* p. 353, Los Angeles, Calif., IEEE Publications (1986).

M. I. NATHAN, W. P. DUMKE, G. BURNS, F. J. DILL, JR., and G. J. LASHER, "Stimulated Emission of Radiation from GaAs p-n Junction," *Appl. Phys. Lett.,* 1, p. 62 (1962).

Optoelectronics Designer's Catalog, Hewlett Packard, (1985).

H. NOBUHARA, T. SANADA, M. KUNO, M. MAKIUCHI, T. FUJII, and O. WADA, *Extended Abstracts of the 18th (1986 International) Conference on Solid State Materials and Devices,* Tokyo, pp. 185–188 (1986).

R. H. PAREKH and A. M. BARNETT, *IEEE Trans. Electron Devices,* ED-31, no. 5, p. 689 (1984).

T. M. QUIST, R. H. REDIKER, R. J., KEYES, W. E. KRAG, B. LAX, A. L. MCWHORTER, and H. J. ZEIGLER, "Semiconductor Maser of GaAs," *Appl. Phys. Lett.,* 1, p. 91 (1962).

P. RAPPOPORT and J. J. WYSOCKI, "The Photovoltaic Effect in GaAs, CdS, and Other Compound Semiconductors," *Acta Electron.,* 5, p. 364 (1961).

H. J. ROUND, "A Note on Carborundum," *Electron World,* 19, p. 309 (1907).

H. SAKAI, K. MARUYAMA, T. YOSHIDA, Y. ICHIKAWA, T. HAMA, M. UENO, M. KAMIYAMA, and Y. UCHIDA, *Technical Digest of the Internatonal PVSEC-1,* Kobe, Japan, p. 591 (1984).

A. L. SCHAWLOW and C. H. TOWNES, "Infrared and Optical Masers," *Phys. Rev.,* 112, p. 1940 (1958).

K. SENDA, E. FUJII, Y. HIROSHIMA, and T. TAKAMURA, *IEDM Technical Digest,* p. 369, Los Angeles, Calif., IEEE Publications (1986).

I. SHIMIZU, *Proceedings of the Eleventh International Conference on Amorphous and Liquid Semiconductors,* Rome, ed. F. Evangelisti and J. Stuke, pp. 1363–1372, North-Holland, (1986).

R. A. SINTON, Y. KWARK, J. Y. GAN, and R. M. SWANSON, *IEEE Electron Device Lett.,* EDL-7, no. 7, p. 567 (1986).

W. E. Spear and P. G. LeComber, *J. Non-Crystal. Solids,* 8-10, p. 727 (1972).

W. E. Spear, P. G. LeComber, S. Kinmond, and M. H. Brodsky, *Appl. Phys. Lett.,* 28, p. 105 (1976).

D. L. Stabler and C. R. Wronski, *Appl. Phys. Lett.,* 31, p. 292 (1977).

T. Sugino, A. Yoshikawa, A. Yamamoto, M. Hirose, G. Kano, and I. Teramoto, *IEDM Technical Digest,* p. 618, Los Angeles, Calif., IEEE Publications (1986).

A. Susuki, T. Uji, Y. Inomoto, J. Hayashi, Y. Isoda, and H. Nomura, *IEEE Trans. Electron Devices,* ED-32, no. 12, p. 2609 (1985).

S. Sze, *Physics of Semiconductor Devices,* John Wiley & Sons, New York (1981).

K. Taguchi, T. Torikai, Y. Sugimoto, K. Makita, H. Ishihara, S. Fujita, and K. Minemura, *IEEE Electron Device Lett.,* EDL-7, no. 4, p. 257 (1986).

L. E. Tannas, Jr., *Spectrum,* p. 37, October (1986).

Y. Tawada, M. Kondo, H. Okamoto, and Y. Hamakawa, *Sol. Energy Mat.,* 6, p. 237 (1982).

M. Thayer, W. A. Anderson, and B. B. Rao, *IEEE Trans. Electron Devices,* ED-31, no. 5, p. 619 (1984).

K. Vahala , C. Chiu, S. Margalit, and A. Yariv, *Appl. Phys. Lett.,* 42, p. 631 (1983).

M. Yamano and H. Takesada, *Proceedings of the Eleventh International Conference on Amorphous and Liquid Semiconductors,* Rome, ed. F. Evangelisti and J. Stuke, pp. 1383–1388, North-Holland, (1986).

A. Yariv, *Quantum Electronics,* 2d ed., New York, John Wiley & Sons (1975).

A. Yariv, *Optical Electronics,* 3d ed., New York, Holt, Rinehart and Winston (1985).

A. van der Ziel, *Solid State Electronics,* Prentice-Hall, Englewood Cliffs, N.J. (1976).

PROBLEMS

5-1-1. Derive eqs. (5-1-8) to (5-1-10),

$$\varepsilon = \varepsilon_0 n_r^2(1 - \chi^2) \tag{5-1-8}$$

$$\sigma = 2n_r^2\chi\omega\varepsilon_0 \tag{5-1-9}$$

$$\alpha = \sigma/(n_r^2 c\varepsilon_0) \tag{5-1-10}$$

from eqs. (5-1-1), (5-1-6), and (5-1-7),

$$n_r^* = n_r(1 - \chi)^{1/2} \tag{5-1-1}$$

$$n_r^* = (\varepsilon^*/\varepsilon_0)^{1/2} \tag{5-1-6}$$

$$\varepsilon^* = \varepsilon - i\sigma/\omega \tag{5-1-7}$$

5-2-1. Derive an expression for the light-generated current, I_L, for a short p-n diode where the length of the n section, W_n, is much smaller than the diffusion length of holes, L_p

and the length of the p section, W_p, is much smaller than the diffusion length of electrons, L_n. The generation rate is equal to G_L. The device cross section is equal to S.

5-2-2. Assume that the refraction indexes of two media are 3.3 and 1. Find the reflection coefficients with and without the antireflection coating with the refraction index 1.83.

5-2-3. Equation (5-2-7) for the light-generated current was obtained assuming a uniform generation rate of electron-hole pairs in the sample. A more realistic approach should account for the nonuniform generation rate. If we consider a p^+-n solar cell with a narrow p^+ region, neglecting the light absorption in the p^+ region and possible light reflection from the back side of the cell, the generation rate for a monochromatic light is given by

$$G_L = \alpha Q_{eff} F_{ph} \exp(-\alpha x) = \alpha N_{eh} \exp(-\alpha x) \tag{P5-2-3-1}$$

where F_{ph} is the incident photon flux, Q_{eff} is the quantum efficiency (i.e., the number of electron-hole pairs generated per each photon), and $N_{eh} = Q_{eff} F_{ph}$. Substituting this expression into the continuity equation for holes (see eq. (5-2-1)), assuming the appropriate boundary conditions and choosing the bias voltage $V = 0$, we can calculate the light-generated current. If we assume the constant surface recombination velocity S_{sf} at the back contact, the boundary conditions are

$$S_{sf} \Delta p_n = -D_p \, dp_n/dx \quad \text{at } x = W \tag{P5-2-3-2}$$

$$\Delta p_n \, (x = 0) = 0 \tag{P5-2-3-3}$$

(a) Derive the expression for the hole profile, $\Delta p_n(x)$.

(b) Derive the expression for the light-generated current,

$$I_{pL} = qD_p S \left. \frac{d\Delta p_n}{dx} \right|_{x=0} \tag{P5-2-3-4}$$

5-2-4. Solve the continuity equation for electrons in the p section of the n^+-p solar cell,

$$\frac{d^2 \Delta n_p}{dx^2} - \frac{\Delta n_p}{L_p^2} = 0$$

using the boundary conditions

$$\Delta n_p(W) = 0$$

$$\Delta n_p(0) = n_{po}[\exp(V/V_T) - 1]$$

and derive the expression for the short-circuit current. Compare the result with eqs. (5-2-11) and (5-2-12).

5-2-5. The current-voltage characteristics of two solar cells constituting one tandem solar cell are shown in Fig. P5-2-5 (for AM1 conditions).

(a) Which cell (A or B) corresponds to the higher energy-gap material?

(b) What will be the upper bound for the light-generated short-circuit current for the tandem cell?

(c) What will be the upper bound for the open-circuit voltage for the tandem cell?

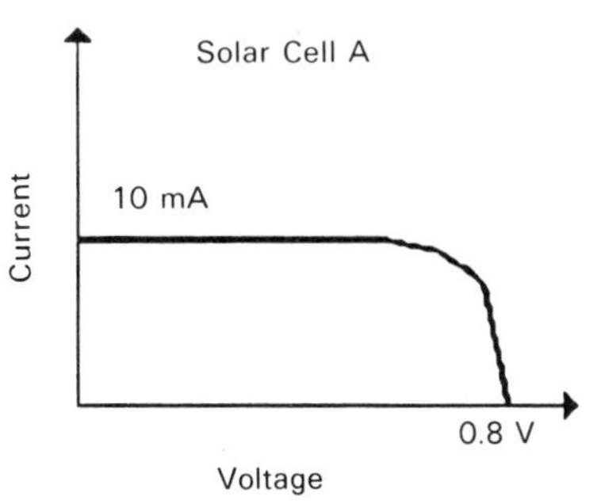

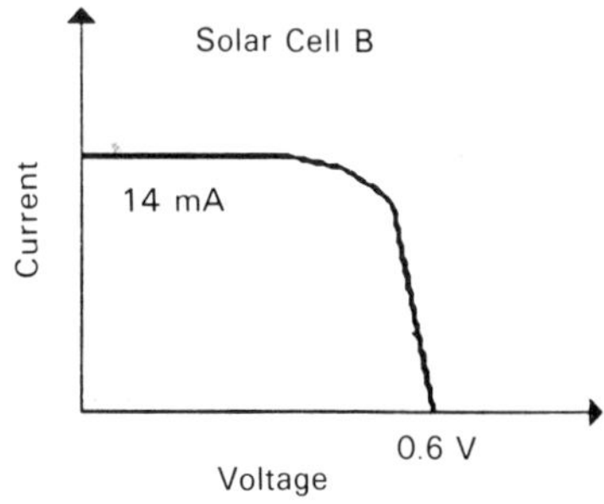

Fig. P5-2-5

5-3-1. Using a simplified distribution of donorlike and acceptorlike states in the energy gap of a-Si,

$$g_d(E) \approx g_v \exp[(E_v - E)/E_d]$$
$$g_a(E)\, dE \approx g_c \exp[(E - E_c)/E_a]$$

give expressions for the positions of electron and hole quasi-Fermi levels, E_{Fn} and E_{Fp}, as functions of the electron-hole concentration, P, created by light. Assume that the generation rate $G = P/\tau$, where P is the density of electron-hole pairs and τ is the lifetime. Furthermore, assume that practically all light-generated carriers are trapped by localized states. (Holes are trapped by donorlike states and electrons are trapped by acceptorlike states.)

Assuming $E_g = 1.7$ eV, $E_a = 0.086$ eV, $E_d = 0.13$ eV, $g_v = 6 \times 10^{18}$ cm^{-3}ev^{-1}, $g_c = 5 \times 10^{18}$ cm^{-3}ev^{-1}, $G = 10^{23}$ cm^{-3}s^{-1}, and $\tau = 10^{-6}$ s, calculate the positions of E_{Fn} and E_{Fp} with respect to the conduction band edge, E_c.

Hint: Use eqs. (4-12-2) and (4-12-3) and a zero-temperature limit for the Fermi-Dirac occupation function.

$$Q_d = q\int_{E_F}^{E_c} g_d(E)\, dE \approx qg_v \int_{E_F}^{\infty} \exp[(E_v - E)/E_d]\, dE = qg_vE_d \exp[(E_v - E_F)/E_d] \qquad (4\text{-}12\text{-}2)$$

$$Q_a = -q\int^{E_F} g_a(E)\, dE \approx qg_c \int^{E_F} \exp[(E - E_c)/E_a]\, dE = qg_cE_a \exp[(E_F - E_c)/E_a] \qquad (4\text{-}12\text{-}3)$$

5-3-2. Assume that the tandem cell shown in Fig. 5-3-3 may be modeled as a series combination of two cells with the current-voltage characteristics decribed by the following equations:

$$I_1 = I_{L1} - I_{s1}[\exp(qV_1/k_BT) - 1]$$
$$I_2 = I_{L2} - I_{s2}[\exp(qV_2/k_BT) - 1]$$

Assume that the light generated current, I_{L2}, for the second narrow-band-gap cell, is given by

$$I_{L2} = I_{Lm2} - I_{L1}$$

where I_{Lm2} is the maximum light-generated current that can be produced by this narrow-band-gap material, and

$$I_{L1} \approx I_{Lm1}[1 - \exp(-\alpha_{eff} L_1)]$$

where I_{Lm1} is the maximum light-generated current that can be produced by this wide-band-gap material, L_1 is the the thickness of the wide-band-gap active layer, and α_{eff} is the effective absorption coefficient of high-energy photons in the wide-band-gap material.

Assume $\alpha_{eff} = 10^5$ cm^{-1}, $I_{s1} = 10^{-11}$ A/cm^2, $I_{s2} = 10^{-9}$ A/cm^2, $I_{Lm1} = 10^{-2}$ A/cm^2, and $I_{Lm2} = 1.5 \times 10^{-2}$ A/cm^2.

(a) Calculate the current-voltage characteristics of the tandem cell for $L_1 = 0.1$ μm, 0.2 μm, and 0.4 μm.

(b) What value of L_1 would correspond to the highest efficiency of the tandem cell?

5-4-1. Solve the continuity equation for holes in the n$^-$ region of an p$^+$-n$^-$-n$^+$ device. Assume that the length of the n$^-$ section, L, is much greater than the diffusion length of holes, and use boundary conditions

$$p_n = 0 \qquad \text{for } x = 0$$

and

$$p_n = p_{no} \qquad \text{for } x >> L_p$$

Then find the diffusion current

$$I_{diff} = -qD_p S \partial p_n / \partial x \qquad \text{at } x = 0$$

where x corresponds to the boundary between the p$^+$ region and the n region.

5-4-2. Calculate the quantum efficiency–transit frequency product vs. length, L, for a p-i-n photodetector. Assume the carrier velocity 10^5 m/s, the diffusion length $L_p = 100$ μm, and the reflection coefficient $R = 0.3$. Vary L from 1 to 300 μm.

5-5-1. In a GaP red LED, the radiation mechanism corresponds to the transition from a donor to an acceptor level. The photon frequency is given by

$$h\nu = E_g - E_d - E_a + q^2/(\varepsilon_s r)$$

where E_g is the energy gap, E_d is the the donor energy (the energy difference between the donor level and the bottom of the conduction band), E_a is the acceptor energy (the energy difference between the acceptor level and the ceiling of the valence band), and r is the average separation between the donor and acceptor. The energy gap $E_g = 2.3$ eV, the donor is oxygen ($E_d = 0.803$ eV), and the acceptor is zinc ($E_a = 0.04$ eV). The wavelength of the emitted light is 7000 Å. What is the average separation between the donor and acceptor atoms?

5-6-1. Derive eq. (5-6-43) from eq. (5-6-42).

5-6-2. Calculate the threshold concentration of electron-hole pairs needed for lasing in GaAs as a function of temperature.

Effective density of states in the conduction band $= 4.7 \times 10^{17}\ \mathrm{cm}^{-3}$ (at 300 K)
Effective density of states in the valence band $= 7.0 \times 10^{18}\ \mathrm{cm}^{-3}$ (at 300 K)
Energy gap $= 1.424$ eV

5-6-3. Calculate the threshold concentration of electron-hole pairs needed for lasing in $Al_xGa_{1-x}As$ at room temperature as a function of composition x for $0 < x < 0.45$. The effective density of states mass in the conduction band $m_c = (0.067 + 0.083x)m$. The effective density of states mass in the valence band $m_v = (0.62 + 0.14x)m$. Energy gap $E_g = 1.424 + 1.247x$ (eV).

6

Transferred-Electron Devices and Avalanche Diodes

6-1. INTRODUCTION

In this Chapter we consider two-terminal microwave semiconductor devices of two types: transferred-electron devices utilizing the bulk negative resistance of gallium arsenide or related compounds, and avalanche devices that use impact ionization in high electric fields. At the present time, these devices are the most powerful solid-state sources of microwave energy. In Section 6-2 we consider the mechanism of negative differential resistance in GaAs that is related to the electron transfer from the central valley of the conduction band of GaAs into the satellite valleys. Transferred Electron Oscillators (TEOs), amplifiers, and logic and functional devices are discussed in Section 6-2. In Section 6-3 we introduce a concept of the phase delays related to avalanche breakdown, carrier injection in a p-n junction, and carrier drift in a semiconductor diode. We then describe different devices that utilize these phase delays for obtaining a 180° phase shift between voltage and current waveforms. These phase shifts between the device current and applied voltage create a dynamic negative differential resistance that is used to generate microwave power.

6-2. RIDLEY-WATKINS-HILSUM-GUNN EFFECT

The dependence of the electron velocity on the electric field, $v(F)$, in GaAs (and some other compound semiconductors) has a region where the differential mobility,

$$\mu_d = \frac{dv}{dF} \tag{6-2-1}$$

is negative (see Section 1-9). The mechanism of this negative differential mobility and, hence, negative differential conductivity in GaAs,

$$\sigma_d = qn\mu_d \tag{6-2-2}$$

was first proposed by Ridley and Watkins (1961) and, independently, by Hilsum (1962). Gunn (1963) was first to discover microwave oscillations of current in GaAs and InP samples related to instabilities caused by the negative differential conductance.

The negative slope of the v vs. F characteristic develops as a consequence of the intervalley transition of electrons from the central Γ valley of the conduction band into the satellite valleys. When the electric field is low, electrons are primarily located in the central valley of the conduction band (see Fig. 6-2-1a). As the electric field increases, many electrons gain enough energy from the electric field for the intervalley transition into the satellite valleys. The electron effective masses in the L and X valleys of the conduction band are much greater than in the Γ valley. Also, the intervalley transition is accompanied by an increased intervalley scattering. These factors result in the decrease of the electron velocity in high electric fields (see Fig. 1-9-1).

This mechanism of negative differential conductance proposed by Ridley, Watkins, and Hilsum may be explained using a simple two-valley model (see Fig. 6-2-2). In this model all X and L minima of the conduction band are represented by one equivalent minimum (one "upper valley") of the conduction band. The low "valley" is the lowest minimum (Γ minimum) of the conduction band. When the electric field is low, practically all electrons are in the lowest minimum of the conduction band, and the electron drift velocity v is given by

$$v_1 = \mu F \tag{6-2-3}$$

where μ is the low-field mobility and F is the electric field. In a higher electric field, electrons are "heated" by the field, and some carriers may have enough energy to transfer into upper valleys, where the electron velocity is

$$v_2 \simeq v_s \tag{6-2-4}$$

Here v_s is the saturation velocity. Strictly speaking, the velocity of electrons in upper valleys is not constant and may depend on the electric field. However, the assumption that $v_2 \approx v_s$ allows us to reproduce correctly this velocity saturation in high electric fields.

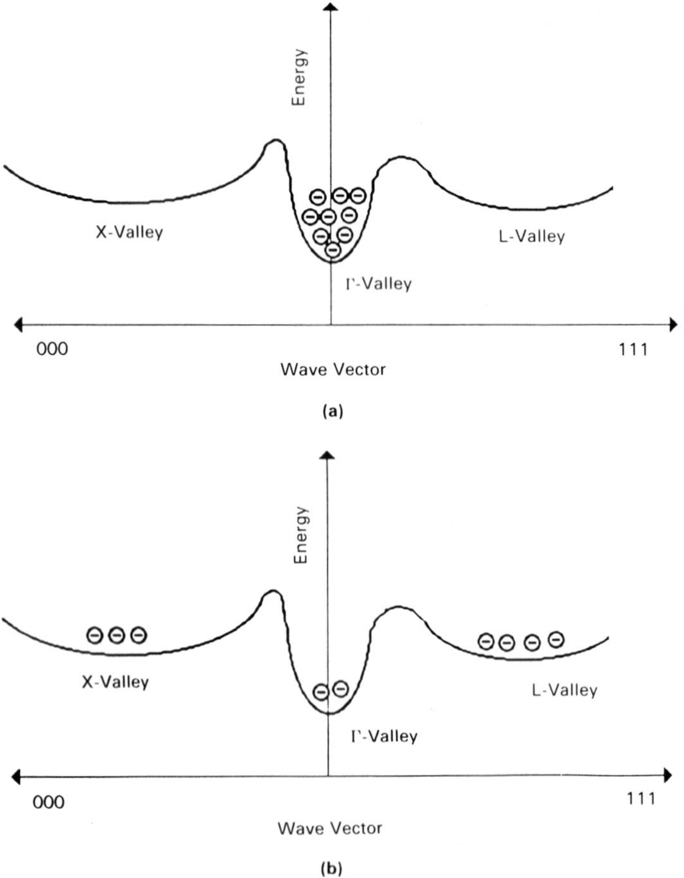

Fig. 6-2-1. Schematic illustration of intervalley transition in GaAs. (a) Electrons in the conduction band in low electric field. (b) Electrons in the conduction band in high electric field.

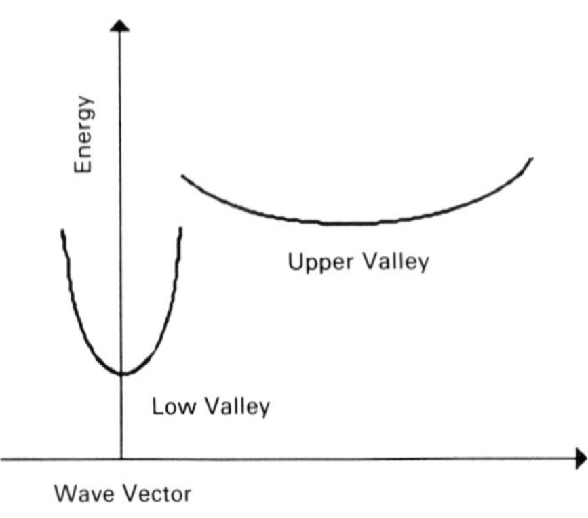

Fig. 6-2-2. Simplified two-valley model. In this model all X and L minima of the conduction band are represented by one equivalent minimum (upper valley).

The current density is given by

$$j = qv_1(F)n_1 + qv_2(F)n_2 \tag{6-2-5}$$

where n_1 is the electron concentration in the lowest valley and n_2 is the electron concentration in the upper valley;

$$n_1 + n_2 = n_o \tag{6-2-6}$$

where $n_o \simeq N_D$.

Using the results of a Monte Carlo calculation (see Section 1-9), one can check that this simplified model is quite adequate for GaAs at 300 K if we assume that the fraction of electrons in the upper valleys, $p = n_2/n_o$ is given by

$$p = \frac{A(F/F_s)^t}{1 + A(F/F_s)^t} \tag{6-2-7}$$

where $F_s = v_s/\mu$. Equation (6-2-5) can then be rewritten as

$$\begin{aligned} v(F) &= j/(qn_o) = \mu(1 - p)F + v_s p \\ &= v_s\left[1 + \frac{F/F_s - 1}{1 + A(F/F_s)^t}\right] \end{aligned} \tag{6-2-8}$$

Comparison with the results of the Monte Carlo calculation shows that such an interpolation is accurate within 10% to 15% (see Fig. 6-2-3).

From eq. (6-2-8) we find that

$$\frac{dv}{dF} = \mu(1 - p) + (v_s - \mu F)\frac{dp}{dF} \tag{6-2-9}$$

As can be seen from eq. (6-2-9), when dp/dF exceeds a critical value,

$$\frac{dp}{dF} > \frac{1 - p}{F - F_s} \tag{6-2-10}$$

the differential mobility becomes negative.

As will be shown presently, the uniform field distribution becomes unstable when F exceeds the peak field, F_p. This instability may lead to the formation of high-field domains and to the effect first observed by J. B. Gunn, as well as other phenomena that may be used to generate or amplify microwave signals.

Indeed, the negative differential resistance may lead to a growth of small fluctuations in the space charge in a sample. A simplified equivalent circuit may be presented as a parallel combination of the differential resistance,

$$R_d = \frac{L}{qn\mu_d S} \tag{6-2-11}$$

and the differential capacitance,

$$C_d = \frac{\varepsilon S}{L} \tag{6-2-12}$$

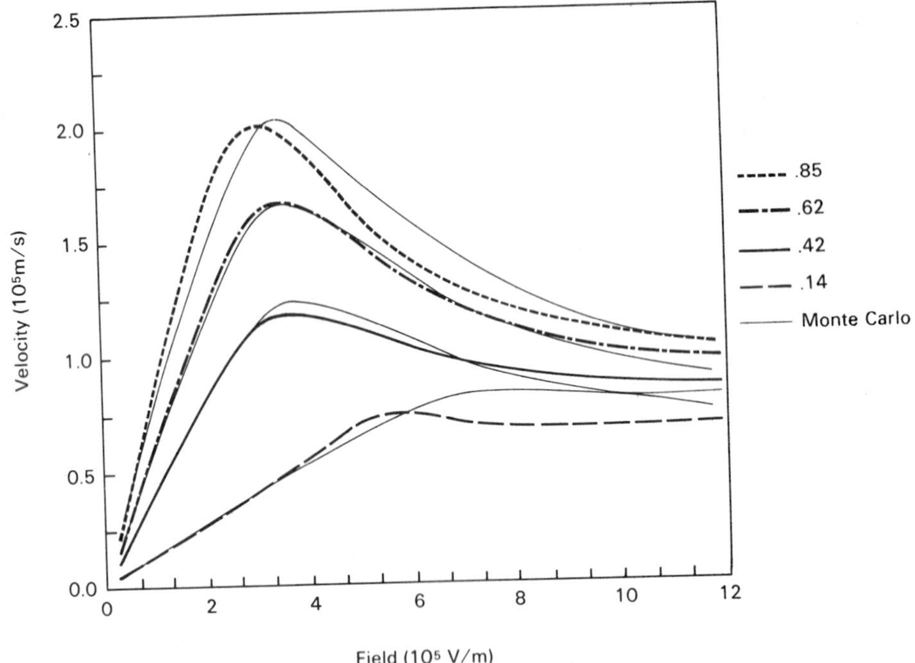

Fig. 6-2-3. Curves v and F in GaAs given by the interpolation formula (6-2-8) for different values of μ (in m²/Vs). (from J. Xu and M. Shur, *IEEE Trans. Electron Devices*, ED-34, no. 8, pp. 1831–1832 (1987). © 1987 IEEE). The curves obtained by Monte Carlo calculations from J. G. Ruch and W. Fawcett, *J. Appl. Phys.*, 41, no. 9, pp. 3843–3849 (1970). © 1970 IEEE). Y. K. Pozela and A. Reklaitis, *Solid State Electronics*, 23, pp. 927–933 (1980). © 1980 IEEE)., and J. Xu, B. A. Bernhardt, M. Shur, C. H. Chen, and A. Peczalski, *Appl. Phys. Lett.*, 49, no. 6, pp. 342–344 (1986). © 1986 IEEE) are shown by comparison. Parameters A, t, and v_s were chosen as the following functions of μ: $A = 0.6[e^{10(\mu-0.2)} + e^{-35(\mu-0.2)}]^{-1} + 0.01$, $t = 4[1 + 320/\sin h(40\mu)]$, and $v_s(10^5 \text{ m/s}) = 0.6 + 0.6\mu - 0.2\mu^2$ (where μ is in m²/Vs).

Here S is the cross section of the sample and L is the sample length. Hence the equivalent RC time constant determining the evolution of the space charge is given by

$$R_d C_d = \frac{\varepsilon}{q n_d \mu_d S} \tag{6-2-13}$$

This time constant is called the *Maxwell differential dielectric relaxation time*. In a material with a positive differential conductivity, a space-charge fluctuation, ΔQ, decays exponentially with time:

$$\Delta Q = \Delta Q(0) \cdot \exp(-t/\tau_d) \tag{6-2-14}$$

where $\Delta Q(0)$ is the magnitude of the fluctuation at $t = 0$. When the differential conductivity is negative, the space-charge fluctuation may actually grow with time:

$$\Delta Q = \Delta Q(0)\exp(t/\tau_d) \tag{6-2-15}$$

This process can be analyzed more accurately based on the Poisson equation,

$$\frac{\partial F}{\partial x} = -\frac{q(n - N_D)}{\varepsilon} \tag{6-2-16}$$

and the equation for the total current,

$$I = q\mu nF + qDn\frac{\partial n}{\partial x} + \varepsilon\frac{\partial F}{\partial t} \tag{6-2-17}$$

Here F is the electric field, x is the space coordinate, n is the electron concentration, N_D is the concentration of shallow ionized donors, ε is the dielectric permittivity, q is the electronic charge, I is the total current density, and μ is the low-field mobility. In the simplest case of a uniform sample in a constant electric field (F = const), with constant total current density, I, the solution of eqs. (6-2-16) and (6-2-17) is given by

$$n^{(o)} = N_D \tag{6-2-18}$$

$$F^{(o)} = \frac{I}{q\mu N_D} \tag{6-2-19}$$

Small-signal analysis is based on seeking the solutions of eqs. (6-2-16) and (6-2-17) in the following form:

$$n = n^{(o)} + n_1 \cdot \exp[i(\omega t - kx)] \tag{6-2-20}$$

$$F = F^{(o)} + F_1 \cdot \exp[i(\omega t - kx)] \tag{6-2-21}$$

where $n_1 << n^{(o)}$ and $F_1 << F^{(o)}$. Substitution of eqs. (6-2-20) and (6-2-21) into eqs. (6-2-16) and (6-2-17) yields two uniform algebraic equations for n_1 and F_1, which have a nonzero solution if and only if the system determinant is zero. This leads to the following equation:

$$\omega = -kv(F^{(o)}) - i\left[\frac{1}{\tau_{md}(F^{(o)})} + D_n k^2\right] \tag{6-2-22}$$

where

$$\tau_{md} = \frac{\varepsilon}{qn_o\, dv/dF} \tag{6-2-23}$$

is the differential dielectric relaxation time.

Equation (6-2-22), called the *dispersion equation,* may be used to describe the evolution of the sinusoidal fluctuation of the electron concentration, which is related to the space-charge and electric-field fluctuations (called the *space-charge waves*).

The wavelength (the characteristic size of the fluctuation) λ is related to the wave vector:

$$k = 2\pi/\lambda \tag{6-2-24}$$

The real part of ω,

$$\mathrm{Re}(\omega) = -kv(F^{(o)}) \tag{6-2-25}$$

determines the frequency, and the imaginary part of ω,

$$\mathrm{Im}(\omega) = -\frac{1}{\tau_{\mathrm{md}}} - D_n k^2 \tag{6-2-26}$$

determines the attenuation or growth time constant of the fluctuation.

If $\mathrm{Im}(\omega) < 0$, the fluctuation decays, and if $\mathrm{Im}(\omega) > 0$, the fluctuation grows. This means that

$$\tau_{\mathrm{md}} < 0 \tag{6-2-27}$$

is the condition of instability. The values of k for unstable fluctuations are limited by

$$0 \leq k < \frac{1}{(D_n|\tau_{\mathrm{md}}|)^{1/2}} \tag{6-2-28}$$

Equation (6-2-22) can also be used to analyze how the fluctuation induced at a given frequency varies in space. In this case we consider ω real and k complex. Neglecting diffusion, we find from eq. (6-2-22) that

$$\mathrm{Re}(k) = -\frac{\omega}{v} \tag{6-2-29}$$

$$\mathrm{Im}(k) = -\frac{1}{\tau_{\mathrm{md}} v} \tag{6-2-30}$$

As can be seen from eqs. (6-2-29) and (6-2-30), the wave grows in the direction of propagation, i.e., from the cathode to the anode when $\tau_{\mathrm{md}} < 0$. The magnitude of this increase is proportional to $\exp[\mathrm{Im}(k)L]$. Hence, two possible limiting cases are $\mathrm{Im}(k)L << 1$ and $\mathrm{Im}(k)L >> 1$.

In the first case, fluctuation growth over the length, equal to the wavelength of the fluctuation, $\lambda = 2\pi/k$, is small. Under such conditions the device can behave as an amplifier. In one possible scheme, such an amplification is achieved using a sample with two coupling probes—one near the cathode and one near the anode (see Fig. 6-2-4). The input signal is applied to the probe close to the cathode contact. It excites a space-charge wave with a wave vector, k, related to the frequency of the input signal by eq. (6-2-22). The exited space-charge wave grows and induces a larger (amplified) output signal at the output probe. In a practical

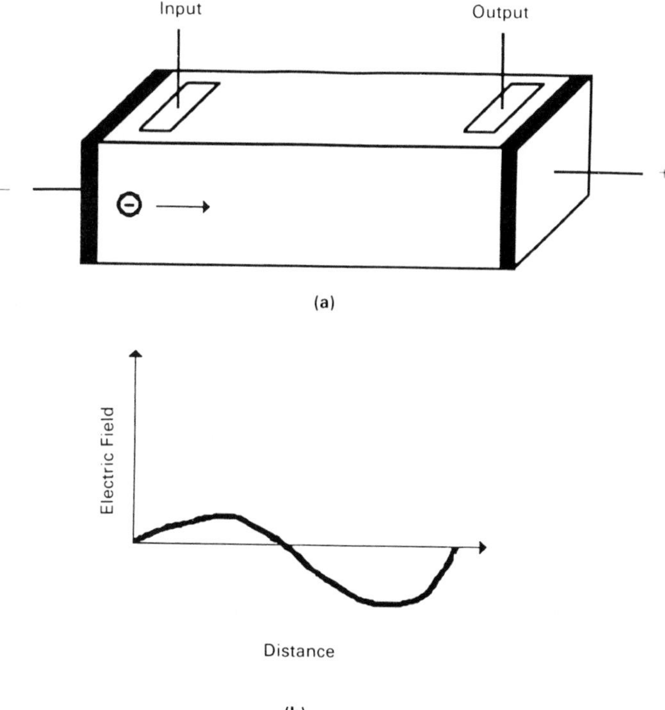

Fig. 6-2-4. Travelling wave amplifier.

device, such an amplification can only happen if an integer number of wavelengths, $\lambda = 2\pi/k$, of the space-charge wave can fit inside the sample. As can be seen from eq. (6-2-22), this happens when the frequency ω is equal to the transit-time frequency, $\omega_T = 2\pi f_{TR}$, or its harmonics. Here

$$f_{TR} = \frac{v(F^{(o)})}{L} = \frac{1}{T} \tag{6-2-31}$$

where T is the transit time. This is illustrated by Fig. 6-2-5, which shows the conductance of a GaAs sample as a function of frequency. Negative conductance corresponds to the amplification. The amplification at harmonics of the transit frequency is suppressed by diffusion. This is because higher frequencies correspond to higher values of k (see eq. (6-2-22)), and hence, the diffusion term, which is proportional to $D_n Dk^2$, becomes more important.

From the condition $\mathrm{Im}(k)L << 1$ and eqs. (6-2-30) and (6-2-23) we find that such an amplification regime takes place when

$$n_o L < \frac{\varepsilon v}{q|\mu_d|} \tag{6-2-32}$$

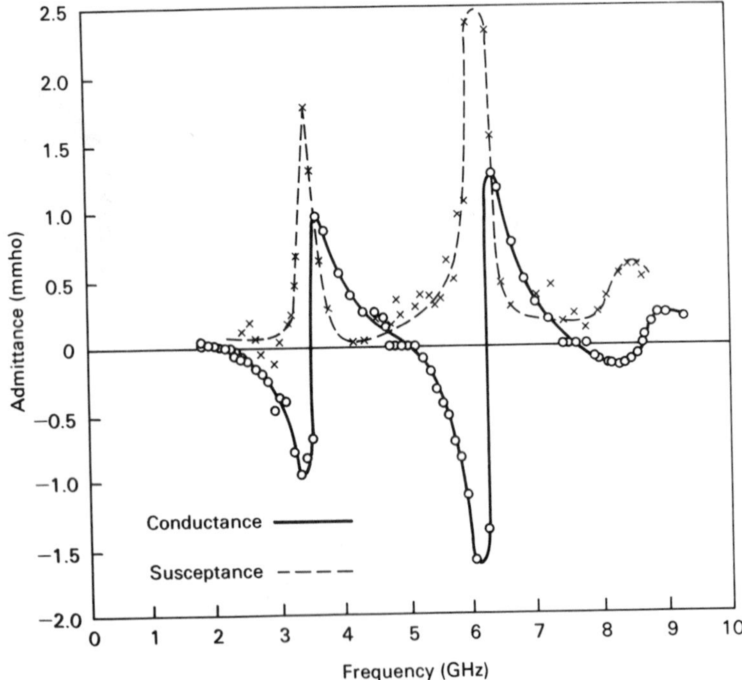

Fig. 6-2-5. Conductance of transferred electron device vs. frequency (from B. W. Hakki, *J. Appl. Phys.*, 38, p. 808 (1967)). Bias electric field 4.8 kV/cm, electron concentration 3×10^{13} cm^{-3}, device length 70 μm, room temperature.

Assuming $v \simeq 10^5$ m/s, $|\mu_d| \simeq 0.05$ m^2/Vs, and $\varepsilon = 1.14 \times 10^{-10}$ F/m, we obtain $n_o L < 1.5 \times 10^{11}$ cm^{-2}. More accurate calculations yield $n_o L < 10^{12}$ cm^{-2}. This condition is called the *Kroemer criterion* because it was first introduced by Kroemer (1965).

When the fluctuation growth, caused by the negative differential conductivity, is large (i.e., $\text{Im}(k)L >> 1$), the fluctuation grows very rapidly over a small fraction of the total sample length. As John Gunn proved in 1963, making elegant measurements of time-dependent potential distributions in his samples (using capacitive probes), propagating regions of high electric field are formed under such conditions (see Fig. 6-2-6).

The mechanism leading to the formation of high-field domains may be explained as follows. Let us assume that the average electric field in the sample is higher than the peak field, F_p, and that a small fluctuation of the electric field is present near the cathode (see Fig. 6-2-7). Such a fluctuation may be present, for example, as a consequence of a slightly smaller doping level.

The higher field in this region leads to a smaller electron velocity because of the negative slope of the v vs. F curve. Hence, the electrons in front of the high-

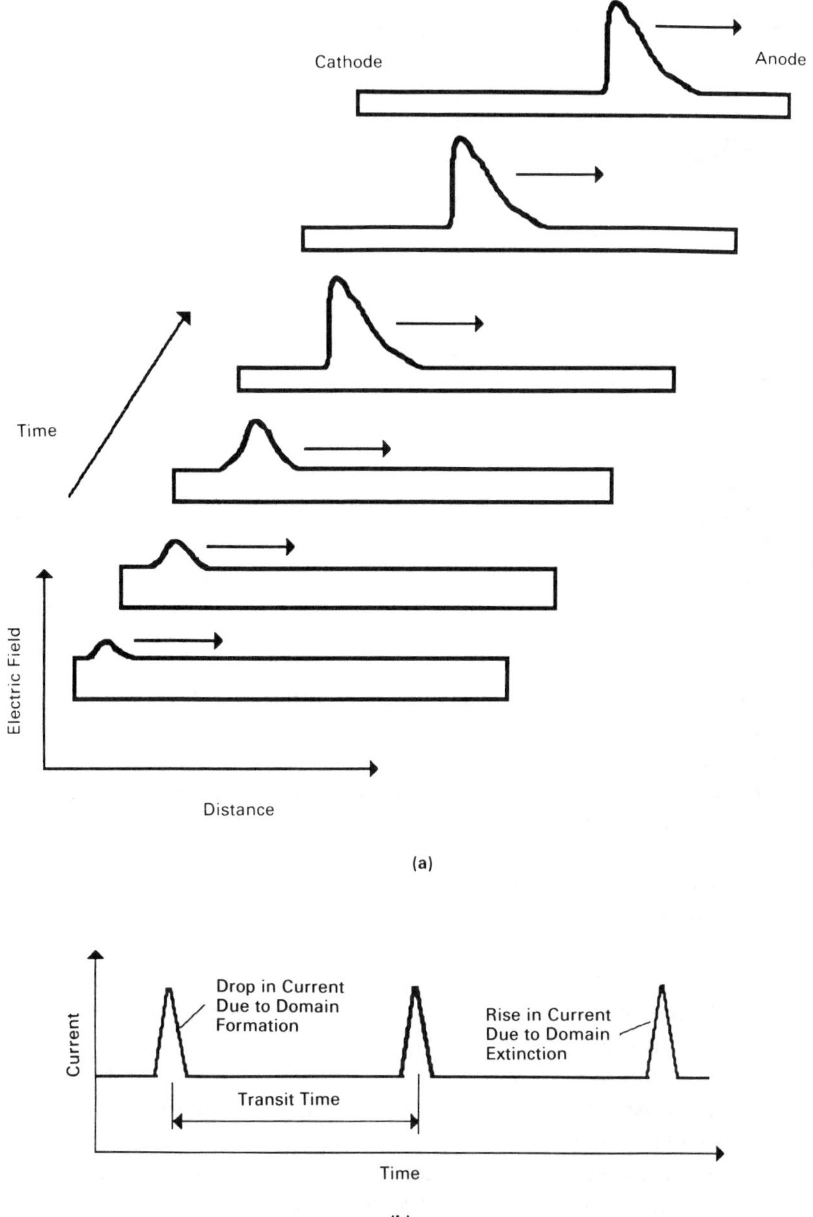

Fig. 6-2-6. Propagation of high field domain (a) and current waveform for sample with propagating domains (b).

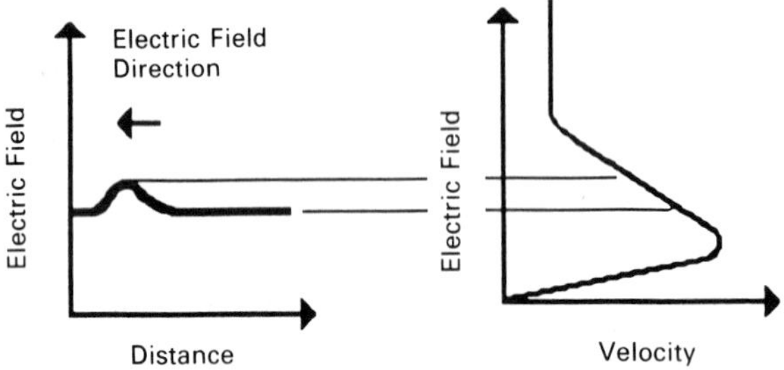

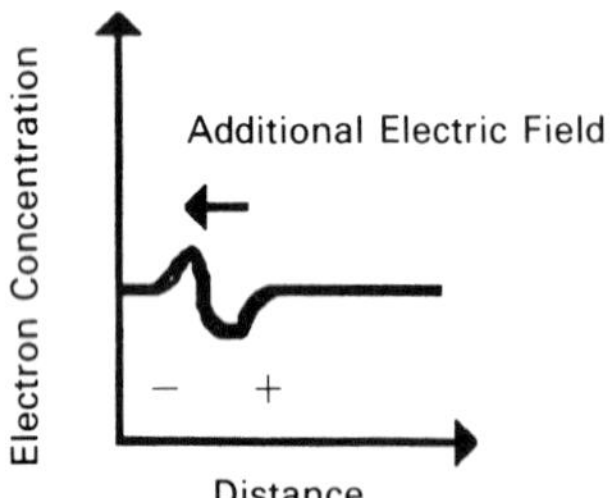

Fig. 6-2-7. Mechanism of domain formation.

field region and behind the high-field region move faster than in the region. This leads to the depletion of electrons in the leading edge and to the accumulation of electrons in the trailing edge. The resulting dipole layer of charge increases the electric field in the region, intensifying the growth of the fluctuation, which propagates toward the anode. If the bias voltage is kept constant, the growth of the fluctuation comes at the expense of the electric field, F_r, outside the domain. The decrease of F_r leads to a current drop during high-field-domain formation. The annihilation of the high-field domain when it reaches the anode causes a temporary increase in the current. Then the next domain nucleates near the cathode and the process repeats itself.

Most often, just one high-field domain forms in the sample because the voltage drop across the domain results in a field smaller than F_p everywhere except within the domain.

In most cases (but not always) the domain forms near the cathode and propagates through the sample with a velocity close to the saturation velocity, v_s. The electrons outside the domain also move with the saturation velocity. Hence, the current density in a sample with a domain is given by

$$j_s = qn_o v_s \tag{6-2-33}$$

When the domain is extinct, the current density is close to the peak current density,

$$j_p = qn_o v_p \tag{6-2-34}$$

where v_p is the peak velocity. During the domain annihilation and the formation of a new domain the current density rises from j_s to j_p. The domain transit time is given by

$$T \simeq L/v_s \tag{6-2-35}$$

The time of the initial growth of a small fluctuation, t_{gr}, is determined by the differential dielectric relaxation time, τ_{md}:

$$t_{gr} \simeq 3|\tau_{md}| = \frac{3\varepsilon}{q|\mu_d|n_o} \tag{6-2-36}$$

For formation of a stable propagating domain, t_{gr} should be considerably smaller than the transit time:

$$\frac{3\varepsilon}{q|\mu_d|n_o} < \frac{L}{v_s} \tag{6-2-37}$$

This condition coincides with the Kroemer criterion discussed previously (see eq. (6-2-32)):

$$n_oL > (n_oL)_l = \frac{3\varepsilon v_s}{q|\mu_d|} \tag{6-2-38}$$

When the n_oL product is close to 10^{12} cm^{-2} or above, the formation of stable propagating domains take place when F exceeds the velocity peak field, F_p. When $(n_oL)_l < 10^{12}$ cm^{-2}, the threshold field for domain formation, F_t, exceeds F_p. The value of F_t depends on n_oL and on the shape of the v vs. F curve.

Once formed, the domain does not disappear if the applied voltage is decreased below the threshold voltage during the domain propagation. The reason is that the field within the domain remains higher than the peak field, F_p, even though the average field may be below the peak field unless the voltage drop is too big. For large values of $n_oL \sim 10^{12}$ cm^{-2} the domain-sustaining field is close to F_s.

The difference between F_t and F_s leads to a hysteresis in the domain current-voltage characteristics. It can be utilized for applications of transferred-electron devices in logic circuits.

For $n_oL >> (n_oL)_l$ the electric field outside the stable domain is close to F_s. The domain field depends on n_oL and the applied voltage and may vary from approximately 30 to 200 kV/cm. It is limited by the critical field for avalanche breakdown.

The time of stable domain formation may be considerably larger than the initial growth time, $3|\tau_{md}|$. It decreases with an increase in n_o and may vary from a few picoseconds to a hundred picoseconds.

Domain shape is also dependent on the doping density. When the doping density is large (much larger than $\sim 10^{15}$ cm^{-3} for GaAs) a small relative change in the equilibrium electron concentration is sufficient to produce a large space charge, supporting a high domain field. In this case the domain shape is nearly symmetrical. In the opposite limiting case ($n_o << 10^{15}$ cm^{-3}), the leading edge is totally depleted of carriers, creating a positive space-charge density $qn_o = q(N_D - N_A)$. At the same time, the density of electrons in the accumulation layer (i.e., in the trailing domain edge) is limited only by the diffusion processes and can exceed n_o many times. The resulting field distribution is very similar to the field distribution in a p$^+$-n junction (see Fig. 6-2-8) moving with a domain velocity that is close to the saturation velocity, v_s. In the nearly totally depleted leading edge the drift and diffusion currents are small, and the current density is determined by the displacement current,

$$j = \varepsilon \frac{\partial F}{\partial t} \simeq \varepsilon v_s \frac{qn_o}{\varepsilon} \tag{6-2-39}$$

so that the total current is

$$I = qv_s n_o S \tag{6-2-40}$$

Equations (6-2-39) and (6-2-40) are valid even when n_o or the device cross section S depend on the space coordinate x, i.e., are valid for a device with a nonuniform

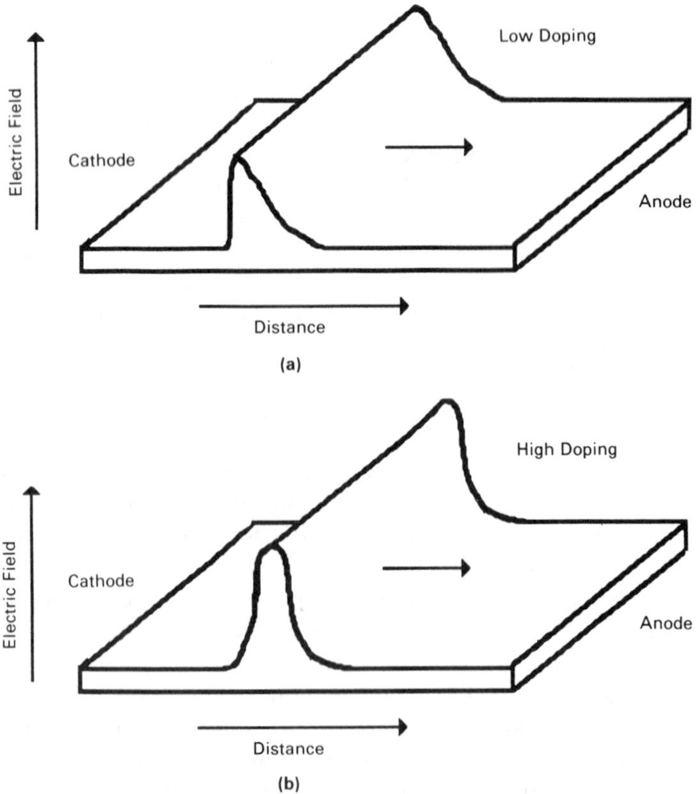

Fig. 6-2-8. Domain shape in low doped (a) and highly doped (b) samples.

doping profile or a complex shape. This means that the current waveform should reproduce the $n_o(x)S(x)$ profile:

$$I(t) = qn_o(v_s t)S(v_s t)v_s \tag{6-2-41}$$

where t is counted from the time of domain formation (see Fig. 6-2-9). This domain property may be utilized for implementing numerous analog and logic functions (see, for example, Shoji 1967 and Engelbrecht 1967).

Although we derived eq. (6-2-41) for the limiting case of relatively low-doped samples when the leading edge is totally depleted, it is approximately valid for an arbitrary doping level.

Stable propagating domains exist in relatively long samples, which are used as microwave oscillators at relatively low frequencies (10 to 20 GHz or less). In

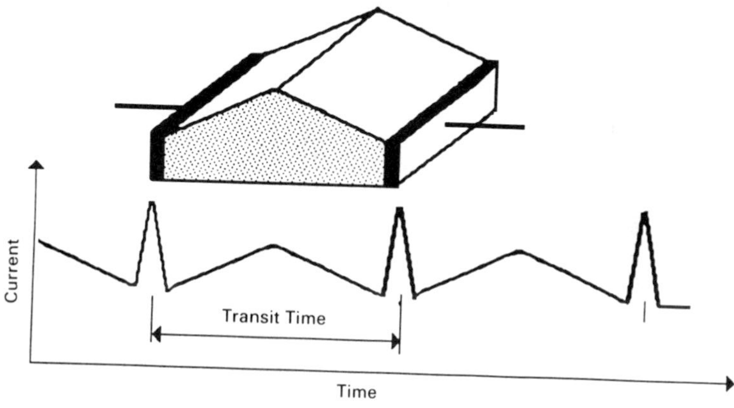

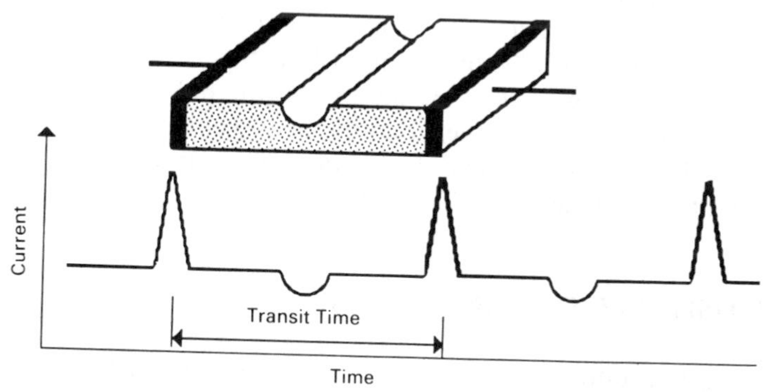

Fig. 6-2-9. Current waveform in samples with nonuniform cross-sections. Spikes in the current waveform correspond to domain annihilation and to formation of a new domain. The part of the waveform between the spikes reproduces the shape of the device.

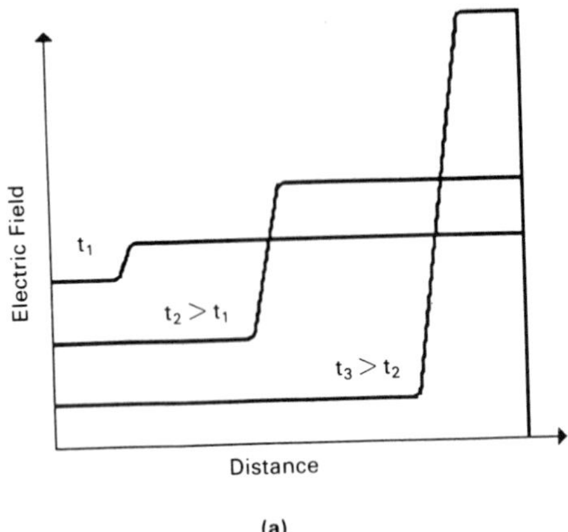

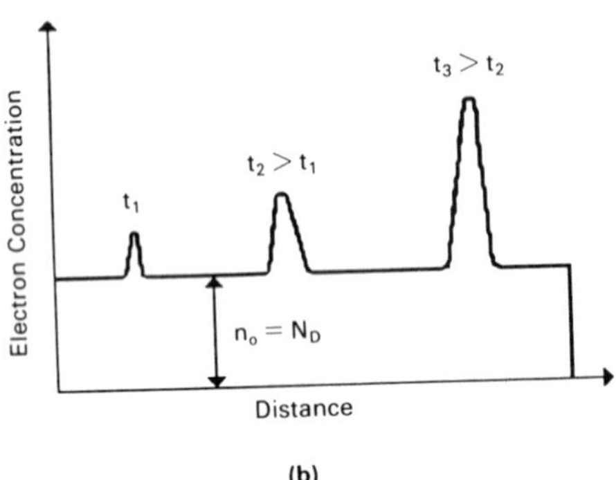

Fig. 6-2-10. Propagation of accumulation layer in GaAs sample: (a) electric field distributions at different times, *t* and (b) electron concentration at different times, *t*.

short samples, operating as microwave oscillators at higher frequencies (50 GHz and above), a typical form of instability is the propagation of *accumulation layers* (see Fig. 6-2-10). The theory describing this mode of operation was developed by D'yakonov et al. (1981). It shows that the propagation velocity of accumulation layers may be even larger than the peak velocity in GaAs.

For a more detailed description of the Ridley-Watkins-Hilsum-Gunn effect see, for example, Shur 1987, Chapter 4.

6-3. TRANSFERRED-ELECTRON DEVICES

Negative differential resistance in GaAs and other compound semiconductors (such as InP) can be utilized for generation or amplification of microwave signals. Relatively low-doped or short devices can operate as microwave amplifiers, whereas highly doped or long samples can be used as oscillators operating at microwave frequencies (see eq. (6-2-32)).

Indeed, when the bias voltage corresponding to the negative differential resistance region of the velocity vs. electric-field curve is applied to a GaAs sample with $n_o L \geq 10^{12}$ cm^{-2}, the current waveform has spikes, resulting in the generation of microwave power at the transit frequency (see Fig. 6-2-6b). However, the efficiency of such an oscillator is very low.

Much higher efficiencies can be obtained when the device is placed in a microwave cavity. In a high-Q parallel LRC circuit (see Fig. 6-3-1a) the voltage across the device is equal to

$$U = U_o + U_1 \cos \omega t \tag{6-3-1}$$

where U_o is the dc bias voltage, U_1 is the ac voltage amplitude, and ω is the resonance frequency of the circuit (including the device).

In a series LRC circuit (see Fig. 6-3-1b), the current I through the device is controlled by the external circuit,

$$I = I_o + I_1 \cos \omega t \tag{6-3-2}$$

where I_o is the dc current component and I_1 is the ac current amplitude.

In a practical microwave circuit the voltage or current waveforms may include the second and higher harmonics of the resonance frequency.

A microwave cavity is usually used to provide an equivalent parallel or series LRC circuit. The amplitude of the ac voltage in the parallel LRC circuit is usually chosen in such a way that

$$U_o - U_1 < F_s L \tag{6-3-3}$$

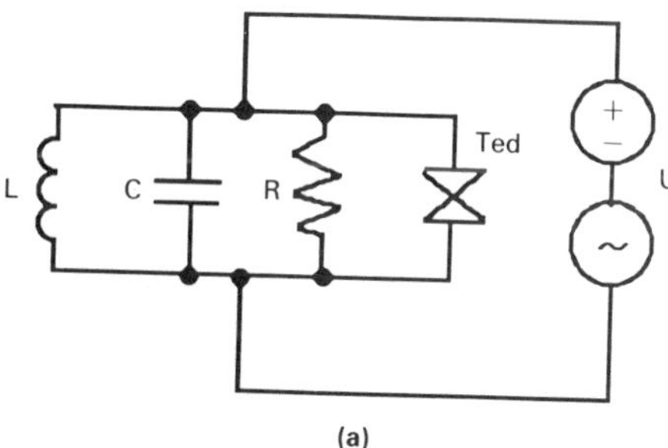

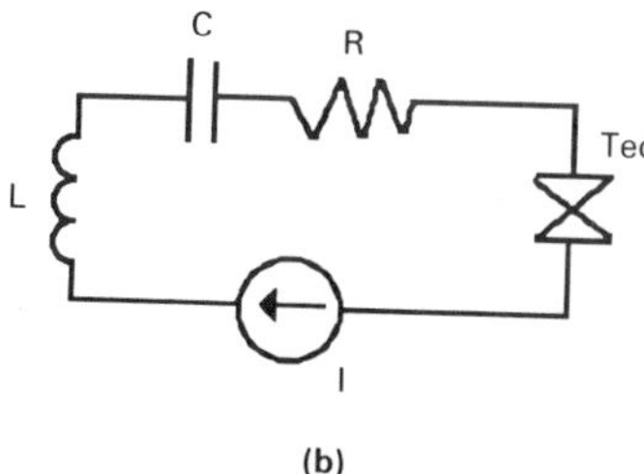

Fig. 6-3-1. (a) Parallel and (b) series resonant LRC circuits for transferred electron devices.

where F_s is the domain-sustaining field, so that the device is biased below the peak field, F_p, for a fraction of each period. During this fraction of the period, the space charge in the device decays, and hence, propagation of high-field domains or other instabilities are controlled by the applied voltage.

As was first pointed out by Kroemer (1965), when the frequency ω is large enough so that

$$\omega\tau_f >> 1 \tag{6-3-4}$$

where τ_f is the domain formation time, the high-field domains do not form. In this case the current is determined by the electron drift velocity,

$$i = qn_o v[F(t)] \tag{6-3-5}$$

where

$$F(t) = \frac{U(t)}{L} \tag{6-3-6}$$

is the bias electric field, L is the sample length, and n_o is the electron concentration in the sample. (As was discussed in Section 1-11, at very high frequencies the electron drift velocity cannot be considered as an instantaneous function of the bias. Retardation effects related to a finite energy relaxation time have to be included in a more realistic analysis of this mode of operation.) Such a regime is called the *Limited Space charge Accumulation mode* (LSA; see Copeland 1966 and 1967). In many cases an accumulation layer (see Section 6-1) propagates in the device during a part of the cycle, but the accumulated charge disappears when the voltage drops below the threshold (see Eastman 1978).

When

$$\omega\tau_f \sim 1 \tag{6-3-7}$$

the domains start to form, but the domain formation cannot be completed during the rf cycle. This mode of operation, is called the *hybrid mode*. The hybrid mode has been demonstrated to have high efficiencies at frequencies close to 10 GHz (see Dizhur et al. 1968, Huang and MacKenzie 1968, and Monroe and Briggs 1968).

If the frequency of the microwave circuit is low enough,

$$\omega\tau_f << 1 \tag{6-3-8}$$

so that the stable domain formation and propagation take place, and the device operates in the *transit mode* of operation. A waveform in the transit regime depends on the ratio ω/ω_T, where $\omega_T = 2\pi f_{TR}$,

$$f_{TR} = \frac{v_s}{L} \tag{6-3-9}$$

is the transit frequency, and v_s is the electron saturation velocity that is approximately equal to the domain propagation velocity (see Section 6-1). If the frequency $f > f_{TR}$, stable domains are quenched before they reach the anode. The current waveform for this mode of operation is shown in Fig. 6-3-2. This transit mode is called the *quenched mode* of operation.

At frequencies smaller than the transit frequency, the domains travel across the diode and disappear at the anode. If the voltage across the device at the end of the domain transit is smaller than the threshold voltage, a new domain nucleates only after the delay that is necessary to reach the threshold voltage (see Fig. 6-3-3). This transit mode is called the *delayed mode* of operation.

As was discussed in Chapter 1 (see Fig. 1-11-1), electron negative differential mobility in GaAs decreases with an increase in the operating frequency. This limits the frequency range of transferred-electron devices. However, the frequency range may be extended by extracting power at higher harmonics of the fundamental frequency. This can be achieved by utilizing a microwave cavity that has some impedance at higher harmonics, so that the voltage waveform becomes

$$U = U_o + U_1 \cos \omega t + U_2 \cos 2\omega t + \ldots \quad (6\text{-}3\text{-}10)$$

This mode of operation, which is sometimes called the *harmonic extraction mode*, was studied, for example, by Eddison and Brookbanks (1981) and Haydl (1981).

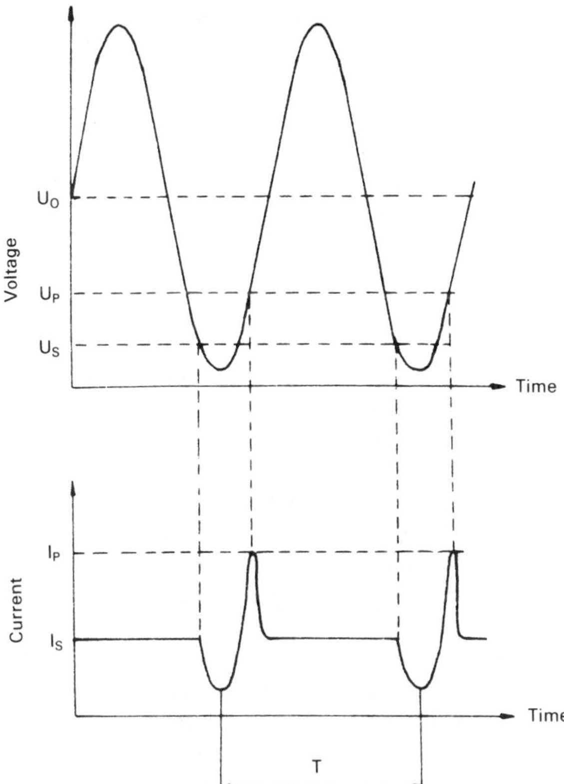

Fig. 6-3-2. Voltage and current waveforms for the quenched mode of operation (from M. Shur, *GaAs Devices and Circuits*, Plenum Publishing, New York (1987)).

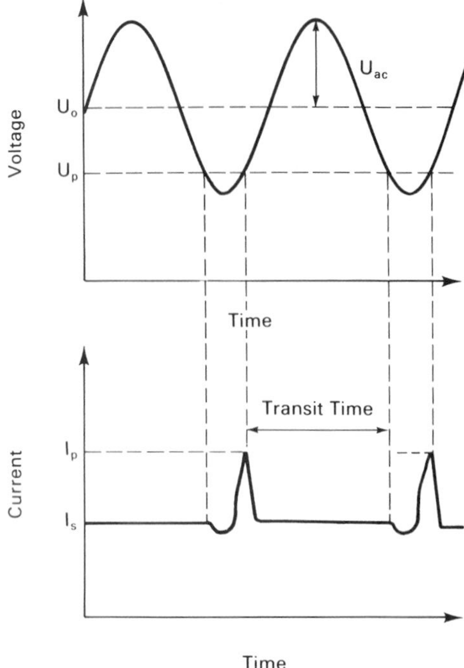

Fig. 6-3-3. Voltage and current waveforms for the delayed mode of operation (from M. Shur, *GaAs Devices and Circuits*, Plenum Publishing, New York (1987)).

Haydl (1981) showed that 1.8- to 2-μm-long devices, operating at the second harmonic, and 2.2- to 2.4-μm devices, operating at the third harmonic, can provide microwave power at frequencies close to 100 GHz. A 1.3- to 1.5-μm-long device, operating at the second harmonic, or a 1.7- to 2-μm device, operating at the third harmonic, may provide power up to 140 GHz.

At frequencies much smaller than the transit frequency, a transferred-electron device may still exhibit a negative differential resistance. Also, the dc I-V characteristic may have a current drop at the threshold voltage due to domain formation. As a result-low-frequency sinusoidal or relaxation oscillations may occur in resonance circuits connected to the device at frequencies from 1 kHz to the transit frequency.

We might also mention the *multidomain* regime of operation, which occurs if the voltage across the diode with doping fluctuations changes very fast (faster than $U_D/\tau_f \sim 10^{12}$ V/s, where U_D is the domain voltage and τ_f is the domain formation time), so that one domain cannot absorb the increasing voltage (see Thim 1968 and Slater and Harrison 1976).

Thus, transferred-electron devices operating in different modes (low-frequency oscillations, transit mode, hybrid mode, and LSA mode) may oscillate in a wide frequency range, from several kilohertz to a few hundred gigahertz. CW powers from a few milliwatts to a few watts have been generated with efficiencies of up to 15% to 20%. In the pulsed mode of operation, output powers have reached several kilowatts, with efficiencies up to 30%. Low cost, long times before failure (up to several decades), acceptable noise figures, and wide tunability have led to many applications of transferred-electron oscillators in microwave systems.

A crude estimate of device efficiency may be obtained by assuming that both voltage and current waveforms are sinusoidal and out of phase:

$$U = U_o + U_1 \cos \omega t \tag{6-3-11}$$

$$I = I_o - I_1 \cos \omega t \tag{6-3-12}$$

Then the output rf power is

$$P_\sim = (\tfrac{1}{2})U_1 I_1 \tag{6-3-13}$$

and the efficiency is

$$\eta = \frac{P_\sim}{P_o} = \frac{U_1 I_1}{2U_o I_o} \tag{6-3-14}$$

For large bias voltages, $U_o \simeq U_1 >> U_p$ (where U_p is the peak voltage), and

$$\frac{I_1}{I_o} \leq \frac{I_p - I_s}{I_p + I_s} \tag{6-3-15}$$

leading to the following estimate for the efficiency, η_{max}, and output power, $P_\sim$:

$$\eta_{max} \leq \frac{1 - I_s/I_p}{2(1 + I_s/I_p)} \tag{6-3-16}$$

$$P_\sim \leq q v_s F_p \eta_{max} \frac{U_o}{U_p} n_o LS \tag{6-3-17}$$

where $U_p = F_p L$, F_p is the peak electric field, q is the electronic charge, v_s is the electron saturation velocity, n_o is the carrier concentration, L is the device length, and S is the device cross section. Assuming that the peak-to-valley ratio for GaAs $I_p/I_s \sim 2.75$, we obtain $\eta_{max} \leq 23.3\%$. The efficiency may be even higher if the current and voltage waveforms have a rectangular (not a sinusoidal) shape. In this case we find that the efficiency at the fundamental frequency is given by

$$\eta_{max} \leq \frac{8}{\pi^2} \frac{1 - I_s/I_p}{1 + I_s/I_p} \tag{6-3-18}$$

(see Kino and Kuru 1969). Thus, for GaAs, $\eta_{max} < 37.8\%$. The actual efficiency value depends on device parameters, bias, and the external circuit. A complete solution of the transport equations for the device, together with the Kirchhoff equations describing the external circuit, is required to simulate the device behavior (see, for example, Shaw et al. 1979).

A simpler approach is based on the assumption that the voltage waveform for the parallel RLC circuit (or the current waveform for the series RLC circuit) is given. This assumption makes it possible to calculate the current waveform and calculate the load resistance at the fundamental frequency. The impedances at

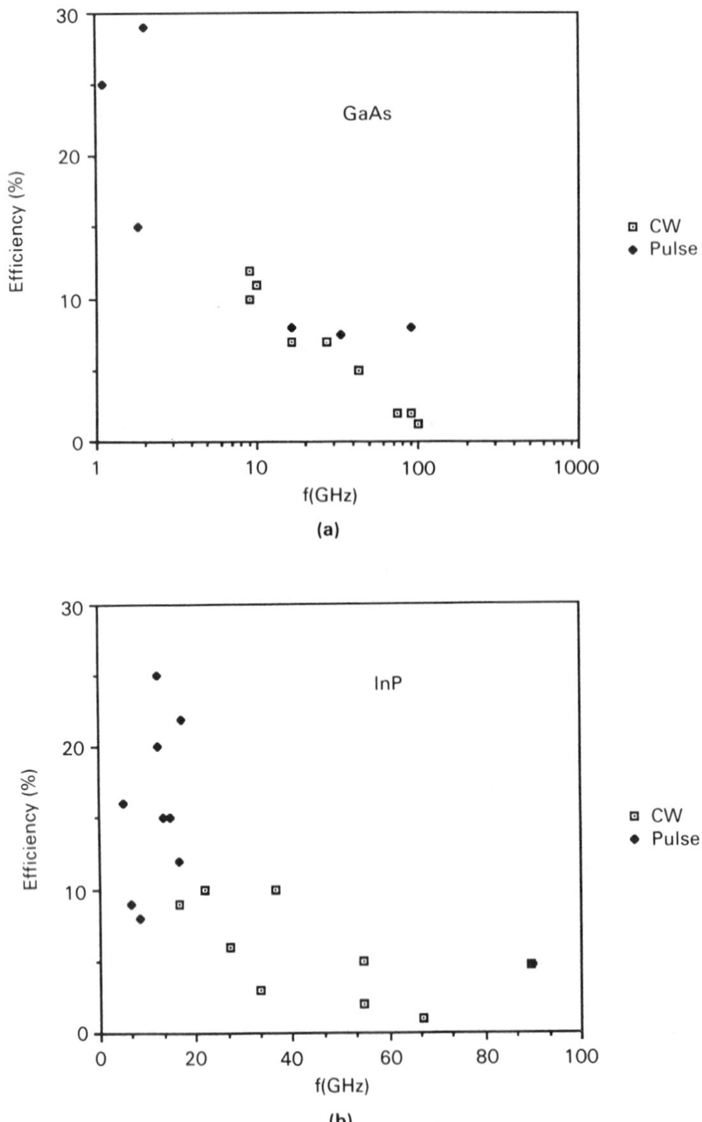

Fig. 6-3-4. Efficiency of (a) GaAs and (b) InP transferred electron oscillators versus frequency (from M. Shur, *GaAs Devices and Circuits*, Plenum Publishing, New York (1987)).

harmonics can also be calculated if the applied voltage has a complex nonsinusoidal shape (see, for example, Levinshtein and Shur 1968).

Efficiency and output microwave power of transferred-electron devices vs. frequency are shown in Figs. 6-3-4 and 6-3-5, respectively. The data points used in these figures are taken from Sze 1981 (Chapter 11, Fig. 30), Eddison and

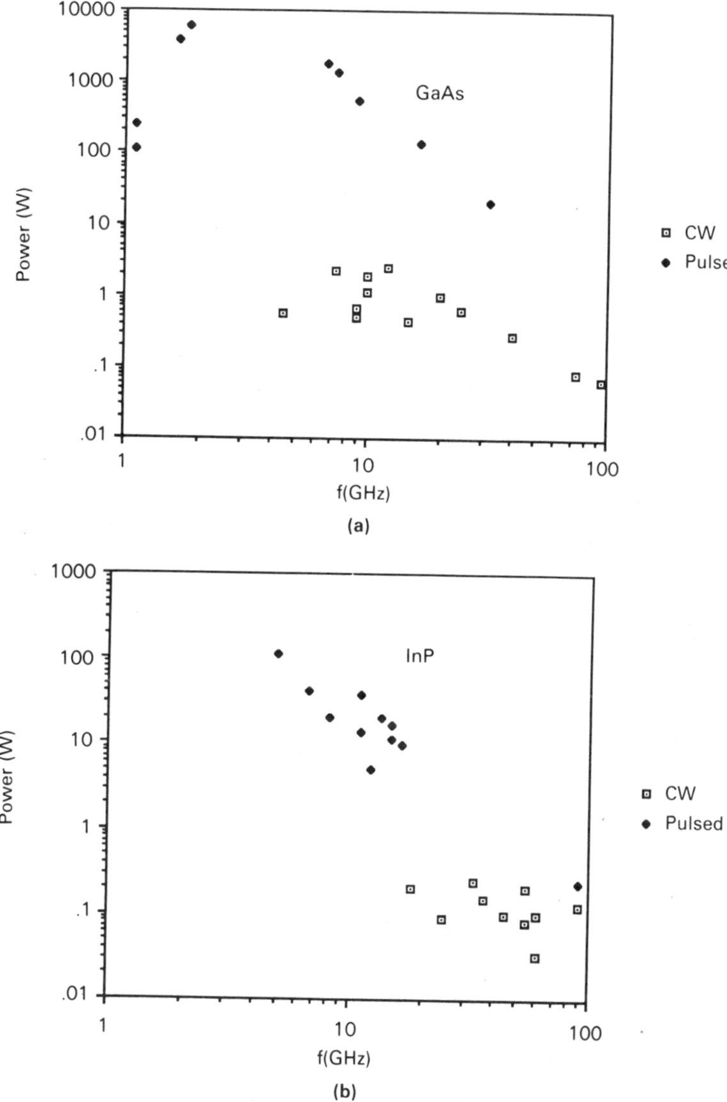

Fig. 6-3-5. Output power of (a) GaAs and (b) InP transferred electron oscillators versus frequency (from M. Shur, *GaAs Devices and Circuits*, Plenum Publishing, New York (1987)).

Brookbanks 1981, Haydl 1981, and Crowley et al. 1980. Recently, higher output power (up to 150 W) in the microwave C-band (3.9 to 6.2 GHz) was reported by Signon and Ayyagari (1987). They achieved such high power by combining several transferred-electron devices (Gunn diodes) operating in pulsed mode (see Fig. 6-3-6).

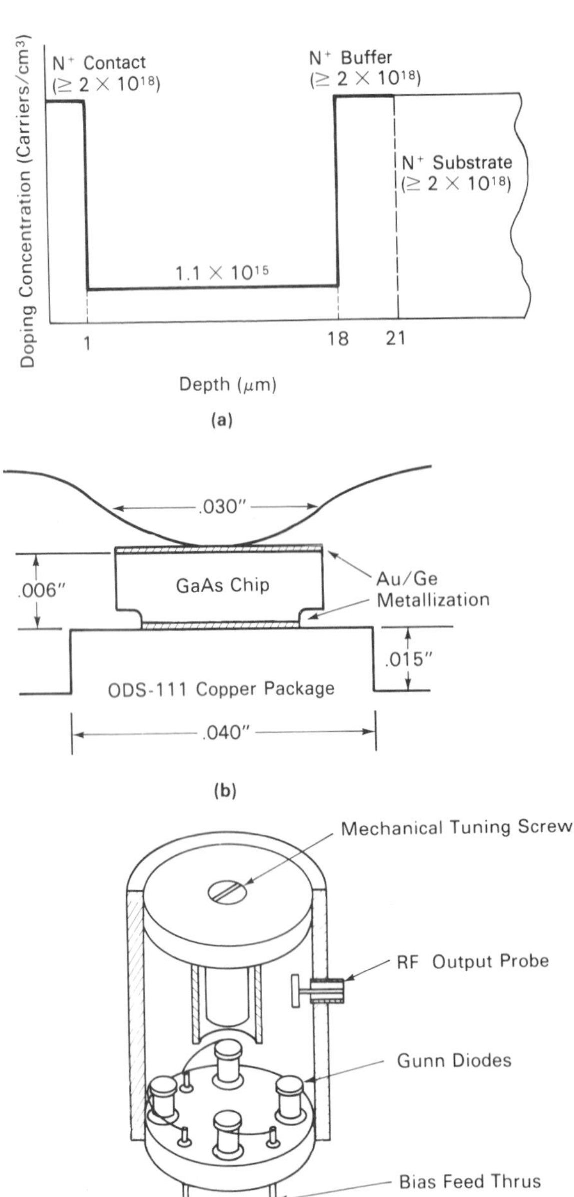

Fig. 6-3-6. Gunn devices operating in microwave C-band (3.9 GHz-6.2 Ghz) in pulsed mode (from B. E. Signon and M. Avvagari, *1987 IEEE MTT Symposium Digest*, pp. 871 (1987). © 1987 IEEE). (a) doping profile, (b) device design, and (c) cut-away view of power combiner with several Gunn diodes.

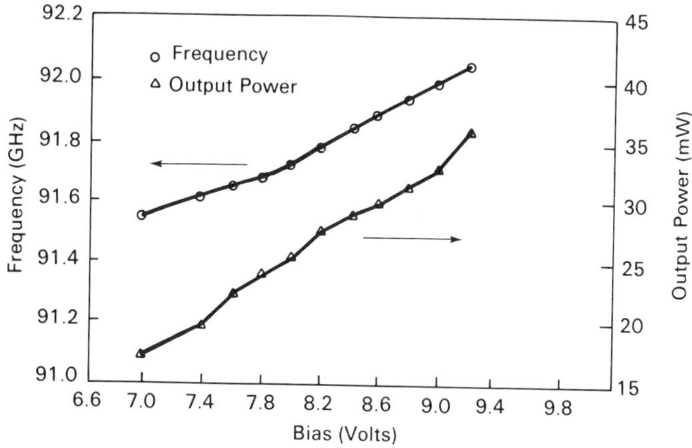

Fig. 6-3-7. Output power and frequency of InP Gunn diode vs. bias (from D. R. Singh, *1987 IEEE MTT Symposium Digest*, pp. 971–972 (1987). © 1987 IEEE).

Fig. 6-3-7 shows the frequency and output power of an InP Gunn oscillator vs. bias (from Singh 1987). The output power of 35 mW at 92 GHz was achieved for this device operating in a microstrip circuit, with a bias voltage tuning range of 400 MHz.

As was mentioned in Section 6-2, negative differential resistance in GaAs, InP, and some other compound semiconductors may also be utilized in microwave amplifiers. This can be done using samples with low values of the n_oL product (see eq. (6-2-32)), where high-field domains do not develop. Such devices are called *subcritically doped amplifiers*. Alternatively, samples with larger values of the n_oL product may be used if the domains are either suppressed or if the amplification occurs at a frequency different from the transit frequency, so that the microwave signal generated by the propagating domains does not interfere with the output signal of the transferred-electron amplifier. Amplifiers of this type are called *supercritically doped amplifiers*. A major drawback of these amplifiers is their relatively large noise, considerably higher than what can be achieved with GaAs or AlGaAs/GaAs field-effect transistors (see Chapter 4).

A more detailed discussion of different transferred-electron devices (including functional devices utilizing nonuniform samples; see Fig. 6-2-9) and their characteristics was given by Shur (1987, Chapters 5 and 6).

6-4. IMPATT, TRAPATT, AND BARITT DIODES

The transferred-electron devices (TEDs) described in Sections 6-1 and 6-2 utilize bulk negative differential resistance related to the intervalley transition in GaAs and other compound semiconductors. In another family of devices, represented by IMPATT, TRAPATT, and BARITT diodes, the phase shifts between the device current and applied voltage create a dynamic negative differential resistance that is used to generate microwave power.

In an IMPact ionization Avalanche Transit Time (IMPATT) diode, two mechanisms create additive phase delays: avalanche breakdown (which is used to generate carriers) and carrier drift across the special drift section of the device.

This principle of operation is illustrated by Fig. 6-4-1, which shows sinusoidal voltage and current waveforms with phase shifts of 90° and 180°. Avalanche breakdown creates a phase delay of 90° between the ac voltage and current, as explained presently. An additional 90° (or so) phase delay is created by utilizing carrier drift across the drift region. As a consequence, the total delay is close to 180°. As can be seen from the figure, when the phase shift is 180°, the current decreases when the voltage increases. This corresponds to a negative differential resistance (i_ω is proportional to $(-V_\omega)$).

A schematic structure of an IMPATT diode, along with doping and field profiles, is shown in Fig. 6-4-2. When a large enough voltage is applied to the device, carriers are generated by the avalanche breakdown process in the high-field region at the p$^+$-n interface. These carriers move in the electric field, creating the electric current at the p$^+$-n interface.

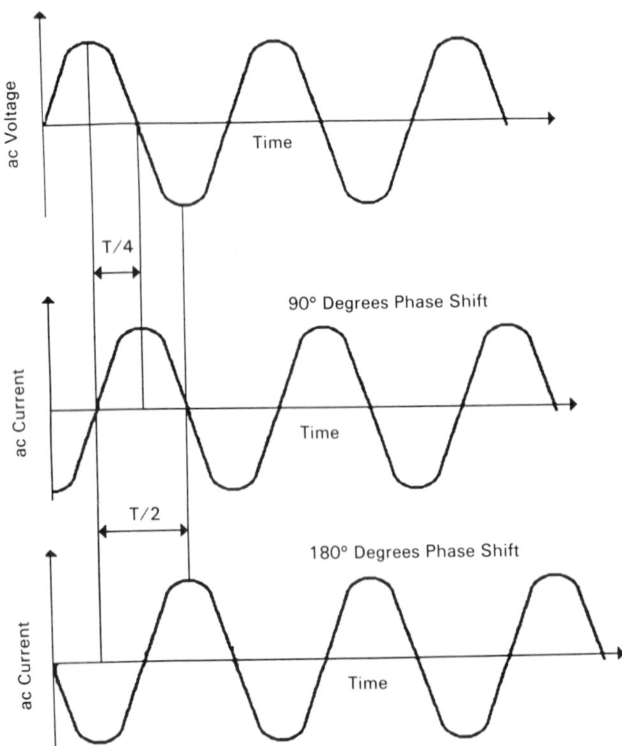

Fig. 6-4-1. Sinusoidal voltage and current waveforms with phase shifts of 90° and 180°.

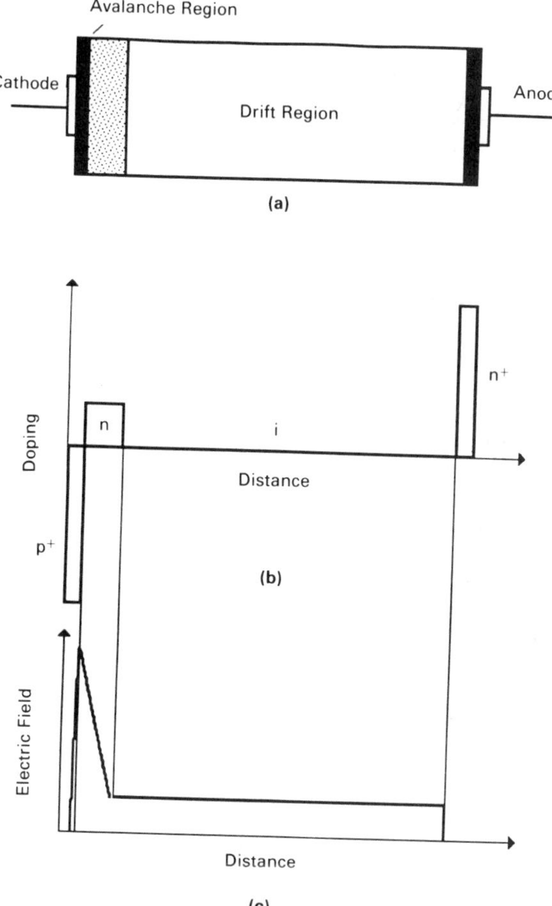

Fig. 6-4-2. Schematic diagram of IMPATT diode.

Let us consider the situation when the sum of the dc bias and ac voltage is applied to the diode (as shown in Fig. 6-4-3). This is a typical situation for a microwave device, such as an IMPATT diode, when it operates in a microwave cavity (see Section 6-3). Let us further assume that the dc voltage applied is equal to the critical voltage needed to initiate avalanche breakdown. The impact ionization generation rate is a very strong (exponential) function of the maximum electric field. Also, the generation rate is proportional to the number of carriers. Hence, when the ac voltage increases the total voltage applied to the device, the generation rate rises sharply. When the ac voltage decreases from the peak value, the generation rate decreases. The rate of current increase is proportional to the generation rate. Hence, the current continues to rise, reaching the peak when the ac voltage drops to zero. This means that the phase of the generated current is delayed by 90° with respect to the ac voltage.

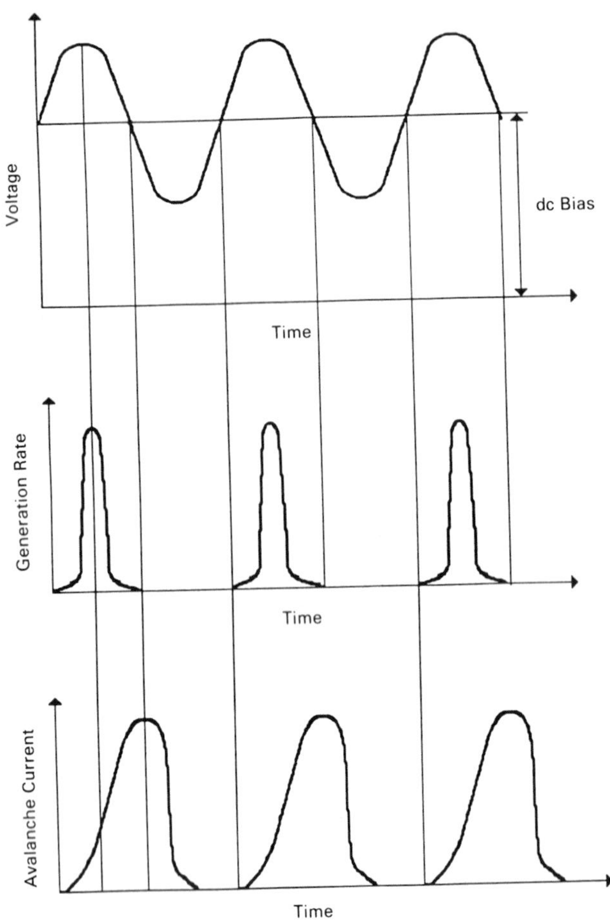

Fig. 6-4-3. Qualitative voltage, generation rate, and current waveforms for IMPATT diode.

To obtain dynamic negative resistance, we need an additional 90° phase delay between the terminal current and the ac voltage. This is achieved by utilizing the drift region, shown in Fig. 6-4-2. This phase shift is equal to ωT_{tr}, where T_{tr} is the transit time of carriers across the drift region and ω is the oscillation frequency.

The idea of a p^+-n-i-n^+ IMPATT device was first proposed by Read (1958), and therefore this device is frequently called the *Read diode*. Johnston et al. (1965) were first to report IMPATT oscillations in an avalanching silicon diode in a microwave cavity. Different modifications of the Read diode have been proposed to optimize the generated current pulse and increase the breakdown voltage, and hence, the generated microwave power. In practice, almost any reverse biased p-n junction may operate in the IMPATT regime. Misawa (1966) showed that even a p-i-n structure, in which the electric field in the undoped region is nearly constant, may operate as an IMPATT diode.

In the device shown in Fig. 6-4-2 electrons generated by avalanche breakdown travel across the drift region, and holes are collected by the cathode contact. An alternative n^+-p-i-p^+ structure in which holes are the carriers propagating through the drift region toward the cathode and electrons are collected by the anode (next to the avalanche region) is also possible. Many practical devices utilize *double-drift IMPATT* diodes, which have two drift regions for electrons and holes back to back with the avalanche region in between. Doping and field profiles for a typical double-drift IMPATT diode are shown in Fig. 6-4-4 (from Vasudev 1984). This particular device was made of GaAs and delivered over 40 W of output power with 20% efficiency in the microwave X-band (8–12 GHz).

In a millimeter wave band (94 GHz and above), silicon IMPATTs have higher power and efficiency than GaAs or InP diodes. This is primarily due to the higher thermal conductivity of silicon (1.5 W/cm-°C) compared with that of GaAs (0.46 W/cm-°C). Output power close to 1 W was obtained for 90 GHz silicon IMPATT diodes (see Midford and Bernick 1979).

Prager et al. (1967) discovered a completely new and more efficient operation regime of IMPATT diodes at relatively low frequencies. Computer simulations of Scharfetter et al. (1968) showed that in this new regime, during part of the cycle, the drift region of the diode is filled by a high-density electron-hole plasma that is "trapped" within the device. In this state, the device impedance is very small, and hence, the voltage across the device is low. The electrons and holes are removed from the device via electron and hole transit to the contacts. Therefore, this new regime is called the TRApped Plasma Avalanche Triggered Transit (TRAPATT) regime, and avalanche diodes operating in this regime are called TRAPATT diodes.

The physics of this new regime may be understood by considering the electric current in the undoped (drift) region. The total current through the device should be continuous. The avalanche current in the avalanche region is carried by

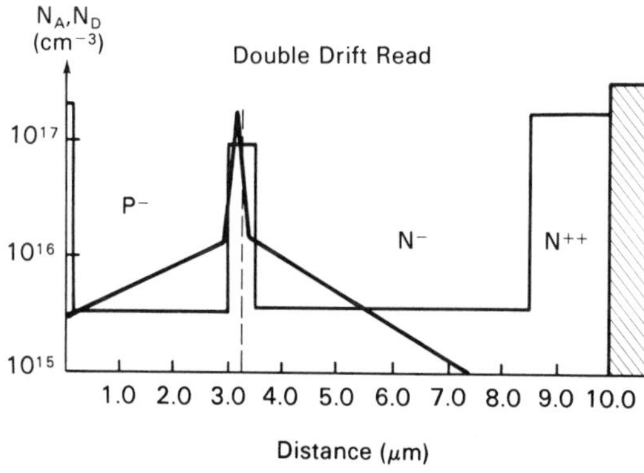

Fig. 6-4-4. Doping and field profiles for GaAs Double Drift IMPATT diode (from P. K. Vasudev, *IEEE Trans. Electron Devices*, ED-31, no. 8, pp. 1044–1050 (1984). © 1984 IEEE).

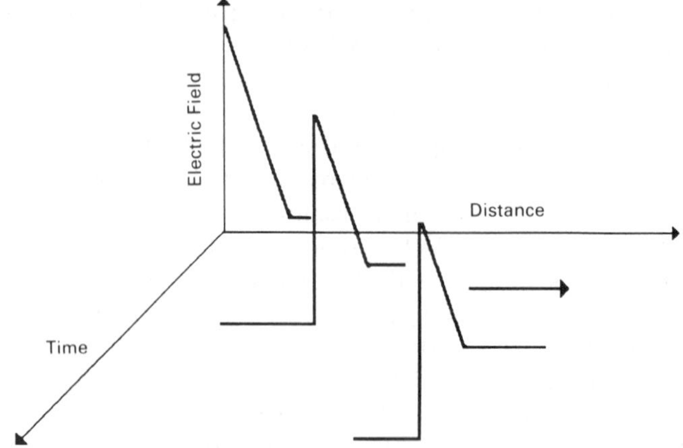

Fig. 6-4-5. Propagation of avalanche breakdown region in TRAPATT diode.

electrons and holes. However, there are no carriers in the drift region, and hence, the current in this region must be the displacement current with the current density $\varepsilon \partial F/\partial t = \varepsilon \omega F$, where F is the electric field. The maximum value of the displacement current is $\varepsilon \omega F_{max}$, where F_{max} is the maximum field in the structure. But what happens if the avalanche current is greater than $\varepsilon \omega F_{max}$? Then the current continuity is maintained by the electric field redistribution in such a way that the avalanche region moves toward the anode, leaving high-density electron-hole plasma behind. This process is schematically illustrated by Fig. 6-4-5.

An additional phase delay between current and voltage waveforms may also be created by utilizing carrier thermionic injection over a Schottky barrier, rather

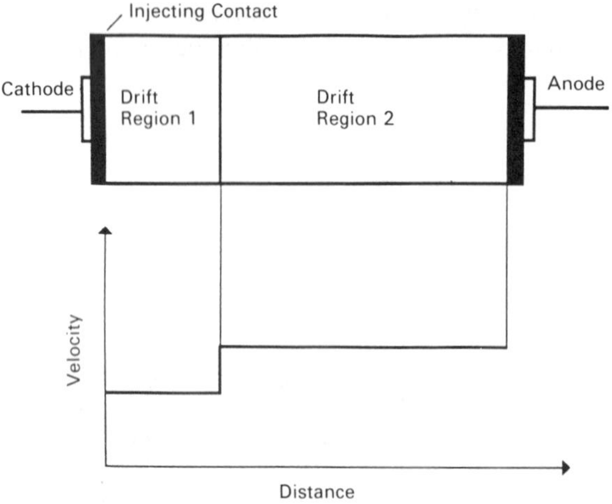

Fig. 6-4-6. Schematic velocity profile for DOuble VElocity Transit Time (DOVETT) diode.

than avalanche breakdown. Such a mechanism is used in Barrier Injection Transit Time (BARITT) diodes. Such structures were first proposed by Ruegg (1968) and by Wright (1968). The operation of a BARITT device was reported by Coleman and Sze (1981).

A possible modification of a BARITT diode is a DOuble VElocity Transit Time (DOVETT) diode. In this device, there are two drift regions and the electron drift velocity in the region near the injecting contact is substantially smaller (see Fig. 6-4-6).

Typical output powers and efficiencies of IMPATT, BARITT, and TRAPATT diodes are compared in Fig. 6-4-7. The main problem with avalanche diodes, however, is the large noise associated with the avalanche-breakdown proc-

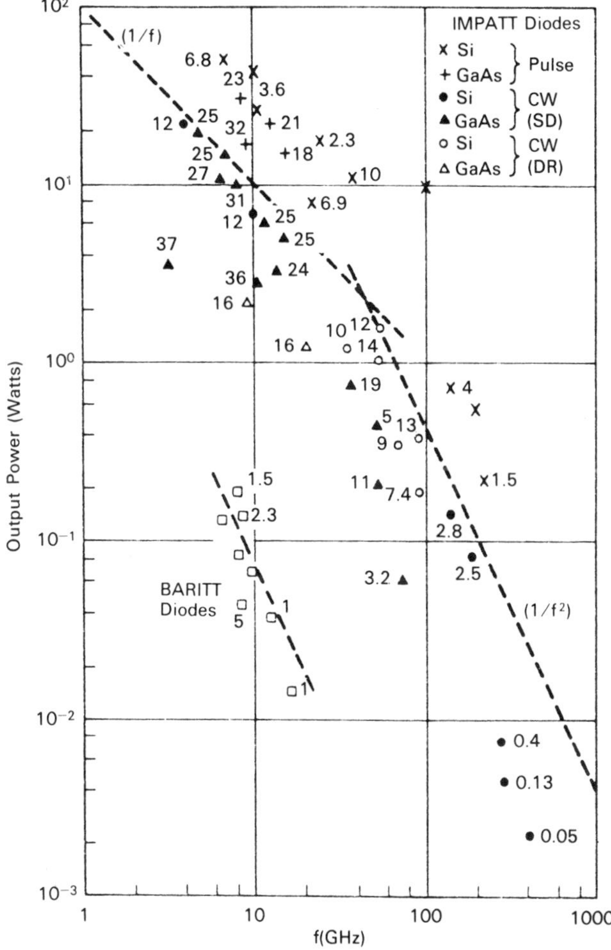

Fig. 6-4-7. Typical output powers of IMPATT, BARITT, and TRAPATT diodes (from S. M. Sze, *Physics of Semiconductor Devices*, John Wiley & Sons, New York (1981)). Numbers near experimental points indicate efficiencies (in percent). SD are single drift structures and DR are double drift structures.

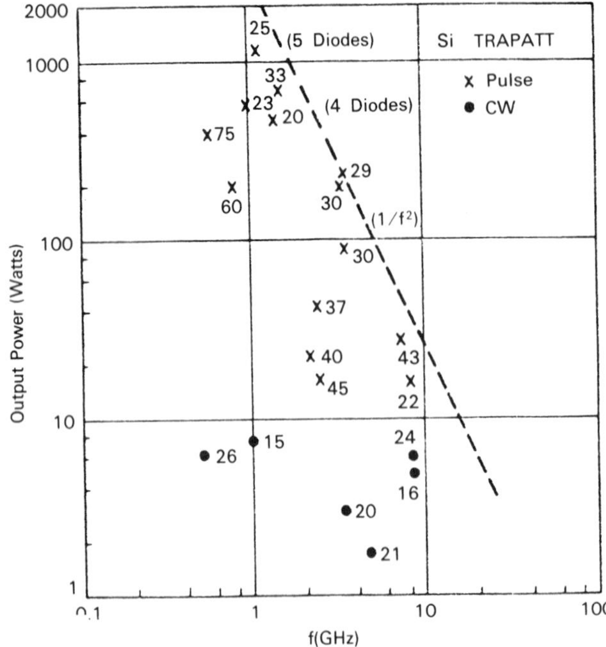

Fig. 6-4-7. Cont.

ess. In this respect GaAs IMPATT diodes have a clear advantage compared with Si IMPATT diodes (see, for example, Sze 1981, p. 603), and transferred-electron devices have the advantage over all IMPATTs (see Carrol 1970 for comparison.) The technology that has the lowest noise at microwave frequencies is compound semiconductor field-effect transistor technology (see Sections 4-10 and 4-11).

REFERENCES

J. Carrol, *Hot Electron Microwave Generators,* Edward Arnolds, London (1970).

D. J. Coleman and S. M. Sze, *Bell Syst. Tech. J.*, 50, p. 1695 (1981).

J. A. Copeland, *Proc. IEEE,* 54, no. 10, pp. 1479–1480 (1966).

J. A. Copeland, *J. Appl. Phys,* 38, no. 8, pp. 3096–3101 (1967).

J. D. Crowley, J. J. Sowers, B. A. Janis, and F. B. Fank, "High Efficiency 90 GHz InP Gunn Oscillators," *Electronics Lett.,* 16, no. 18, pp. 705–706 (1980).

D. P. Dizhur, M. E. Levinstein, and M. S. Shur, *Electronics Lett.*, 4, no. 21, pp. 444–446 (1968).

M. I. D'yakonov, M. E. Levinstein, and G. S. Simin, *Soviet Phys. Semicond.*, 13, no. 7, pp. 1365–1368 (1981).

L. F. EASTMAN, "Transferred-Electron Devices," in *Microwave Devices: Device-Circuit Interaction,* ed. M. J. Howes and D. V. Morgan, Wiley, Chichester (1978).

R. S. ENGELBRECHT, "Bulk Effect Devices for Future Transmission Systems," *Bell Lab. Rec.,* 45, no. 6, pp. 192–201 (1967).

J. B. GUNN, "Microwave Oscillations of Current in III-V Semiconductors," Solid State Communications, 1, no. 4, pp. 88–91 (1963).

B. W. HAKKI, *J. Appl. Phys.,* 38, p. 808 (1967).

C. HILSUM, "Transferred Electron Amplifiers and Oscillators," *Proc. IRE,* 50, 2, pp. 185–189 (1962).

H. C. HUANG and L. A. MACKENZIE, *Proc. IEEE,* 56, no. 7, pp. 1232–1233 (1968).

R. L. JOHNSTON, B. C. DELOACH, JR., and B. G. COHEN, *Bell Syst. Tech. J.,* 44, p. 369 (1965).

G. S. KINO and I. KURU, *IEEE Trans. Electron Devices,* ED-16, no. 9, pp. 735–748 (1969).

H. KROEMER, *Proc. IEEE,* 53, no. 9, p. 1246 (1965).

M. E. LEVINSHTEIN and M. SHUR, *Electronics Lett.,* 4, no. 11, pp. 233–235 (1968).

T. A. MIDFORD and R. L. BERNICK, *IEEE Trans. Microwave Theory and Technique,* MTT-27, pp. 483–492 (1979).

T. MISAWA, *IEEE Trans. Electron Devices,* ED-13, no. 9, p. 137 (1966).

I. W. MONROE and F. M. BRIGGS, *Proc. IEEE,* 56, no. 11, pp. 2066–2068 (1968).

Y. K. POZELA and A. REKLAITIS, *Solid State Electronics,* 23, pp. 927–933 (1980).

H. J. PRAGER, K. K. N. CHANG, and S. WEISBROD, *Proc. IEEE,* 55, p. 586 (1967).

W. T. READ, *Bell Syst. Tech. J.,* 37, p. 401 (1958).

B. K. RIDLEY and T. B. WATKINS, "The Possibility of Negative Resistance," *Proc. Phys. Soc.,* 78, no. 8, pp. 293–304 (1961).

J. G. RUCH and W. FAWCETT, *J. Appl. Phys.,* 41, no. 9, pp. 3843–3849 (1970).

H. W. RUEGG, *IEEE Trans. Electron Devices,* ED-15, p. 577 (1968).

D. L. SCHARFETTER, D. J. BARTELINL, H. K. GUMMEL, and R. L. JOHNSON, *IEEE Trans. Electron Devices,* ED-15, p. 691 (1968).

M. P. SHAW, H. GRUBIN, and P. SOLOMON, *The Gunn-Hilsum Effect,* Academic Press, New York (1979).

M. SHOJI, "Functional Bulk Semiconductor Oscillators," *IEEE Trans. Electron Devices,* ED-14, no. 9, pp. 533–546 (1967).

M. SHUR, *GaAs Devices and Circuits,* Plenum Publishing, New York (1987).

B. E. SIGNON and M. AYYAGARI, *1987 IEEE MTT Symposium Digest,* p. 871 (1987).

D. R. SINGH, *1987 IEEE MTT Symposium Digest,* pp. 971–972 (1987).

S. M. SZE, *Physics of Semiconductor Devices,* John Wiley & Sons, New York (1981).

P. K. VASUDEV, *IEEE Trans. Electron Devices,* ED-31, no. 8, pp. 1044–1050 (1984).

G. T. WRIGHT, *Electronics Lett.,* 4, p. 453 (1968).

J. XU, B. A. BERNHARDT, M. SHUR, C. H. CHEN, and A. PECZALSKI, *Appl. Phys. Lett.,* 49, no. 6, pp. 342–344 (1986).

J. XU and M. SHUR, *IEEE Trans. Electron Devices,* ED-34, no. 8, pp. 1831–1832 (1987).

PROBLEMS

6-2-1. Using eq. (6-2-8),

$$v(F) = v_s\left[1 + \frac{F/F_s - 1}{1 + A(F/F_s)^t}\right] \tag{6-2-8}$$

where

$$A = 0.6[e^{10(\mu-0.2)} + e^{-35(\mu-0.2)}]^{-1} + 0.01$$

$$t = 4[1 + 320/\sinh(40\mu)]$$

$$v_s\ (10^5\ \text{m/s}) = 0.6 + 0.6\mu - 0.2\mu^2$$

(here μ is in m²/Vs), calculate and plot the value of

$$(n_o L)_{cr} = \frac{\varepsilon_V}{q|\mu_d|}$$

(see eq. (6-2-32)) vs. electric field, F, for electric fields larger than the peak electric field. Assume $\mu = 6000$ cm²/Vs, $\varepsilon = 1.14 \times 10^{-10}$ F/m.

6-3-1. The output power, $P_\sim$, of a transferred-electron oscillator can be found as

$$P_\sim \approx q v_s F_p \eta (U_o/U_p) n_o L S \tag{P6-3-1}$$

where $U_p = F_p L$ is the peak voltage, F_p is the peak electric field, η is efficiency, q is the electronic charge, v_p is the electron peak velocity, v_s is the electron saturation velocity, n_o is the carrier concentration, S is the device cross section, and (U_o/U_p) is the ratio of the bias voltage over the peak voltage (compare with eq. (6-3-17)). Calculate the output power as a function of frequency, f, for GaAs devices and compare with data in Fig. 6-3-5a. Assume that $\eta \approx 7\%$, $v_s = 10^5$ m/s, $n_o L = 5 \times 10^{12}$ cm^{-2}, $F_p = 3.5$ kV/cm, $U_o/U_p = 5$, $S = K_s L^2$ where K_s is a constant, and $L = K_l/f$, where K_l is a constant. What values of $K_{sl} = K_s K_l^2$ give the best fit with the experimental data shown in Fig. 6-3-5a?

Hint: Fit data for CW and pulsed oscillators separately.

6-4-1. Calculate the critical avalanche current density, J_{cr}, required for the propagation of the avalanche region in TRAPATT diodes as a function of frequency, f. Assume that the dielectric permittivity, ε, is equal to 1.14×10^{-10} F/m and that the maximum field in the structure $F_{max} = 500$ kV/cm.

7

Novel Transistor Structures

7-1. ELECTRON TRANSPORT IN SHORT DEVICES AND COMPOUND SEMICONDUCTOR TECHNOLOGY

In this chapter we will consider several novel devices that have the potential of achieving ultra-high-speed operation. These devices include the Permeable Base Transistor (PBT), the Vertical Ballistic Transistor (VBT), Planar Doped Barrier (PDB) devices, Hot Electron Transistors (HETs) based on the effect of real-space hot-electron transfer in heterostructures, superlattice devices, and resonant tunneling structures.

Most of these new transistors utilize extremely short dimensions and are based on compound semiconductor technology. The increase in operating frequency with the scaling down of device size was already noticed in the seventeenth century when Marin Mersenne of Paris (1588–1648) stated: "I inferred that the number of vibrations of the shorter string must also be twenty times as great" (quoted by J. A. Zahm, *Sound and Music,* Chicago, McClurg, 1892).

In shorter devices, carrier transit times are reduced, leading to shorter propagation delays and higher operating frequencies. In addition, a smaller active device volume means lower power required for a switching event.

Silicon devices dominate conventional electronics. However, submicron devices made of compound semiconductors successfully compete in the area of microwave and ultrafast digital circuits. By combining elements from columns III and V of the periodic table (such as Ga and As, for example) one creates compound semiconductors with the same number of valence electrons per atom as in Si (four). Compounds such as GaAs, InP, InAs, InSb, and AlAs have semiconducting properties and band structures somewhat similar to those of classic elemental semiconductors, such as silicon or germanium. At the same time, they provide a range of materials with different band gaps (direct and indirect), different lattice constants, and other physical properties for the semiconductor device designer. Moreover, some of these elements can be combined to form solid-state solutions such as $Al_xGa_{1-x}As$, where the composition, x, may vary continuously from 0 to 1, with a corresponding change in physical properties from those of GaAs into those of AlAs.

Gallium arsenide is the most studied and understood compound semiconductor material. It has proven indispensable for many device applications—from ultra-high-speed transistors to lasers and solar cells. Its lattice constant (5.6533 Å) is very close to that of AlAs (5.6611 Å), and the heterointerface between the two materials has a very small density of interface states. Newer technologies, such as Molecular Beam Epitaxy (MBE) and Metal Organic Chemical Vapor Deposition (MOCVD), allow us to grow these materials with very sharp and clean heterointerfaces and to have very precise control over doping and composition profiles, literally (in the case of MBE) changing these parameters within an atomic distance.

Other compound semiconductors important for applications in ultra-high-speed submicron devices include InP, $In_xGa_{1-x}As$, AlN, GaP, and $Ga_xIn_{1-x}As$.

There are several advantages of GaAs and related compound semiconductors for applications in submicron devices. First of all, the effective mass of electrons in GaAs is much smaller than in Si ($0.067m$ in GaAs compared with $0.98m$ longitudinal effective mass and $0.19m$ transverse effective mass in Si, where m is the free electron mass). This leads to much higher electron mobility in GaAs—approximately 8500 cm^2/Vs in pure GaAs at room temperature compared with 1500 cm^2/Vs in silicon. Moreover, in high electric fields, electron velocity in GaAs is also larger (see Fig. 1-9-1), which is even more important in submicron devices, in which electric fields are high. Light electrons in GaAs are much more likely to experience ballistic transport, i.e., to move through the short active region of a high-speed device without having any collisions (with lattice imperfections or phonons). This may boost their velocity far beyond the values expected for long devices. Ballistic transport may become important in very short devices with sizes on the order of 0.1 μm or less. This was first predicted by Shur and Eastman (1979) and was recently observed by Heiblum et al.(1985) and Levy et al. (1985).

A schematic band diagram of the hot-electron transistor used by Heiblum et al. (1985) is shown in Fig. 7-1-1. Electrons tunnel through the emitter-base barrier

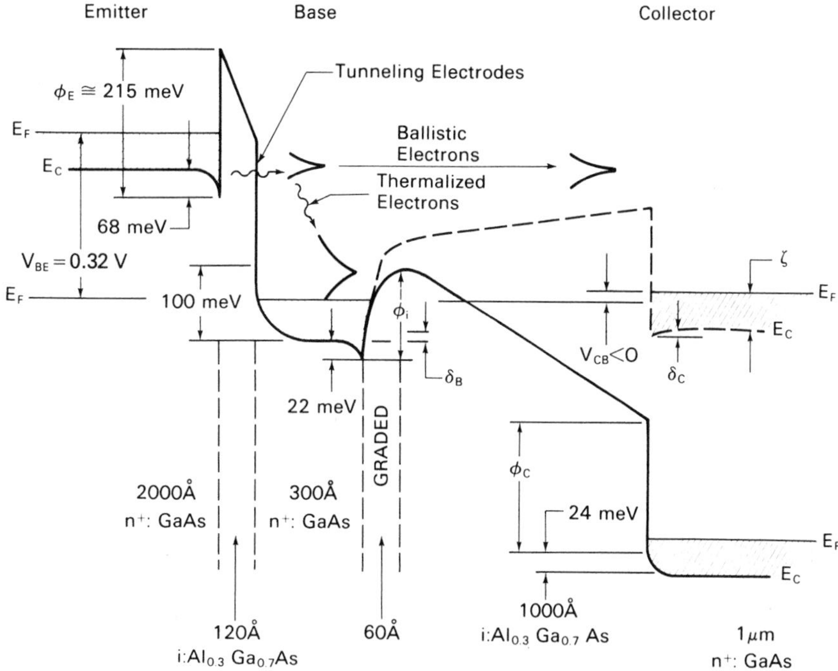

Fig. 7-1-1. Schematic band diagram of hot electron transistor for two different collector-base voltages (from M. Heiblum, M. I. Nathan, D. C. Thomas and C. M. Knoedler, "Direct observation of Ballistic Transportation GaAs," *Phys. Rev. Lett., 55,* p. 2200 (1985)).

and travel across the base. Electrons with energies higher than the collector barrier are collected. By changing the collector voltage, one can change the number of the collected electrons so that the energy distribution of the collected electrons is proportional to the derivative of collector current with respect to collector-base voltage. This measured derivative is shown in Fig. 7-1-2. The sharp peak that increases with emitter current corresponds to ballistic electrons that travel across the base without collisions. At the present time, this device is primarily used as a hot-electron spectrometer that yields information about the energy distribution of hot electrons. However, if parasitics are minimized and base spreading resistance is kept low, it may also be used for ultra-high-frequency applications.

Xu and Shur (1987) proposed a different version of this device—a Double Base Hot Electron Transistor (DBHET), in which the first (doped and/or graded) base region acts as an electron gun, accelerating electrons, and as a lens, providing a better-focused ballistic electron beam that is injected into the second base, where an input signal is applied. The same concept may be applied to form a beam of ballistic electrons in other structures, including a field-effect transistor (see Shur 1989).

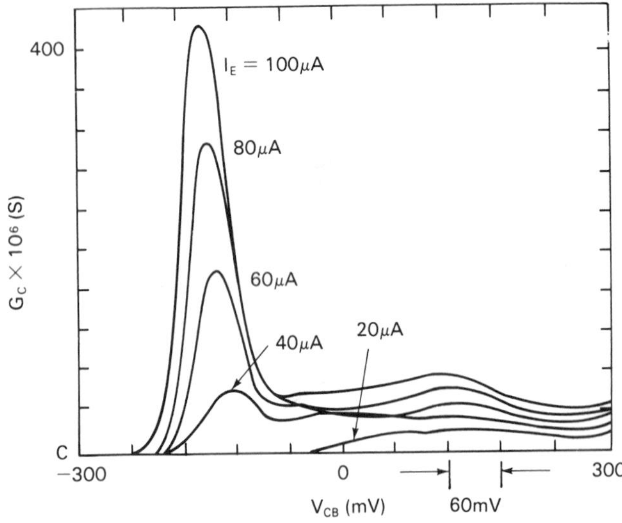

Fig. 7-1-2. Energy distribution of collected electrons in hot electron transistor for different emitter currents (from M. Heiblum, M. I. Nathan, D. C. Thomas and C. M. Knoedler, "Direct observation of Ballistic Transportation GaAs," *Phys. Rev. Lett.*, *55*, p. 2200 (1985)). $G_c = \partial I_c/\partial V_{cb}$ where I_c is the collector current and V_{cb} is the base-collector voltage.

In somewhat longer GaAs devices (with dimensions between 0.1 and 1.5 μm or so) *overshoot* effects are important. These effects (first predicted by the computer simulations of Ruch [1972]) are related to the finite time that it takes for an electron to change its energy, and they may also result in boosting the electron velocity to considerably higher levels than the equilibrium values observed in long-channel devices. The increase of electron velocity in short samples is illustrated by Fig. 7-1-3 (from Shur 1976), which shows the evolution of electron velocity as a function of distance in a constant electric field. Very close to the injecting contact, electrons are moving ballistically and electron velocity is proportional to time. Further from the contact the velocity reaches a peak value and then decreases. However, owing to overshoot effects, the peak value of velocity is far higher than the equilibrium value reached as the distance increases further.

More information about compound semiconductor technology and short-channel submicron devices may be found in books by Shur (1987a) and Watts (1988).

In silicon, ballistic and overshoot effects may also be important. However, they are much less pronounced because of a larger electron effective mass. To summarize, electrons in GaAs, and many related compounds, are faster than in silicon. This increase of effective velocity in short-channel field-effect transistors is illustrated by Fig. 4-10-4 (from Cappy et al. 1980), which shows computed electron velocities vs. distance for field-effect transistors fabricated from different materials. We also show, for comparison, the values of peak velocities in long-channel devices fabricated from the same materials.

Devices such as the hot-electron transistor shown in Fig. 7-1-1, heterostructure FETs, and HBTs (considered in Chapters 3 and 4) are possible because of the availability of excellent compound heterostructure systems, such as

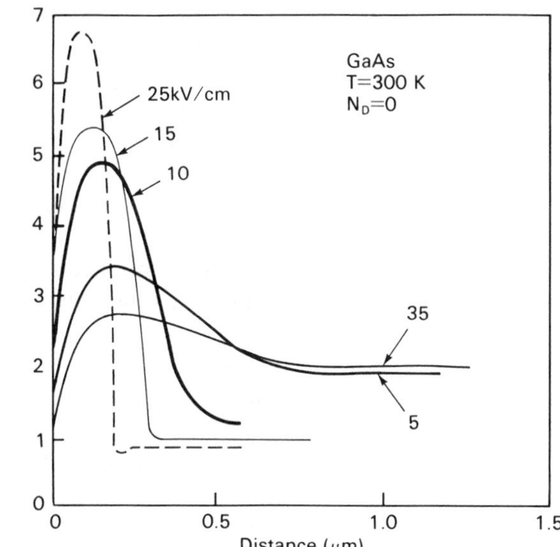

Fig. 7-1-3. Electron velocity vs. distance for constant electric field (from M. Shur, "Influence of the Non-Uniform Field Distribution in the Channel on the Frequency Performance of GaAs FETs," *Electronics Lett.*, 12, pp. 615–616 (1976). © 1976 IEE). The curves were calculated by solving the equations describing the electron transport using effective energy-dependent momentum and energy relaxation times.

AlGaAs/GaAs, GaInAs/InP, and InGaAs/AlGaAs, and new technologies, such as MBE and MOCVD. These technologies allow us to obtain extremely sharp heterointerfaces with low densities of interface states as well as to change the composition of a compound, such as $Al_xGa_{1-x}As$, as a function of distance in a controlled fashion. This approach, which is sometimes called energy-band engineering, opens unlimited opportunities for experimentation with new device structures.

Another important advantage of compound semiconductors such as GaAs and InP for applications in high-speed submicron devices is the availability of semi-insulating substrates that eliminate parasitic capacitances related to junction isolation in silicon circuits and allow us to fabricate microstrip lines with small losses (the latter being especially important for applications in microwave monolithic integrated circuits). We may speculate that with the development of high-T_c superconductors, which may be used in such microstrip lines, this advantage will become especially important.

Also, as was mentioned in Chapter 5, GaAs and related compounds are widely used in optoelectronic applications. The direct gap makes possible a monolithic integration of ultra-high-speed submicron transistors and lasers or light-emitting diodes on the same chip for applications in optical communications. This and the resulting faster recombination rates may also lead to better radiation hardness.

There are also drawbacks of compound semiconductor technology. Silicon is blessed with an excellent native oxide. Silicon nitride is also used as an excellent insulator in silicon field-effect transistors. A poor quality of oxide on GaAs and a corresponding high density of interface states make it difficult to fabricate a

GaAs metal oxide semiconductor field-effect transistor. A wide-band-gap AlGaAs and, more recently, AlN may substitute for such an insulator, but only in a limited way. Only very recently, the new approach of oxidizing a thin silicon layer grown epitaxially on the GaAs surface has offered hope for the development of viable GaAs MOSFET technology (see Tiwari et al. 1988).

In a compound semiconductor, such as GaAs, the material composition is of the utmost importance. For example, defects may be caused by a deficiency of arsenic atoms. Silicon is an elemental semiconductor. Of course, silicon purity determines the device quality and one does not have to worry about composition.

Silicon is abundantly available in nature. Gallium is a relatively rare element. Arsenic is toxic. As a consequence, GaAs wafers cost many times more than comparable silicon wafers and may be in short supply.

GaAs thermal conductivity at 300 K is three times less than that for silicon (0.46 W/cm-°C for GaAs compared with 1.5 W/cm-°C for silicon).

Compound semiconductor transistors are widely used for microwave and ultra-high-speed applications, in which their high-speed properties are the most important. Other possible uses include high-temperature electronics (because of the larger energy gap in GaAs), power devices (because of a higher breakdown field and an ability to speed up their turn-on by light), optoelectronic applications, and radiation-hard electronics (because of the direct band gap). However, microwave and ultra-high-speed applications are the mainstream of compound semiconductor technology. Hence, scaling down the device sizes to enhance the electron velocity by ballistic or overshoot effects and reduce the transit time is especially important for compound semiconductor transistors.

For all novel devices, an accurate control of the vertical device dimensions and doping profile is needed to maintain device uniformity across the wafer, which is a necessary precondition for achieving high yield or large scale integration. The necessity of achieving a very low contact resistance in vertical devices is another practical limitation. Finally, parasitics and interconnects play an important role in practical devices. All these factors make it difficult to reach, and in some cases even to estimate, the potential performance limits of novel devices. But technological difficulties have never stopped the emergence of new ideas, new materials, and new device concepts. Some of these concepts are considered in the following sections.

7-2. PERMEABLE BASE TRANSISTORS

In a Permeable Base Transistor (PBT) an ultrafine metal grid is incorporated into an epitaxial GaAs film (see Fig. 7-2-1; from Bozler et al. 1979). The device consists of four layers: an n^+ substrate, an n-type emitter layer, a thin-film tungsten grid, and an n-type collector layer. The doping concentration, N_D, in the n-layer is such that the zero bias depletion width around the Schottky barrier is larger than the space between the tungsten strips. As a result, the current is zero at zero

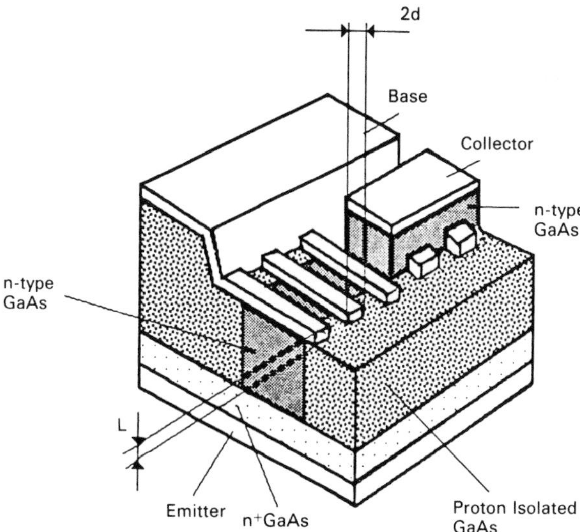

Fig. 7-2-1. Permeable base transistor. A tungsten grating with 300 Å fingers is embedded into a single crystal of GaAs. The electrons flow from emitter to collector through the slits in the grating. (From Bozler et al. (1979), reprinted with permission of Lincoln Laboratory, Massachusetts Institute of Technology, Lexington, Massachusetts.)

emitter-base voltage. When a positive bias is applied to the base the depletion layer shrinks and a conductive path may form between the collector and the emitter. In fact, numerical simulations show that the collector current starts flowing even when the openings between the fingers of the grid are still depleted (see Bozler and Alley 1980). A positive bias applied to the grid decreases the potential barrier for the electrons. In this sense, the operation of a PBT has some similarity to a bipolar junction transistor, in which the emitter-base voltage changes the potential barrier for the emitter-base junction.

The PBT fabrication sequence includes epitaxial growth of a GaAs layer, electron beam evaporation of the tungsten grating using X-ray lithography, and GaAs epitaxial growth over the tungsten grating. Standard nickel-gold-germanium ohmic contacts are used. The devices are isolated using proton bombardment. Such a bombardment creates heavy damage in GaAs, producing deep traps that make GaAs semi-insulating.

The effect of doping density and device dimensions on the cutoff frequency, f_T, is shown in Fig. 7-2-2 (from Bozler and Alley 1980). In this calculation, the device dimensions were scaled down with N_D so that the threshold voltage, V_T, remains nearly constant. As can be seen from the figure, f_T increases rapidly with an increase in doping and with a commensurate decrease in device size.

The current-voltage characteristics of a PBT are shown in Fig. 7-2-3 (from Bozler and Alley 1980). The small-signal equivalent circuit of the PBT is shown in Fig. 7-2-4. For this particular device the cutoff frequency $f_T \approx 37$ GHz (maximum oscillation frequency $f_{max} \approx 10.4$ GHz). Much higher values of the maximum oscillation frequency ($f_{max} > 100$ GHz) have been predicted (see Alley et al. 1982). The application of PBTs to logic circuits has also been proposed (Bozler

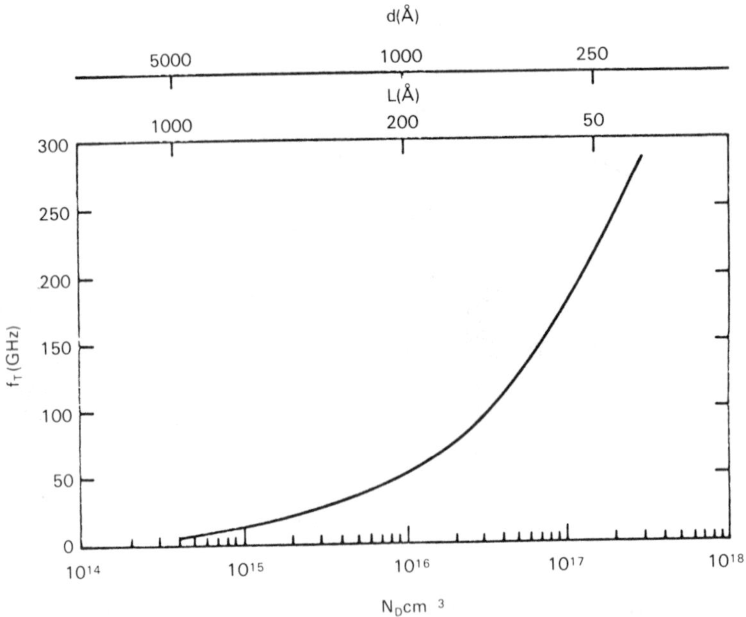

Fig. 7-2-2. PBT cutoff frequency f_T vs. doping. Device dimensions are scaled down with the increase in doping as shown in the figure. (From Bozler and Allen (1980), reprinted with permission of Lincoln Laboratory, Massachusetts Institute of Technology, Lexington, Massachusetts.)

and Alley 1982). Vojak et al. (1984) fabricated a dual self-aligned PBT. The advantage of this transistor is the possibility of four-terminal device operation.

Recently, a bipolar version of a PBT in which one of the injecting contacts is p-type and the other is n-type was proposed (see Fig. 7-2-5). In this device, the electron-hole plasma is injected into the spacings between the isolated fingers of

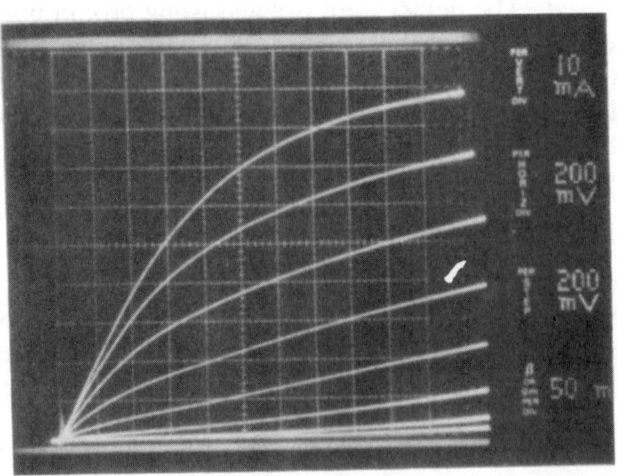

Fig. 7-2-3. PBT collector current vs. emitter-collector voltage. The top curve corresponds to the emitter-base voltage $V_{BE} = 0.5$ V. (From Bozler and Allen (1980), reprinted with permission of Lincoln Laboratory, Massachusetts Institute of Technology, Lexington, Massachusetts.)

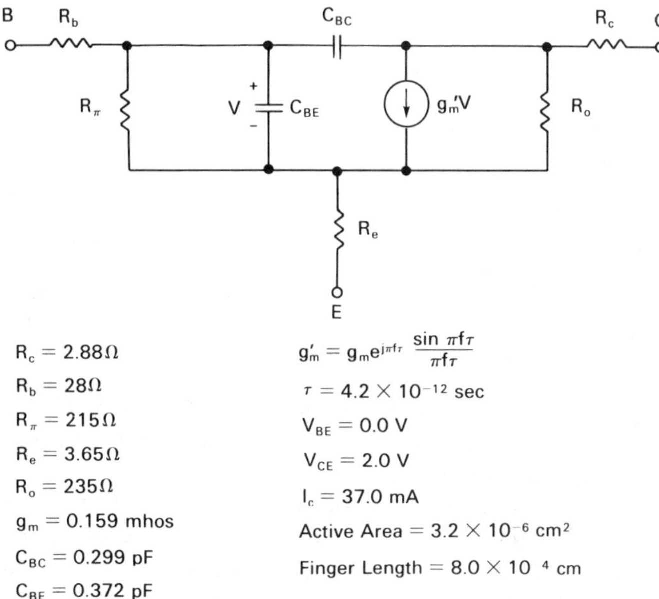

Fig. 7-2-4. PBT equivalent circuit and element values derived from measured data. Bias voltages are $V_{CE} = 2.0$ V and $V_{BE} = 0.0$ V. (From Bozler and Allen (1980), reprinted with permission of Lincoln Laboratory, Massachusetts Institute of Technology, Lexington, Massachusetts.)

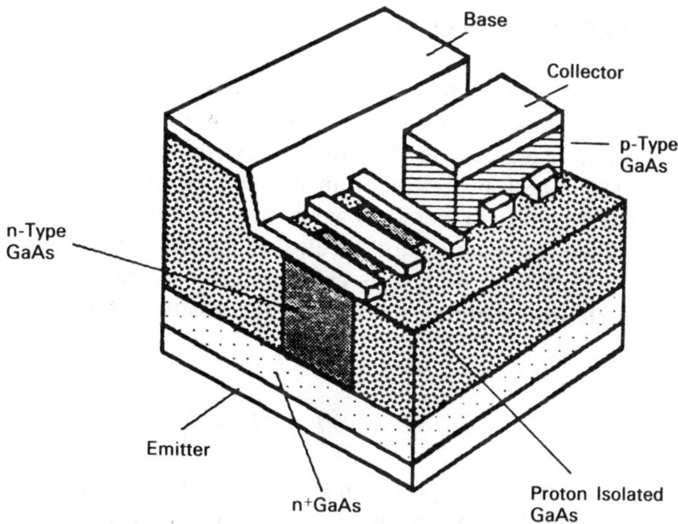

Fig. 7-2-5. Bipolar permeable base transistor (from M. Shur, "Bipolar Permeable Base Transistor, in *Proceedings of IEEE 1978 Bipolar Circuits and Technology Meeting,* Minneapolis, p. 37 (1987). © 1987 IEEE).

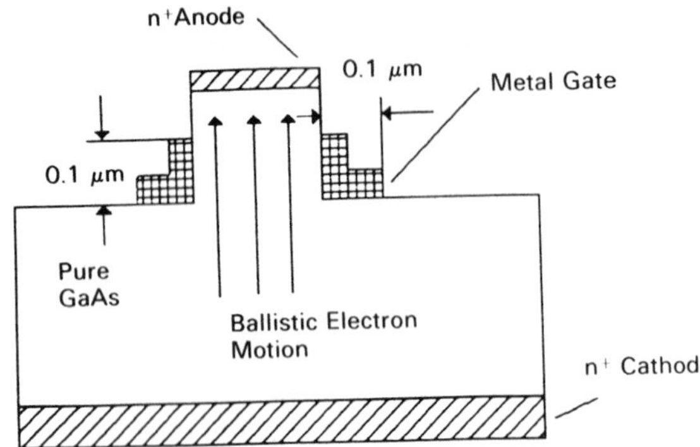

Fig. 7-2-6. Vertical Ballistic Transistor (VBT) (from L. F. Eastman, R. Stall, D. Woodard, N. Dandekar, C. Wood, M. S. Shur, and K. Board, "Ballistic Electron Motion in GaAs at Room Temperature," *Electron. Lett.*, 16, pp. 524–525 (1980). © 1980 IEEE).

the grid. The density of the plasma is modulated by changing the grid potential, similar to such modulation in amorphous silicon double-injection field-effect transistors (see Section 4-14).

Eastman et al. (1980) proposed a device that they called a Vertical Ballistic Transistor (VBT; see Fig. 7-2-6). The principle of operation of this device is similar to that of a PBT. The gate electrodes deplete the conduction path by fringing fields. The main advantage of this structure is that the gate electrodes are on the surface of the device, not embedded in the device as in a permeable base transistor. This makes the structure much easier to fabricate. The device performance should be close to the predicted performance of a permeable base transistor.

7-3. PLANAR DOPED BARRIER DEVICES

The planar doped barrier structure utilizes a plane of a p-type dopant positioned within an undoped region sandwiched between two n^+ regions (see Fig. 7-3-1a). An alternative planar doped barrier structure is a p^+-i-n^+-i-p^+ structure (see Fig. 7-3-1b). The p-region doping in an n^+-i-p^+-i-n^+ planar doped barrier structure is such that it is fully depleted. Two narrow positively charged depletion regions in the n^+ layers maintain charge neutrality for the entire device. As a result of this charge distribution the field is constant between the n^+ layers and the p^+ plane, with a field discontinuity induced by the positive charge of the p^+ plane,

$$F_1 d_1 = F_2 d_2 \tag{7-3-1}$$

$$F_2 - F_1 = \frac{Q_p}{\varepsilon} \tag{7-3-2}$$

leading to a triangular potential barrier (see Fig. 7-3-1).

The zero-bias barrier height can be found from eqs. (7-3-1) and (7-3-2):

$$\phi_o = \frac{d_1 d_2 Q_p}{(d_1 + d_2)\varepsilon} \tag{7-3-3}$$

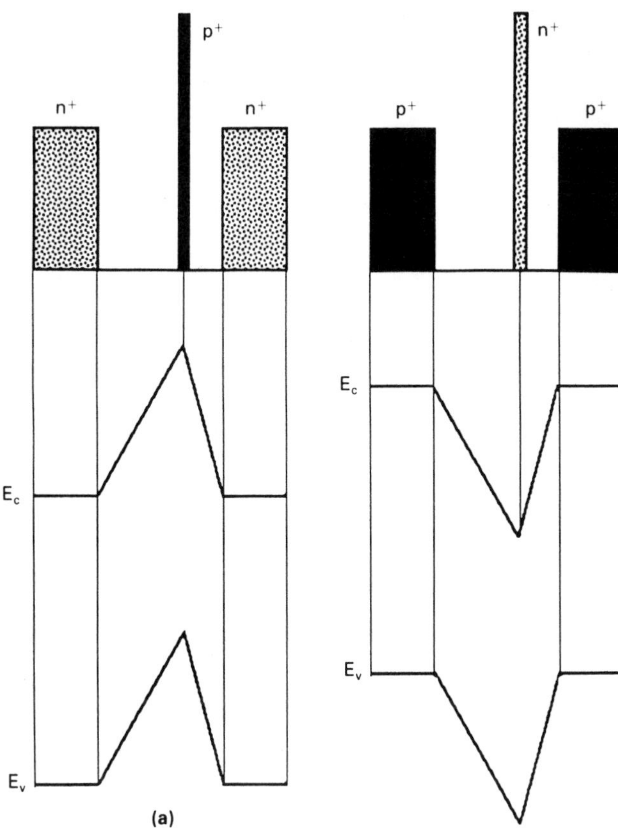

Fig. 7-3-1. Schematic structure of planar doped barrier diode and energy band edge diagrams under applied bias: (a) n^+-i-p^+-i-n^+ structure; (b) p^+-i-n^+-i-p^+ structure.

where Q_p is the total charge per unit area in the p plane. Thus, the barrier height can be modified by changing the doping in the p plane (Q_p) or the device geometry (d_1 and d_2). In a sense, the basic idea of this device has some similarity to attempts to change the effective barrier height of Schottky barriers by implanting n and p regions near the surface (see Sze 1981).

Under applied bias, the barrier becomes asymmetrical (see Fig. 7-3-1b). The voltage drop, V, is related to the electric fields F_1 and F_2 as follows:

$$F_1 d_1 - F_2 d_2 = -V \tag{7-3-4}$$

with eq. (7-3-2) still valid. From eqs. (7-3-2) and (7-3-4) we find that

$$\phi_1 = \phi_0 + \frac{d_1}{d_1 + d_2} V \tag{7-3-5}$$

$$\phi_2 = \phi_0 - \frac{d_2}{d_1 + d_2} V \tag{7-3-6}$$

If the distances d_1 and d_2 are such that the voltage drop across d_1 and d_2 is much larger than $k_B T/q$, the current over the barrier in both directions will be determined by thermionic emission. A crude estimate of the current may be obtained using eqs. (7-3-5) and (7-3-6) (see Malik et al. 1980):

$$J = A^* T^2 \exp(-q\phi_0/k_B T)[\exp(\alpha_2 V) - \exp(\alpha_1 V)] \tag{7-3-7}$$

where J is the current density, A^* is the Richardson constant, T is the lattice temperature, and

$$\alpha_1 = \frac{qd_1}{(d_1 + d_2)k_B T} \tag{7-3-8}$$

$$\alpha_2 = \frac{qd_2}{(d_1 + d_2)k_B T} \tag{7-3-9}$$

The capacitance of the PDB device is approximately independent of V and given by

$$C = \frac{\varepsilon S}{d_1 + d_2} \tag{7-3-10}$$

where S is the device cross section. As can be seen from Fig. 7-3-2 (from Malik et al. 1980), these predictions agree with the experimental results.

A more realistic model describing the PDB structure should account for the effect of the mobile carrier spillover from the n^+ regions into the i regions. This

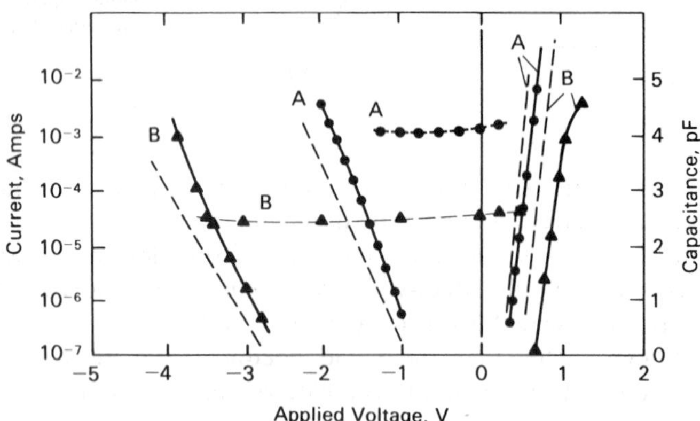

Fig. 7-3-2. Current and capacitance voltage characteristics of a planar doped barrier diode (from R. J. Malik, T. R. AuCoin, R. L. Ross, K. Board, C. E. C. Wood, and L. F. Eastman, "Planar Doped Barriers in GaAs by Molecular Beam Epitaxy," *Electronics Lett.*, 16, pp. 836–837 (1980). © 1980 IEE). Device A: d_1 = 500 Å; d_2 = 2000 Å; $N_p = 10^{12}$ cm^{-2}. Device B: d_1 = 250 Å; d_2 = 2000 Å; $N_p = 2 \times 10^{12}$ cm^{-2}.
—— current (measured)
-·-·- current (calculated)
----- capacitance (measured)

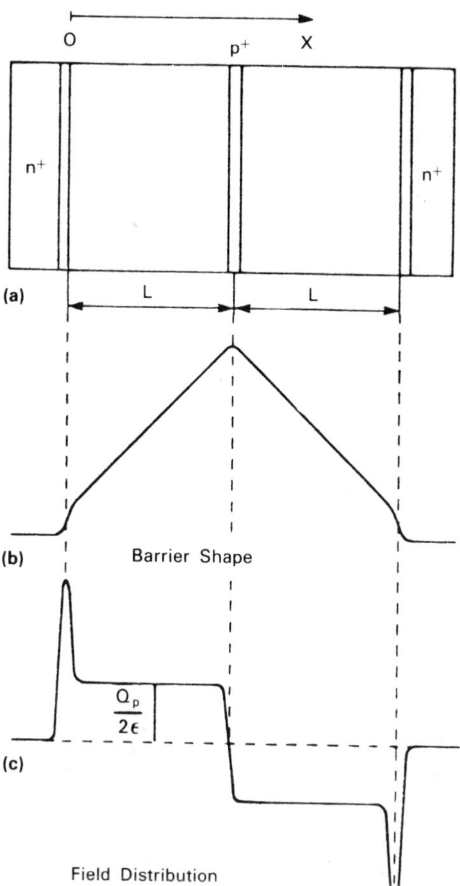

Fig. 7-3-3. Symmetrical planar doped barrier diode (from M. S. Shur, "Spill-over Effects in Planar Doped Barrier Structures," *Appl. Phys. Lett.*, 47, no. 8, pp. 869–871, October (1985)). (a) Schematic structure. (b) Conduction band edge diagram. (c) Electric field distribution.

spillover leads to two effects. First, the shape of the barrier is different because of high-field regions near the contacts (see Fig. 7-3-3). (A theory of PDB diodes that takes this effect into account was developed by Luryi and Kazarinov [1981]). Second, the barrier height in short structures may be substantially different from that predicted by eq. (7-3-3), as shown in Fig. 7-3-4 (from Shur 1985).

Malik et al. (1980) used two planar doped barrier diodes back-to-back to build a planar doped transistor. The doping profile for such a structure is shown in Fig. 7-3-5. Malik et al. (1980) suggested that an opportunity to tailor the barrier height provided by the structure may be used for launching fast ballistic or near-ballistic electrons into a submicron structure. As in a wide-gap emitter HBT, the hot electrons are injected into the base with very high velocity (up to 6×10^5 m/s). Levi et al. (1985) used a planar doped barrier transistor as a hot-electron spectrometer and observed ballistic transport of electrons across the narrow base region. The cutoff frequency of the structure is limited by the RC constants of the

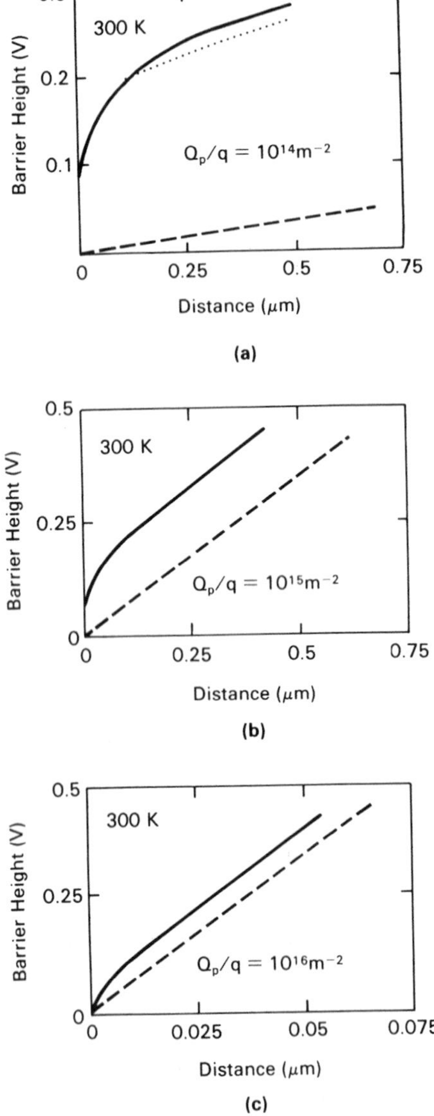

Fig. 7-3-4. Effective barrier height for a symmetrical planar doped barrier diode vs. half of device length, L (from M. S. Shur, "Spill-over Effects in Planar Doped Barrier Structures," *Appl. Phys. Lett.*, 47, no. 8, pp. 869–871, October (1985)). Q_p is the total concentration of acceptors in the p^+ plane. Solid lines correspond to exact calculation and dashed lines correspond to geometric model given by eq. (7-3-3). The dotted line in Fig. 7-3-4a is calculated for an *n-i-n* structure, neglecting the charges in the p^+ plane and using the theory developed by van der Ziel et al. (1983).

emitter and collector circuits. Simple estimates show that a cutoff frequency higher 100 GHz may be achieved (see Malik et al. 1980).

Two terminal switching devices utilizing the planar doped barrier have also been proposed (see Board et al. 1981). Other devices utilizing planar doped barrier structures include high-speed logic switches (see Wood et al. 1982), mixers (see Malik and Dixon 1982), fast photodiodes (see Chen et al. 1981), and BARITT

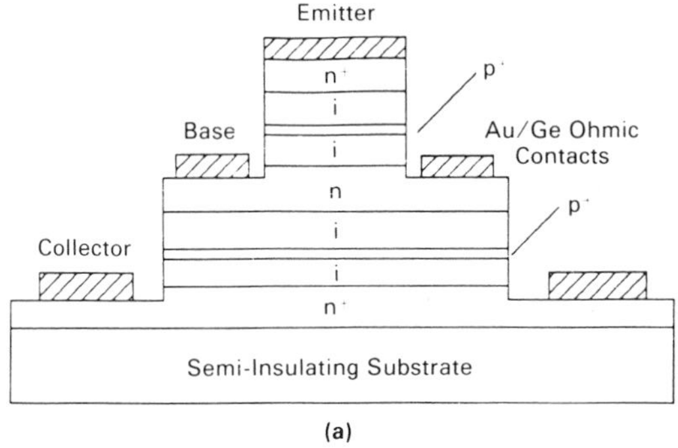

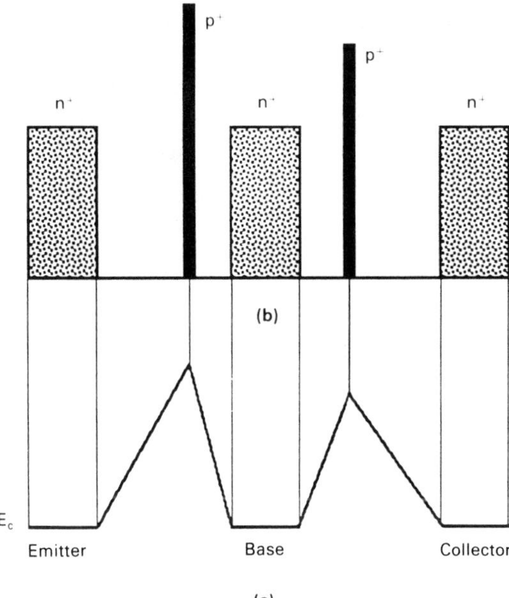

Fig. 7-3-5. Planar doped barrier transistor (from R. J. Malik, T. R. AuCoin, R. L. Ross, K. Board, C. E. C. Wood, and L. F. Eastman, "Planar Doped Barriers in GaAs by Molecular Beam Epitaxy," *Electronics Lett.*, 16, pp. 836–837 (1980). © 1980 IEE). (a) Schematic cross-section. (b) Doping profile. (c) Band diagram.

diodes (see Luryi and Kazarinov 1982). Planar doped barrier devices have been made from gallium arsenide, silicon, and amorphous silicon (see Chang et al. 1986).

7-4. REAL SPACE TRANSFER AND HOT-ELECTRON INJECTION TRANSISTORS

Hess et al. (1979) and Keever et al. (1979) predicted and then observed the *real-space* transfer effect in AlGaAs/GaAs heterostructures. In a typical real-space transfer diode electrons are moving in a GaAs layer sandwiched between AlGaAs layers. In high electric fields the electron temperature increases, and they are transferred by thermionic emission into AlGaAs regions where the electron velocity is low (see Fig. 7-4-1). This mechanism may lead to a negative differential resistance.

Kastalsky and Luryi (1983) proposed a three-terminal device—the NEgative Resistance Field-Effect Transistor (NERFET)—based on a similar concept. Again, the increase in electron temperature with the increase in the drain-to-source voltage leads to an increase in the substrate current and, as a consequence, a drop in the drain-to-source current, i.e., to negative differential resistance. The basic idea of hot-electron injection devices may be illustrated by comparing the structure shown in Fig. 7-4-2a to a vacuum-tube diode (see Fig.7-4-2b; from Luryi and Kastalsky 1984). In a vacuum-tube diode, the anode current may be controlled by varying the cathode temperature. In a hot-electron injection device, the current, I_{HOT}, over the barrier separating the channel from the substrate may be controlled by varying the electron temperature in the channel.

Three new devices—the CHarge Injection Transistor (CHINT), the NERFET, and the hot-electron erasable programmable random access memory—

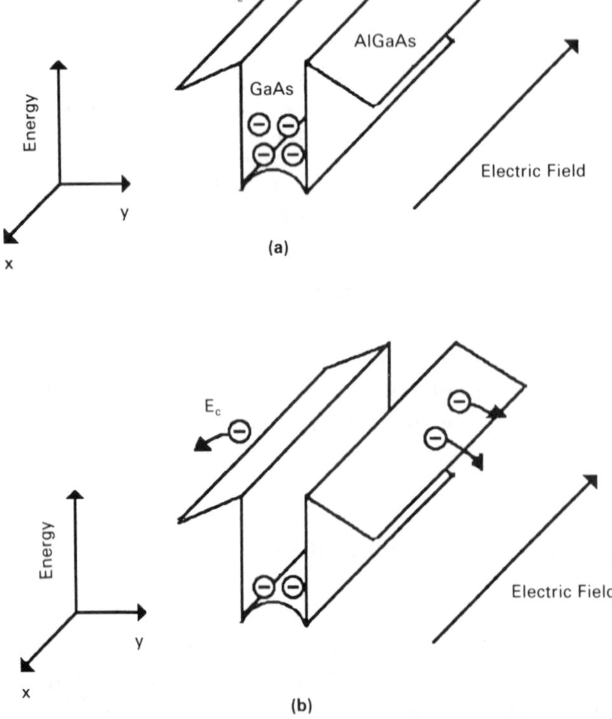

Fig. 7-4-1. Illustration of real space transfer. (a) In low electric field electrons are localized in a two-dimensional quantum well. (b) In high electric field many electrons gain enough energy from the field to transfer over the barrier.

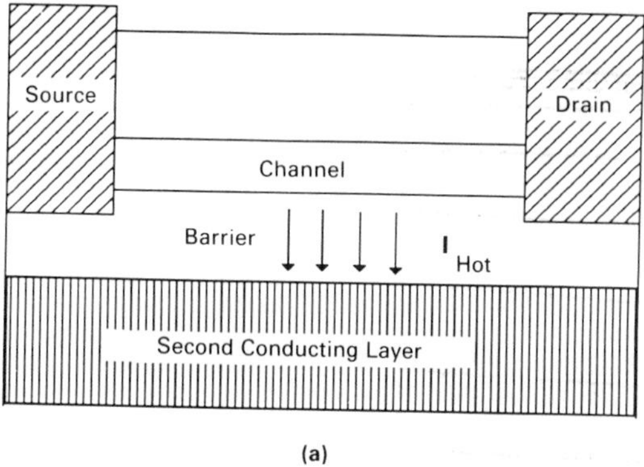

(a)

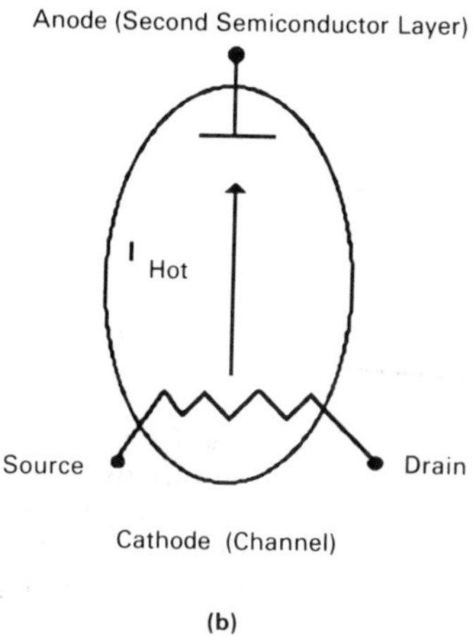

(b)

Fig. 7-4-2. Principle of operation of hot-electron injection transistors (after S. Luryi, A. Kastalsky, A. C. Gossard, and R. Hendel, *IEEE Trans. Electron Devices*, ED-31, p. 832 (1984). © 1984 IEEE). (a) Schematic device structure. (b) vacuum tube analogy.

have been proposed based on this principle (see Kastalsky and Luryi 1983, Luryi et al. 1984, and Kastalsky et al. 1984).

The primary advantage of this device is the possibility of achieving a higher speed of operation because the modulation of the drain-to-source current in this regime is related to the electron heating. The change in electron temperature is limited by the longest of the two time constants: the energy relaxation time (of the order of 1 ps or so for GaAs) and the time constant of the electric field variation. The latter time is determined by the electron transit time across the high-field region near the drain and may be several times shorter than the transit time of electrons across the gate, which sets a limit for the intrinsic speed of conventional field-effect transistors.

Shur et al. (1986) observed negative differential resistance in n-channel Heterostructure Insulated Gate Field Effect Transistors (HIGFETs) at high gate voltages (see Fig. 7-4-3a). This negative resistance may be explained by a mecha-

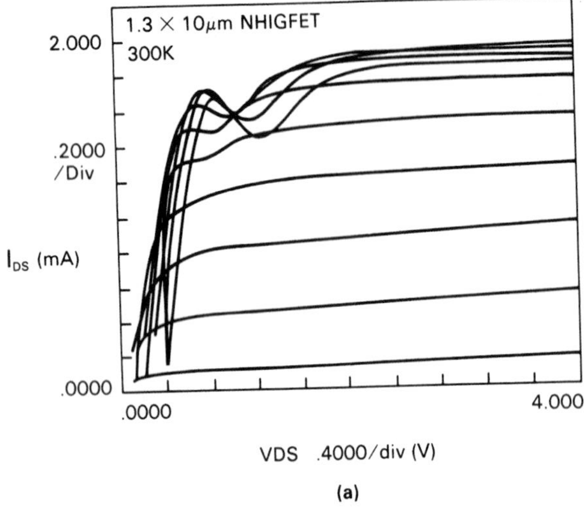

(a)

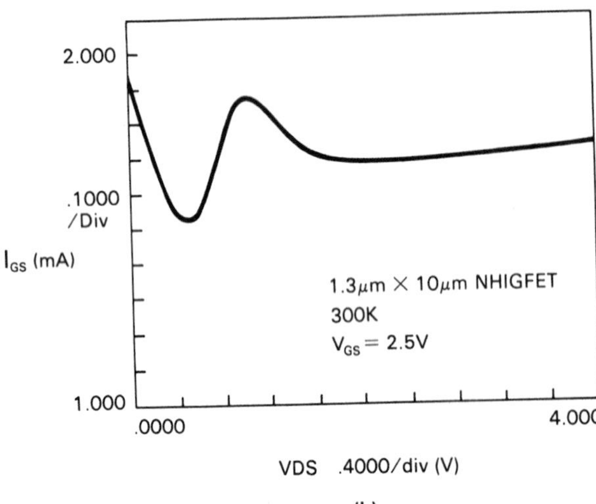

(b)

Fig. 7-4-3. Current-voltage characteristics of heterostructure insulated gate field-effect transistor exhibiting negative differential resistance at high gate voltages (a) and dependence of gate current on drain-to-source voltage (b) (from M. Shur, D. K. Arch, R. R. Daniels, and J. K. Abrokwah, *IEEE Electron Device Lett.*, EDL-7, pp. 78–80 (1986). © 1986 IEEE).

nism similar to that causing the negative differential resistance in NERFETs. Indeed, as can be seen from the comparison of Fig. 7-4-3a with Fig. 7-4-3b, which shows the measured dependence of the gate current on the drain-to-source voltage, the decrease in the drain current is commensurate with the increase in the gate current caused by the hot-electron current. Similar negative differential resistance was observed in modulation doped field-effect transistors by Chen et al. (1987), who utilized this effect for microwave amplification at frequencies up to 26 GHz. These results show that the real-space transfer effect in heterostructure FETs opens up an opportunity to operate these devices as fast hot-electron transistors.

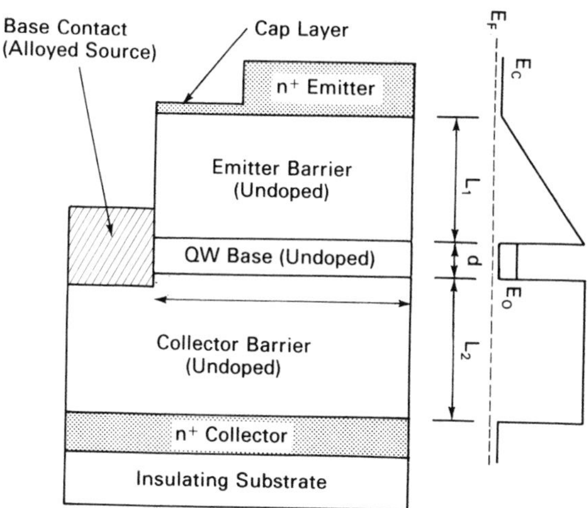

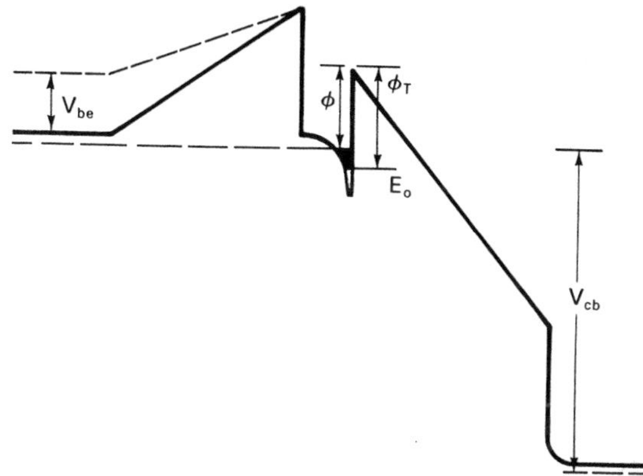

Fig. 7-4-4. Schematic structure and band diagram of induced base transistor (from S. Luryi, "An Induced Base Hot-Electron Transistor," *IEEE Electron Device Lett.*, EDL-6, pp. 178–180 (1985). © 1985 IEEE) and C. H. Chung, W. C. Liu, M. S. Jame, Y. H. Wang, S. Lury, and S. M. Sze, "Induced Base Transistor Fabricated by Molecular Beam Epitaxy," *IEEE Electron Device Lett.*, EDL-7, pp. 497–499 (1986). © 1986 IEEE). (a) Under equilibrium conditions and (b) under bias V_{cb} applied to the collector.

We should also mention a new and interesting device concept proposed by Luryi (1985a). In this device, called the Induced Base Transistor (IBT), base conduction occurs in a two-dimensional gas induced by the collector voltage (see Fig. 7-4-4). Chung et al. (1986) fabricated IBTs and obtained common-base current gain, α, as high as 0.96.

7-5. SUPERLATTICE DEVICES

Esaki and Tsu (1969, 1970) were first to propose a semiconductor superlattice based on a periodic structure of alternating layers of GaAs and $Al_xGa_{1-x}As$. Layers of semiconductor material with a more narrow energy gap form potential wells for electrons and holes (see Fig. 7-5-1). If the layer thickness is small

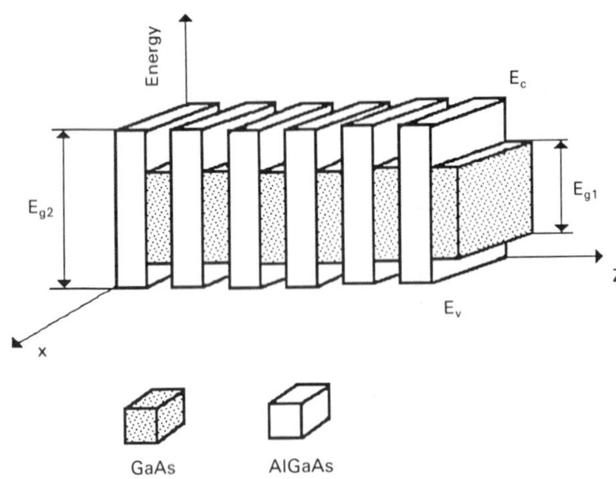

Fig. 7-5-1. Band structure of semiconductor superlattice consisting of alternating layers of GaAs and AlGaAs.

enough, electronic motion perpendicular to the superlattice plane (i.e., heterointerface plane) is quantized (see Section 1-2 and Fig. 1-2-2). The carrier motion parallel to the heterointerfaces is not quantized, so that the wave function is proportional to $\exp[i(k_x x + k_y y)]$. Here k_x and k_y are components of the wave vector in the plane of the superlattice, x and y are coordinates in the superlattice plane, and z is the coordinate perpendicular to the superlattice plane (see Problem 1-6-4). In other words, each energy level found for such a quantum well from the solution of the one-dimensional Schrödinger equation (as was done in Section 1-2 for an infinitely deep quantum well) corresponds to a subband of states. The density of states, D, in each subband is constant and is given by

$$D = m^*/(\pi\hbar^2) \tag{7-5-1}$$

(see eq. (1-6-51) and Fig. 7-5-2.

If the thickness of the wide-band-gap layers is such that carriers may tunnel through, the individual levels corresponding to quantized motion perpendicular to the heterointerface are split into bands (called minibands). Figure 7-5-3 shows how individual levels corresponding to the bottom of a subband in a superlattice, with quantum wells separated by wide potential barriers, are split into such minibands when the potential wells are brought closer to each other.

Esaki and Tsu (1970) proposed another approach toward creation of a man-made periodic potential in a semiconductor crystal, namely periodic n- and p-type doping. (Somewhat later, but independently, a similar idea was proposed by Ovsyannikov et al. [1971]). Such a superlattice, called a *doping superlattice* or an *n-i-p-i superlattice*, is shown in Fig. 7-5-4. This superlattice has very interesting and unusual properties. When electron-hole pairs are generated in a doping su-

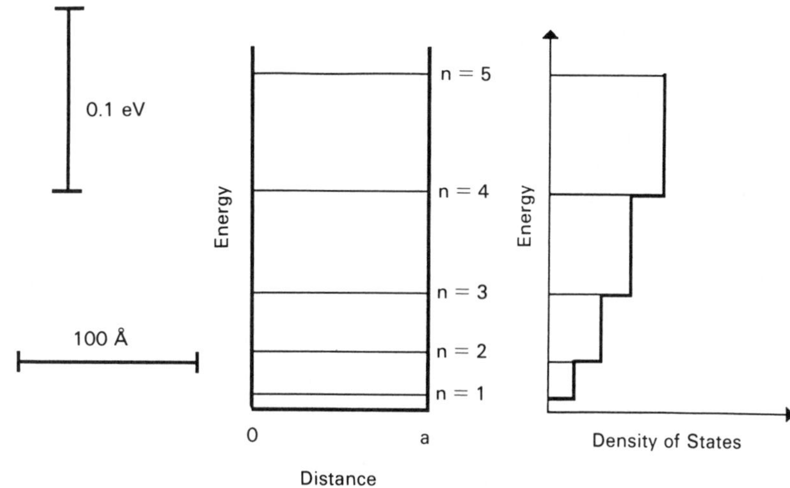

Fig. 7-5-2. Energy levels and density of states in two-dimensional quantum well.

perlattice (for example, by light) electrons and holes are separated by the space-charge electric fields. Electrons accumulate in the minima of the conduction band, and holes accumulate in the maxima of the valence band. As a consequence of this spatial separation, the effective carrier lifetime may be very long (and dependent on applied electric fields). This property allows one to design various

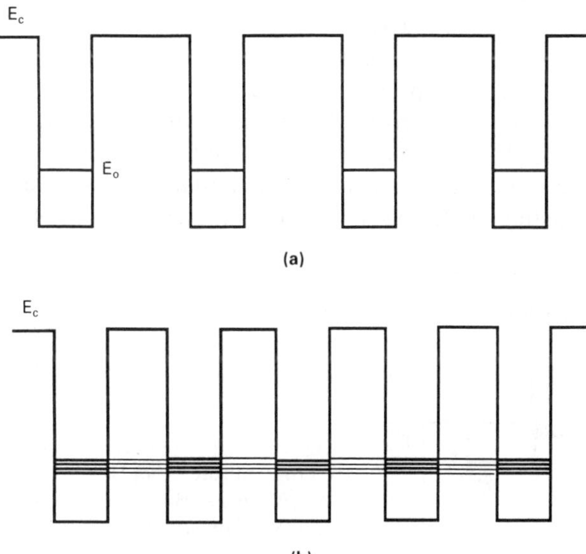

Fig. 7-5-3. Formation of minibands in superlattice: (a) large separation between quantum wells and (b) small separation between quantum wells.

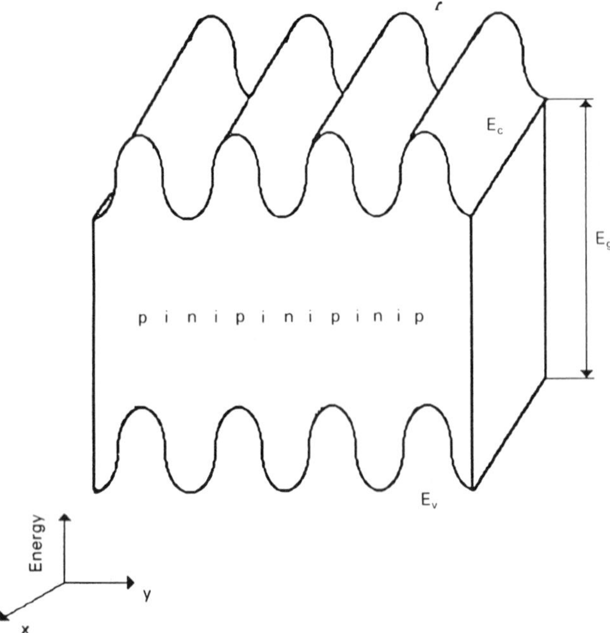

Fig. 7-5-4. Doping (*n-i-p-i*) superlattice.

optoelectronic devices, such as tunable light sources, optical amplifiers, modulators, and photodetectors (see, for example, Döhler 1986).

Other, less conventional ways to create a periodic superlattice potential include metal-dielectric superlattices and superlattices in which periodic potential is created by a standing light or acoustic wave.

Sakaki et al. (1975) discussed the superlattice transistor structure shown in Fig. 7-5-5. In this structure, the periodic potential of the "corrugated" gate interacts with the electron gas induced in a narrow channel near the semiconductor–insulator interface. This may lead to the creation of minibands and, as was shown by Sakaki (1975), to negative differential resistance. A similar idea may be realized using a superlattice consisting of alternating layers of materials with different band gaps (see, for example, Sakaki 1986).

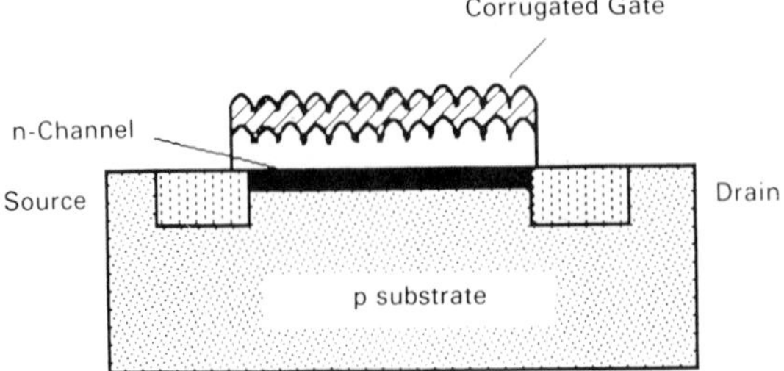

Fig. 7-5-5. Field effect transistor with corrugated gate (from H. Sakaki, K. Wagatsuma, J. Hamasaki, and S. Saito, *Thin Solid Films*, 36, p. 497 (1976). © 1976 IEEE).

Sweeny et al. (1987) proposed devices containing superlattices of one-dimensional quantum wires connecting the source and the drain of a field-effect transistor. In particular, they considered the properties of holes in such structures and found that hole mobility may be greatly enhanced compared with that in conventional three-dimensional devices. This opens up an opportunity of developing faster low-power complementary technology.

Baba et al. (1983), Tu et al. (1986), and Arch et al. (1987) used superlattice structures in which the doping is located in thin GaAs layers bounded by AlGaAs layers in heterostructure field-effect transistors. The advantage of such an approach is that it eliminates doping in AlGaAs (otherwise needed to control the device threshold voltage; see Section 4-11). Doping in the AlGaAs layer may lead to many undesirable effects, such as threshold voltage shift with temperature and persistent photoconductivity (see, for example, Valois et al. 1983).

The schematic structure of a superlattice HFET is shown in Fig. 7-5-6 (from Arch et al. 1987). In this device, the undoped 0.5-μm GaAs layer is grown by molecular beam epitaxy (MBE) on a semi-insulating GaAs substrate. This is followed by the deposition of 95 Å of an undoped p-$Al_{0.5}Ga_{0.5}$ insulator layer, five periods of the superlattice (with each period including 17 Å of doped GaAs and 24 Å of undoped AlGaAs), and finally 95 Å of an undoped p-$Al_{0.5}Ga_{0.5}$ insulator layer. The doping level in the GaAs (doped n-type by silicon) was estimated to be close to 2×10^{18} cm^{-3}. Calculated band diagrams of such a structure at two different gate voltages are shown in Fig. 7-5-7.

The typical measured transconductance in the saturation regime and drain-to-source saturation current vs. gate-to-source voltage V_{gs} are shown in Fig. 7-5-8. As can be seen from Fig. 7-5-8a, the transconductance at 300 K nearly saturates at gate voltages of the order of 0.8 V and then peaks sharply with further increase in the gate voltage. The maximum transconductance 310 mS/mm is reached at $V_{gs} = 1.08$ V. At 77 K the measured transconductance has two peaks, and the gate voltage swing, $V_{gs} - V_T$, corresponding to the maximum transconductance is larger (see Fig. 7-5-8b). These results may be explained by the effects caused by parallel conduction paths in the superlattice at high gate voltages. As

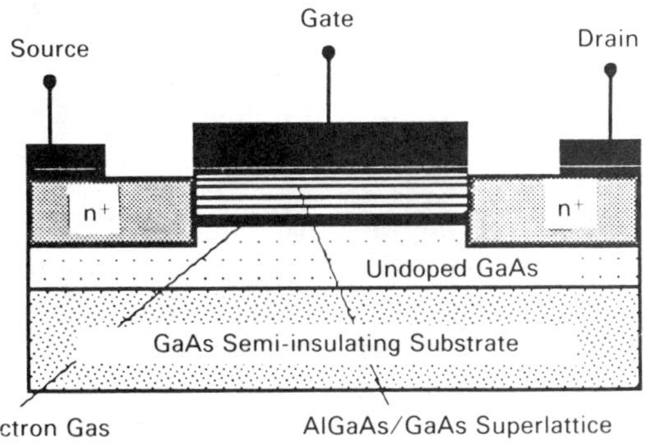

Fig. 7-5-6. Schematic structure of superlattice HFET (from Arch et al. [1987]).

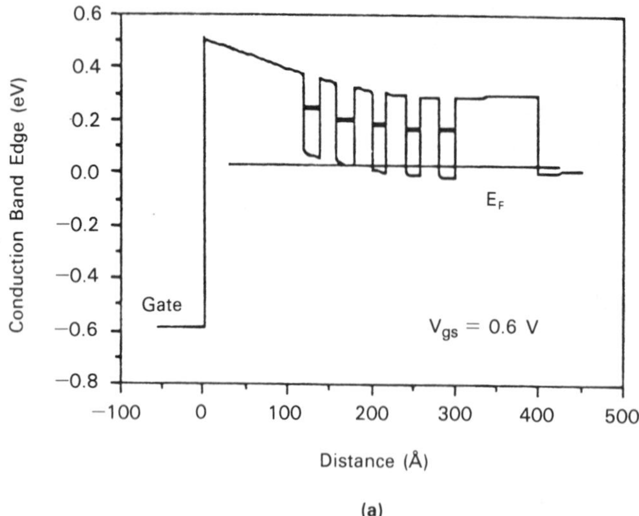

(a)

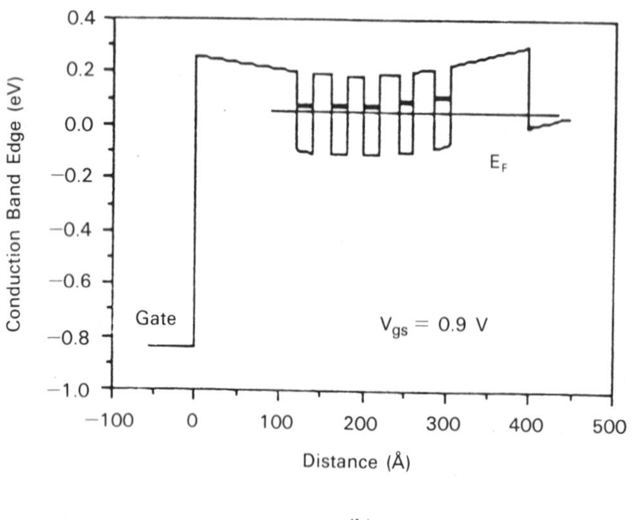

(b)

Fig. 7-5-7. Calculated band diagrams of superlattice HFET for two different gate voltages (from D. K. Arch, M. Shur, J. Abrokwah, and R. R. Daniels, *J. Appl. Phys.*, 61, no. 4, pp. 1503–1509 (1987).

was shown by Shur et al. (1987), electron mobility in narrow highly doped GaAs quantum wells may be considerably increased when electrons are induced in the quantum well by the field effect. This mobility enhancement is a consequence of a higher Fermi energy and more effective screening of the ionized impurity potential by conducting electrons induced in the quantum well.

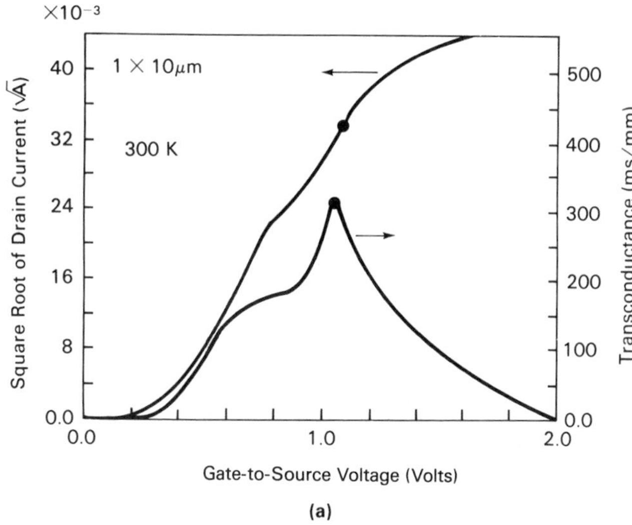

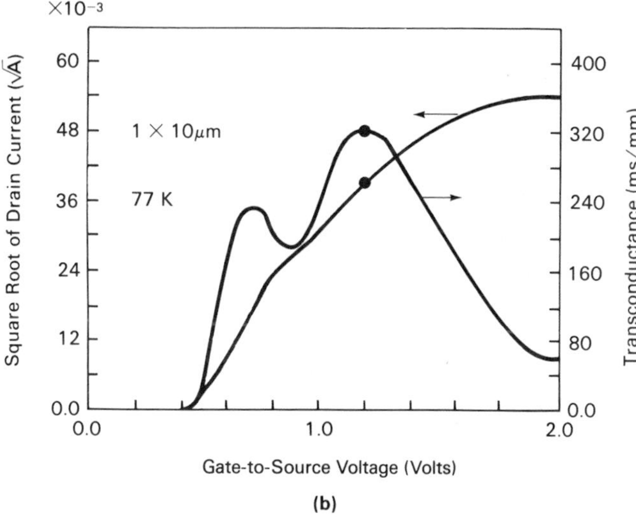

Fig. 7-5-8. Transconductance and drain-to-source saturation current vs. gate voltage (from D. K. Arch, M. Shur, J. Abrokwah, and R. R. Daniels, *J. Appl. Phys.*, 61, no. 4, pp. 1503–1509 (1987). (a) 300 K. (b) 77 K.

7-6. RESONANT TUNNELING DEVICES

Tunneling—the finite probability of a particle's penetrating a potential barrier—is one of the quantum mechanical phenomena that defy our everyday experience. It is illustrated by Fig. 2-7-2, which shows a fish trying to "tunnel" through the fishbowl wall. Of course, such a feat is not possible for a fish, but for a particle light enough (like an electron in GaAs) and a barrier narrow enough (with a thickness comparable to or smaller than the de Broglie wavelength) tunneling is quite possible and even probable.

The idea of *resonant tunneling* in multibarrier semiconductor structures (which has greatly enhanced tunneling probability when subband energies in adjacent quantum wells coincide, i.e., are "in resonance") was first considered by Kazarinov and Suris (1971). Chang et al. (1974) were first to observe resonant tunneling in double barrier semiconductor structures (see Fig. 7-6-1a).

The idea of a resonant tunneling device is illustrated by Fig. 7-6-1a, which shows a quantum well separated from two n^+ contacts by symmetrical barriers.

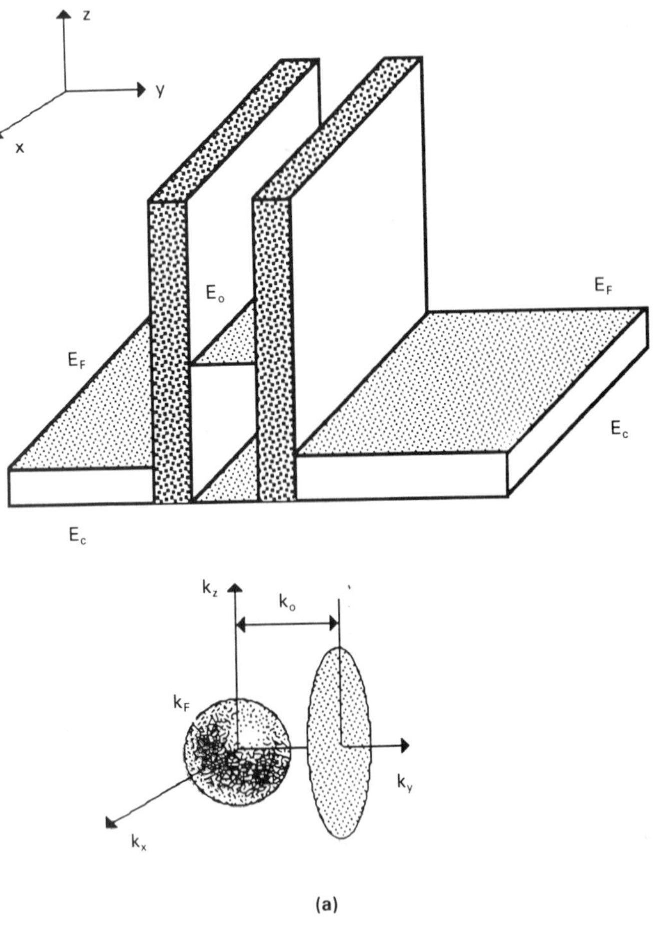

Fig. 7-6-1. Double barrier structure: (a) zero bias voltage and (b) bias voltage corresponding to E_o greater than the bottom of the conduction band in the emitter region and smaller than the Fermi energy of the emitter electrons. Also shown is the Fermi sphere of the emitter electrons in the k space and plane $k_y = k_o$ where $k_o = [2m_n(E_o - E_c)]^{1/2}/\hbar$.

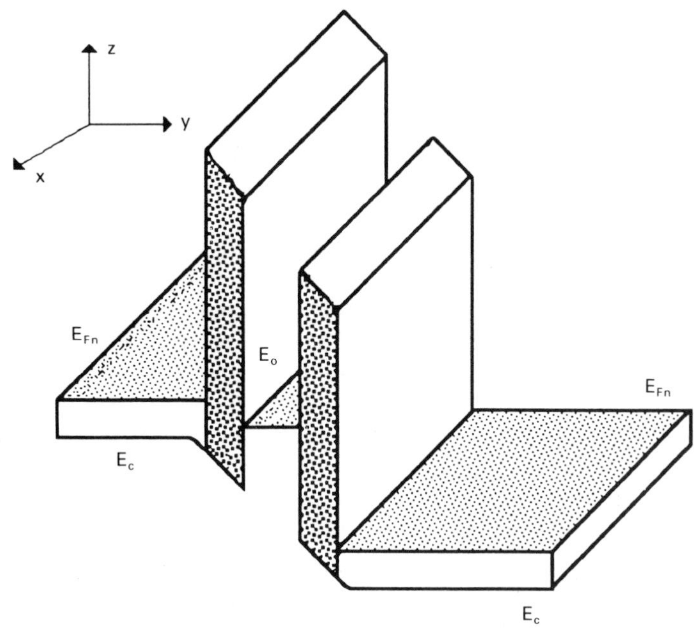

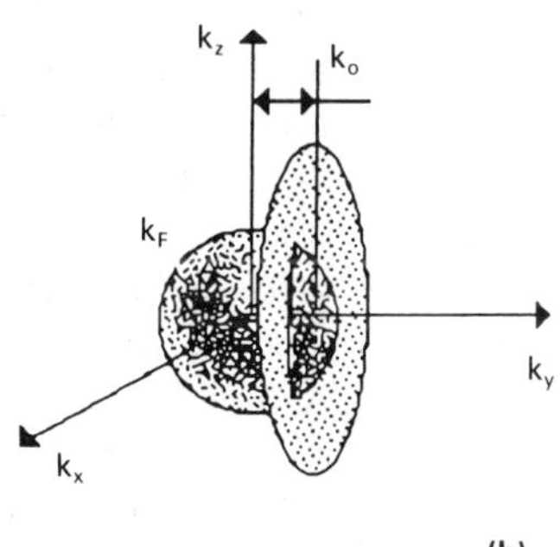

(b)

Fig. 7-6-1. Cont.

Conduction through such a structure is determined by electron tunneling through the barriers. Following Luryi (1985a), let us first consider the situation depicted in Fig. 7-6-1a when the wave vector

$$k_o = [2m_n(E_o - E_c)]^{1/2}/\hbar \tag{7-6-1}$$

is greater than the Fermi wave vector of electrons, k_F, in the emitter region:

$$k_F = [2m_n(E_F - E_c)]^{1/2} / \hbar \tag{7-6-2}$$

Here m_n is the electron effective mass and E_o is the subband energy. In this case, no electrons in the emitter region with energies corresponding to the energy level in the quantum well, E_o, are available for tunneling, and the tunneling probability is very low. The tunneling probability is greatly enhanced, however, when there are energy states available such that both energy and momentum can be conserved during the tunneling process. This situation is illustrated by Fig. 7-6-1b. Now voltage applied to the structure is such that E_o is below the Fermi energy in the emitter region but higher than the bottom of the conduction band, E_c, in the emitter region. The momentum is conserved for states that have the component of the wave vector k_y (in the direction perpendicular to the barriers) equal to k_o. As was pointed out by Luryi (1985b) the wave vectors of these electrons must lie in the disk formed by the intersection of the Fermi sphere of the emitter electrons with the plane where $k_y = k_o$. As the bias is increased, this disk radius first increases from zero to k_F and then decreases from k_F to zero (when the bottom of the conduction band in the emitter region exceeds E_o). As a consequence, the current through this structure exhibits a peak for bias voltages corresponding to the situation when E_o is greater than the bottom of the conduction band in the emitter region and smaller than the Fermi energy of the emitter electrons. At higher bias voltages the current through the device decreases and the device may exhibit negative differential resistance.

This explanation shows that tunneling into a two-dimensional electron gas leads to negative differential resistance. Barrier symmetry is not required for observing such a negative resistance, as was first pointed out by Luryi (1985b). However, if the barriers are symmetrical, interference of the wave functions leads to a substantial increase in the magnitude of the wave function in the quantum well and, as a consequence, to a much larger peak current. The interference pattern of the electron wave functions transmitted through and reflected from two identical barriers is very similar to the light interference pattern in the *Fabry-Perot effect*. (The Fabry-Perot effect is the effect of light interference between two parallel partially reflective and partially transmitting plates.) The physics of resonant tunneling in symmetrical structures was considered by Ricco and Azbel (1984).

Tsu and Esaki (1973) and Vassell et al. (1983) calculated the current-voltage characteristics of multibarrier structures. The resulting current-voltage characteristics have several peaks (see Fig. 7-6-2; from Vassell et al. 1983). These peaks correspond to resonance between the energies of the emitter electrons and different subbands in the quantum well between the barriers. Unfortunately, it is extremely difficult to predict the exact values of the peak and valley currents. These values may be strongly affected by other mechanisms, such as phonon-assisted tunneling, impurity-assisted tunneling, and scattering. Measured negative resistance caused by tunneling in a multibarrier structure looks more like the curves shown in Fig. 7-6-3 (from Sollner et al. 1989). The positions of the current peaks are, however, easier to establish, since they are related to energy levels of the subbands. This brings one to a classic quantum mechanical problem of calculating quantum levels in a finite potential well (the infinitely deep potential well was

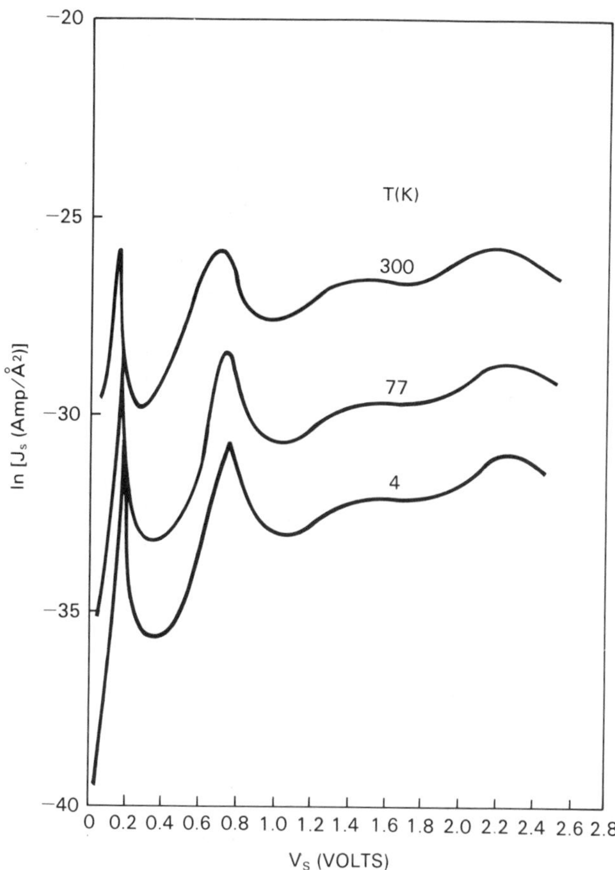

Fig. 7-6-2. Calculated current-voltage characteristics of double barrier structure (from M. O. Vassel, J. Lee, and H. F. Lockwood, *J. Appl. Phys.*, 54, no. 9, pp. 5206–5213 (1983).

considered in Section 1-2.) These levels for the finite potential well under bias are given by the roots, E, of the following equation (see Vassell et al. 1983 and Problem 1-2-4):

$$kd_o = n\pi - \arcsin\{\gamma E/[V_o + (\gamma - 1)E]\}^{1/2} - \arcsin\{\gamma E/[V_o + V_a + (\gamma - 1)E]\}^{1/2} \quad (7\text{-}6\text{-}3)$$

where $E = E_o - E_c$,

$$k = [2m_n(E_o - E_c)]^{1/2}/\hbar \quad (7\text{-}6\text{-}4)$$

d_o is the width of the quantum well, $n = 1,2,3, \ldots$ (all integer values for which solutions of eq. (7-6-3) exist), V_o is the depth of the potential well (under zero bias), V_a is the voltage drop across the potential well (see Fig. 7-6-4), and γ is the ratio of the effective mass in the barriers and in the potential well. For the case of

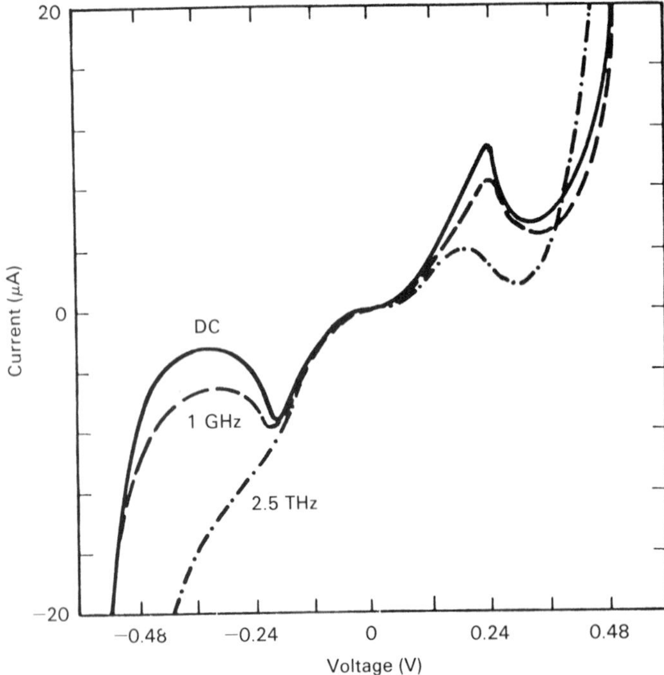

Fig. 7-6-3. Current-voltage characteristics of resonant tunneling diode (from T. C. L. G. Sollner, E. R. Brown, and H. Q. Le, "Microwave and Millimeter-Wave Resonant-Tunneling Devices," in *Physics of Quantum Devices,* ed. F. Capasso, Springer-Verlag (1989). The solid curve shows measured dc characteristics and the dashed and dotted-dashed curves show I-V characteristics expected at high frequencies.

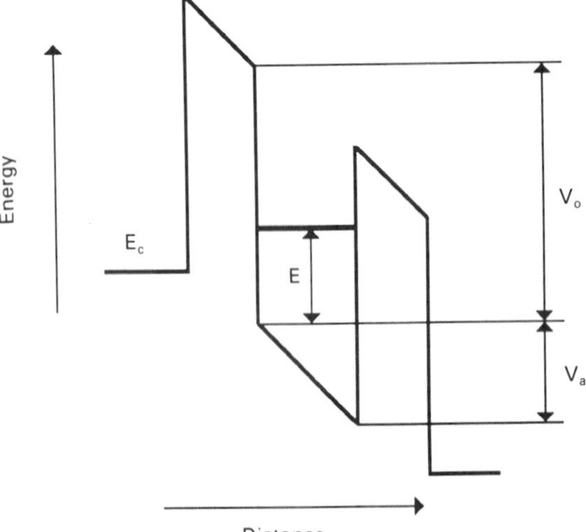

Fig. 7-6-4. Band diagram of double barrier structure. Notation in the figure corresponds to eq. 7-6-3.

Al_xGa_{1-x} As barriers and a GaAs quantum well,

$$\gamma \approx 1 + 1.239x \tag{7-6-5}$$

Numerous novel devices utilizing resonant tunneling structures have been proposed (see, for example, Capasso 1986 and Sollner et al. 1987). One possible device structure is a Resonant tunneling Bipolar Transistor (RBT; see Yokoyma 1986 and Futatsugi et al. 1987). Schematic band diagrams of this device (with a regular collector and with a heterostructure collector) are shown in Fig. 7-6-5.

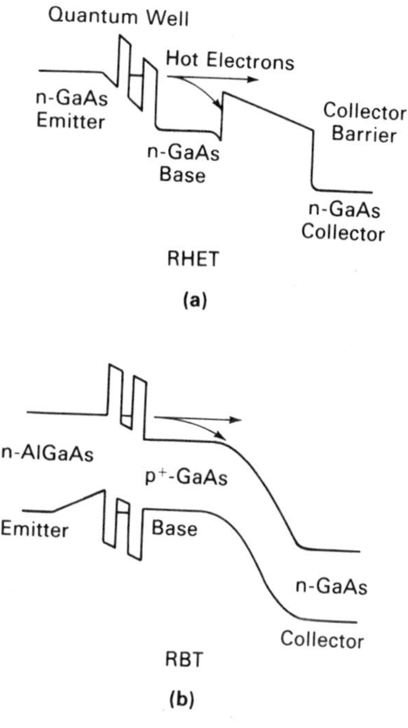

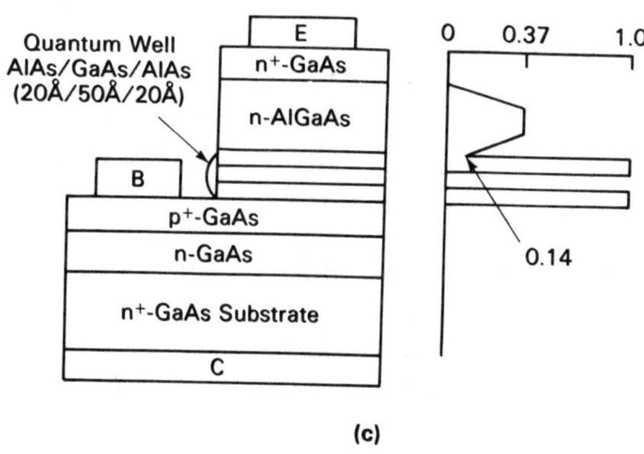

Fig. 7-6-5. Schematic band diagrams of resonance bipolar transistor (from T. Futatsugi, Y. Yamaguchi, K. Imamura, S. Muto, N. Yokoyama, and A. Shibatomi, *J. J. Appl. Phys.*, 26, no. 2, pp. L131–L133, February (1987)). (a) regular collector, (b) heterostructure collector, (c) schematic cross-section and composition profile, and (d) current-voltage characteristics.

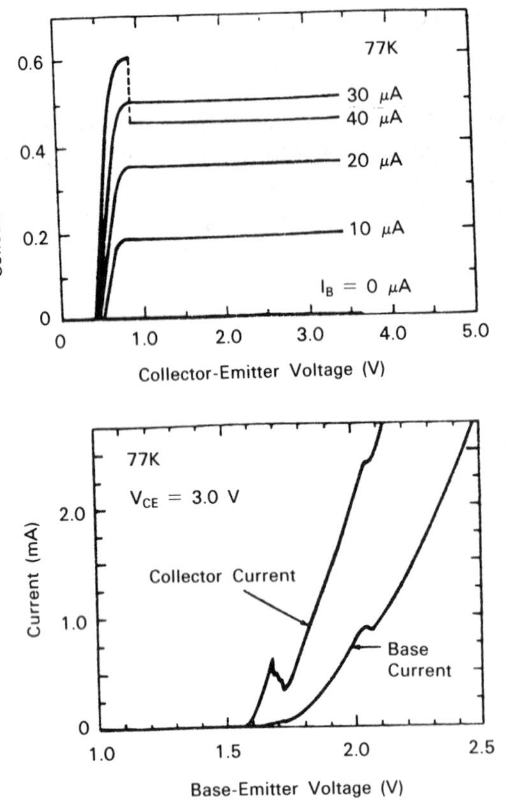

Fig. 7-6-5. Cont.

Fig. 7-6-5c shows a schematic cross section and composition profile for this device. As may be expected, the current-voltage characteristics of this device exhibit negative differential resistance under certain bias conditions (see Fig. 7-6-5d). This new shape of current-voltage characteristic may be adopted for realizing various logic functions.

Capasso and Kiehl (1985) proposed a Resonant Tunneling Transistor (RTT) utilizing ballistic electron injection into the base, which contains a double barrier or even a multibarrier structure (see Fig. 7-6-6). The collector current of such a device should exhibit a series of peaks corresponding to resonant tunneling through different energy states.

Luryi and Capasso (1985) proposed a new surface-resonant-tunneling transistor utilizing tunneling of two-dimensional electrons into a one-dimensional electron gas (see Fig. 7-6-7). This structure may develop into a very-low-power device. (Earlier, a version of such a device with multiple barriers was proposed by Sakaki 1983.)

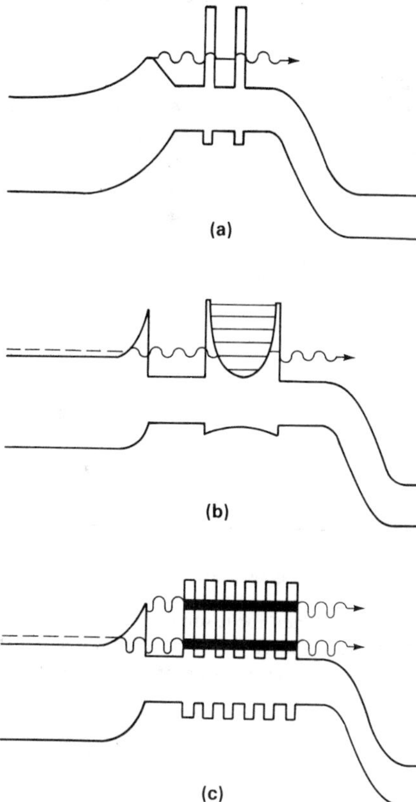

Fig. 7-6-6. Band energy diagrams and schematic structure of a Resonant Tunneling Transistor (RTT) (from F. Capasso, S. Sen, A. C. Gossard, A. L. Hutchinson, and J. H. English, *IEEE Electron Device Lett.*, EDL-7, no. 10, pp. 573–575 (1986). © 1986 IEEE). Inserts show dependencies of the collector current on the base current. At small base currents, the device behavior is similar to that of a conventional BJT (Fig. 7-6-6a). At large base currents, the potential difference between across the AlAs layer raises the conduction band edge in the emitter region with respect to the base. This terminates the resonance tunneling and leads to the drop in the collector current (see Fig. 7-6-6c). Fig. 7-6-6a shows the band diagram at the threshold.

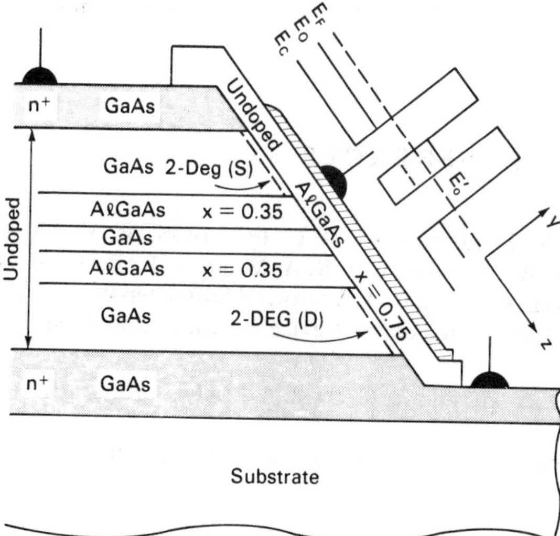

Fig. 7-6-7. Schematic structure and band diagram of surface resonant tunneling transistor (from S. Luryi and F. Capasso, *Appl. Phys. Lett.*, 47, pp. 1347–1349 (1985). (a) Structure with compositionally graded emitter, (b) structure with graded (parabolic) quantum well in the base, and (c) multibarrier structure.

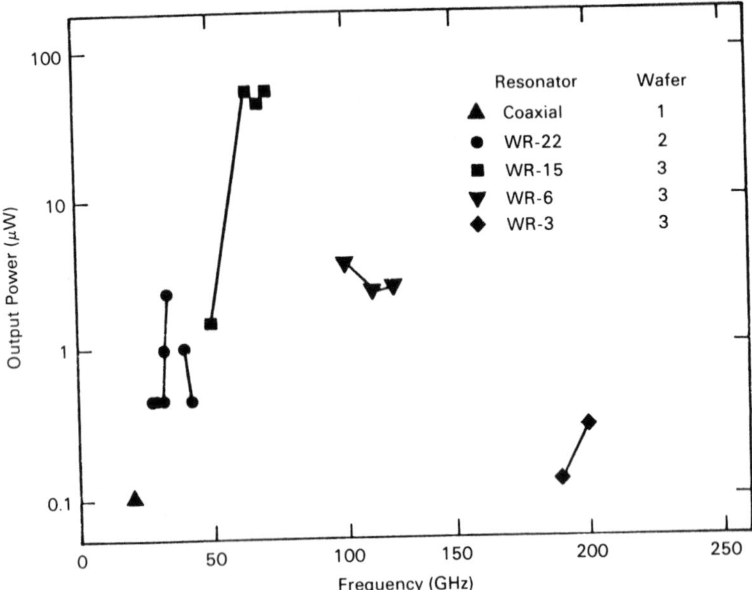

Fig. 7-6-8. Output power vs. frequency for resonant tunneling oscillators (from T. C. L. G. Sollner, E. R. Brown, and H. Q. Le, "Microwave and Millimeter-Wave Resonant-Tunneling Devices," in *Physics of Quantum Devices,* ed. T. Capasso, Springer-Verlag (1989)).

Millimeter-wave and microwave applications of resonant tunneling diodes were discussed by Sollner et al. (1987). These devices operate as oscillators (owing to their negative differential resistance) up to very high frequencies (see Fig. 7-6-8).

REFERENCES

G. D. ALLEY, C. O. BOZLER, N. P. ECONOMOU, D. C. FLANDERS, M. W. GEIS, G. A. LINCOLN, W. T. LINDLEY, R. W. MCCLELLAND, R. A. MURPHY, K. B. NICHOLS, W. J. PIACENTINI, S. RABE, J. P. SALERNO, and B. A. VOJAK, "Multimeter-Wavelength GaAs Permeable Base Transistors," presented at the Device Research Conference, Ft. Collins, Colo. (1982).

D. K. ARCH, M. SHUR, J. ABROKWAH, and R. R. DANIELS, *J. Appl. Phys.,* 61, no. 4, pp. 1503–1509 (1987).

T. BABA, T. MUZUTANI, M. OGAWA, and K. OHATA, *Jpn. J. Appl. Phys.,* 23, p. L164 (1983).

K. Board, K. Singer, R. Malik, C. E. C. Wood, and L. F. Eastman, "A Planar Doped Barrier Switching Device," in *Proc. 8th Bien.* Cornell Electrical Engineering Conference, Cornell University Press, Ithaca, N.Y., pp. 115–124, August (1981).

C. O. Bozler, G. D. Alley, R. A. Murphy, D. C. Flanders, and W. T. Lindley, "Permeable Base Transistor," in *Proc. 7th Bien. Cornell Conf. on Active Microwave Semiconductor Devices,* Cornell University Press, Ithaca, N.Y., August (1979).

C. O. Bozler and G. D. Alley, "Fabrication and Numerical Simulation of the Permeable Base Transistor," *IEEE Trans. on Electron Devices,* ED-27, no. 6, pp. 1128–1141 (1980).

C. O. Bozler and G. D. Alley, "The Permeable Base Transistor and Its Application to Logic Circuits," *Proc. IEEE,* 70, no. 1, January (1982).

F. Capasso and R. A. Kiehl, *J. Appl. Phys.,* 58, pp. 1366–1368 (1985).

F. Capasso, S. Sen, A. C. Gossard, A. L. Hutchinson, and J. H. English, *IEEE Electron Device Lett.,* EDL-7, no. 10, pp. 573–575 (1986).

A. Cappy, B. Carnes, R. Fauquembergues, G. Salmer, and E. Constant, "Comparative Potential Performance of Si, GaAs, GaInAs, InAs Submicrometer-gate FETs," *IEEE Trans. Electron Devices,* ED-27, pp. 2158–2168 (1980).

L. L. Chang, L. Esaki, and R. Tsu, *Appl. Phys. Lett.,* 24, p. 593 (1974).

Y. K. Chen, D. C. Radulescu, G. W. Wang, A. N. Lepore, P. J. Tasker, L. F. Eastman, and E. Strid, "Bias-Dependent Microwave Characteristics of an Atomic Planar-Doped AlGaAs/InGaAs/GaAs Double Heterojunction MODFET," in *Proc. IEEE MTT Symposium,* p. 871, Las Vegas, June (1987).

C. H. Chung, W. C. Liu, M. S. Jame, Y. H. Wang, S. Luryi, and S. M. Sze, "Induced Base Transistor Fabricated by Molecular Beam Epitaxy," *IEEE Electron Device Lett.,* EDL-7, pp. 497–499 (1986).

G. H. Döhler , *IEEE J. Quant. Electron.,* QE-22, no. 9, pp. 1682–1695 (1986).

L. F. Eastman, "Ballistic Transistors," presented at WOCSEMMAD 81.

L. F. Eastman, R. Stall, D. Woodard, N. Dandekar, C. Wood, M. S. Shur, and K. Board, "Ballistic Electron Motion in GaAs at Room Temperature," *Electron. Lett.,* 16, pp. 524–525 (1980).

L. Esaki and R. Tsu, "Superlattice and Negative Conductivity in Semiconductors," *IBM Res. Note,* RC-2418, March (1969).

L. Esaki and R. Tsu, "Superlattice and Negative Differential Conductivity in Semiconductors," *IBM J. Res. Develop.,* pp. 61–65, January (1970).

T. Futatsugi, Y. Yamaguchi, K. Imamura, S. Muto, N. Yokoyama, and A. Shibatomi, *J. J. Appl. Phys.,* 26, no. 2, pp. L131–L133, February (1987).

M. Heiblum, M. I. Nathan, D. C. Thomas, and C.M. Knoedler, "Direct Observation of Ballistic Transport in GaAs," *Phys. Rev. Lett.,* 55, p. 2200 (1985).

K. Hess, H. Morkoç, H. Shichijo, and B. G. Streetman, *Appl. Phys. Lett.,* 35, p. 469 (1979).

A. Kastalsky and S. Luryi, *IEEE Electron Device Lett.,* EDL-4, p. 334 (1983).

A. Kastalsky, S. Luryi, A. C. Gossard, and R. Hendel, *IEEE Electron Device Lett.,* EDL-5, p. 57 (1984).

R. F. Kazarinov and R. A. Suris, *Sov. Phys. Semicond.,* 5, pp. 707–709 (1971).

M. KEEVER, H. SHICHIJO, K. HESS, S. BANERJEE, L. WITKOWSKI, H. MORKOÇ, and B. G. STREETMAN, *Appl. Phys. Lett.*, 35, p. 459 (1979).

A. F. J. LEVI, J. R. HAYES, P. M. PLATZMAN, and W. WIEGMANN, "Injected Hot Electron Transport in GaAs," *Phys. Rev. Lett.*, 55, pp. 2071–2073 (1985).

S. LURYI, An Induced Base Hot-Electron Transistor, *IEEE Electron Device Lett.*, EDL-6, pp. 178–180 (1985a).

S. LURYI, *Appl. Phys. Lett.*, 47, pp. 167–168 (1985b).

S. LURYI, A. KASTALSKY, A. C. GOSSARD, and R. HENDEL, *IEEE Trans. Electron Devices,* ED-31, p. 832 (1984).

S. LURYI and A. KASTALSKY, "Hot Electron Injection Devices," presented at International Conference on Superlattices, Microstructures and Microdevices, Champaign, Illinois (1984).

S. LURYI and F. CAPASSO, *Appl. Phys. Lett.*, 47, pp. 1347–1349 (1985).

R. J. MALIK, T. R. AUCOIN, R. L. ROSS, K. BOARD, C. E. C. WOOD, and L. F. EASTMAN, "Planar Doped Barriers in GaAs by Molecular Beam Epitaxy," *Electronics Lett.*, 16, pp. 836–837 (1980).

M. I. OVSYANNIKOV, YU. A. ROMANOV, and V. N. SHABANOV, *Proc. Int. Conf. Phys. Chem. Semicond. Heterojunctions, Layer Structures,* VI, Budapest, Hungary, pp. 205-211 (1971).

B. RICCO and M. YA. AZBEL, *Phys. Rev., B* 29, pp. 1970–1981 (1984).

J. G. RUCH, "Electronics Dynamics in Short Channel Field-Effect Transistors," *IEEE Trans. Electron Devices,* ED-19, pp. 652–654 (1972).

H. SAKAKI, K. WAGATSUMA, J. HAMASAKI, and S. SAITO, *Thin Solid Films,* 36, p. 497 (1976).

H. SAKAKI, *Proceedings of 15th International Conference on Solid State Devices and Materials,* Tokyo, Japan, pp. 3–6 (1983).

H. SAKAKI, *IEEE J. Quant. Electron.*, QE-22, no. 9, pp. 1845–1852 (1986).

M. SHUR and L. F. EASTMAN, "Ballistic Transport in Semiconductors at Low Temperatures for Low Power High Speed Logic," *IEEE Trans. Electron Devices,* ED-26, pp. 1677–1683 (1979).

M. SHUR, "Influence of the Non-Uniform Field Distribution in the Channel on the Frequency Performance of GaAs FETs," *Electronics Lett.*, 12, pp. 615–616 (1976).

M. SHUR, D. K. ARCH, R.R. DANIELS, and J. K. ABROKWAH, *IEEE Electron Device Lett.*, EDL-7, pp. 78–80 (1986).

M. SHUR, *GaAs Devices and Circuits,* Plenum Publishing, New York (1987a).

M. SHUR, "Bipolar Permeable Base Transistor," in *Proceedings of IEEE 1987 Bipolar Circuits and Technology Meeting,* Minneapolis, p. 37 (1987b).

M. SHUR, J. K. ABROKWAH, R. R. DANIELS, and D. K. ARCH, *J. Appl. Phys.*, 61, no. 4, pp. 1643–1645 (1987).

M. SHUR, "Split-Gate Field-Effect Transistor," Appl. Phys. Lett., 542, pp. 162–164, January (1989).

T. C. L. G. SOLLNER, E. R. BROWN, and H. Q. LE, "Microwave and Millimeter-Wave Resonant-Tunneling Devices," in *Physics of Quantum Devices,* ed. F. Capasso, Springer-Verlag (1989).

M. SWEENY, J. XU, and M. SHUR, "Light and Heavy Holes in One-Dimensional Systems," presented at International Conference on Superlattices and Microstructures, Chicago, August (1987).

S. TIWARI, S. L. WRIGHT, and J. BATEY, "Unpinned GaAs MOS Capacitors and Transistors," *IEEE Electron Device Lett.*, 9, no. 9, pp. 488–489 (1988).

R. TSU and L. ESAKI, *Appl. Phys. Lett.*, 22, p. 562 (1973).

C. W.TU, W. L. JONES, R. F. KOPF, L. D. URBANEK, and S. S. PEI, *IEEE Electron Device Lett.*, EDL-7, no. 9, p. 552 (1986).

A. J. VALOIS, G. Y. ROBINSON, K. LEE and M. S. SHUR, "Temperature Dependence of the I-V Characteristics of Modulation-doped FETs," *J. Vac. Sci. Tech.*, B1, no. 2, pp. 190–195, April-June (1983).

M. O. VASSELL, J. LEE, and H. F. LOCKWOOD, *J. Appl. Phys.*, 54, no. 9, pp. 5206–5213 (1983).

B. A. VOJAK, R. W. MCCLELLAND, G. A. LINCOLN, A. R. CALAWA, D. C. FLANDERS,and M. W. GEIS, "A Self-Aligned Dual-Grating GaAs Permeable Base Transistor," *IEEE Electron Device Lett.*, EDL-5, no. 7, pp. 270–272, July (1984).

R. K. WATTS, ED., *Integrated Circuits in Dimensional Range 0.5 μm to 0.05 μm*, John Wiley & Sons, New York (1988).

N. YOKOYMA, *Extended Abstracts of the 18th (1986 International) Conference on Solid State Devices and Materials*, pp. 347–350, Tokyo (1986).

PROBLEMS

7-1-1. Assume that the cutoff frequency, f_T, of a field-effect transistor is determined by the following capacitance:

$$C_{total} = C_{gate} + C_{int} + C_L$$

where $C_{gate} = \varepsilon WL/d$ is the gate capacitance, ε is the dielectric permittivity, W is the gate width, L is the gate length, d is the distance between the gate and the channel, C_{int} is the interconnect capacitance, and C_L is the effective load capacitance. Assume that the intrinsic device transconductance, g_{mo}, is proportional to the gate capacitance and to the electron velocity in the channel, v, and inversely proportional to the gate length, L,

$$g_{mo} \approx \frac{C_{gate}v}{L}$$

Calculate the cutoff frequency for $v = 2.5 \times 10^5$ m/s and $L = 1$ μm. Assume $C_{int} =$ 0.2 pF, $W = 20$ μm, $C_L = 0.5C_{gate}$, $d = 300$ Å, $\varepsilon = 1 \times 10^{-10}$ F/m.

7-1-2. Consider that a hot-electron transistor can be modeled by a structure that includes two impenetrable barriers (emitter-base and collector-base) separated by a highly doped base (see Fig. P7-1-2).

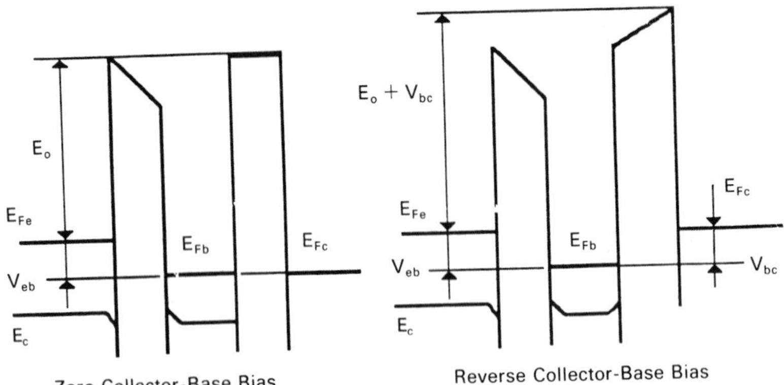

Fig. P7-1-2

Calculate and plot the collector current as a function of the reverse collector-base bias ($0 < V_{ce} < 0.3$ V). Use the following device parameters: effective electron velocity $v_s = 2 \times 10^5$ m/s, device cross-section $S = 10^{-6}$ cm^2, mean free path $\lambda = 1000$ Å, $E_o = 0.2$ eV, $T = 300$ K, and the total structure thickness $d = 0.3$ μm. Assume that at high energies the energy distribution of electrons in the emitter region may be approximated as $N(E) = N_o \exp[q(E_{Fe} - E)/k_B T]$ where E_{Fe} is the electron quasi-Fermi level in the emitter region, E is energy, and $N_o = 2 \times 10^{18}$ cm$^{-3}eV^{-1}$.

Hint: Assume that the electron current over the collector barrier (the collector) current) is proportional to the effective saturation velocity, v_s, and to the number of electrons, n, reaching the top of the collector barrier without collisions. This number is proportional, in turn, to the number of electrons emitted with energies higher than the top of the collector barrier and to $1 - \exp(-\lambda/d)$. (This is a fairly crude approach because it assumes that any electron that experiences a collision does not reach a collector contact and contributes to the base current. It also neglects the possible acceleration of electrons in the base region).

7-3-1. Calculate and plot the current-voltage characteristics of a planar-doped barrier diode. Assume $A^* = 4.4$ A/cm^{2}K^2, $d_1 = 0.1$ μm, $d_2 = 0.2$ μm for $T = 300$ K and $T = 77$ K, the charge concentration in the doped plane $Q_p = 10^{12}$ cm^{-2}, and cross section $S = 10^{-2}$ cm^2.

Appendixes

INTRODUCTION

In addition to information about units, the Greek alphabet, important physical constants, etc., these appendixes contain information about material properties of many important semiconductor compounds. Even though silicon is the semiconductor material dominating modern semiconductor electronics, compound semiconductors and amorphous silicon are becoming increasingly important, and a brief summary of their properties may be useful for the reader. Additional information can be found in the Landolt-Börnstein book *Numerical Data and Functional Relationships in Science and Technology,* New Series, group 3, volume 22, subvolume a, *Intrinsic Properties of Group IV Elements and III-V, II-VI, and I-VII Compounds,* and subvolume b, *Impurities and Defects in Group IV Elements and III-V Compounds,* ed. O. Madelung, Springer-Verlag, Berlin.

APPENDIX 1. SOME IMPORTANT PHYSICAL CONSTANTS

Quantity	Symbol	Value
Avogadro number	N_{AV}	6.0221367×10^{23} 1/mol
Bohr energy	E_B	13.6060 eV
Bohr magneton	μ_B	5.78832×10^{-5} eV/T
Bohr radius	a_B	0.52917 Å
Boltzmann constant	k_B	1.38066×10^{-23} J/K
Boltzmann constant/q	k_B/q	8.61738×10^{-5} eV/K
Electronic charge	q	1.60218×10^{-19} C
Electronvolt	eV	1.60218×10^{-19} J
Fine structure constant	α	0.00729735308 ($\approx$ 1/137)
Gas constant	R	1.98719 cal mol^{-1} K^{-1}
Gravitational constant	γ	6.67259×10^{-11} $\text{m}^3/(\text{kg s}^2)$
Impedance of free space	$1/c\varepsilon_o = \mu_o c$	376.732 Ω
Mass of electron at rest	m_e	$0.91093897 \times 10^{-30}$ kg
Mass of proton at rest	M_p	$1.6726231 \times 10^{-27}$ kg
Permeability in vacuum	μ_o	1.26231×10^{-8} H/cm ($4\pi \times 10^{-9}$)
Permittivity in vacuum	ε_o	8.85418×10^{-12} F/m ($1/\mu_o c^2$)
Planck constant	h	$6.6260755 \times 10^{-34}$ J-s
Reduced Planck constant	$\hbar = h/(2\pi)$	$1.0545727 \times 10^{-34}$ J-s
Speed of light in vacuum	c	2.99792458×10^8 m/s
Standard atmosphere		1.01325×10^5 N/m^2
Thermal voltage at 300 K	$k_B T/q$	0.025860 V
Thermal voltage at 293 K	$k_B T/q$	0.025256 V
Thermal voltage at 77 K	$k_B T/q$	0.006354 V
Wavelengths of visible light	λ	0.4 to 0.7 μm

APPENDIX 2. GREEK ALPHABET

Letter	Lowercase	Capital
Alpha	α	A
Beta	β	B
Gamma	γ	Γ
Delta	δ	Δ
Epsilon	ε	E
Zeta	ζ	Z
Eta	η	H
Theta	θ	Θ
Iota	ι	I
Kappa	κ	K
Lambda	λ	Λ
Mu	μ	M
Nu	ν	N
Xi	ζ	Ξ
Omicron	o	O
Pi	π	Π
Rho	ρ	P
Sigma	σ	Σ
Tau	τ	T
Upsilon	υ	Υ
Phi	ϕ	Φ
Chi	χ	X
Psi	ψ	Ψ
Omega	ω	Ω

APPENDIX 3. UNITS

Quantity	Unit	Symbol	
Fundamental units			
Length	meter		m
Time	second		s
Mass	kilogram		kg
Temperature	degree Kelvin		K
Current	ampere		A
Light intensity	candela		Cd
Additional units			
Angle	radian		rad
Solid angle	radian		rad
Other named units			
Frequency	Hertz	Hz	1/s
Force	Newton	N	kg m/s^2
Energy	Joule	J	Nm
Pressure	Pascal	Pa	N/m^2
Power	Watt	W	J/s
Electric charge	Coulomb	C	A s
Potential	Volt	V	J/C
Resistance	Ohm	Ω	V/A
Conductance	Siemens	S	A/V
Capacitance	Farad	F	C/V
Magnetic flux	Weber	Wb	V s
Inductance	Henry	H	Wb/A
Magnetic induction	Tesla	T	Wb/m^2
Light flux	Lumen	Lm	Cd rad

APPENDIX 4. MAGNITUDE PREFIXES

Magnitude prefix	Multiple factor	Symbol
exa	10^{18}	E
peta	10^{15}	P
tera	10^{12}	T
giga	10^{9}	G
mega	10^{6}	m
kilo	10^{3}	k
hecto	10^{2}	h
deka	10	da
deci	10^{-1}	d
centi	10^{-2}	c
milli	10^{-3}	m
micro	10^{-6}	μ
nano	10^{-9}	n
pico	10^{-12}	p
femto	10^{-15}	f
atto	10^{-18}	a

APPENDIX 5. UNIT CONVERSION FACTORS

$$1\ \text{Å} = 1 \times 10^{-10}\ \text{m}$$

$$1\ \text{mil} = 25.4\ \mu\text{m}$$

$$1\ \text{eV} = 1.60218 \times 10^{-19}\ \text{J}$$

$$1\ \text{erg} = 10^{-7}\ \text{J} = 6.242 \times 10^{11}\ \text{eV} = 2.389 \times 10^{-8}\ \text{cal}$$

$$1\ \text{degree} = 0.01745\ \text{rad}$$

$$1\ \text{rad} = 57.3\ \text{degree}$$

$$1\ \text{dyne} = 10^{-5}\ \text{N}$$

$$1\ \text{barye} = 10^{5}\ \text{Pa}$$

$$1\ \text{kW hour} = 3.6 \times 10^{6}\ \text{J}$$

APPENDIX 6. PROPERTIES OF SILICON (Si)

Atomic number	14
Atoms/cm^3	5.02×10^{22}
Electronic shell configuration	$1s^2 2s^2 2p^6 3s^2 3p^2$
Atomic weight	28.09
Crystal Structure	diamond
Breakdown field (V/cm)	$\sim 3.0 \times 10^5$
Density (g/cm^3)	2.329 (at 298 K)
Dielectric constant	11.7
Diffusion constant (cm^2/s)	37.5 (electrons) (at 300 K) 13 (holes) (at 300 K)
Effective density of states in the conduction band (cm^{-3})	2.8×10^{19} (at 300 K)
Effective density of states in the valence band (cm^{-3})	1.04×10^{19} (at 300 K)
Effective electron mass (in units of m_e)	longitudinal : 0.92 (at 1.26 K) transverse : 0.19 (at 1.26 K) density of states : 1.28 (at 600 K) : 1.18 (at 300 K) : 1.08 (at 77 K) : 1.026 (at 4.2 K)
Effective hole mass (in units of m_e)	heavy hole : 0.537 (at 4.2 K) heavy hole : 0.49 (at 300 K) light hole : 0.153 (at 4.2 K) light hole : 0.16 (at 300 K) density of states : 0.591 (at 4.2 K) : 0.62 (at 77 K) : 0.81 (at 300 K)
Electron affinity (V)	4.05
Energy gap (eV)	1.12 (at 300 K) 1.17 (at 77 K)
Index of refraction	3.42
Intrinsic carrier concentration (cm^{-3})	1.02×10^{10} cm^{-3} (at 300 K)
Intrinsic Debye length (μm)	24
Intrinsic resistivity (Ω-cm)	3.16×10^5 (at 300 K)
Lattice constant (Å)	5.43107 (at 298.2 K)
Melting point (°C)	1412
Mobility (cm^2/V-s)	1450 (electron, at 300 K) 500 (hole, at 300 K)
Optical phonon energy (eV)	0.063
Specific heat (J/g-°C)	0.7
Thermal conductivity (W/cm-°C)	1.31 (at 300 K)
Thermal diffusivity (cm^2/s)	0.9
Thermal expansion, linear (°C^{-1})	2.6×10^{-6} (at 300 K)
Young's modulus (dyn/cm^2)	1.9×10^{12} (111) direction

APPENDIX 7. PROPERTIES OF GERMANIUM (Ge)

Atomic number	32
Atoms/cm^3	4.42×10^{22}
Electronic shell configuration	$1s^2 2s^2 2p^6 3s^2 3p^6 3d^{10} 4s^2 4p^2$
Atomic weight	72.6
Crystal structure	diamond
Breakdown field (V/cm)	$\sim 10^5$
Density (g/cm^3)	5.3234 (at 298 K)
Dielectric constant	16.2 (at 300 K) 15.8 (at 77 K)
Diffusion constant (cm^2/s)	100 (electrons) (at 300 K) 49.1 (holes) (at 300 K)
Effective density of states in the conduction band (cm^{-3})	1.04×10^{19} (at 300 K)
Effective density of states in the valence band (cm^{-3})	6.0×10^{18} (at 300 K)
Effective electron mass (in units of m_e)	longitudinal : 1.64 (at 300 K) transverse : 0.082 (at 300 K)
Effective hole mass (in units of m_e)	heavy hole : 0.28 (at 300 K) light hole : 0.044 (at 300 K)
Electron affinity (V)	4.0 (at 300 K)
Energy gap (eV)	0.664 (at 291 K) 0.741 (at 4.2 K)
Index of refraction	3.98 (300 K)
Intrinsic carrier concentration (cm^{-3})	2.33×10^{13} (at 300 K)
Intrinsic Debye length (μm)	0.68 (at 300 K)
Intrinsic resistivity (Ω-cm)	47.62 (at 300 K)
Lattice constant (Å)	5.65791 (at 298.15 K)
Melting point (°C)	937.4
Mobility (cm^2/V-s)	electron : 3900 (at 300 K) hole : 1900 (at 300 K)
Optical phonon energy (eV)	0.037
Specific heat (J/g-°C)	0.31
Thermal conductivity (W/cm-°C)	0.6 (at 300 K)
Thermal diffusivity (cm^2/s)	0.36
Thermal expansion, linear (K^{-1})	2.2×10^{-6} (at 100 K) 5.9×10^{-6} (at 300 K)

APPENDIX 8. PROPERTIES OF GALLIUM ARSENIDE (GaAs)

Crystal structure	zinc blende
Breakdown field (V/cm)	$\sim 4.0 \times 10^5$
Density (g/cm^3)	5.3176 (at 298 K)
Dielectric constant (κ_s)	12.9 (at 300 K)
(κ_o)	10.89 (at 300 K)
Diffusion constant (cm^2/s)	207 (electrons, at 300 K)
	10 (holes, at 300 K)
Effective density of states in the conduction band (cm^{-3})	4.7×10^{17} (at 300 K)
Effective density of states in the valence band (cm^{-3})	7.0×10^{18} (at 300 K)
Effective electron mass (in units of m_e)	0.067 (0 K)
	0.063 (300 K)
Effective hole mass (in units of m_e)	heavy hole : 0.51 (at $T < 100$ K)
	: 0.50 (at 300 K)
	light hole : 0.084 (at $T < 100$ K)
	: 0.076 (at 300 K)
	density of states : 0.53
Electron affinity (V)	4.07
Energy gap (eV)	1.424 (at 300 K)
	1.507 (at 77 K)
	1.519 (at 0 K)
Index of refraction	3.3
Intrinsic carrier concentration (cm^{-3})	2.1×10^6 (at 300 K)
Intrinsic Debye length (μm)	2250 (at 300 K)
Intrinsic resistivity (Ω-cm)	10^8 (at 300 K)
Lattice constant (Å)	5.6533 (at 300 K)
Melting point (°C)	1240
Mobility (cm^2/V-s)	8500 (electron, at 300 K)
	400 (hole, at 300 K)
Optical phonon energy (eV)	0.035
Specific heat (J/g-°C)	0.35
Thermal conductivity (W/cm-°C)	0.46
Thermal diffusivity (cm^2/s)	0.44
Thermal expansion, linear (°C^{-1})	6.86×10^{-6} (at 300 K)

APPENDIX 9. PROPERTIES OF ALUMINUM GALLIUM ARSENIDE ($Al_xGa_{1-x}As$)

Crystal structure	zinc blende
Density (g/cm^3)	$5.36 - 1.6x$
Dielectric constant (κ_s)	$13.18 - 3.12x$ (at 300 K)
(κ_o)	$10.89 - 2.78x$ (at 300 K)
Effective electron mass (in units of m_e)	$0.067 + 0.083x$ (Γ minimum, density of states) $0.85 - 0.14x$ (X minimum, density of states) $0.56 + 0.10x$ (L minimum, density of states) $0.067 + 0.083x$ (G minimum, conductivity) $0.32 - 0.06x$ (X minimum, conductivity) $0.11 + 0.03x$ (L minimum, conductivity)
Effective hole mass (in units of m_e)	heavy hole : $0.62 + 0.14x$ (density of states) light hole : $0.087 + 0.063x$ (density of states) split-off band : $0.15 + 0.09x$ (density of states)
Electron affinity (V)	$4.07 - 1.1x$ ($x < 0.45$) $3.64 - 0.14x$ ($0.45 < x < 1.0$)
Energy gap (eV)	$1.424 + 1.247x$ ($x < 0.45$) $1.9 + 0.125x + 0.143x^2$ ($0.45 < x < 1.0$)
Conduction band discontinuity	$\Delta E_c = \Delta E_g - \Delta E_v$
Lattice constant (Å)	$5.6533 + 0.0078x$
Melting point (K)	$1511 - 58x + 560x^2$ (solidus curve) $1511 + 1082x - 580x^2$ (liquidus curve)
Mobility (cm^2/V-s), electrons	$\approx 8000 - 22000x + 10000x^2$ (for $x < 0.45$) $\approx -255 + 1160x - 720x^2$ (for $x < 0.45$)
Mobility (cm^2/V-s), holes	$\approx 370 - 970x + 740x^2$
Thermal resistivity (cm-K/W)	$2.27 + 20.83x - 30x^2$
Thermal expansion, linear (in 10^{-6} K^{-1})	$6.4 - 1.2x$
Young's modulus (in 10^{11} dyn/cm^2)	$8.53 - 0.18x$
Band discontinuities at the $Al_XGa_{1-X}As$/$In_YGa_{1-Y}As$ heterointerface	
Valence band discontinuity	$\Delta E_v = 0.4\Delta E_{gg}$ where ΔE_{gg} (eV) $= 1.247X + 1.5Y - 0.4Y^2$ is the difference between Γ valleys in $Al_XGa_{1-X}As$ and $In_YGa_{1-Y}As$
Energy gap discontinuity	$\Delta E_g = \Delta E_{gg}$ for $X < 0.45$ $\Delta E_g = 0.476 + 0.125X + 0.143X^2 + 1.5Y - 0.4Y^2$ for $X \geq 0.45$ (eV)
Conduction band discontinuity	$\Delta E_c = \Delta E_g - \Delta E_v$

APPENDIX 10. PROPERTIES OF INDIUM PHOSPHIDE (InP)

Crystal structure	zinc blende
Density (g/cm^3)	4.81
Dielectric constant (κ_s)	12.56 (at 300 K)
	11.93 (at 77 K)
(κ_o)	9.61
Diffusion constant (cm^2/s)	118 (electrons at 300 K)
	4 (holes at 300 K)
Effective electron mass (in units of m_e)	0.077 (at 300 K Γ valley)
	0.068 (at 500 K Γ valley)
	0.325 (at 300 K L valley)
	0.26 (at 300 K X valley)
Effective hole mass (in units of m_e)	heavy hole : 0.60 (at 110 K)
	light hole : 0.12 (at 4.2 K)
Energy gap (eV)	1.34 (at 300 K)
	1.414 (at 77K)
Intrinsic carrier concentration (cm^{-3})	1.2×10^8 (at 293 K)
Index of refraction	3.1
Lattice constant (Å)	5.8687 (at 291.15 K)
Melting point (°C)	1062
Mobility (cm^2/V-s)	4600 (electron, at 300 K)
	40,000–60,000 (at 77 K)
	150 (hole, at 300 K)
	1200 (hole, at 77 K)
Peak velocity (10^5 m/s)	2.3 (at 300 K)
Saturation velocity (10^5 m/s)	0.92 (at 300 K)
Sound velocity (cm/s)	5.13×10^5
Thermal expansion, linear (K^{-1})	4.75×10^{-6} (at 298.15 K)

APPENDIX 11 PROPERTIES OF INDIUM ARSENIDE (InAs)

Crystal structure	zinc blende
Density (g/cm^3)	5.667 (at 300 K)
Dielectric constant	κ_s : 15.15 (at 300 K)
	κ_o : 12.3 (at 300 K)
Effective electron mass (in units of m_e)	0.0231 (at 150 K)
	0.0219 (at 250 K)
Effective hole mass (in units of m_e)	heavy hole : 0.43 (along [111])
	: 0.35 (along [100])
	light hole : 0.026 (at 20 K)
Energy gap (eV)	0.354 (at 295 K)
	0.414 (at 77 K)
	0.418 (at 4.2 K)
Index of refraction	3.51
Intrinsic carrier concentration (cm^{-3})	1.3×10^{15} (at 300 K)
Lattice constant (A)	6.0583 (at 298.15 K)
Melting point (°C)	942
Mobility (cm^2/V-s)	electron: 1×10^5 (at 77 K)
	3.3×10^4 (at 300 K)
	hole : 100–450 (at 300 K)
Thermal expansion, linear (K^{-1})	4.52×10^{-6}

APPENDIX 12. PROPERTIES OF INDIUM GALLIUM ARSENIDE ($In_xGa_{1-x}As$)

Crystal structure	zinc blende
Effective electron mass (in units of m_e)	0.041 (at T → 0 K) for $x = 0.53$
Effective hole mass (in units of m_e) for $x = 0.53$	heavy hole 0.465(along [001]) 0.56 (along [110]) 0.60 (along [111]) light hole 0.0503
Energy gap (eV)	0.75 (at 300 K for $x = 0.53$)
Lattice constant (Å)	$6.058-0.405x$
Mobility (cm^2/V-s) for $x = 0.53$	electron: 13800 (at 300 K) 70000 (at 77 K)

Band discontinuities at the $Al_XGa_{1-X}As/In_YGa_{1-Y}As$ heterointerface

Valence band discontinuity	$\Delta E_v = 0.4\Delta E_{gg}$ where ΔE_{gg} (eV) $= 1.247X + 1.5Y - 0.4Y^2$ is the difference between Γ valleys in $Al_XGa_{1-X}As$ and $In_YGa_{1-Y}As$
Energy gap discontinuity	$\Delta E_g = \Delta E_{gg}$ for $X < 0.45$ $\Delta E_g = 0.476 + 0.125X + 0.143X^2 + 1.5Y - 0.4Y^2$ for $X \geq 0.45$ (eV)
Conduction band discontinuity	$\Delta E_c = \Delta E_g - \Delta E_v$

APPENDIX 13. PROPERTIES OF DIAMOND

Atomic number	6
Electronic shell configuration	$1s^2 2s^2 2p^2$
Atomic weight	12
Crystal structure	diamond
Density (g/cm^3)	3.52
Dielectric constant	5.70 (at 300 K)
Effective electron mass (in units of m_e)	longitudinal 1.4
	transverse 0.36
Effective hole mass (in units of m_e)	heavy hole 2.18 (at 1.2 K)
	light hole 0.7 (at 1.2 K)
	density of states 0.75 (at 300 K)
Energy gap (eV)	5.5 (at 300 K)
Index of refraction	2.39
Lattice constant (Å)	3.56683 (at 298 K)
Melting point (°C)	3827
Mobility ($cm^2/V\text{-}s$)	electron: ~ 2000 (at 300 K)
	hole : 2100 (at 300 K)
Thermal conductivity (W/cm-K)	6–10 (at 293 K)
Thermal expansion, linear (K^{-1})	1.0×10^{-6} (at 300 K)

APPENDIX 14. PROPERTIES OF SILICON CARBIDE (SiC)

Breakdown field (V/cm)	2–3 $\times$ 10^6 (at 300 K)
Density (g/cm^3)	3.21 (at 300 K)
Dielectric constant (κ_o)	6.52 (at 300 K)
(κ_s)	9.72 (at 300 K)
Effective electron mass (in units of m_e)	1.5 (300 K) longitudinal
	0.25 (300 K) transverse
Effective hole mass (in units of m_e)	1.0 (300 K)
Energy gap (eV)	2.86 (300 K)
Index of refraction	2.55
Lattice constant (Å)	a = 3.0806 (297 K)
	c = 15.1173 (297 K)
Melting point (°C)	2830
Mobility (cm^2/V-s)	900 (300 K) electron
	50 (300 K) hole
Optical phonon energy (meV)	104.2 longitudinal
	95.6 transverse
Thermal expansion, linear (K^{-1})	2.9 $\times$ 10^{-6} (at 300 K)

APPENDIX 15. PROPERTIES OF ZINC SELENIDE (ZnSe)

Density (g/cm^3)	5.28 (at 4 K)
Dielectric constant (κ_s)	9.1 (at 300 K)
(κ_o)	6.3 (at 300 K)
Effective electron mass (in units of m_e)	0.21 (at 300 K)
Effective hole mass (in units of m_e)	0.6 (at 300 K)
Energy gap(eV)	2.70 (at 295 K)
Index of refraction	2.5
Lattice constant (Å)	5.6676
Melting point (°C)	1520
Mobility (cm^2/V-s)	electron: 500 (at 300 K)
	hole : 30 (at 300 K)
Optical phonon energy (meV)	longitudinal : 31.7 (at 90–250 K)
	longitudinal : 63.5 (at 90–250 K)
	transverse : 26.0 (at 90–250 K)
Thermal conductivity (W/cm-K)	0.19

APPENDIX 16. PROPERTIES OF ZINC TELLURIDE (ZnTe)

Density (g/cm^3)	5.636 (at 298 K)
Dielectric constant	7.4 (at 300 K)
Effective electron mass (in units of m_e)	0.20 (at 300 K)
Effective hole mass (in units of m_e)	0.1–0.3 (at 300 K)
Energy gap (eV)	2.26 (at 300 K)
Index of refraction	2.72
Lattice constant (Å)	6.1037
Melting point (°C)	1295
Mobility (cm^2/V-s)	electron: 340 (at 300 K) hole : 100 (at 300 K)
Thermal expansion, linear (K^{-1})	819×10^{-8} (at 283 K) 294×10^{-8} (at 75 K)

APPENDIX 17. PROPERTIES OF AMORPHOUS Si

Fermi level in intrinsic material ($E_c - E_{Fo}$, eV)	~ 0.6–0.7
Deep localized states density at $E = E_{FO}$ ($cm^{-3}eV^{-1}$)	10^{15}–10^{16}
Deep states characteristic energy (meV)	86
Tail localized states density at $E = E_C$ ($cm^{-3}eV^{-1}$)	~ 2×10^{21}
Tail states characteristic energy (meV)	23
Width of tail states band (meV)	~ 150
Dielectric constant	~ 11
Diffusion constant (cm^2/s)	0.26–0.52 (electrons in conduction band at 300 K) 0.13–0.26 (holes in valence band at 300 K)
Effective density of states in conduction band (cm^{-3})	~ 10^{19} (at 300 K)
Effective density of states in valence band (cm^{-3})	~ 10^{19} (at 300 K)
Energy gap (eV)	1.72 (at 300 K)
Index of refraction	3.32
Mobility (cm^2/V-s)	10–20 (electrons in conduction band at 300 K) 5–10 (holes in valence band at 300 K)

APPENDIX 18. PROPERTIES OF SiO_2

Breakdown field (V/cm)	10^7
Density (g/cm^3)	2.2
Dielectric constant (κ_s)	3.9
Dielectric constant (κ_o)	2.13
Energy gap (eV)	9
Index of refraction	1.46
Resistivity (Ω-cm)	10^{14}–10^{16} (at 300 K)
Thermal conductivity (W/cm-K)	0.014
Thermal expansion, linear ($°C^{-1}$)	5×10^{-7} (at 300 K)

APPENDIX 19. PROPERTIES OF Si_3N_4

Breakdown field (V/cm)	10^7
Density (g/cm^3)	3.1
Dielectric constant (κ_s)	7.5
Dielectric constant (κ_o)	4.2
Energy gap (eV)	~ 5
Index of refraction	2.05
Resistivity (Ω-cm)	10^{14} (at 300 K)
Resistivity (Ω-cm)	2×10^{13} (at 500 K)

APPENDIX 20. HYPERBOLIC FUNCTIONS

$$\sinh(x) = [\exp(x) - \exp(-x)]/2$$

$$\cosh(x) = [\exp(x) + \exp(-x)]/2$$

$$\tanh(x) = [\exp(x) - \exp(-x)]/[\exp(x) + \exp(-x)]$$

$$\coth(x) = [\exp(x) + \exp(-x)]/[\exp(x) - \exp(-x)]$$

$$\mathrm{sech}(x) = 2/[\exp(x) + \exp(-x)]$$

$$\mathrm{csch}(x) = 2/[\exp(x) - \exp(-x)]$$

$$d[\sinh(x)]/dx = \cosh(x)$$

$$d[\cosh(x)]/dx = \sinh(x)$$

$$d[\tanh(x)]/dx = 1/\cosh^2(x)$$

$$d[\coth(x)]/dx = -\ 1/\sinh^2(x)$$

APPENDIX 21. FERMI INTEGRALS

Fermi integrals, $F_n(\eta)$, arise in the calculation of different averages using the Fermi-Dirac distribution function,

$$F_n(\eta) = \frac{2}{\sqrt{\pi}} \int_o^\infty \frac{x^n \, dx}{1 + \exp(x - \eta)} \tag{A21-1}$$

The Fermi integral,

$$F_{1/2}(\eta) = \frac{2}{\sqrt{\pi}} \int_o^\infty \frac{x^{1/2} \, dx}{[1 + \exp(x - \eta)]} \tag{A21-2}$$

can be evaluated analytically in two important limiting cases:

1. When $\eta << -1$

$$F_{1/2}(\eta) \approx \exp(\eta) \tag{A21-3}$$

2. When $\eta >> 1$

$$F_{1/2}(\eta) \approx \frac{4\eta^{3/2}}{3\sqrt{\pi}} \tag{A21-4}$$

For $-10 < \eta < 10$ the Fermi integral $F_{1/2}(\eta)$ can be interpolated by the following expression (see Fig. A21-1):

$$F_{1/2}(\eta) = \exp(-0.32881 + 0.74041\eta - 0.045417\eta^2 - 8.797 \times 10^{-4}\eta^3 + 1.5117 \times 10^{-4}\eta^4)$$

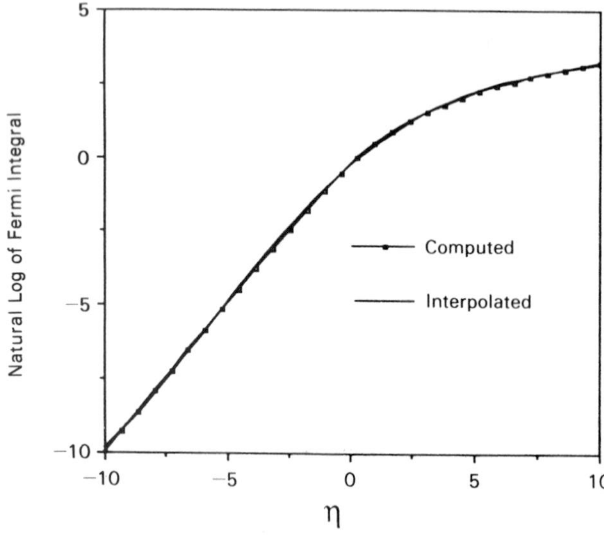

Fig. A21-1. Ln ($F_{1/2}$) vs. η.

APPENDIX 22. OVERLAP FACTOR AND SCATTERING RATES

Overlap Factor

For the electrons in the central valley Γ of the conduction band the difference of overlap factor, G, in eq. (1-13-12) from unity is related to the nonparabolicity. As shown by Fawcett et al. (1970),

$$G(k, k') = \frac{\{[1 + \alpha E(k)]^{1/2}[1 + \alpha E(k)]^{1/2} + \alpha[E(k)E(k')]^{1/2}\cos\theta\}^2}{[1 + 2\alpha E(k)][1 + 2\alpha E(k')]} \qquad \text{(A22-1)}$$

where θ is the angle between k and k'. For the intraband transitions within heavy or light hole bands we have

$$G_{h,l}(k,k') = \frac{3}{4}(1 + 3\cos^2\theta) \qquad \text{(A22-2)}$$

(see Ruch and Fawcett 1970) and for interband hole transitions we have

$$G_{h,l}(k,k') = \frac{3}{4}\sin^2\theta \qquad \text{(A22-3)}$$

Expressions for the electron scattering rates, obtained using eqs. (1-13-11) to (1-13-14) are as follows (see Ruch and Fawcett 1970):

Polar Optical Scattering

$$\lambda_o(E) = 5.616 \times 10^{15}\left(\frac{m^*}{m_e}\right)^{1/2} E_o\left(\frac{1}{\kappa_o} - \frac{1}{\kappa_s}\right)\frac{1 + 2\alpha E'}{\gamma^{1/2}(E)} F_o(E, E') \times \begin{cases} N_o & \text{(absorption)} \\ N_o + 1 & \text{(emission)} \end{cases} \quad (\text{sec}^{-1}) \qquad \text{(A22-4)}$$

where $E_o = \hbar\omega_o/q$ (i.e., the energy of the optical phonon in eV),

$$E' = \begin{cases} E + E_o & \text{(absorption)} \\ E - E_o & \text{(emission)} \end{cases}$$

$$\gamma(E) = E(1 + \alpha E)$$

$$F_o(E, E') = C^{-1}\left(A\,\ell n\left|\frac{\gamma^{1/2}(E) + \gamma^{1/2}(E')}{\gamma^{1/2}(E) - \gamma^{1/2}(E')}\right| + B\right)$$

$$N_o = \frac{1}{\exp(\hbar\omega_o/k_B T) - 1}$$

$$A = \{2(1 + \alpha E)(1 + \alpha E') + \alpha[\gamma(E) + \gamma(E')]\}^2$$

$$B = -2\alpha\gamma^{1/2}(E)\gamma^{1/2}(E')\{4(1 + \alpha E)(1 + \alpha E') + \alpha[\gamma(E) + \gamma(E')]\}$$

$$C = 4(1 + \alpha E)(1 + \alpha E')(1 + 2\alpha E)(1 + 2\alpha E')$$

Here E is the electron energy (in eV), m^*/m_e is the relative electron effective mass, and κ_o and κ_s are high- and low-frequency dielectric constants (10.92 and 12.9 for GaAs).

For the polar optical scattering with the emission of a phonon, eq. (A22-4) is valid only for $E' = E - \hbar\omega_o > 0$. For $E < \hbar\omega_o$ the scattering rate, λ, corresponding to the phonon emission is zero.

For the parabolic bands $\alpha = 0$ and

$$A = C = 4$$
$$B = 0$$

so that eq. (A22-4) can be simplified.

The energy dependencies of the polar optical scattering rate in GaAs, computed from eq. (A22-1), are shown in Figs. 1-13-1 and 1-13-2 for 300 K and 77 K, respectively (see Fawcett et al. 1970).

The values of the scattering parameters for GaAs used in Pozhela and Reklaitis 1980 are given in Table A22-1.

The energy of the final state in polar optical scattering is equal to $E + \hbar\omega_o$ for the emission of, and to $E - \hbar\omega_o$ for the absorption of, a phonon. The angular probability distribution of wave vectors of the final states can be found from

$$P(\theta)\, d\theta \cong \frac{[\gamma^{1/2}(E)\gamma^{1/2}(E') + \alpha EE' \cos\theta]^2 \sin\theta\, d\theta}{[\gamma(E) + \gamma(E') - 2\gamma^{1/2}(E)\gamma^{1/2}(E')\cos\theta]} \tag{A22-5}$$

(see Fawcett et al. 1970 and Fig. 1-13-3). Here θ is the angle between **k** and **k**$'$. As can be seen from Fig. 1-13-3, at high electron energies the scattering in the same direction is more probable.

TABLE A22-1. PARAMETERS CHARACTERIZING ELECTRON SCATTERING IN GaAs (after Yu. K. Pozhela, and A. Reklaitis, "Electron Transport Properties in GaAs at High Electric Fields," *Solid-State Electronics,* 23, pp. 927–933 (1980). Pergamon Press plc).

Energy separation between valleys (eV)	Γ–L	0.33
	Γ–X	0.52
Effective mass (m^*/m_e)	Γ	0.063
	L	0.17
	X	0.58
Intervalley coupling constant (10^9 eV/cm)	Γ–L	0.18
	Γ–X	1
	L–X	0.1
	L–L	0.5
	X–X	1
Nonparabolicity (eV^{-1})	Γ	0.62
	L	0.5
	X	0.3
Intervalley phonon energy (eV)		0.0299
Acoustic deformation potential (eV)		7
LO phonon energy (eV)		0.0362
Static dielectric constant		12.9
Optical dielectric constant		10.92
Sound velocity (cm/sec)		5.2×10^5
Density (g/cm^3)		5.37

Nonpolar Optical Scattering (Intravalley Scattering).

$$\lambda_{on}(E) = 1.129 \times 10^{-5}\left(\frac{m^*}{m_e}\right)^{3/2} \frac{D_o^2}{\rho E_{on}} \gamma^{1/2}(E')(1 + 2\alpha E')\, F_{on}(E, E') \times \begin{Bmatrix} N_{on} & \text{(absorption)} \\ N_{on} + 1 & \text{(emission)} \end{Bmatrix} \tag{A22-6}$$

Here

$$N_{on} = \frac{1}{\exp\left(\dfrac{qE_{on}}{k_B T}\right) - 1}$$

$$F_{on} = \frac{(1 + \alpha E)(1 + \alpha E')}{(1 + 2\alpha E)(1 + 2\alpha E')}$$

All energies are measured in electronvolts and are counted from the bottom of the conduction band minimum,

$$E' = E \pm E_{on}$$

where $E_{on} = \hbar\omega_{on}/q$ is the energy of the nonpolar optical phonon. The plus sign in the last equation corresponds to phonon absorption, the minus sign to phonon emission. D_o is the deformation potential constant (in eV/cm), and ρ is the density in g/cm^3.

Acoustic Scattering

The scattering rate is given by

$$\lambda_a(E) = \frac{0.449 \times 10^{18}\left(\dfrac{m^*}{m_e}\right)^{3/2} T E_1^2}{\rho u^2} \gamma^{1/2}(E)(1 + 2\alpha E)F_a(E)(\text{sec}^{-1}) \tag{A22-7}$$

where

$$F_a(E) = \frac{(1 + \alpha E)^2 + 1/3(\alpha E)^2}{(1 + 2\alpha E)^2}$$

Here ρ is the crystal density in g/cm^3, E_1 is the acoustic deformation potential (in eV), T is the lattice temperature (K), and u is the sound velocity (cm/s). The acoustic scattering is practically elastic because the energy of the acoustic phonons involved is small compared with $k_B T$.

The angular probability distribution for the acoustic scattering is given by

$$P(\theta)\, d\theta \cong [1 + \alpha E(1 + \cos\theta)^2] \sin\theta\, d\theta \tag{A22-8}$$

which reduces to

$$P(\theta)\, d\theta \cong \sin\theta\, d\theta \tag{A22-9}$$

for $\alpha = 0$ (i.e., for the parabolic band). Equation (A22-9) is equivalent to the statement that all final states are equally probable, and hence, the probability of the angle between $\mathbf{k}'$ and $\mathbf{k}$ being between θ and $\theta + d\theta$ is proportional to the number of states on a circle with the radius $|\mathbf{k}| \sin \theta$. In other words, the acoustic scattering processes are randomizing.

Impurity Scattering

The ionized impurity scattering rate is given by

$$\lambda_i(E) = 4.84 \times 10^{11}\left(\frac{m^*}{m_e}\right)^{1/2} \cdot \frac{T}{\kappa_o} \frac{1 + 2\alpha E}{[E(1 + \alpha E)]^{1/2}} \ (\text{sec}^{-1}) \qquad \text{(A22-10)}$$

(see Pozhela and Reklaitis 1980).

The scattering rate given by eq. (A22-10) is independent of N_D. This result is the consequence of the long-range Coulomb potential leading to the divergence of the scattering cross section for small scattering angles. The scattering cross section is limited only by the average distance between impurities. The magnitude of the electron concentration affects the transport properties by changing the angular distribution of scattering, because as N_D increases, the electrons are on average scattered through to a larger angle θ. The angular probability $P(\theta)d\theta$ is found to be

$$P(\theta)\, d\theta \cong \left[\frac{1 + \alpha E(1 + \cos\theta)}{2k^2(1 - \cos\theta) + \beta^2}\right]^2 \sin\theta \, d\theta$$

where

$$\beta = \left(\frac{q^2 N_D}{\varepsilon_o k_B T}\right)^{1/2}$$

is the inverse screening length. When $N_D \rightarrow 0$, $\beta \rightarrow 0$ and $P(\theta)$ diverges as $\theta \rightarrow 0$, so that only zero-angle scattering remains and impurity scattering has no effect.

Intervalley Scattering between Nonequivalent Valleys

In high electric fields the electron energy in GaAs may become sufficiently high to enable electrons to scatter into L and X valleys. Then, in addition to polar optical scattering, the intervalley scattering becomes important. The intervalley scattering rate for the scattering between the nonequivalent (000) and upper valleys is given by

$$\lambda_{ij}(E) = 1.129 \times 10^{-5} \cdot Z_j\left(\frac{m_j^*}{m_e}\right)^{3/2} \frac{D_{ij}^2}{\rho E_{ij}} \gamma_j^{1/2}(1 + 2\alpha_j E')F_{ij}(E, E') \times$$

$$\begin{Bmatrix} N_{ij} & \text{(absorption)} \\ N_{ij} + 1 & \text{(emission)} \end{Bmatrix} \ (\text{sec}^{-1}) \qquad \text{(A22-11)}$$

(see Fawcett et al. 1970). Here

$$N_{ij} = \frac{1}{\exp\left(\frac{\hbar\omega_{ij}}{k_B T}\right) - 1}$$

$$F_{ij}(E, E') = \frac{(1 + \alpha_i E)(1 + \alpha_j E')}{(1 + 2\alpha_i E)(1 + 2\alpha_j E')}$$

The energies are measured in eV and are counted from the bottoms of the minima of the conduction band,

$$E' = E \pm E_{ij}$$

$E_{ij} = \hbar\omega_{ij}/q$ is the energy of the intervalley scattering phonon in eV, Z_j is the number of the equivalent valleys, and D_{ij} is the deformation potential constant in eV/cm. This scattering is caused by the nonpolar optical phonons. It is randomizing, and the angular probability distribution is given by eq. (A22-5).

Intervalley Scattering between Equivalent Valleys

This scattering is also caused by the nonpolar optical phonons. The scattering rate is given by

$$\lambda_e(k) = 1.129 \times 10^{-5}(Z_e - 1)\left(\frac{m^*}{m_e}\right)^{3/2} \frac{D_e^2}{\rho E_e} (E')^{1/2} \times \begin{cases} N_e & \text{(absorption)} \\ N_e + 1 & \text{(emission)} \end{cases} \quad (\text{sec}^{-1}) \tag{A22-12}$$

where Z_e is the number of the equivalent valleys,

$$N_e = \frac{1}{\exp\left(\frac{\hbar\omega_e}{k_B T}\right) - 1}$$

E_e is the energy of the equivalent intervalley scattering phonon (in eV; $E_e = \hbar\omega_e/q$), D_e is the equivalent intervalley scattering constant (in eV/cm), and nonparabolicity is neglected.

This scattering is also randomizing.

APPENDIX 23. MOMENTUM RELAXATION TIMES AND LOW–FIELD MOBILITIES

Ionized Impurity Scattering

The calculation of the impurity relaxation time including the effect of the screening of the Coulomb potential of an impurity ion by free carriers leads to the following expression for the ionized impurity scattering time:

$$\tau_{ii} = \frac{2^{9/2}\pi E^{3/2}(m^*)^{1/2}(\kappa_s\varepsilon_o)^2}{N_I Z^2 q^4[\ln(1+\beta^2) - \beta^2/(1+\beta^2)]} \tag{A23-1}$$

where E is the carrier (electron or hole) energy, m^* is the effective mass, κ_s is the static dielectric constant, N_I is the impurity concentration, Z is the impurity charge in units of the electronic charge ($Z = 1$ for a hydrogenlike shallow donor or acceptor), q is the electronic charge, $\beta = 2kL_D$, k is the electronic (or hole) wave vector, and

$$L_D = \left(\frac{q^2 n}{\kappa_s\varepsilon_o k_B T}\right)^{-1/2} \tag{A23-2}$$

is the Debye radius. Here n is the electron concentration (it should be replaced by the hole concentration p for a p-type semiconductor).

For a semiconductor with ellipsoidal surfaces of equal energy (such as Si or Ge),

$$m^* = \frac{9}{\left(\dfrac{1}{m_t^{1/2}} + \dfrac{2}{m_l^{1/2}}\right)^2} \tag{A23-3}$$

Substituting eq. (A23-1) into eq. (1-13-37) and performing the integration, we obtain

$$\langle\tau_{ii}\rangle = \frac{2^{15/2}\pi^{1/2}(m^*)^{1/2}(\kappa_s\varepsilon_o)^2(k_B T)^{3/2}}{Z^2 q^4 N_I A} \tag{A23-4}$$

where

$$A = \ln(1+\beta_B^2) - \frac{\beta_B^2}{1+\beta_B^2} \tag{A23-5}$$

and

$$\beta_B = \frac{(24 m^* k_B T)^{1/2} L_D}{\hbar} = \frac{k_B T}{\hbar q}\left(\frac{24 m^*\kappa_s\varepsilon_o}{n}\right)^{1/2} \tag{A23-6}$$

Substituting constants into eqs. (A23-4) and (A23-6), we find that

$$\langle \tau_{ii} \rangle (s) = \frac{9.7 \times 10^9}{AN_I(m^{-3})} \kappa_s^2 \frac{1}{Z^2} \left(\frac{m^*}{m_e} \right)^{1/2} \left(\frac{T}{300} \right)^{3/2} \tag{A23-7}$$

and

$$\beta_B = 3.41 \times 10^9 \left(\frac{T}{300} \right) \left(\frac{m^*}{m_e} \right)^{1/2} \kappa_s^{1/2} \frac{1}{[n(\text{cm}^{-3})]^{1/2}} \tag{A23-8}$$

Equation (A23-4) is called the *Brooks-Herring formula* (see Brooks 1955).

The first calculation of the ionized impurity relaxation time was done by Conwell and Weisskopf (1950), who estimated the radius of the effective cross section as an average distance $N_I^{-1/3}$ between the ionized impurities. Their result is similar to the Brooks-Herring formula except for the logarithmic term A, which should be replaced by

$$A_{cw} = \ln(1 + \beta_c^2) \tag{A23-9}$$

where

$$\beta_c = \frac{12\pi \kappa_s \varepsilon_0 k_B T}{q^2 n^{1/3}} = 5.38 \times 10^5 \kappa_s \frac{T}{300} \frac{1}{[n(\text{cm}^{-3})]^{1/3}} \tag{A23-10}$$

The results obtained for Si using these formulas are compared in Fig. A23-1, in which mobility

$$\mu_{ii} = \frac{q\langle \tau_{ii} \rangle}{m^*} \tag{A23-11}$$

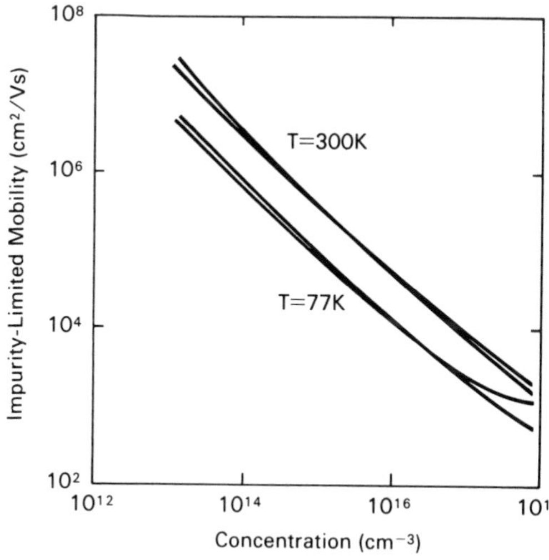

Fig. A23-1. Mobility μ_{imp} limited by ionized impurity scattering vs. ionized impurity concentration for n-type silicon. Curves which flatten out at large values of N_I are calculated using the Brooks-Herring formula. Curves that are closer to straight lines are calculated using the Conwell-Weisskopf formula. The curves are calculated using program PLOTF with subroutines PMBIMP (see Program Description Section).

is shown as a function of N_I (assuming $n = N_I$). The differences become important for large values of N_I, for which both formulas may become inaccurate because the degeneracy of the electron gas has not been taken into account. We should notice that for gallium arsenide the Brooks-Herring formula actually predicts an increase in η_{ii} at 77 K, with an increase in N_I for very large values of N_I. Again, this is a consequence of not accounting for the electron degeneracy. The Conwell-Weisskopf formula may still be used for large N_I as a crude but meaningful interpolation.

Acoustic (Deformation Potential) Scattering

The momentum relaxation time for deformation potential scattering is given by Rode (1970):

$$\frac{1}{\tau_a(E)} = 4.15 \times 10^{19} \frac{(m^*/m_e)^{3/2}T^{3/2}(E_1)^2X^{1/2}}{\rho u^2} \tag{A23-12}$$

where ρ is the density in g/cm^3, E_1 is the acoustic deformation potential in eV, u is the sound velocity in cm/s, $X = E/k_BT$, and E is the electron energy.

In most cases only the scattering by the longitudinal acoustic waves is taken into account ($u = u_c$), even though the scattering by the transverse acoustic phonons may also be important.

Substitution of eq. (A23-12) into eq. (1-13-37) and substitution of the result of the integration into the equation for the mobility,

$$\mu_a = \frac{q\langle\tau_a\rangle}{m^*} \tag{A23-13}$$

yields

$$\mu_a = \frac{2\sqrt{2\pi}\hbar^4 c_{11}}{3(m^*)^{5/2}E_1^2(k_BT/q)^{3/2}q^{1/2}} \tag{A23-14}$$

In the case of ellipsoidal minima, the effective mass, m^*, in eq. (A23-14) should be replaced by

$$m_a^* = \left(\frac{\frac{3}{2} m_t m_l^{1/2}}{\frac{2}{m_t} + \frac{1}{m_l}}\right)^{2/5} \tag{A23-15}$$

Equation (A23-14) may be rewritten as

$$\mu_a(m^2/V\cdot s) = 0.589 \times 10^{-12} \frac{c_{11}(\text{dyn/cm}^2)}{\left(\frac{m^*}{m_e}\right)^{5/2}\left[\frac{T(K)}{300}\right]^{3/2} E_1^2(\text{eV})} \tag{A23-16}$$

Polar Optical Scattering

Polar optical scattering is a dominant scattering mechanism in lightly or moderately doped GaAs samples at room temperature (see Fig. 1-9-2). It is inelastic scattering because the optical phonon energy, $\hbar\omega_o/q$, is comparable to the thermal energy at room temperature ($\hbar\omega_o/q = 0.0362$ eV for GaAs). Hence, the correction to the distribution function depends not only on the energy of the electron prior to the scattering but also on the energy of the final state, and strictly speaking, the relaxation time approximation cannot be used. The Boltzmann equation may be solved by the variation method or by the iteration technique as discussed by Rode (1970). However, a simple and useful semiempirical relaxation time has been introduced by Harrison and Hauser (1976):

$$1/\tau_{op} = [1/X^{1/2} + \exp(z)/(X - z)^{1/2}]/\tau_o \tag{A23-17}$$

where

$$\tau_o = 1.93 \times 10^{-14} \frac{T^{1/2}}{T_{po}(m^*/m_e)^{1/2}\,(1/\kappa_o - 1/\kappa_s)} [\exp(z) - 1] \tag{A23-18}$$

Here T_{po} is the Einstein temperature ($T_{po} = \hbar\omega_o/k_B$; 420 K for GaAs), $z = T_{po}/T$, κ_o is the high-frequency dielectric constant (10.92 for GaAs), and κ_s is the static dielectric constant (13.1 for GaAs).

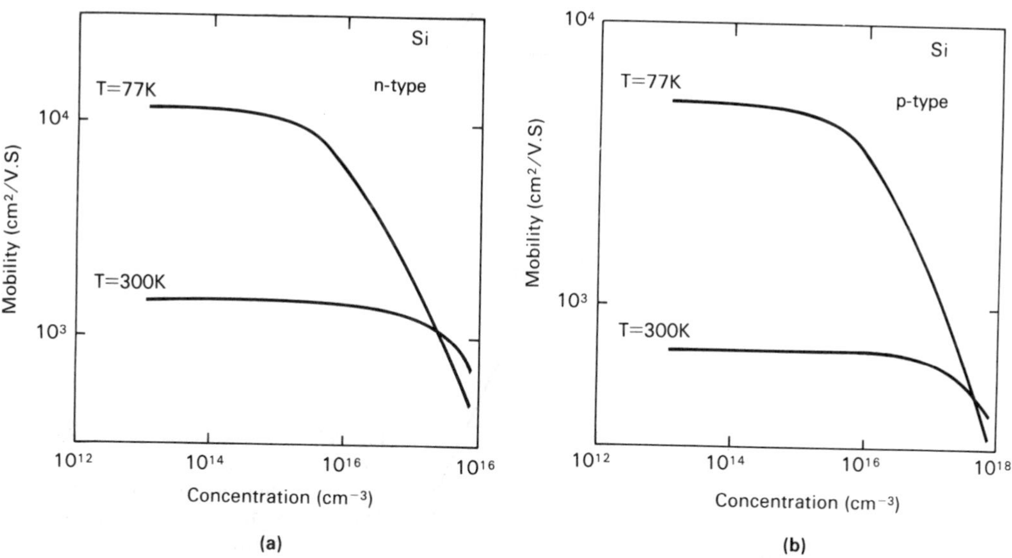

Fig. A23-2. Electron (a) and hole (b) mobility in silicon vs. concentration of ionized impurities. This approximate calculation takes into account acoustic scattering by longitudinal phonons and ionized impurity scattering. The curves were generated using program PLOTF with subroutines PMOBIL (see program description section).

For an analytical calculation, the following empirical formula may be used for the mobility limited by the polar optical scattering (see Lee et al. 1983):

$$\mu_{po} = \frac{A}{T^2} + \frac{B}{T^6} \tag{A23-19}$$

where for GaAs $A = 7.42 \times 10^4$ m^2 (degree K)^2/Vs and $B = 1.18 \times 10^{13}$ m^2 (degree K)^6/Vs. Figures A-23-2 and A-23-3 show the calculated electron and hole mobilities in silicon and gallium arsenide vs. the concentration of ionized impurities.

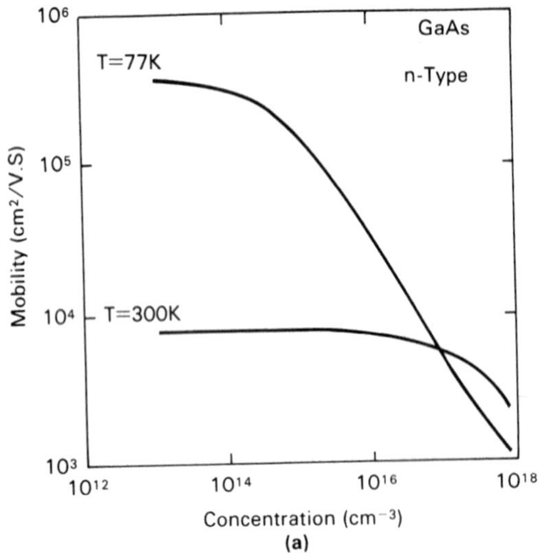

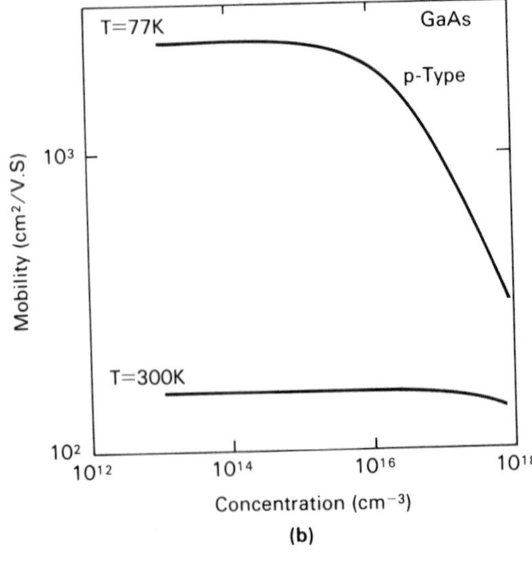

Fig. A23-3. (a) Electron and (b) hole mobility in gallium arsenide vs. concentration of ionized impurities. This approximate calculation takes into account acoustic scattering by longitudinal phonons, by polar optical phonons and ionized impurity scattering. The curves were generated using program PLOTF with subroutines PMOBIL (see program description section).

APPENDIX 24. DEVICE AND CIRCUIT SIMULATION PROGRAMS

Introduction

Computer simulation of device fabrication and device characteristics as well as computer-aided design of semiconductor devices and circuits have become important steps in developing semiconductor technologies. A typical development sequence is illustrated by Fig. A24-1. Nearly all development steps rely on computer software.

In this appendix we briefly describe some of the computer programs that are available for semiconductor device simulation and modeling. Most of these programs are distributed to the public by universities (most notably by Berkeley and Stanford) for a charge ranging from several hundred to several thousand dollars.

We first briefly describe SUPREME, SIMPL, and SAMPLE, which are programs for the simulation of fabrication processes of silicon and GaAs devices. We then discuss programs for the simulation of semiconductor devices, such as SEDAN, BIPOLE, PISCES, and BAMBI. Finally, we describe SPICE, which is the most popular semiconductor circuit simulator today.

SUPREME, SIMPL, and SAMPLE: Computer Programs for Fabrication-Process Modeling

SUPREME is a very powerful program for simulating fabrication processes. It has been developed and distributed by Stanford University (see Ho et al. 1983 and Plummer 1986). It includes models for simulating diffusion, ion implantation, and oxidation of silicon. At the present time, version 3 is available and version 4 is under development. A typical application of the program is illustrated in Fig. A24-2 (from Plummer 1986), which shows an NMOS Si field-effect transistor and simulated phosphorus, arsenic, and boron concentration profiles in two device cross sections. These dopant profiles may be directly used as an input to device simulation programs SEDAN and PISCES, described in the next section. This

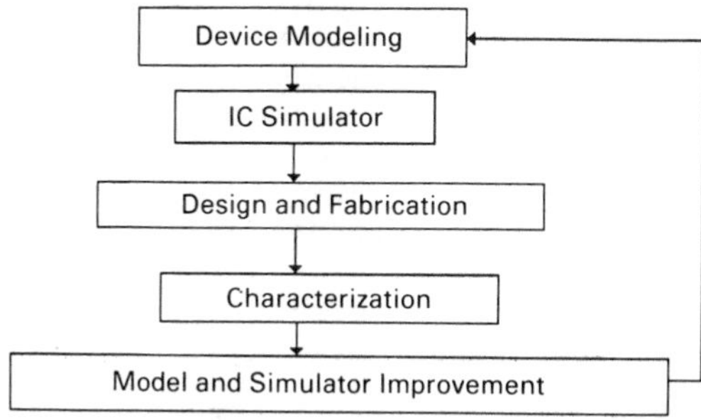

Fig. A24-1. Typical development sequence for semiconductor devices and integrated circuits.

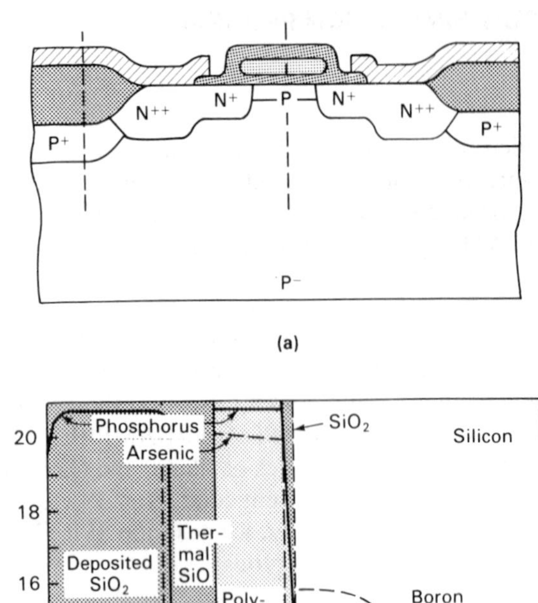

(a)

(b)

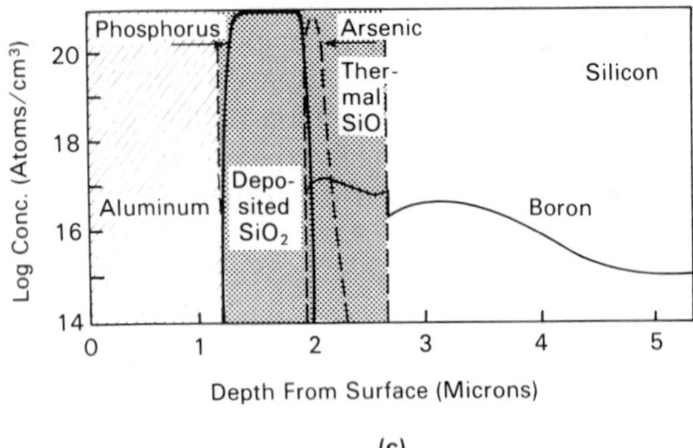

(c)

Fig. A24-2. SUPREM-3 simulation. (a) cross-section of NMOS Si field-effect transistor and (b) and (c) impurity profiles in cross-sections shown by dashed lines in Fig. A24-2-1a. (From Plummer (1986), reprinted with the permission of Solid State Technology).

simulation is based on one-dimensional models. However, SUPREME-4 will include two-dimensional models as well that will be especially useful for modeling the fabrication processes for short-channel devices.

SIMPL (SIMulated Profiles from the Layout) is a computer program that generates device cross sections from the circuit layout (see Grimm et al. 1983 and

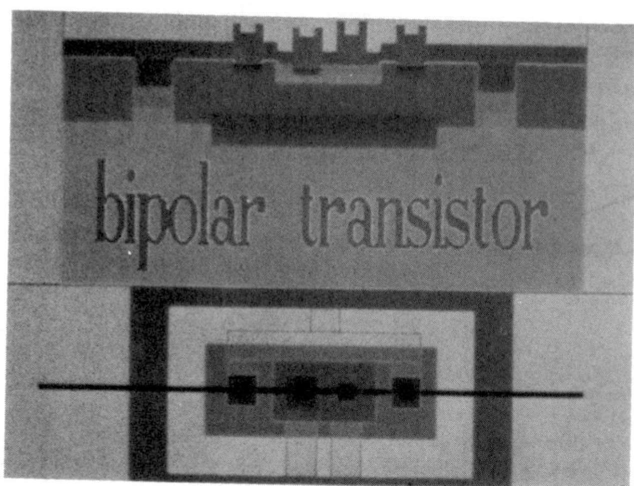

Fig. A24-3. SIMPL output: Bipolar transistor layout and device cross-section, shown on the layout by a black line. (From Lee and Neureutner (1985), reprinted with permission of Solid State Technology."

Neureuther 1986). The input data to the program include process and mask information. Figure A24-3 shows a typical program output—a bipolar transistor cross section and device cross-section. A newer version of this program, SIMPL-2, (see Lee and Neureuther 1985), allows for a more accurate simulation of two-dimensional process effects. SIMPL is written in C computer language and requires a color terminal.

SAMPLE (Simulation and Modeling of Profiles in Lithography and Etching) simulates lithography, etching, and deposition processes. It includes approximately fifteen-thousand lines of FORTRAN and requires a few minutes of VAX11/780 computer time per processing step (see Neureuther 1986). It can also

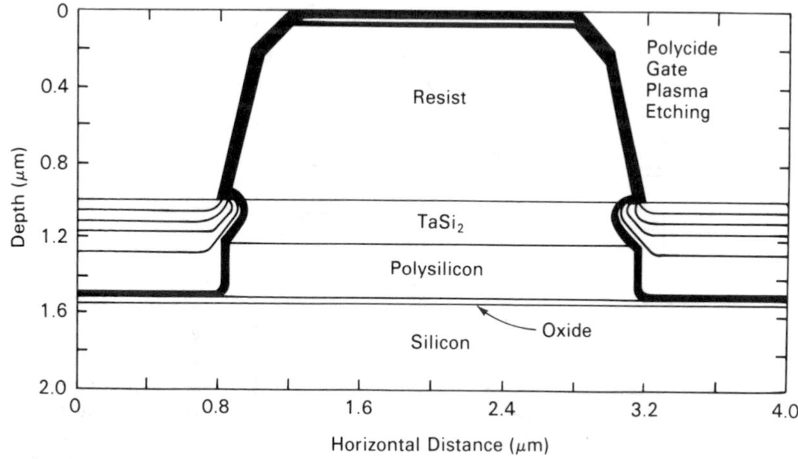

Fig. A24-4. SAMPLE output: simulated edge profile of polycide ($TaSi_2$ on polysilicon) gate obtained by plasma etching. (from Neureuther 1986)

run on IBM PC-XT (with a floating point 8087 chip) and compatible computers. As an example, Fig. A24-4 shows the simulated edge profile of a polycide ($TaSi_2$ on polysilicon) gate obtained by plasma etching (from Neureuther 1986).

Both SIMPL and SAMPLE were developed at Berkeley, and more information may be obtained from EECS Industrial Liaison Program, 457 Cory Hall, University of California, Berkeley, CA 94720.

SEDAN, BIPOLE, PISCES, and BAMBI: Computer Programs for Device Modeling

The SEDAN (SEmiconductor Device ANalysis) program has been developed and distributed by the Integrated Circuit Laboratory at Stanford University (see Yu and Dutton 1985). It computes potential, electron, and hole concentration profiles as functions of time for one-dimensional device structures. The SEDAN III version simulates both silicon (including transistors with polysilicon emitters) and GaAs (or AlGaAs/GaAs) devices. The program solves basic semiconductor equations based on field-dependent electron and hole velocities (see Section 1-11). The manual includes a detailed description of the material and device models and lists all material parameters used in the simulation. The program is written in FORTRAN 77 and is run under the UNIX or the HP-UX operating system. As an example of the program output, Fig. A24-5 shows the computed energy-band diagram of a tunneling emitter bipolar transistor (from Xu and Shur 1986). Another example—carrier concentrations in a bipolar junction transistor—is shown in Fig. 3-1-3a (from Yu and Dutton 1985).

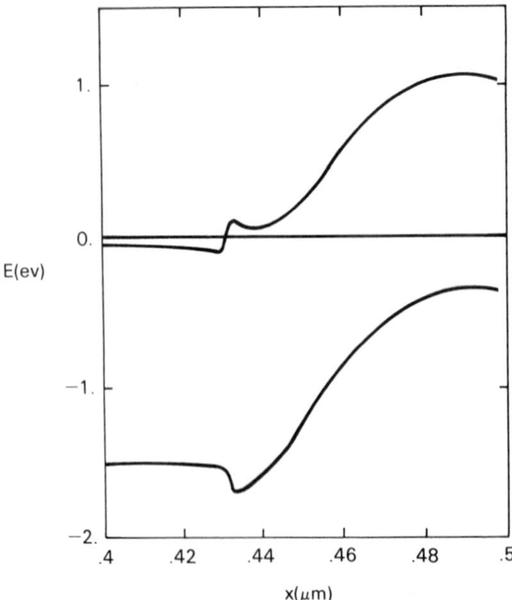

Fig. A24-5. Example of SEDAN-3 simulation, the computed energy band diagram of a tunneling emitter bipolar transistor (from J. Xu and M. Shur, "Tunneling Emitter Bipolar Junction Transistor," *IEEE Electron Device Lett.*, EDL-7, pp. 416–418 (1986). © 1986 IEEE).

BIPOLE is a program that computes the characteristics of bipolar junction transistors from the information about the fabrication process and mask dimensions. It was developed at the Department of Electrical Engineering of the University of Waterloo, Canada, and is distributed by Waterloo Engineering Software, 180 Columbia Street, West, Waterloo, Ontario, Canada N2L 3L3. The program takes into account band-gap narrowing in heavily doped emitter regions, the dependence of the lifetime on doping, temperature, electric field, and doping dependencies of electron and hole mobilities. It is applicable for simulation of integrated structures and GaAs and AlGaAs/GaAs heterostructure bipolar transistors.

PISCES is a two-dimensional device simulator developed by Stanford Electronics Laboratories, Department of Electrical Engineering, Stanford University, Stanford, CA 94305 (see Pinto et al. 1985). The program solves basic semiconductor equations based on field-dependent electron and hole velocities (see Section 1-11). It handles four types of boundary conditions: ohmic contacts, Schottky barriers, insulator interfaces, and reflective boundaries. It includes models for Shockley-Read-Hall recombination, Auger recombination, and surface recombination (see Section 1-12). The program requires several megabytes of storage. The program is written in ANSI FORTRAN 77 and runs under the UNIX operating system. Graphics output requires a Hewlett-Packard HP2648 terminal or a Tektronix 4107 terminal. As an example of the program output, Fig. A24-6 shows the computed electron and hole concentration profiles in a double-injection transistor (from Xu and Shur 1987).

BAMBI (Basic Analyzer of MOS and BIpolar DEvices) is a two-dimensional device simulator developed by Franz and Franz (1985) and distributed by Dr. Siegfrid Selbergeher, Institut fur Allgemeine Electrotechnik und Elektronik, Technische Universitatet Wien, Gusshausstraße 27-29, A-1040 Wien, Austria. It handles both dc and transient problems. Arbitrary shapes can be simulated. Ohmic and Schottky barrier contacts may be specified. The device may include boundaries between a semiconductor and dielectric as well as surface charges at any interface boundary. The program solves basic semiconductor equations based on field-dependent electron and hole velocities (see Section 1-11). The process of grid generation and management in BAMBI is automated. The program also includes an automatic time-step control algorithm (see Franz and Franz 1984). As an example of BAMBI output, the results of the simulation of a power silicon MOS field, effect transistor are shown in Fig. A24-7. BAMBI is written in ANSI FORTRAN. BAMBI can run on CDC computers (NOS operating system), on VAX (under the VMS operating system), and other computers.

Programs FISHID, PUPHS1D, PUPHS2D, DEMON, and SEQUAL have been developed by Professor Mark Lundstrom of Purdue University. FISHID solves Poisson's equation in compositionally nonuniform semiconductors (such as GaAs/AlGaAs device structures), both under equilibrium conditions and under bias (assuming zero current densities). The program allows you to plot energy-

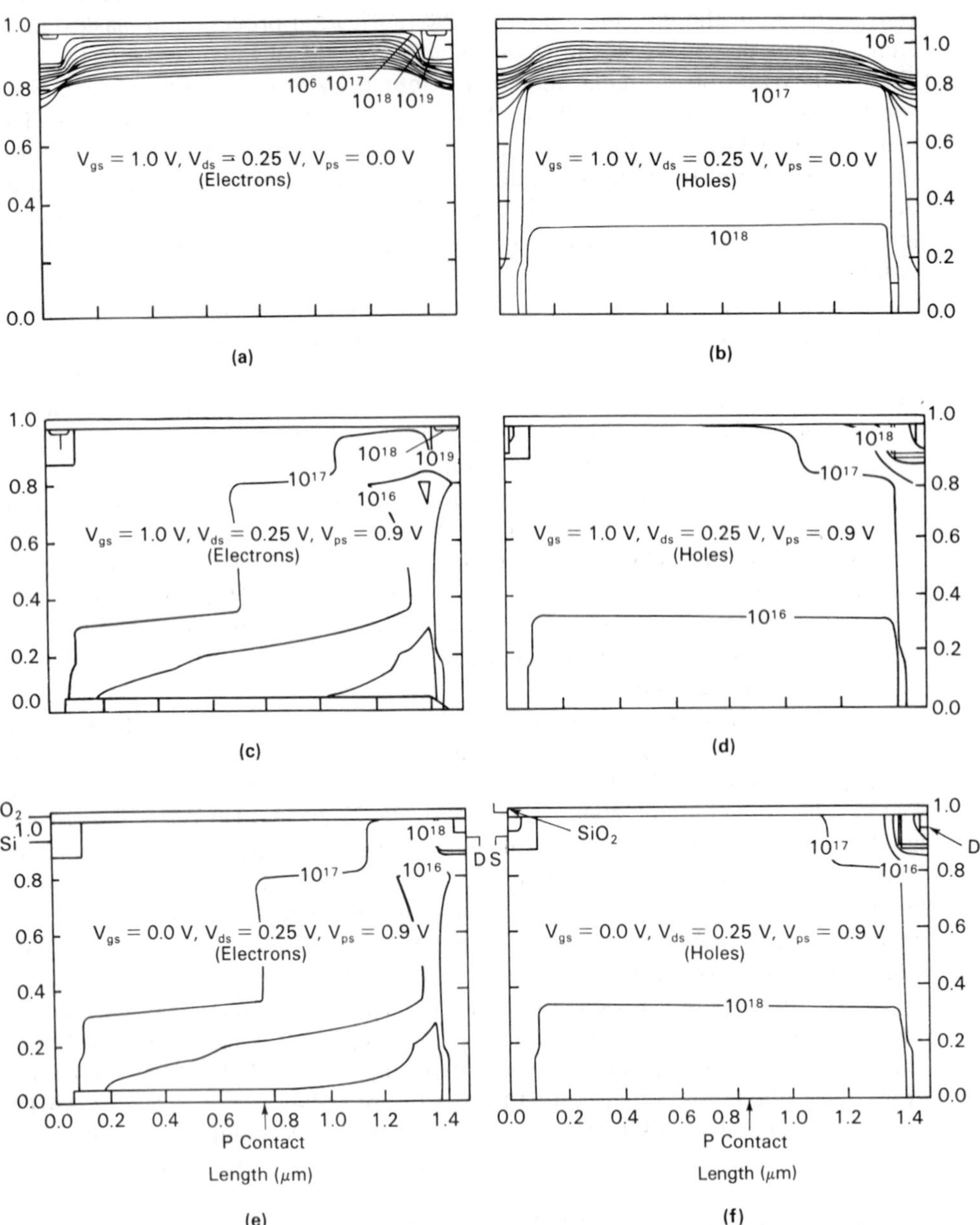

Fig. A24-6. Example of PISCES simulation, the computed electron and hole concentration profiles in a double-injection transistor (from J. Xu and M. Shur, *J. Appl. Phys.*, 63, no. 3, pp. 1108–1111 (1987)).

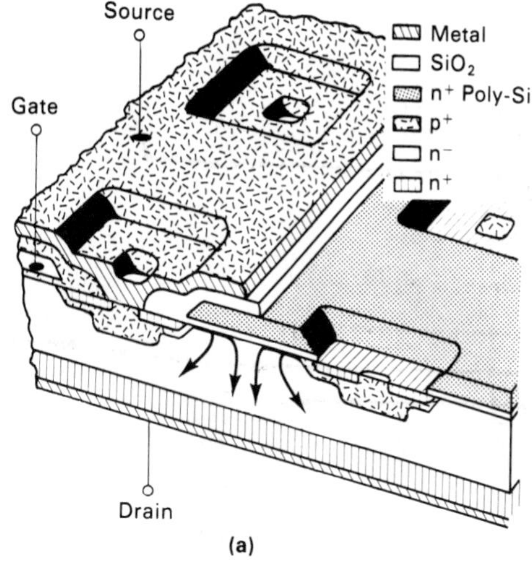

(a)

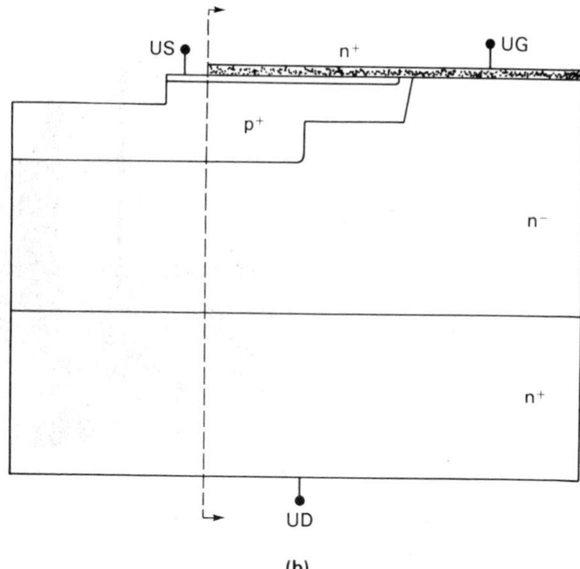

(b)

Fig. A24-7. Example of BAMBI output, results of the simulation of power silicon MOS field-effect transistor (from A. F. Franz and G. A. Franz, *IEEE Trans. Computer-Aided Design,* CAD-4, no. 3, pp. 177–189 (1985). © 1985 IEEE). (a) Schematic device structure, (b) simulation geometry, (c) computer output characteristics, and (d) computed electron concentration for drain voltage of 3 V, gate voltage of 10 V.

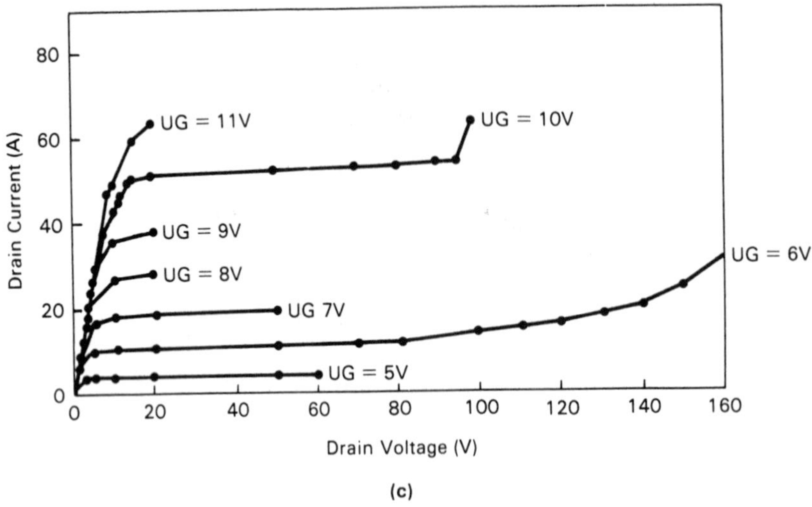

(c)

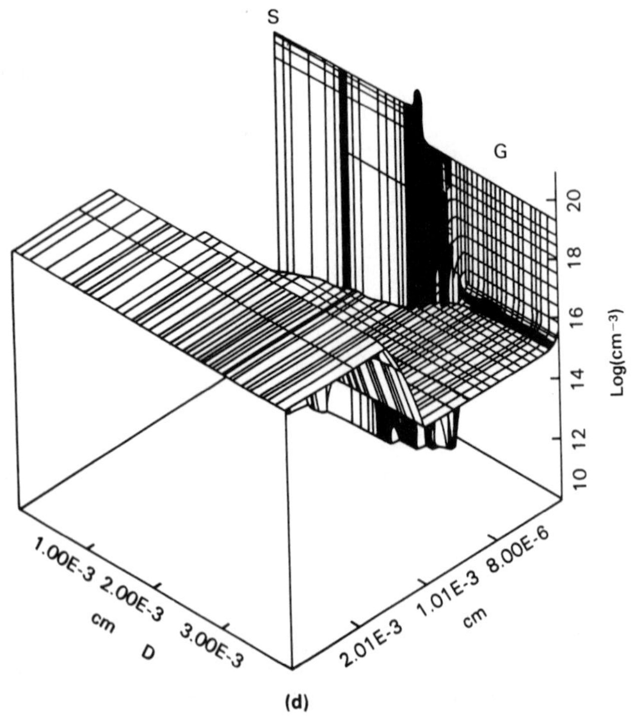

(d)

Fig. A24-7. Cont.

band diagrams, potential, field, and carrier concentration profiles, and to compute C-V characteristics (see Gray and Lundstrom 1985).

PUPHS1D and PUPHS2D are one-dimensional and two-dimensional versions of a device simulator, respectively. These programs solve the electron and hole continuity equations for compositionally nonuniform semiconductors in equilibrium and under bias. These programs have been used to simulate solar cells, diodes, field-effect transistors, bipolar junction transistors, and heterostructure bipolar junction transistors (see Lundstrom and Schulke 1983).

DEMON simulates electron transport in compositionally nonuniform semiconductors using one-dimensional Monte Carlo simulation (see Maziar et al. 1986 and Klausmeier-Brown 1986).

SEQUAL solves a one-dimensional Schrödinger equation for a given one-dimensional band diagram (obtained, for example, using FISH1D or PUPHS1D). This program was described by Cahay et al. (1987) and by McLennan (1987).

SPICE: Computer Program for Simulating Semiconductor Circuits

SPICE is the most popular circuit simulator today. It has been developed and supported by the Electronic Research Laboratory of the College of Engineering of the University of California, Berkeley (see Nagel 1975). It is written in FORTRAN (except the most recent version, which is written in C). It includes models for linear elements (resistors, capacitors, inductors, coupled inductors, transmission lines, voltage sources, current sources, and voltage-controlled current sources) and nonlinear elements (nonlinear voltage-controlled current sources,

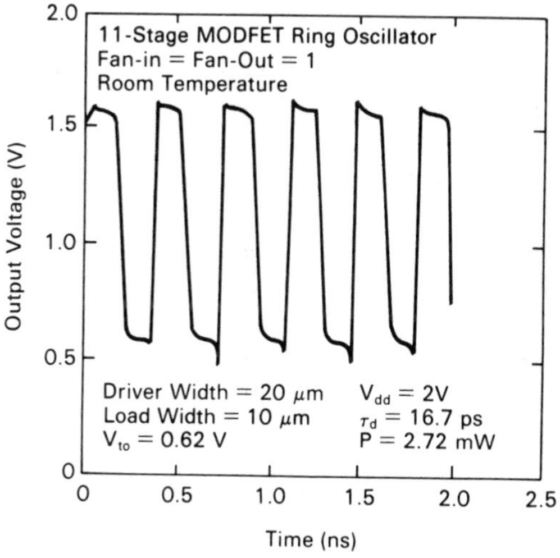

Fig. A24-8. Example of UM-SPICE output, the waveform of AlGaAs-GaAs MODFET ring oscillator (from C. H. Hyun, M. Shur, and A. Peczalski, *IEEE Trans. Electron Devices,* ED-33, no. 10, pp. (1986). © 1986 IEEE).

diodes, bipolar junction transistors, junction field-effect transistors, and metal oxide field-effect transistors). Information about the most recent releases of the program may be obtained from EECS Industrial Liaison Program, 457 Cory Hall, University of California, Berkeley, CA 94720.

A popular commercial version, PSPICE for IBM personal computers and compatibles, also includes models for GaAs MESFETs. This version is distributed by MicroSim Corporation, P.O. Box 2055-233, Tustin, CA 92680. Another commercial version of SPICE, Interactive-Graphics SPICE (I-G SPICE), is available from AB Associates, Inc., P.O. Box 82215, Tampa, FL 33682.

The version of SPICE developed at the University of Minnesota (see Hyun et al. 1986) includes models for GaAs MESFETs, ungated GaAs FETs, and heterostructure FETs (such as MODFETs and HIGFETs; see Section 4-11). An example of UM-SPICE output—the waveform of an AlGaAs/GaAs MODFET ring oscillator—is shown in Fig. A24-8.

REFERENCES

CAHAY M., M. MCLENNAN, S. DATTA, and M. LUNDSTROM, "Importance of Space-Charge Effects in Resonant Tunneling Devices," *Appl. Phys. Lett.*, 50, pp. 612–614 (1987).

A. F. FRANZ and G. A. FRANZ, *IEEE Trans. Computer-Aided Design,* CAD-4, no. 3, pp. 177–189 (1985).

A. F. FRANZ and G. A. FRANZ, in *Simulation of Semiconductor Devices and Processes,* Swansea, Pineridge, pp. 204–218 (1984).

J. L. GRAY and M. S. LUNDSTROM, "Solution to Poisson's Equation with Application to C-V Analysis of Heterostructure Capacitors," *IEEE Trans. Electron Devices,* ED-32, pp. 2102–2109 (1985).

M. A. GRIMM, K. LEE, A. R. NEUREUTHER, *IEDM Technical Digest,* pp. 255–258 (1983).

C. P. HO, J. D. PLUMMER, S. E. HANSEN, and R. W. DUTTON, *IEEE Trans. Electron Devices,* ED-30, no. 11, pp. 1438–1453, November (1983).

C. H. HYUN, M. SHUR, and A. PECZALSKI, *IEEE Trans. Electron Devices,* ED-33, no. 10, pp. (1986a).

C. H. HYUN, M. SHUR, and N. C. CIRILLO, JR., *IEEE Trans. Computer-Aided Design,* CAD-5, no. 2, pp. 284–292 (1986b).

M. E. KLAUSMEIER-BROWN, "Monte Carlo Studies of Electron Transport in III-V Heterostructures," M.S.E.E. thesis, Purdue University, May (1986).

M. S. LUNDSTROM and R. J. SCHULKE, "Numerical Simulation of Heterostructure Devices," *IEEE Trans. Electron Devices,* ED-30, no. 11, pp. 1151–1159 (1983).

C. MAZIAR, M. E. KLAUSMEIER-BROWN, and M. S. LUNDSTROM, "Proposed Structure for Transit Time Reduction in AlGaAs/GaAs Bipolar Transistors," *IEEE Electron Dev. Lett.,* EDL-8, p. 486 (1986).

M. McLennan, "Quantum Ballistic Transport in Semiconductor Heterostructures," M.S.E.E. thesis, Purdue University, May (1987).

L. W. Nagel, "SPICE 2: A Computer Program to Simulate Semiconductor Circuits," Memorandum No. ERL-M520, Electronic Research Laboratory, College of Engineering, University of California, Berkeley (1975).

A. R. Neureuther, *Solid State Technology,* 29, no. 3, pp. 71–75 (1986).

M. R. Pinto, C. S. Rafferty, H. R. Yeager, and R. W. Dutton, "PISCES-IIB," Supplementary Report, Stanford Electronics Laboratories, Department of Electrical Engineering, Stanford University, Stanford, CA 94305 (1985).

J. D. Plummer, *Solid State Technology,* 29, no. 3, pp. 61–66 (1986).

J. Xu and M. Shur, "Tunneling Emitter Bipolar Junction Transistor," *IEEE Electron Device Lett.,* EDL-7, pp. 416–418 (1986).

J. Xu and M. Shur, *J. Appl. Phys.,* 63, no. 3, pp. 1108–1111 (1987).

Z. Yu and R. W. Dutton, *Sedan III: A General Electronic Material Device Analysis Program,* Program Manual, Stanford University, July (1985).

APPENDIX 25. RELATIONSHIP BETWEEN H PARAMETERS AND PARAMETERS OF T EQUIVALENT CIRCUIT AND π EQUIVALENT CIRCUIT

Equations relating the parameters of T and π equivalent circuits and h parameters are as follows (see Casasent 1973, p. 87):

$$r_e = h_{re}/h_{oe} = h_{ib} - h_{rb}(1 + h_{fb})/h_{ob} \tag{A25-1}$$

$$r_{bb'} = h_{ie} - h_{re}(1 + h_{fe})/h_{oe} = h_{rb}/h_{ob} \tag{A25-2}$$

$$r_c = (1 + h_{fe})/h_{oe} = (1 - h_{rb})/h_{ob} \tag{A25-3}$$

$$\alpha = h_{fe}/(1 + h_{fe}) = -h_{fb} \tag{A25-4}$$

$$\beta = h_{fe} = -h_{fb}/(1 + h_{fb}) \tag{A25-5}$$

The h parameters may also be related to the parameters of the hybrid-π equivalent circuit using the circuit analysis.

REFERENCE

D. Casasent, *Electronic Circuits,* Quantum Publishers, New York (1973).

APPENDIX 26. DATA SHEETS FOR MOTOROLA 2N2219A GENERAL–PURPOSE SILICON n-p-n TRANSISTORS (from *Small-Signal Transistor Data,* published by Motorola Semiconductor Products, Inc., Phoenix Ariz., 1983)

Maximum ratings			
Rating	Symbol		Unit
Collector-emitter voltage	V_{CEO}	30	V(dc)
Collector-base voltage	V_{CBO}	60	V(dc)
Collector current—continuous	I_C	800	mA(dc)
Total power dissipation	P_D	0.8	Watt
Total power dissipation @ $T_A = 25°C$	P_D	0.8	Watt
Derate above 25°C		4.57	mW/°C
Operating and storage junction temperature range	T_J, T_{stg}	−65 to 200	°C

Electrical characteristics (T_A = 25°C unless otherwise noted)				
Characteristic	Symbol	Min	Max	Unit
On characteristics				
DC current gain	h_{FE}			
(I_C = 0.1 mA(dc), V_{CE} = 10 V(dc)		35	—	
(I_C = 1 mA(dc), V_{CE} = 10 V(dc)		50	—	
(I_C = 10 mA(dc), V_{CE} = 10 V(dc)		75	—	
(I_C = 10 mA(dc), V_{CE} = 10 V(dc), T_A = −55°C			35	—
(I_C = 150 mA(dc), V_{CE} = 10 V(dc)		100	300	
(I_C = 150 mA(dc), V_{CE} = 1.0 V(dc)		50	—	
(I_C = 500 mA(dc), V_{CE} = 10 V(dc)		30	—	
Small-signal characteristics				
Current gain—bandwidth product	f_T	300		MHz
Output capacitance	C_{obo}		8	pF
Input capacitance	C_{ibo}		30	pF
Input impedance	h_{ie}			kΩ
(I_C = 1 mA(dc), V_{CE} = 10 V(dc), f = 1 kHz)			2	8
(I_C = 10 mA(dc), V_{CE} = 10 V(dc), f = 1 kHz)			0.25	1.25
Voltage feedback ratio	h_{re}			
(I_C = 1 mA(dc), V_{CE} = 10 V(dc), f = 1 kHz)			—	8×10^{-4}
(I_C = 10 mA(dc), V_{CE} = 10 V(dc), f = 1 kHz)			—	4×10^{-4}

Electrical characteristics (T_A = 25°C unless otherwise noted)

Characteristic	Symbol	Min	Max	Unit
Small-signal current gain	h_{fe}			
(I_C = 1 mA(dc), V_{CE} = 10 V(dc), f = 1 kHz)		50	300	
(I_C = 10 mA(dc), V_{CE} = 10 V(dc), f = 1 kHz)		75	375	
Output admittance	h_{oe}			
(I_C = 1 mA(dc), V_{CE} = 10 V(dc), f = 1 kHz)		5	35	μmhos
(I_C = 10 mA(dc), V_{CE} = 10 V(dc), f = 1 kHz)		25	200	
Collector-base time constant (I_E = 20 mA(dc), V_{CE} = 20 V(dc), f = 31.8 MHz)	$r_{b'}C_c$		150	ps
Noise figure (I_C = 100 μA(dc), V_{CE} = 10 V(dc), R_S = 1 kΩ, f = 1.0 kHz)	NF		4	dB
Real part of common-emitter high-frequency input impedance (I_C = 20 mA(dc), V_{CE} = 20 V(dc), f = 300 MHz)	$Re(h_{ie})$		60	Ω
Switching characteristics				
Delay time	t_d		10	ns
Rise time	t_r		25	ns
(V_{CC} = 30 V(dc), $V_{BE(off)}$ = 0.5 V(dc), I_C = 150 mA(dc), I_{B1} = 15 mA(dc))				
Storage time	t_s		225	ns
Fall time	t_f		605	ns
(V_{CC} = 30 V(dc), $V_{BE(off)}$ = 0.5 V(dc), I_C = 150 mA(dc), I_{B1} = I_{B2} = 15 mA(dc))				
Active region time constant (I_C = 150 mA(dc), V_{CC} = 30 V(dc))	t_f		605	ns

Appendix 27. Periodic Table of Elements

Key (example: Polonium):

- atomic number → 84
- 209 → atomic weight based upon carbon 12
- • → Metaloid (elements to the right are non-metals, elements to the left are metals)
- density (g/cubic cm for solids and liquids and g/liter for gases at 273K and 1 atm) → 9.4
- Po → symbol
- Xe$4f^{14}5d^{10}6s^2p^4$ → electron configuration*
- r → radioactive
- solid → solid, liquid, gas at 300 K or synthetically prepared
- name → Polonium

GROUP IA	IIA	IIIA	IVA	VA	VIA	VIIA	VIIIA	VIIIA	VIIIA	IB	IIB	IIIB	IVB	VB	VIB	VIIB	VIIIB
1 1.0079 0.0899 **H** $1s^1$ gas Hydrogen																	2 4.0026 0.1787 **He** $1s^2$ gas Helium
3 6.941 0.53 **Li** $1s^22s^1$ solid Lithium	4 9.0122 1.85 **Be** $1s^22s^2$ solid Beryllium											5 10.81 • 2.34 **B** $1s^22s^2p^1$ solid Boron	6 12.01 2.62 **C** $1s^22s^2p^2$ solid Carbon	7 14.007 1.251 **N** $1s^22s^2p^3$ gas Nitrogen	8 16 1.429 **O** $1s^22s^2p^4$ gas Oxigen	9 19 1.696 **F** $1s^22s^2p^5$ gas Fluorine	10 20.18 0.901 **Ne** $1s^22s^2p^6$ gas Neon
11 22.9898 0.97 **Na** Ne$3s^1$ solid Sodium	12 24.305 1.74 **Mg** Ne$3s^2$ solid Magnesium											13 26.98 2.70 **Al** Ne$3s^2p^1$ solid Aluminum	14 28.09 • 2.33 **Si** Ne$3s^2p^2$ solid Silicon	15 30.97 1.82 **P** Ne$3s^2p^3$ solid Phosphorus	16 32.06 2.07 **S** Ne$3s^2p^4$ solid Sulfur	17 35.45 3.17 **Cl** Ne$3s^2p^5$ gas Chlorine	18 39.95 1.784 **Ar** Ne$3s^2p^6$ gas Argon
19 39.0983 0.86 **K** Ar$4s^1$ solid Potassium	20 40.08 1.55 **Ca** Ar$4s^2$ solid Calcium	21 44.9559 3.0 **Sc** Ar$3d^14s^2$ solid Scandium	22 47.90 4.5 **Ti** Ar$3d^24s^2$ solid Titanium	23 50.94 5.8 **V** Ar$3d^34s^2$ solid Vanadium	24 51.996 7.19 **Cr** Ar$3d^54s^1$ solid Chromium	25 54.94 7.43 **Mn** Ar$3d^54s^2$ solid Manganese	26 55.85 7.86 **Fe** Ar$3d^64s^2$ solid Iron	27 58.93 8.90 **Co** Ar$3d^74s^2$ solid Cobalt	28 58.70 8.90 **Ni** Ar$3d^84s^2$ solid Nickel	29 63.55 8.96 **Cu** Ar$3d^{10}4s^1$ solid Copper	30 65.38 7.14 **Zn** Ar$3d^{10}4s^2$ solid Zinc	31 69.72 5.91 **Ga** Ar$3d^{10}4s^2p^1$ liquid Gallium	32 72.59 • 5.32 **Ge** Ar$3d^{10}4s^2p^2$ solid Germanium	33 74.92 • 5.72 **As** Ar$3d^{10}4s^2p^3$ solid Arsenic	34 78.96 4.80 **Se** Ar$3d^{10}4s^2p^4$ solid Selenium	35 79.90 3.12 **Br** Ar$3d^{10}4s^2p^5$ liquid Bromine	36 83.80 3.74 **Kr** Ar$3d^{10}4s^2p^6$ gas Krypton
37 85.47 1.53 **Rb** Kr$5s^1$ solid Rubidium	38 87.62 2.6 **Sr** Kr$5s^2$ solid Strontium	39 88.91 4.5 **Y** Kr$4d^15s^2$ solid Yttrium	40 91.22 6.49 **Zr** Kr$4d^25s^2$ solid Zirconium	41 92.91 8.55 **Nb** Kr$4d^45s^1$ solid Niobium	42 95.94 10.2 **Mo** Kr$4d^55s^1$ solid Molibdenum	43 98 11.5 **Tc** Kr$4d^55s^2$ synthetic Technetium	44 101.07 12.2 **Ru** Kr$4d^75s^1$ solid Ruthenium	45 102.91 12.4 **Rh** Kr$4d^85s^1$ solid Rhodium	46 106.4 12 **Pd** Kr$4d^{10}$ solid Palladium	47 107.87 10.5 **Ag** Kr$4d^{10}5s^1$ solid Silver	48 112.41 8.65 **Cd** Kr$4d^{10}5s^2$ solid Cadmium	49 114.82 7.31 **In** Kr$4d^{10}5s^2p^1$ solid Indium	50 118.69 7.30 **Sn** Kr$4d^{10}5s^2p^2$ solid Tin	51 121.75 • 6.88 **Sb** Kr$4d^{10}5s^2p^3$ solid Antimony	52 127.60 • 6.24 **Te** Kr$4d^{10}5s^2p^4$ solid Tellurium	53 126.9 4.92 **I** Kr$4d^{10}5s^2p^5$ solid Iodine	54 131.3 5.89 **Xe** Kr$4d^{10}5s^2p^6$ gas Xenon
55 132.91 1.87 **Cs** Xe$6s^1$ liquid Cesium	56 137.33 3.5 **Ba** Xe$6s^2$ solid Barium	57 138.91 6.7 **La*** Xe$5d^16s^2$ solid Lanthanum	72 178.49 13.1 **Hf** Xe$4f^{14}5d^26s^2$ solid Hafnium	73 180.95 16.6 **Ta** Xe$4f^{14}5d^36s^2$ solid Tantalum	74 183.85 19.3 **W** Xe$4f^{14}5d^46s^2$ solid Tungsten	75 186.21 21.0 **Re** Xe$4f^{14}5d^56s^2$ solid Rhenium	76 190.2 22.4 **Os** Xe$4f^{14}5d^66s^2$ solid Osmium	77 192.22 22.5 **Ir** Xe$4f^{14}5d^76s^2$ solid Iridium	78 195.09 21.4 **Pt** Xe$4f^{14}5d^96s^1$ solid Platinum	79 196.97 19.3 **Au** Xe$4f^{14}5d^{10}6s^1$ solid Gold	80 200.59 13.53 **Hg** Xe$4f^{14}5d^{10}6s^2$ liquid Mercury	81 204.37 11.85 **Tl** Xe$4f^{14}5d^{10}6s^2p^1$ solid Tallium	82 207.2 11.4 **Pb** Xe$4f^{14}5d^{10}6s^2p^2$ solid Lead	83 208.98 9.8 **Bi** Xe$4f^{14}5d^{10}6s^2p^3$ solid Bismuth	84 209 • 9.4 **Po** Xe$4f^{14}5d^{10}6s^2p^4$ solid r Polonium	85 210 - **At** Xe$4f^{14}5d^{10}6s^2p^5$ solid r Astatine	86 222 9.91 **Rn** Xe$4f^{14}5d^{10}6s^2p^6$ gas r Radon
87 223 - **Fr** Rn$7s^1$ liquid r Francium	88 226.03 5 **Ra** Rn$7s^2$ solid r Radium	89 227.03 10.1 **Ac**** Rn$6d^17s^2$ solid r Actinium	104 261 - **Unq** Rn$5f^{14}6d^27s^2$ synthetic r Unnilquadium	105 262 - **Unp** Rn$5f^{14}6d^37s^2$ synthetic r Unnilpentium	106 263 - **Unh** Rn$5f^{14}6d^47s^2$ synthetic r Unnilhexium												

*Lanthanides (Rare Earths)	58 140.12 6.78 **Ce** Xe$4f^15d^16s^2$ solid Cerium	59 140.91 6.77 **Pr** Xe$4f^36s^2$ solid Praseodymium	60 144.24 7.00 **Nd** Xe$4f^46s^2$ solid Neodymium	61 145 6.475 **Pm** Xe$4f^56s^2$ synthetic r Promethium	62 150.4 7.54 **Sm** Xe$4f^66s^2$ solid r Samarium	63 151.96 5.26 **Eu** Xe$4f^76s^2$ solid Europium	64 157.25 7.89 **Gd** Xe$4f^75d^16s^2$ solid Gadolinium	65 158.93 8.27 **Tb** Xe$4f^96s^2$ solid Terbium	66 162.50 8.54 **Dy** Xe$4f^{10}6s^2$ solid Dysprosium	67 164.93 8.80 **Ho** Xe$4f^{11}6s^2$ synthetic Holmium	68 167.26 9.05 **Er** Xe$4f^{12}6s^2$ solid Erbium	69 168.93 9.33 **Tm** Xe$4f^{13}6s^2$ solid Thulium	70 173.04 6.98 **Yb** Xe$4f^{14}6s^2$ solid Ytterbium	71 174.97 9.84 **Lu** Xe$4f^{14}5d^16s^2$ solid Lutetium
**Actinides	90 232.04 11.7 **Th** Rn$6d^27s^2$ solid r Thorium	91 231.04 15.4 **Pa** Rn$5f^26d^17s^2$ solid r Protactinium	92 238.03 18.90 **U** Rn$5f^36d^17s^2$ solid r Uranium	93 237.05 20.4 **Np** Rn$5f^46d^17s^2$ synthetic r Neptunium	94 244 19.8 **Pu** Rn$5f^67s^2$ synthetic r Plutonium	95 243 13.6 **Am** Rn$5f^77s^2$ synthetic r Americium	96 247 13.5 **Cm** Rn$5f^76d^17s^2$ synthetic r Curium	97 247 - **Bk** Rn$5f^97s^2$ synthetic r Berkelium	98 251 - **Cf** Rn$5f^{10}7s^2$ synthetic r Californium	99 252 - **Es** Rn$5f^{11}7s^2$ synthetic r Einsteinium	100 257 - **Fm** Rn$5f^{12}7s^2$ synthetic r Fermium	101 258 - **Md** Rn$5f^{13}7s^2$ synthetic r Mendelevium	102 259 - **No** Rn$5f^{14}7s^2$ synthetic r Nobelium	103 260 - **Lr** Rn$5f^{14}6d^17s^2$ synthetic r Lawrencium

* The following notation is used in the Table in order to save space: $3s^2p^6$ means $3s^23p^6$, etc.

Index

Tear out this card and fill in all necessary information. Then enclose this card with your check or money order *only* in an envelope and mail to:

Book Distribution Center
PRENTICE HALL
Route 59 at Brook Hill Drive
West Nyack, New York 10995

PHYSICS OF SEMICONDUCTOR DEVICES—Shur

Please send the item(s) checked below. PAYMENT ENCLOSED (Check or money order *only*). The Publisher will pay all shipping and handling charges.

_____ Please send Software and Manual to accompany PHYSICS OF SEMICONDUCTOR DEVICES (66658-6)—$18.00

NAME ______________________________

DEPT. ______________________________

SCHOOL ______________________________

CITY ______________ STATE ________ ZIP ________

NOTE: PROFESSIONAL/REFERENCE BOOKS ARE TAX-DEDUCTIBLE.

Prices subject to change without notice. Please add sales tax for your area.

Dept. 1

Tear out this card and fill in all necessary information. Then enclose this card with your check or money order *only* in an envelope and mail to:

Book Distribution Center
PRENTICE HALL
Route 59 at Brook Hill Drive
West Nyack, New York 10995

PHYSICS OF SEMICONDUCTOR DEVICES—Shur

Please send the item(s) checked below. PAYMENT ENCLOSED (Check or money order *only*). The Publisher will pay all shipping and handling charges.

_____ Please send Software and Manual to accompany PHYSICS OF SEMICONDUCTOR DEVICES (66658-6)—$18.00

NAME ______________________________

DEPT. ______________________________

SCHOOL ______________________________

CITY ______________ STATE ________ ZIP ________

NOTE: PROFESSIONAL/REFERENCE BOOKS ARE TAX-DEDUCTIBLE.

Prices subject to change without notice. Please add sales tax for your area.

Dept. 1